GaAs MESFET Circuit Design

The Artech House Materials Science Library

DiLorenzo, James V., and Deen D. Khandelwal, **GaAs FET Principles and Technology**

Gilmour, A.S., Jr., **Microwave Tubes**

Granatstein, Victor L., and Igor Alexeff, eds., **High-Power Microwave Sources**

Robert, Philippe, **Electrical and Magnetic Properties of Materials**

Soares, Robert A., ed., **GaAs MESFET Circuit Design**

Williams, Ralph E., **Gallium Arsenide Processing Techniques**

GaAs MESFET Circuit Design

edited by

Robert Soares

Artech House
Boston and London

British Library Cataloguing in Publication Data
GaAs MESFET circuit design.
1. Field-effect transistors
I. Soares, R.
621.3815′284 TK7871.95

ISBN 0–89006–267–6

Library of Congress Cataloging-in-Publication Data
GaAs MESFET circuit design.

Bibliography: p.
Includes index.
1. Electronic circuit design—Data processing. 2. Gallium arsenide semiconductors. 3. Metal semiconductor field-effect transistors. I. Soares, Robert, 1946– .
TK7867.G33 1988 621.3815′3 88–19597
ISBN 0–89006–267–6

ARTECH HOUSE, INC.
685 Canton Street
Norwood, MA 02062

International Standard Book Number: 0-89006-267-6
Library of Congress Catalog Card Number: 88-19597

10 9 8 7 6 5 4 3 2 1

Contents

Authors' Biographies

Josep **BAUCELLS** Bois was born in Vic, Spain, in 1960. He obtained his degree in telecommunication engineering from the Escola Tècnica Superior de Ingenieros Telecomunicación de Barcelona (ETSETB), Spain, in 1983. He worked on GaAs MESFET mixers in the Electrophysics Department of ETSETB for 2 years. From March 1985 to October 1986 he worked on nonlinear computer-aided design at the Centre National d'Etudes des Télécommunications (CNET), Lannion, France, and since October 1986 he has been working on a Ku band SSPA under a European Space Agency contract at Mier Allende SA, Spain.

Alain **CAPPY** was born in Chalons sur Marne, France, on January 25, 1954. In 1977 he joined the Centre Hyperfréquences et Semiconducteurs at the University of Lille, France, and he received the Docteur en électronique and Docteur ès sciences degrees from the University of Lille. His main research interests are concerned with semiconductor device modelling, noise modelling in submicrometer-gate FETs and TEGFETs, and FET characterization techniques. Dr Cappy is a founder of the International GaAs Simulation Group.

Jean Michel **DUMAS** was born in Charente, France, on November 28, 1948. He received the Diplôme d'Ingénieur degree from the Institut National des Sciences Appliquées, University of Toulouse, and the Doctorat ès sciences physiques degree from the Institut de Recherches en Communications Optiques et Microondes, University of Limoges. He is currently responsible for the microwave device reliability studies at the Centre National d'Etudes des Télécommunications, Lannion, France.

Maurice **GLOANEC** was born in Tremeven, France, on January 4, 1945. He received an electrical engineering degree from Le Conservatoire National des Arts et Métiers, Paris, France, in 1975. In 1969 he joined the Central Research Laboratory of Thomson-CSF, located in Orsay, France, where he first worked on silicon transducer integrated circuits. From 1974 to 1977 he investigated the digital applications of GaAs planar transferred electron devices. Since 1978 he has been involved in GaAs MESFET digital integrated circuit design. He has developed high performance SSI integrated circuits, e.g. a 1.6 GHz D-type flip-flop in 1979, a 5.7 GHz frequency divider in 1981 and a 4 GHz PCM decision circuit in 1983. In 1984 he was promoted to manager of Thomson-Semiconductors GaAs Analog Integrated Circuits Group. He is responsible for the development of a new family of GaAs analog components for high frequency signal processing. He holds several patents in the fields of GaAs planar TED, GaAs MESFET devices and high frequency packaging and is co-author of about 15 papers in scientific journals and conferences.

Christian **KERMARREC** received his engineering degree from Le Conservatoire National des Arts et Métiers in Paris in 1975. First he worked at CIT Alcatel in Paris, France, on the design of active and passive millimeter-wave components for use in an experimental TE_{01} circular waveguide link (30–60 GHz). In 1980 he joined the Laboratoires d'Electronique et de Physique Appliquée (Philips Research Center in France) to design GaAs MMICs, in particular a 12 GHz DBS receiver. From 1985, he was in charge of the implementation of a GaAs MMIC facility at Radiotechnique Compélec (Philips center in France for the development and production of electronic components). In 1986, he joined SPAR Aerospace Ltd, in Montreal, Canada, to introduce miniaturized technologies (thin film and GaAs MMICs) in satellite payload circuitry. He is now at Tachonics Corporation, NJ, USA.

Amarpal S. **KHANNA** received his B.S. degree from Punjab Engineering College, Chandigan, India, in 1971, and his Doctorate of Engineering (microwaves and optical communications) in 1981 from the University of Limoges, France. From 1971 to 1977 he worked for the Government of India on radar and radio navigation aids. He worked for Giga Modules France (1982–1983) and Loral MD, USA (1983–1985), before joining Avantek Inc., Santa Clara, CA, USA, in July 1985. He is at present the senior member of the technical staff in the Modular Components and

Subassemblies Division of Avantek where he is responsible for the advanced development of transistor oscillators and subassemblies in the microwave and millimeter frequency range. Dr Khanna has authored or co-authored more than 25 publications and four books. He holds four US and two French patents on dielectric resonator circuits.

Yves-Michel **MARTI** is the manager of the Computer-Aided Engineering Group for the Monolithic Microwave Integrated Circuits Department at Electronique Serge Dassault (Saint-Cloud, France). He is currently a visiting scientist at Pacific Monolithics (Sunnyvale, CA, USA). He came to Electronique Serge Dassault after being a consultant for computer-aided design at the Inforama Consulting Group (Paris, France). Previously, he started and led the Analog CMOS Integrated Circuits Group at the Société Anonyme des Télécommunications (Paris, France). He received his graduate engineering degree in 1982 from the Ecole Nationale Supérieure des Télécommunications, Brittany, France, and holds a double Master of Sciences in integrated circuits design and integrated circuits technology from the University of Grenoble, France. He has taught classes on computer-aided design at the Universities of Jussieu (Paris, France), Orsay (Paris, France) and Limoges (France). He is currently preparing his own book on CAD/CAM, and a series of classes for the University of California at Berkeley.

Angel **MEDIAVILLA** Sánchez was born in Santander, Spain, in 1955. He graduated (with honors) in 1978 and received the Doctor of Physics degree in 1984, both from the Universidad de Cantabria, Santander, Spain. From 1980 to 1983 he was Ingénieur Stagière at Thomson-CSF, Corbeville, France. Now he is professor at the Department of Electronics of Cantabria University, Spain, and his current fields of research are on active microwave circuits, mainly in the area of nonlinear MESFET modelling and their applications in large-signal computer design.

Marc **PARISOT** was born in October 1942. He received the Ph.D. degree from the University of Orsay, France, in 1967. He then joined the Philips Research Laboratory (LEP) in Limeil-Brevannes where he was involved in the design of infrared light modulators for the Symphonie satellite project. In 1972 he joined the microwave hybrid circuit team to develop YIG-tuned Gunn oscillators and to design one of the first wideband 4–18 GHz sweep generators and power and low noise FET amplifiers.

During this time, he developed a 2–18 GHz automatic test station including a vectorial network analyzer, a load-pull and a noise parameter test system with microstrip de-embedding procedures and designed a microwave circuit simulation program. In 1981 he moved to the DMH department of Thomson-Semiconductors in Orsay where he was in charge of the microwave characterization of GaAs device models and MMICs, and also the design of wideband monolithic amplifiers. He is at present responsible for MMIC products development and the Project Manager of a T/R module study and development program. He has authored a number of papers on microwave characterization and GaAs circuit applications.

Joseph **PAUGAM** was born in Finistère, France, on November 3, 1948. He received the Baccalauréat ès sciences degree from Brest High School and then went to the Telecommunications School in Paris. He is now engaged in microwave device reliability studies at the Centre National d'Etudes de Télécommunications, Lannion, France.

Ernesto H. **PEREA** holds an Ingeniero Electrico–Electronico degree from the Catholic University of Cordoba, Argentina, a Licenciado en Fisica degree from the National University of Cordoba, Argentina, and a Ph.D. in physics from Massachusetts Institute of Technology, Cambridge, MA, USA. He was a Kaiser-Renault Fellow from March 1965 to December 1970 and a Fulbright Fellow from July 1975 to December 1979. From January 1980 to February 1981 he was a member of the technical staff at Bell Laboratories CMOS IC Design Department, in charge of the design and evaluation of medium-size CMOS gate arrays. He was also a member of the Low End Microprocessor Design Group. From February 1981 to October 1983 he was a research engineer at Thomson-CSF's Central Research Laboratory, in charge of the computer-aided design and testing of GaAs digital ICs. At present he manages the GaAs Signal Processing IC Development Group which includes analog and digital IC design and testing. He has published extensively in the fields of compound semiconductor growth and properties, MOS and GaAs digital circuit design and instrumentation. He holds several patents in the field of GaAs devices and is a member of the Institute of Electrical and Electronic Engineers (IEEE), the Electron Device Society, the Solid State Circuit Society and the Computer Society, since 1980.

Christian **RUMELHARD** qualified as electrical engineer, Conservatoire National des Arts et Métiers, in 1966 and received his Docteur Ingénieur degree from the University of Paris VI in 1977. He worked at Thomson-CSF on the design of microwave tubes until 1969 and then on hybrid circuits: low noise and power amplifiers, voltage-controlled oscillators and phase shifters. In 1975 he joined the Computer Science Division to develop computer-aided design algorithms and numerical models for the simulation of microwave circuits and devices. In 1980 he was head of the Microwave Integrated Circuits Laboratory of the Central Research Laboratory of Thomson-CSF. Since 1984 he has been head of the MMIC Group at Thomson Hybride et Microondes.

Robert Alain **SOARES** was born in Bombay, India, and was educated at St Joseph's College, Darjeeling. He received his degrees (B.Sc. (Hons) 1967, Ph.D. 1975) in electronic engineering from the University of London, United Kingdom. His work in microwaves began at the GEC Hirst Research Centre (1970–1973) where he worked on S band low noise bipolar amplifier design. In 1974 he joined the Plessey Allen Clark Research Centre (Caswell) where he worked on X band low noise GaAs MESFET amplifiers. He has been at the Centre National d'Etudes des Télécommunications (CNET), Lannion, France, since November 1976 where he has successively worked on power GaAs MESFET amplifier design and nonlinear modelling, GaAs MESFET dielectric resonator oscillators, GaAs MESFET multipliers and recently low noise GaAs MESFET amplifiers at 20 and 30 GHz. Dr Soares is a member of the IEE and IEEE. He has authored or co-authored over 20 technical articles and holds one patent. He is also the main author and co-editor of *Applications of GaAs MESFETs* and *Applications des Transistors à Effet de Champ en Arseniure de Gallium.*

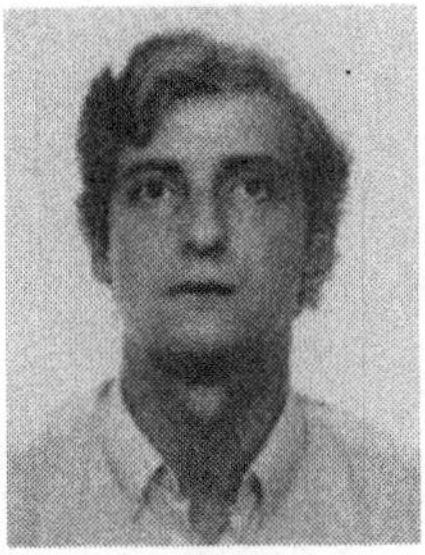

Antonio **TAZON** Puente was born in Santander, Spain, in 1951. He graduated in 1978 and received the Doctor of Physics degree in 1987, both from the Universidad de Cantabria, Santander, Spain. Now he is professor at the Department of Electronics of Cantabria University, and his current fields of research are large-signal MESFET modelling and the use of computer-aided design in the analysis and optimization of nonlinear microwave circuits.

Christian **VUCHENER** was born in Doubs, France, on November 1, 1955. He received the Baccalauréat ès sciences degree from Vezoul High School and then attended the Telecommunications School in Paris. He is now engaged in microwave device reliability studies at the Centre National d'Etudes des Télécommunications, Lannion, France.

Romuald A. **WROBLEWSKI** received his M.Sc. and D.Eng. degrees in electronic engineering from the Technical University of Gdańsk, Poland, in 1968 and 1977 respectively. From 1968 to 1979 he was with the Department of Electronic Engineering at the Technical University of Gdańsk, working on IMPATT and Gunn diode oscillators. From 1979 to 1983 he worked on avalanche phenomena and power limitations in GaAs MESFETs at the Centre Hyperfréquences et Semiconducteurs of Lille University I, Villeneuve d'Ascq, France. From 1983 to 1986 he was with Dateno SA, Dinard, France, where he was responsible for the development of power FET amplifiers, DBS converters for Telecom 1, and millimeter-wave components for 30 GHz digital radio. Since 1986 he has been with SPAR Aerospace Ltd leading a project on low noise, high stability oscillators for space applications.

Foreword

For years now there has been a joke about GaAs being the material of the future and staying that way. The problem with GaAs as a material is that, while it offers greatly improved performance over silicon, it brings difficulties in handling and processing that keep manufacturing engineers on their toes.

For both digital and microwave applications, GaAs MESFETs appear to offer significant advantages over silicon. For digital applications that advantage is of the order of 2:1, whereas in the case of microwave applications the advantage is of the order of 5:1. A unique characteristic of compound semiconductors, of which GaAs is just a single example, is that material properties can be tailored to specific requirements, whereas silicon, being a single material, cannot be mixed with itself in varying quantities to obtain anything but silicon.

This book, composed of 10 chapters devoted to circuit design using GaAs MESFETs, balances theory and applications applicable to monolithic microwave integrated circuits and hybrid circuits.

The early chapters are devoted to textbook basics: the nature and principle of operation of GaAs MESFETs; their high frequency and noise equivalent circuits; the mathematical formulation of *S* parameters and their use in microwave circuit theory. A comprehensive treatment of GaAs MESFET *S* parameter and noise measurement is also presented in Chapters 2 and 3. In addition this book has an important information content enabling the design of low noise and wideband amplifiers, electrically tunable and fixed-frequency dielectric resonator oscillators, and power amplifiers, independently of whether the final circuits are realized in hybrid or monolithic form.

In the microwave business it is not usually acceptable to use integrated circuits with much less than the best achievable performance with discrete devices. In low frequency silicon linear circuits, the transistors have current gain cut-off frequencies perhaps a thousand times higher than the frequency of use. This kind of excess performance is simply not available for microwave monolithic circuits. More typically that number is less than 10:1.

As seen in the chapter on monolithic circuit design, linear amplifiers are not the only components integrated on GaAs. For example, complete MMIC front-ends are available for microwave links. The r.f. amplifier, mixer, i.f.

amplifier, local oscillator and buffer—five different parts of a normal microwave system—are all colocated on a monolithic chip of less than 1 mm on a side.

Small complex circuits with multiple cross-couplings put a special emphasis on MMIC computer-aided design. Programs suitable for hybrid circuits simply do not have the flexibility required for the highly integrated and most cost effective MMICs. Moreover, these MMICs have oscillators, buffers and mixers which require both nonlinear analysis and linear analysis. These analysis programs need to communicate without much human intervention.

Another factor making microwave circuits particularly challenging is the fact that the interconnecting circuitry is as important to the performance of a circuit as the active devices are. Automation of the interconnections between active devices, resistors or diodes is not as easily accomplished as it is with silicon low frequency circuits where the effects of the interconnections do not tune the circuit.

To extract the maximum performance from microwave devices in a complex circuit, accurate device modelling and computer-aided design are essential. The chapter on computer-aided engineering tools (Chapter 4) and nonlinear simulators (Chapter 5) are applicable to both hybrid and monolithic technology. These design aids are useful time-saving devices in hybrid circuit realizations, and they are essential in monolithic circuit design.

GaAs logic circuits have long attracted attention as candidates for very high speed digital integrated circuit (VHSIC) applications. Indeed, the very first GaAs integrated circuits demonstrated were logic circuits. A certain "credibility" gap has persisted, however, with regard to these circuits, and it is only now that GaAs digital integrated circuits are being designed into the next generation of supercomputers. This subject receives the comprehensive treatment it deserves in this text.

The versatility of the GaAs MESFET and its widespread use in microwave equipment have aroused concern as to its reliability. The subject has been extensively studied over the last 10 years and this has led to substantial improvements in fabrication technology. Only a few "gray areas" remain, one of which is surface-induced degradation, and this is treated in Chapter 10.

This is a book covering a wide range of practical problems in a truly original and international collaboration. With this book under one arm and a computer under the other, an enterprising engineer could turn into a giant.

ALLEN PODELL
Co-founder and Vice-President of
Pacific Monolithics Inc.

Acknowledgments

We would like to thank the directors of the Centre National d'Etudes des Télécommunications (CNET) for permission to prepare this book. Our thanks also to the many people who have provided material for the book and support in many ways during its production. Many of the photographs in the book were taken by Thierry Scart for Thomson-CSF. We are especially grateful for the care with which Christine Sharrock of Omega Scientific shepherded the book through from manuscript to typeset pages. Finally, people who have contributed to specific chapters are acknowledged in those chapters.

Chapter 1

Fabrication and Operation of GaAs MESFETs

A. CAPPY

1.1 Fabrication of GaAs MESFETs

Some fundamental concepts and definitions relevant to the fabrication of GaAs MESFETs are reviewed in this section. Various methods of obtaining the active layer are described, details of the specific process steps are given and finally the two main techniques used for GaAs MESFET fabrication are discussed.

1.1.1 Material requirements for MESFET fabrication

A schematic diagram of the active layer of a typical low noise MESFET is shown in Fig. 1.1. First a GaAs buffer layer is grown onto the GaAs semi-insulating (SI) substrate. The growth of this very pure layer before deposition of the active layer itself improves electron mobility near the interface between the active layer and the substrate by reducing the defect density [1.1]. It should be noted that such a layer is not always present, as for example in ion-implanted layers.

After the active layer has been grown, a highly doped top layer is usually introduced in order to reduce the value of access resistances by lowering both the contact resistances and the resistances of the source-to-gate and gate-to-drain regions. This layer is removed from under the gate by a recess process. The most frequently used methods for achieving such a layer are vapor phase epitaxy, molecular beam epitaxy and ion implantation. Basically, epitaxy consists of growing GaAs material on an SI GaAs substrate and ion implantation refers to the implantation of dopant atoms directly into the substrate.

1.1.1.1 Vapor phase epitaxy

Various vapor phase epitaxy techniques have been developed and more exhaustive descriptions can be found in the literature [1.2, 1.3]. The epitaxial process begins with an SI substrate which is a single crystal of GaAs obtained

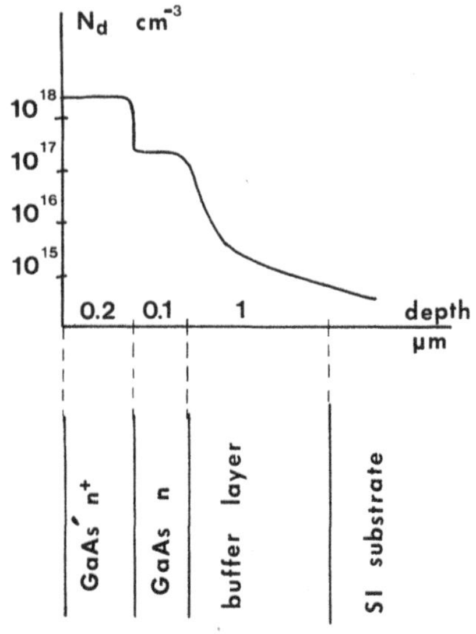

Fig. 1.1 The active layer structure of a low noise MESFET.

using either the Bridgmann process [1.4] or a liquid-encapsulated Czochralski method [1.5, 1.6].

Vapor phase epitaxy with trichloride transport The slice is placed in a quartz slice holder in the reactor tube. At a temperature close to 800 °C, growth is obtained by the reaction of $AsCl_3$ with gallium in the presence of hydrogen to produce GaCl, As_4 and HCl. The main chemical equilibrium is given by [1.2]

$$4GaCl + As_4 + 2H_2 \rightleftharpoons 4GaAs + 4HCl \tag{1.1}$$

Although this method is the main epitaxial technique for the preparation of GaAs device layers, its limitations in terms of uniformity and profile control have led to the development of an alternative vapor system using an organometallic compound.

Metallorganic vapor phase epitaxy In this method trimethylgallium ($Ga(CH_3)_3$) is used as the gallium source and arsine (AsH_3) is used as the arsenic source (Fig. 1.2). The basic reaction, which takes place on a heated substrate, is [1.7]

$$Ga(CH_3)_3 + AsH_3 \rightleftharpoons GaAs + 3CH_4 \tag{1.2}$$

The main advantages of the method are the ability to produce complex heterostructures and the good control of both layer thickness and doping. The method seems to be one of the most promising for achieving high performance MESFETs and two-dimensional electron gas field-effect transistors (TEGFETs).

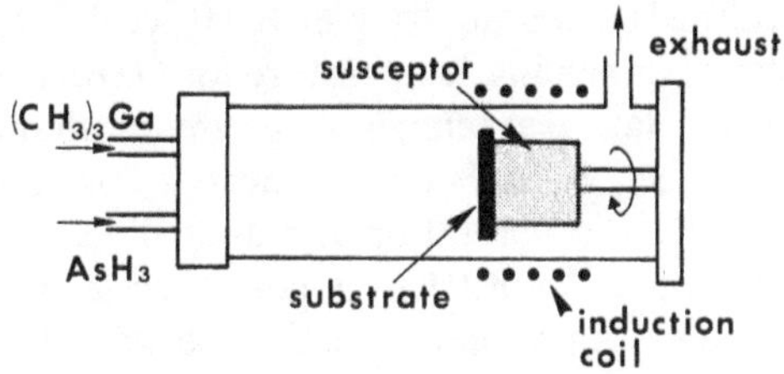

Fig. 1.2 Sketch of a reactor for the epitaxial growth of GaAs by the organometallic process.

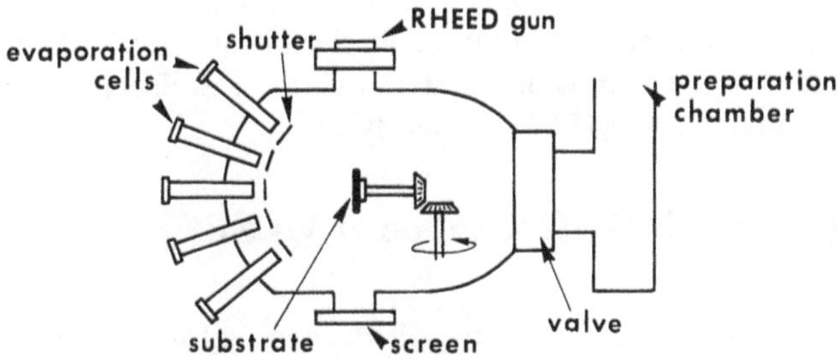

Fig. 1.3 Schematic diagram of a molecular beam epitaxy system (RHEED, reflection high energy electron diffraction).

1.1.1.2 Molecular beam epitaxy

Molecular beam epitaxy is the most recent method developed for epitaxial growth. The substrate is heated in an ultrahigh vacuum system and is rotated to improve uniformity. The elemental species (gallium, arsenic, aluminum, silicon etc.) are placed in effusion cells and evaporated. Each beam is controlled using a shutter placed in front of the effusion cell (Fig. 1.3).

This technique has a number of advantages, in particular a low growth rate that allows the realization of complex heterostructures such as superlattices [1.8]. Despite the high cost of the equipment, molecular beam epitaxy is likely to play an important role in the future development of GaAs integrated circuits (ICs) and discrete devices.

1.1.1.3 Ion implantation

Ion implantation is widely used in the fabrication of both discrete MESFETs and GaAs ICs [1.9]. In this method, dopant atoms are accelerated and introduced into the near-surface regions of the GaAs substrate. The shape of the doping profile can be accurately controlled by the dose (typically 10^{12}–10^{14} atoms cm^{-2}) and the ion energy (typically 100–400 keV for GaAs). Moreover the profile can be kept very uniform over the slice. After implantation, annealing is necessary to remove lattice damage and to activate the dopant. The typical annealing temperature for GaAs is 800–900 °C. Since GaAs is subject to dissociation at temperatures greater than 600 °C, the GaAs surface must be protected by either an encapsulating layer (Si_3N_4) or an arsenic overpressure. After implantation and annealing, the concentration of impurities as a function of depth has a near-Gaussian distribution. This particular profile involves a gradual transition between the active layer and the substrate.

In addition to the precise control of the doping profile, ion implantation has another significant advantage over epitaxial techniques—selective implantation. In this case ions are implanted into localized areas in order to define the active region of devices or to reduce the access resistances [1.10, 1.11]. Finally it should be noted that ion implantation is a very fast technique (a few seconds for a commercial implanter) and thus is well suited for the large-scale production of GaAs MESFETs and ICs.

1.1.2 Basic steps in the fabrication of MESFETs

The fabrication of a field-effect transistor requires four major steps [1.12]: (i) isolation; (ii) ohmic contacts; (iii) Schottky contact; (iv) passivation and metallization. Each step involves localized areas of the slice which are defined by conventional photolithography or electron beam lithography [1.12]. As shown in Fig. 1.4 conventional photolithography consists of three steps: deposition of photoresist; exposure; development. In the case of electron beam lithography, energetic electrons are used instead of light and the writing is performed directly (without any masks). The resolution achievable with contact photolithography is approximately 1 μm, whereas resolutions of better than 1000 Å can be obtained with electron beam lithography.

The most frequently used technique for metallization, which is performed after development of the photoresist, is the lift-off process (Fig. 1.5) in which

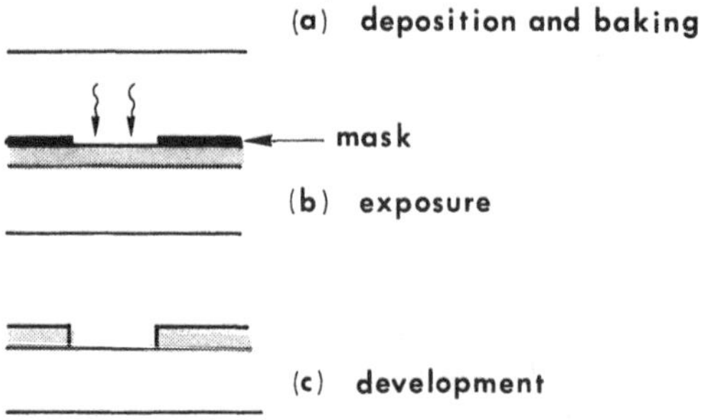

Fig. 1.4 Steps in the conventional photolithography technique.

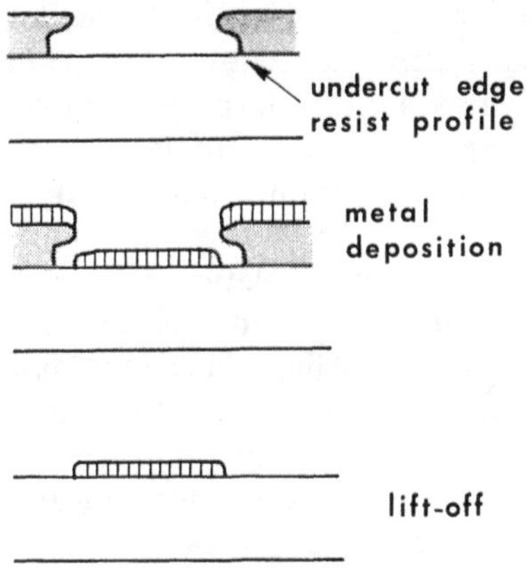

Fig. 1.5 The lift-off process.

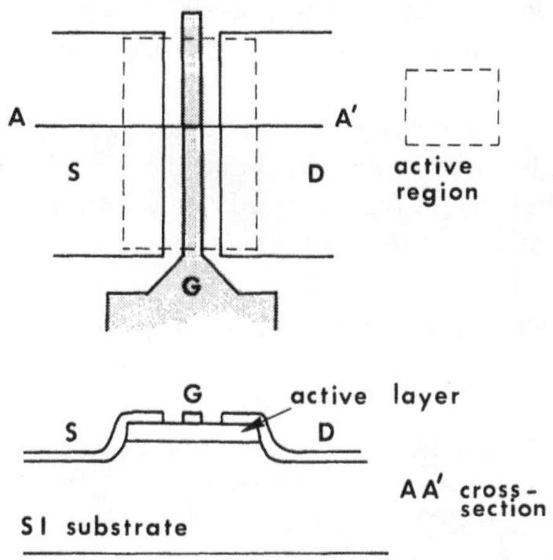

Fig. 1.6 Sketch of a device using mesa isolation.

metal is deposited onto the slice by evaporation or sputtering, the resist is dissolved using a solvent and finally the metal deposited on top of the resist is "lifted off". The main difficulty with this technique is the provision of an undercut resist edge profile as shown in the figure.

1.1.2.1 Isolation

The purpose of the isolation step is to restrict the electrically conductive portion of the slice to the active region of each device. Therefore contact pads are deposited on the insulating material (Fig. 1.6).

The most popular technique for discrete devices is mesa isolation

(Fig. 1.6). In this technique the active region of the device is protected by a photoresist and the slice is etched to remove material in the unmasked areas. Wet etching is generally used for this purpose. The main disadvantage of the technique is the nonplanarity of the slice after mesa formation, which results in difficulties in resist deposition, contact photolithography and metal step coverage.

Another method of achieving isolation is to use ion implantation or selective implantation. In the first method ions are implanted in the nonactive region of the device without annealing. Since ion implantation damages the lattice, the implanted region becomes insulating. In selective implantation, which is often used for ICs, the active layer is formed by implanting ions into the active region of each device and then performing an activating anneal. The main advantage of the two techniques is the conservation of planarity.

1.1.2.2 Ohmic contacts

The source and drain contacts of a GaAs MESFET should have linear I–V characteristics with as low a contact resistance as possible. Measures of device performance such as the noise figure are very sensitive to the value of the contact resistance.

The method most frequently used to achieve good ohmic contacts is to deposit an AuGeNi layer by evaporation and then to alloy this layer with the GaAs [1.13, 1.14]. In fact, the formation of ohmic contacts on GaAs is a very complex process that is not yet completely understood. For planar ohmic contacts in MESFETs the transition between the bulk material and the contact is characterized by a contact resistance R_c that is commonly less than 0.1 Ω mm.

1.1.2.3 Schottky barrier

Owing to its small dimensions, the formation of the Schottky barrier is the critical step in MESFET fabrication. The most common method of realizing Schottky barriers on GaAs is to deposit aluminum or TiPtAu in the desired pattern. This results in I–V characteristics of the type shown in Fig. 1.7. Under forward-biasing conditions the I–V characteristic can be expressed as

$$I = A^* T^2 S \exp\left(\frac{qV_b}{kT}\right) \exp\left(\frac{qV_{GS}}{nkT}\right) \tag{1.3}$$

where A^* is the Richardson constant, T is the temperature, S is the area, V_b is the barrier height, n is the ideality factor and V_{GS} is the gate bias voltage. Usually V_b is close to 0.8 and n is close to 1.1. Some methods of determining these parameters will be given later.

The gate is also characterized by the metallization resistance which is a limiting parameter of the device performance.

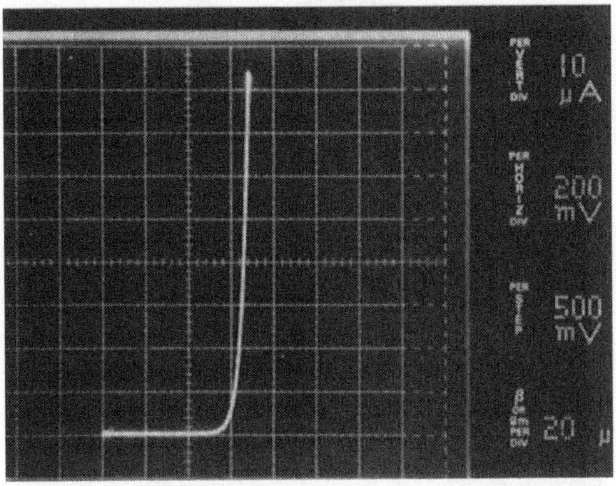

Fig. 1.7 The *I–V* characteristic of a Schottky barrier.

1.1.2.4 Dielectric formation: interconnections

The final steps in MESFET fabrication are dielectric formation and interconnection.

The purpose of the dielectric layer is to protect the surface from chemical or mechanical damage. These layers are generally formed by plasma-assisted deposition or sputtering [1.12]. The dielectric most frequently used on GaAs is silicon nitride (Si_3N_4). In the case of GaAs ICs dielectric layers are also used as isolation between two levels of metallization [1.12].

The interconnection problem is important in both ICs and power MESFETs when it is required to increase the total gate width. Two main techniques are used to combine the elementary cells in parallel: air bridges or dielectric crossovers and via holes (Fig. 1.8). These techniques are characterized by a reduction of electrical parasitics, particularly the source inductance. Air bridges have several advantages over dielectric crossovers, in particular a lower parasitic capacitance. These bridges are fabricated by depositing plated gold onto a thick resist layer that is removed after plating. Via holes are fabricated by etching a previously thinned GaAs slice. The back side of the slice is then metallized and gold is plated on to increase conductivity [1.12].

1.1.3 Techniques for fabricating MESFETs

There are many ways of fabricating MESFETs and the process is always adapted to the application for which they are intended, e.g. low noise transistors, power transistors, digital ICs and monolithic microwave ICs (MMICs). Nevertheless the various methods are derived from two basic processes—the lift-off technique and the self-aligned technique [1.12].

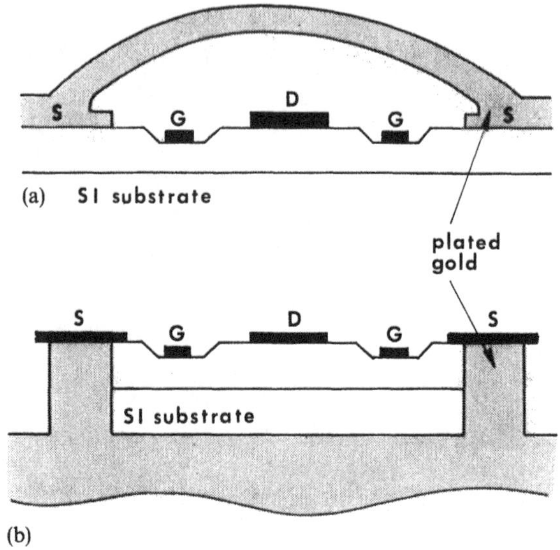

Fig. 1.8 Interconnections using (a) air bridges and (b) via holes.

1.1.3.1 The lift-off technique

The lift-off technique is the most popular for discrete low noise devices. The steps in the process, which are shown in Fig. 1.9, are as follows: (a) isolation by mesa etching; (b) formation of source and drain ohmic contacts; (c) etching of the gate recess and formation of the Schottky barrier; (d) dielectric deposition; (e) deposition of gold on the contact pads. Steps (b), (c) and (e) are performed using the lift-off technique described earlier.

1.1.3.2 The self-aligned technique

The self-aligned method is frequently used for both digital and analog ICs with either epitaxial or implanted active layers. The steps in the process, which are shown in Fig. 1.10, are as follows: (a) formation of the epitaxial layer; (b) deposition of the refractory gate; (c) gate definition; (d) n^+ implantation in access regions; (e) Si_3N_4 deposition and anneal; (f) metallization and alloying of ohmic contacts. This technology requires a gate metal exhibiting stable Schottky characteristics during the high temperature anneal. A TiW Schottky gate that stays unalloyed up to at least 860 °C can be used.

It should be pointed out that the process steps listed above represent only a general description of the method. Several other methods of realizing MESFETs are available, and the interested reader should refer to the literature (e.g. refs 1.2 and 1.12).

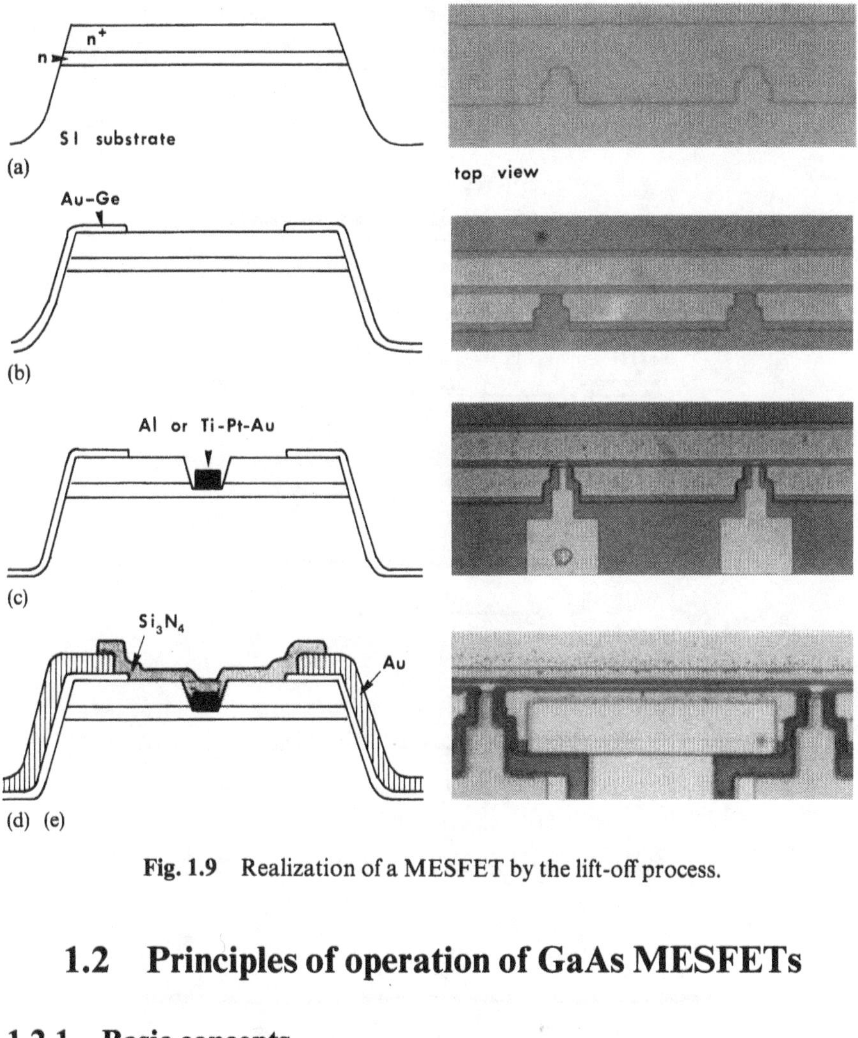

Fig. 1.9 Realization of a MESFET by the lift-off process.

1.2 Principles of operation of GaAs MESFETs

1.2.1 Basic concepts

The cross-section of an idealized planar MESFET is presented in Fig. 1.11. The physical behavior of the device can be described in terms of three regions: (1) the source and drain access regions; (2) the active zone under the gate; (3) the buffer layer or SI substrate.

The electrical field in the source and drain access regions (1) remains low and these two regions behave like perfect resistors. The equivalent resistance is the sum of the source (drain) contact resistance and the resistance of the active layer between the active zone and the source (drain) contact. These two regions are characterized from an electrical point of view by two resistances R_s and R_d.

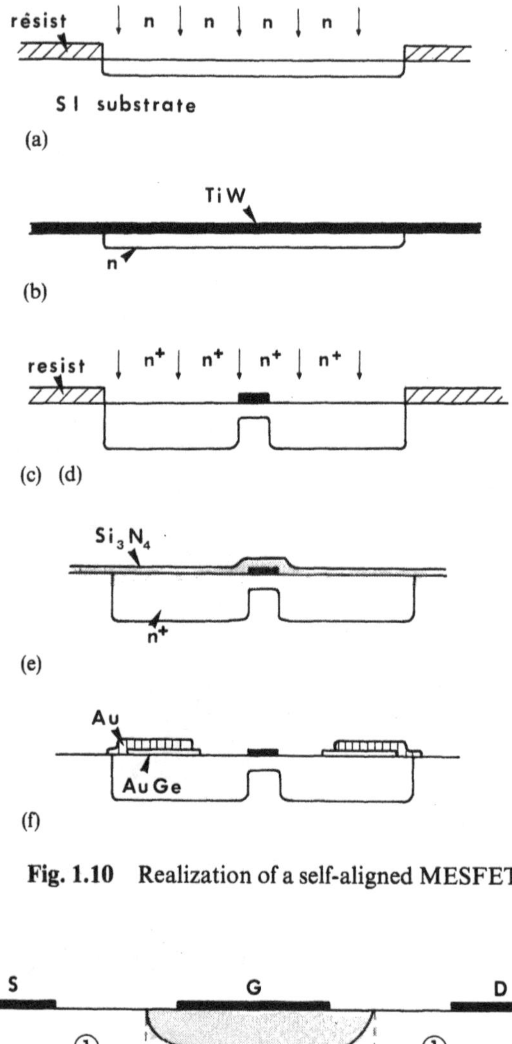

Fig. 1.10 Realization of a self-aligned MESFET.

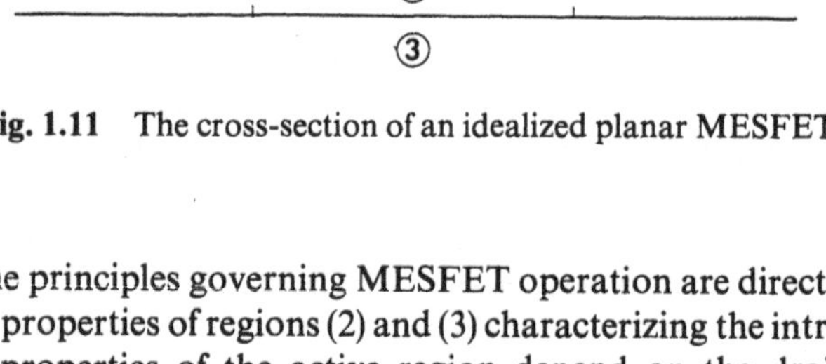

Fig. 1.11 The cross-section of an idealized planar MESFET.

Therefore the principles governing MESFET operation are directly related to the physical properties of regions (2) and (3) characterizing the intrinsic device. In fact the properties of the active region depend on the drain-to-source voltage V_{DS}.

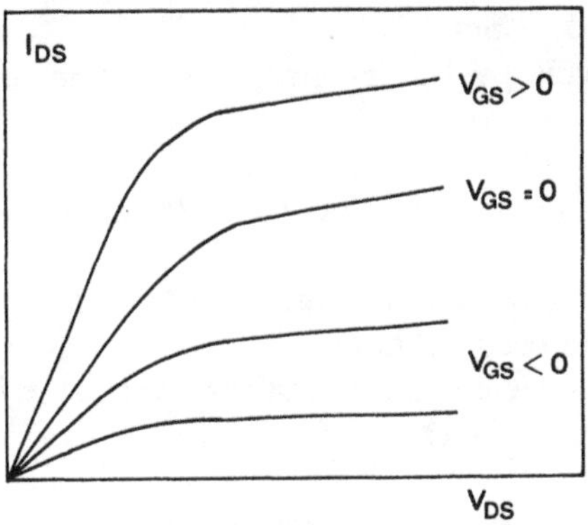

Fig. 1.12 The $I_{DS}(V_{GS}, V_{DS})$ characteristics of a MESFET.

1.2.1.1 Low V_{DS} range

The Schottky contact is usually reverse biased and a depletion region is set up beneath the gate. The thickness y_d of the depletion region increases with the reverse gate voltage. If an active layer with a uniform doping level N_d is assumed, the depletion thickness y_d is given by Shockley's law

$$y_d(V_{GS}) = \left\{\frac{2\varepsilon(V_b - V_{GS})}{qN_d}\right\}^{1/2} \tag{1.4}$$

where ε is the permittivity of GaAs, V_b is the built-in potential and q is the electronic charge.

At low V_{DS} the depletion thickness is roughly constant under the gate and the active zone behaves like a resistor that is controlled by the gate voltage. If the edge effects on each side of the gate are neglected, the equivalent resistance is given by

$$R_c(V_{GS}) = \frac{1}{q\mu_0 N_d} \frac{L_g}{\{A - y_d(V_{GS})\}z} \tag{1.5}$$

where μ_0 is the carrier mobility, A is the thickness of the active layer and z is the width of the device. The drain current between source and drain is then given by (Fig. 1.12)

$$I_{DS} = \frac{V_{DS}}{R_{DS}(V_{GS})}$$

where (1.6)

$$R_{DS}(V_{GS}) = R_s + R_d + R_c(V_{GS})$$

It should be pointed out that in this case there are no carriers in the buffer layer or substrate under the gate (region (3)) and thus it has no major influence on the drain current.

1.2.1.2 High V_{DS} range

As the drain voltage V_{DS} increases three important phenomena are observed.

(i) The drain voltage modifies the shape of the depletion region under the gate.

(ii) The electron velocity is no longer proportional to the electric field and the carriers are said to become "hot".

(iii) Carriers are injected into the buffer layer and yield a parasitic current flowing outside the active layer.

These effects will be described below in more detail.

Two-dimensional effects The two-dimensional effects in MESFETs are due to the planar position of the different electrodes. The drain-to-source voltage tends to create an electric field parallel to the surface (x direction) whereas the gate-to-source voltage induces an electric field perpendicular to the surface (y direction), except in the two quasi-elliptical depleted regions located on each side of the gate (Fig. 1.13).

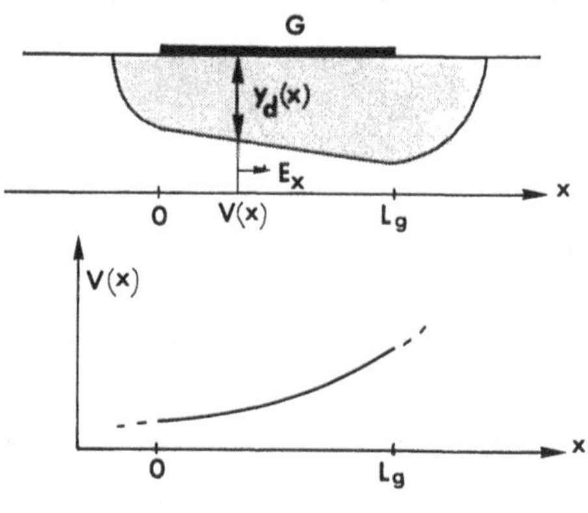

Fig. 1.13 Shape of the depleted layer under the gate for $V_{DS} > 0$.

To a first approximation the active layer under the gate can be divided into two regions (Shockley's gradual-channel approximation) [1.15]: the depleted layer under the gate where E_x can be neglected; the active channel where E_y can be neglected.

Since the electrostatic potential $V(x)$ in the channel increases from the source to the drain, the Schottky barrier becomes more reverse biased as the drain end of the gate is approached and hence the thickness of the conductive channel decreases (Fig. 1.13). Therefore the active channel tends to be "pinched off" at the drain end of the gate. More accurately, the height of the active channel can be expressed as

$$y_c(x) = A - y_d(x) = A - \left[\frac{2\varepsilon\{V_b - V_{GS} + V(x)\}}{qN_d}\right]^{1/2} \tag{1.7}$$

where A is the thickness of the active layer and $V(x)$ is the local electrical potential. This expression shows that, for a sufficiently high drain voltage, the active channel can be completely pinched off. In fact the gradual-channel approximation is now no longer valid for three main reasons: (i) the term $\partial E_x/\partial x$ cannot be neglected in Poisson's equation for the calculation of the channel height; (ii) the two-dimensional fringing effects near the drain end of the gate must be taken into consideration; (iii) carriers are injected into the buffer layer. Consequently the height of the active channel remains finite in practice and the drain current tends to saturate as a function of the drain voltage as shown in Fig. 1.12.

Electron dynamics effects One of the major physical parameters of semiconducting material is the relation between the applied electric field (assumed to be uniform in space and time) and electron velocity. This relation for GaAs is presented in Fig. 1.14. The curve is characterized by three important regions: at electric field strengths below $4\,\mathrm{kV\,cm^{-1}}$ the electron velocity increases linearly with the field ($v = \mu_0 E$); between 5 and $15\,\mathrm{kV\,cm^{-1}}$ the v–E relation shows a negative differential mobility effect which corresponds to electron transfer from a high mobility central valley to a low mobility satellite valley [1.16, 1.17]; above $15\,\mathrm{kV\,cm^{-1}}$ the velocity becomes almost constant as a result of equilibrium between increasing Coulomb force strength qE and an increasing rate of electron scattering from the lattice.

It must be emphasized that the v–E relation shown in Fig. 1.14 corresponds to the case of an electric field which is uniform in space and time.

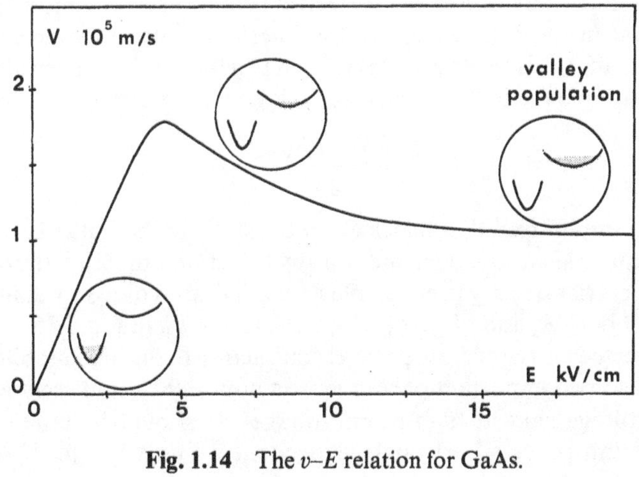

Fig. 1.14 The v–E relation for GaAs.

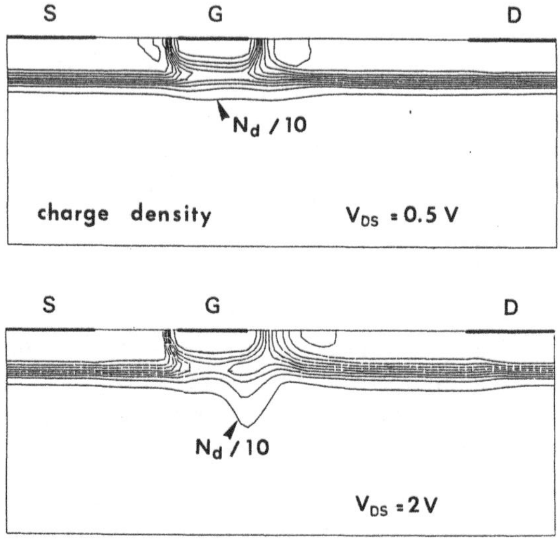

Fig. 1.15 Result of two-dimensional modelling showing the influence of V_{DS} on carrier injection into the buffer layer.

The electric field in the channel of a MESFET with a submicrometer gate length presents strong spatial variations, and consequently the carrier velocity is no longer simply related to the field by the v–E relation shown in Fig. 1.14. These nonstationary effects have to be considered in the physical behavior of transistors with very short gate lengths [1.18].

The buffer layer As indicated earlier, the longitudinal electric field increases strongly in the channel at high drain voltage. Because of the continuity of the electrostatic potential at the active layer–buffer layer interface an increasing longitudinal electric field is also present in the buffer layer, at least in the region close to the interface. Since Poisson's equation holds in the buffer, this electric field is related to a local electron density n. If the impurity density in the buffer is neglected, n is given to a first approximation by

$$n = \frac{\varepsilon}{q}\frac{\partial E_x}{\partial x} \tag{1.8}$$

It should be emphasized that the electron density n in the buffer is controlled by the longitudinal electric field and thus by the drain potential. In particular, the injected carrier density is negligible at small drain voltages because of the low values of both E_x and $\partial E_x/\partial x$ [1.19, 1.20] as shown in Fig. 1.15.

The injected carriers create parallel conduction in the high mobility buffer layer and yield two important phenomena: a non-zero output conductance; a threshold voltage shift at high drain voltages. It should be noted that the carrier injection process is strongly dependent on gate length. If, as a first

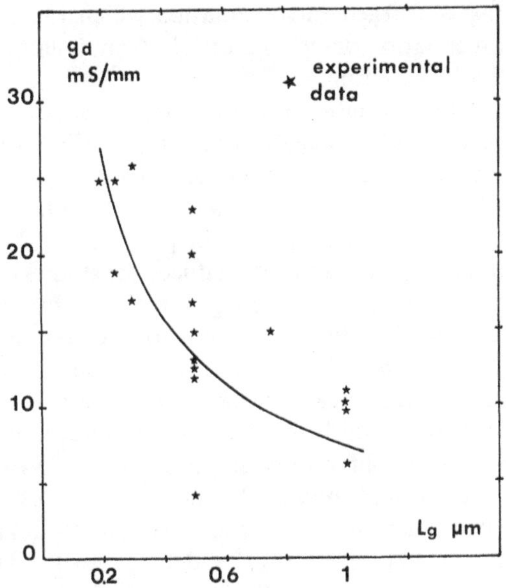

Fig. 1.16 Experimental relation between the output conductance and the gate length: ——, eqn (1.12).

approximation, a constant electric field slope under the gate is assumed, the number of injected electrons becomes proportional to V_{DS}/L_g^2. This indicates that the output conductance increases with decreasing gate length as shown in Fig. 1.16.

In summary, the drain current characteristic $I_{DS}(V_{DS})$ for a given gate voltage can be divided into three regions (Fig. 1.12): the low V_{DS} range where the MESFET behaves like a resistor; the intermediate region where the drain current tends to saturate; the high voltage range or saturation region where the current increases slowly with V_{DS}.

1.2.2 Physical model of a MESFET

There are two approaches to the calculation of the $I_{DS}(V_{GS}, V_{DS})$ characteristics of a MESFET: analytical or numerical one-dimensional modelling; two-dimensional modelling.

1.2.2.1 One-dimensional models

One-dimensional models are derived from Shockley's original theory [1.15]. As indicated above, the device is divided into a fully depleted layer and a conductive channel. These models are based on the self-consistent resolution of three equations: a charge control equation which yields the height of the depleted layer; an equation describing the electron dynamics in the channel; the current continuity equation.

The charge control equation generally used is Shockley's law (eqn (1.7)) for the region located under the gate. The fringing effects on each side of the gate are neglected or are taken into account by using circular or elliptical depleted zones [1.18]. The electron dynamics in the channel are frequently described using an idealized v–E relation [1.21, 1.22]. Nevertheless, nonstationary electron dynamics can be introduced for studies of MESFETs with submicrometer gate lengths by using macroscopic electron transport equations deduced from localized Boltzmann transport equations [1.18, 1.23].

The main advantage of these models is their simplicity. Analytic expressions or simple numerical methods can be used to obtain the steady state characteristics, particularly for the computer-aided design of GaAs ICs. Nevertheless various simplifying assumptions and approximations are required for one-dimensional models. In most cases nonstationary electron dynamics or carrier injection in the buffer layer are not taken into consideration, and this results in difficulties in modelling MESFETs with submicrometer gate lengths and in predicting the output conductance value.

1.2.2.2 Two-dimensional models

A more rigorous approach is to use two-dimensional models which are based on a self-consistent resolution of (i) the two-dimensional Poisson equation, (ii) a two-dimensional electron transport equation and (iii) the continuity of the current density. Either finite-difference or finite-element formulations can be used to solve these equations. Electron transport in devices with long gate lengths is frequently described using the conventional steady state v–E relation, whereas the Monte Carlo or macroscopic transport equation methods (hydrodynamic models) [1.24–1.27] are used for MESFETs with submicrometer gate lengths.

Since these two-dimensional models constitute a more rigorous approach, their contribution to our physical understanding of the device is considerable. The principal effects investigated include the influence of gate fringing effects and/or the shape of the recess [1.28], the influence of the substrate current [1.25], the influence of a large transverse field in the channel which can affect the electron transport [1.19] and the influence of the overshoot effect [1.29]. However, two-dimensional models require a powerful computer and long computation times. Therefore they are not suitable for systematic studies or device optimization but are intended more for physical analysis of phenomena occurring under steady state operating conditions. In addition, both two-dimensional and one-dimensional models can provide very useful information concerning the values of the various elements of the small-signal equivalent circuit as a function of device technological parameters and biasing conditions.

1.2.3 The small-signal equivalent circuit

From a practical point of view a knowledge of the high frequency equivalent circuit is of primary importance. Figure 1.17 shows the equivalent circuit generally used and the location of the various elements in the device. The circuit can be divided into two parts: the extrinsic parameters and the intrinsic parameters. The intrinsic parameters characterize the active region under the gate and are functions of biasing conditions, whereas the extrinsic parameters depend, at least to a first approximation, only on the technological parameters.

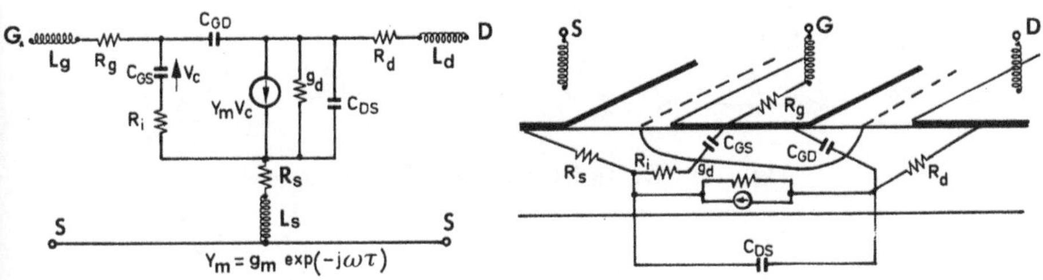

Fig. 1.17 MESFET equivalent circuit and origin of the various elements in the structure.

1.2.3.1 Extrinsic parameters

The most important extrinsic parameters are the gate, source and drain inductances due to the bond wires and the gate, source and drain resistances. As shown previously, the source and drain access resistances can be calculated from the active-layer parameters (contact resistance, doping, thickness, n^+ top layer etc.).

The series resistance of the gate is due to the voltage drop associated with the flow of gate current in the gate metallization. A gate finger can be considered as a distributed *RC* network (Fig. 1.18). It can easily be shown that the equivalent lumped gate resistance of one finger of width z is simply $R_m z/3$, where R_m is the linear resistance of the gate metallization (in $\Omega\,\text{mm}^{-1}$). Therefore for a device of total width z consisting of N fingers the total gate resistance is given by [1.30]

$$R_g = \frac{1}{N^2}\frac{R_m z}{3} \tag{1.9}$$

The gate resistance can be reduced by connecting the gate fingers in parallel. Two structures are generally used for this purpose: the interdigitated gate MESFET and the π gate MESFET (Fig. 1.19).

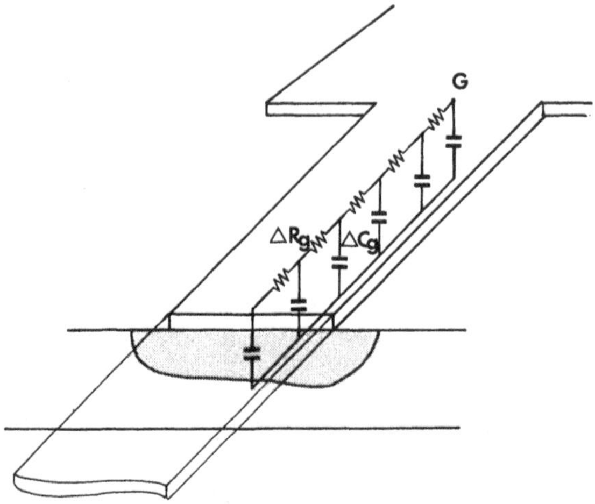

Fig. 1.18 Equivalent network of the *RC* gate.

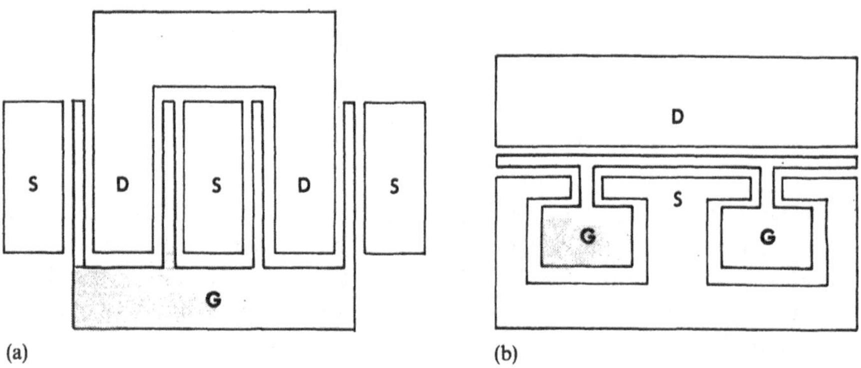

Fig. 1.19 Pattern of (a) interdigitated and (b) π gate MESFETs.

Finally, two other parameters can be considered as extrinsic elements: the capacitance C_{PG} of the gate pad and the drain-to-source capacitance C_{DS} which are both virtually independent of biasing conditions, at least in the case of low noise devices.

1.2.3.2 Intrinsic parameters

The four main intrinsic parameters are the input capacitance C_{GS}, the transconductance g_m, the output conductance g_d and the feedback capacitance C_{GD}. These parameters are simply related to variations in drain current and stored charge in the active region of the device as a function of the gate and drain voltages. In addition, the electrical behavior of the intrinsic device requires the introduction of two more parameters: the intrinsic resistance R_i which can be related to the distributed nature of the RC input network [1.30], and the delay τ, which is introduced in the expression for the current generator and corresponds to the time needed for the carriers to travel under the gate [1.30].

The intrinsic small-signal parameters are a function of the biasing conditions and device technological parameters. The experimental variation in these parameters is shown in Fig. 1.20 as a function of V_{GS} for given V_{DS}. However, the relation between these elements and the technological parameters is not easy to obtain from measurements and therefore is deduced from theoretical work. The following set of analytic expressions, deduced from a numerical model [1.20] with a drain current of 100 mA mm^{-1} and a drain voltage of 3 V, gives results in good agreement with experimental findings:

$$C_{GS} = \frac{\varepsilon z(L_g + 2A)}{A} + C_0 z \quad \text{with} \quad C_0 = 300\,\text{fF mm}^{-1} \tag{1.10}$$

$$g_m = \frac{\varepsilon z v_0}{A} \quad \text{with} \quad v_0 = 1.9 \times 10^5\,\text{m s}^{-1} \tag{1.11}$$

$$g_d = \beta \frac{zA N_d^{1/2}}{L_g + A} \quad \text{with} \quad \beta = 1.8 \times 10^{10}\,\text{USI} \tag{1.12}$$

$$C_{GD} = \gamma z L_g N_d^{1/2} + C_1 z \quad \text{with} \quad \gamma = 3.5 \times 10^{-16}\,\text{USI} \quad \text{and} \quad C_1 = 30\,\text{fF mm}^{-1} \tag{1.13}$$

1.2.4 Typical operating frequencies

Except for the noise performance, which is discussed in the next section, the microwave performance of MESFETs can be related to three typical frequencies: the cut-off frequency f_c of the current gain, the cut-off frequency f_m of the maximum available gain and the maximum frequency f_u of oscillation. These frequencies are related to the small-signal parameters by the following expressions [1.20]:

$$f_c = \frac{g_m}{2\pi C_{GS}} \tag{1.14}$$

$$\begin{aligned} f_m &= f_c \left\{ 4g_d\left(R_0 + \frac{\omega_c L_s}{2}\right) + 2\omega_c C_{GD}(R_0 + R_g + \omega_c L_s) \right\}^{-1/2} \\ &= \frac{f_c}{2g_d^{1/2}} \left\{ \left(R_0 + \frac{\omega_c L_s}{2}\right) + \frac{f_c}{2f_0}(R_0 + R_g + \omega_c L_s) \right\}^{-1/2} \end{aligned} \tag{1.15}$$

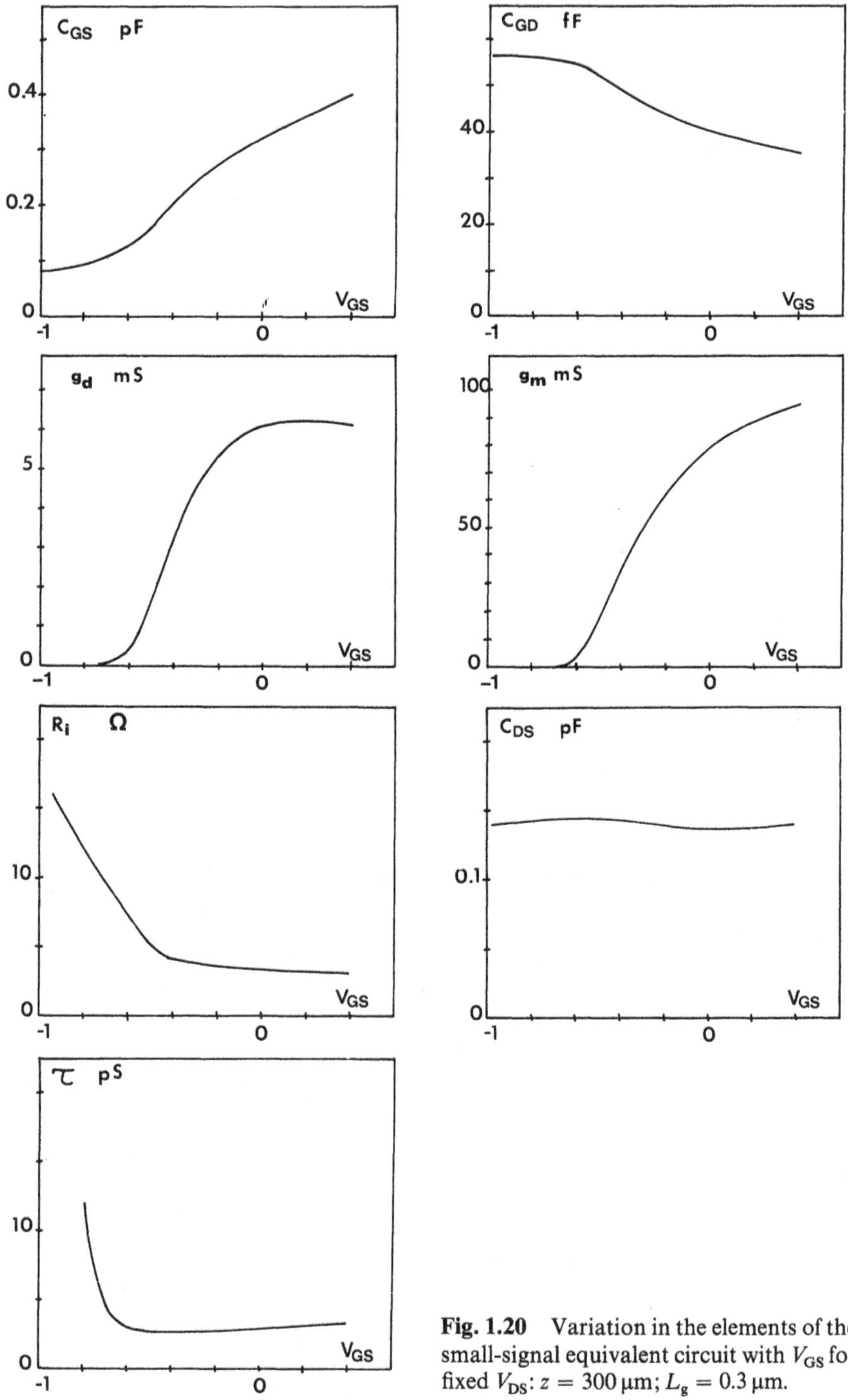

Fig. 1.20 Variation in the elements of the small-signal equivalent circuit with V_{GS} for fixed V_{DS}: $z = 300\,\mu m$; $L_g = 0.3\,\mu m$.

$$f_u = \frac{f_c}{2}\left(R_0 g_d + \omega_c R_g C_{GD}\right)^{-1/2}$$
$$= \frac{f_c}{2 g_d{}^{1/2}}\left(R_0 + R_g \frac{f_c}{f_0}\right)^{-1/2} \qquad (1.16)$$

where $R_0 = R_i + R_g + R_s$, $\omega_c = 2\pi f_c$ and $f_0 = g_d/2\pi C_{GD}$. In order to maximize both f_m and f_u, the following conditions must be satisfied: high f_c and small g_d yielding high $f_c/g_d{}^{1/2}$; low parasitic R_s, R_g and L_s; high $f_0 = g_d/2\pi C_{GD}$ which implies a low feedback capacitance C_{GD}. These requirements can be achieved by reducing the gate length and epilayer thickness whilst increasing the doping level in order to maintain a near-constant pinch-off voltage.

1.2.5 The noise performance of MESFETs

1.2.5.1 Noise sources

A knowledge of the small-signal equivalent circuit is insufficient for low noise amplifier design and it must be supplemented by the incorporation of noise sources [1.31] (Fig. 1.21). Except for the case of the spectral purity of oscillators, which is determined by low frequency noise sources (1/*f* generation–recombination), the noise in the microwave frequency range is due mainly to fluctuations in the carrier velocity. At low electric fields these velocity fluctuations give rise to "thermal noise", whereas at higher fields they result in "diffusion noise" or "hot electron noise".

The analysis of high frequency noise in MESFETs is usually performed in three stages [1.31, 1.32].

(i) First, the noisy device is separated into two parts: the access resistances and the intrinsic MESFET. The noise associated with the access resistances is characterized by thermal noise sources that can be expressed according to the Nyquist formula

$$\overline{e^2} = 4kTR\Delta f \qquad (1.17)$$

where k is the Boltzmann constant, T is the temperature, R is the resistance and Δf is the frequency bandwidth. The noise associated with the intrinsic MESFET is described using two correlated noise sources which represent the effects of the local noise sources distributed along the channel on the drain and gate electrodes. Since they have the same physical origin, the gate and drain noise sources are correlated.

(ii) Two correlated equivalent noise sources preceding the noiseless extrinsic MESFET are then defined using a simple circuit calculation [1.31] (Fig. 1.21).

(iii) These two noise sources are decorrelated by introducing two "correlation admittances" or "correlation impedances" (Fig. 1.21).

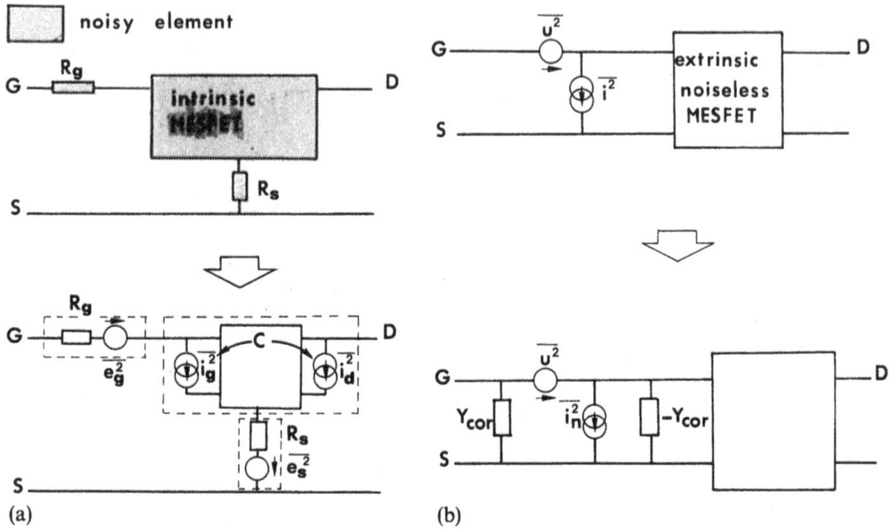

Fig. 1.21 Equivalent circuit of a noisy MESFET and noise source transformation for calculating the noise figure.

At this point the calculation of the minimum noise figure F_{min} is straightforward [1.31]:

$$F_{min} = 1 + 2R_n(G_{opt} + G_{cor}) \tag{1.18}$$

where

$$Y_{cor} = G_{cor} + jB_{cor}$$

and

$$Y_{opt} = G_{opt} + jB_{opt} = \left(G_{cor}{}^2 + \frac{G_n}{R_n}\right)^{1/2} - jB_{cor}$$

This indicates that the main problem in MESFET noise analysis is determination of the gate and drain noise sources of the *intrinsic* device. Following Van der Ziel [1.32], these two noise sources can be written

$$\overline{i_d{}^2} = 4kTg_m P\Delta f \tag{1.19}$$

$$\overline{i_g{}^2} = 4kT\frac{\omega^2 C_{GS}{}^2}{g_m} R\Delta f \tag{1.20}$$

where P and R are two dimensionless coefficients which depend on the biasing conditions and the technological parameters of the device.

In summary, calculation of the noise figure requires a knowledge of the small-signal equivalent circuit and three noise parameters P, R and the correlation coefficient C between the gate and drain noise sources [1.20, 1.31].

1.2.5.2 The minimum noise figure

At sufficiently low operating frequencies, the feedback capacitance C_{GD} can be neglected. The minimum noise figure can then be calculated analytically as follows [1.22]:

$$F_{min}=1+2\{R+P-2C(RP)^{1/2}\}^{1/2}\frac{f}{f_c}\left\{g_m(R_s+R_g)+\frac{RP(1-C^2)}{R+P-2C(RP)^{1/2}}\right\}^{1/2} \quad (1.21)$$

This expression demonstrates two important effects. First, owing to the correlation between the gate and drain noise sources, the drain noise is partially compensated by the gate noise as shown by the expressions $R+P-2C(RP)^{1/2}$ and $RP(1-C^2)$ that are decreasing functions of C for given R and P. Second, the noise figure is due partly to the access resistances and partly to the intrinsic device noise. If the intrinsic noise is neglected, the expression for F_{min} reduces to the well-known Fukui formula [1.34]

$$F_{min} = 1+k_F\frac{f}{f_c}\{g_m(R_s+R_g)\}^{1/2} \quad (1.22)$$

It should be emphasized that for a typical device with a gate length of 0.5 μm ($z = 300$ μm, $g_m = 40$ mS, $R_s+R_g = 4\,\Omega$, $P = 1$, $R = 0.5$ and $C = 0.8$) we have

$$g_m(R_s+R_g) = 0.16$$

$$\frac{RP(1-C^2)}{R+P-2C(RP)^{1/2}} = 0.49$$

Therefore the minimum noise figure for such a device strongly depends on the intrinsic noise which cannot be neglected.

Figure 1.22 shows a typical variation in the minimum noise figure as a function of the d.c. drain current I_{DS}. The noise figure has a minimum value that is usually obtained for a drain current close to 30 mA mm^{-1}. This variation results from two opposing effects: the decrease in the parameter P leads to a decrease in the noise figure with decreasing I_{DS}; at very low values of I_{DS} the transconductance g_m and the cut-off frequency f_c both approach zero, yielding a sharp increase in F_{min} near the threshold. Since the increase in F_{min} at low drain currents is mainly determined by the g_m–V_{GS} relation, high transconductance values are required at low drain currents in order to provide low noise figures [1.35]. Devices with high mobility buffer layers and a sharp decrease in doping level at the active layer–buffer layer interface are suitable for this purpose.

In many applications such as broadband amplifiers the device is not matched to the minimum noise figure. The mismatch effect on the noise figure is then expressed as follows:

$$F = F_{min}+\frac{4R_n}{1-|\Gamma_s|^2}\frac{|\Gamma_s-\Gamma_{opt}|^2}{|1+\Gamma_{opt}|^2} \quad (1.23)$$

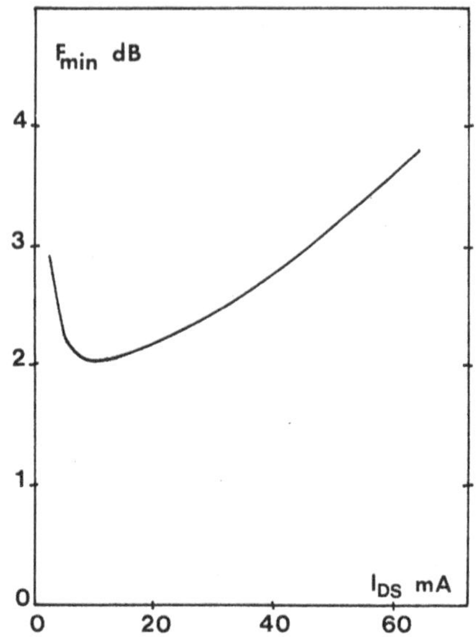

Fig. 1.22 Variation in F_{min} with I_{DS}.

where F_{min} is the minimum noise figure, Γ_{opt} is the optimum reflection coefficient, R_n is the equivalent noise resistance (normalized to 50 Ω) and Γ_s is the actual source reflection coefficient [1.31, 1.32]. This expression shows that a small R_n is essential for reducing the mismatch effect. To a first approximation this resistance can be written (without normalizing to 50 Ω) as

$$R_n = R_s + R_g + \frac{P}{g_m} \tag{1.24}$$

Equation (1.24) shows that low parasitic resistances and high g_m are required to give low R_n.

1.3 Low frequency characterization of MESFETs

The purpose of the low frequency characterization of MESFETs is to determine the main technological device parameters (gate length L_g, epilayer thickness A, doping N_d and mobility μ_0) as well as the values of the various elements of the small-signal equivalent circuit.

1.3.1 Determination of the technological parameters and access resistances R_s and R_d

The determination of both the technological parameters and the access resistances of MESFETs is based on three measurements: the gate-to-source capacitance and the drain-to-source resistance as a function of V_{GS} for $V_{DS} = 0$, and the I–V characteristic of the Schottky barrier.

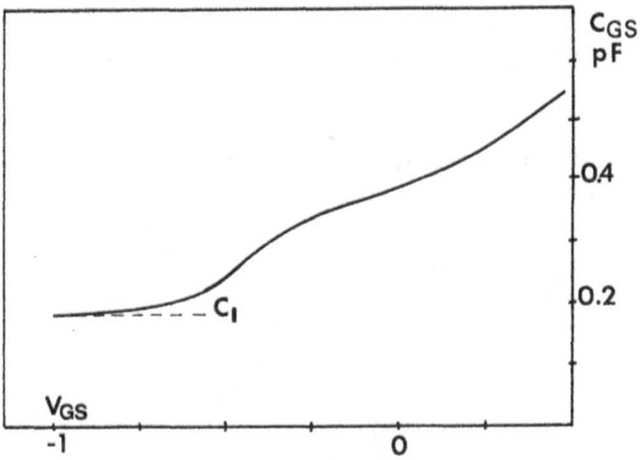

Fig. 1.23 Variation in the gate capacitance with V_{GS} for zero drain bias.

1.3.1.1 $C_{GS}(V_{GS})$

Figure 1.23 shows a typical $C_{GS}(V_{GS})$ curve. Below the pinch-off voltage V_p the capacitance remains almost constant. This residual value C_1 is due to the capacitance of the gate pad and to the electrostatic coupling between the source and gate metallization. Therefore an integration of $C_{GS} - C_1$ over V_{GS} gives the amount of charge $Q(V_{GS})$ under the gate [1.20]:

$$Q(V_{GS}) = \int_{V_p}^{V_{GS}} (C_{GS} - C_1)\, dV_{GS} \tag{1.25}$$

According to Shockley's law, $Q(V_{GS})$ can be expressed as

$$Q(V_{GS}) = qzL_gAN_d\left\{1 - \left(\frac{V_b - V_{GS}}{W_{00}}\right)^{1/2}\right\} \tag{1.26}$$

where $W_{00} = qN_dA^2/2\varepsilon$. As shown in Fig. 1.24 the plot of $Q(V_{GS})$ as a function of $(V_b - V_{GS})^{1/2}$ is linear. The pinch-off potential W_{00} and the total amount of charge $Q_0 = qzN_dL_gA$ under the gate can easily be deduced by extrapolation.

1.3.1.2 $R_{DS}(V_{GS})$

The second basic characterization consists in the measurement of the drain-to-source resistance as a function of V_{GS} for zero drain bias. It can easily

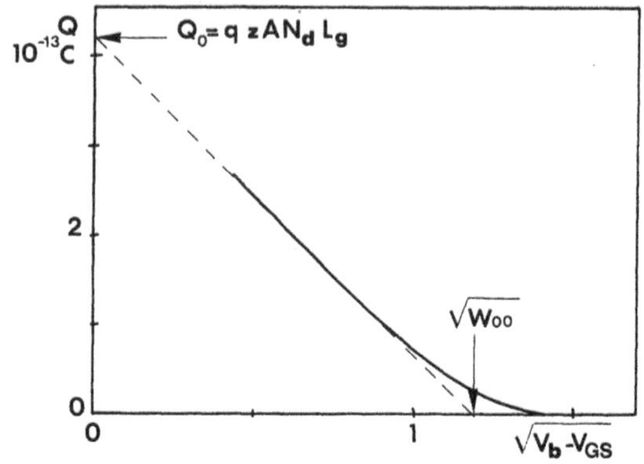

Fig. 1.24 Variation in the amount of charge under the gate as a function of $(V_b - V_{GS})^{1/2}$

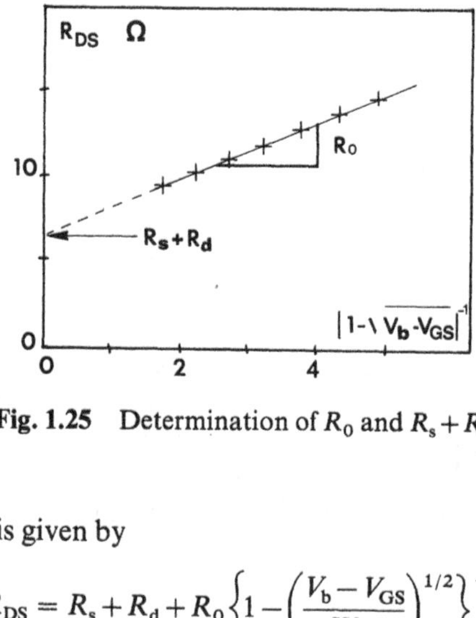

Fig. 1.25 Determination of R_0 and $R_s + R_d$.

be shown that R_{DS} is given by

$$R_{DS} = R_s + R_d + R_0 \left\{ 1 - \left(\frac{V_b - V_{GS}}{W_{00}} \right)^{1/2} \right\}^{-1} \tag{1.27}$$

where $R_0 = L_g / q\mu_0 N_d A z$. Therefore the plot of R_{DS} as a function of $[1 - \{(V_b - V_{GS})/W_{00}\}^{1/2}]^{-1}$ is almost a straight line. R_0 is given by the slope of this line and the sum $R_s + R_d$ is given by the intercept with the y axis (Fig. 1.25).

The $R_{DS}(V_{GS})$ measurement can also be used to determine the mobility profile by means of the geometrical magnetoresistance technique [1.33]. The geometry of a typical MESFET structure—a rectangle with contacts along the two wider edges and a large width-to-length ratio—is similar to that of a Corbino disk. The resistance in a transverse magnetic field is then given approximately by [1.33]

$$R_{DS}(B) = R_{DS}(0)(1 + \mu_0{}^2 B^2) \tag{1.28}$$

This expression shows that the mobility μ_0 can be obtained from R_{DS} measurements performed in the presence and the absence of a magnetic field.

1.3.1.3 $I_{GS}(V_{GS})$

If the Schottky barrier is forward biased, the I–V characteristic is given by [1.34]

$$I_{GS} = A^* T^2 S \exp\left(\frac{qV_b}{kT}\right) \exp\left(\frac{qV_{GS}}{nkT}\right) \tag{1.29}$$

The ideality figure n and the built-in voltage V_b can therefore be obtained from a semilogarithmic plot of the gate current as shown in Fig. 1.26.

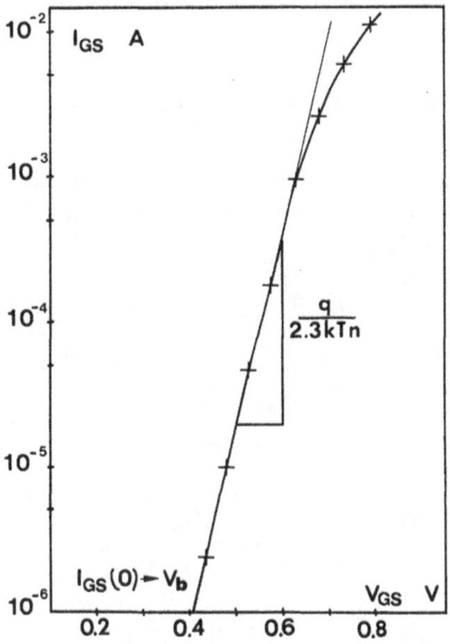

Fig. 1.26 Determination of the ideality factor n and the built-in potential V_b.

At high values of the gate bias the gate current I_{GS} tends to saturate owing to series resistance effects. If the drain-to-source voltage is measured in this current range we obtain [1.36]

$$V_{DS} = (R_s + \alpha R_0) I_{GS} \tag{1.30}$$

where $\alpha = 0.5$ arises from the distributed resistance of the channel under the gate and R_0 is the resistance of the fully open channel. The slope of $V_{DS}(I_{GS})$ provides $R_s + \alpha R_0$ as shown in Fig. 1.27. The same measurement can be performed between gate and drain. In that case we have

$$V_{GD} = (R_d + \alpha R_0) I_{GD} \tag{1.31}$$

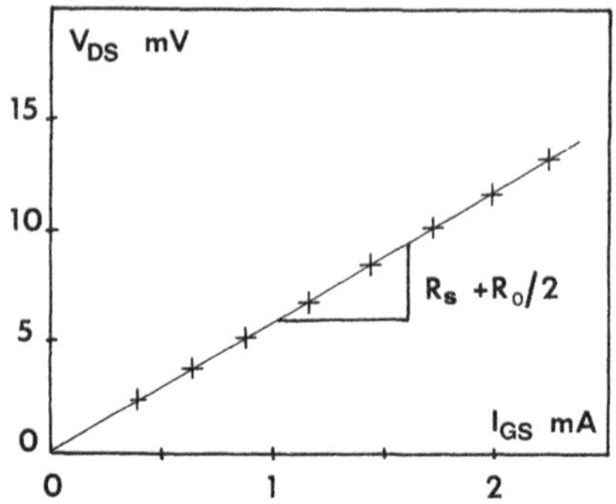

Fig. 1.27 Plot of $V_{DS}(I_{GS})$ giving $R_s + R_0/2$.

The difference $R_d - R_s$ between the drain and source access resistances can be obtained from $R_s + \alpha R_0$ and $R_d + \alpha R_0$.

In summary, the various quantities that can be deduced from d.c. and low frequency measurements are given in Table 1.1. It is easy to deduce the values of the technological parameters and the access resistances from these quantities.

Table 1.1

Measurement	*Extracted parameters*	
$C_{GS}(V_{GS})$	$W_{00} = \dfrac{qN_dA^2}{2\varepsilon}$	$\dfrac{L_g}{A} = \dfrac{2\varepsilon z W_{00}}{Q_0}$
	$Q_0 = qzN_dAL_g$	$N_dA^2 = \dfrac{2\varepsilon W_{00}}{q}$
$R_{DS}(V_{GS})$	$R_s + R_d$	
	$R_0 = \dfrac{L_g}{q\mu_0 N_d A z}$	
	μ_0	
$I_{GS}(V_{GS})$	n V_b	
$I_{GD}(V_{GD})$	$R_s + \dfrac{R_0}{2}$ $R_d + \dfrac{R_0}{2}$	
	$R_d - R_s$	

1.3.2 Determination of the small-signal equivalent circuit

The equivalent circuit of a MESFET is shown in Fig. 1.17. It involves both intrinsic elements (R_i, C_{GS}, C_{GD}, g_m, g_d, τ and C_{DS}) and extrinsic access elements (R_s, R_d, R_g, L_s, L_d and L_g). The purpose of this section is to show how the values of these elements can be obtained from measurements performed in the 100 MHz–4 GHz frequency range. This method is divided into two steps: (i) determination of the extrinsic parameters from measurements performed at $V_{DS} = 0$; (ii) determination of the intrinsic parameters from measurements at $V_{DS} = 0$ after removing the extrinsic parameters.

1.3.2.1 Determination of access elements

Device measurements are generally made using a network analyzer and the parameters obtained are the scattering parameters S_{ij}. Unfortunately these parameters are not simply related to the elements of the small-signal equivalent circuit that have to be determined and the use of impedance or admittance parameters is more convenient.

As originally proposed by Diamant and Laviron [1.37], the elements of a MESFET impedance matrix at zero drain bias can be written as follows:

$$Z_{11} = R_s + R_g + \frac{R_c}{3} + Z_{dy} + j\omega(L_s + L_g) \tag{1.32}$$

$$Z_{12} = Z_{21} = R_s + \frac{R_c}{2} + j\omega L_s \tag{1.33}$$

$$Z_{22} = R_s + R_d + R_c + j\omega(L_s + L_d) \tag{1.34}$$

R_c is the channel resistance and Z_{dy} is the equivalent impedance of the Schottky barrier. For reverse gate voltages $Z_{dy} = 1/jC_{GS}\omega$, whereas in the case of forward gate voltages $Z_{dy} = nkT/qI_{GS}$. Therefore the imaginary part X_{ij} of all the Z_{ij} is proportional to ω for forward gate voltages. The slope of X_{ij} as a function of ω gives the parasitic inductances L_s, L_g and L_d from X_{12}, X_{11} and X_{22} respectively.

The plot of the real part R_{11} of Z_{11} as a function of $1/I_{GS}$ is a straight line and the intercept with the y axis gives the sum $R_s + R_g + R_c/3$. Since R_s and R_d are known from d.c. measurements, the gate resistance R_g can easily be deduced from this plot.

Once the extrinsic parameters are known, the intrinsic parameters are obtained as follows. Both gate and drain bias voltages are applied and the scattering parameters S_{ij} are measured using a network analyzer in the 100 MHz–4 GHz frequency range. The scattering parameters are then transformed into impedance parameters using the classical formulae given in Chapter 2. The extrinsic parameters in series are then subtracted, giving the impedance matrix of the intrinsic device. Finally the impedance matrix is

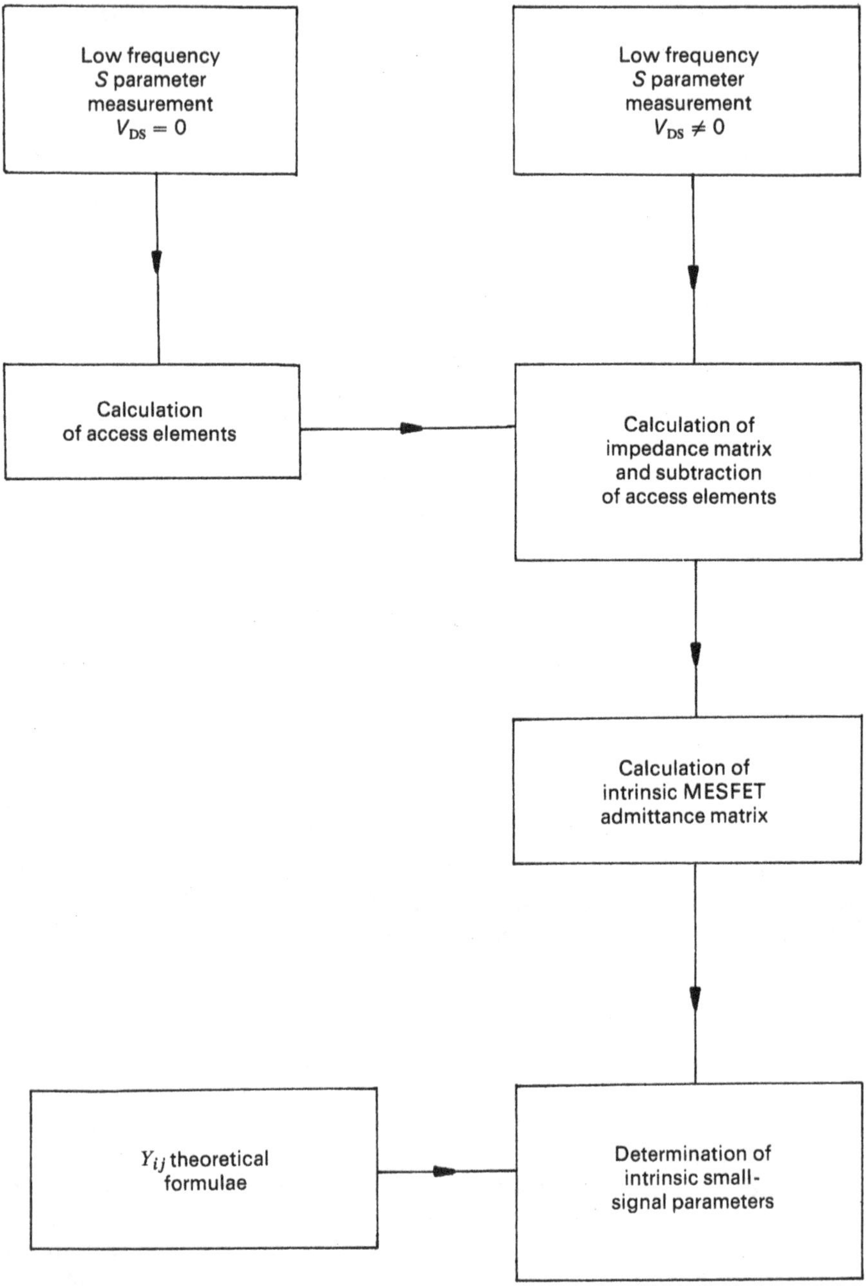

Fig. 1.28 Flowchart of the measurement method giving the values of both the extrinsic and intrinsic elements.

transformed into the admittance matrix because the use of admittance parameters is more convenient for the π topology of the intrinsic device. In fact, the admittance parameters are simply related to the elements of the small-signal equivalent circuit in the frequency range under consideration by eqn (2.33) given in Chapter 2, Section 2.6.1 [1.38]. If the various Y_{ij} are known, C_{GD} can be obtained from Y_{12}, R_i and C_{GS} can be obtained from Y_{11}, g_m and τ can be obtained from Y_{21}, and g_d and C_{DS} can be obtained from Y_{22}. A flowchart of this method is shown in Fig. 1.28. It is fast and accurate and can be applied to both packaged devices and chips by using microwave wafer probing. An example of the results obtained is given in Fig. 1.20.

References

1.1 Ogawa, M. *et al. IEEE Trans. Microwave Theory Tech.*, **24** (6), 300–305, 1976.

1.2 Hollan, L. and Hallais, J. In Dilomenzo, J. V. (ed.), *GaAs FET Principles and Technology*, pp. 67–99, Artech House, Dedham, MA, 1982.

1.3 Cox, H. M. and Di Lorenzo, J.V. *Inst. Phys. Conf. Ser.*, **33b**, 11–22, 1976.

1.4 Ainslie, N. G. *et al. J. Appl. Phys.*, **33**, 2391–2393, 1962.

1.5 Au Coin, T. R. *et al. Solid State Technol.*, **22** (59), 1979.

1.6 Lindquist, P. F. and Ford, W. M. In Dilomenzo, J. V. (ed.), *GaAs FET Principles and Technology*, pp. 1–60, Artech House, Dedham, MA, 1982.

1.7 *J. Cryst. Growth*, **55** (1), October 1981.

1.8 Dingle, R. *et al. J. Appl. Phys.*, **33** (7), 665–667, 1978.

1.9 Eisen, F. *et al.* In Dilomenzo, J. V. (ed.), *GaAs FET Principles and Technology*, pp. 117–146, Artech House, Dedham, MA, 1982.

1.10 Yamasaki, K. *et al. IEEE Trans. Electron Devices*, **29** (11), 1772–1777, 1982.

1.11 Zucca, R. *et al.* In *Semi-insulating III–V Materials*, pp. 335–345, Shiva, Nottingham, 1980.

1.12 Williams, R. E. *Gallium Arsenide Processing Techniques*, Artech House, Dedham, MA, 1984.

1.13 Rideout, V. L. *Solid-State Electron*, **18**, 541, 1975.

1.14 Piotrowska, A. *et al. Solid-State Electron*, **26** (3), 179–197, 1983.

1.15 Shockley, W. *Proc. IRE*, **40**, 1365, 1952.

1.16 Ruch, J. G. and Fawcett, W. *J. Appl. Phys.*, **41** (9), 3843–3849, 1970.

1.17 Fawcett, W. *et al. J. Phys. Chem. Solids*, **31**, 1963–1990, 1970.

1.18 Carnez, B. *et al. J. Appl. Phys.*, **51** (1), 784–790, 1980.

1.19 El Sayed, O. *et al. Solid-State Electron.*, **30** (6), 643–654, 1987.

1.20 Cappy, A. *Thèse d'Etat*, Université de Lille, 1986.

1.21 Turner, J. A. and Wilson, B. L. *Inst. Phys. Conf. Ser.*, **7**, 195–204, 1969.

1.22 Pucel, R. A. *et al. Adv. Electron. Electron Phys.*, **38**, 193–265, 1975.

1.23 Shur, M. *Electron. Lett.*, **12** (23), 615, 1976.

1.24 Yamaguchi, K. *IEEE Trans. Electron Devices*, **23**, 545, 1976.

1.25 Reiser, M. *IEEE Trans. Electron Devices*, **20** (1), 35, 1973.

1.26 Moglestue, C. *Comput. Methods Appl. Mech. Eng.*, **30**, 173–208, 1982.

1.27 Curtice, W. *IEEE Trans. Electron Devices*, **28** (8), 954–962, 1981.

1.28 Moglestue, C. *Proc. IEEE*, **132** (5), 217–223, 1985.

1.29 Buot, F. A. and Frey, J. *Solid-State Electron.*, **26** (7), 617–632, 1983.

1.30 Wolf, P. *IBM J. Res. Dev.*, **14** (2), 125–141, 1970.

1.31 Rothe, H. and Dahlke, W. *Proc. IRE*, **44**, 811–818, 1956.

1.32 Van der Ziel, A. *Proc. IRE*, **51**, 461–467, 1963.
1.33 Sites, J. R. and Wieder, H. H. *IEEE Trans. Electron Devices*, **27** (12), 2277–2281, 1980.
1.34 Fukui, H. *Bell Syst. Tech. J.*, **58** (3), 771–795, 1979.
1.35 Van der Ziel, A. *Proc. IRE*, **50**, 1808–1812, 1962.
1.36 Lee, K. *et al. IEEE Trans. Electron Devices*, **32** (5), 987–992, 1985.
1.37 Diamant, F. and Laviron, M. *Proc. 12th European Microwave Conf.*, pp. 451–456, Microwave Exhibitions and Publishers, Tunbridge Wells, 1982.
1.38 Minasian, R. A. *Electron. Lett.*, **13** (8), 549–551, 1977.

Chapter 2

GaAs MESFETs: *S* Parameter Measurements and their Use in Circuit Design

MARC PARISOT AND ROBERT SOARES

2.1 Introduction

This chapter is the second devoted to the development of basic principles which will be used elsewhere in the book. It includes many disparate elements such as device handling and mounting techniques, definition of the normalized *S* parameter, *S* parameter measurement for both on-wafer and encapsulated devices, various de-embedding schemes and the principles of correcting measurement data. The chapter can be divided into three main areas: (a) basic principles involving device biasing and handling, followed by the elementary theory necessary for *S* parameter measurement; (b) three sections devoted to *S* parameter measurements (characterization of coaxial or waveguide devices, measurement of devices mounted in a test fixture and on-wafer *S* parameter measurements); (c) a section showing how these basic measurements can be exploited, particularly for amplifier design, and used in computer-aided programs.

2.2 Basic principles

2.2.1 MESFET biasing techniques and precautions

As discussed in Chapter 1, the operation of a microwave depletion-mode GaAs MESFET consists in modulating the d.c. flow between source and drain by an r.f. signal applied to a negatively biased gate. If this modulated current is now fed into an appropriate load resistor, the r.f. signal out of the device is considerably larger than the gate input signal and there is amplification. The device has the d.c. characteristics shown in Fig. 2.1. It can be seen intuitively, and confirmed experimentally, that the common source configuration offers the highest amplifier gain. It also offers the best r.f. stability.

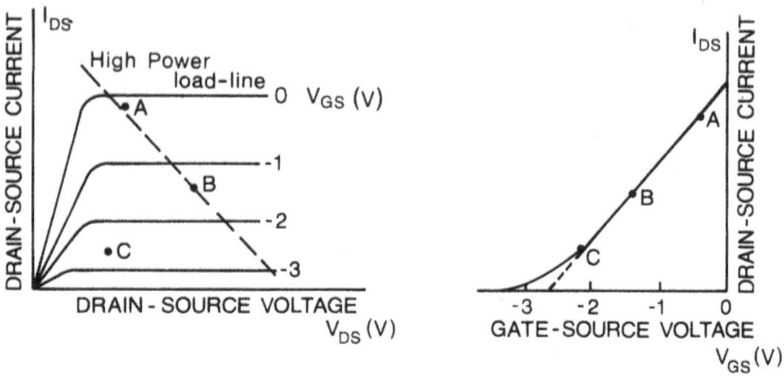

Fig. 2.1 Optimum bias points for various applications.

The current and voltage bias points selected for various applications of a typical GaAs MESFET are also shown in Fig. 2.1. Essentially, the highest gain under small-signal conditions is obtained at point A where the drain current I_{DS} is high (Fig. 2.2) and V_{DS} is maintained between 3 and 4 V to avoid thermal problems. The trade-off between minimum noise figure and acceptable gain is optimum at C, corresponding to a drain current of around $I_{DSS}/8$. Finally, the best linearity for high power operation is obtained at B by biasing the transistor at $I_{DSS}/2$ and $V_{ds\,max}/2$, where $V_{ds\,max}$ is the maximum excursion along the drain–source axis compatible with safe linear operation. In modern devices V_{DS}, corresponding to $V_{ds\,max}/2$, is typically 10 V. A higher r.f./d.c. efficiency and longer transistor lifetime, however, will be obtained by operating at $V_{DS} = 9$ V.

Practical biasing configurations [2.1] are shown in Fig. 2.3. The first circuit requires two power supplies of opposite polarity; the complexity of this arrangement is offset by the fact that the source leads of the MESFET are directly grounded, thereby minimizing source series inductance and, in consequence, maximizing gain. This is particularly important at frequencies above 18 GHz where gain is at a premium.

The circuits shown in Figs 2.3(b) and 2.3(c) permit the use of a single-polarity bias supply (positive in the case of Fig. 2.3(b) and negative for Fig. 2.3(c)). The last two circuits (Figs 2.3(d) and 2.3(e)) include resistors in the source-to-ground connection and are widely used for small-signal broadband applications. As the drain voltage is applied, the gate is simultaneously reverse biased with respect to source by the current flowing through the series resistor. A 50 pF bypass capacitor provides an r.f. ground, and the resistor protects the transistor against both transients and low frequency instability. The source resistor can also be used to stabilize the d.c. drain current, thereby rendering an amplifier less vulnerable to device spreads, temperature effects and bias voltage fluctuations [2.2].

For high power transistors, where the drain–source current is high ($I_{DS} > 0.5$ A), dual-polarity supplies should be used with a sequencing circuit to ensure that the gate–source is reverse biased before the drain voltage is

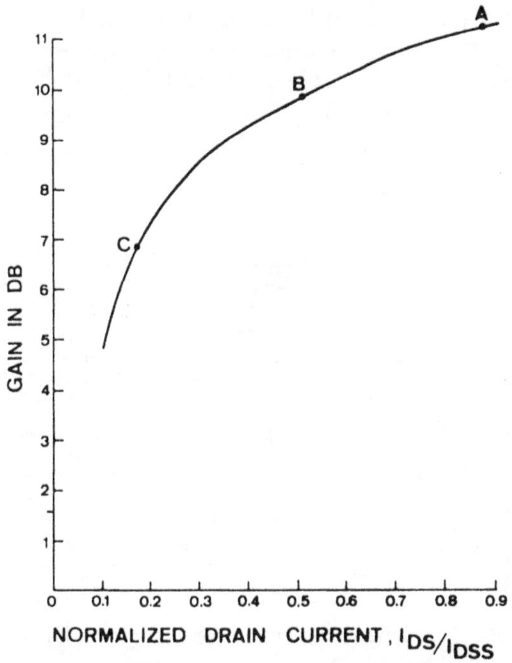

Fig. 2.2 Available power gain versus normalized output current.

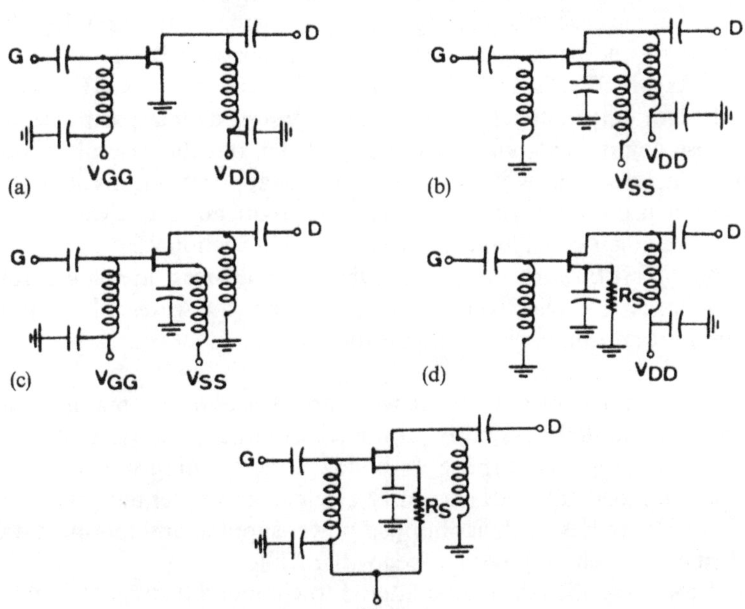

Fig. 2.3 Five practical GaAs MESFET bias configurations [2.1].

applied. A practical power MESFET circuit is given in Fig. 2.4. R.f. chokes and grounding capacitors are selected to provide sufficient low frequency (below 500 MHz) isolation between the GaAs MESFET and the power supply. The 10 kΩ series resistance in the gate bias circuit protects the device from being driven into the forward conduction region.

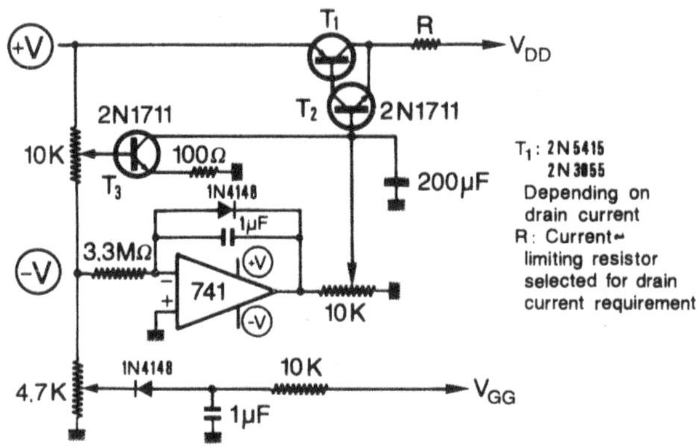

Fig. 2.4 Practical sequenced bias circuit for high power GaAs MESFETs.

GaAs MESFETs are sensitive devices and must be used with care, particularly in the case of unpackaged chip transistors. Subassemblies should be designed around packaged devices whenever possible. Both low noise and power GaAs MESFETs are now available for use up to 15 GHz, often in prematched (or fully matched [2.3]) packages designed for a specific frequency band. These components are often highly priced, but the ease of utilization they afford makes them worthwhile choices for applications involving small and medium production runs over medium frequency bandwidth specifications. Packaging is considered in greater detail in Section 2.2.4.1.

GaAs MESFETs, in particular submicron devices, are vulnerable to static discharge and electromagnetic pulses on the gate. The following elementary precautions save both time and money.

(i) The operator, or at least the work area where device manipulation is being effected, should be properly grounded. Conductive straps and mats are widely available for this purpose. In bonders, the bonding wedge and stage should be grounded, all leads should be shielded and potential generators of electromagnetic pulses, such as chopped power supplies and thermostatically controlled heaters, should be equipped with r.f. filters.

(ii) Power supplies should be filtered to eliminate transients and power surges. Similarly, great care should be taken with d.c. curve tracers and probes; the probes should be shielded as close as possible to the tips.

(iii) The GaAs MESFET is a three-port component which can be used in the reverse-channel configuration, i.e. with a negative drain voltage. In all cases, however, as can be seen from the bias circuit configurations, the gate electrode must be maintained at a zero or negative potential with respect to the most negative of the other two electrodes. In addition, the use of a current-limiting device in the drain bias supply protects the MESFET from accidental drain–source short circuits during tuning.

2.2.2 Normalized S parameters

Instead of defining the behavior of microwave networks in terms of the currents and voltages at their terminals, we can equally well define them in terms of linear combinations of these variables. The variables most commonly used in microwave networks derive from transmission line (and waveguide) concepts of incident and reflected waves normalized to the characteristic impedance of the homogeneous transmission medium.

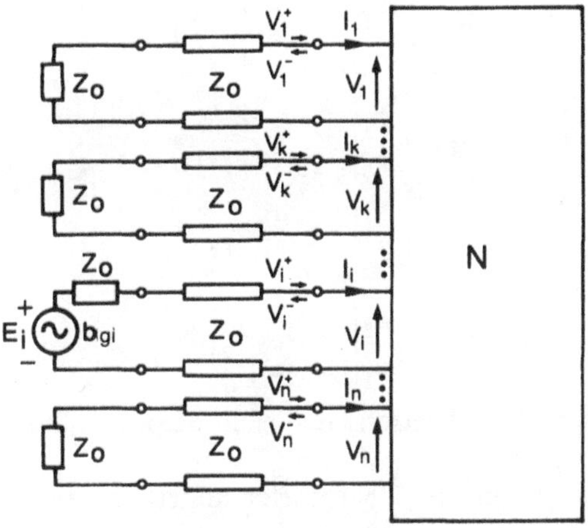

Fig. 2.5 Normalized S parameters for an n-port network.

We can thus define an incident wavevector $\boldsymbol{a}$ and a reflected wavevector $\boldsymbol{b}$ in terms of an incident voltage wavevector $\boldsymbol{V}^+$ and a reflected voltage wavevector $\boldsymbol{V}^-$ respectively normalized to a diagonal impedance matrix, each element of which is a real and positive impedance Z_0 (in practice 50 Ω). At any port i of an n-port network (Fig. 2.5)

$$a_i = \frac{V^+{}_i}{Z_0{}^{1/2}} \qquad i = 1, 2, \ldots, n$$
$$b_i = \frac{V^-{}_i}{Z_0{}^{1/2}} \qquad i = 1, 2, \ldots, n \tag{2.1}$$

or, in terms of the voltage and current at port i,

$$a_i = \tfrac{1}{2}(V_i + I_i Z_0) Z_0^{-1/2} \qquad i = 1, 2, \ldots, n$$
$$b_i = \tfrac{1}{2}(V_i - I_i Z_0) Z_0^{-1/2} \qquad i = 1, 2, \ldots, n \tag{2.2}$$

We can now define a scattering parameter matrix relating the variables a_i and b_i as follows:

$$\boldsymbol{b} = \boldsymbol{S} \cdot \boldsymbol{a} \tag{2.3}$$

where

$$\boldsymbol{b}^{\mathrm{T}} = [b_1, b_2, \ldots, b_i, \ldots, b_n]$$

and

$$\boldsymbol{a}^{\mathrm{T}} = [a_1, a_2, \ldots, a_i, \ldots, a_n]$$

Thus for a two-port network we obtain

$$b_1 = S_{11} a_1 + S_{12} a_2 \tag{2.4a}$$

$$b_2 = S_{21} a_1 + S_{22} a_2 \tag{2.4b}$$

whence

$$S_{11} = \left.\frac{b_1}{a_1}\right|_{a_2 = 0} = \frac{V^{+}{}_1}{V^{-}{}_1} = \Gamma_1$$

$$S_{22} = \left.\frac{b_2}{a_2}\right|_{a_1 = 0} = \frac{V^{+}{}_2}{V^{-}{}_2} = \Gamma_2$$

$$S_{12} = \left.\frac{b_1}{a_2}\right|_{a_1 = 0} = T_{12}$$

$$S_{21} = \left.\frac{b_2}{a_1}\right|_{a_2 = 0} = T_{21}$$

where Γ_i represents the reflection coefficient at port i and T_{ij} represents the transmission from ports j to i.

We can thus obtain the S parameter matrix for an n-port network by successively placing a normalized signal source b_{gi} (operating out of a source impedance Z_0) onto ports $1, 2, \ldots, i, \ldots, n$, terminating all other ports in Z_0 and measuring the response b_i on all ports. S parameter measurements of two-port active networks are described fully in Section 2.3.

In order to apply computer analysis to cascaded two-port networks (e.g. an amplifier made up of input-matching network, transistor and output-matching network), it is convenient to define R or cascading parameters [2.4]. In this case, for a two-port network the independent variables are a_2, b_2 and the dependent variables are b_1, a_1.

Thus, for each of the two-port networks 1 and 2 of Fig. 2.6 we obtain

$$\begin{bmatrix} b_1 \\ a_1 \end{bmatrix}_{\mathrm{A}} = \begin{bmatrix} R_{11} & R_{12} \\ R_{21} & R_{22} \end{bmatrix}_{\mathrm{A}} \begin{bmatrix} a_2 \\ b_2 \end{bmatrix}_{\mathrm{A}} \tag{2.5}$$

$$\begin{bmatrix} b_1 \\ a_1 \end{bmatrix}_B = \begin{bmatrix} R_{11} & R_{12} \\ R_{21} & R_{22} \end{bmatrix}_B \begin{bmatrix} a_2 \\ b_2 \end{bmatrix}_B \tag{2.6}$$

where subscript A refers to network 1 and subscript B to network 2. It can be seen from Fig. 2.6 that $(b_2)_A = (a_1)_B$ and $(a_2)_A = (b_1)_B$. Thus

$$\begin{bmatrix} b_1 \\ a_1 \end{bmatrix}_A = \begin{bmatrix} R_{11} & R_{12} \\ R_{21} & R_{22} \end{bmatrix}_A \begin{bmatrix} R_{11} & R_{12} \\ R_{21} & R_{22} \end{bmatrix}_B \begin{bmatrix} a_2 \\ b_2 \end{bmatrix}_B \tag{2.7}$$

When this procedure is extended to n cascading two-port networks, the relationship between the variables (b_1, a_1) and (a_2, b_2) is given simply by multiplying the individual R matrices in the order $\boldsymbol{R}_1$, $\boldsymbol{R}_2$, ..., $\boldsymbol{R}_{n-1}$, $\boldsymbol{R}_n$. T matrices accomplish the same function, except that in this case the independent and dependent variables are (a_1, b_1) and (b_2, a_2) respectively. The individual T matrices are thus multiplied in the reverse order: $\boldsymbol{T}_n$, $\boldsymbol{T}_{n-1}$, ..., $\boldsymbol{T}_2$, $\boldsymbol{T}_1$.

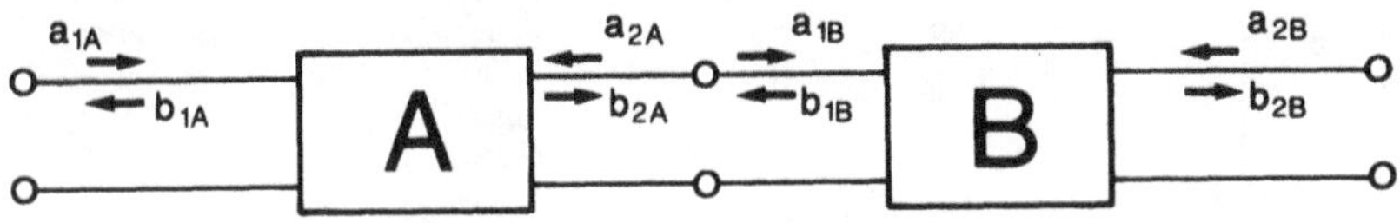

Fig. 2.6 Cascaded two-port matrices.

It should be noted that linear relations exist between S parameters and other sets of parameters: Z (impedance), Y (admittance), $ABCD$ (voltage–current cascading) or R (normalized wave cascading). This permits direct conversion between the different parameter sets, thereby greatly facilitating the integration of S parameters into standard computer optimization programs. The relations between standard parameter sets are given in Appendix 2.1 at the end of this chapter. Table 2.1 meanwhile gives Y, Z, $ABCD$, S and R matrix descriptions of some standard circuit elements.

2.2.3 Change in the reference plane

It is often necessary during S parameter measurements of two-port active or passive devices to measure components physically at a reference plane different from that actually desired. In general there are two possibilities.

(i) The device is characterized via two lossless homogeneous transmission lines at planes 1 and 2, whereas the data are required at planes 1′ and 2′ respectively (Fig. 2.7).

(ii) Alternatively, our correction standards, notably in the through–short–delay (TSD) method (Section 2.3.5), ensure that measurements are effected with respect to a common plane A, whereas we wish to correct for

Table 2.1 Matrix representations for some two-port networks

Element	*Y matrix* *Z matrix*	*ABCD matrix*	*S matrix*	*R matrix*
Homogeneous transmission line matrix $\Gamma = (Z'_0 - Z_0)/(Z'_0 + Z_0)$ $\theta = \beta \ell$ 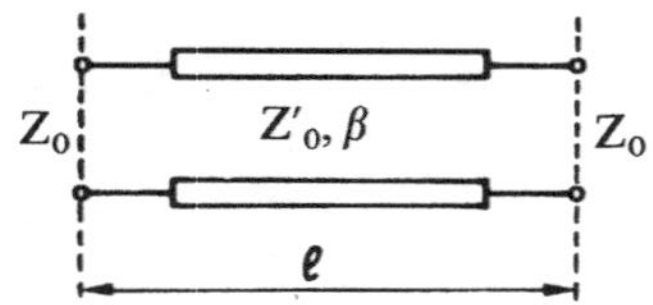	$Y_{11} = Y_{22} = -j(\cot\theta)/Z'_0$ $Y_{21} = Y_{12} = j(\operatorname{cosec}\theta)/Z'_0$ $Z_{11} = Z_{22} = -jZ'_0 \cot\theta$ $Z_{21} = Z_{12} = -jZ'_0 \operatorname{cosec}\theta$	$A = D = \cos\theta$ $B = jZ'_0 \sin\theta$ $C = j(\sin\theta)/Z'_0$	$S_{11} = S_{22} = \dfrac{\Gamma(1 - e^{-2j\theta})}{1 - \Gamma^2 e^{-2j\theta}}$ $S_{21} = S_{12} = \dfrac{(1 - \Gamma^2)e^{-j\theta}}{1 - \Gamma^2 e^{-2j\theta}}$	$R_{11} = \dfrac{e^{-j\theta} - \Gamma^2 e^{j\theta}}{1 - \Gamma^2}$ $R_{22} = \dfrac{e^{j\theta} - \Gamma^2 e^{-j\theta}}{1 - \Gamma^2}$ $-R_{21} = R_{12} = \dfrac{\Gamma(e^{j\theta} - e^{-j\theta})}{1 - \Gamma^2}$
Shunt element 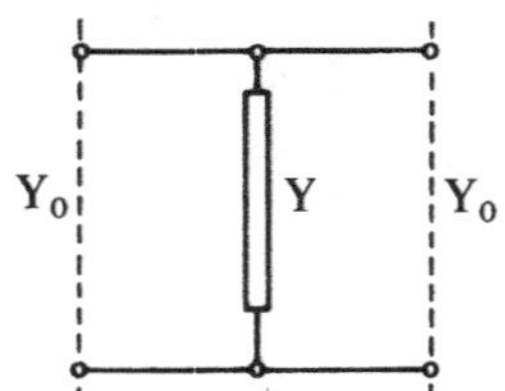 $y = Y/Y_0$	No Y matrix $Z_{11} = Z_{22} = 1/Y$ $Z_{12} = Z_{21} = 1/Y$	$A = D = 1$ $B = 0$ $C = Y$	$S_{11} = S_{22} = -y/(y+2)$ $S_{21} = S_{12} = 2/(y+2)$	$R_{11} = \frac{1}{2}(y+2)$ $R_{22} = \frac{1}{2}(2-y)$ $R_{21} = R_{12} = y/2$
Series element 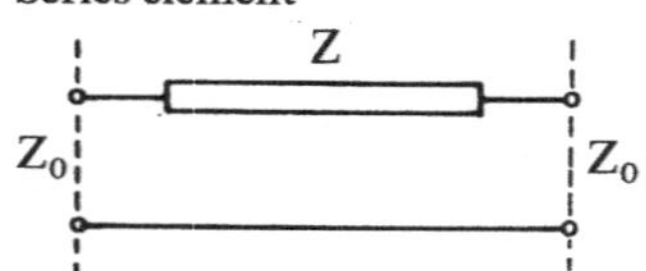 $z = Z/Z_0$	$Y_{11} = Y_{22} = 1/Z$ $Y_{12} = Y_{21} = 1/Z$ No Z matrix	$A = D = 1$ $B = Z$ $C = 0$	$S_{11} = S_{22} = z/(z+2)$ $S_{21} = S_{12} = 2/(z+2)$	$R_{11} = \frac{1}{2}(2-z)$ $R_{22} = \frac{1}{2}(2+z)$ $R_{21} = R_{12} = z/2$

(continued)

π network

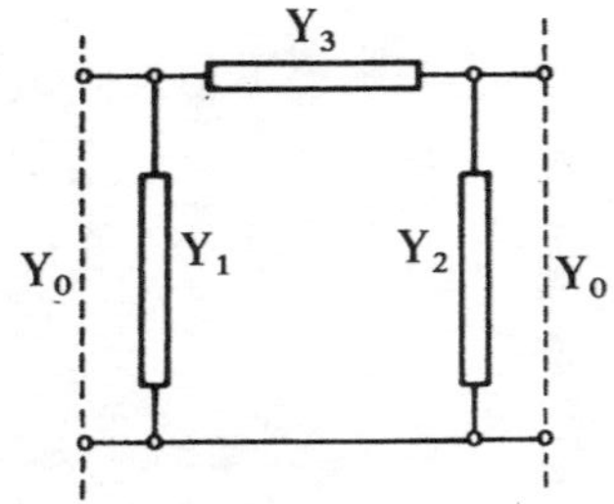

$y_1 = Y_1/Y_0,\ y_2 = Y_2/Y_0$
$y_3 = Y_3/Y_0$

$Y_{11} = Y_1 + Y_3$	$A = 1 + Y_2/Y_3$	$S_{11} = (1 - y_1 + y_2 - \Delta^y)/D_1$	$R_{11} = 1 + (y_1 + y_2 + \Delta^y - 1)/2y_3$
$Y_{22} = Y_2 + Y_3$	$B = 1/Y_3$	$S_{22} = (1 + y_1 - y_2 - \Delta^y)/D_1$	$R_{22} = 1 + (y_1 + y_2 + \Delta^y + 1)/2y_3$
$Y_{12} = Y_{21} = -Y_3$	$C = -\Delta^Y/Y_3$	$S_{12} = S_{21} = -2y_3/D_1$	$R_{12} = (y_2 - y_1 - \Delta^y + 1)/2y_3$
$Z_{11} = Y_2 + Y_3/\Delta^Y$	$D = 1 + Y_1/Y_3$	$D_1 = 1 + y_1 + y_2 + 2y_3 + \Delta^y$	$R_{21} = (y_2 - y_1 + \Delta^y - 1)/2y_3$
$Z_{22} = Y_1 + Y_3/\Delta^Y$			
$Z_{12} = Z_{21} = Y_3/\Delta^Y$			

T network

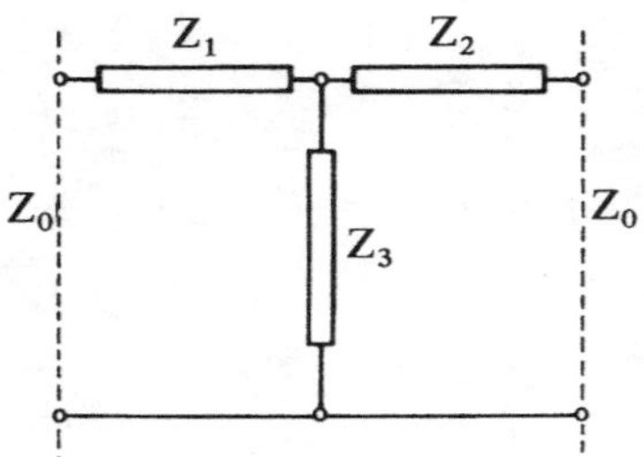

$z_1 = Z_1/Z_0,\ z_2 = Z_2/Z_0$
$z_3 = Z_3/Z_0$

$Y_{11} = Z_2 + Z_3/\Delta^Z$	$A = 1 + Z_1/Z_3$	$S_{11} = (z_1 - z_2 - 1 + \Delta^z)/D_3$	$R_{11} = 1 + (z_1 + z_2 - \Delta^z - 1)/2z_3$
$Y_{22} = Z_1 + Z_3/\Delta^Z$	$B = \Delta^Z/Z_3$	$S_{22} = (z_2 - z_1 - 1 + \Delta^z)/D_3$	$R_{22} = 1 + (z_1 + z_2 + \Delta^z + 1)/2z_3$
$Y_{12} = Y_{21} = -Z_3/\Delta^Z$	$C = 1/Z_3$	$S_{12} = S_{21} = 2z_3/D_3$	$R_{12} = (z_1 - z_2 + \Delta^z - 1)/2z_3$
$Z_{11} = Z_1 + Z_3$	$D = 1 + Z_2/Z_3$	$D_3 = 1 + z_1 + z_2 + 2z_3 + \Delta^z$	$R_{21} = (z_1 - z_2 - \Delta^z + 1)/2z_3$
$Z_{22} = Z_2 + Z_3$			
$Z_{12} = Z_{21} = Z_3$			

$\Delta^K = K_{11}K_{22} - K_{12}K_{21}$

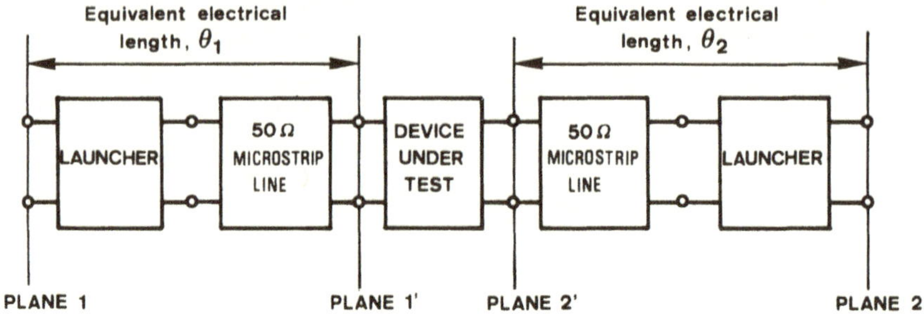

(a)

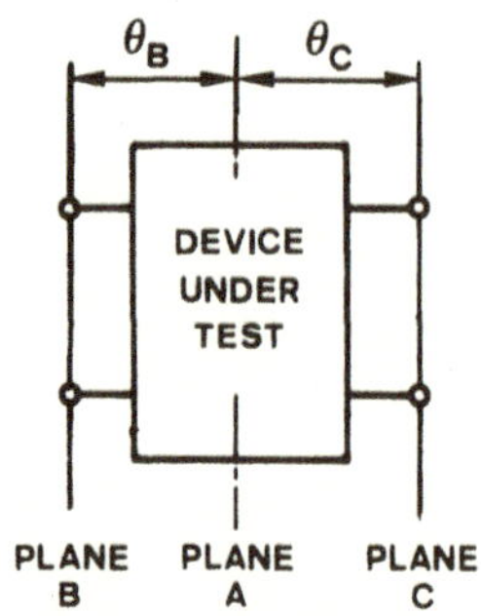

(b)

Fig. 2.7 Two examples of a change in reference plane.

the physical size of the device and offset the results to planes B and C situated at the component edges.

Consider the first case. From Table 2.1 the R matrix description of a homogeneous transmission line of electrical length θ is given by

$$\boldsymbol{R}_{\mathrm{TL}} = \begin{bmatrix} \exp(-\mathrm{j}\theta) & 0 \\ 0 & \exp(\mathrm{j}\theta) \end{bmatrix} \tag{2.8}$$

The R matrix $\boldsymbol{R}'$ of the network between planes 1′ and 2′ in terms of the overall measured chain parameters is given by

$$\boldsymbol{R}' = \boldsymbol{R}_1^{-1}\boldsymbol{R}_{\mathrm{m}}\boldsymbol{R}_2^{-1} \tag{2.9}$$

or

$$\boldsymbol{R}' = \begin{bmatrix} \exp(\mathrm{j}\theta_1) & 0 \\ 0 & \exp(-\mathrm{j}\theta_1) \end{bmatrix} \begin{bmatrix} R_{\mathrm{m11}} & R_{\mathrm{m12}} \\ R_{\mathrm{m21}} & R_{\mathrm{m22}} \end{bmatrix} \begin{bmatrix} \exp(\mathrm{j}\theta_2) & 0 \\ 0 & \exp(-\mathrm{j}\theta_2) \end{bmatrix}$$

$$\boldsymbol{R}' = \begin{bmatrix} \exp\{\mathrm{j}(\theta_1+\theta_2)\}\, R_{\mathrm{m11}} & \exp\{\mathrm{j}(\theta_1-\theta_2)\}\, R_{\mathrm{m12}} \\ \exp\{-\mathrm{j}(\theta_1-\theta_2)\}\, R_{\mathrm{m21}} & \exp\{-\mathrm{j}(\theta_1+\theta_2)\}\, R_{\mathrm{m22}} \end{bmatrix} \tag{2.10}$$

On converting back to S parameters we observe that

$$S' = \begin{bmatrix} S_{11}\exp(2j\theta_1) & S_{12}\exp\{j(\theta_1+\theta_2)\} \\ S_{21}\exp\{j(\theta_1+\theta_2)\} & S_{22}\exp(2j\theta_2) \end{bmatrix} \tag{2.11}$$

This is equivalent to changing the phases of S_{11}, S_{22} and both S_{12} and S_{21} by $-2\theta_1$, $-2\theta_2$ and $-(\theta_1+\theta_2)$ respectively.

The second case is dealt with in a similar manner except that we now *add* $2\theta_B$ and $2\theta_C$ to the phases of the common plane parameters S_{11} and S_{22} respectively, and $\theta_B+\theta_C$ to both S_{12} and S_{21}.

2.2.4 De-embedding

The operation of inverse chain matrix multiplication used to change the reference plane in the previous section is a particular example of "de-embedding". This operation is often necessary at microwave frequencies because, for one or more of the reasons listed below, it is virtually impossible to obtain direct access to the r.f. performance characteristics of a given component, whether it is an active device, a passive network or a complex subassembly.

(a) The component we wish to measure is encapsulated in a microstrip package.

(b) The device to be measured is mounted in a microstrip test fixture furnished with r.f. connectors compatible with the measuring apparatus. This fixture is in general composed of 50 Ω microstrip lines to access the packaged or chip component and microstrip-to-coaxial launchers.

(c) The device has been tuned at the input to give a certain (e.g. low noise) performance and we wish to know the tuning impedance presented.

(d) The device is matched at input *and* output ports to give a certain (low noise, optimum power, doubler etc.) performance and we need to know the passive source and load impedances presented.

We consider each case in turn.

2.2.4.1 Package and mount de-embedding

Two approaches are available to correct for package parasitics (lead inductances and lead-to-ground capacitances (see Section 2.6.1)) and for errors (multiple reflections, microstrip loss and dispersion) introduced by the test fixture: (i) introduce calibration standards at the transistor chip position (Section 2.3.4); (ii) precharacterize the package and/or mount and use de-embedding or peeling to obtain the device parameters [2.5–2.8].

Techniques for precharacterizing transistor and diode ceramic packages are given in ref. 2.5, and techniques for precharacterizing the various

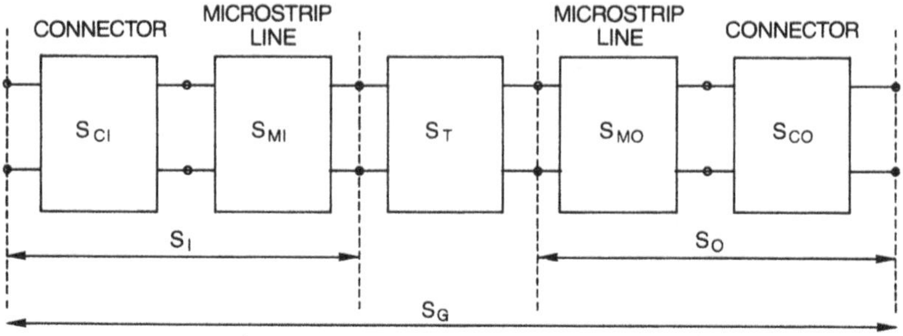

Fig. 2.8 De-embedding for 50 Ω *S* parameter measurements.

components of a test fixture are described in the following section. Having obtained either an electrical equivalent circuit or an *S* parameter representation of the text fixture, we can now apply the peeling procedure (Fig. 2.8). The input and output circuits of the fixture consist of coaxial–microstrip adaptors, defined by the *S* parameters $\boldsymbol{S}_{\mathrm{CI}}$ and $\boldsymbol{S}_{\mathrm{CO}}$ respectively, and 50 Ω microstrip lines, defined by the *S* parameters $\boldsymbol{S}_{\mathrm{MI}}$ and $\boldsymbol{S}_{\mathrm{MO}}$ respectively. All the *S* parameters are converted into their corresponding *R* parameters, and the transistor *S* parameters $\boldsymbol{S}_{\mathrm{T}}$ are obtained from the fixture *S* parameters $\boldsymbol{S}_{\mathrm{G}}$ as follows:

$$\boldsymbol{R}_{\mathrm{T}} = (\boldsymbol{R}_{\mathrm{CI}}\boldsymbol{R}_{\mathrm{MI}})^{-1}\boldsymbol{R}_{\mathrm{G}}(\boldsymbol{R}_{\mathrm{CO}}\boldsymbol{R}_{\mathrm{MO}})^{-1} \tag{2.12}$$

$$\boldsymbol{R}_{\mathrm{T}} \rightleftharpoons \boldsymbol{S}_{\mathrm{T}} \tag{2.13}$$

2.2.4.2 Input and output tuner de-embedding

We consider the case where the transistor is matched by means of a removable tuner at its input and output. The tuner consists of metal disks, dielectric slivers or silver paint applied to the 50 Ω microstrip input and/or output lines of the test fixture in order to obtain a given low noise or other r.f. performance condition. Errors due to loss and dispersion in the microstrip line are minimized by matching as close as is physically possible to the device terminal. The *S* parameters of the terminal matching circuit are then obtained in a manner similar to that described above by applying the following *S* parameter measurement methodology (Fig. 2.9).

(a) Measure the two-port fixture *S* parameters with input and output matching in place. This gives $\boldsymbol{S}_{\mathrm{G1}}$.

(b) Measure the two-port fixture *S* parameters with either input or output matching removed. This gives $\boldsymbol{S}_{\mathrm{G2}}$.

(c) Measure the two-port fixture *S* parameters with all matching removed. This gives $\boldsymbol{S}_{\mathrm{G3}}$.

The input- and/or output-matching network matrices are obtained from these measurements as follows.

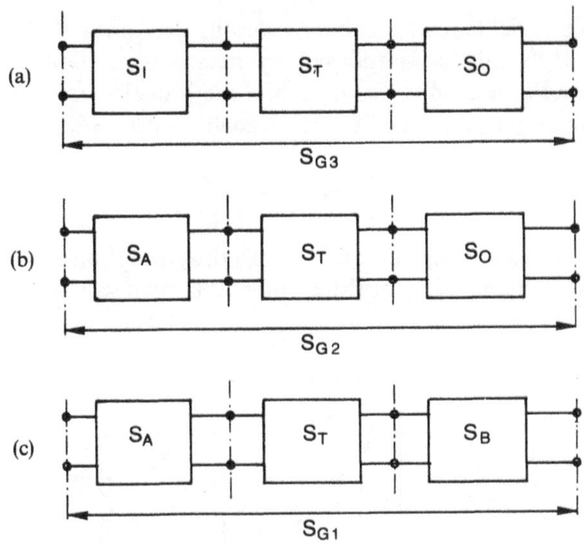

Fig. 2.9 Process used for determining input and output matching network matrices.

(i) Determine the transistor S matrix $\boldsymbol{S}_\mathrm{T}$ from $\boldsymbol{S}_{\mathrm{G3}}$ by applying the inverse R matrix multiplication techniques described in the previous section:

$$\boldsymbol{R}_\mathrm{T} = \boldsymbol{R}_\mathrm{I}^{-1}\boldsymbol{R}_{\mathrm{G3}}\boldsymbol{R}_\mathrm{O}^{-1} \tag{2.14}$$

(ii) Determine the input load matrix $\boldsymbol{S}_\mathrm{A}$ when $\boldsymbol{S}_{\mathrm{G2}}$ corresponds to the case where the output matching has been removed,

$$\boldsymbol{R}_\mathrm{A} = \boldsymbol{R}_{\mathrm{G2}}(\boldsymbol{R}_\mathrm{T}\boldsymbol{R}_\mathrm{O})^{-1} \tag{2.15a}$$

and the output load matrix $\boldsymbol{S}_\mathrm{B}$ when $\boldsymbol{S}_{\mathrm{G2}}$ corresponds to the case where the input matching has been removed,

$$\boldsymbol{R}_\mathrm{B} = (\boldsymbol{R}_\mathrm{I}\boldsymbol{R}_\mathrm{T})^{-1}\boldsymbol{R}_{\mathrm{G2}} \tag{2.15b}$$

(iii) Determine the remaining unknown load matrix from $\boldsymbol{S}_{\mathrm{G1}}$:

$$\boldsymbol{R}_\mathrm{B} = (\boldsymbol{R}_\mathrm{A}\boldsymbol{R}_\mathrm{T})^{-1}\boldsymbol{R}_{\mathrm{G1}} \tag{2.16a}$$

$$\boldsymbol{R}_\mathrm{A} = \boldsymbol{R}_{\mathrm{G1}}(\boldsymbol{R}_\mathrm{T}\boldsymbol{R}_\mathrm{B})^{-1} \tag{2.16b}$$

2.2.5 Launcher and microstrip line characterization

The technique used in launcher and microstrip line characterization is that pioneered by Deschamps [2.9] in which an unknown two-port network is characterized by successive measurement of the reflection coefficient at one of the ports as a function of a varying but well-known load applied to the other.

In general the variable load is of fixed standing wave ratio (SWR) and varying phase (this is usually accomplished by means of a sliding short). The S parameter equation used in the case of a reciprocal two-port network ($S_{12} = S_{21}$) terminated by a load of reflection coefficient $\Gamma_L = a_2/b_2$ is

$$\Gamma_i = \frac{b_1}{a_1} = S_{11} - \frac{S_{12}^2}{S_{22} - 1/\Gamma_L} \tag{2.17}$$

The unknown network parameters can thus be obtained from a minimum of three measurements of Γ_i for three different known values of Γ_L (Fig. 2.10).

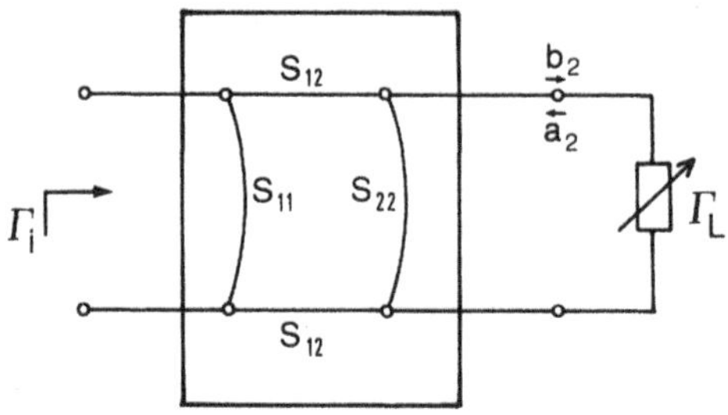

Fig. 2.10 Method of determining the parameters of a reciprocal network.

2.2.5.1 Microstrip line

In the case of a microstrip line we use a variable "open circuit" made up of a 50 Ω microstrip line divided into 2 mm sections separated by 0.1 mm gaps on an alumina substrate 0.635 mm thick [2.10, 2.11].

The S matrix of a homogeneous lossless nonreflecting line of length l is (Table 2.1)

$$\mathbf{S}_{TL} = \begin{bmatrix} 0 & \exp(-\gamma_l) \\ \exp(-\gamma_l) & 0 \end{bmatrix} \tag{2.18}$$

where the propagation constant $\gamma = \alpha + j\beta$, with α the attenuation per unit length and β the phase constant. The attenuation per unit length is principally a function of the width of the line, the thickness of the metallization, the conductivity of the metal, the thickness of the dielectric, the dielectric loss tangent and the operating frequency [2.12]. The phase constant is given by $\omega \varepsilon_{eff}^{1/2}/c$, where ω is the angular frequency of the signal, ε_{eff} is the effective dielectric constant of the microstrip and c is the speed of light.

Initially, all line segments are carefully interconnected with conducting silver paint. The reflection coefficient at the launcher input is then measured at discrete frequency points for line lengths which successively decrease by a fixed decrement d, giving values Γ_0, Γ_1, Γ_2, Γ_3. The propagation constant is then obtained from

$$\{\cosh(\gamma d)\}^2 = \frac{(\Gamma_0 - \Gamma_2)(\Gamma_1 - \Gamma_3)}{4(\Gamma_0 - \Gamma_1)(\Gamma_2 - \Gamma_3)} \tag{2.19}$$

It should be noted that this relationship is independent of both the launchers *and* the characteristic impedance of the microstrip line, provided that the line used is homogeneous.

2.2.5.2 Launcher

Having obtained the propagation constant of the microstrip line, we can now characterize the launcher by successive measurement of Γ_i for various line segment lengths l:

$$\Gamma_i = \frac{\Gamma_L(S_{11}S_{22} - S_{12}{}^2) - S_{11}}{\Gamma_L S_{22} - 1} \tag{2.20}$$

where $\Gamma_L = \Gamma_C \exp(-\gamma_l)$ and Γ_C is the reflection coefficient at plane CC′ of Fig. 2.11. The Γ_0, Γ_1, Γ_2 and Γ_3 measured previously can be used again, and accuracy is enhanced by carrying out at least five measurements.

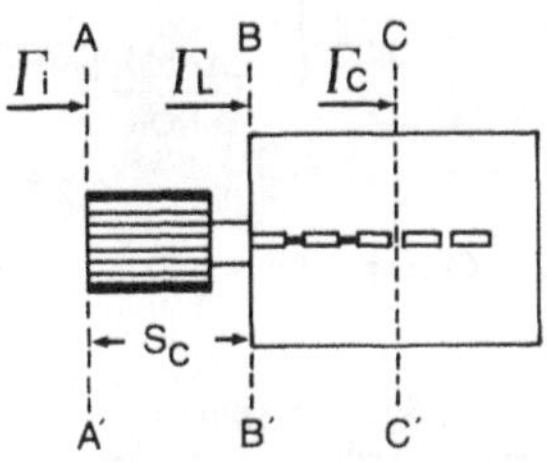

Fig. 2.11 Variable open circuit made up of 50 Ω line segments 2 mm long (0.1 mm gaps).

2.3 *S* parameter measurements—general concepts

A survey of the calibration techniques and calibration standards used most frequently for commercial automatic network analyzers is presented in this section. The concept of systematic error removal using the flowchart approach is analyzed. This concept is extended to fixture and on-wafer device measurements in Sections 2.4 and 2.5.

2.3.1 One-port error modelling

A block diagram of the measurement system, which consists of a sweep oscillator connected to the unknown one-port through a reflectometer, is shown in Fig. 2.12. This imperfect reflectometer can be modelled by taking all the linear errors of the system and combining them into a fictitious two-port error adaptor between the reflectometer and the unknown one-port. This results in an absolutely perfect reflectometer with no loss, infinite directivity, no mismatch and no frequency response errors. The fictitious two-port error adaptor has four error terms which can be reduced to three when ratioed measurements are made [2.13]. The flowchart of the fictitious two-port error adaptor is connected to the unknown reflection coefficient (Fig. 2.13).

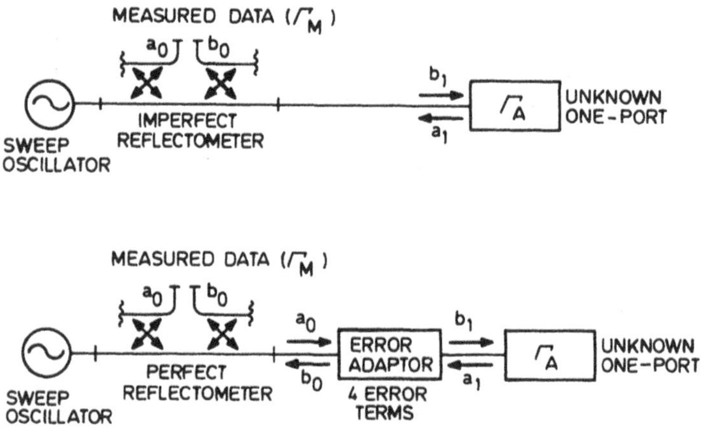

Fig. 2.12 One-port error modelling [2.13].

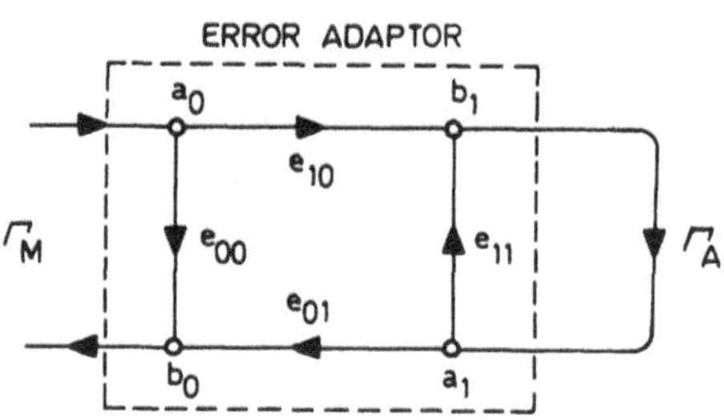

Fig. 2.13 One-port flowchart and solution.

A flowchart is a good method of giving a practical appreciation of how signals are flowing in a microwave network [2.14, 2.15]. For example, suppose that Γ_A is a perfect termination Z_0, equal to the characteristic impedance of the measuring system, but that when Γ_M is measured there are some residual signals caused by the directivity e_{00} of the system. This signal does not flow to the device under test but to the branch of the flowchart labelled e_{00}. The frequency response terms e_{10} and e_{01} are real reproducible frequency-dependent errors due to the couplers, the front-end mixers, the cables and other hardware. The error term e_{11} is the residual reflection coefficient seen by the unknown device when looking into the test port of the measurement system. The relationship between Γ_M and Γ_A can easily be obtained using flowchart reduction techniques [2.16]:

$$\Gamma_M = \frac{b_0}{a_0} = e_{00} + \frac{e_{10}e_{01}\Gamma_A}{1 - e_{11}\Gamma_A} \tag{2.21}$$

$$\Gamma_A = \frac{\Gamma_M - e_{00}}{e_{11}(\Gamma_M - e_{00}) + e_{10}e_{01}} \tag{2.22}$$

2.3.2 One-port calibration techniques

The calibration procedure consists of generating three different measured reflection coefficients Γ_M with three different known loads Γ_A. This yields three equations which can be solved to obtain the error terms e_{00}, $e_{01}e_{10}$ and e_{11}. These loads, which are known as "standards", must meet the following requirements: they must be known exactly over the calibration frequency range; they must be nonredundant; their connection to the test port must be repeatable. The four most popular calibration techniques are discussed below.

2.3.2.1 Three different offset shorts

This technique eliminates the need for a load Z_0 but it is frequency sensitive and bandwidth limited and does not give an accurate value for the error term e_{00}.

2.3.2.2 A load Z_0 (fixed or sliding), a short and an offset short

The offset short is a quarter wavelength at the center of the frequency band. This technique is not broadband but is accurate.

2.3.2.3 A load Z_0 (fixed or sliding), a short and an open circuit

This technique is the most frequently used and gives a good compromise between accuracy, bandwidth and practicality. However, the open circuit has a fringing capacitance that needs to be modelled very precisely. The example in Fig. 2.14 shows the measured reflection coefficient of an APC7 open circuit. If the d.c. capacitance is modelled, the peak-to-peak error is reduced from 8 dB to 1 dB. If a frequency-dependent capacitance term is added, this error is reduced to less than 0.1 dB peak to peak.

2.3.2.4 Two sliding terminations and a short circuit

This technique has the potential of achieving the best performance but is the most time consuming and also the most mathematically complex. It uses a low reflection sliding load and a sliding short. Neither the magnitude and phase of these terminations nor the frequency have to be known exactly. Certain assumptions are made about the sliding terminations: (1) the magnitude of the reflection coefficient does not change as it rotates in phase; (2) airline loss is negligible.

In conclusion, these techniques change an actual reflectometer to a quasi-perfect reflectometer, but error correction does not remove all the errors. Residual errors remaining in the measurement system may be due to imperfect standards, repeatability of switches and cables, noise etc. However, the accuracy of measurements is much improved. In fact the improvement is typically greater than 20 dB (Fig. 2.15).

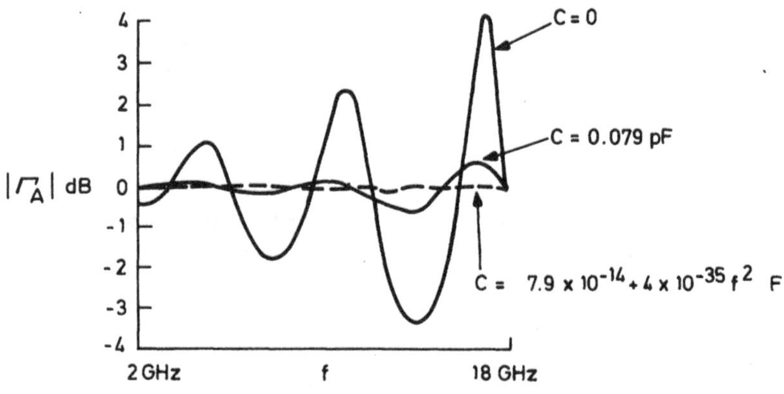

Fig. 2.14 Measured reflection coefficient of an APC7 open circuit as a function of the fringing capacitance model [2.13].

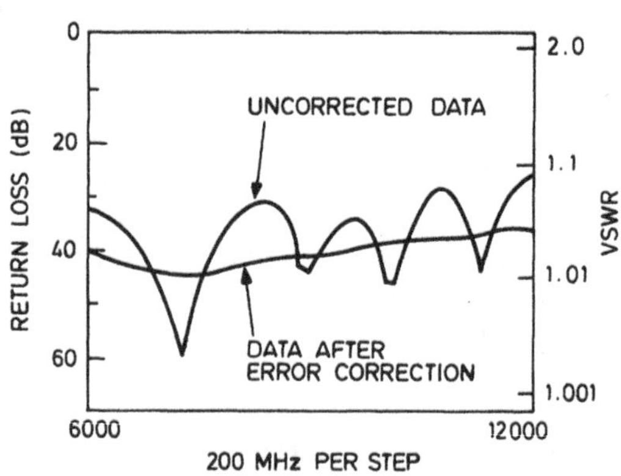

Fig. 2.15 One-port before and after correction [2.13].

2.3.3 Two-port error modelling

The two-port system duplicates part of the one-port measurement system. A dual directional coupler is attached to the input of the unknown two-port and another dual directional coupler is attached to the output. A switch changes the direction of the incident power to the unknown two-port for forward and reverse measurements and terminates the unknown two-port in an impedance Z_0. In this case there are four measurement ports.

As in the previous one-port analysis, all the linear errors of the imperfect reflectometer can be combined into an error adaptor. In this case the fictitious error adaptor has 16 error terms (Fig. 2.16). To reduce the number of errors and take into account the changing port match due to the different switch positions, it is possible to create one fictitious error adaptor for the forward direction and another for the reverse direction.

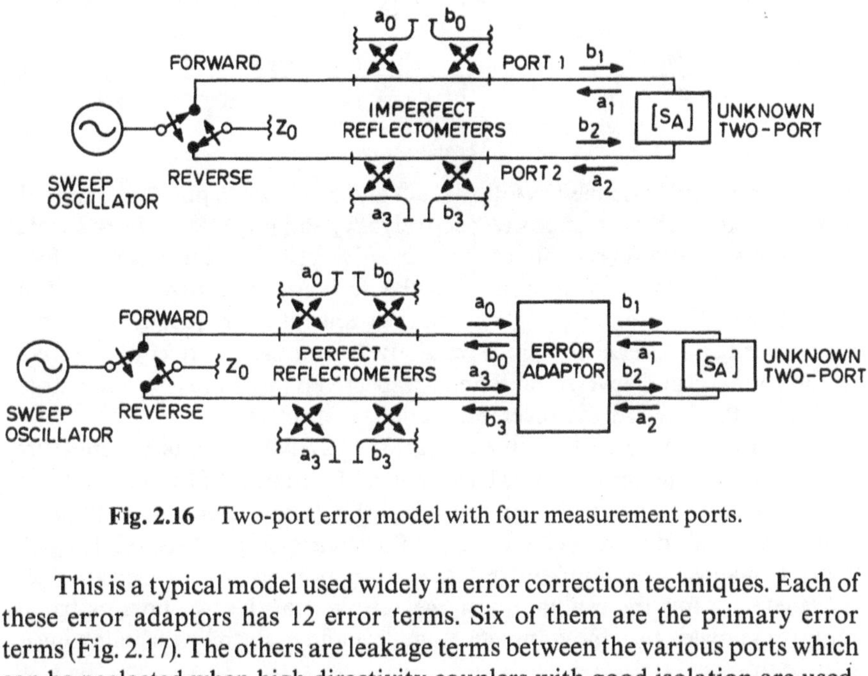

Fig. 2.16 Two-port error model with four measurement ports.

This is a typical model used widely in error correction techniques. Each of these error adaptors has 12 error terms. Six of them are the primary error terms (Fig. 2.17). The others are leakage terms between the various ports which can be neglected when high directivity couplers with good isolation are used. These two sets of six error models lead to the well-known 12-term error model (Figs 2.18 and 2.19) [2.17, 2.18]:

$$S_{11M}=\frac{b_0}{a_0}=e_{00}+e_{10}e_{01}\frac{S_{11A}-e_{22}\Delta^{A}}{1-e_{11}S_{11A}-e_{22}S_{22A}+e_{11}e_{22}\Delta^{A}} \tag{2.23}$$

$$S_{21M}=\frac{b_3}{a_0}=e_{30}+e_{10}e_{32}\frac{S_{21A}}{1-e_{11}S_{11A}-e_{22}S_{22A}+e_{11}e_{22}\Delta^{A}} \tag{2.24}$$

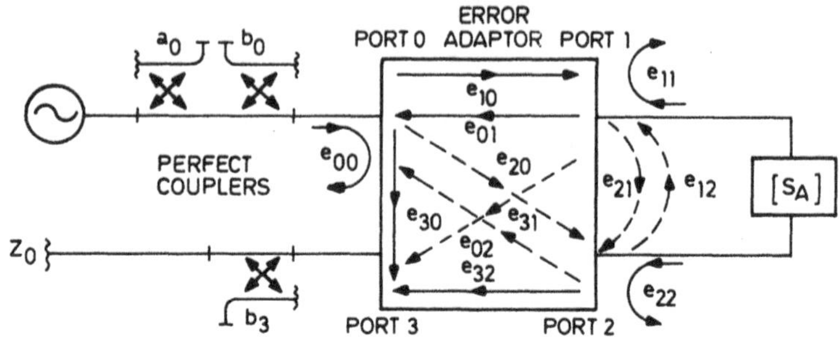

Fig. 2.17 Forward configuration with three measurement ports.

$$S_{22M}=\frac{b'_3}{a'_3}=e'_{33}+e'_{23}e'_{32}\frac{S_{22A}-e'_{11}\Delta^A}{1-e'_{11}S_{11A}-e'_{22}S_{22A}+e'_{11}e'_{22}\Delta^A} \tag{2.25}$$

$$S_{12M}=\frac{b'_0}{a'_3}=e'_{03}+e'_{23}e'_{01}\frac{S_{12A}}{1-e'_{11}S_{11A}-e'_{22}S_{22A}+e'_{11}e'_{22}\Delta^A} \tag{2.26}$$

$$\Delta^A=S_{11A}S_{22A}-S_{21A}S_{12A} \tag{2.27}$$

The error terms are defined as follows: e_{11} and e'_{11} refer to port 1 match in the forward and reverse directions respectively; e_{22} and e'_{22} refer to port 2 match in the forward and reverse directions respectively; $e_{10}e_{01}$ and $e'_{23}e'_{32}$ refer to the reflected frequency response in the forward and reverse directions respectively; $e_{10}e_{32}$ and $e'_{23}e'_{01}$ refer to the transmitted frequency response in the forward and reverse directions respectively; e_{00} and e'_{33} refer to directivity in the forward and reverse directions respectively; e_{30} and e'_{03} refer to the leakage in the forward and reverse directions respectively.

It should be noted that in eqns (2.23)–(2.27) all terms of the measured matrix $\boldsymbol{S}_M$ are functions of error terms *and* all four terms of the actual device-under-test (DUT) matrix $\boldsymbol{S}_A$. Extraction of the various elements of the DUT matrix is achieved by reducing the error flowcharts (Figs 2.18 and 2.19) [2.19, 2.20]. Conversely, each unknown S_{ijA} is a function of the four measured S_{ijM} parameters. Thus the evaluation of *each* element of the two-port matrix $\boldsymbol{S}_A$ requires knowledge of the complete $\boldsymbol{S}_M$ matrix and is given by the following set of equations:

$$S_{21A}=\frac{1}{D}\frac{S_{21M}-e_{30}}{e_{10}e_{32}}\left\{1+\left(\frac{S_{22M}-e'_{33}}{e'_{23}e'_{32}}\right)(e'_{22}-e_{22})\right\} \tag{2.28}$$

$$S_{11A}=\frac{1}{D}\left\{\frac{S_{11M}-e_{00}}{e_{10}e_{01}}\left(1+\frac{S_{22M}-e'_{33}}{e'_{23}e'_{32}}e'_{22}\right)-e_{22}\frac{S_{21M}-e_{30}}{e_{10}e_{32}}\frac{S_{12M}-e'_{03}}{e'_{23}e'_{01}}\right\} \tag{2.29}$$

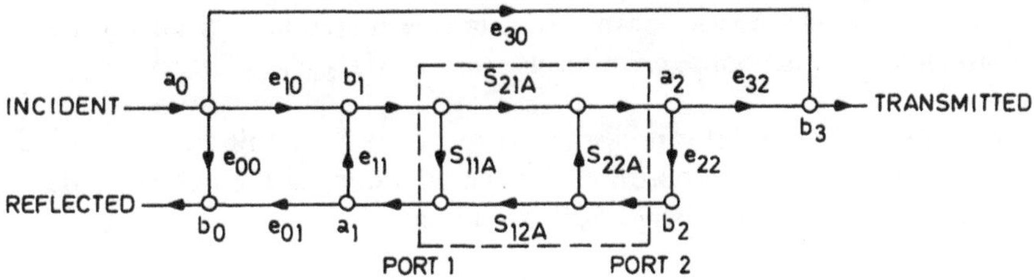

Fig. 2.18 Two-port forward flowchart with three measurement ports.

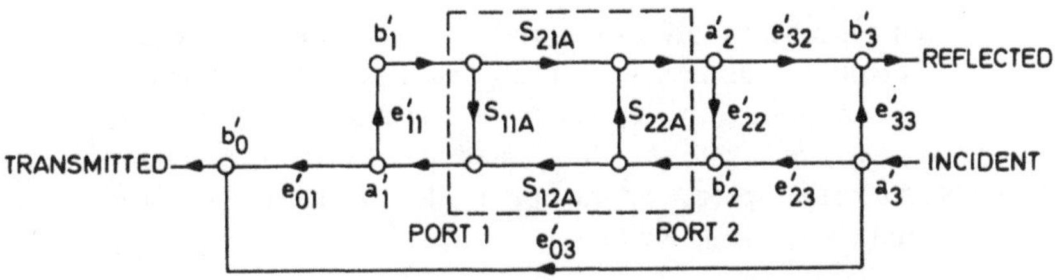

Fig. 2.19 Two-port reverse flowchart with three measurement ports.

$$S_{12A} = \frac{1}{D} \frac{S_{12M} - e'_{03}}{e'_{23}e'_{01}} \left\{1 + \left(\frac{S_{11M} - e_{00}}{e_{10}e_{01}}\right)(e_{11} - e'_{11})\right\} \tag{2.30}$$

$$S_{22A} = \frac{1}{D} \left\{\frac{S_{22M} - e'_{33}}{e'_{23}e'_{32}} \left(1 + \frac{S_{11M} - e_{00}}{e_{10}e_{01}} e_{11}\right) - e'_{11} \frac{S_{21M}e_{30}}{e_{10}e_{32}} \frac{S_{12M} - e'_{03}}{e'_{23}e'_{01}}\right\} \tag{2.31}$$

where

$$D = \left(1 + \frac{S_{11M} - e_{00}}{e_{10}e_{01}} e_{11}\right)\left(1 + \frac{S_{22M} - e'_{33}}{e'_{23}e'_{32}} e'_{22}\right) - \frac{S_{21M} - e_{30}}{e_{10}e_{32}} \frac{S_{12M} - e'_{03}}{e'_{23}e'_{01}} e_{22}e'_{11} \tag{2.32}$$

2.3.4 The two-port standard calibration procedure

The standard calibration procedure comprises three steps.

(1) Ports 1 and 2 are calibrated using the procedure described in Section 2.3.2. The error terms e_{00}, e_{11}, $e_{01}e_{10}$, e'_{33}, e'_{22} and $e'_{32}e'_{23}$ are then determined.

(2) The leakage terms e_{30} and e'_{03} are measured with loads Z_0 connected to ports 1 and 2.

(3) Ports 1 and 2 are connected together and the port match e_{22} and e'_{11} are determined since ports 1 and 2 have been calibrated during step (1). The

two transmission terms S_{21M} and S_{12M} enable the equations for the transmission frequency response error terms $e_{10}e_{32}$ and $e'_{23}e'_{01}$ to be solved.

It should be noted that it is necessary to know the Z_0 termination accurately during this calibration procedure. The actual values of the other standards (shorts, offset shorts and open) are also required.

2.3.5 The two-port self-calibration procedure

The self-calibration procedure requires very little knowledge of the standards. It cannot take into account the leakage errors, which are assumed to have been determined in step (2) of the standard procedure. When these error terms have been removed, the measurement system can be modified so that it has an error adaptor X on the port 1 side and an error adaptor Y on the port 2 side of the DUT (Fig. 2.20).

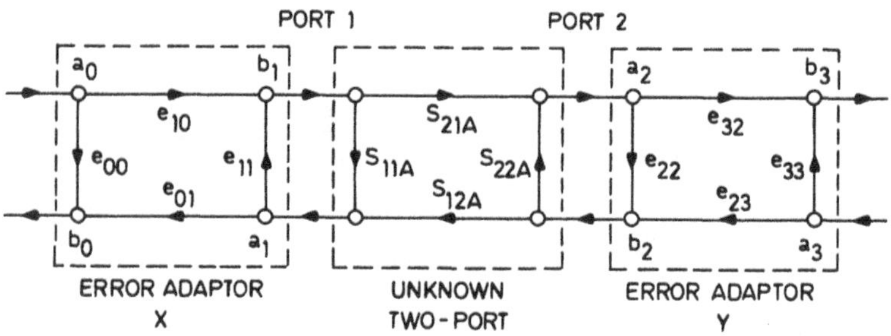

Fig. 2.20 Error model for self-calibration.

The basic idea of this procedure [2.21, 2.22], which is sometimes called the through–short–delay (TSD) procedure, is to replace the load Z_0 of the standard procedure by the S parameter measurement of two transmission lines with the same characteristic impedance but different physical lengths. The three steps in this procedure are summarized in Fig. 2.21. It is not necessary to know the high reflection termination which is connected to the system ports. This is used as a constant to solve the error terms.

A total of 10 measurements are made during the self-calibration procedure. Only eight unknown error terms need to be determined. The redundant information can be used to calculate the propagation constant of the delay line and the reflection coefficient of the high reflection termination that was placed on the test ports. This can be used as a check of the calibration. An example is given in Section 2.5.

Details of the determination of the error terms [2.23] are given in Appendix 2.2 at the end of this chapter.

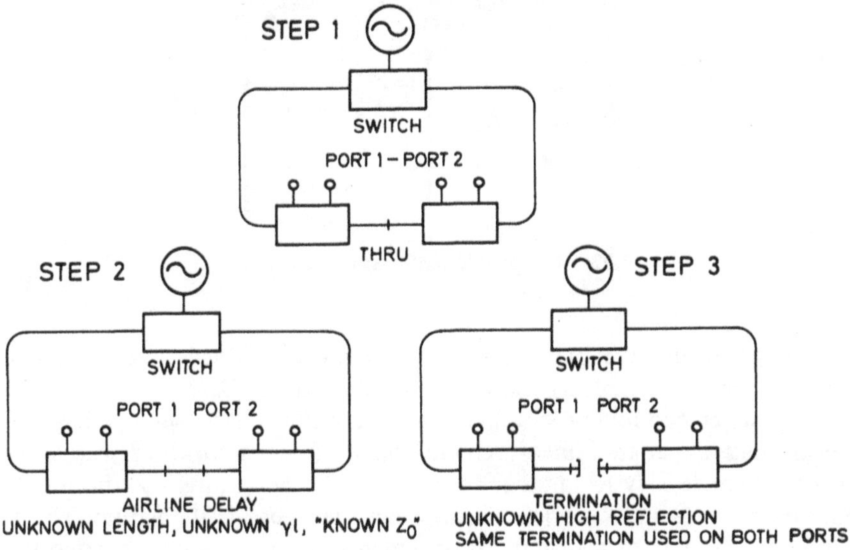

Fig. 2.21 Self-calibration procedure.

2.4 Measurement of the fixture *S* parameters

2.4.1 Statement of the problem: the concept of de-embedding

A large class of network measurements have the property that the actual device to be measured cannot be connected directly to the connection available at the measurement plane of a calibrated automatic network analyzer. This class of DUT measurements can be termed fixture measurements because the DUT is separated from the calibrated automatic network analyzer by some kind of transitional network or fixture (Fig. 2.22).

This is the case for measurement of chip or packaged transistors. As a consequence, the measurement plane which can be obtained through calibration at the coaxial connection is separated from the DUT by a network—the fixture. In general this fixture is not lossless and reflectionless. It may comprise connectors, transitions between different types of transmission lines and the connections to the DUT. Although more restrictive than the original definition [2.24] in one way, the term "de-embedding" can be used to describe error correction by a process of establishing measurement planes different from those provided by calibration. It is important to note that the term de-embedding does not refer to the method used for calibrating the networks between the DUT and the data collection planes but only to the process of

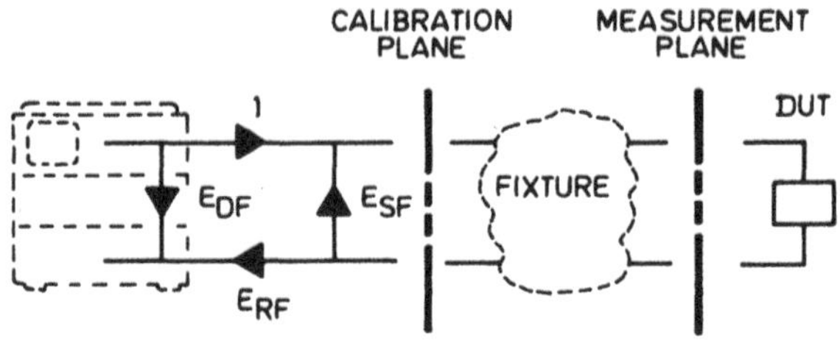

Fig. 2.22 The problem of fixture measurement.

providing measurement planes different from those obtained by conventional automatic network analyzer calibration with standards [2.25].

The de-embedding process can also be used to shift noise impedance or power load impedance measurement planes from a location where the measurement is easy to another location inside the fixture, e.g. the input–output planes of active devices that are matched outside the fixture. This is particularly useful for load-pull measurements or noise parameter determination. Various cases where the de-embedding is used to good effect are described in detail in Sections 2.2.3–2.2.5, and we shall not discuss them further here.

2.4.2 Fixture considerations

A number of issues must be considered in designing and using a fixture for the measurement of a DUT to ensure that optimum results are obtained. Because of the wide variety of devices to be measured, there is no universal solution to the fixture design problem. In fact, in the limit each device is unique and requires its own specific measurement system. However, some general considerations from which to start the design of a test fixture may be useful.

2.4.2.1 Repeatability of the test fixture

Fixture repeatability establishes fundamental bounds on accuracy since nonrepeatable errors cannot be corrected.

The fixture characteristics should not change between the calibration and the measurements or between successive measurements. This problem can generally be solved by a careful mechanical design. The test fixture and, if possible, the DUT chip carrier or package should be designed with tight margins to ensure accurate positioning of the DUT. Special attention has to be given to the problem of grounding. The ground path has to be localized near the input–output r.f. signal paths, particularly at the critical interfaces.

For low power applications (typically below 2 W) the coplanar chip carrier is superior to the microstrip carrier because the r.f. and ground contacts are situated in the same plane. A repeatability of ± 0.04 dB in amplitude and $\pm 0.2°$ in phase over 50 measurements has already been reported [2.26]. The microstrip carrier requires a much more complicated text fixture design and closer tolerance control of the substrate thickness.

2.4.2.2 Independence of the test fixture

The test fixture should be electrically invariant with respect to the DUT. This independence of the fixture from the device being measured is essential if the linear model in the flowcharts described in the previous section is to apply. Each error term has to be independent of the DUT. Dependence can occur when discontinuities or couplings are located near the DUT. Such fixture discontinuities must be made small enough and separated far enough from the measurement plane to allow a linear model to describe the fixture adequately. It is important to minimize leakage terms due to imperfect ground paths. The amplitude of these terms is often dependent on the device topology.

2.4.2.3 Characterization of the test fixture

Since a DUT measurement requires the fixture to be characterized, enough standards should be used to allow this. This can be accomplished by conventional calibration with device plane standards or by measuring the standards or pseudo-standards and using computer optimization to fit the data to a model. Some examples are given in Section 2.4.3.

2.4.2.4 Convenience and practicality of the test fixture

The previous items may be necessary but are not sufficient to provide useful measured data. For this to be the case, operation must be reasonably convenient. It should not be necessary to recalibrate the fixture constantly, and the characterization process needs to be acceptable. Multiple measurements of standards which are hard to connect may be unacceptable. Additionally, the DUT itself should be reasonably easy to install in the fixture.

This list of considerations is not intended to be exhaustive but rather a starting point for fixture design. The question of the electrical quality of the test fixture (low losses, low voltage standing wave ratio (VSWR)) is not critical in the case of *S* parameter measurements because these errors are embedded with those of the network analyzer and associated connecting cables and are easily eliminated through the calibration process. A VSWR of 2.5:1 can be corrected if it is repeatable. It would be unacceptable, however, for noise or power measurements.

The only essential electrical characteristic of the test fixture is isolation between the test ports.

2.4.3 Calibration of the fixture

Calibration of the fixture is not always necessary for network analysis applications. Direct calibration of the automatic network analyzer is faster and in some cases more accurate as will be seen in the next section.

Fixture calibration can be used when the insertion of standards is difficult or time consuming. In this case the automatic network analyzer is first calibrated using one of the methods described in Sections 2.3.4 and 2.3.5, and then the fixture is calibrated from these calibration planes using secondary standards or de-embedding.

The de-embedding can be processed by a computer from the data measured at the automatic network analyzer calibration plane and the error model of the fixture, or in real time by the automatic network analyzer. In the latter case the computer calculates the result of the cascade of the error flowchart of the automatic network analyzer and the fixture and redirects the equivalent error flowchart into the internal memory of the analyzer. If the fixture is rarely dismounted, it is only necessary to repeat the automatic network analyzer calibration and the cascading operation from time to time.

Fixture calibration is also needed when there are no standards available at the chip or package level. A good example of such a calibration is used with the transistor test fixture (TTF) manufactured by the Maury Microwave Corporation (model MT950) or Hewlett-Packard (model HP85841A) (Fig. 2.23). The TTF can be dismounted and calibrated using a coaxial calibration

Fig. 2.23 Transistor test fixture (TTF). (Courtesy of Maury Microwave.)

kit. In fact, when the packaged transistor is placed in the proper insert the calibration planes are not those which interest the designer (Fig. 2.24). The effects of coupling between the different ground planes and also of the transistor leads going across an air gap can be modelled [2.27] by two C_1–L–C_2 networks describing the transition regions between the calibration planes and the DUT planes (Fig. 2.25). The values of the LC components are not known and can be extracted from the measurements of pseudo-standards known as insert calibration devices (Fig. 2.26) using an optimization program. A single set of C_1–L–C_2 values is then used to calculate the equivalent error

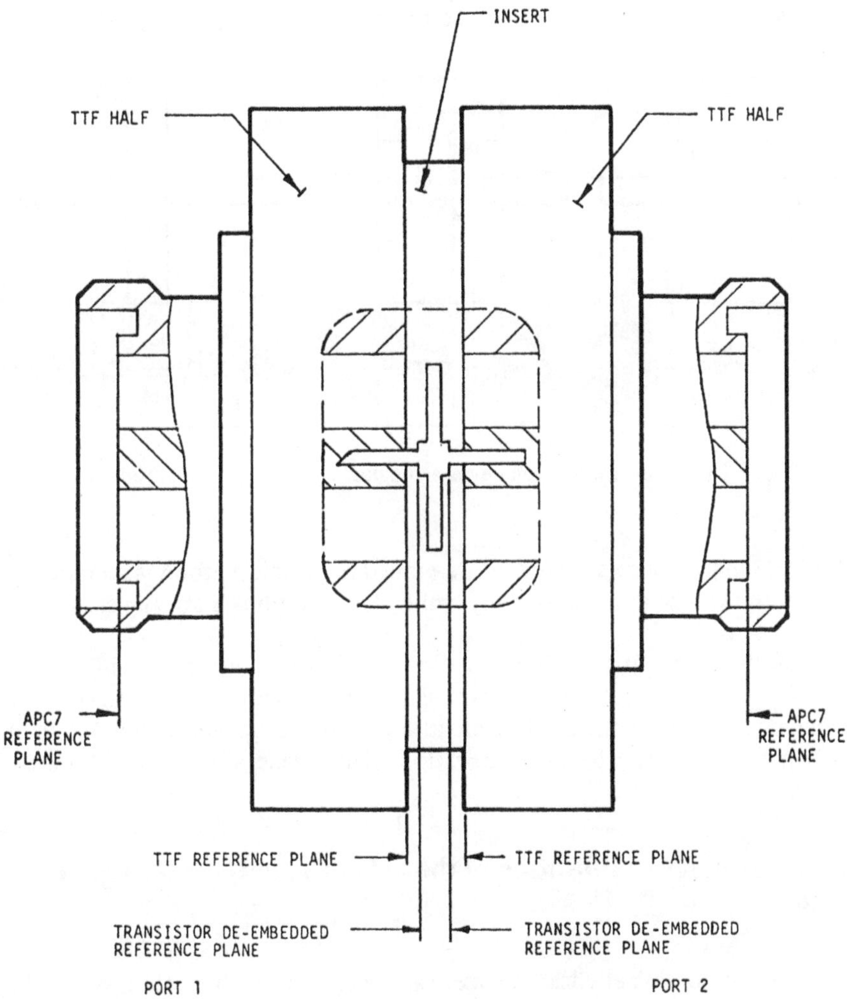

Fig. 2.24 Reference planes of the TTF.

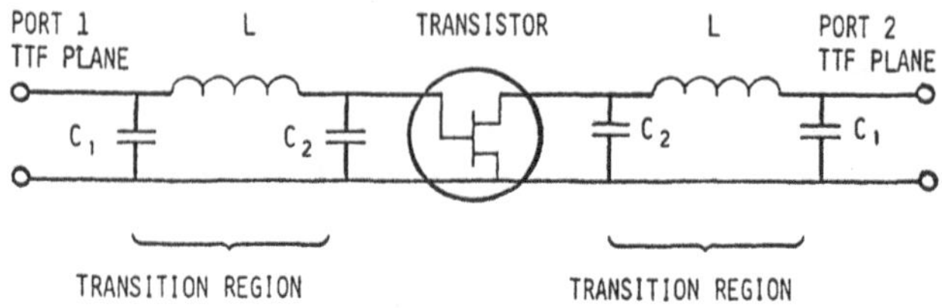

Fig. 2.25 C–L–C π network used to model insert transition regions for de-embedding transistor data.

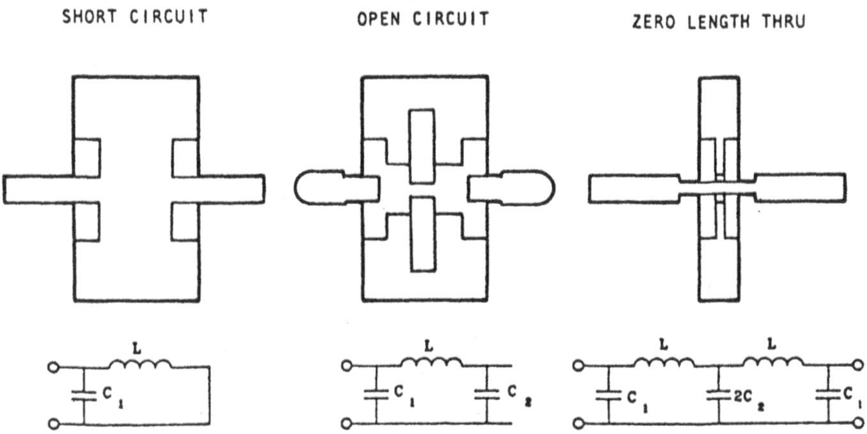

Fig. 2.26 Insert calibration devices and corresponding circuit models.

adaptor S matrix necessary for the de-embedding. This method is completely general and can be applied to many embedded situations if no chip standards are available.

Another application of fixture calibration has been mentioned in Section 2.4.1. In this case the intrinsic characteristics of the fixture are used to shift the noise or power measurements from the automatic network analyzer calibration planes to the fixture calibration plane (generally the DUT plane) [2.28].

2.4.4 Standard calibration of the network analyzer through a fixture

The automatic network analyzer can be calibrated through a fixture if a set of standards as described in Section 2.3.2 is available. Owing to the difficulty of realizing a good Z_0 load in a chip form which is usable at high frequency, it is simpler to design a set of open and/or short circuits in coplanar or microstrip configuration.

The coplanar calibration kit is simple to design because the short circuit has a low parasitic inductance and the open circuit can be made at the end of a tapered 50 Ω transmission line, thus minimizing the fringing end effect.

The microstrip calibration kit can be composed of a set of four different open-circuit lines as already described in previous work [2.10, 2.11]. In all these applications a good knowledge of the effective dielectric constant in the transmission medium is necessary. In addition, the series inductance of the grounding via holes and the shunt capacitance of the open-circuit lines should be carefully modelled.

2.4.5 Self-calibration of the network analyzer through a fixture

As already seen in Section 2.3.5 self-calibration does not require critical standards. Three standards are used with a total of four connections, whereas the standard calibration needs a minimum of four standards and seven connections. The following considerations may be useful in designing these standards.

2.4.5.1 The through line

The through line is simple to realize in coplanar, microstrip or slotline configurations. The only critical parameter is the characteristic impedance Z_0, which can be checked by time domain reflectometry measurements. Line losses are not critical and are perfectly corrected.

The reference planes are given by the zero length through line. Since zero length is not physically realizable, it is interesting to use a nonzero length section l of transmission line. In this case the reference planes are located at the middle of this section. Each error adaptor X and Y (see Section 2.3.5) includes a half-length of transmission line of characteristic impedance Z_0. When this technique is used, the delay line can become very similar to the through line; this important point will be discussed when the delay line is described. The length l of the through line should be greater than the longest DUT to be measured to ensure the independence of the fixture (see Section 2.4.2).

By using the effective propagation constant given by the self-calibration, it is very simple to shift the center calibration planes to the actual location of the DUT (Section 2.2.3).

2.4.5.2 The delay line

The delay line is simply a through line of length l with an excess length dl. Great care should be taken to avoid redundancy with the transmission coefficient of the through line. The transmission phase shift of the dl section must be less than 180° at the upper frequency. If this condition cannot be met physically, the designer should carefully choose the frequency band and the frequency steps to eliminate anomalous results.

A second length limitation occurs if the delay line is slightly more reflective than the through line. This may be the case if the test fixture is designed for a constant-length DUT to avoid excessive dismounting. The delay line has to be folded (Fig. 2.27). The 90° angles create discontinuities which cannot be completely compensated by chamfering. Simulations and experimental results show that the induced errors mainly affect the S_{21} parameter of the MESFET. This effect increases rapidly with the amplitude of the discontinuities and is very obvious at low frequencies where the phase shift of the dl section is a minimum. Since it is related to the quality of the delay line, it is difficult to fix a minimum value for dl. The following solutions are possible: use of an adjustable test fixture with straight through and delay lines; use of two similarly shaped delay lines (see Fig. 2.27); limitation of the minimum measurement frequency; use of a large dl and elimination of the aberrant points. The effective propagation constant is also better determined using two identically shaped delay lines. The effect of the discontinuities on the true length of the lines is thus removed [2.26].

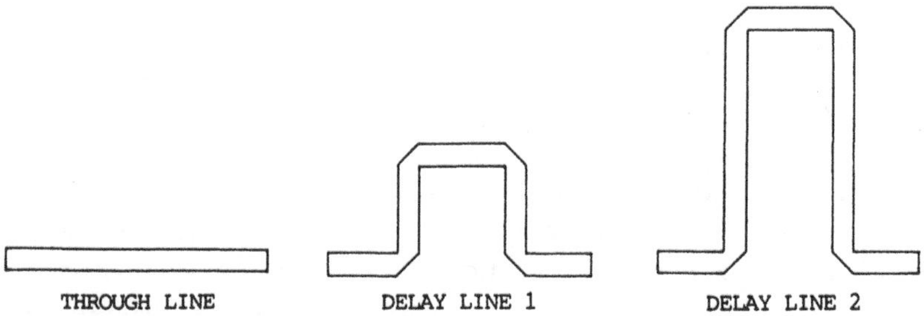

Fig. 2.27 Microstrip through and delay lines for self-calibration.

2.4.5.3 The high reflection termination

The high reflection termination can be one of those discussed in the previous section. The difference is that the parasitics do not need to be modelled because their value is not required. An approximate knowledge of the phase of the standard is a good aid to the calibration program (see Appendix 2.2).

In conclusion the self-calibration method is well adapted to the problem of fixture measurements. A drawback is that it needs a computer program to solve the error flowchart and load it into the network analyzer. Nevertheless good results can be obtained and examples are given in Section 2.5.

2.5 On-wafer *S* parameter measurements

The possibility of measuring microwave *S* parameters directly on semiconductor wafers was first suggested in 1981 [2.29] and was demonstrated experimentally in 1982 [2.30]. A measuring system working up to 26.5 GHz is now commercially available. An example of a two-port wafer-probing system is shown in Fig. 2.28. A complete on-wafer microwave measurement station can be realized using an automatic probe station and an automatic network analyzer (Fig. 2.29).

For this application the DUT must have an access cell compatible with the topology of the probe tips. Since these tips are coplanar, the DUT must provide ground planes at the top of the substrate. Figure 2.30 shows the two

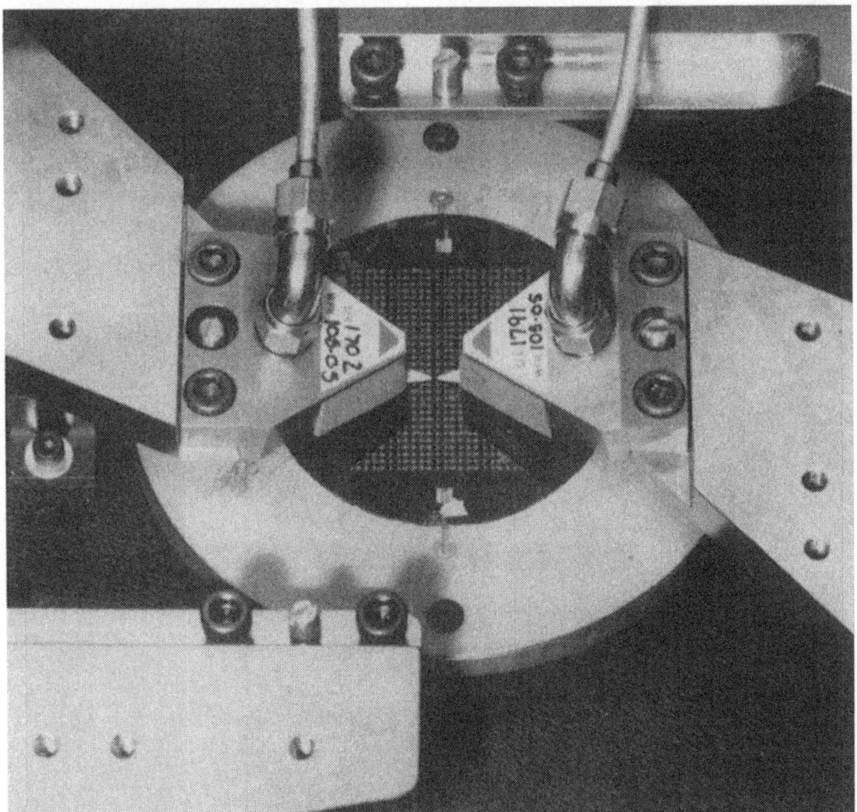

Fig. 2.28 Two-port microwave wafer probe. (Courtesy of Cascade Microtech.)

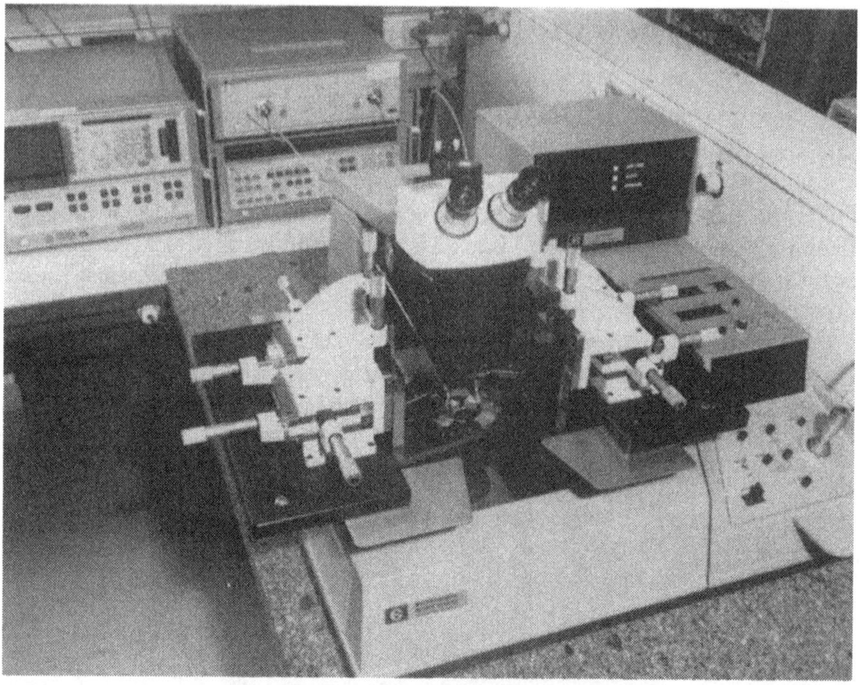

Fig. 2.29 On-wafer microwave measurement station. (Courtesy of Thomson-SC.)

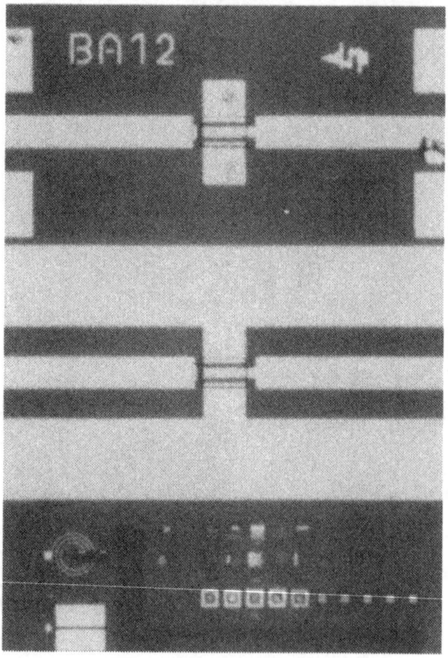

Fig. 2.30 Test pattern for MESFET microwave probing. (Courtesy of Thomson-SC.)

possible patterns. Above, the GaAs MESFET with sources grounded with via holes is connected to the input–output planes by two 50 Ω transmission lines. The access pads are also grounded with via holes at each side of the transmission line. Below, the GaAs MESFET is connected in a 50 Ω coplanar configuration.

A network analyzer calibration kit can be defined for these two different configurations. For instance, a coplanar calibration kit is shown in Fig. 2.31. All the standards necessary for self-calibration or standard calibration can be used: open and short circuits, through and delay lines, 50 Ω reflection or transmission loads etc.

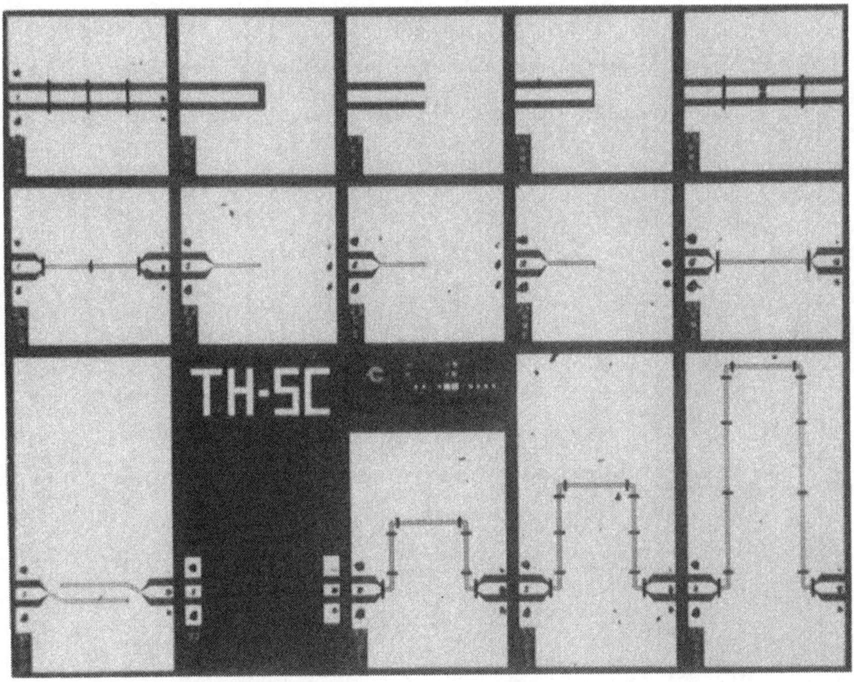

Fig. 2.31 Coplanar on-wafer calibration kit. (Courtesy of Thomson-SC.)

Figure 2.32 shows the experimental result that can be achieved using direct self-calibration of the network analyzer. The determination of the open circuit using the procedure described in Appendix 2.2 is accurate to ± 0.08 dB in amplitude and $\pm 0.7°$ in phase up to 26 GHz. Another typical result is given in Fig. 2.33. The *S* parameters measured for 20 coplanar 150 μm MESFETs of the type shown in Fig. 2.30 exhibit a very narrow spread which could not be obtained with conventional fixtures.

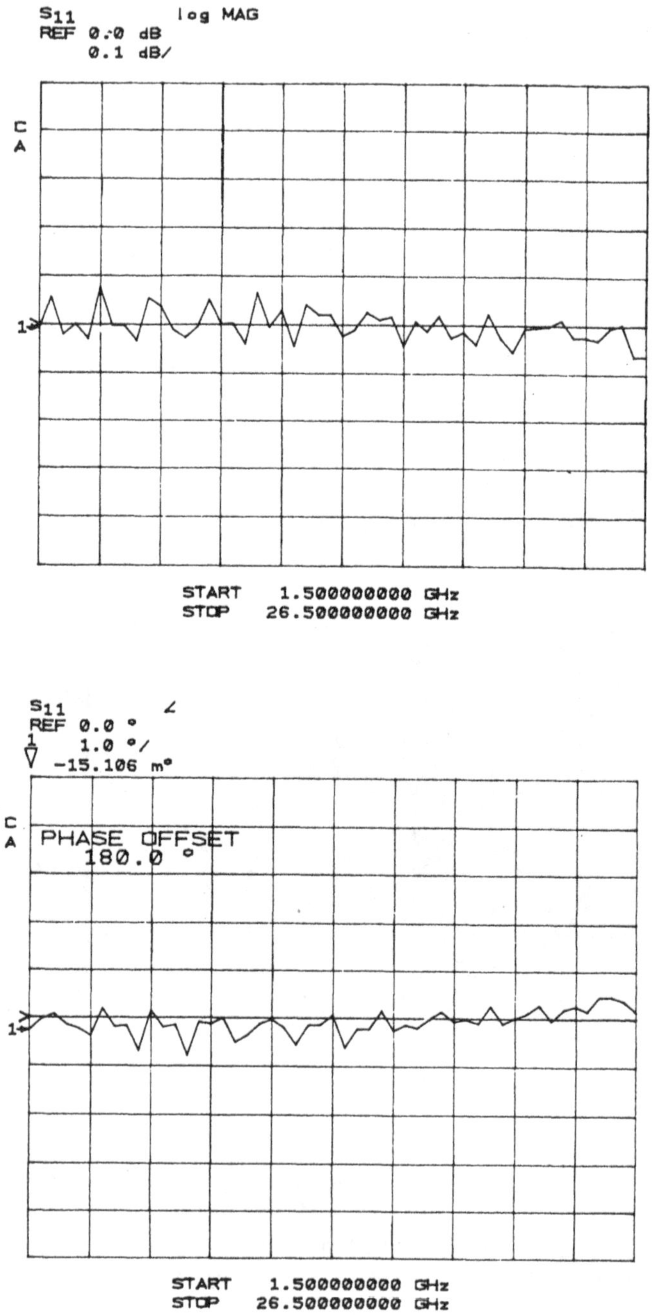

Fig. 2.32 Determination of the open circuit by self-calibration.

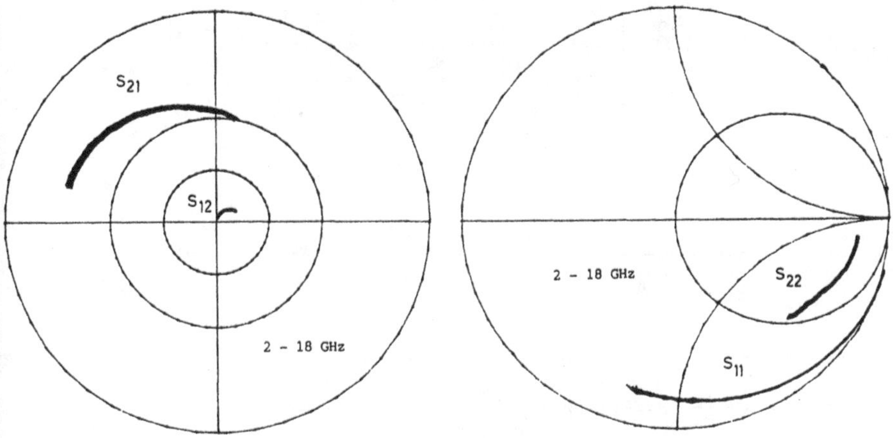

Fig. 2.33 On-wafer S parameter measurement of 20 coplanar 150 μm GaAs MESFETs.

2.6 Utilization of *S* parameters

2.6.1 GaAs MESFET equivalent circuit

When the S parameters of the GaAs MESFET have been measured, it is often convenient to represent its r.f. behavior by an equivalent circuit. Such a circuit must simultaneously have a physical significance and accurately simulate the S parameter performance of the transistor over a wide range of frequencies. A frequently used equivalent circuit is that given in Fig. 2.34. This circuit will be referred to many times in other sections of this book, notably in Chapter 3 on the design of low noise amplifiers.

The circuit (Fig. 2.34(a)) comprises two parts, one representing the intrinsic, or fundamental, behavior of the device and the other representing certain parasitic, or extrinsic, effects. The circuit within the broken lines is often taken as the intrinsic circuit, although it includes many parasitic effects such as the access resistances R_s and R_d and the gate metallization resistance R_g. These effects are often measured under d.c. conditions using the techniques described in Chapter 1. The reactive elements of the "intrinsic" circuit can be obtained using 10 MHz LC bridge measurements [2.31] or S parameter measurements made between 100 MHz and 2 GHz [2.32], or by fitting a simplified "passive device" equivalent circuit (obtained by setting the drain–source voltage V_{DS} to zero and carrying out measurements at several gate voltage values) to S parameter measurements carried out in the range 4–18 GHz [2.33]. External package elements (Fig. 2.34(b)) can be obtained in a

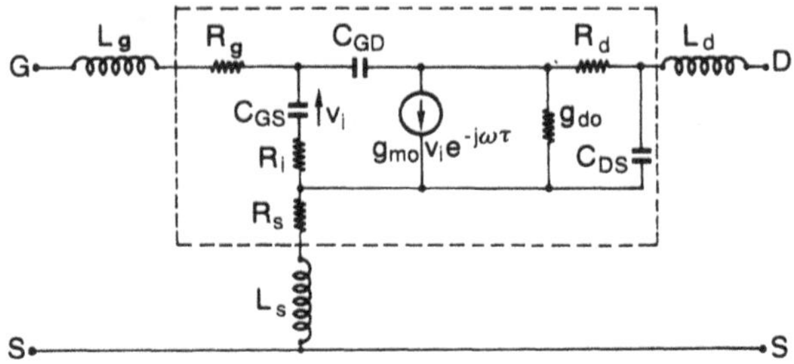

(a)

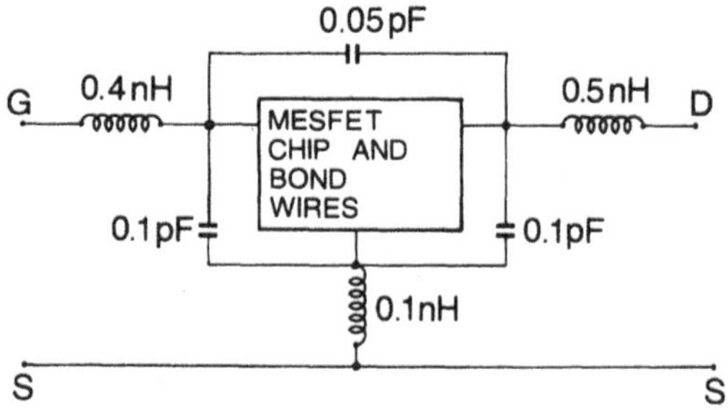

(b)

Fig. 2.34 (a) Transistor equivalent circuit; (b) package equivalent circuit.

similar manner. When the largest number of equivalent circuit elements has been fixed, values for the remainder are obtained by measuring the GaAs MESFET at frequencies from 2 to 26 GHz and using an optimization program, preferably with a least squares error function algorithm, to ensure that the equivalent circuit reflects r.f. behavior over this frequency range. An example of an equivalent circuit fitted to measured *S* parameter data is given in Fig. 2.35.

In order to be able to switch from a common source to a common gate or common drain representation of the GaAs MESFET, it is often useful to express the three-port equivalent circuit of the device directly in *Y* parameters. This can be done with respect to the simplified equivalent circuit of Fig. 2.36 by applying the following rules [2.34].

(i) The number of rows (and columns) in the indefinite matrix is equal to the number of network ports.

(ii) Each term Y_{ii} on the main diagonal associated with a particular terminal i is the sum of all the admittances incident at this terminal.

(iii) Each off-diagonal term Y_{ik} associated with the ports i and k is the negative of all the admittances connected directly between these ports.

(iv) The sum of each row and each column of the indefinite impedance matrix must be zero if the circuit is to obey Kirchhoff's laws.

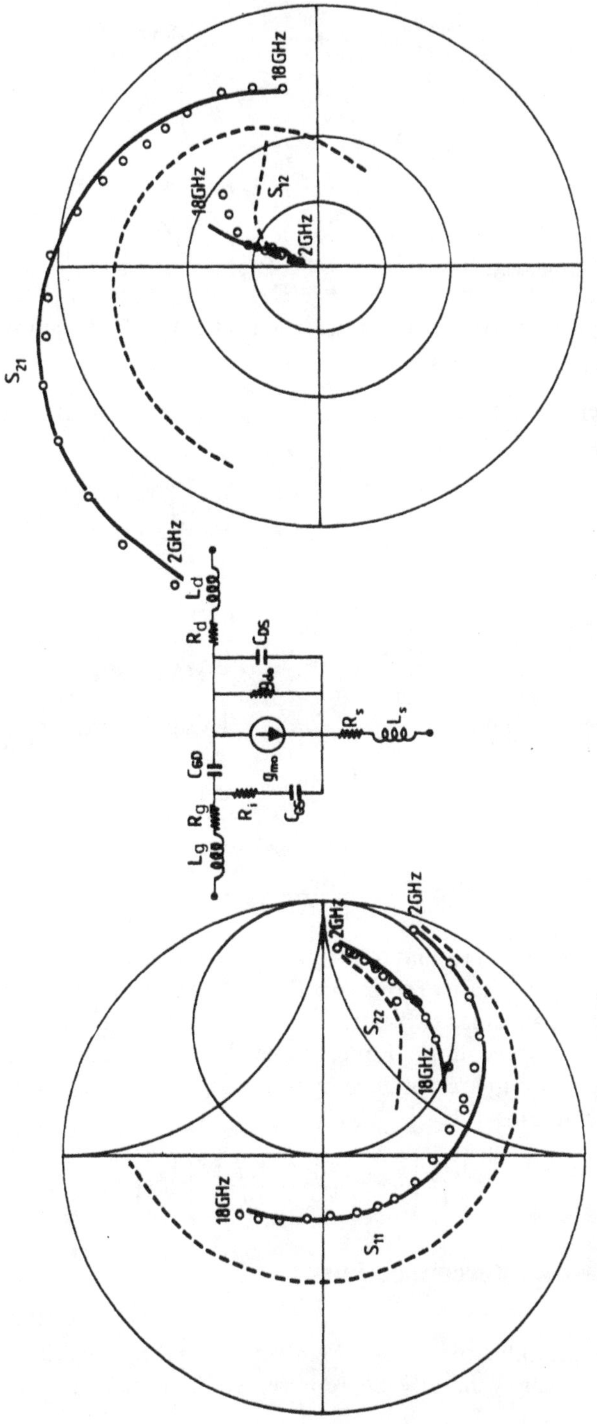

Fig. 2.35 Fit of modelled to measured S parameters: - - -, before optimization; ——, after optimization.

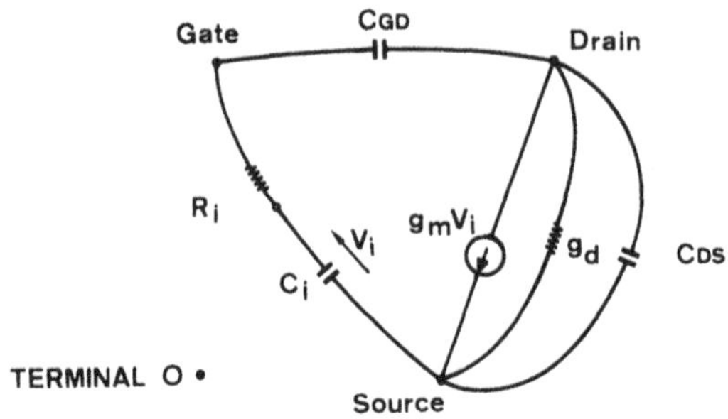

Fig. 2.36 Simplified Y parameter representation of the MESFET equivalent circuit.

We can thus write the Yparameter indefinite impedance matrix as follows:

$$\begin{bmatrix} I_g \\ I_d \\ I_s \end{bmatrix} = \begin{bmatrix} Y_{GG} & Y_{GD} & Y_{GS} \\ Y_{DG} & Y_{DD} & Y_{DS} \\ Y_{SG} & Y_{SD} & Y_{SS} \end{bmatrix} \begin{bmatrix} V_g \\ V_d \\ V_s \end{bmatrix} \tag{2.33}$$

where

$$Y_{GG} = \frac{sC_i}{1+sC_iR_i} + sC_{GD} \qquad Y_{DG} = \frac{g_m}{1+sC_iR_i} - sC_{GD}$$

$$Y_{GD} = -sC_{GD} \qquad Y_{DD} = g_d + sC_{DS} + sC_{GD}$$

$$Y_{GS} = \frac{-sC_i}{1+sC_iR_i} \qquad Y_{DS} = -\left(g_d + sC_{DS} + \frac{g_m}{1+sC_iR_i}\right)$$

$$Y_{SG} = -\frac{sC_i + g_m}{1+sC_iR_i}$$

$$Y_{SD} = -(g_d + sC_{DS})$$

$$Y_{SS} = \frac{sC_i + g_m}{1+sC_iR_i} + g_d + sC_{DS}$$

The complex variable s is equal to $\sigma + j\omega$ and is limited in this case to $j\omega$, and $g_m = g_{mo}\exp(-j\omega\tau)$ where τ represents the electron transit time delay.

To obtain a particular two-port configuration, it is sufficient to eliminate the unwanted row and column. For example, the common drain S parameters of the equivalent circuit are obtained by removing the second row and column of eqn (2.33) to obtain

$$\begin{bmatrix} I_g \\ I_s \end{bmatrix} = \begin{bmatrix} Y_{GG} & Y_{GS} \\ Y_{SG} & Y_{SS} \end{bmatrix} \begin{bmatrix} V_g \\ V_s \end{bmatrix} \tag{2.34}$$

2.6.2 Network interconnections

It is often desirable to "dress" or embed a circuit by the addition of extra elements which can generally be represented in terms of a Y parameter configuration for elements to be added in parallel, a Z parameter configuration for elements to be added in series, or R or $ABCD$ parameters for elements to be cascaded. The various conversions between parameter sets in the case of two-port networks are given in Appendix 2.1. We shall describe certain n-port network interconnections in this section.

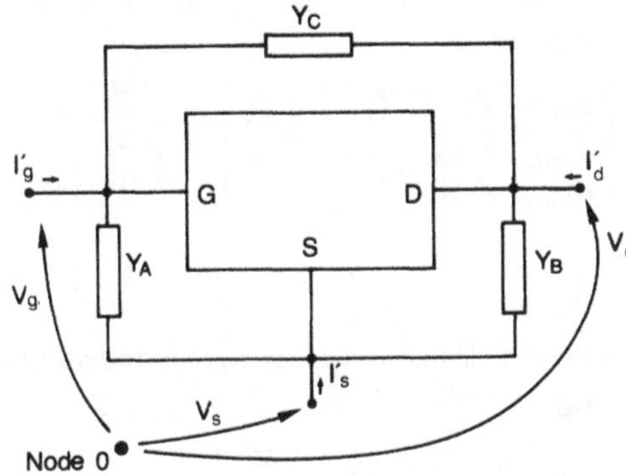

(a)

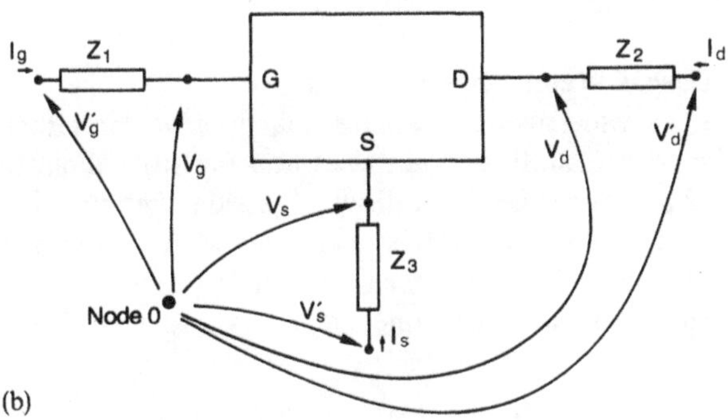

(b)

Fig. 2.37 (a) Parallel embedding of an n-port network; (b) series interconnections.

2.6.2.1 Parallel and series connections

Consider first the case shown in Fig. 2.37(a) in which we seek to embed a three-port Y parameter network, such as the GaAs MESFET equivalent circuit, in a parallel element network configuration, e.g. package capacitances. First we obtain the Y parameter matrix of the embedding network by applying the rules outlined in the preceding section:

$$\begin{bmatrix} I'_g \\ I'_d \\ I'_s \end{bmatrix} = \begin{bmatrix} Y_A + Y_C & -Y_C & -Y_A \\ -Y_C & Y_B + Y_C & -Y_B \\ -Y_A & -Y_B & Y_A + Y_B \end{bmatrix} \begin{bmatrix} V_g \\ V_d \\ V_s \end{bmatrix} \tag{2.35}$$

or

$$\boldsymbol{I'} = \boldsymbol{Y'V}$$

The Y matrix of the two networks in parallel is now simply

$$Y_T = Y' + Y \tag{2.36}$$

In the case of series-connected elements (Fig. 2.37(b)) we can invert the equivalent circuit from Y matrix form to Z matrix form when the matrices are well conditioned (the determinant is not zero) and add the corresponding series element Z matrix. Thus

$$\begin{bmatrix} V'_g \\ V'_d \\ V'_s \end{bmatrix} = \begin{bmatrix} Z_{GG} & Z_{GD} & Z_{GS} \\ Z_{DG} & Z_{DD} & Z_{DS} \\ Z_{SG} & Z_{SD} & Z_{SS} \end{bmatrix} \begin{bmatrix} I_g \\ I_d \\ I_s \end{bmatrix} + \begin{bmatrix} Z_1 & 0 & 0 \\ 0 & Z_2 & 0 \\ 0 & 0 & Z_3 \end{bmatrix} \begin{bmatrix} I_g \\ I_d \\ I_s \end{bmatrix} \tag{2.37}$$

or

$$V' = (Z + Z')I$$

In the case of ill-conditioned matrices we can arrive at the same result by applying the following formulae [2.35]:

$$Y_T = (U + YZ')^{-1} Y \tag{2.38}$$

where U represents the unitary matrix.

2.6.2.2 Cascaded connections

In order to cascade two S parameter matrices we can either convert the matrices to the corresponding R matrices and multiply them in the correct order (Section 2.2.2) or proceed as follows. Consider two matrices, one a $k \times l$ matrix and the other an $l \times m$ matrix, which we wish to join at the common port l (Fig. 2.38). The easiest manner in which to proceed is to partition the matrices according to the interconnected ports and the "free" ports. Thus [2.36]

$$\begin{bmatrix} b_k \\ \cdots \\ b_l \end{bmatrix} = \begin{bmatrix} S_{kk} & \vdots & S_{kl} \\ \cdots & \vdots & \cdots \\ S_{lk} & \vdots & S_{ll} \end{bmatrix} \begin{bmatrix} a_k \\ \cdots \\ a_l \end{bmatrix} \tag{2.39}$$

$$\begin{bmatrix} b'_l \\ \cdots \\ b'_m \end{bmatrix} = \begin{bmatrix} S'_{ll} & \vdots & S'_{lm} \\ \cdots & \vdots & \cdots \\ S'_{ml} & \vdots & S'_{mm} \end{bmatrix} \begin{bmatrix} a'_l \\ \cdots \\ a'_m \end{bmatrix} \tag{2.40}$$

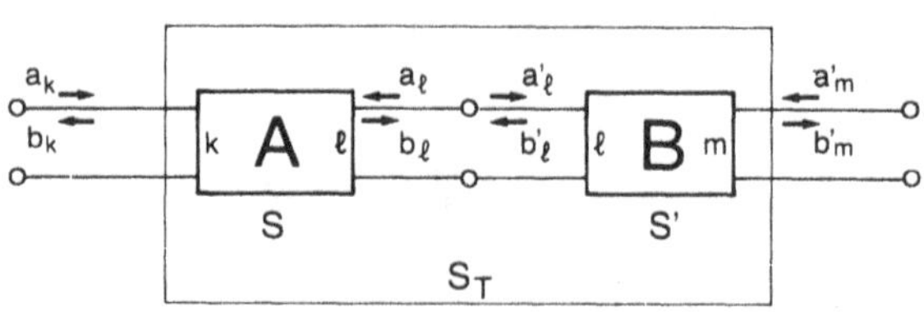

Fig. 2.38 S parameter representation of cascaded matrices.

By applying the joining equations $a_l = b'_l$ and $b_l = a'_l$ we arrive at the following result [2.36]:

$$S_T = \begin{bmatrix} (S_T)_{kk} & \vdots & (S_T)_{km} \\ \cdots\cdots & \vdots & \cdots\cdots \\ (S_T)_{mk} & \vdots & (S_T)_{mm} \end{bmatrix} \tag{2.41}$$

where

$$(S_T)_{kk} = S_{kk} + S_{kl}(U - S'_{ll}S_{ll})^{-1}S'_{ll}S_{lk} \tag{2.42}$$

$$(S_T)_{km} = S_{kl}(U - S'_{ll}S_{ll})^{-1}S'_{lm} \tag{2.43}$$

$$(S_T)_{mk} = S'_{ml}(U - S_{ll}S'_{ll})^{-1}S_{lk} \tag{2.44}$$

$$(S_T)_{mm} = S'_{mm} + S'_{ml}(U - S_{ll}S'_{ll})^{-1}S_{ll}S'_{lm} \tag{2.45}$$

2.6.3 Terminal impedances and power transfer

In this section we shall review the fundamental concepts of impedance matching and power transfer which apply in general to all microwave devices, both active and passive.

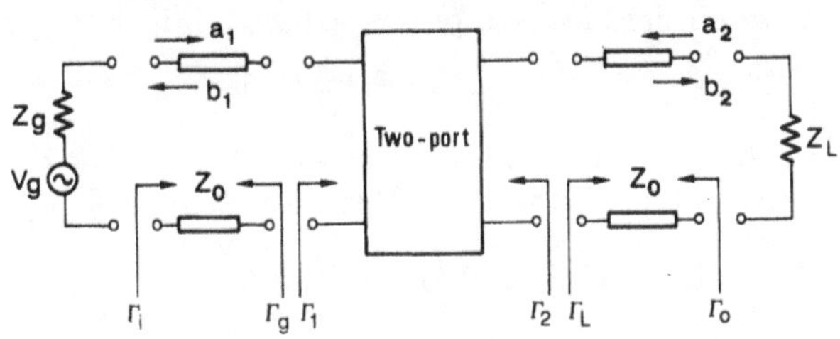

Fig. 2.39 The two-port matching problem.

Consider the problem represented in Fig. 2.39 in which we seek to connect a generator and a load to a two-port network via two transmission line sections of characteristic impedance Z_0. In the case of a nonunilateral component such as the GaAs MESFET, the load impedance will have a direct bearing on the input impedance and the source impedance will directly affect the output impedance. It can easily be shown that the corresponding reflection coefficients Γ_1 and Γ_2 at the input and output respectively of the device are given by [2.37]

$$\Gamma_1 = \frac{b_1}{a_1} = S_{11} + \frac{S_{12}S_{21}\Gamma_L}{1 - S_{22}\Gamma_L} \tag{2.46}$$

$$\Gamma_2 = \frac{b_2}{a_2} = S_{22} + \frac{S_{12}S_{21}\Gamma_g}{1 - S_{11}\Gamma_g} \tag{2.47}$$

Using the result of eqn (2.46), we can describe the behavior of the two-port terminated on port 2 by Γ_L using the simple equivalent impedance Z_1, as shown in Fig. 2.40(a), where

$$Z_1 = Z_0 \frac{1-\Gamma_1}{1+\Gamma_1} \tag{2.48}$$

and

$$Z_L = Z_0 \frac{1-\Gamma_L}{1+\Gamma_L} \tag{2.49}$$

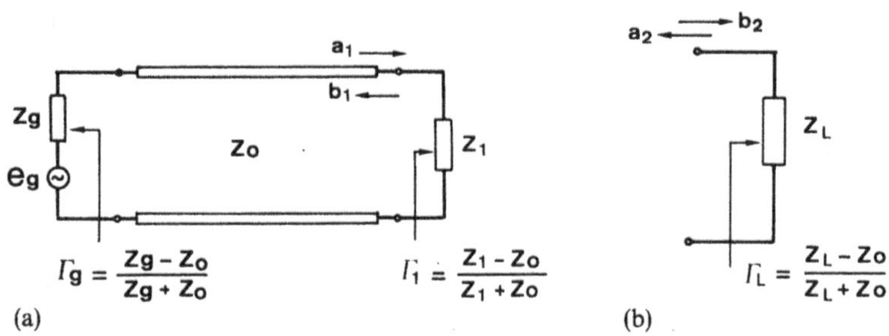

Fig. 2.40 (a) Power delivered to the device input port; (b) power available from the output port.

The power that is delivered to the input impedance Z_1 is thus

$$P_i = |a_1|^2(1-|\Gamma_1|^2) \tag{2.50}$$

When the generator is connected to the transmission line Z_0, the incident wave b_g gives rise to a reflected wave b_1. It is readily shown [2.4] that

$$a_1 = b_g + b_1\Gamma_g \tag{2.51}$$

where

$$\Gamma_g = \frac{Z_g - Z_0}{Z_g + Z_0} \tag{2.52}$$

and

$$b_g = \frac{V_g Z_0^{1/2}}{Z_g + Z_0} \tag{2.53}$$

Substituting eqns (2.46) and (2.51) into eqn (2.50) we obtain

$$P_i = |b_g|^2 \frac{1-|\Gamma_1|^2}{|1-\Gamma_1\Gamma_g|^2} \tag{2.54}$$

Therefore we see that the maximum power that the generator is able to deliver to a load, i.e. the generator available power P_A, occurs when $\Gamma_g = \Gamma_1^*$ or equivalently when $Z_g = Z_1^*$. Thus

$$P_A = \frac{|b_g|^2}{1-|\Gamma_1|^2} \tag{2.55}$$

Alternatively, by substituting eqns (2.52) and (2.53) in eqn (2.55) we obtain

$$P_A = \frac{|V_g|^2}{4\mathrm{Re}(Z_g)} \tag{2.56}$$

which is readily recognizable from circuit theory as the maximum available generator power.

We could have arrived at this result more easily had we chosen to define a_1 and b_1 directly in terms of the source impedance Z_g rather than define *both* Γ_g and Γ_1 with respect to a common real impedance Z_0. We shall see the advantages of generalized impedance scattering parameters, particularly with respect to amplifier design, in Section 2.6.4.

The power P_L delivered to the output load impedance is given in a similar manner (Fig. 2.40(b)) by

$$P_L = |a_2|^2(1-|\Gamma_L|^2) \tag{2.57}$$

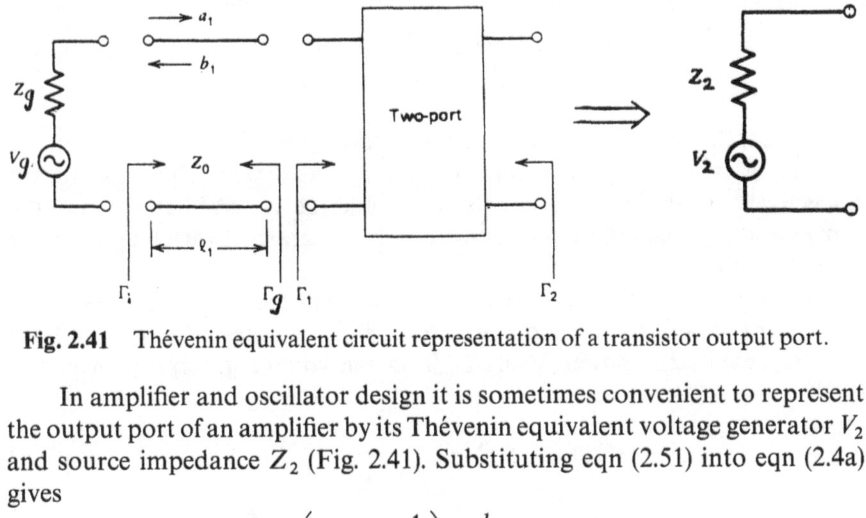

Fig. 2.41 Thévenin equivalent circuit representation of a transistor output port.

In amplifier and oscillator design it is sometimes convenient to represent the output port of an amplifier by its Thévenin equivalent voltage generator V_2 and source impedance Z_2 (Fig. 2.41). Substituting eqn (2.51) into eqn (2.4a) gives

$$a_1\left(-S_{11}+\frac{1}{\Gamma_g}\right) = \frac{b_g}{\Gamma_g}+S_{12}a_2 \tag{2.58}$$

whence

$$a_1 = \frac{b_g+S_{12}\Gamma_g a_2}{1-\Gamma_g S_{11}} \tag{2.59}$$

Substituting eqn (2.59) into eqn (2.4b) gives

$$\begin{aligned} b_2 &= (b_g+S_{12}\Gamma_g a_2)\frac{S_{21}}{1-\Gamma_g S_{11}}+S_{22}a_2 \\ &= \frac{b_g S_{21}}{1-\Gamma_g S_{11}}+\left(S_{22}+\frac{S_{12}S_{21}\Gamma_g}{1-\Gamma_g S_{11}}\right)a_2 \\ &= b_{2g}+\Gamma_2 a_2 \end{aligned} \tag{2.60}$$

Equation (2.60) is formally identical with eqn (2.51) which governs the input to the amplifier. As in that case, we can obtain the conditions for maximum power transfer at the output, which are shown to occur when $\Gamma_2 = \Gamma_L^*$ or $Z_2 = Z_L^*$ where

$$\Gamma_2 = \frac{Z_2 - Z_0}{Z_2 + Z_0} \tag{2.61}$$

2.6.4 The generalized scattering matrix

Instead of normalizing our scattering parameter variables with respect to 50 Ω on all ports, let us consider a more general definition in which the variables at a given port are defined with respect to the Thévenin impedance (generator or load) seen at that port (Fig. 2.42). We thus redefine *generalized* incident and reflected waves a'_k and b'_k at port k as follows:

$$a'_k = \tfrac{1}{2}\{\text{Re}(Z_{sk})\}^{-1/2}(V_k + I_k Z_{sk}) \tag{2.62}$$

$$b'_k = \tfrac{1}{2}\{\text{Re}(Z_{sk})\}^{-1/2}(V_k - I_k Z_{sk}^*) \tag{2.63}$$

where Z_{sk} is the complex Thévenin impedance presented by the *external* source or load attached to port k, and V_k and I_k are the voltage and current respectively at the port terminals. It should be noted that we use the conventional prime notation to designate the generalized variables and their related parameters.

Generalized scattering variables have many interesting properties which make them suitable for use in microwave amplifier synthesis (Chapter 3). Thus $|a'_k|^2$ represents the power available from the voltage generator on port k [2.38]:

$$|a'_k|^2 = \frac{|E_k|^2}{4\text{Re}(Z_{sk})} \tag{2.64}$$

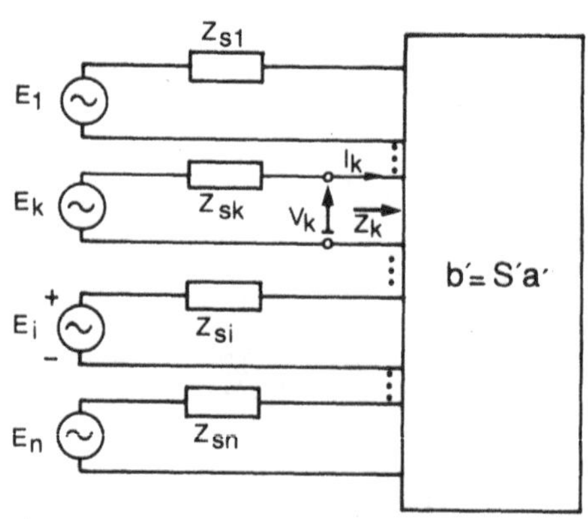

Fig. 2.42 Generalized S parameters.

and

$$|a'_k|^2 - |b'_k|^2 = \frac{(V_k + I_k Z_{sk})(V_k^* + I_k^* Z_{sk}^*) - (V_k - I_k Z_{sk}^*)(V_k^* - I_k^* Z_{sk})}{4R_{sk}}$$

$$= \tfrac{1}{4}(R_{sk})^{-1}(Z_{sk} + Z_{sk}^*)(V_k I_k^* + V_k^* I_k)$$

$$= \mathrm{Re}(V_k^* I_k)$$

$$= |a'_k|^2(1 - |\Gamma'_k|^2) = \text{power absorbed at port } k \tag{2.65}$$

where the generalized reflection coefficient Γ'_k is defined as

$$\Gamma'_k = \frac{b'_k}{a'_k} = \frac{V_k - I_k Z_{sk}^*}{V_k + I_k Z_{sk}} = \frac{Z_k - Z_{sk}^*}{Z_k + Z_{sk}} \tag{2.66}$$

and

$$R_{sk} = \mathrm{Re}(Z_{sk})$$

Thus we see that, while the generalized S parameters are closely related to their normalized counterparts and maintain many of the same properties, they are far more suited to amplifier design in view of the fact that they can be used to include the complex impedances at the transistor terminals. Thus, for the n-port network of Fig. 2.42, we can define the vectors $\boldsymbol{a}'$ and $\boldsymbol{b}'$ as follows:

$$\boldsymbol{a}' = \boldsymbol{R}_s \cdot (\boldsymbol{V} + \boldsymbol{Z}_s \boldsymbol{I}) \tag{2.67}$$

$$\boldsymbol{b}' = \boldsymbol{R}_s \cdot (\boldsymbol{V} - \boldsymbol{Z}_s^{\mathrm{T}*} \boldsymbol{I}) \tag{2.68}$$

where $\boldsymbol{R}_s$ and $\boldsymbol{Z}_s$ are diagonal matrices whose ith components are given by $\frac{1}{2}\{\mathrm{Re}(Z_{si})\}^{-1/2}$ and Z_{si} respectively. The generalized S parameter matrix relating the variable vectors $\boldsymbol{a}'$ and $\boldsymbol{b}'$ is now given by

$$\boldsymbol{b}' = \boldsymbol{S}'\boldsymbol{a}' \tag{2.69}$$

The relationship $\boldsymbol{V} = \boldsymbol{ZI}$ between the S' matrix and the impedance matrix of the n-port network (Fig. 2.42) can be obtained directly from eqns (2.67)–(2.69):

$$\boldsymbol{S}' = \boldsymbol{R}_s(\boldsymbol{Z} - \boldsymbol{Z}_s^{\mathrm{T}*})(\boldsymbol{Z} + \boldsymbol{Z}_s)^{-1}\boldsymbol{R}_s^{-1} \tag{2.70}$$

and reciprocally

$$\boldsymbol{Z} = \boldsymbol{R}_s^{-1}(\boldsymbol{U} - \boldsymbol{S}')^{-1}(\boldsymbol{S}'\boldsymbol{Z}_s + \boldsymbol{Z}_s^{\mathrm{T}*})\boldsymbol{R}_s \tag{2.71}$$

In order to understand the physical significance of the generalized S parameters, consider Fig. 2.43 in which all ports of the network except the ith (excited by source E_i) are terminated by their normalizing impedances. By applying the relations

$$V_k = -Z_{sk} I_k \qquad k = 1, 2, \ldots, n \qquad k \neq i \tag{2.72}$$

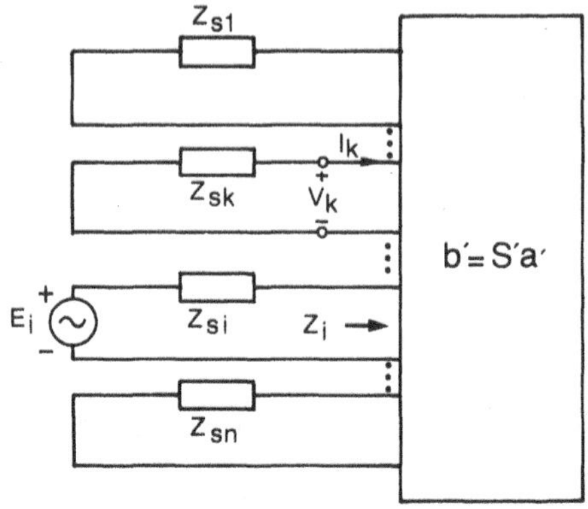

Fig. 2.43 Physical significance of the generalized S parameters.

and

$$E_{si} = V_i + I_i Z_{si} \tag{2.73}$$

to eqns (2.67)–(2.69), it can be shown that [2.36]

$$|S'_{ki}|^2 = \frac{\mathrm{Re}(Z_{sk})|I_k|^2}{|E_{si}|^2/4\mathrm{Re}(Z_{si})} = \frac{|b'_k|^2}{|a'_i|^2} = \frac{P_k}{P_{Ai}} \tag{2.74}$$

Thus, as will be seen in Section 2.6.5, $|S'_{ki}|$ represents the transducer gain between ports i and k. Similarly it can be shown that

$$S'_{ii} = \frac{b'_i}{a'_i} = \frac{V_i - Z_{si}^* I_i}{V_i + Z_{si} I_i} = \Gamma'_i \tag{2.75}$$

S'_{11} represents the reflection coefficient looking into port i when all the other ports are terminated in their conjugate match impedances. In the case of the two-port amplifier described in the previous section, the input and output reflection coefficient relations described by eqns (2.46) and (2.47) are given directly by S'_{11} and S'_{22} respectively.

Finally, it is easily shown [2.39] that the generalized scattering matrix retains the properties of the normalized matrix with regard to lossless passive networks. Thus

$$U - S'^{\mathrm{T}*} \cdot S' = 0 \tag{2.76}$$

which gives the following standard results in the case of a two-port network:

$$|S'_{11}|^2+|S'_{21}|^2 = 1 \tag{2.77}$$

$$S'_{12}S'_{11}{}^*+S'_{22}S'_{21}{}^* = 0 \tag{2.78}$$

$$S'_{12}{}^*S'_{11}+S'_{22}{}^*S'_{21} = 0 \tag{2.79}$$

$$|S'_{12}|^2+|S'_{22}|^2 = 1 \tag{2.80}$$

It is often desirable to be able to convert between measured S parameters normalized to 50 Ω and scattering parameters generalized to specific load and source impedances. In this case it is necessary to renormalize at each device port using the following general formula [2.36]:

$$\boldsymbol{S}' = \boldsymbol{A}^{-1}(\boldsymbol{S}-\boldsymbol{\Gamma}^{\mathrm{T}*})(\boldsymbol{U}-\boldsymbol{\Gamma}\boldsymbol{S})^{-1}\boldsymbol{A}^{\mathrm{T}*} \tag{2.81}$$

where $\boldsymbol{A}$ and $\boldsymbol{\Gamma}$ are diagonal matrices whose kth elements are given by

$$A_k = |1-\Gamma_k|^{-1}(1-\Gamma_k{}^*)(1-|\Gamma_k|^2)^{1/2} \tag{2.82}$$

$$\Gamma_k = \frac{Z'_{sk}-Z_{sk}}{Z'_{sk}+Z_{sk}{}^*} \tag{2.83}$$

Z_{sk} is the old normalizing impedance (Z_0 in the case of an impedance normalized to 50 Ω) and Z'_{sk} is the new normalizing impedance.

2.6.5 Stability

The first and most important step in amplifier design is to ensure that the active device is stable: it will not oscillate under certain load terminations. The problem of stability is as old as that of amplifier design. Essentially it can be defined as ensuring that the real parts of the input and output impedances are simultaneously positive when the device is terminated in certain impedances [2.40]. We can thus identify two possibilities, which in generalized S parameter notation can be written as follows:

Unconditional stability

$$|S'_{11}|<1 \qquad \forall|\Gamma_{\mathrm{L}}|<1 \tag{2.84}$$

$$|S'_{22}|<1 \qquad \forall|\Gamma_{\mathrm{g}}|<1 \tag{2.85}$$

Conditional stability

$$|S'_{11}|<1 \qquad \text{for certain } |\Gamma_{\mathrm{L}}|<1 \tag{2.86}$$

$$|S'_{22}|<1 \qquad \text{for certain } |\Gamma_{\mathrm{g}}|<1 \tag{2.87}$$

We consider the second case first. If eqn (2.46) is used for S'_{11}

$$S'_{11} = \frac{S_{11}-\Delta\Gamma_{\mathrm{L}}}{1-S_{22}\Gamma_{\mathrm{L}}} \tag{2.88}$$

where

$$\Delta = S_{11}S_{22} - S_{12}S_{21} \tag{2.89}$$

As we are in the situation in which the transistor is unstable and thus likely to oscillate under certain load impedances, it is important to be able to determine clearly the terminating reflection coefficients Γ_g and Γ_L which cause instability.

We consider Γ_L first. In order to present the results in a simple form which is easy to understand we shall map onto a Smith chart representing the load reflection coefficient those impedances which render the real part of the input impedance negative. In order to be able to do this we rewrite eqn (2.88):

$$\Gamma_L = \frac{S_{11} - S'_{11}}{\Delta - S_{22}S'_{11}} \tag{2.90}$$

Equation (2.90) is a bilinear equation and the locus of the circle $|S'_{11}| = 1$, which is the limit condition for stability ($\mathrm{Re}(Z_1) = 0$), will also be a circle. The center C_{stL} and the radius R_{stL} of the circle are obtained by applying the well-known bilinear formulae [2.4]:

$$C_{stL} = \frac{S_{11}\Delta^* - S_{22}^*}{|\Delta|^2 - |S_{22}|^2} \tag{2.91}$$

$$|R_{stL}| = \frac{|S_{12}S_{21}|}{|\Delta|^2 - |S_{22}|^2} \tag{2.92}$$

Using these parameters we are able to trace the circle limiting the unstable zone onto the Smith chart representing the output plane (Fig. 2.44).

Having taken the boundary condition $|S_{11}| = 1$ for the stability circle, there is an ambiguity concerning the unstable zone: is it inside or outside the circle? The problem is solved in a simple but elegant manner by calculating S'_{11} for a known load reflection coefficient $\Gamma_L = 0$, i.e. the center of the Smith chart. If this point is outside the circle and $|S'_{11}|$ (which reduces to $|S_{11}|$ in this case) is greater than unity, the unstable area is outside the circle (Fig. 2.45(b)). If the contrary is true, the unstable impedances (the shaded area in the figure) are confined to the interior of the circle (Fig. 2.45(a)).

In the case of an amplifier design using a conditionally stable transistor over a frequency band, it is advisable to perform the stability circle construction at a minimum of three frequencies: the center frequency and the band edges. In this case the impedance values to be avoided in the design of the matching circuit are clearly represented by the shaded area (Fig. 2.46).

The same approach is followed with respect to the generator impedance. In this case the relevant formulae are

$$C_{stg} = \frac{S_{22}\Delta^* - S_{11}}{|\Delta|^2 - |S_{11}|^2} \tag{2.93}$$

$$|R_{stg}| = \frac{|S_{12}S_{21}|}{|\Delta|^2 - |S_{11}|^2} \tag{2.94}$$

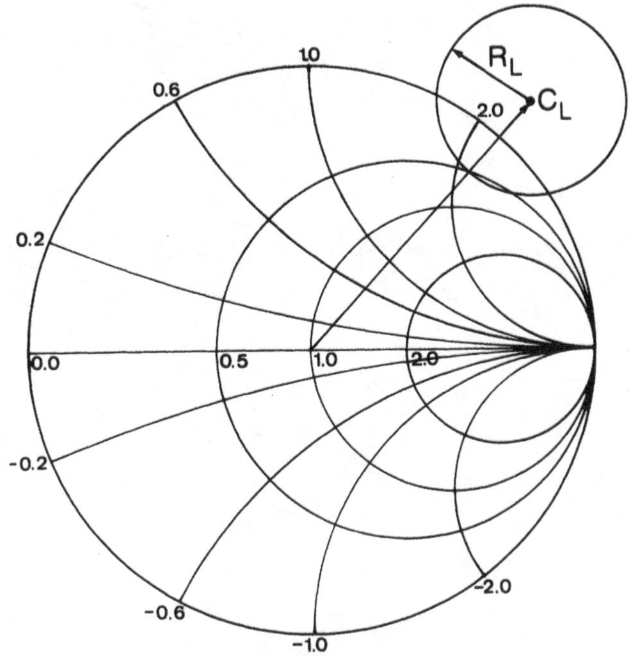

Fig. 2.44 Stability circles.

Finally, let us consider the case when the transistor is unconditionally stable. For this to obtain it is necessary for the inequalities (2.80) and (2.81) to be satisfied simultaneously. It can be shown that the necessary and sufficient conditions for unconditional stability are [2.41]

$$K = \frac{1-|S_{11}|^2-|S_{22}|^2+|\Delta|^2}{2|S_{12}S_{21}|} > 1 \tag{2.95}$$

$$|\Delta| = |S_{11}S_{22}-S_{12}S_{21}| < 1 \tag{2.96}$$

2.6.6 Power gains

An important point to establish when discussing amplifier gain is which definition is being applied. The various possibilities are clearly seen by referring to Fig. 2.47.

On the source side we distinguish between the power P_A available from the generator and the power P_i absorbed by the transistor. Similarly, using the Thévenin equivalent representation established earlier (eqn (2.60)), we can consider the power P_{avo} available at the output of the transistor and the power

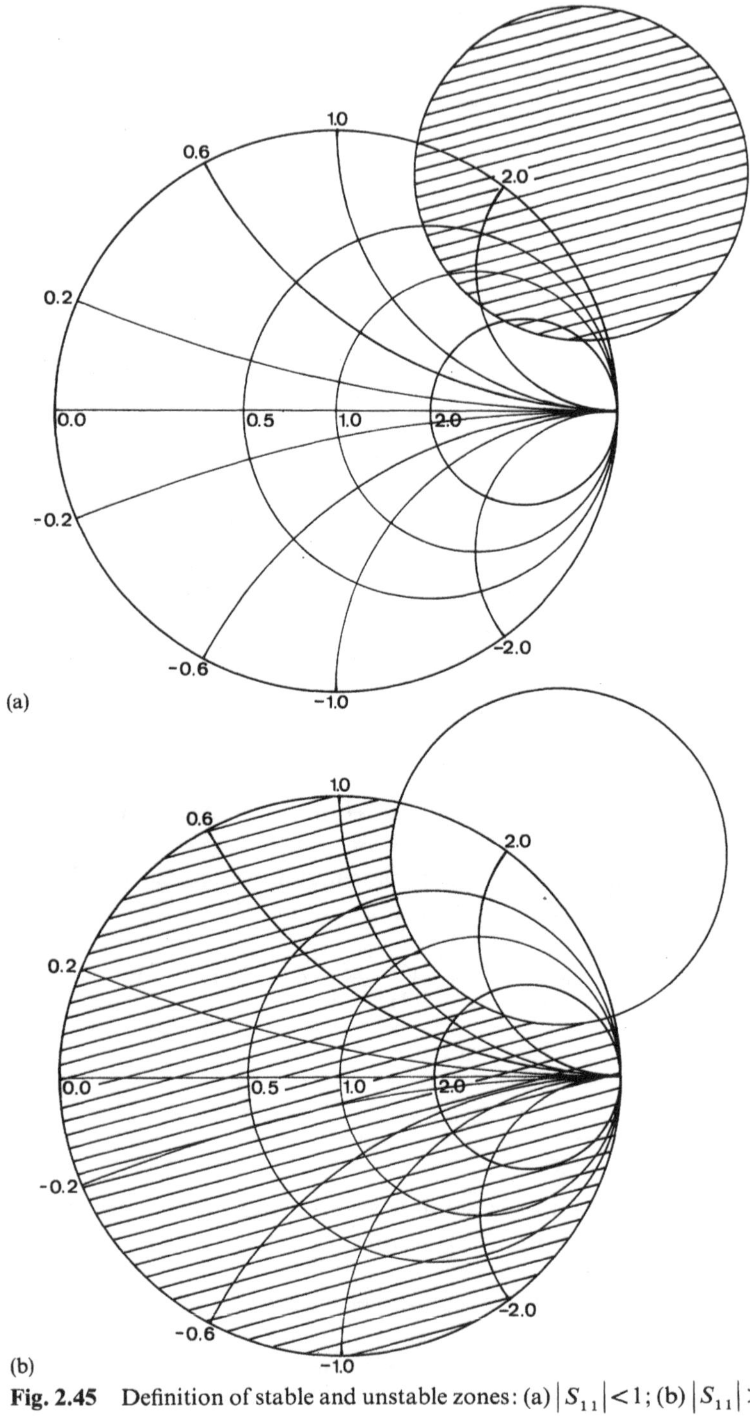

Fig. 2.45 Definition of stable and unstable zones: (a) $|S_{11}|<1$; (b) $|S_{11}|>1$.

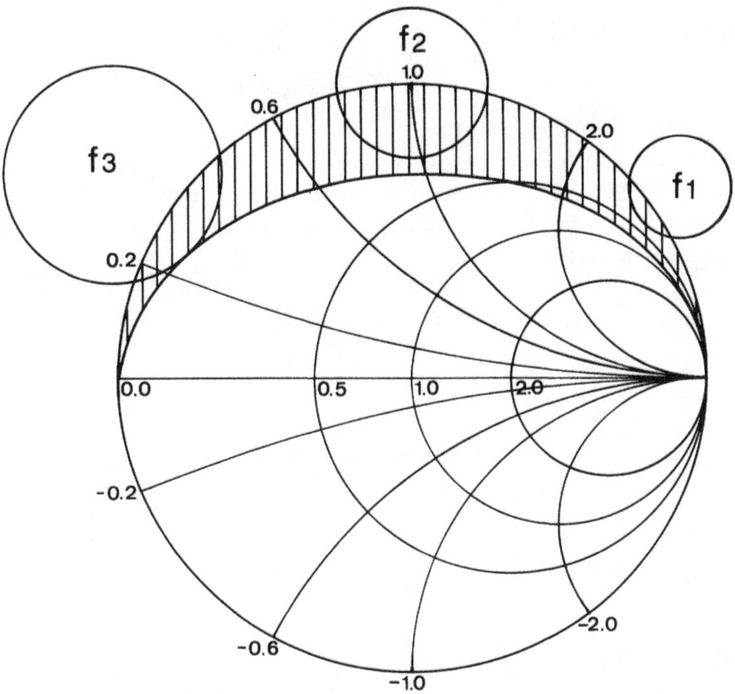

Fig. 2.46 Stable operating zone over a frequency bandwidth.

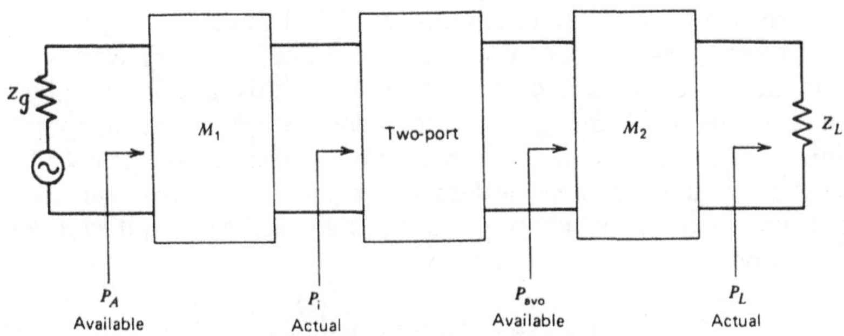

Fig. 2.47 Various parameters used in gain definitions.

P_L actually fed into the load. There are thus four possibilities available. In general, however, we limit ourselves to three: the operating power gain

$$G = P_L/P_i = f(S, \Gamma_L) \quad (2.97a)$$

the available power gain

$$G_A = P_{avo}/P_A = f(\Gamma_g, S) \quad (2.97b)$$

and the transducer gain

$$G_{\mathrm{T}} = P_{\mathrm{L}}/P_{\mathrm{A}} = f(\Gamma_{\mathrm{g}}, S, \Gamma_{\mathrm{L}}) \tag{2.97c}$$

Let us consider each case in turn.

The operating power gain, as its name implies, is an actual power gain [2.42] in so far as it measures the ratio of the power delivered to the load to the power absorbed by the device. Using the formulae developed in Section 2.6.3 it is readily shown that

$$G = \frac{P_{\mathrm{L}}}{P_i} = \frac{|b_2|^2(1-|\Gamma_{\mathrm{L}}|^2)|1-S'_{11}\Gamma_{\mathrm{g}}|^2}{|b_{\mathrm{g}}|^2(1-|S'_{11}|^2)} \tag{2.98}$$

but

$$\left|\frac{b_2}{a_1}\right|^2 = \frac{|S_{21}|^2}{|1-S_{22}\Gamma_{\mathrm{L}}|^2} \tag{2.99}$$

and

$$\left|\frac{a_1}{b_{\mathrm{g}}}\right|^2 = \frac{1}{|1-S'_{11}\Gamma_{\mathrm{g}}|^2} \tag{2.100}$$

whence

$$G = \frac{(1-|\Gamma_{\mathrm{L}}|^2)|S_{21}|^2}{|1-S_{22}\Gamma_{\mathrm{L}}|^2(1-|S'_{11}|^2)} \tag{2.101}$$

The available power gain is the gain obtainable from the transistor when it is matched at the input and output ports. This definition is very convenient when discussing system noise, as we shall see in Chapter 3, for it avoids having to take mismatch effects into account. However, this necessarily obliges us, when applying available gain or loss concepts in noise measurement equations, to ensure that each component in the test bench is matched at all ports. An alternative is to use isolators with good return loss characteristics throughout the measurement network. By applying the relevant equations to (2.97b) we obtain

$$G_{\mathrm{A}} = \frac{(1-|\Gamma_{\mathrm{g}}|^2)|S_{21}|^2}{|1-S_{11}\Gamma_{\mathrm{g}}|^2(1-|S'_{22}|^2)} \tag{2.102}$$

The transducer power gain provides a measure of the advantage of using a transistor over simply matching the generator to the output load [2.42]. Applying the relevant equations to the definition (2.97c) enables us to obtain, in a manner similar to that employed in the previous case,

$$G_{\mathrm{T}} = \frac{(1-|\Gamma_{\mathrm{L}}|^2)|S_{21}|^2(1-|\Gamma_{\mathrm{g}}|^2)}{|1-S'_{22}\Gamma_{\mathrm{L}}|^2|1-S_{11}\Gamma_{\mathrm{g}}|^2} \tag{2.103}$$

The definition of transducer gain is the most complete in that it is the only one in which both input *and* output match appear. It is for this reason that it is extensively used in amplifier analysis and synthesis. It is given directly by $|S'_{21}|^2$ when this is normalized to Γ_g and Γ_L at the input and output respectively. In fact, on carefully examining eqn (2.103) it is seen to be made up of three components, one linked to the input match, one linked to $|S_{21}|^2$ and one linked to the output match:

$$G_T = G_g G_0 G_L \tag{2.104}$$

where

$$G_g = \frac{1-|\Gamma_g|^2}{|1-S_{11}\Gamma_g|^2} \tag{2.105}$$

$$G_0 = |S_{21}|^2 \tag{2.106}$$

$$G_L = \frac{1-|\Gamma_L|^2}{|1-S'_{22}\Gamma_L|^2} \tag{2.107}$$

Thus transducer gain is a function of the generator and load impedances as well as the transistor two-port S parameters.

2.6.6.1 Unilateral transducer gain

Equation (2.103) can be simplified by using the approximation $S_{12} = 0$, so that in this case it becomes

$$G_{Tu} = \frac{(1-|\Gamma_L|^2)|S_{21}|^2(1-|\Gamma_g|^2)}{|1-S_{22}\Gamma_L|^2|1-S_{11}\Gamma_g|^2} \tag{2.108}$$

Maximum unilateral gain occurs when $S_{11} = \Gamma_g^*$ *and* $S_{22} = \Gamma_L^*$, and is given by

$$G_{Tu\,max} = \frac{|S_{21}|^2}{(1-|S_{22}|^2)(1-|S_{11}|^2)} \tag{2.109}$$

with

$$G_{gu\,max} = \frac{1}{1-|S_{11}|^2} \tag{2.110}$$

$$G_{Lu\,max} = \frac{1}{1-|S_{22}|^2} \tag{2.111}$$

G_0 is unchanged, and in this case the matching problem reduces to designing circuits at transistor input and output ports which satisfy the conditions $S_{11} = \Gamma_g^*$ and $S_{22} = \Gamma_L^*$.

Above a certain frequency threshold (around 1 GHz for 0.5 μm devices) the transistor intrinsic gain $G_0 = |S_{21}|^2$ rolls off with frequency. If the amplifier is to satisfy a gain flatness specification over a frequency band exceeding 25%, it is necessary to design matching networks that compensate for this roll-off. This is generally done by matching the device for optimum gain at the highest frequency point and mismatching at input and/or output ports at other frequencies in order to obtain a flat gain response. This work will be greatly facilitated if we can obtain the locus of mismatch impedances corresponding to a given loss in gain. In the case of the input port this is obtained from eqn (2.105) by rearranging the terms as follows:

$$G_g = \frac{1-|\Gamma_g|^2}{(1-S_{11}\Gamma_g)(1-S_{11}{}^*\Gamma_g{}^*)} \tag{2.112}$$

Thus

$$|\Gamma_g|^2(1+G_g|S_{11}|^2)-G_g(S_{11}{}^*\Gamma_g{}^*+S_{11}\Gamma_g)+G_g-1=0 \tag{2.113}$$

which describes the locus of a circle whose radius R_g and center C_g are functions of G_g:

$$C_g = \frac{G_g S_{11}{}^*}{1+G_g|S_{11}|^2} \tag{2.114}$$

$$|R_g| = \frac{\{1-G_g(1-|S_{11}|^2)\}^{1/2}}{1+G_g|S_{11}|^2} \tag{2.115}$$

where G_g is a real number (not in decibels) greater than unity which gives the desired matching gain.

We can select different values of G_g at different design band frequencies for which we have S parameters so as to equalize amplifier gain over the band. We next trace on the Smith chart representing the input match plane at each frequency a set of circles corresponding to the different values chosen for G_g. From these constant-gain circles we are in a position to design a physically realizable input matching circuit which fits the locus of source impedance versus frequency. The same approach can be applied to the output load impedance plane. In this case the relevant constant-gain circle formulae are

$$C_L = \frac{G_L S_{22}{}^*}{1+G_L|S_{22}|^2} \tag{2.116}$$

$$|R_L| = \frac{\{1-G_L(1-|S_{22}|^2)\}^{1/2}}{1+G_L|S_{22}|^2} \tag{2.117}$$

It is equally possible to separate the mismatch loss between the input and output matching networks in order to facilitate network synthesis.

2.6.6.2 Maximum transducer gain

Let us now consider the case when S_{12} is no longer negligible. In order to obtain maximum transducer gain, we must simultaneously match input and output ports of the transistor. Of course, this is only possible if the device is unconditionally stable. Accepting this condition, we must solve eqns (2.46) and (2.47) simultaneously, since Γ_1 is affected by Γ_L and Γ_2 is conditioned by Γ_g. We rewrite the equations as follows:

$$\Gamma_{g\,opt}{}^* = \Gamma_1 = S_{11} + \frac{S_{12}S_{21}\Gamma_{L\,opt}}{1 - S_{22}\Gamma_{L\,opt}} \tag{2.118}$$

$$\Gamma_{L\,opt}{}^* = \Gamma_2 = S_{22} + \frac{S_{12}S_{21}\Gamma_{g\,opt}}{1 - S_{11}\Gamma_{g\,opt}} \tag{2.119}$$

where $\Gamma_{g\,opt}$ and $\Gamma_{L\,opt}$ are the source and load reflection coefficients which permit maximum transducer gain to be obtained. By substituting for one or other of these parameters we obtain an equation of the form

$$\Gamma_i^2 - \Gamma_i \frac{B_i}{C_i} + \frac{C_i^*}{C_i} = 0 \qquad i = 1, 2 \tag{2.120}$$

which has a solution

$$\Gamma_i = \frac{C_i^*}{2|C_i|^2}\{B_i \pm (B_i^2 - 4|C_i|^2)^{1/2}\} \qquad i = 1, 2 \tag{2.121}$$

where the subscripts 1 and 2 represent $\Gamma_{g\,opt}$ and $\Gamma_{L\,opt}$ respectively and

$$B_1 = 1 + |S_{11}|^2 - |S_{22}|^2 - |\Delta|^2 \tag{2.122}$$

$$B_2 = 1 - |S_{11}|^2 + |S_{22}|^2 - |\Delta|^2 \tag{2.123}$$

$$C_1 = S_{11} - \Delta S_{22}{}^* \tag{2.124}$$

$$C_2 = S_{22} - \Delta S_{11}{}^* \tag{2.125}$$

$$\Delta = S_{11}S_{22} - S_{12}S_{21} \tag{2.126}$$

Examination of eqn (2.121) shows that only one solution is viable [2.42], i.e. that given by B_i positive, $B_i/2|C_i| > 1$, and taking the negative sign of the square root to ensure that Γ_i is less than unity. B_i negative indicates potential instability, $B_i/2|C_i| \leqslant 1$ gives the trivial solution $|\Gamma_i| = 1$ and taking the positive sign for B_i gives a nonrealizable $|\Gamma_i| > 1$.

Several approaches are possible to find the maximum transducer gain. The simplest is to substitute the derivations for $\Gamma_{g\,opt}$ and $\Gamma_{L\,opt}$ into eqn (2.103), remembering that $S'_{22} = \Gamma_{L\,opt}{}^*$. After considerable manipulation we obtain

$$G_{T\,max} = \left|\frac{S_{21}}{S_{12}}\right|\{K \pm (K^2 - 1)^{1/2}\} \tag{2.127}$$

(the negative sign is used for $B_1 > 0$). In this case there is maximum power transfer from generator to transistor input and from transistor output to load. Thus

$$G_{max} = G_{T\,max} = G_{A\,max} \tag{2.128}$$

Finally, in the case of conditionally stable transistors maximum gain is obtainable by improving the input and output match from values giving $K > 1$ until the boundary stability condition $K = 1$ is attained, when from eqn (2.127) it is seen that the maximum stable gain is

$$G_{\text{st max}} = \left|\frac{S_{21}}{S_{12}}\right| \tag{2.129}$$

2.7 Conclusion

In this chapter we have considered three essential themes: an introduction to the manipulation and biasing of the GaAs MESFET and to basic microwave S parameter concepts, a thorough treatment of S parameter measurement techniques applicable to encapsulated devices, transistor chips and components on-wafer, and a discussion of the use of the measured S parameters, particularly in small-signal amplifier design (Chapter 3).

Acknowledgments

One of the authors (RS) would like to thank J.P. Le Rolland, B. Mathieu, J. Guilcher and M. Bouguennec for doing the artwork, and the directors of the Centre National d'Etudes des Télécommunications for permission to published the material presented in this chapter.

References

2.1 Vendelin, G. *Design of Amplifiers and Oscillators by the S-Parameter Method*, Wiley, New York, 1982.

2.2 Arden, J. A. GaAs FET circuits begin with biasing choices, *Microwaves*, **21**, 89–91, June 1982.

2.3 Manufacturer's data sheets (e.g. NEC, Fujitsu, MSC etc.).

2.4 Kerns, D. M. and Beatty, R. W. *Basic Theory of Waveguide Junctions and Introductory Microwave Network Analysis*, Pergamon, Oxford, 1967.

2.5 Owens, R. Mount independent equivalent circuit of the S4 diode package, *Electron. Lett.*, **7** (19), 580–582, 1971.

2.6 Slaymaker, N. A., Soares, R. A. and Turner, J. A. GaAs MESFET small-signal X-band amplifiers, *IEEE Trans. Microwave Theory Tech.*, **24** (6), 329–337, 1976.

2.7 Soares, R. A and Cripps, S. C. Some novel characterization techniques used in the design of X-band GaAs FET amplifiers, *Proc. European Microwave Conf., Rome*, pp. 19–23, Microwave Exhibitions and Publishers, Tunbridge Wells, 1976.

2.8 Soares, R. A. Novel large signal S-parameter measurement technique aids GaAs power amplifier design, *Proc. European Microwave Conf., Copenhagen*, pp. 113–117, Microwave Exhibitions and Publishers, Tunbridge Wells, 1977.
2.9 Deschamps, G. A. Determination of reflection coefficients and insertion loss of wave-guide junction, *J. Appl. Phys.*, **24** (8), 1046–1050, 1953.
2.10 Parisot, M., Binet, M. and Rabier, A. Caractérisation automatique en hyperfréquences du transistor à effet de champ, *Acta Electron.*, **23** (2), 137–149, 1980.
2.11 Bianco, B. and Parodi, M. Measurement of the effective relative permittivities of microstrip, *Electron. Lett.*, **11**, 71–72, 1975.
2.12 Hammarstad, E. and Jensen, O. Accurate models for microstrip computer design, *IEEE Microwave Theory Tech. Symp. Dig.*, pp. 407–409, IEEE, New York, 1980.
2.13 Rytting, D. An analysis of vector measurement accuracy techniques, *Hewlett-Packard RF and Microwave Symp.*, 1980.
2.14 Mason, S. J. Feedback theory—some properties of signal flow graphs, *Proc. IRE*, **41**, 1144–1156, 1953.
2.15 Chow, Y. and Cassignol, E. *Linear Signal Flow Graphs and Applications*, Wiley, New York, 1962.
2.16 Kuhn, N. Simplified signal flow graph analysis, *Microwave J.*, **6**, 59–66, November 1963.
2.17 Hackborn, R. A. An automatic network analyzer system, *Microwave J.*, **11** (5), 45–52, May 1968.
2.18 Hand, B. P. Developing accuracy specifications for network analyzer systems, *Hewlett Packard J.*, **21** (6), 16–19, February 1970.
2.19 Kruppa, W. and Sodomsky, K. F. An explicit solution for the scattering parameters of a linear two-port measured with an imperfect test set, *IEEE Trans. Microwave Theory Tech.*, **19**, 122–123, 1971.
2.20 Rehnmark, S. On the calibration process of automatic network analyzer systems, *IEEE Trans. Microwave Theory Tech.*, **22**, 457–458, 1974.
2.21 Franzen, N. R. and Speciale, R. A. A new procedure for system calibration and error removal in automated S parameter measurements, *Proc. European Microwave Conf., Hamburg*, pp. 67–73, Microwave Exhibitions and Publishers, Tunbridge Wells, 1975.
2.22 Speciale, R. A. A generalization of the TSD network analyzer calibration procedure, covering *n*-port scattering parameter measurements, affected by leakage errors, *IEEE Trans. Microwave Theory Tech.*, **25**, 1100–1115, 1977.
2.23 Rytting, D. Appendix to an analysis of vector measurement accuracy techniques, *Hewlett-Packard RF and Microwave Symp.*, 1980.
2.24 Bauer, R. F. and Penfield, P., Jr. De-embedding and unterminating, *IEEE Trans. Microwave Theory Tech.*, **22**, 282–288, 1974.
2.25 Elmore, G. De-embedded measurements using H.P. 8510 microwave network analyzer, *Hewlett-Packard RF and Microwave Symp.*, 1985.
2.26 Parisot, M., Archambault, Y., Pavlidis, D. and Magarshack, J. Highly accurate design of spiral inductors for MMIC's with small size and high cut-off frequency characteristics, *Proc. IEEE Microwave and Millimeter Wave Monolithic Circuits Symp.*, pp. 91–95, IEEE, New York, 1984.
2.27 MT950 transistor test fixture operating manual, Maury Microwave.
2.28 Chusseau, L., Parisot, M. and Jousseaume, N. Automatic full noise characterization of microwave GaAs FETS, *Proc. European Microwave Conf., Rome*, pp. 628–634, Microwave Exhibitions and Publishers, Tunbridge Wells, 1987.
2.29 Strid, E. and Gleason, K. A microstrip probe for microwave measurements on

GaAs FET and IC wafers, *GaAs IC Symp., Las Vegas, AZ*, Paper 31, November 1980.

2.30 Strid, E. and Gleason, K. A DC–12 GHz monolithic GaAs FET distributed amplifier, *IEEE Trans. Microwave Theory Tech.*, **30**, 969–975, 1982.

2.31 Borrego, J., Gutmann, R., Chudzicki, M. and Moghe, S. Determination of GaAs MESFET equivalent circuit parameters from *I*/*V* and 10 MHz measurements, *Electron. Lett.*, **13** (24), 756–757, 1977.

2.32 Perichon, R. A. In Soares, R., Graffeuil, J. and Obregon, J. (eds), *Applications des Transistors à Effet de Champ en Arseniure de Gallium, Collection Scientifique et Technique des Télécommunications*, pp. 69–73, Eyrolles, Paris, 1984.

2.33 Curtice, W. and Camisa, R. Self-consistent FET models for amplifier design and device diagnostics, *IEEE Trans. Microwave Theory Tech.*, **32**, 1573–1578, 1984.

2.34 Spence, R. *Linear Active Networks*, Wiley, New York, 1970.

2.35 Boulouard, A. Personal communication, April 1987.

2.36 Ha, T. T. *Solid-State Microwave Amplifier Design*, Wiley, New York, 1981.

2.37 Anderson, R. *S*-parameter techniques for faster more accurate network design, *Hewlett-Packard J.*, **18**, 13–24, 1967.

2.38 Kurokawa, K. Power waves and the scattering matrix, *IEEE Trans. Microwave Theory Tech.*, **13**, 194–202, 1965.

2.39 Collins, R. *Foundations for Microwave Engineering*, McGraw-Hill, New York, 1966.

2.40 Lewellyn, F. Some fundamental properties of transmission systems, *Proc. IRE*, **40**, 271–283, 1952.

2.41 Bodway, G. Two port power flow analysis using generalized scattering parameters, *Microwave J.*, **10** (6), 61–69, May 1967.

2.42 Carson, R. *High Frequency Amplifiers*, Wiley, New York, 1975.

Appendix 2.1 Conversions between two-port parameters normalized to Z_0

This appendix presents in tabular form simple formulae permitting conversion between different two-port parameter sets. The parameter sets described can be separated into two groups: *R* and *S* parameters defined with respect to normalizing impedances (or admittances) at input and output ports, and *Z*, *Y* and *ABCD* parameters defined independently of terminating impedances. Some conversion equations in which the complex impedances are specified as Z_1 on port 1 and Z_2 on port 2 are given in ref. 2.4. In this appendix we restrict ourselves to the case when $Z_1 = Z_2 = Z_0$. This case covers most applications and permits a major simplification of the formulae.

	S	Z	Y	R	ABCD
S		$S_{11}=\frac{(z_{11}-1)(z_{22}+1)-z_{12}z_{21}}{(z_{11}+1)(z_{22}+1)-z_{12}z_{21}}$	$S_{11}=\frac{(1-y_{11})(1+y_{22})+y_{12}y_{21}}{(1+y_{11})(1+y_{22})-y_{12}y_{21}}$	$S_{11}=\frac{R_{12}}{R_{22}}$	$S_{11}=\frac{A+BY_0-CZ_0-D}{A+BY_0+CZ_0+D}$
	$\begin{bmatrix}b_1\\b_2\end{bmatrix}=\begin{bmatrix}S_{11}&S_{12}\\S_{21}&S_{22}\end{bmatrix}\begin{bmatrix}a_1\\a_2\end{bmatrix}$	$S_{12}=\frac{2z_{12}}{(z_{11}+1)(z_{22}+1)-z_{12}z_{21}}$	$S_{12}=\frac{-2y_{12}}{(1+y_{11})(1+y_{22})-y_{12}y_{21}}$	$S_{12}=\frac{\Delta^R}{R_{22}}$	$S_{12}=\frac{2(AD-BC)}{A+BY_0+CZ_0+D}$
		$S_{21}=\frac{2z_{21}}{(z_{11}+1)(z_{22}+1)-z_{12}z_{21}}$	$S_{21}=\frac{-2y_{21}}{(1+y_{11})(1+y_{22})-y_{12}y_{21}}$	$S_{21}=\frac{1}{R_{22}}$	$S_{21}=\frac{2}{A+BY_0+CZ_0+D}$
		$S_{22}=\frac{(z_{11}+1)(z_{22}-1)-z_{12}z_{21}}{(z_{11}+1)(z_{22}+1)-z_{12}z_{21}}$	$S_{22}=\frac{(1+y_{11})(1-y_{22})+y_{12}y_{21}}{(1+y_{11})(1+y_{22})-y_{12}y_{21}}$	$S_{22}=\frac{-R_{21}}{R_{22}}$	$S_{22}=\frac{-A+BY_0-CZ_0+D}{A+BY_0+CZ_0+D}$
Z	$z_{11}=\frac{(1+S_{11})(1-S_{22})+S_{12}S_{21}}{(1-S_{11})(1-S_{22})-S_{12}S_{21}}$		$z_{11}=\frac{Y_{22}}{\Delta^Y}$	$z_{11}=\frac{R_{22}+R_{11}+R_{12}+R_{21}}{D_1}$	$z_{11}=\frac{A}{C}$
	$z_{12}=\frac{2S_{12}}{(1-S_{11})(1-S_{22})-S_{12}S_{21}}$	$\begin{bmatrix}V_1\\V_2\end{bmatrix}=\begin{bmatrix}z_{11}&z_{12}\\z_{21}&z_{22}\end{bmatrix}\begin{bmatrix}I_1\\I_2\end{bmatrix}$	$z_{12}=\frac{-Y_{12}}{\Delta^Y}$	$z_{12}=2\frac{\Delta^R}{D_1}\quad z_{21}=\frac{2}{D_1}$	$z_{12}=\frac{\Delta^A}{C}$
	$z_{21}=\frac{2S_{21}}{(1-S_{11})(1-S_{22})-S_{12}S_{21}}$		$z_{21}=\frac{-Y_{21}}{\Delta^Y}$	$z_{22}=\frac{R_{22}+R_{11}-R_{12}-R_{21}}{D_1}$	$z_{21}=\frac{1}{C}$
	$z_{22}=\frac{(1-S_{11})(1+S_{22})+S_{12}S_{21}}{(1-S_{11})(1-S_{22})-S_{12}S_{21}}$		$z_{22}=\frac{Y_{11}}{\Delta^Y}$	$D_1=Y_0(R_{22}-R_{11}-R_{12}+R_{21})$	$z_{22}=\frac{D}{C}$
Y	$y_{11}=\frac{(1-S_{11})(1+S_{22})+S_{12}S_{21}}{(1+S_{11})(1+S_{22})-S_{12}S_{21}}$	$y_{11}=\frac{Z_{12}}{\Delta^Z}$		$Y_{11}=\frac{R_{22}+R_{11}-R_{12}-R_{21}}{D_2}$	$Y_{11}=\frac{D}{B}$
	$y_{12}=\frac{-2S_{12}}{(1+S_{11})(1+S_{22})-S_{12}S_{21}}$	$y_{12}=\frac{-Z_{12}}{\Delta^Z}$	$\begin{bmatrix}I_1\\I_2\end{bmatrix}=\begin{bmatrix}Y_{11}&Y_{12}\\Y_{21}&Y_{22}\end{bmatrix}\begin{bmatrix}V_1\\V_2\end{bmatrix}$	$Y_{12}=-2\frac{\Delta^R}{D_2}\quad Y_{21}=\frac{-2}{D_2}$	$Y_{12}=\frac{-\Delta^A}{B}$
	$y_{21}=\frac{-2S_{21}}{(1+S_{11})(1+S_{22})-S_{12}S_{21}}$	$y_{21}=\frac{-Z_{21}}{\Delta^Z}$		$Y_{22}=\frac{R_{22}+R_{11}+R_{12}+R_{21}}{D_2}$	$Y_{21}=\frac{-1}{B}$
	$y_{22}=\frac{(1+S_{11})(1-S_{22})+S_{12}S_{21}}{(1+S_{11})(1+S_{22})-S_{12}S_{21}}$	$y_{22}=\frac{Z_{11}}{\Delta^Z}$		$D_2=Z_0(R_{22}-R_{11}+R_{12}-R_{21})$	$Y_{22}=\frac{A}{B}$

R	$R_{11}=\frac{-\Delta^S}{S_{21}}$	$R_{11}=\frac{z_{11}+z_{22}-\Delta^z-1}{2z_{21}}$	$R_{11}=\frac{y_{22}+y_{11}-\Delta^y-1}{(-2y_{21})}$		$R_{11}=\frac{1}{2}(A-BY_0-CZ_0+D)$
	$R_{12}=\frac{S_{11}}{S_{21}}$	$R_{12}=\frac{z_{11}-z_{22}+\Delta^z-1}{2z_{21}}$	$R_{12}=\frac{y_{22}-y_{11}-\Delta^y+1}{(-2y_{21})}$	$\begin{bmatrix} b_1 \\ a_1 \end{bmatrix}=\begin{bmatrix} R_{11} & R_{12} \\ R_{21} & R_{22} \end{bmatrix}\begin{bmatrix} a_2 \\ b_2 \end{bmatrix}$	$R_{12}=\frac{1}{2}(A+BY_0-CZ_0-D)$
	$R_{21}=\frac{-S_{22}}{S_{21}}$	$R_{21}=\frac{z_{11}-z_{22}-\Delta^z+1}{2z_{21}}$	$R_{21}=\frac{y_{22}-y_{11}+\Delta^y-1}{(-2y_{21})}$		$R_{21}=\frac{1}{2}(A-BY_0+CZ_0-D)$
	$R_{22}=\frac{1}{S_{21}}$	$R_{22}=\frac{z_{11}+z_{22}+\Delta^z+1}{2z_{21}}$	$R_{22}=\frac{y_{22}+y_{11}+\Delta^y+1}{(-2y_{21})}$		$R_{22}=\frac{1}{2}(A+BY_0+CZ_0+D)$
ABCD	$A=\frac{(1+S_{11})(1-S_{22})+S_{12}S_{21}}{2S_{21}}$	$A=\frac{Z_{11}}{Z_{21}}$	$A=\frac{-Y_{22}}{Y_{21}}$	$A=\frac{1}{2}(R_{22}+R_{11}+R_{21}+R_{12})$	
	$Y_0B=\frac{(1+S_{11})(1+S_{22})-S_{12}S_{21}}{2S_{21}}$	$B=\frac{\Delta^Z}{Z_{21}}$	$B=\frac{-1}{Y_{21}}$	$B=\frac{1}{2}Z_0(R_{22}-R_{11}-R_{21}+R_{12})$	$\begin{bmatrix} V_1 \\ I_1 \end{bmatrix}=\begin{bmatrix} A & B \\ C & D \end{bmatrix}\begin{bmatrix} V_2 \\ -I_2 \end{bmatrix}$
	$Z_0C=\frac{(1-S_{11})(1-S_{22})-S_{12}S_{21}}{2S_{21}}$	$C=\frac{1}{Z_{21}}$	$C=\frac{-\Delta^Y}{Y_{21}}$	$C=\frac{1}{2}Y_0(R_{22}-R_{11}+R_{21}-R_{12})$	
	$D=\frac{(1-S_{11})(1+S_{22})+S_{12}S_{21}}{2S_{21}}$	$D=\frac{Z_{22}}{Z_{21}}$	$D=\frac{-Y_{11}}{Y_{21}}$	$D=\frac{1}{2}(R_{22}+R_{11}-R_{21}-R_{12})$	

$\Delta^K = K_{11}K_{22}-K_{12}K_{21}$
$z_{ik} = Z_{ik}/Z_0$
$Y_0 = 1/Z_0$
$y_{ik} = Y_{ik}/Y_0$

Appendix 2.2 Self-calibration procedure: error term determination

In the flowchart shown in Fig. 2.20 it is assumed that there are two unknown error adaptors X and Y between the actual two-port which is described by the unknown matrix $\boldsymbol{S}_A$ and the measured matrix $\boldsymbol{S}_m$. It is assumed that the leakage errors have been removed previously (see Section 2.3.5). The eight error terms to be solved by this procedure are

e_{00} and e_{33}	the directivity on ports 1 and 2
e_{11} and e_{22}	the port match on ports 1 and 2
$e_{10}e_{01}$ and $e_{23}e_{32}$	the reflection frequency response
$e_{10}e_{32}$ and $e_{23}e_{01}$	the transmission frequency response

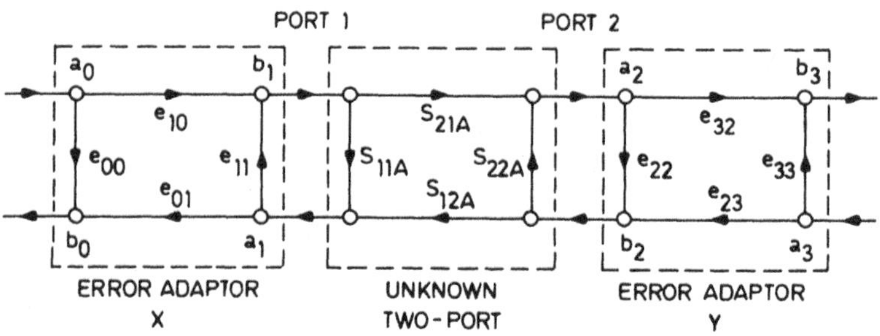

Fig. 2.20 Error model for self-calibration.

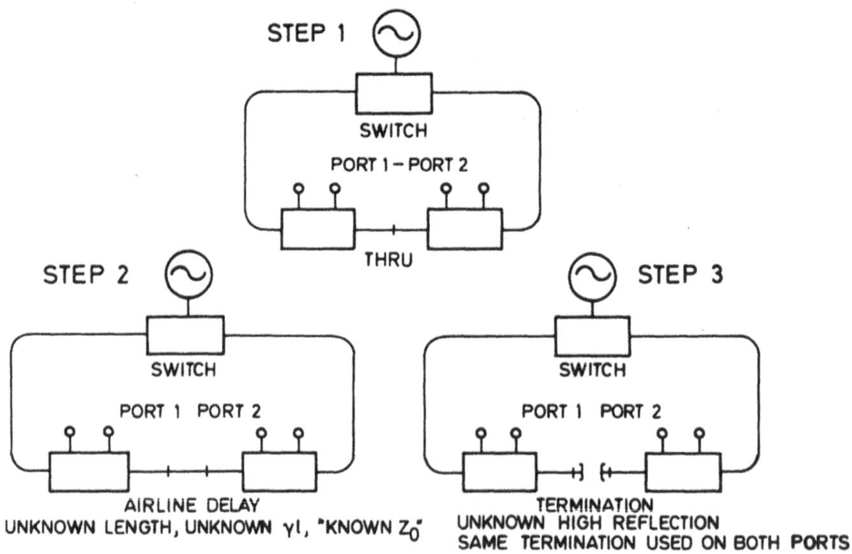

Fig. 2.21 Self-calibration procedure.

In the self-calibration procedure we make use of a through line, an airline with known delay and a termination with a high reflection coefficient to solve for the error terms (Fig. 2.21). The error terms are described using an S matrix to ensure compatibility with the notation used in Sections 2.3.2 and 2.3.3. In fact the resolution of these error terms is simpler if an R matrix representation is used.

An R matrix is related to its equivalent S matrix by the well-known relation (Appendix 2.1)

$$\boldsymbol{R} = \begin{bmatrix} R_{11} & R_{12} \\ R_{21} & R_{22} \end{bmatrix} = \frac{1}{S_{21}} \begin{bmatrix} S_{12}S_{21} - S_{11}S_{22} & S_{11} \\ -S_{22} & 1 \end{bmatrix}$$

The inverse transformation may also be useful:

$$\boldsymbol{S} = \begin{bmatrix} S_{11} & S_{12} \\ S_{21} & S_{22} \end{bmatrix} = \frac{1}{R_{22}} \begin{bmatrix} R_{12} & R_{11}R_{22} - R_{12}R_{21} \\ 1 & -R_{21} \end{bmatrix}$$

The error terms x_{ij} and y_{ij} of the error adaptors $\boldsymbol{R}_x$ and $\boldsymbol{R}_y$ can be expressed as

$$\boldsymbol{R}_x = \begin{bmatrix} x_{11} & x_{12} \\ x_{21} & x_{22} \end{bmatrix} = \frac{1}{e_{10}} \begin{bmatrix} e_{10}e_{01} - e_{00}e_{11} & e_{00} \\ -e_{11} & 1 \end{bmatrix} \tag{1}$$

$$\boldsymbol{R}_y = \begin{bmatrix} y_{11} & y_{12} \\ y_{21} & y_{22} \end{bmatrix} = \frac{1}{e_{32}} \begin{bmatrix} e_{32}e_{23} - e_{22}e_{33} & e_{22} \\ -e_{33} & 1 \end{bmatrix} \tag{2}$$

The $\boldsymbol{R}_m$ and $\boldsymbol{R}_a$ matrices equivalent to the measured $\boldsymbol{S}_m$ and actual $\boldsymbol{S}_a$ matrices can be written

$$\boldsymbol{R}_m = \begin{bmatrix} R_{11m} & R_{12m} \\ R_{21m} & R_{22m} \end{bmatrix} = \frac{1}{S_{21m}} \begin{bmatrix} S_{12m}S_{21m} - S_{11m}S_{22m} & S_{11m} \\ -S_{22m} & 1 \end{bmatrix} \tag{3}$$

$$\boldsymbol{R}_a = \begin{bmatrix} R_{11a} & R_{12a} \\ R_{21a} & R_{22a} \end{bmatrix} = \frac{1}{S_{21a}} \begin{bmatrix} S_{12a}S_{21a} - S_{11a}S_{22a} & S_{11a} \\ -S_{22a} & 1 \end{bmatrix} \tag{4}$$

The actual R matrix value of a zero-length reflectionless through line is the unit matrix

$$\boldsymbol{R}_{at} = \boldsymbol{U} = \begin{bmatrix} 1 & 0 \\ 0 & 1 \end{bmatrix} \tag{5}$$

Similarly the R matrix of a reflectionless delay line is

$$\boldsymbol{R}_{ad} = \begin{bmatrix} e^{-\gamma l} & 0 \\ 0 & 1/e^{-\gamma l} \end{bmatrix} \tag{6}$$

where γ is the complex propagation constant and l is the excess delay line length.

With the R matrix representation the measured matrix $\boldsymbol{R}_{\mathrm{m}}$ is simply expressed by

$$\boldsymbol{R}_{\mathrm{m}} = \boldsymbol{R}_{\mathrm{x}}\boldsymbol{R}_{\mathrm{a}}\boldsymbol{R}_{\mathrm{y}} \tag{7}$$

When the through line is connected we measure

$$\boldsymbol{R}_{\mathrm{mt}} = \boldsymbol{R}_{\mathrm{x}}\boldsymbol{R}_{\mathrm{at}}\boldsymbol{R}_{\mathrm{y}} = \boldsymbol{R}_{\mathrm{x}}\boldsymbol{R}_{\mathrm{y}} \tag{8}$$

Then with the delay line we obtain

$$\boldsymbol{R}_{\mathrm{md}} = \boldsymbol{R}_{\mathrm{x}}\boldsymbol{R}_{\mathrm{ad}}\boldsymbol{R}_{\mathrm{y}} \tag{9}$$

From eqn (8) we obtain

$$\boldsymbol{R}_{\mathrm{y}} = \boldsymbol{R}_{\mathrm{x}}^{-1}\boldsymbol{R}_{\mathrm{mt}} \tag{10}$$

and substituting $\boldsymbol{R}_{\mathrm{y}}$ in eqn (9) gives

$$\boldsymbol{M}\boldsymbol{R}_{\mathrm{x}} = \boldsymbol{R}_{\mathrm{x}}\boldsymbol{R}_{\mathrm{ad}} \tag{11}$$

where

$$\boldsymbol{M} = \boldsymbol{R}_{\mathrm{md}}\boldsymbol{R}_{\mathrm{mt}}^{-1} = \begin{bmatrix} m_{11} & m_{12} \\ m_{21} & m_{22} \end{bmatrix} \tag{12}$$

Using eqns (1), (6) and (12), we can rewrite eqn (11) as follows:

$$\begin{bmatrix} m_{11} & m_{12} \\ m_{21} & m_{22} \end{bmatrix}\begin{bmatrix} x_{11} & x_{12} \\ x_{21} & x_{22} \end{bmatrix} = \begin{bmatrix} x_{11} & x_{12} \\ x_{21} & x_{22} \end{bmatrix}\begin{bmatrix} \mathrm{e}^{-\gamma l} & 0 \\ 0 & 1/\mathrm{e}^{-\gamma l} \end{bmatrix} \tag{13}$$

$$m_{11}x_{11} + m_{12}x_{21} = x_{11}\mathrm{e}^{-\gamma l} \tag{14}$$

$$m_{21}x_{11} + m_{22}x_{21} = x_{21}\mathrm{e}^{-\gamma l} \tag{15}$$

$$m_{11}x_{12} + m_{12}x_{22} = \frac{x_{12}}{\mathrm{e}^{-\gamma l}} \tag{16}$$

$$m_{21}x_{12} + m_{22}x_{22} = \frac{x_{22}}{\mathrm{e}^{-\gamma l}} \tag{17}$$

Elimination of $\mathrm{e}^{-\gamma l}$ between eqns (14) and (15) and also between eqns (16) and (17) gives two equations with the same constant terms and therefore the same roots:

$$m_{21}\left(\frac{x_{11}}{x_{21}}\right)^2 + (m_{22} - m_{11})\frac{x_{11}}{x_{21}} - m_{12} = 0 \tag{18}$$

$$m_{21}\left(\frac{x_{12}}{x_{22}}\right)^2 + (m_{22} - m_{11})\frac{x_{12}}{x_{22}} - m_{12} = 0 \tag{19}$$

Let these roots be a and b. If we look at the physical meaning of these solutions we can make a suitable choice. From eqn (1)

$$a = \frac{x_{11}}{x_{21}} = e_{00} - \frac{e_{10}e_{01}}{e_{11}} \tag{20}$$

$$b = \frac{x_{12}}{x_{22}} = e_{00} \tag{21}$$

For a good reflectometer the term $e_{10}e_{01}$ is almost equal to unity and the term e_{00} is very small. Therefore the choice of roots is dictated by the condition $a \gg b$. Thus the first error term is solved:

$$e_{00} = b \tag{22}$$

In addition, eliminating e_{00} from eqns (20) and (21) yields an interesting relationship between e_{11} and $e_{01}e_{10}$:

$$\frac{e_{01}e_{10}}{e_{11}} = b - a \tag{23}$$

The same approach can be used to solve for the directivity error term e_{33}. As in eqn (10) we extract $\boldsymbol{R}_x$ from eqn (8) and substitute it in eqn (9):

$$\boldsymbol{R}_y N = \boldsymbol{R}_{ad}\boldsymbol{R}_y \tag{24}$$

where

$$N = \boldsymbol{R}_{mt}{}^{-1}\boldsymbol{R}_{md} = \begin{bmatrix} n_{11} & n_{12} \\ n_{21} & n_{22} \end{bmatrix} \tag{25}$$

Using eqns (2), (6) and (25) we can rewrite eqn (24) as

$$\begin{bmatrix} y_{11} & y_{12} \\ y_{21} & y_{22} \end{bmatrix}\begin{bmatrix} n_{11} & n_{12} \\ n_{21} & n_{22} \end{bmatrix} = \begin{bmatrix} e^{-\gamma l} & 0 \\ 0 & 1/e^{-\gamma l} \end{bmatrix}\begin{bmatrix} y_{11} & y_{12} \\ y_{21} & y_{22} \end{bmatrix} \tag{26}$$

$$y_{11}n_{11} + y_{12}n_{21} = y_{11}e^{-\gamma l} \tag{27}$$

$$y_{21}n_{11} + y_{22}n_{21} = y_{21}e^{-\gamma l} \tag{28}$$

$$y_{11}n_{12} + y_{12}n_{22} = \frac{y_{12}}{e^{-\gamma l}} \tag{29}$$

$$y_{21}n_{12} + y_{22}n_{22} = \frac{y_{22}}{e^{-\gamma l}} \tag{30}$$

Finally, the elimination of $e^{-\gamma l}$ between eqns (27) and (28) and between eqns (29) and (30) also yields two equations:

$$n_{12}\left(\frac{y_{11}}{y_{12}}\right)^2 + (n_{22} - n_{11})\frac{y_{11}}{y_{12}} - n_{21} = 0 \tag{31}$$

$$n_{12}\left(\frac{y_{21}}{y_{22}}\right)^2 + (n_{22} - n_{11})\frac{y_{21}}{y_{22}} - n_{21} = 0 \tag{32}$$

Let the roots of these equations be c and d. We obtain from eqn (2)

$$c = \frac{y_{11}}{y_{12}} = -e_{33} + \frac{e_{23}e_{32}}{e_{22}} \tag{33}$$

$$d = \frac{y_{21}}{y_{22}} = -e_{33} \tag{34}$$

As in the previous case the appropriate choice of roots is given by the condition

$$c \gg d$$

and the second error term is solved:

$$e_{33} = -d \tag{35}$$

Eliminating e_{33} from eqns (33) and (34) yields a relationship between e_{22} and $e_{23}e_{32}$:

$$\frac{e_{23}e_{32}}{e_{22}} = c - d \tag{36}$$

The third step of the self-calibration procedure is now used to solve for the other error terms.

We measure G_{mx} with the high reflection termination G_a connected to port 1 of the error adaptor X:

$$G_{mx} = e_{00} + \frac{e_{10}e_{01}G_a}{1 - e_{11}G_a} \tag{37}$$

G_a can be extracted and, using eqns (22) and (23), we obtain

$$G_a = \frac{1}{e_{11}}\frac{b - G_{mx}}{a - G_{mx}} \tag{38}$$

If the termination G_a is now connected to port 2 of Y, we obtain

$$G_{my} = e_{33} + \frac{e_{23}e_{32}G_a}{1 - e_{22}G_a} \tag{39}$$

As before, G_a can be extracted and, using eqns (35) and (36), we obtain

$$G_a = \frac{1}{e_{22}}\frac{d + G_{my}}{c + G_{my}} \tag{40}$$

The elimination of G_a between eqns (38) and (39) yields a relationship between e_{11} and e_{22}:

$$\frac{1}{e_{22}} = \frac{1}{e_{11}}\frac{b - G_{mx}}{a - G_{mx}}\frac{c + G_{my}}{d + G_{my}} \tag{41}$$

We need another relationship between e_{11} and e_{22}. In the first step of the calibration, the error term e_{22} is measured through the error adaptor X as in the standard calibration. The measured value G_{ml} is

$$G_{ml} = e_{00} + \frac{e_{10}e_{01}e_{22}}{1-e_{11}e_{22}} \tag{42}$$

If eqn (42) is solved for e_{11} we obtain

$$e_{11} = \frac{1}{e_{22}} \frac{b-G_{ml}}{a-G_{ml}} \tag{43}$$

The value of e_{11} can now be found by substituting $1/e_{22}$ from eqn (41) into eqn (43):

$$e_{11}{}^2 = \frac{b-G_{mx}}{a-G_{mx}} \frac{c+G_{my}}{d+G_{my}} \frac{b-G_{ml}}{a-G_{ml}} \tag{44}$$

The choice of the appropriate sign to extract the square root of e_{11} will be discussed later.

We now solve eqn (38) for e_{22}:

$$e_{22} = \frac{1}{e_{11}} \frac{b-G_{ml}}{a-G_{ml}} \tag{45}$$

From eqn (23)

$$e_{10}e_{01} = (b-a)e_{11} \tag{46}$$

and from eqn (36)

$$e_{23}e_{32} = (c-d)e_{22} \tag{47}$$

The solution for the frequency response error terms can be obtained from measurement of the through line. The measured transmission terms S_{21m} and S_{12m} are equal to the transmission terms of the cascaded $\boldsymbol{S}_x$ and $\boldsymbol{S}_y$ matrices:

$$S_{21m} = \frac{e_{10}e_{32}}{1-e_{11}e_{22}} \tag{48}$$

$$S_{12m} = \frac{e_{23}e_{01}}{1-e_{11}e_{22}} \tag{49}$$

Finally, the two remaining error terms are determined:

$$e_{10}e_{32} = S_{21m}(1-e_{11}e_{22}) \tag{50}$$

$$e_{23}e_{01} = S_{12m}(1-e_{11}e_{22}) \tag{51}$$

After this step the eight error terms have been solved using 10 measured complex terms: the *S* matrices of the through and delay lines and the high reflection termination G_a on ports 1 and 2. The two remaining terms can be used to solve for G_a and $e^{-\gamma l}$.

The value of G_a given in eqn (38) can be used to validate the self-calibration procedure, as a final check or as an absolute measure of unknown reflection standards. If G_a is known approximately before the calibration is started, eqn (38) can be used to extract the appropriate root of eqn (44) by a simple test of the phase of G_a.

Any of eqns (14)–(17) or (27)–(30) can be used to determine $e^{-\gamma l}$. For example, with eqn (14)

$$e^{-\gamma l} = m_{11} + \frac{m_{12}}{a} \tag{52}$$

The value of $e^{-\gamma l}$ enables the propagation constant γ to be determined in the medium under consideration if l is known accurately.

Chapter 3

Synthesis and Design of Small-signal and Low Noise GaAs MESFET Amplifiers

ROMUALD WROBLEWSKI

3.1 Introduction

From the beginning of the 1970s, when it became apparent that the GaAs MESFET was capable of low noise amplification, the device has undergone a great improvement in performance. In particular, the demand for the lowest possible noise figure at ever higher frequencies has stimulated research on the physics of the device and has had a great impact on its technology. Current GaAs MESFET applications range from v.h.f. to 40 GHz, and some laboratory results indicate its usefulness at 60 GHz and beyond. Applications of MESFET amplifiers include medium-band and broadband (in excess of 2 octaves) bandpass amplifiers, ultrabroadband lowpass amplifiers, very low noise medium-bandwidth amplifiers and ultralow noise cryogenically cooled amplifiers.

Broadband amplifiers with a flat gain and high dynamics are required in electronic countermeasure (ECM) subsystems and command, control, communications and intelligence (C^3I) systems. These amplifiers cover bandwidths of typically 2–6 GHz, 6–18 GHz and 18–40 GHz. The aim for future systems is to cover the whole 2–40 GHz band with a single amplifier. In contrast, in communications and radiometer receivers the lowest possible noise figure is required over rather narrow bandwidths (typically 10%–40%). Lowpass decade-bandwidth amplifiers are required in data transmission systems to amplify pulses with short rise times.

All these applications are treated in this chapter. The design principles are outlined for low noise or broadband amplifiers, and techniques for synthesizing matching networks are reviewed. The following matching principles are treated in detail: reactive matching, feedback, lossy matching and active matching. Ultrabroadband amplifiers using these matching principles as well as distributed amplification are discussed. In the section on low noise amplifiers we describe how accurate noise figure measurements can be made and how the resulting data can be exploited in low noise amplifier design. The chapter ends with a review of ultralow noise cooled GaAs MESFET amplifiers. State-of-the-art noise performances are given and compared with those of the high electron mobility transistor (HEMT).

3.2 Design considerations for small-signal MESFET amplifiers

The design process for microwave MESFET amplifiers can be represented schematically as in Fig. 3.1. It consists essentially of three critical steps where the choices made are of prime importance for performance and cost. These are the selection of the transistor, the choice of circuit configuration and the choice of the technology used for realization of the amplifier. Unfortunately, these are not independent choices. In addition, there is no single solution for each design step. Computer analysis and optimization must be used in an interactive and iterative way in order to obtain an optimum solution. Thus the design process is quite complex and requires a wide background in active device physics, network synthesis and microwave technology.

The GaAs MESFET exhibits an inherent gain roll-off with frequency (typically −6 dB per octave*) and its noise figure increases with frequency. In order to achieve an amplifier stage with flat gain, low noise and low voltage standing wave ratio (VSWR) the active device must be provided with appropriate source and load terminations. The terminations are synthesized using matching networks which are also called equalizers. The synthesis of matching networks is a difficult problem which involves broadband matching as well as impedance transformation and gain roll-off compensation. The problem may be additionally complicated by the potential instability of the device.

The following four definitions of gain are useful in the design of matching networks: transducer gain, unilateral transducer gain, available gain and power gain. These four gains are treated in detail in Chapter 2 together with stability considerations. They enable the amplifier performance to be evaluated under various matching conditions. The definitions are repeated here for completeness.

---

* The number m of octaves for the frequency band from ω_1 to ω_2 is defined as $\omega_2/\omega_1 = 2^m$. The gain slope in decibels per octave is computed from $G_{dB}(\omega)$ as slope $= \{G_{dB}(\omega_2) - G_{dB}(\omega_1)\}/m$.

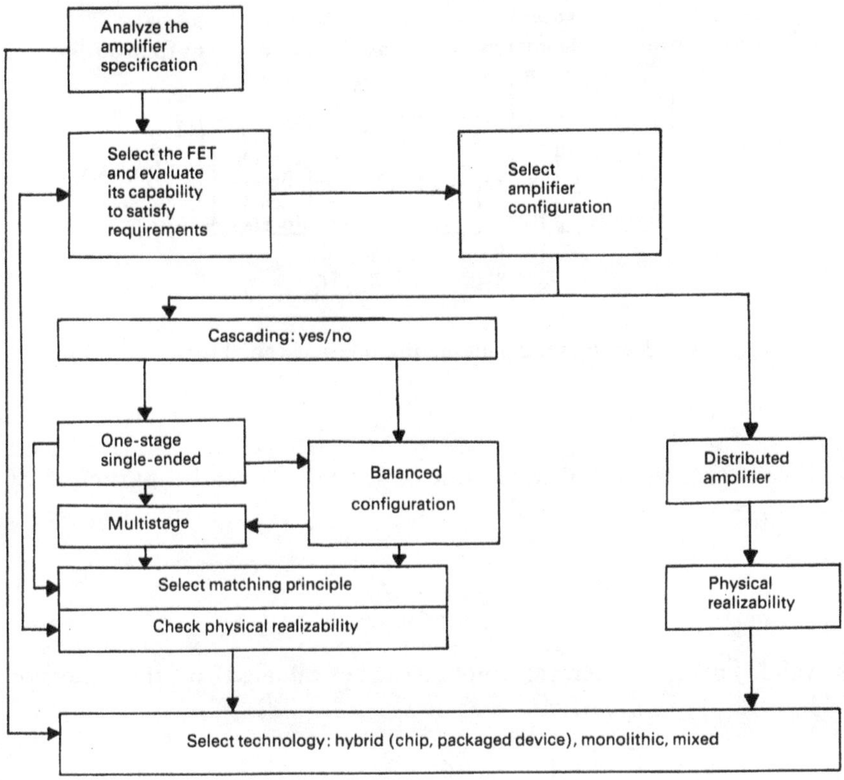

Fig. 3.1 Scheme for the design of microwave amplifiers.

When the S parameters and reflection coefficients of the active device are normalized to the same reference ($Z_0 = 50\,\Omega$) (Fig. 3.2), the transducer power gain is given by [3.1–3.3]

$$G_T = \frac{P_L}{P_A} = G_T(\Gamma_g, \Gamma_L, S) \tag{3.1a}$$

$$G_T = \frac{(1-|\Gamma_g|^2)\,|S_{21}|^2\,(1-|\Gamma_L|^2)}{|(1-\Gamma_g S_{11})(1-S'_{22}\Gamma_L)|^2}$$

where

$$S'_{22} = S_{22} + \frac{S_{12}S_{21}\Gamma_g}{1-S_{11}\Gamma_g} \tag{3.1b}$$

A convenient approximation is to assume that the transistor is unilateral, i.e. $S_{12} = 0$. Then (3.1a) becomes

$$G_{Tu} = \frac{(1-|\Gamma_g|^2)\,|S_{21}|^2\,(1-|\Gamma_L|^2)}{|1-\Gamma_g S_{11}|^2\,|1-\Gamma_L S_{22}|^2} \tag{3.2}$$

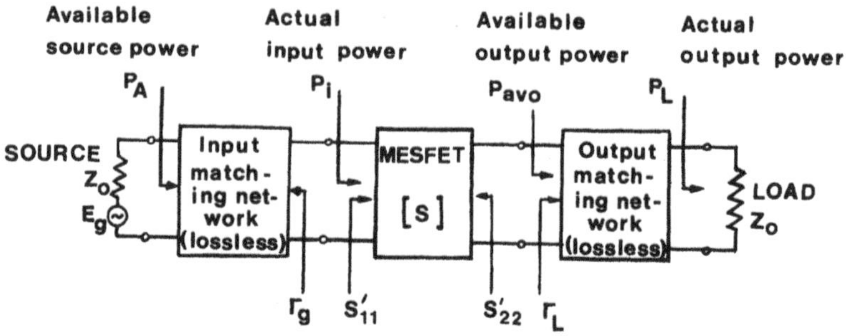

Fig. 3.2 Power definitions in a one-stage amplifier.

G_{Tu} is called the unilateral transducer power gain. It can be maximized by setting $\Gamma_g = S_{11}^*$ and $\Gamma_L = S_{22}^*$:

$$G_{Tu\,max} = \frac{|S_{21}|^2}{(1-|S_{11}|^2)(1-|S_{22}|^2)} \tag{3.3}$$

The validity of the unilateral assumption can be evaluated from the expression [3.1]

$$\frac{1}{(1+u)^2} < \frac{G_T}{G_{Tu\,max}} < \frac{1}{(1-u)^2} \tag{3.4a}$$

where

$$u = \frac{|S_{11}S_{22}S_{12}S_{21}|}{(1-|S_{11}|^2)(1-|S_{22}|^2)} \tag{3.4b}$$

The approximation is acceptable if the variation in $G_T/G_{Tu\,max}$ from the maximum to the minimum is not too large, say ± 0.5 dB.

Another useful definition is the available power gain when the transistor output is conjugate matched:

$$G_A = \frac{P_{avo}}{P_A} = G_A(\Gamma_g, S)$$
$$= \frac{(1-|\Gamma_g|^2)\,|S_{21}|^2}{|1-S_{11}\Gamma_g|^2\,(1-|S'_{22}|^2)} \tag{3.5}$$

The available gain can be maximized for an unconditionally stable transistor.

When the transistor is potentially unstable (the Rollet stability factor K is less than unity (Chapter 2, Section 2.6)), the power gain is the most useful definition:

$$G_p = \frac{P_L}{P_i} = G_p(\Gamma_L, S)$$
$$= \frac{|S_{21}|^2(1-|\Gamma_L|^2)}{|1-\Gamma_L S_{22}|^2(1-|S'_{11}|^2)} \tag{3.6a}$$

where

$$S'_{11} = S_{11} + \frac{S_{12}S_{21}\Gamma_L}{1-S_{22}\Gamma_L} \tag{3.6b}$$

The transducer gain is the most significant parameter because it includes the insertion effect of the active device as well as that of the amplifier. It is generally less than the available gain and the power gain: $G_T \leqslant G_A$, where equality occurs for $\Gamma_L = S'_{22}{}^*$; $G_T \leqslant G_p$, where equality occurs for $\Gamma_g = S'_{11}{}^*$. G_A and G_p are very useful in the synthesis of matching networks since they allow a sequential design to be adopted [3.4–3.6]. When the available power gain definition is used the amplifier is designed proceeding from the source to the load (low noise and/or high gain amplifier design; unconditionally stable transistor). When the power gain definition is used, the amplifier is designed proceeding from the load to the source (potentially unstable transistor). This allows the stability conditions to be controlled by selection of an appropriate load impedance. The sequential design has an advantage over the unilateral approach (see Section 3.3.3) since at each step the source or the load impedance can be specified as well as the amount of gain compensation required from the equalizer.

Usually more than one active device has to be employed to achieve the required gain. This can be done using either multiplicative or additive (distributed) amplification. The majority of amplifiers make use of the multiplicative process through cascading of either active devices with interstage matching networks or matched modules. The input and output VSWRs of the modules should be kept low when cascading because of their impact on the gain flatness, stability, VSWR and group delay of the overall amplifier.

The modules can be built in either single-ended or balanced configurations, as shown in Fig. 3.3. The balanced amplifier employs two identical single-ended amplifiers, which are usually single stage, combined with two 3 dB hybrid couplers [3.7]. This configuration enables the input and output VSWRs to be minimized even when poorly matched stages are used. Isolators are also used between stages to improve cascadability or when very low input and output VSWRs are required, particularly in communications amplifiers. The requirements for flat gain, low noise figure and low VSWR cannot be met simultaneously, and some trade-off between parameters is usually accepted.

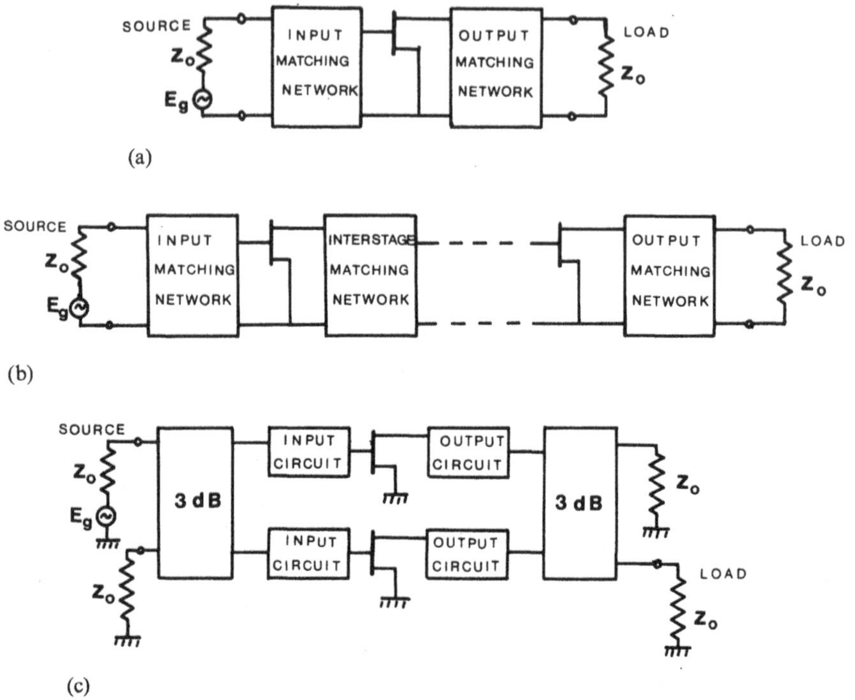

Fig. 3.3 Basic amplifier configurations: (a) single-ended one-stage; (b) single-ended multistage; (c) balanced configuration.

Several matching principles are available for this purpose (reactive matching, negative feedback, lossy matching and active matching) and can be employed either exclusively or in some combination.

The principle of distributed amplification is used for the realization of ultrabroadband amplifiers.

The performances of microwave amplifiers designed using the above principles are summarized in Table 3.1.

A comparative study [3.8] of the gain and VSWR characteristics obtained for single-stage single-ended amplifiers using various matching principles did not identify the optimum circuit type. The choice should be made on the basis of the gain, bandwidth and maximum operating frequency. However, when the gain specifications require the cascading of more than three stages, the distributed amplifier will exhibit a superior performance.

Advanced synthesis methods, both analytical and iterative, have been developed for reactive matching networks. The synthesis of other types of matching circuits starts with simplified formulae (low frequency) and yields a good initial solution for further numerical optimization. Whichever synthesis

Table 3.1 Characteristics of small-signal single-ended broadband microwave amplifiers (1987)

Amplification principle	Multiplicative amplification				Additive amplification
Matching principle	Reactive matching	Negative feedback	Lossy matching	Active matching	Distributed amplifier
Bandwidth	Narrow to large (more than two octaves)	Multi-octave			
Gain per stage	High to small depending on the bandwidth and operating frequency	Small	Small	Very small to small	Small
Noise figure	Very low to fair depending on bandwidth	Fair to poor	Poor	Fair to poor	Poor
Input VSWR	Poor	Fair to good	Fair to good	Very good (common gate)	Good
Output VSWR	Fair to poor	Good	Good	Very good (common drain)	Good
Operating frequencies (GHz)	1–60	0.1–18	0.1–18	0.1–10	1–40
Technology	Hybrid (packaged FET, chip), monolithic	Hybrid or monolithic	Hybrid or monolithic	Monolithic	Hybrid, monolithic preferred
Applications	Bandpass low noise high gain medium-band and wideband amplifiers	Lowpass or bandpass ultrabroadband amplifiers	Lowpass or bandpass ultrabroadband amplifiers	Lowpass ultrabroadband amplifiers	Ultrabroadband amplifiers

method is applied, it has to be completed by analysis using the S parameters of the device, models of matching networks and numerical optimization. The equivalent circuit of the device can sometimes be used in place of its S parameters. The advantage of using the equivalent circuit in computer-aided design is the possibility of device dimension scaling and the ease with which the sensitivity of the amplifier to variations in the physical parameters of the device can be studied. It is particularly useful in the design of monolithic amplifiers. In addition, the S parameters of the device can be extrapolated from the model for the frequency range where measurements are not available.

Several excellent books are devoted to the design of microwave amplifiers and contain very useful guidelines [3.1, 3.2, 3.6, 3.9]. However, amplifier design remains a highly interactive and iterative process even though steps have recently been taken toward fully automated synthesis of reactive matching networks [3.10, 3.11].

3.3 Bandpass amplifiers using the reactive matching principle

Medium and wideband bandpass amplifiers, which are also known as broadband amplifiers, are dealt with in this section. These amplifiers typically have a low noise figure and a frequency bandwidth ranging from 25% up to more than two octaves [3.12–3.14]. The principle of reactive matching (lossless) is preferred when low noise and high gain have to be obtained from the active device. Currently, both analytical (direct) [3.1, 3.10, 3.15–3.21] and iterative (indirect) [3.4, 3.5, 3.22–3.27] approaches are used to synthesize the reactive matching networks.

The analytical approach based on the theory of gain–bandwidth limitation was developed successively by Bode [3.28], Fano [3.29] and Youla [3.30]. Numerous techniques based on this approach have been proposed [3.15–3.17, 3.31–3.38]. They allow synthesis of matching networks with a suitable gain slope which are capable of absorbing the model elements and of realizing the required impedance transformations. All the techniques require a circuit model for the transistor and the assumption of an analytical transfer function for the matching network. This approach has been extensively applied to MESFET amplifiers by Ha [3.1].

The iterative approach is based only on the measured S parameters of the transistor or the impedance data [3.22]. Carlin and coworkers [3.23–3.27] developed a very efficient technique—the real frequency method—which eliminates the need for an analytical representation of the equalizer transfer function. This technique was further simplified by Yarman and Carlin [3.4] who synthesized the equalizer reflection coefficient directly in rational form. This approach gives generally simpler topologies and superior gain performance than those of structures obtained using the analytical approach. Another interesting technique—the transformation Q approach—which makes use of the transforming properties of ladder reactance networks has been proposed by Abrie [3.6].

3.3.1 Definition of the broadband matching problem

In order to achieve either the maximum value of flat gain or a low noise figure together with a flat gain, the transistor must be provided with appropriate input and output terminating impedances (here, flat gain has the meaning zero slope response). However, in many cases it would be impractical to start the synthesis by determining these impedances simultaneously. Instead, the current procedure is to specify the input and output impedances for the transistor (usually taken as corresponding to S_{11} and S_{22} if the unilateral assumption is valid) or to employ a sequential procedure (see Sections 3.2 and 3.3.3). The lossless equalizer, which gives the prescribed transducer gain when it is inserted between specified ports (e.g. between the 50 Ω source and the transistor input), is then synthesized.

This general broadband matching problem was defined in refs 3.19 and 3.25 and is illustrated in Fig. 3.4. The reflection coefficients $S_{11}(j\omega)$ and $S_{22}(j\omega)$ at the input and output ports, normalized to the terminating impedances $Z_g(j\omega)$ and $Z_L(j\omega)$, can be expressed as

$$S_{11}(j\omega) = \frac{Z_1(j\omega) - Z_g^*(j\omega)}{Z_1(j\omega) + Z_g(j\omega)} \tag{3.7a}$$

$$S_{22}(j\omega) = \frac{Z_2(j\omega) - Z_L^*(j\omega)}{Z_2(j\omega) + Z_L(j\omega)} \tag{3.7b}$$

where $Z_1(j\omega)$ and $Z_2(j\omega)$ are the driving-point impedances when the output and input ports are terminated in $Z_L(j\omega)$ and $Z_g(j\omega)$ respectively. Since the equalizer is lossless the transducer power gain $G(\omega^2)$ can be expressed as follows:

$$\begin{aligned} G(\omega^2) &= 1 - |S_{11}(j\omega)|^2 \\ &= 1 - |S_{22}(j\omega)|^2 \\ &= 1 - |\Gamma(j\omega)|^2 \end{aligned} \tag{3.8}$$

The broadband matching problem is to determine the equalizer which achieves a flat $G(\omega^2)$ as high as possible over a prescribed frequency band $\omega_1 \leqslant \omega \leqslant \omega_2$. In multistage amplifiers this requirement is applied to front-end and back-end equalizers, and the transistor gain roll-off is compensated using interstage equalizers. In the latter case $G(\omega^2)$ has a positive slope response with a maximum value at the upper band edge and provides an increasing mismatch with decreasing frequency. Since the large reactance versus

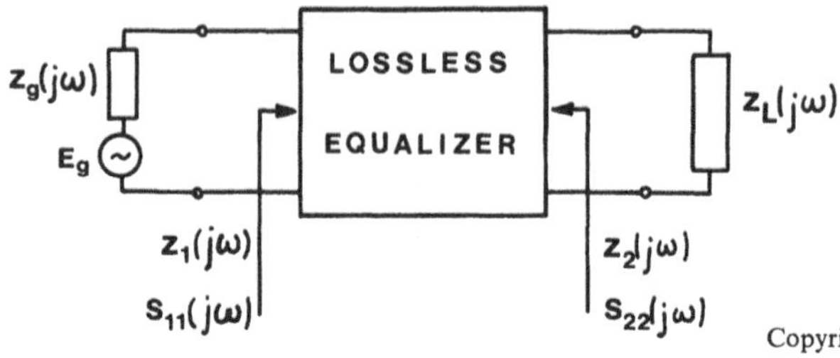

Fig. 3.4 Broadband matching problem (double matching [3.27]).

frequency slope of interstage equalizers causes high group-delay variations and enhances the gain ripple, the multistage configuration with reactive interstage couplings becomes impractical for more than three stages unless a lossy interstage network is used [3.39]. In single-stage amplifiers the gain slope compensation is usually achieved in the input equalizer and results in a high input VSWR at the low frequency edge. Thus the division of the slope compensation between the input and output equalizers may offer advantages [3.40]. Gain slope compensation can also be achieved by using a lossy equalizer in the output [3.12] or by cascading the stages with a network with constant input and output resistances which exhibits variable attenuation with frequency [3.41–3.43].

3.3.2 Synthesis of lossless equalizers using the analytical approach

The analytical approach requires that the load impedance to be matched is first approximated by a circuit model and that an analytical expression is assumed for the transfer function of the equalizer. This enables optimum gain–bandwidth limitations to be derived which determine the capability of reactive element absorption into the equalizer network. In fact the reactive elements of the load must be absorbed into the equalizer and therefore they constrain the gain bandwidth as well as the available topologies.

The transistor circuit model is generally approximated as a unilateral device [3.12, 3.17]. This allows the equivalent circuit to be derived from the S_{11} and S_{22} parameters by numerically fitting pre-assigned topologies to these data. Approximate closed-form expressions for lumped-element models are also available [3.10, 3.44]. Alternatively, the commensurate-distributed model can be used in the synthesis of distributed matching networks [3.40]. The unilateral MESFET model used for the synthesis of matching networks is shown in Fig. 3.5.

The values of the circuit elements and S parameters for the NEC packaged device NE710.84 [3.45] are given in Tables 3.2 and 3.3 for illustration. As can be seen from Fig. 3.5(b), the most suitable topologies for

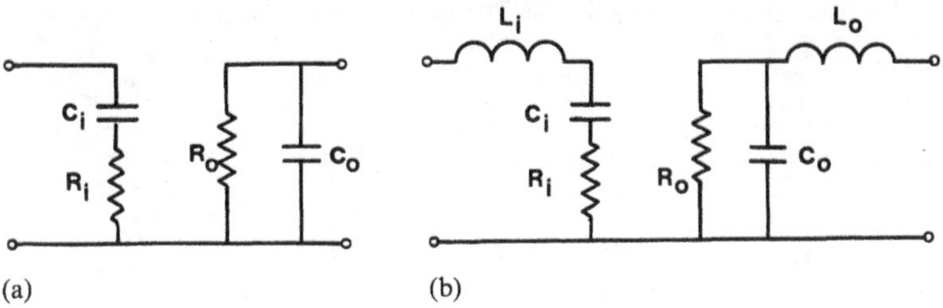

Fig. 3.5 Unilateral MESFET model: (a) chip; (b) packaged device.

input and output equalizers are bandpass and lowpass respectively. The latter can also be transformed into the bandpass topology by using the lowpass-to-bandpass frequency transformation (eqn (3.14)) and Norton transformer equivalences [3.1, 3.46] (Fig. 3.6).

A more elaborate procedure must be employed for potentially unstable devices because an attempt to match S_{11} and S_{22} usually makes the amplifier oscillate. The power gain definition (eqn (3.6)) is used in this case and acceptable load or source impedances are found from constant-gain circles taking into account instability regions in the device source and load plane (Chapter 2) [3.1, 3.2]. Alternatively, some negative feedback (reactive or lossy) can be employed prior to the synthesis procedure [3.6, 3.47].

Table 3.2 Scattering parameters (magnitude and angle) of the NE710.84 device [3.45] ($V_{DS} = 3$ V, $I_{DS} = 10$ mA)

Frequency (GHz)	S_{11}		S_{21}		S_{12}		S_{22}	
2	0.944	−49°	3.483	136°	0.040	59°	0.653	−31°
3	0.942	−70°	3.364	118°	0.057	45°	0.642	−45°
4	0.862	−92°	2.986	97°	0.066	33°	0.607	−61°
5	0.808	−108°	2.707	82°	0.072	25°	0.571	−73°
6	0.789	−126°	2.563	64°	0.077	15°	0.568	−86°
7	0.712	−142°	2.402	48°	0.078	7°	0.532	−100°
8	0.656	−155°	2.281	36°	0.080	4°	0.514	−110°
9	0.608	−171°	2.194	23°	0.085	2°	0.488	−123°
10	0.591	172°	2.169	8°	0.088	−3°	0.482	−139°
11	0.568	154°	2.077	−9°	0.097	−11°	0.494	−157°
12	0.545	137°	1.940	−23°	0.098	−11°	0.505	−173°
13	0.505	117°	1.860	−40°	0.105	−19°	0.515	173°
14	0.474	98°	1.732	−55°	0.114	−27°	0.515	159°
15	0.512	77°	1.689	−70°	0.126	−34°	0.560	142°
16	0.552	57°	1.526	−87°	0.127	−47°	0.606	123°
17	0.579	42°	1.356	−100°	0.127	−53°	0.646	109°
18	0.624	32°	1.270	−111°	0.128	−65°	0.708	100°

Circuit model in 4–8 GHz band.
S_{11}: series RLC with $R = 8.6\,\Omega$, $L = 0.36$ nH, $C = 0.69$ pF.
S_{22}: parallel RC–series L with $R = 166.2\,\Omega$, $C = 0.35$ pF, $L = 0.42$ nH.

Table 3.3 Noise parameters of the NE710.84 device [3.45] ($V_{DS} = 3\,\text{V}$, $I_{DS} = 10\,\text{mA}$)

Frequency (GHz)	F_{min} (dB)	Γ_{opt}		$R_n/50$	Γ^*_{opt} *model series RLC*
1	0.50	0.90	17°	0.50	*4–8 GHz*
2	0.55	0.85	37°	0.47	$R = 15.2\,\Omega$
4	0.60	0.71	85°	0.43	$L = 0.52\,\text{nH}$
6	0.8	0.61	127°	0.35	$C = 0.62\,\text{pF}$
8	1.0	0.54	165°	0.30	
10	1.3	0.52	−158°	0.25	*4–12 GHz*
12	1.6	0.51	−124°	0.20	
14	1.9	0.50	−93°	0.15	$R = 16.2\,\Omega$
16	2.25	0.49	−68°	0.12	$L = 0.59\,\text{nH}$
18	2.55	0.50	−50°	0.09	$C = 0.58\,\text{pF}$

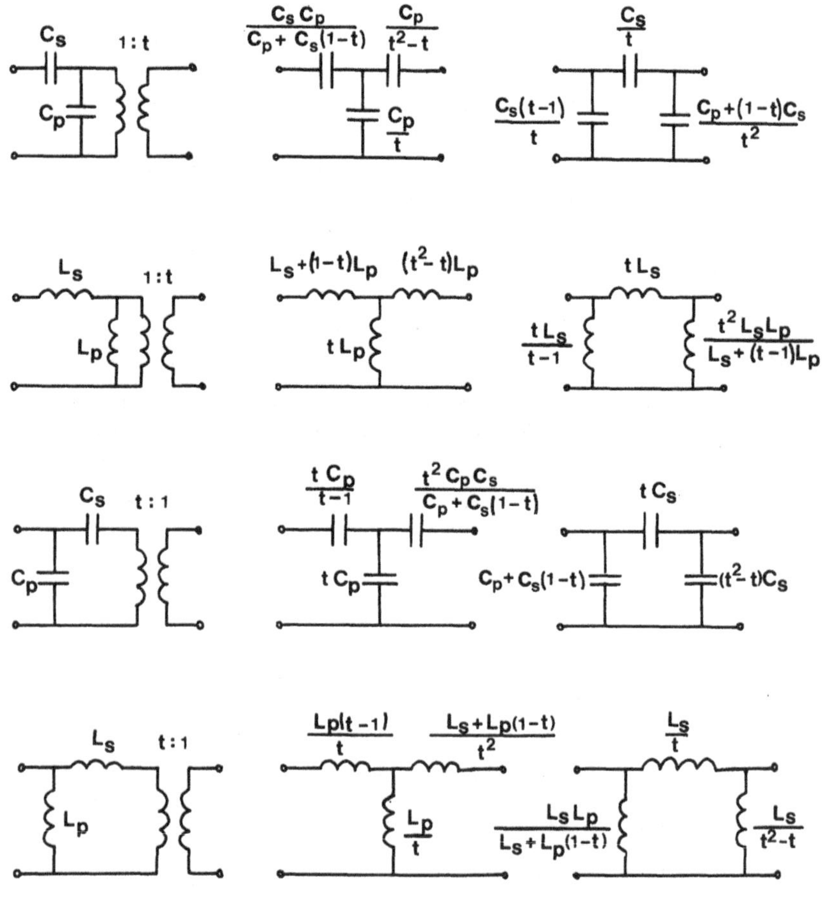

Fig. 3.6 Norton transformer equivalences.

In order to determine an upper bound on the capability of reactive element absorption for a specific bandwidth, gain slope, maximum gain and ripple specification, the optimum gain–bandwidth limitations can be derived from Fano's integral [3.29] assuming an idealized square or tapered shape of gain response. For flat gain the problem is defined [3.16] as in Fig. 3.7. The optimum matching minimizes the value of the reflection coefficient inside the passband and keeps it as high as possible (ideally unity) outside the passband.

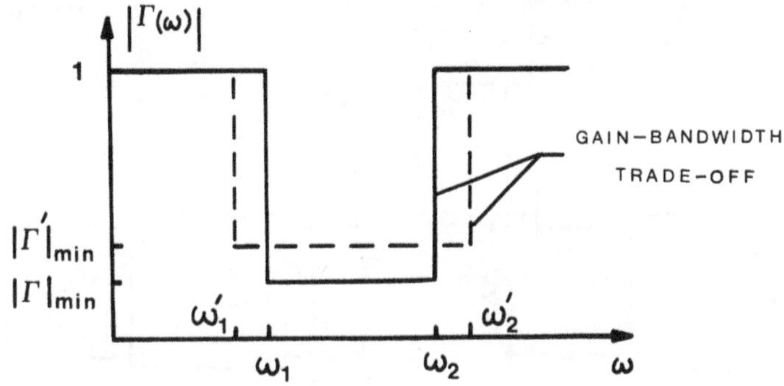

Fig. 3.7 Ideal equalizer response for flat insertion loss.

If we define the fractional bandwidth as

$$\frac{\omega_2-\omega_1}{\omega_0}=\frac{1}{Q_T} \tag{3.9}$$

where $\omega_0=(\omega_1\omega_2)^{1/2}$, and assume a single constraint load Q factor at ω_0 as shown in Fig. 3.8, the following simple expressions for $|\Gamma|_{min}$ and G_{max} are obtained:

$$|\Gamma|_{min}=\exp\left(-\frac{\pi Q_T}{Q}\right) \tag{3.10}$$

$$G_{max}=1-|\Gamma|_{min}{}^2=1-\exp\left(-\frac{2\pi Q_T}{Q}\right) \tag{3.11}$$

Ku and Petersen [3.17] derived the optimum gain–bandwidth limitations using the following expression for $|\Gamma(\omega)|$:

$$|\Gamma(\omega)|^2=1-K_n\left(\frac{\omega}{\omega_2}\right)^{\alpha} \tag{3.12}$$

where K_n is the equalizer gain at $\omega=\omega_2$ ($K_n\leqslant 1$) and

$$\alpha=\frac{x}{10\log 2}\approx\frac{x}{3}$$

where x denotes the required slope (positive) in decibels per octave (on a log–log scale).

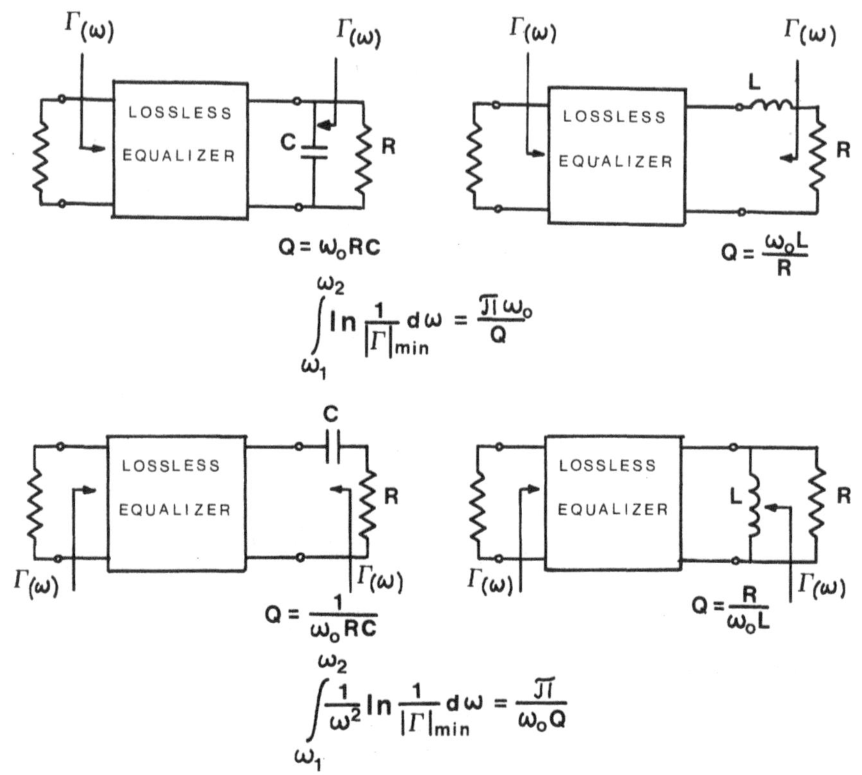

Fig. 3.8 Theoretical gain–bandwidth limitations for single-constraint loads and ideal flat gain response.

The above ideal shapes can only be approximated. The preferred approximation is the Chebyshev polynomial because it offers the best phase response and sharpest transition from $|\Gamma|_{\min}$ to a total reflection, thus allowing the best use of the gain–bandwidth limitation. The flat gain response has the form

$$G(\omega^2) = \frac{K_n}{1+\varepsilon_n^{\,2}\, T_n^{\,2}(\omega')} \tag{3.13}$$

where ω' is related to ω by the bandpass-to-lowpass frequency transformation:

$$\omega' = \frac{\omega_0}{\omega_2-\omega_1}\left(\frac{\omega}{\omega_0}-\frac{\omega_0}{\omega}\right) = Q_{\mathrm{T}}\left(\frac{\omega}{\omega_0}-\frac{\omega_0}{\omega}\right) \tag{3.14}$$

T_n is the Chebyshev polynomial of nth order and ε_n is a passband ripple factor*.

* The peak-to-peak ripple factor is expressed as $r = 10\log(1+\varepsilon_n^{\,2})$ dB.

The maximum and minimum gains in the passband are K_n and $K_n/(1+\varepsilon_n^2)$ respectively. The $+6$ dB per octave gain slope expression is

$$G(\omega^2) = \frac{K_n(\omega/\omega_2)^2}{1+\varepsilon_n^2\, T_n^2(\omega')} \tag{3.15}$$

In most problems n values of 2 or 3 (fourth- or sixth-order bandpass network) give satisfactory results. In order to obtain $G(\omega^2)$ bounded by unity for all ω (physical realizability) the frequency $\omega_m \geqslant \omega_2$ should be found by inspection of eqn (3.15) where $G(\omega^2)$ attains its maximum and K_n should be adjusted to give $G(\omega_m{}^2) \leqslant 1$.

Several analytical techniques are available for the synthesis of an equalizer. The choice depends on the gain characteristics required. Three of these techniques are reviewed briefly below and others are discussed.

3.3.2.1 Levy technique for flat gain response [3.16]

The Levy technique can be applied to synthesize the flat insertion gain equalizer either between a complex load and a resistive source or between two complex impedances. The elements of the lowpass Chebyshev prototype are obtained as closed-form expressions. Following Levy's notation, the formulae are given below for the typical case of $n = 2$:

$$\sinh a = \frac{C_2+(C_2{}^2+2)^{1/2}}{2} \tag{3.16a}$$

$$\sinh b = \sinh a - C_2 \tag{3.16b}$$

where

$$C_2 = \frac{2}{g_1}\sin\left(\frac{\pi}{4}\right) \tag{3.16c}$$

Then

$$K_n = 1-\frac{\sinh^2(2b)}{\sinh^2(2a)} \tag{3.17a}$$

$$\varepsilon_n = \frac{1}{\sinh(2a)} \tag{3.17b}$$

$$g_1 g_2 = \frac{2}{\sinh^2 a+\sinh^2 b+1} \tag{3.18}$$

$$g_3 = \frac{1}{g_2}\frac{2\sin(\pi/4)}{\sinh a+\sinh b} \tag{3.19}$$

The synthesis proceeds as follows.

(a) Determine g_1 from the load model using the required frequency transformation. This automatically guarantees the absorption of the constraint and defines the ripple and minimum insertion loss which must be accepted in the passband.

(b) Find sinh a and sinh b, and compute $g_2, \ldots, g_{n+1}$.

(c) Apply the frequency transformation to the prototype network and denormalize the elements.

(d) Add the ideal transformer to establish the required impedance level.

(e) Apply the Norton transformation to obtain a physically realizable network.

Example The Levy technique is used to synthesize the source impedance that tracks the optimum noise impedance (see Section 3.6.3.4) of the NE710.84 device in the 5.5–7.5 GHz band (31% fractional bandwidth, $Q_T = 3.21$). For this purpose the complex conjugate of this impedance is substituted for the load, i.e. the input impedance of the device. The load has the series RLC topology shown in Fig. 3.9(a). By applying the lowpass-to-bandpass transformation and denormalizing (Fig. 3.9(b)) the bandpass network elements can be identified as (the dual topology is used in the case of parallel RC–series L load)

$$\omega_0 L_s = g_1 Q_T R_i \tag{3.20a}$$

$$\omega_0 C_s = \frac{1}{\omega_0 L_s} \tag{3.20b}$$

$$\omega_0 C_p = \frac{g_2 Q_T}{R_i} \tag{3.21a}$$

$$\omega_0 L_p = \frac{1}{\omega_0 C_p} \tag{3.21b}$$

$$R_s = g_3 R_i \tag{3.22}$$

The transformer ratio is

$$t = \left(\frac{Z_0}{g_3 R_i}\right)^{1/2} \tag{3.23}$$

The Q factors at ω_0 show that the capacitance C_i ($Q = 2.63$) is a stronger constraint than the inductance L_i ($Q = 1.38$); thus $C_i = C_s$ (total absorption) and from eqns (3.20)

$$g_1 = \frac{1}{\omega_0 C_i R_i} \frac{1}{Q_T} = 0.8195$$

The remaining parameters are determined from eqns (3.16a)–(3.19):

$$\sinh a = 1.9784 \qquad \sinh b = 0.2527$$
$$K_n = 0.9965 \qquad \varepsilon_n = 0.1140$$
$$G_{\max} = -0.015\,\text{dB} \qquad G_{\min} = -0.071\,\text{dB}$$

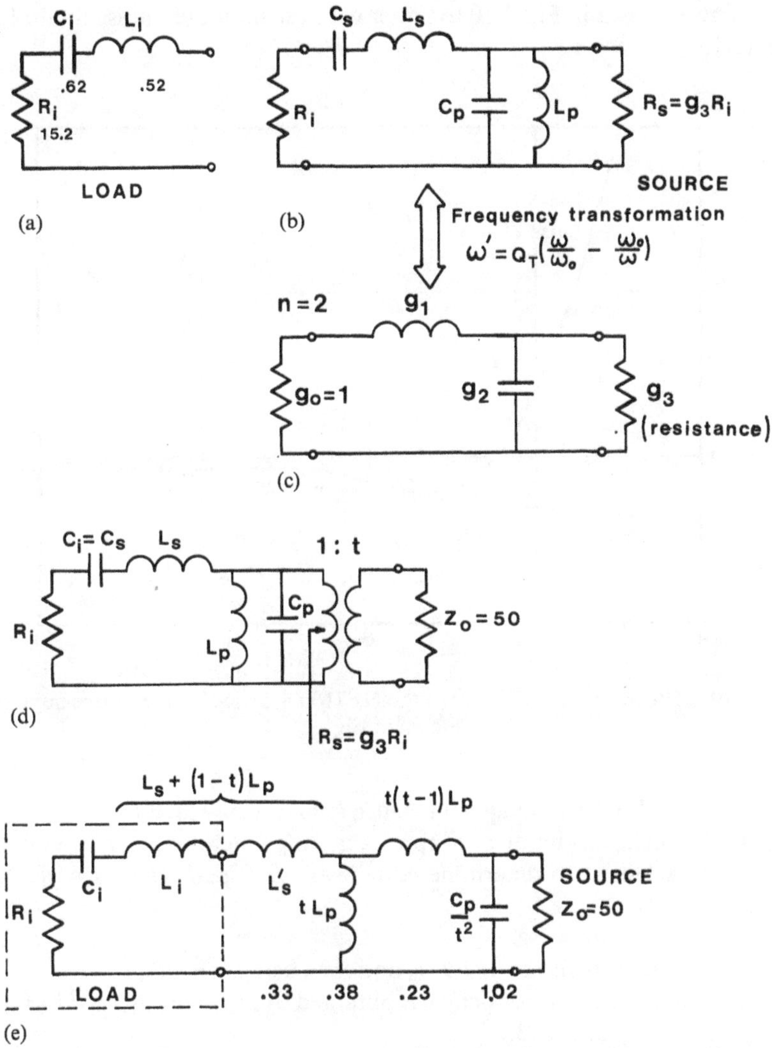

Fig. 3.9 Design example of a flat gain equalizer: (a) load model; (b) bandpass unnormalized network; (c) lowpass prototype; (d) network with ideal transformer added; (e) final network (units nH, pF).

$$
\begin{aligned}
r &= 0.056\,\text{dB} & t &= 1.5951 \\
g_2 &= 0.4903 & g_3 &= 1.2928 \\
C_s &= C_i = 0.62\,\text{pF} & L_s &= 0.99\,\text{nH} \\
C_p &= 2.57\,\text{pF} & L_p &= 0.239\,\text{nH}
\end{aligned}
$$

The final network is obtained (Fig. 3.9(e)) by applying the Norton transformer equivalence from Fig. 3.6. The transistor noise figure corresponding to this matching is shown in Fig. 3.10 together with the minimum noise figure F_{min} of the device.

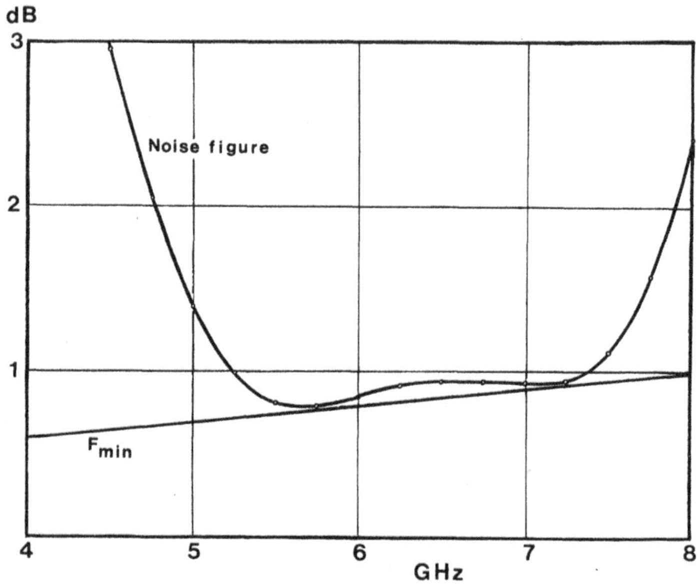

Fig. 3.10 Theoretical noise figure of the NE710.84 device for the source impedance shown in Fig. 3.9.

3.3.2.2 Mellor technique for sloped gain response [3.20]

In the Mellor technique the approximating functions given by eqns (3.13) and (3.15) are used to determine equalizers with gain responses of various slopes.

(a) A slope of 6 dB per octave is obtained using eqn (3.15).

(b) A slope of 12 dB per octave is obtained by replacing $(\omega/\omega_2)^2$ by $(\omega/\omega_2)^4$ in the numerator of eqn (3.15).

(c) A slope of 4 dB per octave is obtained by combining eqns (3.13) and (3.15) using appropriate weighting factors.

Since the ripple factor and the polynomial order are specified, the gain constant K_n has to be adjusted by taking gain–bandwidth limitations into account [3.17]. This guarantees the absorption of reactances associated with the load and source. Once $G(\omega^2)$ is determined, the synthesis proceeds as follows.

(a) Substitute $\omega = s/j$ in $G(\omega^2)$ and determine $\Gamma(s)\,\Gamma(-s)$ from

$$\Gamma(s)\,\Gamma(-s) = 1 - G(-s^2) \tag{3.24}$$

(b) Find $\Gamma(s)$ by using the procedure known as "spectral factorization"; $\Gamma(s)$ is determined from

$$\Gamma(s) = \pm\frac{\prod_{i=1}^{n}(s-Z_i)}{\prod_{i=1}^{n}(s-P_i)} \tag{3.25}$$

where n is the order of the network (for a network of order n the order of the polynomial is $2n$). In order to guarantee that the denominator of $\Gamma(s)$ is a Hurwitz polynomial, the left-hand plane (LHP) poles P_i of $\Gamma(s)\Gamma(-s)$ are assigned to $\Gamma(s)$. Both the LHP and right-hand plane (RHP) zeros Z_i of $\Gamma(s)\Gamma(-s)$ can be assigned to $\Gamma(s)$. With LHP zeros maximum absorption occurs at the load side (R_L), whereas with RHP zeros maximum absorption occurs at the source side. Any combination of zeros can be assigned to $\Gamma(s)$ provided that they are assigned in conjugate pairs. Usually this must be done in order to realize the absorption of the reactances of both load and source.

(c) Select the sign of $\Gamma(s)$ as a function of the topology of the complex load in order to guarantee reactive element absorption. This can be done by inspection of the sign of Γ(s) at the origin and at infinity [3.6].

(d) Find the driving-point impedance $Z(s)$ from $\Gamma(s)$ (Fig. 3.11):

$$Z(s) = R_L\frac{1+\Gamma(s)}{1-\Gamma(s)} \tag{3.26}$$

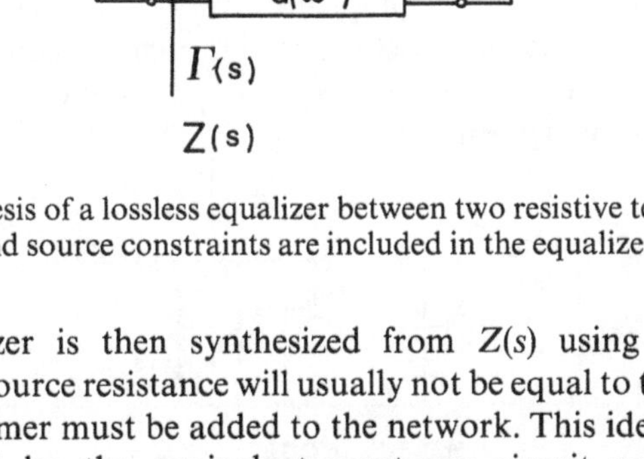

Fig. 3.11 Synthesis of a lossless equalizer between two resistive terminations (load and source constraints are included in the equalizer).

The equalizer is then synthesized from $Z(s)$ using the Darlington procedure. The source resistance will usually not be equal to that specified and an ideal transformer must be added to the network. This ideal transformer is then substituted by the equivalent reactance circuit using the Norton equivalences (Fig. 3.6). Numerical examples of this technique can be found in refs 3.1, 3.6, 3.46 and 3.48. Design tables for lowpass equalizers of specified gain slopes and transforming capabilities are given in refs 3.17 and 3.18.

Many topologies which provide the required absorption of reactive constraints can be obtained using the technique outlined above. Some are less sensitive than others to realization or transistor parameter tolerances [3.49]. The topologies vary in their impedance-transforming capability. Mellor

[3.10] found that this capability depends strongly on the topology itself and is much less dependent on the frequency response (slope, bandwidth, ripple). This opens the way to fully automated synthesis of reactive matching networks.

Young and Scanlan [3.40] have shown that, in contrast with iteratively designed networks, a strict application of the analytical gain–bandwidth approach may yield suboptimal (more complicated) networks. They also found that distributed matching networks have gain ripple properties which are somewhat inferior to those achievable using lumped elements.

The analytical approach was next extended to the synthesis of lumped-lossy-element matching networks and applied to the design of monolithic amplifiers [3.21]. In this case the approximating polynomial in the form of eqn (3.27) is synthesized iteratively by using, for example, the Remez algorithm [3.40] or other numerical optimization methods [3.21]:

$$G(\omega^2) = \frac{K_n \prod_{j=1}^{m} (\omega^2 + \alpha_j^2)}{P_n(\omega^2)} \tag{3.27}$$

where m is the number of highpass elements, $P_n(\omega^2)$ is the nth-order polynomial and α_j is the loss factor of the jth highpass element.

3.3.2.3 Simple analytical technique using a pre-assigned topology [3.38, 3.50]

The eight-element unilateral MESFET model and second-order matching networks are used in this technique. It gives satisfactory results over wide bandwidths and may be very useful in the design of monolithic amplifiers where space-saving topologies are required. The equivalent circuit of the amplifier stage is shown in Fig. 3.12.

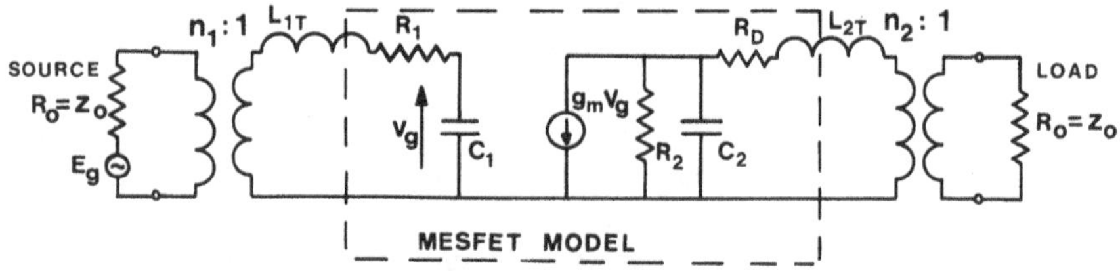

Fig. 3.12 Equivalent circuit of the amplifier stage using an eight-element unilateral MESFET model.

The input and output inductances of the transistor model are included in the total inductances L_{1T} and L_{2T} respectively. In order to increase the operating frequencies the series capacitances can also be included in the tuning circuits. The transducer power gain of the stage is given by

$$G_T(\omega^2) = G_1(\omega^2)G_F G_2(\omega^2) \tag{3.28}$$

where G_F is a frequency-independent term and G_1 and G_2 are the input and output equalizer transducer gains respectively.

$$G_F = 4g_m{}^2 R_0 \tag{3.29a}$$

$$G_1(\omega^2) = \frac{1}{n_1{}^2} \frac{1}{\{1-(\omega/\omega_1)^2\}^2+(1/Q_1{}^2)(\omega/\omega_1)^2} \tag{3.29b}$$

$$G_2(\omega^2) = \frac{n_2{}^2}{\alpha} \frac{1}{\{1-(\omega/\omega_2)^2\}^2+(1/Q_2{}^2)(\omega/\omega_2)^2} \tag{3.29c}$$

where

$$\alpha = 1+\frac{R_D+n_2{}^2R_0}{R_2} \tag{3.30}$$

$$\omega_1{}^2 = \frac{1}{L_{1T}C_1} \tag{3.31}$$

$$\omega_2{}^2 = \frac{\alpha}{L_{2T}C_2} \tag{3.32}$$

$$Q_1 = \frac{1}{\omega_1 C_1(R_1+R_0/n_1{}^2)} \tag{3.33}$$

$$Q_2 = \frac{1}{(1/\omega_2R_2C_2)+\{(\alpha-1)/\alpha\}\omega_2R_2C_2} \tag{3.34}$$

In order to obtain the flat gain G_T, both G_1 and G_2 are designed to have flat responses.

By inspection of (3.29b) we find that $G_1(\omega^2)$ attains its maximum value $G_{1\max}$ under conditions of perfect input match: $\omega = \omega_1$, where ω_1 is the input tuning frequency and $n_1{}^2 = R_0/R_1$. The maximum value decreases at a rate of 6 dB per octave according to

$$G_{1\max} = \frac{1}{4R_0R_1(\omega_1C_1)^2} \tag{3.35}$$

In order to ensure that $G_1(\omega^2)$ is as flat as possible, Q_1 should be equal to unity at ω_1. Then from (3.33) we obtain the condition

$$n_1{}^2 = \frac{\omega_1R_0C_1}{1-\omega_1R_1C_1} \tag{3.36}$$

Thus $G_1(0) = G_1(\omega_1{}^2)$ becomes

$$G_1(0) = \frac{1-\omega_1R_1C_1}{\omega_1C_1R_0} \tag{3.37}$$

and $G(\omega^2)$ remains almost constant between $\omega = 0$ and $\omega = \omega_1$. $G_1(\omega_1{}^2)$ for the flat gain condition (eqn (3.36)) is smaller than $G_{1\max}$ obtained from eqn (3.35) under the perfect match condition unless $\omega_1 \geqslant \omega_{1\mathrm{m}}$, where

$$\omega_{1\mathrm{m}} = \frac{1}{2R_1C_1}$$

Below $\omega_{1\mathrm{m}}$ the input reflection coefficient increases with decreasing frequency:

$$|\Gamma(\omega)|^2 = 1 - \frac{G_1(\omega^2)}{G_{1\max}(\omega_1 = \omega)} \tag{3.38}$$

The output equalizer gain $G_2(\omega^2)$ exhibits similar behavior. Its value can be maximized for perfect match conditions:

$$n_2{}^2 = \frac{1}{R_0}\left\{R_\mathrm{D} + \frac{R_0R_2}{1+(\omega_0R_2C_2)^2}\right\}$$

at (3.39)

$$\omega_0{}^2 = \frac{1}{L_{2\mathrm{T}}C_2} - \frac{1}{(R_2C_2)^2}$$

where ω_0 is the series resonant frequency of the output circuit. ω_0 and ω_2 are related by the expression

$$\omega_0{}^2 = \frac{\omega_2{}^2}{\alpha} - \frac{1}{(R_2C_2)^2} \tag{3.40}$$

If Q_2 is set equal to unity at ω_2, $G_2(\omega^2)$ takes the same value at $\omega = 0$ and $\omega = \omega_2$ and remains almost constant between these frequencies. The conditions for flat G_2 gain are obtained from eqns (3.34) and (3.39):

$$\omega_0 = \omega_2\left(1 - \frac{1}{\omega_2R_2C_2}\right)^{1/2} \tag{3.41}$$

and

$$\frac{1}{\alpha_{\mathrm{flat}}} = 1 - \frac{1}{\omega_2R_2C_2} + \frac{1}{(\omega_2R_2C_2)^2} \tag{3.42}$$

The output transformer ratio is then found from eqn (3.30):

$$n_2{}^2 = \frac{R_2(\alpha_{\mathrm{flat}} - 1) - R_\mathrm{D}}{R_0} \tag{3.43}$$

From the above analysis the design procedure can be summarized as follows.

(a) Set the maximum operating frequency:

$$\omega_{\max} = \omega_1 = \omega_2$$

(b) Compute the resonant frequency ω_0 from eqn (3.41) and the tuning inductance L_{2T} from eqn (3.39):

$$L_{2T} = \frac{C_2 R_2^2}{1 + (\omega_0 C_2 R_2)^2}$$

If L_{2T} is smaller than the transistor inductance, a series capacitance must be added.

(c) Compute from eqn (3.42) the value of α_{flat} which corresponds to the flat gain $G_2(\omega^2)$. The required output transformer ratio n_2 is then found from eqn (3.43).

(d) Compute L_{1T} from eqn (3.31) and n_1 from eqn (3.36).

The gain–bandwidth limitations can be evaluated from eqns (3.37) and (3.29c) with $\omega = 0$ using α_{flat} and n_2 found in step (c). The ideal transformers can be realized as quarter-wavelength transmission lines at ω_{max}. Next, the overall stage is computer optimized taking into account the full S parameters of the transistor.

In addition to the techniques described above there are many other analytical methods which differ essentially in the realization of bandpass circuits and the impedance transformers required. Immittance inverters can be used for bandwidths up to one octave [3.12, 3.42, 3.43]. Semi-analytical techniques may be very useful for medium bandwidths because they do not involve complicated computer routines. Among these, the reactance slope compensation technique (see Chapter 6) [3.51] and reactance transformation and tuning [3.52] (similar to Abrie's Q transformation approach) are of practical use. Medley and Allen [3.53] have proposed an original technique in which the negative-image device model is used to determine the matching network. Practical realization of lumped-element networks using transmission lines is discussed in detail by Abrie (ref. 3.6, Chapter 7).

3.3.3 Synthesis of lossless equalizers using the iterative approach

Iterative techniques have an advantage over the analytical approach in that circuit models for the load and source impedances and an analytical expression for the transducer power gain of equalizers are not required. The networks synthesized using iterative techniques are generally simpler in form with superior gain properties [3.4, 3.5, 3.23–3.27]. Among these, the real frequency method [3.23] and the simplified real frequency technique [3.4] are described in detail below. Other techniques are also discussed briefly.

3.3.3.1 Carlin real frequency technique [3.23]

The resistively terminated equalizer is described by its positive real input impedance as defined in Fig. 3.4 where we set $Z_G(j\omega) = R_G$ and $Z_2(j\omega) = Z_Q(j\omega)$:

$$Z_Q(j\omega) = R_Q(\omega) + jX_Q(\omega) \tag{3.44}$$

Once the real frequency data $R_L(\omega)$ and $X_L(\omega)$ are given, the transducer gain of the equalizer is determined completely by Z_Q (see Section 3.3.1):

$$G(\omega^2) = \frac{4R_L(\omega)R_Q(\omega)}{\{R_L(\omega)+R_Q(\omega)\}^2 + \{X_L(\omega)+X_Q(\omega)\}^2} \tag{3.45}$$

It should be noted that eqn (3.45) has the same form if the admittance parameters Y_L and Y_Q are used. The direct use of eqn (3.45) is the basis of the real frequency technique.

$Z_Q(j\omega)$ is assumed to be a minimum-impedance (admittance) function. Thus $X_Q(\omega)$ can be found using a Hilbert transform once $R_Q(\omega)$ is known (uniquely for the minimum-reactance case). The input equalizer resistance $R_Q(\omega)$ is determined by minimizing the mean square deviation between the desired and the actual transducer gain. The following procedure is carried out.

(a) Choose an ideal gain for $G(\omega^2)$, e.g. a constant value as high as possible over the passband and zero elsewhere.

(b) Determine $R_Q(\omega)$ and its associated reactance $X_Q(\omega)$ to find the "best" approximation of the ideal $G(\omega^2)$ by numerical optimization, e.g. the least-squares technique.

The unknown resistance $R_Q(\omega)$ is represented as a number of linear segments, i.e. semi-infinite slopes with frequency breakpoints at $0 < \omega_1 < \omega_2 < \ldots < \omega_n$ (Fig. 3.13):

$$R_Q(\omega) = r_0 + \sum_{k=1}^{n} a_k(\omega) r_k \tag{3.46}$$

where

$$a_k(\omega) = \begin{cases} 1 & \text{for } \omega_k \leqslant \omega \\ \dfrac{\omega - \omega_{k-1}}{\omega_k - \omega_{k-1}} & \text{for } \omega_{k-1} \leqslant \omega \leqslant \omega_k \\ 0 & \text{for } \omega \leqslant \omega_{k-1} \end{cases} \tag{3.47}$$

$r_0 = R_Q(0)$ and r_k, which is any positive or negative real number, is the unknown resistance excursion of the kth linear segment between the breakpoints ω_{k-1} and ω_k. Since $R_Q(\omega) = 0$ for $\omega > \omega_n$,

$$r_0 + \sum_{k=1}^{n} r_k = 0 \tag{3.48}$$

so that, since r_0 is known ($r_0 = Z_Q(0) = R_G$ for a lowpass network and is zero

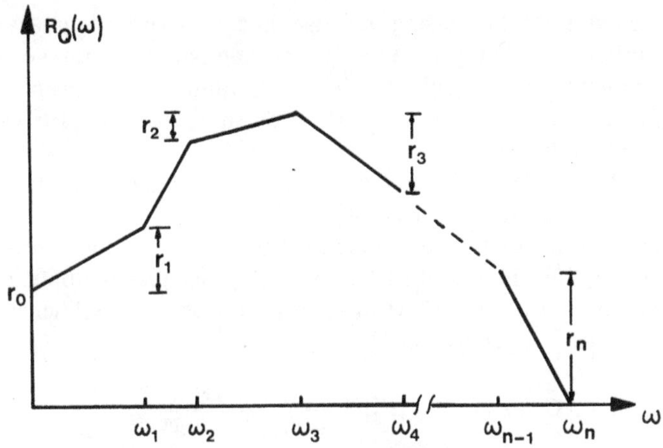

Fig. 3.13 Line-segment representation of $R_Q(\omega)$.

otherwise), there are $n-1$ unknowns, i.e. the r_k with $k = 1, \ldots, n-1$. Now, since the r_k are real, $Z_Q(j\omega)$ is passively realizable provided that $R_Q(\omega) \geqslant 0$.

The crux of this approach is that, as a result of this representation of $R_Q(\omega)$, the reactance $X_Q(\omega)$ is also specified as a linear combination of the same resistive excursions r_k. Thus the exact value of the reactance corresponding to $R_Q(\omega)$ is given by

$$X_Q(\omega) = \sum_{k=1}^{n} b_k(\omega) r_k \tag{3.49}$$

For any frequency, the coefficients $b_k(\omega)$ are immediately known in terms of ω_k and are independent of the r_k. By applying the Hilbert transform we obtain

$$b_k(\omega) = \frac{1}{\pi(\omega_k - \omega_{k-1})} \int_{\omega_{k-1}}^{\omega_k} \ln\left|\frac{y+\omega}{y-\omega}\right| \mathrm{d}y \tag{3.50}$$

The integral (3.50) has a simple closed-form evaluation.

Once the linear segments describing $R_Q(\omega)$ have been found, the appropriate rational approximation $\hat{R}_Q(\omega) \approx R_Q(\omega)$ is determined by numerical optimization (e.g. the least-squares method) and $Z_Q(j\omega)$ is realized as a Darlington driving-point impedance with a resistive termination (the source resistance). The initial estimate of the resistance excursions r_k can be made by assuming that

$$R_Q(\omega_k) = R_L(\omega_k)$$

Alternatively, we can choose to deal with $Y_Q(j\omega)$ as the minimum susceptance (i.e. there are no susceptance poles at real frequencies). The choice is made by considering the load data; for example, a capacitive load implying a series inductance at the back end of the equalizer would be most conveniently handled using $Y_Q(j\omega)$.

In some cases, e.g. for biasing purposes and/or in order to guarantee low frequency stability, it is convenient for the first element seen from the load to be a parallel inductor. This would result in a nonminimum susceptance. The problem can be overcome by associating the inductor with the load data [3.24]. The remaining equalizer is then treated as the one with minimum susceptance. For practical implementation of this technique the reader is referred to ref. 3.24 or the book by Abrie [3.6].

The application of the real frequency technique to the synthesis of a single-stage amplifier is described below. The procedure is applied sequentially as shown in Fig. 3.14. The transducer gain of the amplifier with both equalizers in place is given by [3.24]

$$G_T(\omega^2) = (1-|\Gamma_1|^2)\frac{|S_{21}|^2}{1-|S_{11}|^2}\frac{1-|\Gamma_2|^2}{1-|S'_{22}|^2} \tag{3.51}$$

where Γ_1 and Γ_2 are complex normalized reflection coefficients defined as

$$\Gamma_1(\omega) = \frac{Z_{Q1} - Z^*_{L1}}{Z_{Q1} + Z_{L1}} \tag{3.52}$$

$$\Gamma_2(\omega) = \frac{Z_{Q2} - Z^*_{L2}}{Z_{Q2} + Z_{L2}} \tag{3.53}$$

and S'_{22} is the real normalized reflection coefficient seen at the device output with the input equalizer in place:

$$S'_{22} = S_{22} + \frac{S_{12}S_{21}S_{Q1}}{1 - S_{11}S_{Q1}} \tag{3.54}$$

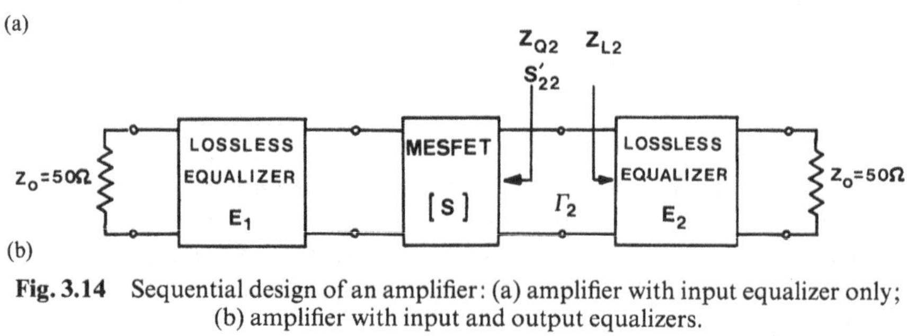

Fig. 3.14 Sequential design of an amplifier: (a) amplifier with input equalizer only; (b) amplifier with input and output equalizers.

where

$$S_{Q1} = \frac{Z_{Q1} - Z_0}{Z_{Q1} + Z_0} \tag{3.55}$$

From eqn (3.51) it can be seen that the term $|S_{21}|^2/(1-|S_{11}|^2)$ is determined solely by the active device, while $1-|S'_{22}|^2$ depends on both the device and the S_{Q1} of the input equalizer. These terms can be used as frequency-dependent weighting factors in several ways in order to obtain an acceptable trade-off between intput–output VSWR, noise figure and flat gain.

First, the input equalizer is synthesized by terminating the active device directly in Z_0 (Fig. 3.14(a)). Thus $|\Gamma_2| = |S'_{22}|$ and eqn (3.51) becomes the transducer gain $G_1(\omega^2)$ of the input equalizer plus the active device:

$$G_1(\omega^2) = (1-|\Gamma_1|^2)\frac{|S_{21}|^2}{1-|S_{11}|^2} \tag{3.56}$$

We can now choose design goals for $G_1(\omega^2)$ as follows.

(a) Z_{Q1} is optimized for a flat gain $G_1(\omega^2)$ as high as possible. In this case the input equalizer will compensate for the amplifier gain roll-off due to the term $|S_{21}|^2/(1-|S_{11}|^2)$. In the second step the output equalizer is added and Z_{Q2} is optimized so that the overall amplifier has a flat gain. Then the weighting factor becomes

$$W(\omega^2) = G_1(\omega^2)\frac{1}{1-|S'_{22}|^2} \tag{3.57}$$

and

$$G_T(\omega^2) = W(\omega^2)(1-|\Gamma_2|^2) \tag{3.58}$$

In this way the role of gain slope compensation is divided between the input and output equalizers.

(b) Z_{Q1} is optimized for either the low input VSWR of the amplifier or the optimum source impedance which corresponds to the minimum noise figure of the transistor (see also Section 3.6.3.4 and the example in Section 3.3.2.1). In this case the term $1-|\Gamma_1|^2$ is determined from the above requirements only. In the next step the output equalizer is added and Z_{Q2} is optimized for a flat gain $G_T(\omega^2)$. In this way the output equalizer compensates for the gain roll-off resulting from $W(\omega^2)$ given by eqn (3.57). The output equalizer can also be synthesized for minimum output VSWR. In this case the sequential design should start from the load side and Z_{Q2} should be optimized first for minimum $|\Gamma_2|$.

3.3.3.2 Yarman–Carlin simplified real frequency technique [3.4]

The simplified real frequency technique eliminates the need for representing the resistance function $R_Q(\omega)$. Instead, the input reflection coefficient of the equalizer is directly computed in rational form.

The equalizer is assumed to be a minimum phase network with transmission zeros only at $\omega = 0$ and $\omega = \infty$. Its real normalized scattering parameters in the complex frequency plane are given by

$$e_{11}(s) = \frac{h(s)}{g(s)} \tag{3.59}$$

$$e_{21}(s) = e_{12}(s) = \pm\frac{s^k}{g(s)} \tag{3.60}$$

$$e_{22}(s) = -(-1)^k\frac{h(s)}{g(s)} \tag{3.61}$$

where k is an integer specifying the order of the zero transmission at $s = \infty$ ($k = 0$ for lowpass networks) and $h(s)$ and $g(s)$ are polynomials. The coefficients of the numerator polynomial $h(s)$ are chosen as unknowns:

$$h(s) = h_0 + h_1 s + \ldots + h_n s^n$$

where $n \geqslant 0$ and $h_0 = 0$ if the ideal transformer is to be avoided. Since the equalizer is lossless, it follows that

$$g(s)g(-s) = h(s)h(-s) + (-1)^k s^{2k} \tag{3.62}$$

Once $h(s)$ is known and $g(s)$ is generated from it by explicit factorization of eqn (3.62), $e_{11}(s)$ is generated from eqn (3.59) and the equalizer input impedance

$$Z(s) = R_L\frac{1+e_{11}(s)}{1-e_{11}(s)}$$

is synthesized using the Darlington procedure.

The amplifier is synthesized sequentially as described in the preceding section by optimizing $h(s)$ at each step. First the input equalizer is synthesized with the transistor terminated into a 50 Ω load. The corresponding transducer gain of the equalizer plus the transistor is given by

$$G_1(\omega^2) = \frac{|e_{21a}|^2|S_{21}|^2}{|1-e_{22a}S_{11}|^2} \tag{3.63}$$

where the subscript a refers to the input. In the next step the output equalizer is added and the overall transducer gain is computed as

$$G_T(\omega^2) = G_1(\omega^2)\frac{|e_{21b}|^2}{|1-e_{11b}S'_{22}|^2} \tag{3.64}$$

where the subscript b refers to the output.

Yarman and Carlin [3.4] also give general formulae for the sequential design of multistage amplifiers.

Among other iterative techniques, the transformation Q approach proposed by Abrie [3.6] uses the transforming properties of reactance ladder networks and automatically guarantees the absorption of reactive constraints. The gain–bandwidth limitations are taken into account iteratively by adjusting the number of reactive series/parallel branches. The advantage of this technique is that it allows for a convenient choice between many available topologies and can easily be used for automated synthesis purposes. The synthesis starts by computing the network elements at the highest frequency of the band. The optimization routine is then applied. Abrie provides an algorithm for the synthesis.

3.3.4 Balanced amplifier stage

The balanced configuration shown in Fig. 3.3(c) is widely used in bandpass amplifiers to improve their cascadability. A detailed analysis of this circuit is given in ref. 3.7, and we summarize its basic characteristics here.

If two single-ended amplifiers are described by their matrices $[S_a]$ and $[S_b]$ and two 3 dB 90° hybrid couplers are used, the overall S matrix of the balanced stage is given by

$$[S_T] = \begin{bmatrix} S_{11T} & S_{12T} \\ S_{21T} & S_{22T} \end{bmatrix}$$

where

$$S_{11T} = \tfrac{1}{2}(S_{11a} - S_{11b}) \tag{3.65a}$$

$$S_{12T} = \tfrac{1}{2}\mathrm{j}(S_{12a} + S_{12b}) \tag{3.65b}$$

$$S_{21T} = \tfrac{1}{2}\mathrm{j}(S_{21a} + S_{21b}) \tag{3.65c}$$

$$S_{22T} = \tfrac{1}{2}(S_{22a} - S_{22b}) \tag{3.65d}$$

If we set

$$S_{21a} = G_1^{1/2} \exp(\mathrm{j}\varphi_1)$$

$$S_{21b} = G_2^{1/2} \exp(\mathrm{j}\varphi_2)$$

the overall gain of the balanced stage can be expressed as

$$G_T = |S_{21T}|^2 = \frac{G_1 + G_2 + 2(G_1 G_2)^{1/2} \cos(\varphi_1 - \varphi_2)}{4} \tag{3.66}$$

It can be seen from eqn (3.66) that $G_T = G_1$ if $G_1 = G_2$ and $\varphi_1 = \varphi_2$. When one of the amplifiers fails, i.e. either $G_1 = 0$ or $G_2 = 0$, G_T is reduced by a factor of 4 (-6 dB).

Similarly, the overall noise figure of the balanced stage is given by

$$F_T = (\tfrac{1}{2}|S_{21a}|^2 F_a + \tfrac{1}{2}|S_{21b}|^2 F_b - \tfrac{1}{2}|S_{21a}|^2 - \\ - \tfrac{1}{2}|S_{21b}|^2 + \tfrac{1}{4}|S_{21b} - S_{21a}|^2 + \tfrac{1}{4}|S_{21a} + S_{21b}|^2)(4|S_{21a} + S_{21b}|^{-2}) \tag{3.67}$$

where F_a and F_b are the noise figures of the single-ended stages. Then, if $S_{21a} = S_{21b}$, then $F_T = (F_a + F_b)/2$, if $S_{21a} = -S_{21b}$, then $F_T = \infty$, and if $S_{21a} = 0$, then $F_T = 2F_b$.

The output power of the balanced configuration is twice that of the single-ended amplifier. However, balanced amplifier modules are more complex, more difficult to fabricate and use twice the number of active devices, resulting in higher power dissipation. 3 dB hybrid couplers are very difficult to manufacture for bandwidths above one octave [3.47], input–output VSWRs of better than 1.5:1 are difficult to achieve and coupler losses increase to 0.8 dB in the e.h.f. range [3.14]. For this reason, cascading of single-ended matched modules is preferred whenever possible, and the balanced configuration is used mainly for power combination in output stages.

3.3.5 Examples of broadband amplifiers using reactive matching

A number of papers on broadband amplifiers have been published. However, the synthesis method used is not always indicated in these publications. In many designs remarkable achievements have been obtained using an experimental approach, e.g. the 2–18 GHz amplifier of Tserng *et al.* [3.13]. Indeed, some matching networks which are "known to be practical" can be optimized in the hope of finding an acceptable performance. However, this approach is time consuming and frustrating even when powerful optimizer routines such as Touchstone or Super Compact are used. Several examples are presented below, with particular emphasis on cases where the researchers have specified the synthesis methods. An example of a monolithic implementation is also given.

Figures 3.15 and 3.16 show the results obtained for 4–8 GHz amplifiers designed using the real frequency approach of Carlin and Komiak [3.24].

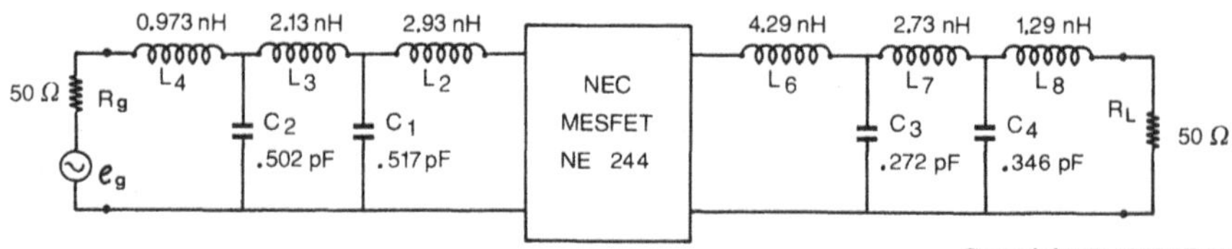

Fig. 3.15 Equalized MESFET amplifier designed using the real frequency approach. (From Carlin and Komiak [3.24].)

Figures 3.17–3.19 show the results obtained for 2.5–8 GHz and 6–18 GHz amplifiers designed using the simple analytical method of Degout *et al.* [3.38]. Figure 3.20 illustrates the result obtained by Pauker [3.33] using the analytical

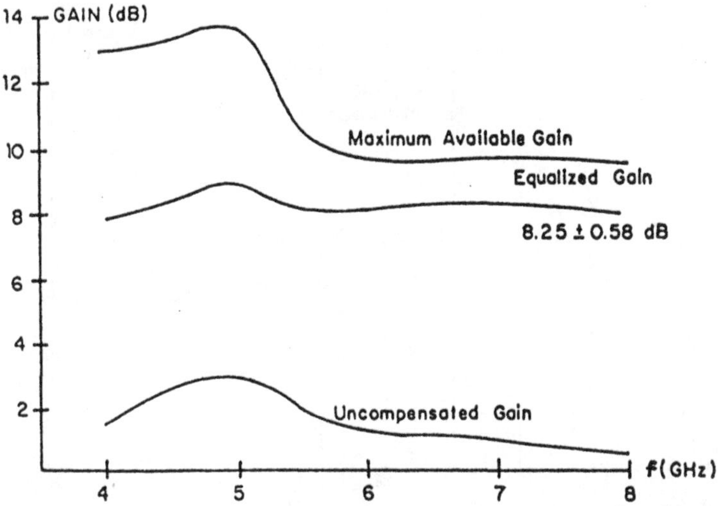

Fig. 3.16 Gain response of an equalized MESFET amplifier. (From Carlin and Komiak [3.24].)

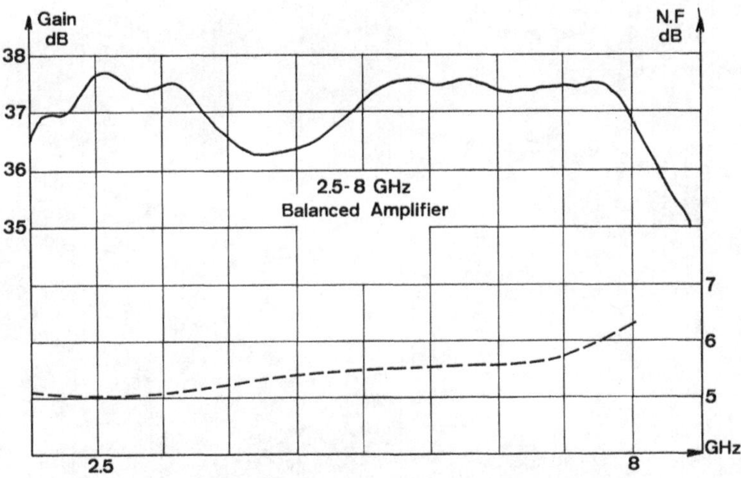

Fig. 3.17 Experimental results for a 2.5–8 GHz amplifier.

approach. A photograph of a three-stage balanced amplifier produced by MITEQ Inc. is shown in Fig. 3.21. It exhibits typically a noise figure of 2 dB with 24 dB gain across the 4–8 GHz band. Figures 3.22 and 3.23 show a commercially available two-stage GaAs monolithic medium power amplifier (Texas Instruments TGA 8014) [3.54, 3.55]. Reactance-matched 900 μm and 1200 μm gate periphery MESFETs provide a nominal small-signal gain of

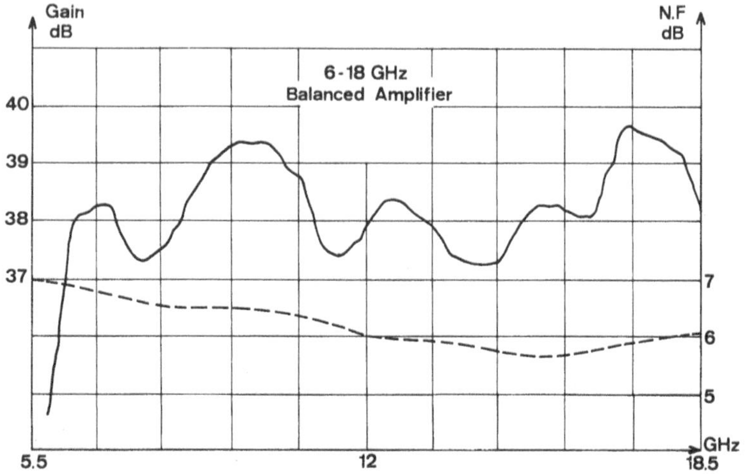

Fig. 3.18 Experimental results for a 6–18 GHz amplifier.

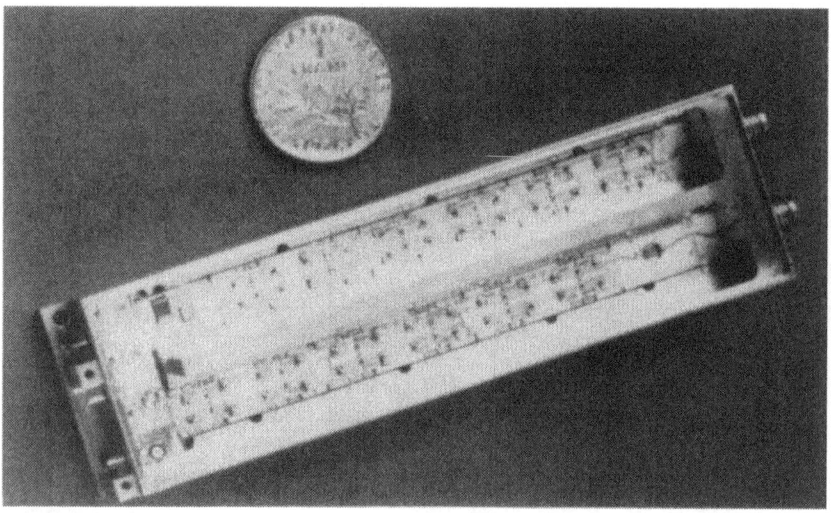

Fig. 3.19 Photograph of a 6–18 GHz amplifier. (Courtesy of Mr S. Bearzotti, Thomson-CSF.)

8 dB in the 6–18 GHz band. The amplifier can deliver typically 0.5 W of midband output power at 1 dB gain compression. The chip dimensions are 3.6 mm × 1.9 mm × 0.1 mm. The lossy matching elements added to the input and interstage matching networks to improve the VSWR and stability should be noted.

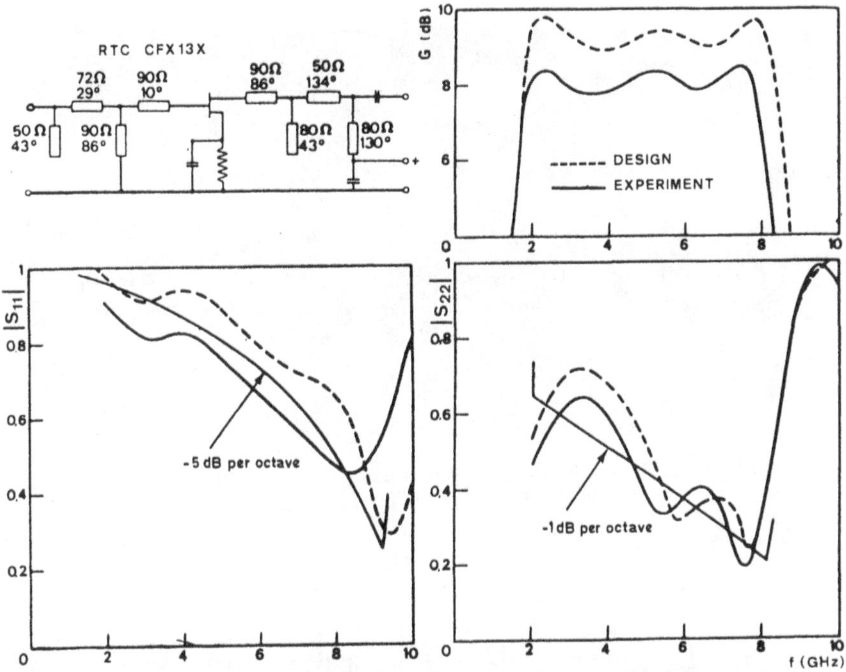

Fig. 3.20 Design and experimental results obtained using the analytical method [3.33].

Fig. 3.21 Three-stage balanced amplifier for the 4–8 GHz band. (Courtesy of Mr A. Kiiss, MITEQ Inc.)

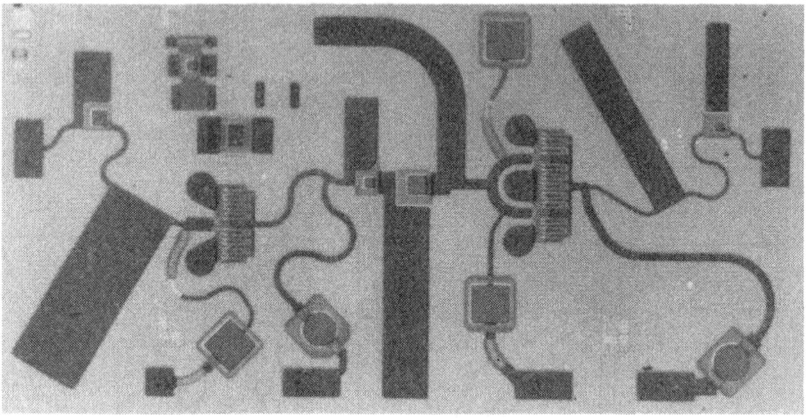

Fig. 3.22 6–18 GHz monolithic amplifier using reactance matching. (Courtesy of Mr R. Toole, Texas Instruments Inc.)

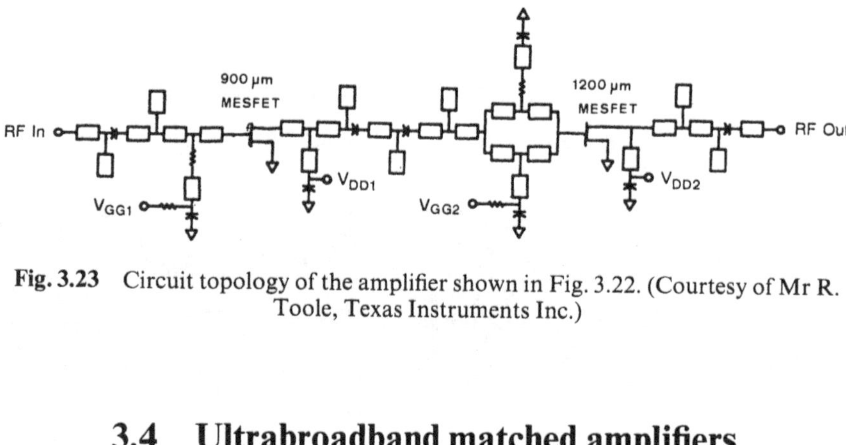

Fig. 3.23 Circuit topology of the amplifier shown in Fig. 3.22. (Courtesy of Mr R. Toole, Texas Instruments Inc.)

3.4 Ultrabroadband matched amplifiers

The conventional microwave amplifier synthesis techniques described previously do not allow simultaneous matching of GaAs MESFETs at low and high frequencies. Therefore they cannot be applied to the synthesis of amplifiers designed for frequency ranges from hundreds of megahertz to several gigahertz. The matching principles that are most suitable for this purpose are lossy matching [3.8, 3.56, 3.57], parallel feedback [3.8, 3.58–3.64] and active matching [3.8, 3.65]. The last principle is used mainly in monolithic technology and will be discussed separately in Section 3.4.4.

The structures of the first two types of amplifier are shown in Fig. 3.24. It can be seen that resistances are fundamental elements of the circuit in both cases. Their purpose is to match the device at low frequency and, indirectly, to

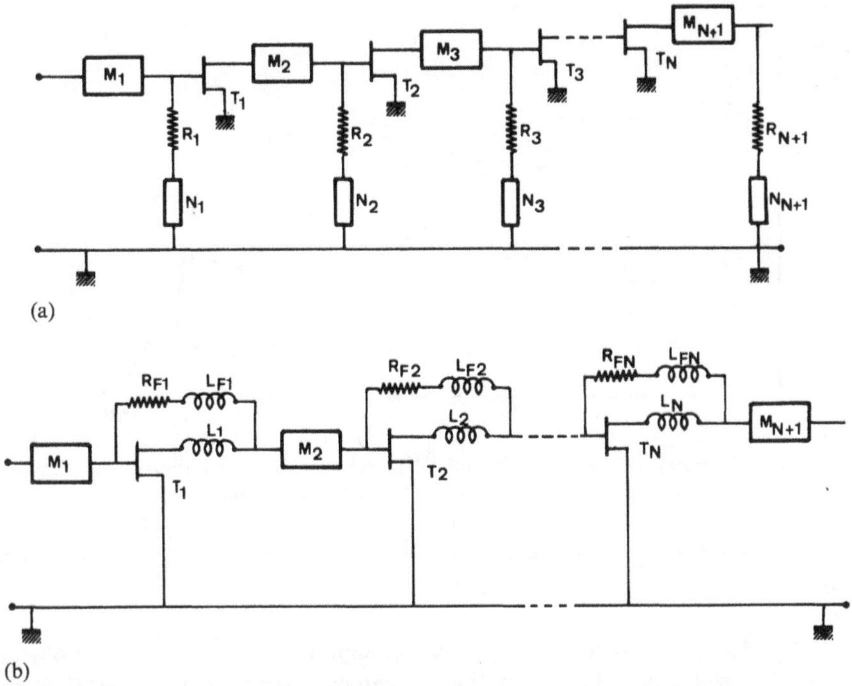

Fig. 3.24 Ultrabroadband matched amplifiers: (a) lossy match amplifier; (b) feedback amplifier.

determine the gain of the amplifier in this part of the band and consequently over the whole frequency range. Lossless elements d.c. couple the resistances at high frequency and match the device ports.

The design procedure for an ultrabroadband matched amplifier (Fig. 3.25) comprises three steps.

(a) At low frequency the only significant parts of the circuits are the resistances and the active devices excluding the microwave parasitic reactances. The values of the resistances that match the device and the power gain obtained can be calculated for this simple circuit. If the transition frequency f_T of the transistor is much higher than the highest frequency of the band, the final circuit can be composed of only the resistances mentioned above and GaAs MESFETs.

(b) For economic reasons it may be necessary to use transistors with an f_T only a few octaves above the highest frequency of the band. This results in a drop in the gain and a mismatch at the ports. To prevent this it is necessary to include frequency-selective lossless elements. This allows gain peaking and good port matching to be obtained toward the upper band edge.

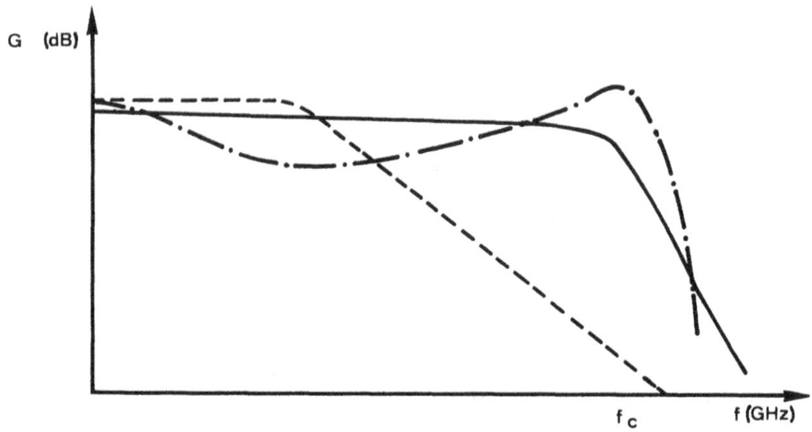

Fig. 3.25 Design procedure for ultrabroadband microwave amplifiers: - - -, device plus resistance; -·-, peak at high frequency; —, final optimization.

(c) Finally, flat gain and good VSWR at the amplifier ports over the entire band are obtained by circuit optimization.

The duration and cost of step (c) can be significantly reduced if a graphical method is used which allows simultaneous analysis over the whole band. Such a method will be described in Section 3.4.3.

3.4.1 Low frequency limitations

However paradoxical it may seem, the *low frequency* behavior of MESFET devices can limit the power gain of the amplifier. This is due to the fact that the only way of matching the high impedance of the ports at these frequencies is by means of resistances that dissipate a large part of the available power.

By using an equivalent circuit model of the MESFET (Fig. 3.26) it is possible to analyze various circuits and to calculate the resistance values and the resulting power gain. When identical matched stages are cascaded, the first two structures give the results shown in Table 3.4 (input and output ports cannot be simultaneously matched in feedback topology). Expressions are calculated for input matching because the output VSWR is rather small for normal values of g_d. The gain provided by a distributed amplifier with N devices is shown in the same table.

The power gain per stage is plotted against the device transconductance in Fig. 3.27. If the devices have a low g_m ($g_m < 25$ mS), it is clear that only distributed amplification can be used and a low gain is obtained: 1.5 dB per transistor [3.66] (many commercially available MESFETs for linear applications in the S, C and X bands have transconductances of about 25 mS). If high

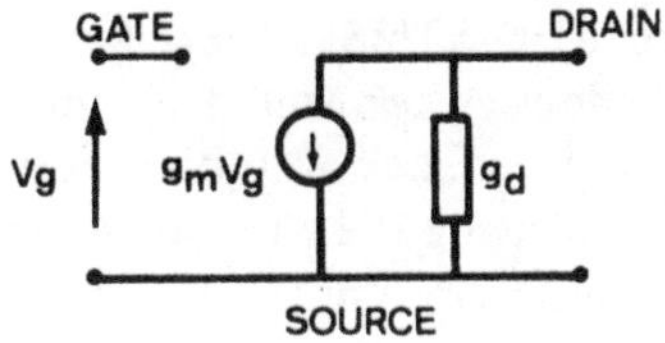

Fig. 3.26 Simplified MESFET model.

Table 3.4

Structure	*Matching resistance*	*Power gain*
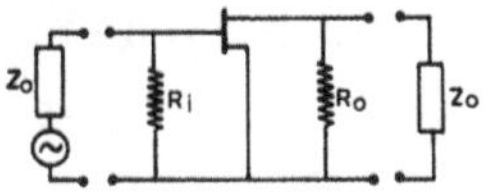	$R_i = Z_0 = \dfrac{1}{G_0}$ $R_o = \dfrac{1}{G_0 - g_d}$	$G = \left(\dfrac{g_m}{2G_0}\right)^2$
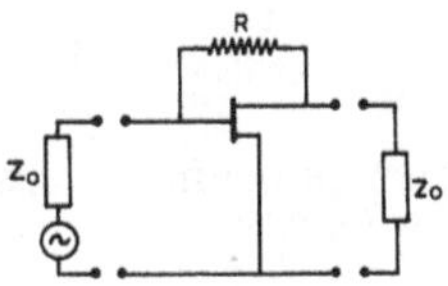	$R = \dfrac{g_m - g_d}{G_0(G_0 + g_d)}$	$G = \left(\dfrac{G_0 - g_m}{G_0 + g_d}\right)^2$
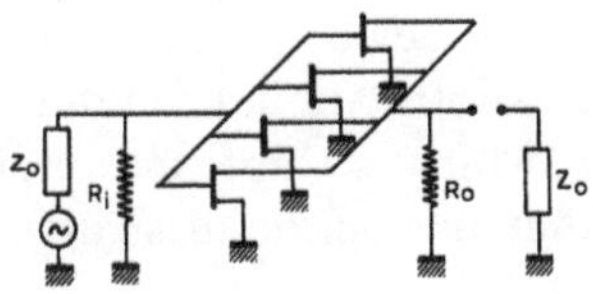	$R_i = Z_0$ $R_o = \dfrac{1}{G_0 - Ng_d}$	$G = \left(\dfrac{N g_m}{2G_0}\right)^2$

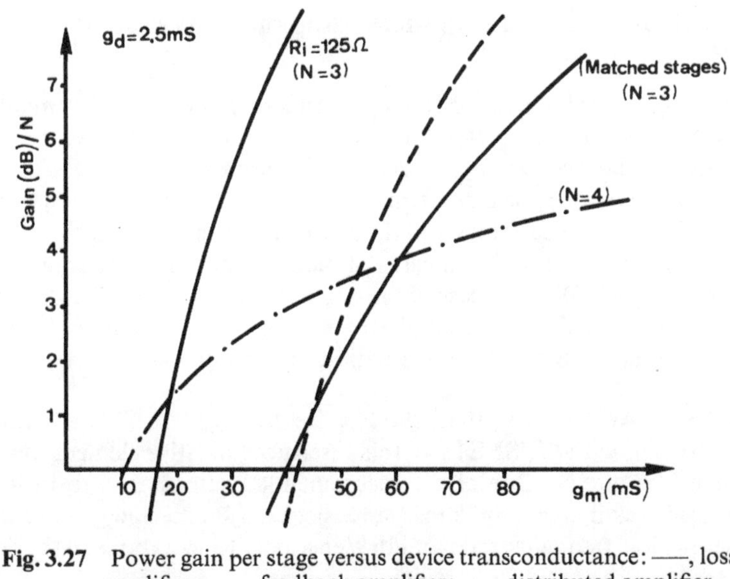

Fig. 3.27 Power gain per stage versus device transconductance: ——, lossy match amplifier; – – –, feedback amplifier; –·–, distributed amplifier.

transconductance devices ($g_m > 50$ mS) are used, which can be either special devices [3.59] or several devices connected in parallel [3.60, 3.66–3.68], a higher gain is obtained in the feedback amplifier configuration.

Finally, one way of increasing the gain per stage with low transconductance devices is by cascading several unmatched stages and matching only the first and last stages. For example, the gain of the lossy match amplifier in Fig. 3.24(a) is, in this case,

$$G = \frac{(g_m{}^N)^2}{\{2G_0(G_2+g_d)(G_3+g_d)\dots(G_N+g_d)\}^2} \tag{3.68}$$

where $G_i = 1/R_i\,(i = 2, \dots, N)$ and the input and output resistance values are

$$R_1 = Z_0 \tag{3.69}$$

$$R_{N+1} = \frac{1}{G_0 - g_d} \tag{3.70}$$

This gain is plotted against g_m in Fig. 3.27. It is clear that a significant increase is achieved compared with matched stage amplifiers [3.69, 3.70].

In summary, the maximum power gain of an ultrabroadband amplifier is limited by the maximum available gain (MAG) of the devices at the highest frequency. Nevertheless, it is necessary to consider another limitation due to the low g_m of commercially available MESFETs, which is sometimes more relevant. The analysis presented above proves that the optimum structure depends on the transconductance of the transistor [3.71].

3.4.2 Ultrabroadband amplifiers using the lossy matching principle

The structure of ultrabroadband amplifiers using the lossy matching principle is shown in Fig. 3.24(a): lossless networks M_i and lossy networks (represented in the diagram by a resistance in series with a lossless network which acts as a frequency-selective equalizer) are connected in cascade with the transistor. The resistance values and the power gain are obtained by using the expressions given in Table 3.4 in cases of matched intermediate stages or by using eqns (3.68)–(3.70) in the case of the unmatched stage amplifier. After this, the MAG of the transistor at the highest frequency can be compared with the low frequency gain. Three different situations can occur.

(a) The MAG is lower than the low frequency gain. This situation is unlikely to occur with MESFETs. In this case, more resistive elements must be included in the circuit. One of the recommended solutions is to include a negative feedback by means of a resistance between source and ground which decreases the low frequency gain and the sensitivity to variations in the device parameters.

(b) Both gains are equal. In this event the resistances must be decoupled at the highest frequency. This can be achieved by quarter-wave impedance transformers with a high characteristic impedance as well as by simple inductors implemented using very fine wire. However, it must be remembered that the thickness of the wire on the drain side must be capable of supporting the maximum d.c. rating I_{DS} of the transistor.

(c) The MAG is higher than the low frequency gain. In this case the resistances must be partially decoupled either by decreasing the inductor values or by designing the quarter-wave transformers at a higher frequency.

Once the equalizers have been designed, it is necessary to calculate the matching networks M_i. Their design is usually very simple because the equalizers also match the device, as in Fig. 3.28 where a design example is shown.

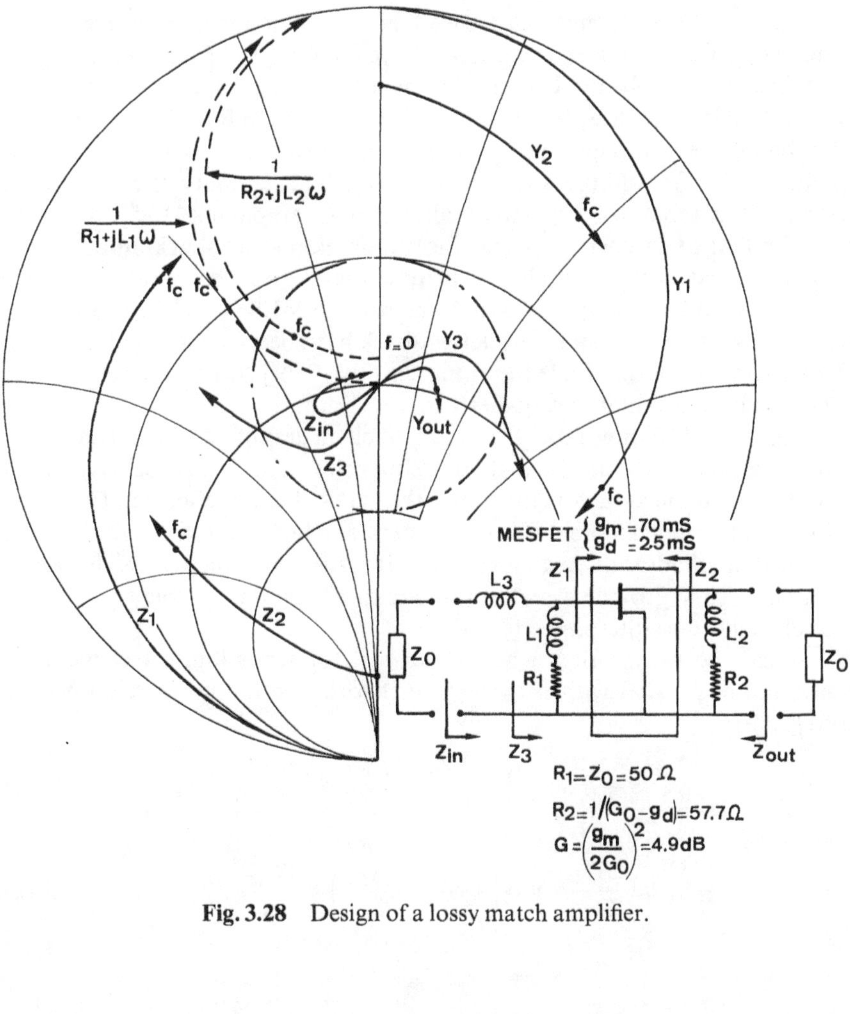

Fig. 3.28 Design of a lossy match amplifier.

The most obvious advantage of this structure is the ease of conception and design as well as the simplicity of the bias network, since the r.f. circuit resistances can be used for this purpose. In fact, these amplifiers have the largest bandwidth potential. This is because the lowest frequency in the band is limited only by the coupling capacitors.

Experimental results support the possibilities of this topology. Amplifiers with bandwidths of up to 16 GHz and power gains of about $6(N-1)$ dB, where N is the number of stages, have been obtained. Amplifiers with similar characteristics and variable gain have also been built using dual-gate MESFETs [3.72]. The main disadvantage is that variations in the MESFET transconductances produce immediate variations in the power gain which is very sensitive to device spread.

3.4.3 Ultrabroadband feedback amplifiers

The use of two strong negative feedbacks to match amplifiers over larger bandwidths was proposed more than 20 years ago [3.73]. These techniques were first extensively used in the design of amplifiers for small television distribution systems. Stages with flat gains and low VSWRs covering v.h.f. and u.h.f. bands were obtained using bipolar transistors with f_T values of a few gigahertz [3.74]. Additionally, the resulting high stability factor allowed several of them to be cascaded to obtain high gain amplifiers [3.75].

The first researcher to extend the application of feedback amplifiers to microwave bands was Ulrich. He proved that series and parallel resistive feedbacks could not be applied simultaneously to MESFETs [3.76]. In recent years many amplifiers using parallel feedback have been constructed in hybrid [3.77], semimonolithic [3.78] and monolithic [3.79] technologies, covering bands from a few hundred megahertz to 20 GHz.

Figure 3.24(b) shows the structure of such an amplifier. The resistance R_F allows the device to be matched at low frequencies, as in the case of lossy match amplifiers, and the inductors maximize the gain at high frequencies. However, in this case it is much more difficult to understand how this is achieved. A set of graphical methods has been proposed for studying microwave feedback amplifiers. They provide a method of circuit design and enable the effect of each element to be interpreted [3.80].

It can be shown that when a network of series-lumped elements is connected in parallel with a two-port network, the following relationships hold [3.60]:

$$\frac{2}{S_{21}{}^{T}} \approx \frac{2}{S_{21}} + y'_{12}\left(1 + \frac{2 + y_{11} + y_{22}}{y_{21}}\right) \approx \frac{2}{S_{21}} + y'_{12} \tag{3.71}$$

$$\frac{2S_{11}{}^{T}}{S_{21}{}^{T}} \approx \frac{2S_{11}}{S_{21}} - y'_{12}\left(1 - \frac{y_{11} + y_{22}}{y_{21}}\right) \approx \frac{2S_{11}}{S_{21}} - y'_{12} \tag{3.72}$$

$$\frac{2S_{22}^{\mathrm{T}}}{S_{21}^{\mathrm{T}}} \approx \frac{2S_{22}}{S_{21}} - y'_{12}\left(1 - \frac{y_{11}+y_{22}}{y_{21}}\right) \approx \frac{2S_{22}}{S_{21}} - y'_{12} \tag{3.73}$$

where the S_{ij} are the S parameters of the two-port network, the S_{ij}^{T} are the S parameters of the feedback two-port network, the y_{ij} are the Y parameters of the two-port network (normalized to Y_0) and the y'_{ij} are the Y parameters of the feedback network (normalized to Y_0).

The functions $2/S_{21}$, $2S_{11}/S_{21}$ and $2S_{22}/S_{21}$ of a typical MESFET are plotted in Fig. 3.29. Expressions (3.71)–(3.73) show that the effect of the feedback network is to translate $2S_{21}$ and $2S_{11}/S_{21}$ by the $\pm y'_{12}$ vector respectively. In an actual design it is necessary to consider the expressions (3.71)–(3.73) in their more complex form. However, for simplicity we use the approximate expressions. Obviously, the feedback network that matches and equalizes the device is that which translates the $2S_{11}/S_{21}$ curves to the origin and the $2/S_{21}$ curve to the constant-S_{21} circle with its center at the origin and a radius of $2/S_{21}$.

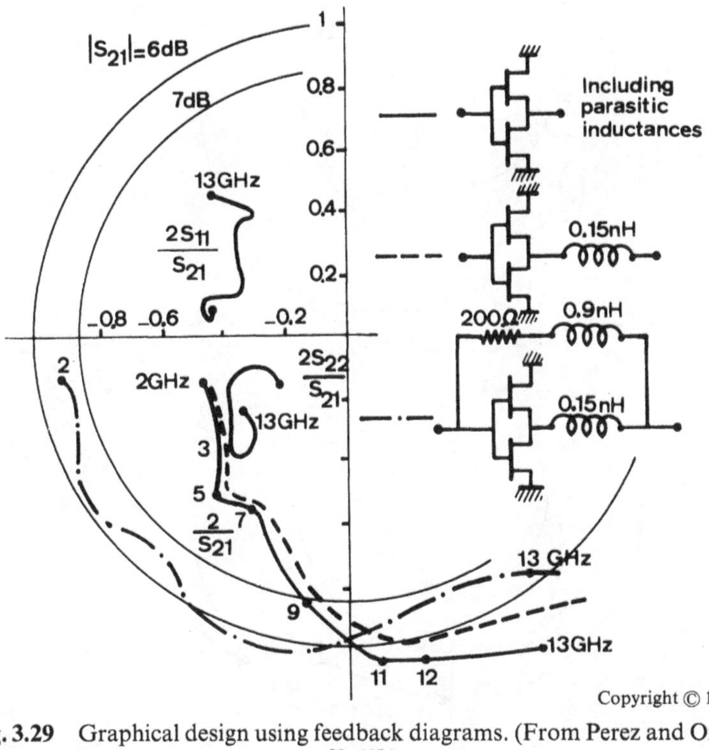

Fig. 3.29 Graphical design using feedback diagrams. (From Perez and Ortega [3.60].)

Figure 3.30(a) shows that, at low frequency, only one resistance ($y'_{12} = 1/R_p$) translates the $2S_{11}/S_{21}$ curves to the origin, thus matching the device.

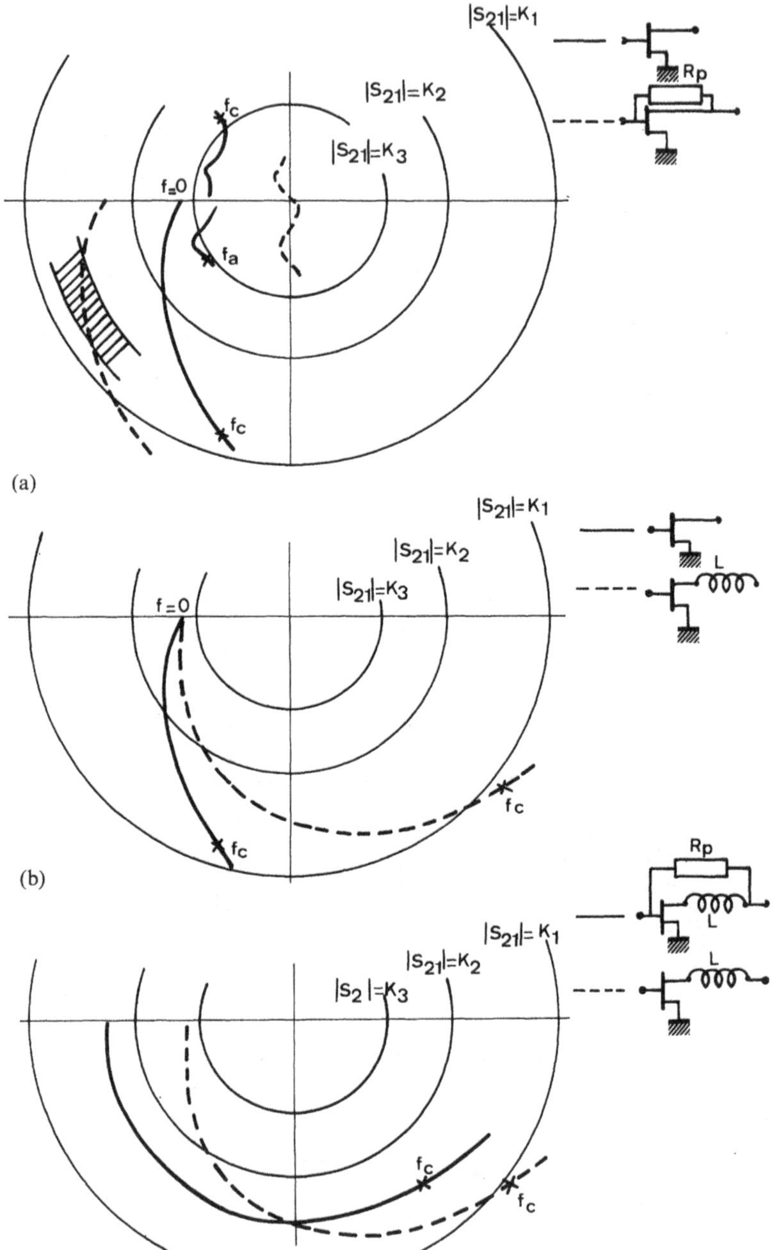

Fig. 3.30 Graphical design of a 0.15–12 GHz feedback amplifier: (a) transistor cell with parallel feedback; (b) cell plus drain series inductance; (c) complete feedback configuration.

This resistance produces a translation in the opposite direction in the $2/S_{21}$ curve, thus showing that a two-port with this feedback has less gain and is equalized in one part of the band. However, the diagram also shows that the effect of the resistance is different for points in different quadrants. If the $2S_{21}$ curve is in the second or third quadrant a feedback resistance degrades the gain, but for points in the first and fourth quadrants the same resistance produces an increase in the S_{21} of the two-port. In other words, the resistance creates a negative feedback at low frequency which could be positive in other parts of the band. This is the case if the phase difference between the input and output of the initial two-port network is adequate and the $2/S_{21}$ points are in the first or fourth quadrants. Thus one method of including a peak at high frequency is to translate this part of the $2/S_{21}$ curve to the fourth quadrant and then add feedback by means of a resistance. A counterclockwise rotation in $2S_{21}$ is produced by an inductor or by a section of transmission line cascaded at the output or a two-port network, thus placing the points of the high part of the band in the said quadrant as shown in Fig. 3.30(b). Finally, an inductor connected in series with the feedback resistance is usually added, thus producing a peak in the response. However, its effect is much less important than that of the inductor connected in series with the device, and its primary application is to simplify the practical realization of the circuit.

Figure 3.31 illustrates the design of an amplifier using two NE-388 transistors in parallel. Figures 3.31 and 3.32 show the final circuit, the experimental results and a photograph of the amplifier built on an alumina substrate 0.6 mm thick. The line sections were printed using thin film technology and the inductors were made using 0.15 mm gold wire. The series inductor has been included in the bond wires connecting the transistor to the main circuit.

In summary, the proposed graphical construction allows the values of R_p and L needed to obtain a two-port equalized over the whole band and matched in its lower part to be determined in such a way that the required result can be calculated using simple networks (directly obtainable from the Smith chart) and a final optimization process. As the values obtained graphically are very close to the final values, the optimization is rapid.

Parallel feedback is increasingly employed in the design of ultrabroadband microwave amplifiers, particularly in monolithic technology [3.64, 3.81, 3.82]. When combined with reactance matching it enables a large gain–bandwidth product and good noise performance to be achieved [3.8, 3.83].

3.4.4 Ultrabroadband amplifiers using the active matching principle

Active matching consists of using MESFETs in configurations that are able to provide an acceptable input or output VSWR. The common-gate (CG) and common-drain (CD) configurations are used as input- and output-

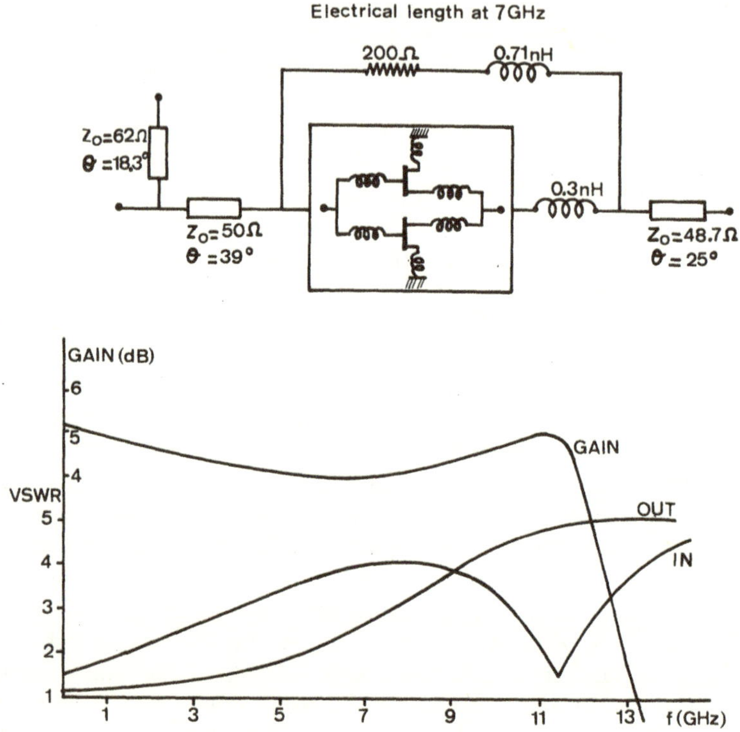

Fig. 3.31 Final circuit and experimental results.

matching devices respectively. An example of an amplifier stage using active matching is shown in Fig. 3.33. The amplifier gain is achieved essentially in a common-source MESFET. The resistors R_1 and R_2 are used as explained below to improve the output and input VSWR of the CG and CD stages respectively.

The CG configuration S parameters at low frequencies for the simple MESFET model in Fig. 3.26 are [3.8]

$$S_{11} = \frac{1-g_m Z_1 + g_d(Z_2 - Z_1)}{1+g_m Z_1 + g_d(Z_2 + Z_1)} \tag{3.74a}$$

$$S_{12} = \frac{2g_d(Z_1 Z_2)^{1/2}}{1+g_m Z_1 + g_d(Z_2 + Z_1)} \tag{3.74b}$$

$$S_{21} = \frac{2(g_m + g_d)(Z_1 Z_2)^{1/2}}{1+g_m Z_1 + g_d(Z_2 + Z_1)} \tag{3.74c}$$

$$S_{22} = \frac{1+g_m Z_1 - g_d(Z_2 - Z_1)}{1+g_m Z_1 + g_d(Z_2 + Z_1)} \tag{3.74d}$$

Fig. 3.32 Photograph of a 0.15–12 GHz feedback amplifier.

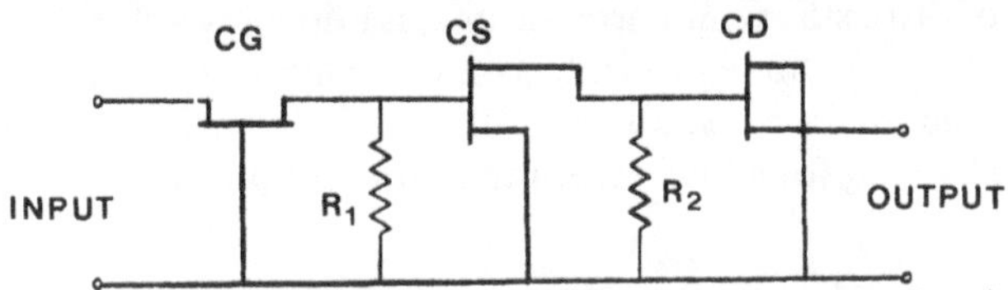

Fig. 3.33 Schematic diagram of a simple amplifier stage using active matching at the input and output. (From Niclas [3.8].)

where Z_1 and Z_2 are the source and load impedances respectively. The condition for perfect input matching is

$$g_m Z_1 = 1 + g_d(Z_2 - Z_1) \tag{3.75}$$

When $Z_1 = Z_2 = Z_0$ and condition (3.75) holds, eqns (3.74a)–(3.74d) become

$$S_{11} = 0 \tag{3.76a}$$

$$S_{12} = \frac{g_d Z_0}{1 + g_d Z_0} \tag{3.76b}$$

$$S_{21} = 1 \tag{3.76c}$$

$$S_{22} = \frac{1}{1 + g_d Z_0} \tag{3.76d}$$

Similarly, the low frequency S parameters for the CD configuration are [3.8]

$$S_{11} = 1 \tag{3.77a}$$

$$S_{12} = 0 \tag{3.77b}$$

$$S_{21} = \frac{2g_m(Z_1Z_2)^{1/2}}{1+(g_m+g_d)Z_2} \tag{3.77c}$$

$$S_{22} = \frac{1-(g_m+g_d)Z_2}{1+(g_m+g_d)Z_2} \tag{3.77d}$$

The condition for a perfect match at the output becomes

$$(g_m+g_d)Z_2 = 1 \tag{3.78}$$

and for $Z_1 = Z_2 = Z_0$ eqn (3.77c) becomes

$$S_{21} = g_m Z_0 \tag{3.79}$$

For practical devices and at higher frequencies the reflection coefficients $|S_{21}|$ for the CG configuration and $|S_{11}|$ for the CD configuration can become greater than unity and thus present unacceptable terminating conditions to the amplifying stage to be placed between the CG and CD stages. The practical solution to this problem is to use shunt resistors as shown in Fig. 3.33 or to employ parallel and series feedback in order to decrease the reflection coefficients of these stages and provide unconditional stability.

The active matching principle is used mainly in lowpass monolithic amplifiers. The highest bandwidth (0.1–10 GHz) has been reported by Petersen *et al.* [3.65] for a four-stage monolithic amplifier.

3.4.5 Biasing circuits for ultrabroadband amplifiers

Bias circuits must be carefully designed in order to ensure the correct performance of an ultrabroadband matched microwave amplifier, particularly in applications such as pulse linear amplification in which the highest potential bandwidth is demanded.

Figures 3.34(a) and 3.34(b) show two conventional bias circuits using two power supplies and one power supply respectively. If the circuit is to amplify the lowest possible frequency signals, the capacitors and inductors must have high values. In that case, however, they will also be physically large and consequently their maximum operating frequency will be low, so that a trade-off will be necessary. The most significant restriction arises from the inductors, as the capacitors have much wider bandwidths.

In lossy match amplifiers the inductors can be eliminated by using the r.f. circuit's own resistances to introduce the biasing (Fig. 3.34(c) and 3.34(d)). This is why they are potentially wide-bandwidth amplifiers. It is also possible to use

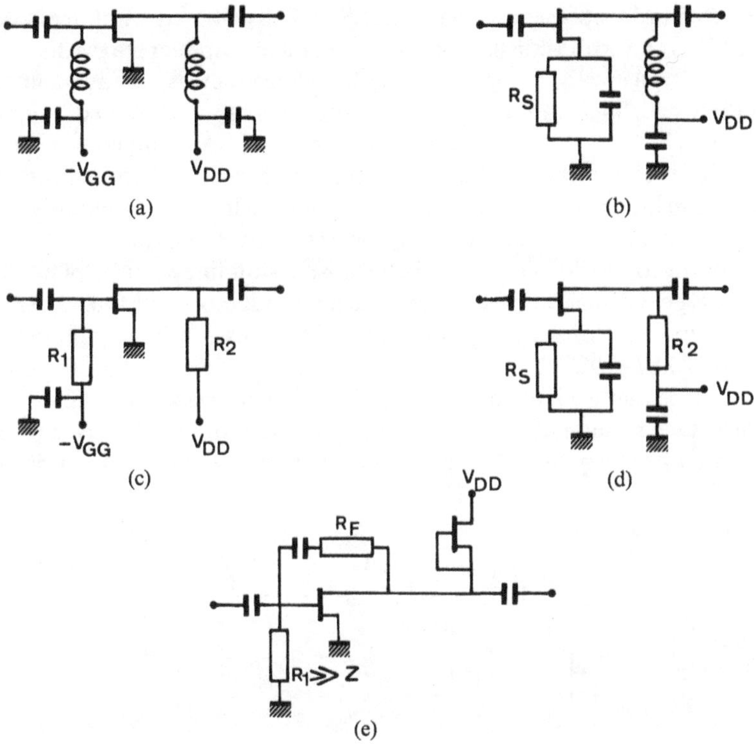

Fig. 3.34 Bias circuits for ultrabroadband matched amplifiers.

two additional resistances instead of inductors in feedback and distributed amplifiers. However, to avoid loading the r.f. circuit the condition $R_1, R_2 \gg Z_0$ must be met. In many cases this prevents the substitution of drain inductors since R_2 would dissipate a large amount of power. The use of an active load (Fig. 3.34(e)) overcomes this difficulty and this solution is frequently adopted in monolithic circuits [3.84].

3.5 Distributed amplifiers

The invention of the distributed amplifier was patented in 1937 by Percival. The concept is based on combining the input and output capacitances of tubes with inductors in such a way that two artificial transmission lines are obtained. The input and output capacitance of each device becomes the capacitance per unit section for these lines and the lines are coupled by the transconductance of the active device. As a result, it is possible to obtain amplification over much wider bandwidths than is achievable with conventional amplifiers.

A schematic representation of a MESFET distributed amplifier is shown in Fig. 3.35. It basically consists of a set of identical amplifier stages distributed along two transmission lines. Ideally, the stages act as voltage-controlled generators with high input and output impedances. A simplified equivalent circuit which takes into account the gate and drain impedances of the MESFETs is shown in Fig. 3.36. These impedances are absorbed into lossy transmission line sections. The unit sections of both lines are essentially loaded constant-K lines in which the parasitic resistances provide the dominant contribution to line losses [3.85]. The transmission lines can be of either the artificial type, i.e. made up of lumped-element inductors [3.66, 3.85, 3.86], or the real type, i.e. consisting of sections of transmission lines (microstrip or coplanar) [3.87–3.92]. In the latter case, some versions make use of transmission lines with identical characteristic impedances and different line lengths between the MESFETs [3.91, 3.92], whereas others are based on equal line lengths between the transistors and different characteristic impedances [3.87, 3.91].

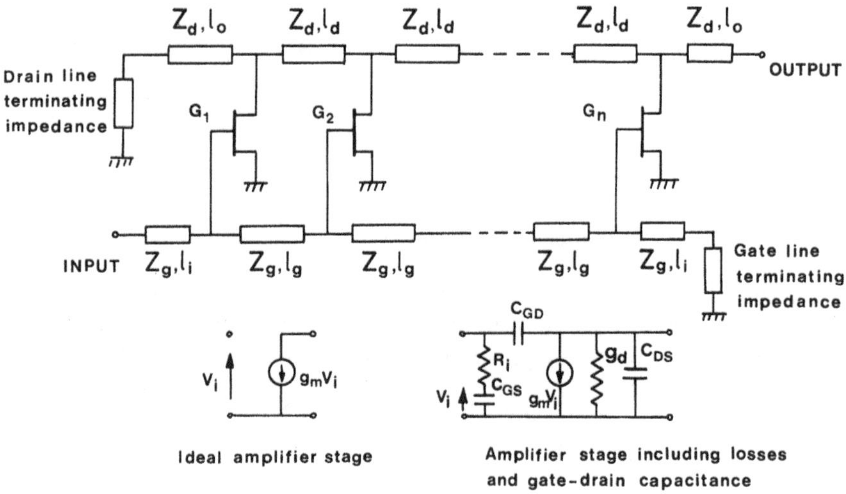

Fig. 3.35 Schematic representation of an n-stage MESFET distributed amplifier.

The operation of the distributed amplifier under the above conditions is easy to explain. An r.f. signal applied at the input port of the gate line travels down the line to the termination where it is absorbed. The travelling signal is picked up by the gates of the individual MESFETs and transferred to the drain line via their transconductances. If the phase velocities on the gate and drain lines are identical, the signals on the drain line add in the forward direction. Any signal which travels backwards and is not entirely cancelled by the out-of-phase additions will be absorbed by the drain line termination. In the ideal

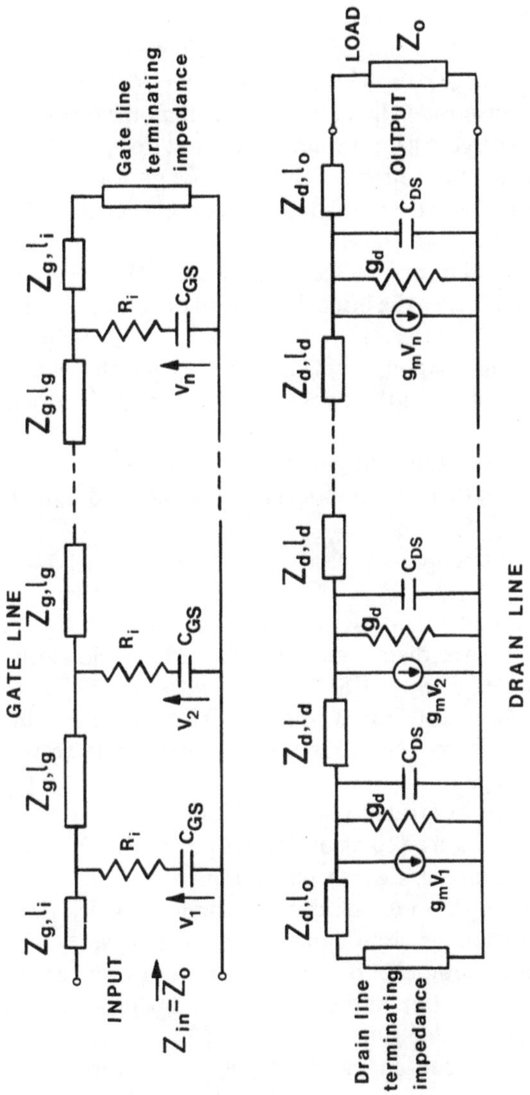

Fig. 3.36 Simplified equivalent circuit of a MESFET distributed amplifier (the parasitic inductors and gate–drain capacitance of the device have been omitted).

case, the small-signal gain expression for the amplifier can be approximated as [3.93]

$$G = \frac{(N g_m Z_0)^2}{4} \tag{3.80}$$

where g_m is the transconductance of each MESFET, N is the number of MESFETs and the Z_0 are the input and output characteristic impedances.

In theory a very large bandwidth can be achieved by making the cut-off frequency f_c of the transmission line much higher than the transition frequency f_T of the MESFET. Thus the gain should remain flat up to f_T. However, several parasitic effects degrade the flatness of the amplifier gain and cause it to fall off at higher frequencies. The most important of these effects are as follows:

(1) increasing loading of line sections by the MESFET parasitic resistances, which makes the intensity $g_m V_i$ of the current sources decrease with distance from the input;
(2) the drain–gate coupling capacitance C_{GD} which enhances the gain ripple by transferring the backward wave from the drain line to the gate line;
(3) mismatch in the drain and gate terminating loads;
(4) series feedback through the source resistances and inductances of the MESFETs;
(5) effects of bias circuits.

Taking into account drain–gate coupling capacitances, Jutzi [3.66] built one of the first microwave distributed amplifiers. Its bandwidth was 2 GHz and it was implemented using MESFET chips with $f_T \approx 6$ GHz and artificial transmission lines. Many distributed amplifiers have been built since then, and bandwidth coverages from 2–18 GHz up to 2–40 GHz have been achieved [3.65].

Designers have concentrated mainly on increasing the gain–bandwidth product and gain flatness, as well as on output power capabilities. Beyer *et al.* [3.85] analyzed the gain–bandwidth limitations of the distributed amplifier and developed a closed-form expression for the gain which depends on the propagation constants of the circuit and the cut-off frequency of the gate line. This approach allows computation of the optimum number of MESFETs [3.85] as well as determination of the physical parameters of the device which give the maximum achievable gain–bandwidth [3.94]. The maximum operating frequency must be chosen appropriately with respect to the cut-off frequency of the transmission lines in order to avoid excessive gain ripple towards the upper band edge.

Gamand [3.92] proposed an improved design in which the terminating loads of transmission lines are synthesized so that they match variations in the characteristic impedance of the line (image impedance of the constant-K

sections) with frequency. The hybrid configuration of this amplifier is shown in Fig. 3.37. The amplifier exhibits a remarkably flat gain of 4.8 ± 0.4 dB over a bandwidth of 2–30 GHz. The very broadband bias circuits which are combined with matched terminating impedances should also be noted.

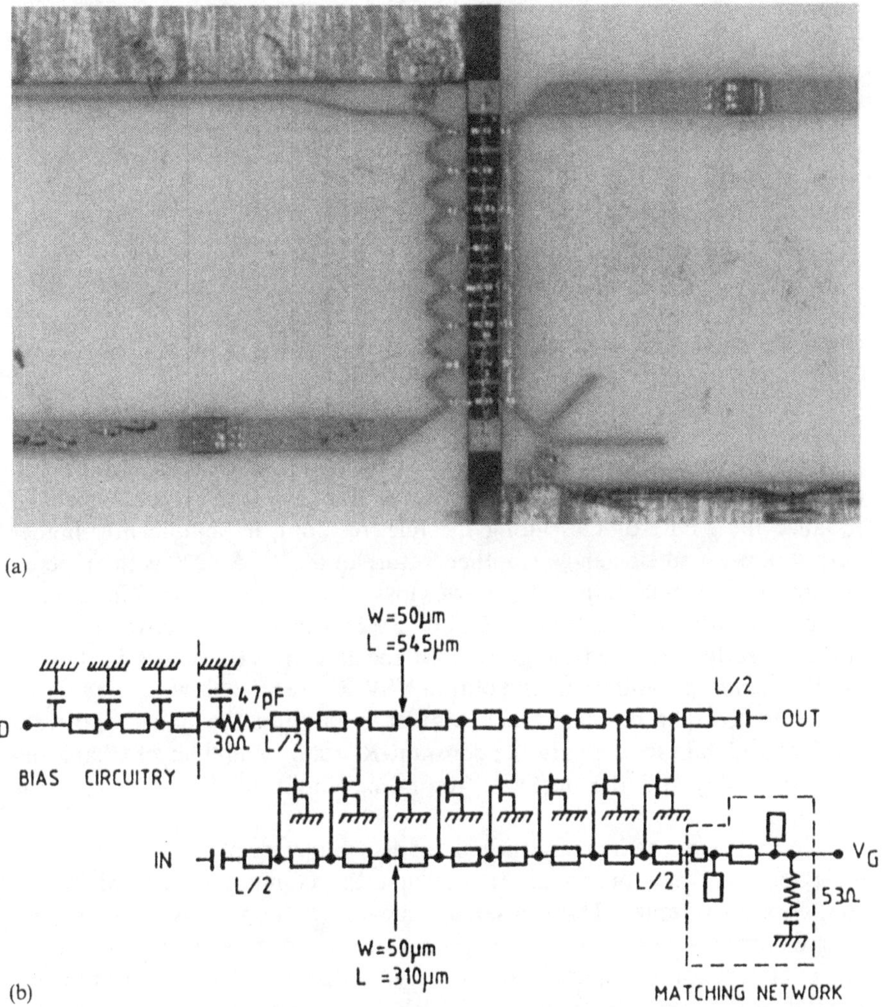

Fig. 3.37 Hybrid distributed amplifier [3.92]: (a) photograph of the complete amplifier; (b) equivalent circuit of the amplifier. (Courtesy of Mr P. Gamand, Laboratoires d'Electronique et de Physique Appliquée, France.)

Figure 3.38 shows a 2–20 GHz distributed amplifier (Texas Instruments TGA 8220) [3.55] which uses six 0.5 μm × 335 μm FETs and achieves 0.4 W with 5 dB gain and 15% power-added efficiency. The chip size is

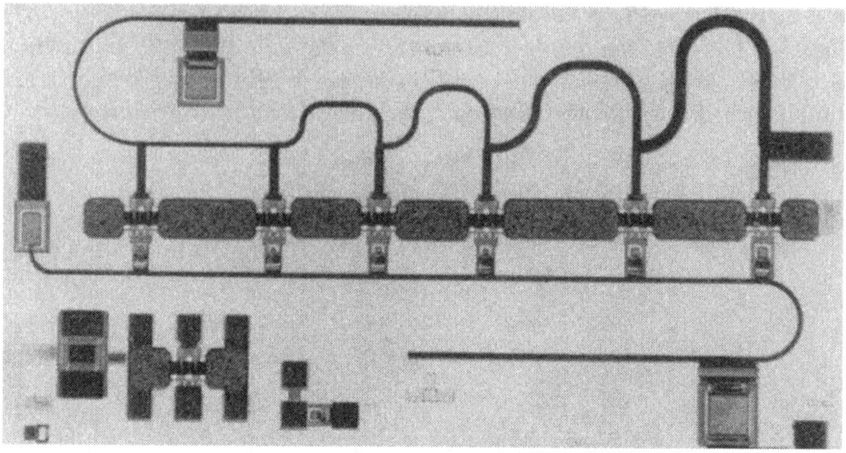

Fig. 3.38 Monolithic distributed amplifier [3.55]. (Courtesy of Mr T. Kilgo, Texas Instruments Inc.)

3.33 mm × 1.88 mm. MIM capacitors were inserted in series with each FET in order to increase the cut-off frequency of the gate transmission line and equalize the FET voltages along the line, resulting in significantly higher output power and efficiency. Another feature of the TGA 8220 is the tapered line approach which reduces the power loss in the drain line termination. In comparison with TGA 8014 (see Section 3.3.5) which uses reactive matching and has nearly the same total gate width, the gain of TGA 8220 is 3 dB lower but the gain ripple and input and output VSWRs are much lower.

Among other designs which aim at improving the gain–bandwidth product and gain flatness are the constant-*R*-section amplifier of Chase and Kennan [3.86] and the declining drain-line-length circuit of Niclas *et al.* [3.90].

In an attempt to increase the gain of distributed amplifiers two solutions have recently been proposed. In the first the common-source MESFET configuration is replaced by a cascode element [3.95] consisting of a common-source first stage followed by a common-gate second stage. The advantage of this configuration is a marked reduction in output conductance. It allows a significant gain improvement by using shorter gate lengths (which means a higher g_m but also higher g_d) and reduces the gate and drain line losses associated with the active devices.

In the second modification a matrix configuration has been proposed [3.96] in which the processes of additive and multiplicative amplification are combined in the same module. The new amplifier configuration consists of an array of m rows and n columns of active devices forming an $m \times n$ rectangular array. The advantages of the new configuration include higher gain, better reverse isolation and reduced size. An experimental 4×2 array amplifier in hybrid configuration achieved $G = 13.8 \pm 0.8$ dB between 2 and 21.5 GHz.

The distributed amplifier is well suited to ultrabroadband power amplification [3.88, 3.97]. Ayasli *et al.* [3.88] studied power-limiting mechanisms in distributed amplifiers and developed a parallel configuration by establishing a common drain line. At present, the main drawback of distributed amplifiers is their relatively poor noise performance.

3.6 Low noise MESFET amplifiers

3.6.1 Introduction

Amplifiers using low noise GaAs MESFETs have had a profound impact on microwave systems over the last decade. The best noise performances obtained approach those of parametric amplifiers, and GaAs MESFET amplifiers are increasingly used in electronic warfare and communications systems.

The primary role of a low noise preamplifier is to improve the operating margin of the system. However, depending on the application envisaged—radio astronomy, radar or microwave radio relay—the noise performance to be attained is different and must take into account sources of noise, external to the system, which accompany the signal coming into the receiver. Such external noise sources, which perturb the useful signal level, can be included in the antenna noise term. This term includes noise sources of such diverse provenance as galactic noise, man-made noise and atmospheric absorption and is expressed in kelvins (K). The noise temperature will depend on the orientation of the antenna relative to the horizon. Additional sources of noise due to molecular absorption will also appear at discrete frequencies, e.g. the water vapor line at 22 GHz and the oxygen line around 60 GHz.

Figure 3.39 gives the antenna temperature for various orientations, and it can easily be seen that it varies from 300 K, when the horizontal antenna is perturbed by the temperature of the Earth's surface, to about 4 K when it is pointing to a cool area of the Galaxy. A study of this curve reveals two types of applications.

(a) Radio astronomy and satellite communications: in both these cases the antenna noise temperature is low, of the order of tens of kelvins, and the Earth station receiver must be optimized to this noise temperature. (It is evident that the radio astronomy receiver, which is essentially a radiometer operating at night, demands the use of very low noise preamplifiers and imposes the more difficult challenge.) In this case a cooled GaAs MESFET amplifier will follow the parametric amplifier stage, although in the case of

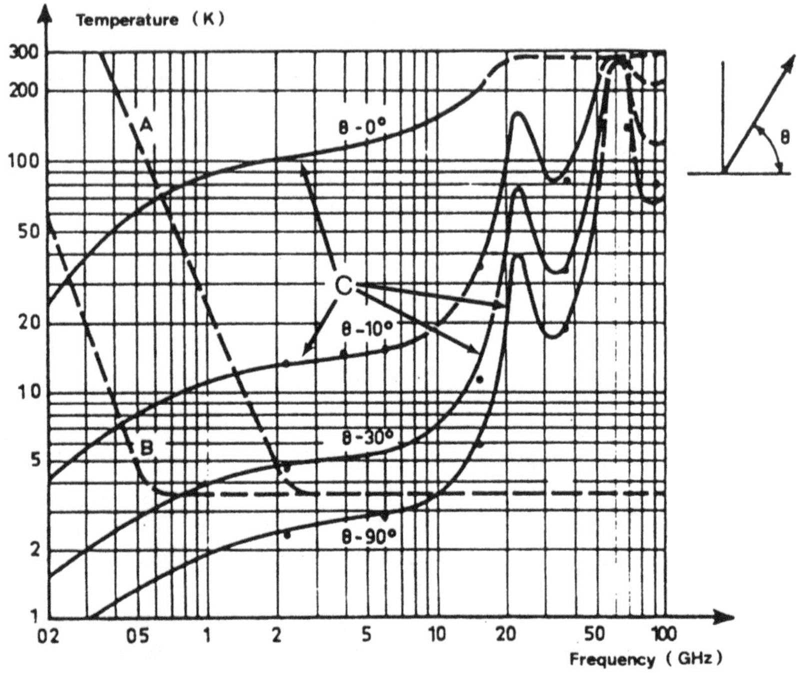

Fig. 3.39 Apparent noise temperature: curve A, galactic noise only; curve B, cosmic noise at 3.4 K only; curves C, with atmospheric absorption and versus antenna elevation.

small Earth stations only the GaAs MESFET amplifier at 100 K will be used directly behind the antenna. Activities of this type led to optimization of the noise performance of transistors and amplifiers, in particular in the frequency bands 3.7–4.2 GHz and 11.7–12.5 GHz. (It should be noted that the antenna of an on-board receiver in a space communications link is pointed at the Earth and as a consequence sees a background noise temperature of 300 K. The on-board front-end thus has a much easier noise specification to meet.)

(b) Microwave radio relay and radar: the horizontally pointed antenna will see a background noise temperature of several hundreds of kelvins. Thus an amplifier noise performance of the same order is sufficient and other considerations will predominate: dynamic range, linearity, bandwidth, overload performance, price and reliability. In Fig. 3.40 we have juxtaposed 1987 state-of-the-art GaAs MESFET noise performance against external noise sources as well as the noise performance of other devices [3.98–3.100]. It is clear from the curves in this figure that, up to X band, low noise GaAs MESFET amplifiers, whether cooled or not, are a useful alternative to parametric amplifiers and that a MESFET amplifier at ambient temperatures can significantly improve the sensitivity of a receiver chain, even with respect to the best mixers.

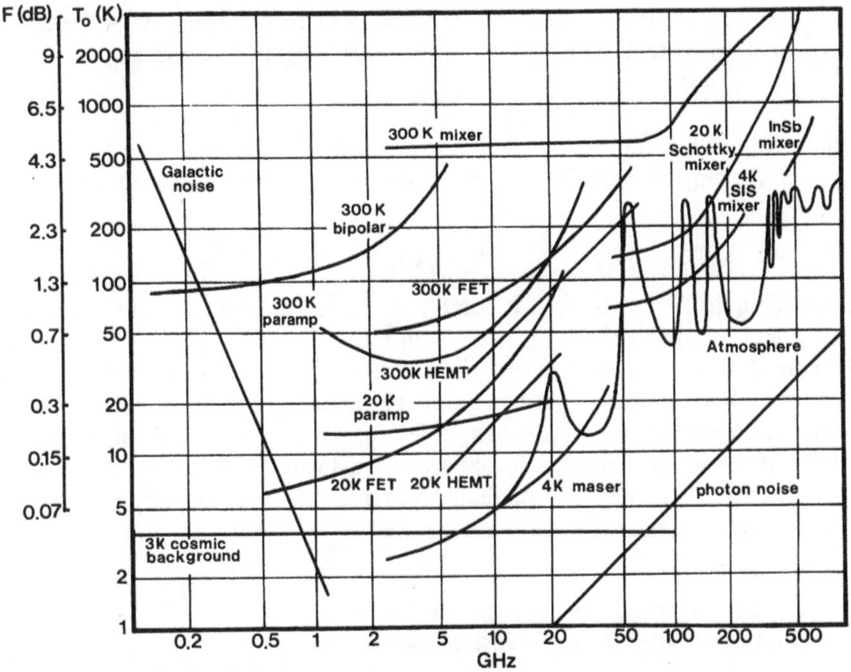

Fig. 3.40 Noise performance of 1987 state-of-the-art low noise GaAs MESFETs compared with other devices and external noise contributions.

In this section we shall treat the design of GaAs MESFET amplifiers with a view to making full use of their low noise potential. In order to do this, we shall briefly review some of the basic concepts of noise measurement before dealing in detail with the measurement of the noise figure of a GaAs MESFET. Finally, we shall show how the cooling of the device results in a noticeable improvement in its noise performance.

3.6.2 Noise measurement—fundamental concepts

3.6.2.1 Noise temperature

In an active two-port network such as a GaAs MESFET, internal noise sources will generate noise power at the device output (see Chapter 1, Section 1.2.5). This noise power will add to noise input to the device from external sources (antenna noise, input load noise etc.) and will be amplified by the transistor. In order to be able to distinguish between internal and external noise sources it is useful to define the noise generated within the two-port network in terms of an equivalent noise source placed at the device input followed by a noiseless network with the same power gain [3.101]. In this case the noise temperature is that of the resistance which, when placed at the device input, will generate the same noise power N_o at the transistor output:

$$N_o = G(N_i + N_a) \tag{3.81}$$

where $N_i = kT_0B$ represents the noise power (from external sources) at the input to the amplifier and N_a is the noise power added by the amplifier which is given by

$$N_a = kT_eB \tag{3.82}$$

k is Boltzmann's constant, B is the amplifier bandwidth, G is the available power gain of the amplifier, T_0 is the standard noise temperature (290 K) and T_e is the effective source noise temperature representing the noise power added by the amplifier and referred to its input. The gain G is the available gain of the device, which is the ratio of the available power at the output to the available power at the input (Chapter 2). Thus we transform our device to a noiseless two-port network which is perfectly matched in order to define its noise behavior and eliminate the noise resistance R.

3.6.2.2 Average noise figure (noise factor)

Two equivalent definitions are available.

(1) The average noise factor of a linear two-port network is the ratio of the total noise power N_o delivered by the network into its output termination when the noise temperature of its input termination is standard (290 K) at all frequencies to the fraction N_iG of the noise power due to the input termination [3.102].

(2) The noise factor of the linear two-port network is the ratio of the signal-to-noise ratio of an ideal network to the signal-to-noise ratio of the actual network [3.103].

From the first definition and using eqns (3.81) and (3.82) the noise factor F is given by

$$F = \frac{kGB(T_0 + T_e)}{kGBT_0} = 1 + \frac{T_e}{T_0} \tag{3.83}$$

This formula links the noise factor to T_e. To avoid confusion we refer to the noise figure as the noise factor (in decibels)

$$F = 10\log_{10}\left(1 + \frac{T_e}{T_0}\right)\text{dB}$$

In practical systems the external noise at the input to the device is often different from the standard noise temperature T_0, i.e. the noise will be generated in an antenna. Equally, noise is generated after the amplifier, in the receiver chain. Therefore it is instructive to consider the contributions of various noise sources in a cascaded two-port network chain.

3.6.2.3 Cascaded noise figure

A receiver chain is in general made up of a cascade of two-ports—preamplifier, amplifier, mixer and i.f. amplifier—matched to one another. Figure 3.41 shows two of these cascaded networks with effective noise temperatures T_1 and T_2, noise figures F_1 and F_2, and available power gains G_1 and G_2 respectively. It is easily shown [3.104] that the overall noise temperature T_{12} at the input of the two-stage and the overall noise factor F_{12} are given by

$$T_{12} = T_1 + \frac{T_2}{G_1} \tag{3.84}$$

and

$$F_{12} = F_1 + \frac{F_2 - 1}{G_1} \tag{3.85}$$

Equations (3.84) and (3.85) can be generalized to the case of n cascaded networks:

$$T_{1n} = T_1 + \frac{T_2}{G_1} + \frac{T_3}{G_1 G_2} + \dots + \frac{T_n}{G_1 G_2 \dots G_{n-1}} \tag{3.86}$$

$$F_{1n} = F_1 + \frac{F_2 - 1}{G_1} + \frac{F_3 - 1}{G_1 G_2} + \dots + \frac{F_n - 1}{G_1 G_2 \dots G_{n-1}} \tag{3.87}$$

Equations (3.86) and (3.87) are known as the Friis formulae. It should be noted that the formulae are equally valid if the two-port network has a gain smaller than unity, i.e. introduces a loss.

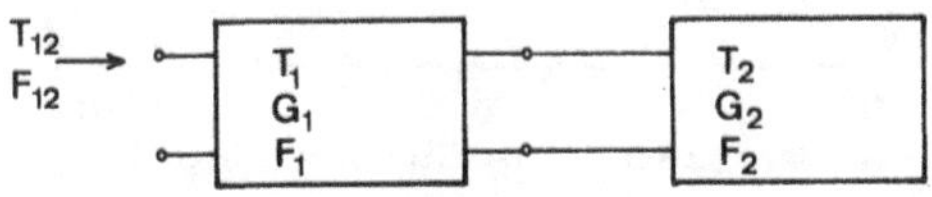

Fig. 3.41 Equivalent model of two cascaded amplifiers.

These formulae are particularly relevant in the case of low noise GaAs MESFET applications since the low noise figure and high gain of the device in a receiver front-end are used to mask the noise contribution of following stages. It is seen from eqn (3.85) that, depending on the noise figure to be masked, the gain in a low noise transistor may be as important a parameter as the noise figure. In Chapter 2, Section 2.8.3, it was shown that, under optimum noise figure bias and matching conditions, as the noise figure of a GaAs MESFET decreases below a certain level the power gain associated with that noise figure also decreases. Consequently, a widely used figure of merit for active low noise devices is the noise measure M. This parameter is defined with respect to the noise figure of an infinite series of devices having identical noise figure F and gain G. The substitution of F and G in eqn (3.87) results in an infinite geometric progression, so that the noise measure is given by

$$M = F_{\infty} - 1 = (F-1)\left(1 + \frac{1}{G} + \frac{1}{G^2} + \ldots\right)$$

or

$$M = \frac{F-1}{1-1/G} \tag{3.88}$$

The optimum value M_{min} of the noise measure depends on a trade-off between F and G. Although M_{min} may not always be the best operating condition to aim for, it provides a useful means of comparing the noise performances of different transistors and is the lowest noise figure one can hope for from an amplifier using identical cascaded low noise transistor stages. Thus the overall noise figure of a multistage amplifier with high gain (more than 15 dB) can be approximated by

$$F_{\infty} = \frac{FG-1}{G-1} \tag{3.89}$$

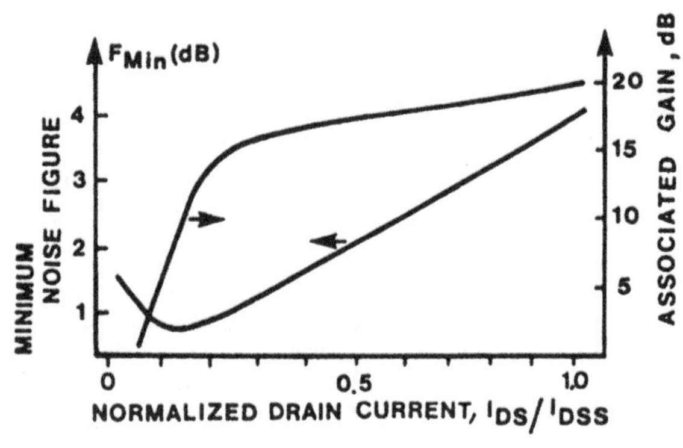

Fig. 3.42 Typical trade-off between the optimum noise figure and the associated gain for a MESFET amplifier versus I_{DS}.

M_{min} has the interesting property of being invariant to lossless feedback [3.105]. A combination of parallel and series lossless feedback is used in the design of low noise amplifiers in hybrid technology for frequency bands up to 4 GHz [3.106]. This approach allows the input and output VSWRs to be improved while keeping the noise figure close to the device minimum value.

3.6.2.4 Minimum noise impedance

The minimum noise figure F_{min} of a GaAs MESFET at any given frequency will be obtained if and only if the noise source presents a complex admittance Y_{min} to the transistor (see Chapter 1). This admittance, which is independent of the output load of the transistor, will generally be different from the conjugate of the MESFET input admittance. This state of affairs presents problems in both the measurement of Y_{min} and the design of ultralow noise amplifiers, as will be seen later. Thus the noise figure F as a function of a given source impedance $Y_s = G_s + jB_s$ is

$$F = F_{min} + \frac{R_n}{G_s}|Y_s - Y_{min}|^2 \tag{3.90}$$

where R_n is the noise resistance and indicates the rate of increase in the noise figure with detuning. Alternatively, the effective noise temperature of the device can be written in a similar form:

$$T_e = T_{min} + T_0\frac{R_n}{G_s}|Y_s - Y_{min}|^2$$

Both F_{min} (or T_{min}) and $N = R_nG_{min}$ are invariant under transformation through lossless reciprocal two-ports connected to the input and/or output of the transistor. F_{min} is also invariant to the common lead inductance in the transistor source. It has been shown that, for F_{min}, T_{min} and N to represent a physical two-port, the following inequality has to be satisfied [3.107]:

$$F_{min} \leqslant 1 + 4NT_0 \qquad (\text{or } T_{min} \leqslant 4NT_0)$$

3.6.2.5 Influence of losses

Losses are of two types: those due to lossy matched structures (isolators, connectors, cables, attenuators etc.) and those introduced by matching the source impedance (generally 50 Ω) to Y_{min}.

In the former case application of the Friis formula enables us to determine the noise figure of the device itself by simply subtracting the losses (in decibels) of the matched components between the noise reference plane and the matched transistor input port.

The matching loss due to transforming the source admittance Y_s to the optimum source noise admittance Y_{min} is given by

$$L = A\frac{1 - |\Gamma_{min}|^2}{1 - A^2|\Gamma_{min}|^2} \tag{3.91}$$

where

$$\Gamma_{min} = \frac{Y_s - Y_{min}}{Y_s + Y_{min}} \tag{3.92}$$

and A is the insertion loss of the 50 Ω line between the tuner and the device.

Since A is a function of line length and loss per unit length, it is important in order to improve measurement accuracy to match as close to the GaAs MESFET as is physically possible and also to use a low loss line, e.g. a line with an air dielectric, between the tuner and the device. L can be estimated from the curves in Fig. 3.43, which show the variation in L as a function of the VSWR $S = (1+|\Gamma_{min}|)/(1-|\Gamma_{min}|)$ for various values of A. Losses will also occur within the tuner itself, which if represented by an S matrix can be described by [3.108]

$$\eta = \frac{|S_{21}|^2}{1-|S_{22}|^2} \tag{3.93}$$

This loss can be measured as follows. The tuner is connected to the transistor and is used to transform Y_s to Y_{min} so as to obtain the minimum noise figure from the MESFET. The MESFET is then disconnected and a second tuner is connected to the first and used to transform Y_{min} back to 50 Ω. The insertion loss of the two tuners is then measured and divided by 2 to give the matching loss. However, this correction does not account for the loss in the 50 Ω line between the tuner and the transistor.

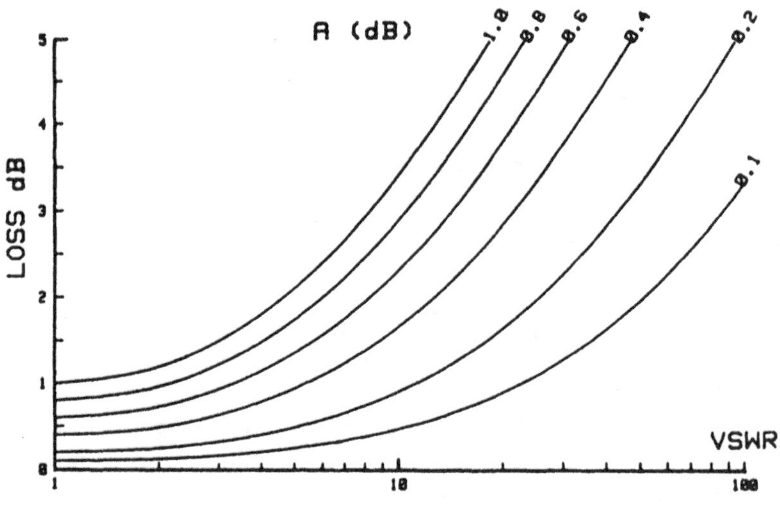

Fig. 3.43 Overall loss (in dB) when a high VSWR is matched through a lossy A (dB) transmission line.

It should be noted that, for very accurate measurements, eqn (3.93) can also be used to correct for mismatch losses in imperfect isolators, cables etc. It also shows the importance of careful selection of the tuning to be used in matching the noise generator impedance to the optimum value Y_{min}.

3.6.2.6 Noise sources

Hot and cold loads Ever since J. B. Johnson discovered that a resistance acts as a noise generator and that the power available from the resistance at a temperature T is given by $P = kTB$ hot and cold noise loads have been used in noise measurement. A resistance at a known temperature is a primary noise power standard, and in order to measure a low noise temperature accurately noise resistances at two different but precisely known temperatures, in general a cryogenic load and a heated load, are used. However, input power changes of about 10:1 are required for the accurate measurement of noise factors in the range 5–15 dB [3.102]. Thus in practice either a gas discharge source or a solid state noise source which has been calibrated against a set of hot and cold loads is generally used.

When hot and cold noise sources are used, two terminations with different available noise powers kT_1B and kT_2B are connected to the input of the device under test. The ratio of the powers delivered to the load is called the Y factor:

$$Y = \frac{T_2 + T_e}{T_1 + T_e} \tag{3.94}$$

It is possible to compute the effective noise temperature of the transistor and hence its noise factor from Y, T_2 and T_1.

Solid state noise source Gas discharge noise sources should be avoided in GaAs MESFET measurement set-ups as the voltage spike which fires the tube can damage or degrade the gate of the transistor. For this reason solid state avalanche diode noise sources are invariably used in practical set-ups. The noise power generated by such a source is defined in terms of the excess noise ratio (ENR) which for an ideal source is defined by

$$\text{ENR} = E = \frac{T_2}{290} - 1 \tag{3.95}$$

where E is the available excess noise ratio. For a source with a reflection coefficient Γ_s

$$E = \left(\frac{T_2}{290} - 1\right)(1 - |\Gamma_s|^2) \tag{3.96}$$

3.6.2.7 Effect of mismatch

So far we have defined all noise parameters in terms of available power, blithely ignoring the effects of mismatch. This is one of the bugbears of noise measurement, and affects the noise power delivered, the measured device noise figure and the available transistor gain. However, this problem is treated thoroughly in refs 3.109–3.112, and so will not be considered further here.

3.6.3 Noise measurement—experimental set-ups

Accurate experimental characterization is the key to low noise amplifier design. In the design process, noise measurement is performed at the start, in order to select the best device and measure its optimum noise impedance, and at the end, in order to check the noise figure obtained from the amplifier. To introduce practical noise-measurement benches, we shall invert the procedure and start with amplifier characterization since the difficulty of measurement increases substantially when a low noise device is being characterized.

3.6.3.1 Noise figure measurement for a MESFET amplifier

A typical low noise MESFET amplifier/device noise-measurement test bench is shown in Fig. 3.44, and the critical points that must be checked for accurate noise figure characterization of state-of-the-art amplifiers are discussed in refs 3.110 and 3.112. Solid state noise sources with an ENR of 15–16 dB are used to measure F values of 3 dB or above, while for F lower than 1.5 dB (T_e < 120 K) a hot–cold noise source should be used. Such sources are commercially available for use at frequencies up to 26 GHz.

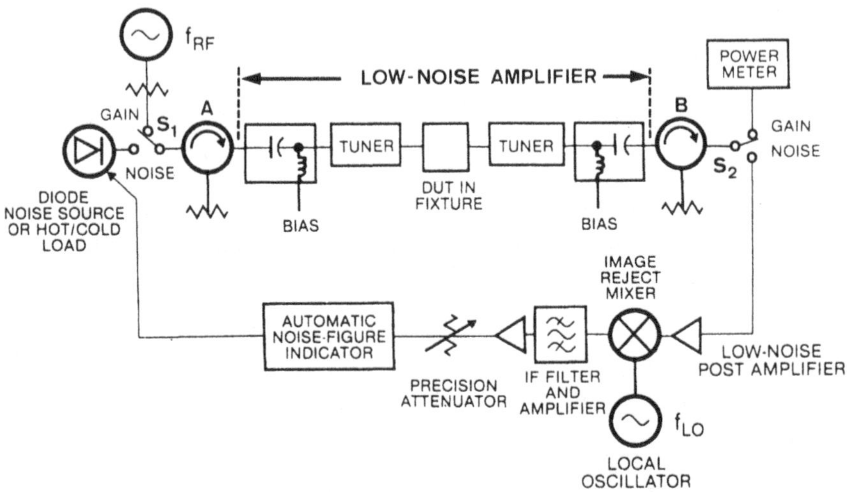

Fig. 3.44 Classical test bench for MESFET noise figure measurement.

The key elements in the test bench are the high quality low loss isolators located before and after the amplifier under test and the low noise post-amplifier. The role of the post-amplifier is to reduce the effect of the noise figure of the receiver chain and thereby reduce the correction necessary to the overall measured noise figure by the application of eqn (3.85). The high quality isolators, which have low terminal VSWRs, reduce the effects of mismatch errors during both the calibration procedure and the actual measurement (Section 3.6.2.7): the mixer always sees the same impedance in the calibration–measurement sequence. This sequence is as follows.

The level of the local oscillator drive to the mixer is optimized for minimum noise. A double sideband mixer can be used for measurements well inside the amplifier bandwidth, provided that the i.f. selected ensures identical gain and noise power distribution at the signal and image sidebands. For enhanced accuracy an i.f. of 10 MHz should be used, particularly at the band edges of the amplifier where there may be a rapid fall-off of gain with frequency. Such a low i.f. is even more important in the case of device matching with external tuners in which the farthest tuning element is a long way from the device, which is undesirable but sometimes unavoidable. Alternatively, a single sideband response can be achieved with an image-reject mixer or an image filter in front of the mixer [3.110]. Calibration is effected by measuring the noise figure F_c of the measurement chain following isolator B. The ENR in this measurement is that of the noise source plus the input isolator loss in decibels (the effects of the isolator VSWR are neglected). The amplifier is now inserted between the isolators and the total noise figure is read. Finally, the gain of the amplifier is measured by switching to the signal generator and power meter via switches S_1 and S_2. Application of eqn (3.85) yields the amplifier noise factor.

The gain and the noise figure can also be obtained directly, without using the signal generator and the power meter, by following the Y factor procedure outlined in Table 3.5 [3.113]. The latest commercial microprocessor-based noise figure set-ups incorporate this procedure automatically.

Table 3.5

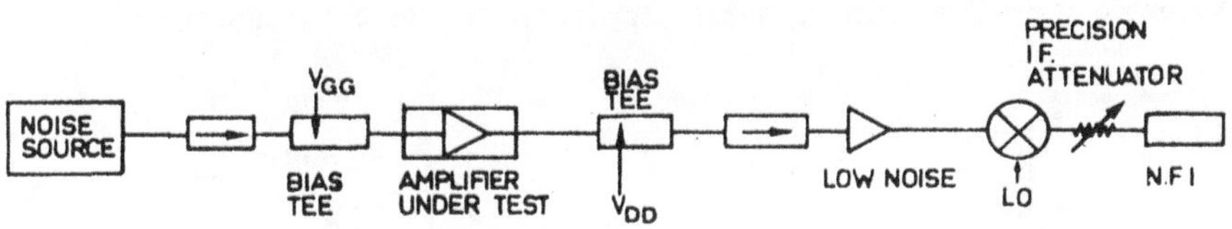

(1) Measure the mixer chain noise figure F_c.

(2) Set the gain on the noise figure indicator (NFI) to an arbitrary level using the i.f. attenuator. *Noise source off.* Note attenuator setting R_1.

(3) Insert the amplifier under test. Turn on the amplifier and adjust the i.f. attenuator to give the previous NFI reading. New attenuator setting R_2. *Noise source off.*

(4) *Noise source on.* Re-adjust i.f. attenuator to give the previous NFI reading. New attenuator setting R_3.

(5) Check for drift by turning amplifier off. Conditions (2) should return: i.f. attenuator setting R_1:

$$Y_0 = (R_2 - R_1)\ \text{dB} \qquad Y_1 = (R_3 - R_1)\ \text{dB}$$

Convert all dBs to ratios:

$$\text{amplifier gain } G = F_c \frac{Y_1 - Y_0}{\text{ENR}} \tag{3.97}$$

$$\text{noise figure } F = F_c \frac{(Y_0 - 1) + 1}{G} \tag{3.98}$$

The noise source ENR should be corrected for ambient temperature variation about 290 K (∓ 0.13 dB for ± 10 K variation).

3.6.3.2 Noise figure measurement for a MESFET device

Although the basic measurement technique is the same as that for low noise amplifiers and the test bench shown in Fig. 3.44 can be used, several additional complications arise in noise figure measurements on transistors.

(1) In addition to the minimum noise figure, we need to know the optimum noise source impedance Y_{min}.

(2) The available gain of the device is lower and is more sensitive to variation in the VSWR of the noise source. Thus a low loss isolator with good isolation and low terminal VSWR must be incorporated into the test bench input port. In addition, the effect of the second-stage noise figure becomes critical. The effect of the second-stage noise figure on the measurement of the optimum source noise impedance is neatly illustrated in ref. 3.114 from which Fig. 3.45 is reproduced. In this figure a typical 1 μm MESFET is tuned at a single frequency by a fixed-value capacitor placed on a 50 Ω line at varying electrical lengths θ from the gate of the device. It can be seen that the value of optimum θ and the noise figure indicated are critically dependent on the device gain and the second-stage noise figure. Application of Friis formula can correct for the second-stage noise figure but not for the error in the optimum source impedance Y_{min}. The use of a low noise, high gain post-amplifier is therefore of primary importance in very low noise device measurement.

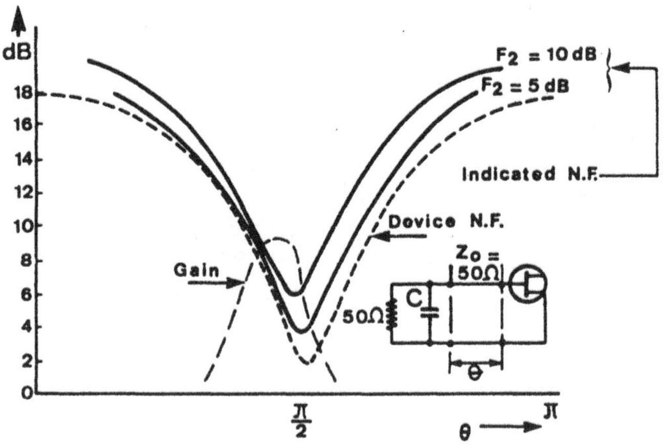

Fig. 3.45 Optimization of the noise figure by input impedance matching depends strongly on the equivalent noise figure F_2 of the subsequent stages.

(3) The insertion loss and the matching loss pose great problems with regard to transitions and tuners respectively. Although the noise figure contribution of adaptors and transitions can be allowed for by adding their loss (in decibels) to the ENR of the noise source, it is first necessary to measure

their loss and to ensure low terminal VSWRs (better than 1.1:1 for precision noise measurements). Transitions and adaptors can be characterized with great accuracy using the techniques described in Chapter 2, Section 2.2.5. Matching losses are minimized by using low loss tuners and by matching as close to the device as possible. One technique consists in using a transistor packed device mounted in a microstrip test fixture with either a substrate of low dielectric constant or air dielectric transmission lines. If the lengths of these 50 Ω lines at the input and output MESFET ports are slightly longer than half a wavelength at the lowest measurement frequency, matching can be effected directly in the test fixture by using metal disks (or slivers of dielectric) on (or under) the lines so as to obtain the optimum noise figure and associated gain. The noise figure and available gain of the matched device, corrected for losses in the coaxial–microstrip transitions, is now measured as for an amplifier. The matching elements are fixed, and the input and output impedances they present to the GaAs MESFET are obtained by following the matching network peeling procedure outlined in Chapter 2, Section 2.2.4, using a network analyzer. An alternative test fixture allows the transistor to be mounted directly in a transmission line fixture incorporating tuning slugs, thereby avoiding unnecessary transitions. It should be noted that a cover must always be used for air-line or low dielectric constant microstrip fixtures so that the device is not perturbed by stray ground planes, e.g. the operator's hand. It is also important to check that the MESFET is stable before proceeding to measure its noise figure.

(4) Finally, to avoid the effects of spurious signals on noise measurement it is important to eliminate sources of spurious signals (such as digital voltmeters), to work in a Faraday cage and to filter mains supplies. For very low noise measurement it is also necessary to correct the noise source ENR for the difference between ambient temperature and 290 K (the ENR change is typically 0.01 dB K^{-1} for a solid state noise source).

3.6.3.3 Noise parameters

As indicated in the previous section, the overall noise figure is measured in all device noise figure measurements, and in the best practical case the minimum noise figure obtained is M_{min}. This corresponds to a second-stage noise figure identical with that of the first stage. In order to measure F_{min} of a transistor directly it is necessary to use a noiseless infinite-gain post-amplifier, which is clearly impossible. The only way to correct rigorously for the second-stage noise figure and to obtain F_{min} is to carry out successive noise figure and noise admittance measurements in order to characterize the GaAs MESFET completely and obtain the four noise parameters F_{min}, R_n, G_{min} and B_{min} (eqn (3.90)). In practice the noise figure corresponding to at least seven different values of the source admittance Y_s is measured on an automated test bench (Fig. 3.46) so as to reduce measurement time and tedium. Such a bench requires the following components:

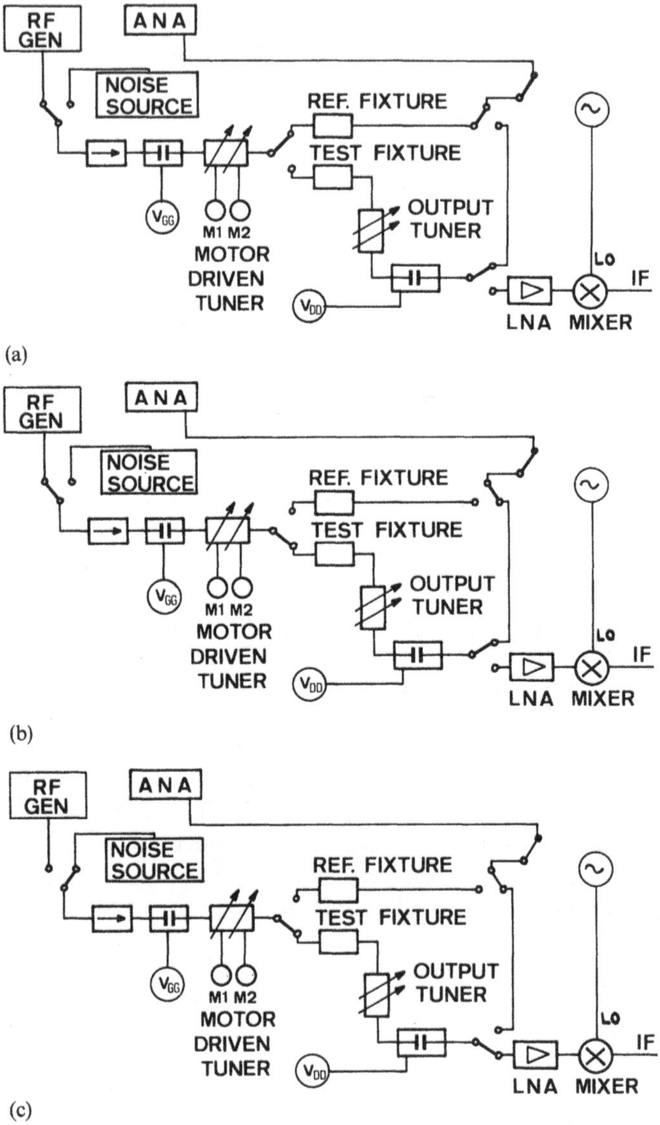

Fig. 3.46 Automatic measurement test bench for noise parameters: (a) impedance measurement; (b) gain measurement; (c) noise measurement. The automatic network analyzer and computer control switches acquire data and compute noise parameters.

(1) two frequency-synthesized r.f. sources, one for amplifier gain and tuner impedance measurements and the other for the local oscillator drive into the mixer; an automatic network analyzer for gain and noise source impedance measurement;

(2) a noise figure indicator;
(3) a programmable tuning device—a motor-driven double-slug tuner [3.115], an automatic slotline [3.116], a varactor tuner which can be tuned electronically [3.117] or more simply by physically connecting in lands to a 50 Ω microstrip line [3.118];
(4) electrically operated single-pole double-throw switches to permit a measurement sequence to be effected which includes noise figure measurement, matched device gain measurement and noise source impedance measurement;
(5) a desk-top calculator to pilot the operations, effect the readings and solve the noise figure equations by either an optimization subroutine or a least-squares fitting procedure.

3.6.3.4 Constant-noise circles

Once the four noise parameters F_{min}, R_n, G_{min} and B_{min} are obtained, the noise figure of the device can be predicted for any source reflection coefficient Γ_s by a simple rearrangement of eqn (3.90):

$$F = F_{min} + 4r_n \frac{|\Gamma_s - \Gamma_{min}|^2}{(1-|\Gamma_s|^2)|1+\Gamma_{min}|^2} \tag{3.99}$$

where

$$r_n = \frac{R_n}{Z_0} = \frac{R_n}{50}$$

$$\Gamma_s = \frac{Y_0 - Y_s}{Y_0 + Y_s}$$

$$\Gamma_{min} = \frac{Y_0 - Y_{min}}{Y_0 + Y_{min}}$$

$$Y_0 = \frac{1}{Z_0} = 20\,\text{mS}$$

Equation (3.99) defines a set of circles for which an analytical expression can be obtained by defining a parameter N_i:

$$N_i = (F_i - F_{min}) \frac{|1+\Gamma_{min}|^2}{4r_n} \tag{3.100}$$

where F_i is an arbitrary noise figure greater than F_{min}.

The centers of the constant-noise-figure circles will all be aligned on a straight line drawn from the center of the Smith chart ($\Gamma = 0$). Their distance from the center is

$$C_{F_i} = \frac{\Gamma_{min}}{1+N_i} \tag{3.101}$$

and their radii are given by

$$R_{F_i} = \frac{1}{1+N_i}\{N_i^2 + N_i(1-|\Gamma_{\min}|^2)\}^{1/2} \tag{3.102}$$

Typical noise figure circles for a low noise MESFET are given in Fig. 3.47. A similar procedure can be carried out with the noise measure M. Formulae for the centers and radii of constant-M circles are given in ref. 3.119. For very low noise amplifiers, in which at least two identical minimum noise measure amplifiers are to be cascaded, M circles, which include the effect of the second-stage noise figure, are more useful than constant-F circles [3.120].

As the noise figure and the noise measure both increase with frequency the amplifier designer must judiciously select a matching circuit to give $M_{\min}$ at the highest frequency and be within the same value of the M circle at lower frequencies in order to obtain a flat noise figure across the design bandwidth.

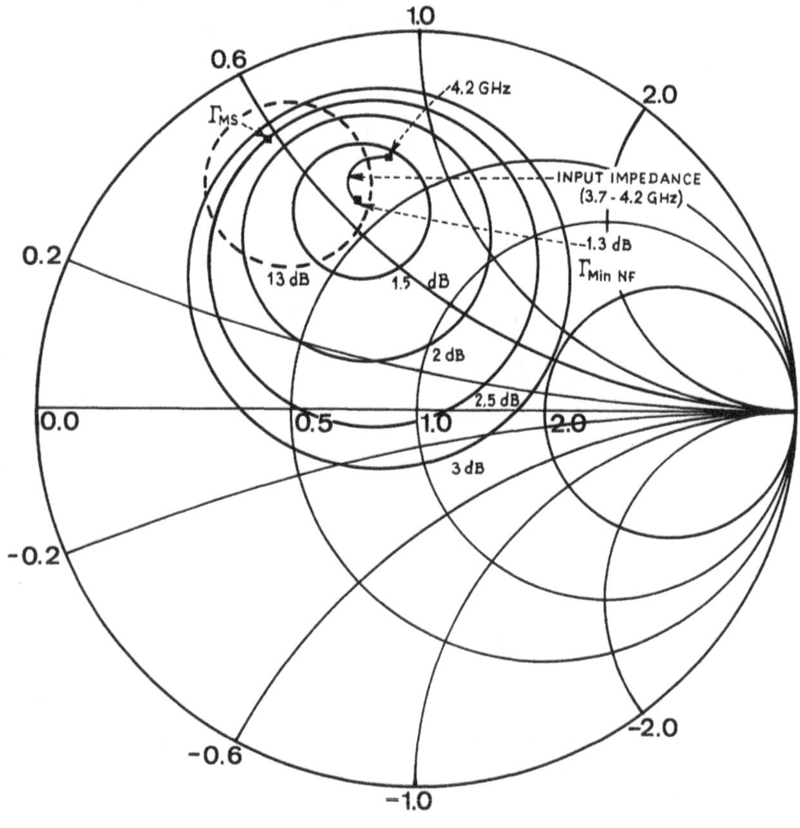

Fig. 3.47 Typical noise circles at $F = 3.7$ GHz. Such circles at different frequencies help the designer to optimize the noise impedance over a given bandwidth.

3.6.4 Examples of low noise amplifiers

We present three types of practical low noise amplifiers in this section. The first makes use of packaged low noise MESFETs covering narrow bands (typically 10%–15%). The second type makes use of transistor chips and is generally dependent on computer-aided design techniques and sensitivity analysis to achieve a repeatable low noise performance with devices that have parameter spreads. The last type is a monolithic implementation which makes use of series feedback to achieve a low input VSWR. In the first two cases the best results are obtained after a final *in situ* retuning, and in the third case they are obtained after measurement and reverse design.

The following standard procedures must always be carried out to exploit the device to its full potential:

(1) the best implementation of the transistor, e.g. low loss grounding of source contacts, launchers, microstrip ground planes etc.;

(2) the use of boxes of reduced dimensions in order to eliminate spurious propagation modes;

(3) for best low noise performance it is possible (and sometimes preferable) to mix dielectrics and to use a substrate with a low dielectric constant ε_r at the input and a substrate with higher ε_r at the output and for succeeding stages.

(4) careful selection of low loss input capacitors;

(5) the use of low loss prematching circuits as close as possible to the transistor input and output.

3.6.4.1 L band GaAs MESFET amplifier

For very low noise applications, either as a preamplifier or an i.f. amplifier, it is desirable to use GaAs MESFETs rather than bipolar transistors, despite the difficulties imposed by high input and output VSWRs and the tendency for packaged devices to oscillate at out-of-band frequencies. The addition of a lossless feedback in the form of a series source inductance reduces the input VSWR and can improve the noise performance of the GaAs MESFET leading to a wider bandwidth design [3.121, 3.122] at the expense of gain.

The noise figure performance of a MESFET can be simulated by using either a lumped electrical equivalent circuit (see Chapter 1) or a two-port model. The value of the source impedance at the required frequency can thus be optimized to give the best compromise between the parameters of gain, bandwidth, input VSWR, stability and noise figure [3.123]. In practice, with a packaged device the necessary source inductance is introduced via the source leads by grounding them at an appropriate distance from the package. This distance is easily adjustable. The input impedance, modified by the series feedback, remains an equivalent series *RC* network (*R* is increased by the feedback), and so a series inductor is required to induce resonance at the center frequency before matching. The output, which can be represented by a parallel *RC* network, is resonated by a shunt inductor which can be a coil of a few turns placed well above the ground plane (Fig. 3.48). These simple inductors present a minimum loss (ohmic and radiation) at low frequencies and are easily tunable. Protection against parasitic oscillation is effected at u.h.f. by chokes wound on a ferrite bead and at high microwave frequencies (15 GHz and above) by an *RC* ladder network in the bias circuit. Amplifiers of this type have been constructed at the Observatoire de Meudon, France, and at the National Astronomy Observatory, USA. Noise figures of less than 70 K at room temperature were obtained for a two-stage amplifier with 24 dB gain at 1.3 GHz.

Fig. 3.48 L band low noise MESFET amplifier. (Courtesy of Mr Beaudin and Mr Demirm, Observatoire de Meudon, France.)

3.6.4.2 Low noise design at 4 GHz

For ground stations, parametric preamplifiers are now being replaced by a new generation of very low noise thermoelectrically cooled HEMT and MESFET amplifiers. The cooling is limited to the first- and second-stage transistors only which are thermally insulated from the amplifier casing. Continuity of the ground plane is ensured by capacitive coupling. Such amplifiers are now capable of ensuring a maximum noise temperature below 50 K over the whole 3.7–4.2 GHz band.

The transistor is stable at 4 GHz and the source needs good grounding. At the input the best prematching circuit remains a series inductance, which reduces the magnitude of the reflection coefficient by increasing the real part of the transistor input impedance while cancelling out its imaginary component. A double-section $\lambda/4$ transmission line circuit is then calculated as in Chapter 6, Section 6.4.2.4, and is computer optimized to give a flat gain response with frequency. For practical realization the inductance is made up of the gate lead of the package, about 4.5 mm long, above a recessed ground milled into the amplifier box (Fig. 3.49). Thus a high impedance transformer with minimal losses is obtained. The double $\lambda/4$ section is realized in microstrip form. To reduce the insertion losses of the gate bias tee, d.c. bias is supplied through the isolated arm of the low loss input circulator.

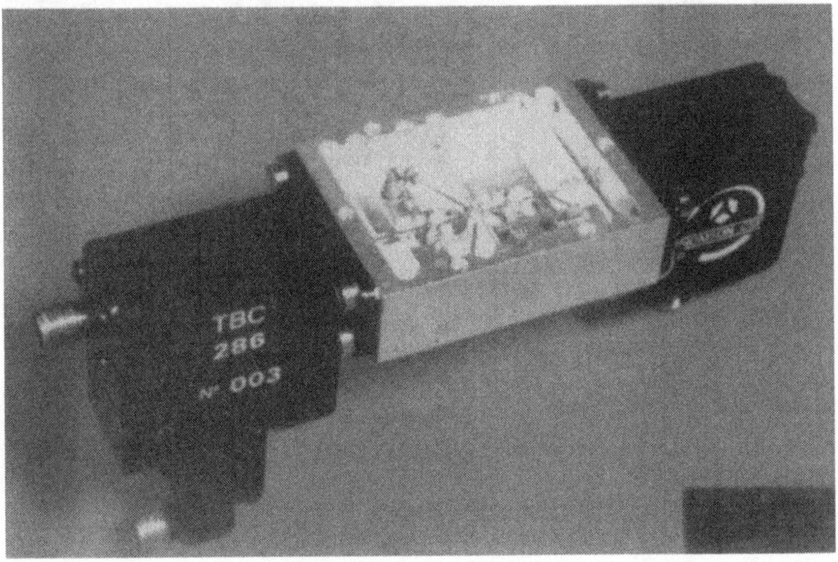

Fig. 3.49 Low noise 3.7–4.2 GHz amplifier. (Courtesy of Mr Deprez, Thomson-CSF; photograph by M.Mathieu.)

3.6.4.3 Low noise X band monolithic amplifier using series feedback

Lehmann and Heston [3.124] reported the first use of series feedback in a monolithic technology. Inductive reactance in the source lead increases the real part of the input impedance. With proper impedance loading at the output of the MESFET (see also ref. 3.125) the conjugate of the MESFET input impedance and the optimum noise match impedance coincide. The reactive series feedback decreases the transistor gain and, since the minimum noise measure is invariant to a lossless feedback, the resulting minimum noise figure also decreases [3.124].

Figure 3.50 shows this monolithic three-stage low noise amplifier which is commercially available from Texas Instruments (model TGA 8021). The circuit topology is shown in Fig. 3.51. The amplifier uses GaAs ion-implanted MESFETs with a gate length of 0.5 μm and a gate width of 300 μm and typically achieves a noise figure of 2.4 dB with a gain of over 24 dB across the 9–10 GHz band. The input and output VSWRs are better than 1.4:1 over the band. The chip dimensions are 3.1 mm × 2.4 mm.

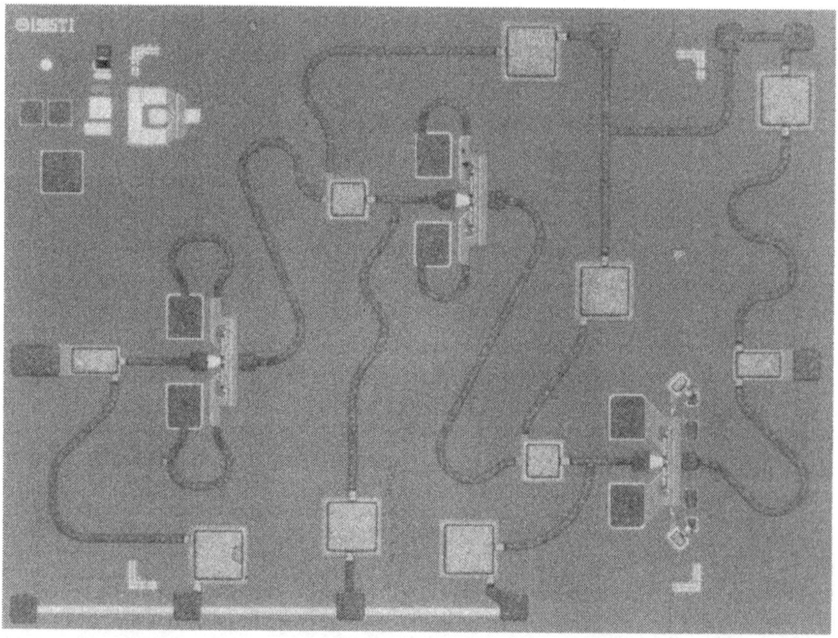

Fig. 3.50 Monolithic three-stage X band low noise amplifier. (Courtesy of Mr R. Toole, Texas Instruments Inc.)

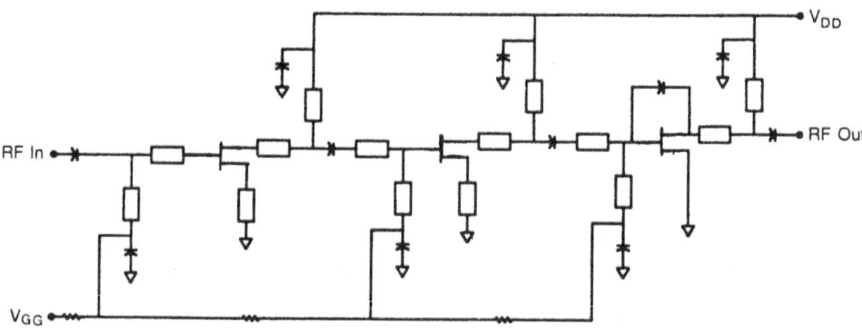

Fig. 3.51 Circuit topology of the low noise amplifier shown in Fig. 3.50. (Courtesy of Mr R. Toole, Texas Instruments Inc.)

3.6.5 Cooling of MESFET amplifiers

Let us consider the effect of temperature on the main parameters of the GaAs MESFET [3.126, 3.127].

(a) Transconductance variation: the mobility in the active channel of GaAs MESFETs increases as the temperature decreases, but the transconductance g_m of *short*-gate-length devices with high doping concentrations is inversely proportional to the square root of the temperature in kelvins because in this case the transconductance depends primarily on velocity saturation.

(b) Input capacitance: this is a second-order variation because the gate capacitance C_{GS} depends, amongst other factors, on the built-in Schottky potential V_{bi} which increases as the temperature falls, thus causing a slight decrease in C_{GS}.

(c) Resistances: in the resistance we have to separate effects due to the resistance in the semiconductor material from those due to the ohmic contact, but at low temperatures attention must be paid to resistance variations which depend on the metallurgical nature of the contacts and the type of metallization used (resistivity and thickness of the metal sandwich).

The GaAs MESFET is an ideal component for cooling: its gain increases and its noise temperature decreases as it is cooled. The effect of cooling is shown in Fig. 3.52. However, some parameters moderate this ideal tendency.

(1) The device ambient temperature is different from the internal temperature of the channel. There is a thermal resistance between the channel and the transistor casing, and this must be taken into account.

(2) Owing to high electrical fields in the channel, it must be assumed that the electron temperature is different from the real temperature of the channel.

(3) Owing to the band structure of GaAs and the high electrical fields, some electrons are transferred to the upper valley. This transfer causes random fluctuations which can be expressed as generation–recombination noise [3.128]. This effect can be modified for very short channels and with devices made up of alternating GaAs–AlGaAs structures (HEMT). As already mentioned there is only a slight variation in the input capacitance, so that the tuning at room temperature for input and output circuits is very nearly the same as at low temperatures. The increasing gain requires application of severe damping to prevent out-of-band oscillation.

The best results to date are summarized in Table 3.6. The first prize goes to the radio astronomy application which reaches an ultimate performance of 7.2 ± 2 K for a single-stage amplifier cooled to 15 K at 1.2 GHz. The results for recent applications using cooled packaged devices (with 0.5 μm gate length) at higher frequencies (up to 24 GHz) are also included in Table 3.6.

Table 3.6

Frequency (GHz)	*Stage no*	*Room temperature*		*Cooled operation*			*Reference*
		Gain (dB)	*Noise temperature* (K)	*Gain* (dB)	*Noise temperature* (K)	*Cooling temperature* (K)	
1.5	2	25	75	30	8	15	3.127
4.7	2	25	100	27	10	17	3.129
10.7	1	10	250	15	40	20	3.130
15.0	1	8	250	9.7	50	20	3.130
24.0	1	5	400	5	70	77	3.130

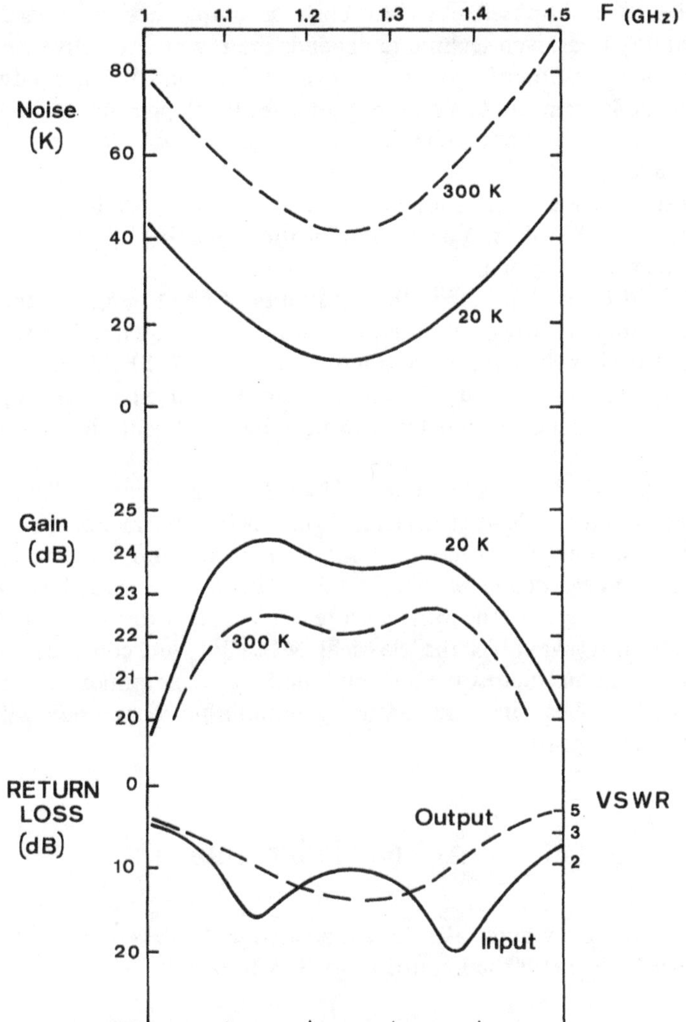

Fig. 3.52 Noise, gain and VSWR of the amplifier shown in Fig. 3.48. The effect of cooling at 20 K is shown.

3.7 Conclusions

Broadband and ultrabroadband MESFET amplifiers can be built for frequency bands extending from 10 MHz to 60 GHz. Various matching principles have been thoroughly analyzed and design approaches have been reviewed. It is impossible to establish a unique design procedure. The applicability of a given technique depends mainly on the active device to be used but also on other factors such as operating frequencies, bandwidth and technology. It seems that a combination of feedback principles with reactance matching is the best approach to obtain high gain, low VSWR and good noise performance.

Distributed amplification appears as a very promising solution for ultrabroadband medium power amplification, opening up new prospects for electronic warfare systems.

Hybrid technology offers the possibility of fine tuning the amplifier for best performance. However, monolithic technology potentially enables a true optimum design through an optimum choice of MESFET parameters and circuit topology. The circuit dimensions can be reduced as well as parasitics, and better control of parameters can be achieved during the manufacturing process.

A GaAs MESFET noise figure of 3.4 dB with an associated gain of 3.8 dB is currently achievable at 60 GHz, and further improvements are expected as well as an extension of the operating frequencies into the 90 GHz region. Laboratory devices have reached $f_T = 55$ GHz and $f_{max} = 180$ GHz. Although new devices, especially the most mature of its competitors—the HEMT—are now becoming available, the classical MESFET will continue to play an important role in microwave and millimetre-wave amplifiers. An increasing number of applications and maturing monolithic technology will help in maintaining its position.

Acknowledgments

The author would like to acknowledge Professor J. Obregon and Professor F. Perez for their contribution to Chapter 3.

References

3.1 Ha, T. T. *Solid State Microwave Amplifier Design*, Wiley, New York, 1981.

3.2 Vendelin, G. D. *Design of Amplifiers and Oscillators by the S-parameter Method*, Wiley, New York, 1982.

3.3 Bodway, G. E. Two-port power flow analysis using generalized scattering parameters, *Microwave J.*, **10**, 61, 1967.

3.4 Yarman, B. S. and Carlin, H. J. A simplified "real frequency" technique applied to broad-band multistage microwave amplifiers, *IEEE Trans. Microwave Theory Tech.*, **30**, 2216, 1982.

3.5 Yarman, B. S. A dynamic CAD technique for designing broadband microwave amplifiers, *RCA Rev.*, **44**, 551, 1983.

3.6 Abrie, P. L. D. *The Design of Impedance-matching Networks for Radio-frequency and Microwave Amplifiers*, Artech House, Dedham, MA, 1985.
3.7 Kurokawa, K. Design theory of balanced transistor amplifiers, *Bell Syst. Tech. J.*, **44** (10), 1675, 1965.
3.8 Niclas, K. B. Multi-octave performance of single-ended microwave solid-state amplifiers, *IEEE Trans. Microwave Theory Tech.*, **32**, 896, 1984.
3.9 Gonzalez, G. *Microwave Transistor Amplifiers, Analysis and Design*, Prentice-Hall, Englewood Cliffs, NJ, 1984.
3.10 Mellor, D. J. Improved computer-aided synthesis tools for the design of matching networks for wide-band microwave amplifiers, *IEEE Trans. Microwave Theory Tech.*, **34**, 1276, 1986.
3.11 Anvari, K. and Fuller, B. A. J. Computer-aided design of microwave amplifiers by the random addition of new components, *Proc. Inst. Electr. Eng., Part H*, **133**, 395, 1986.
3.12 Liechti, C. A. and Tillman, R. L. Design and performance of microwave amplifiers with GaAs Schottky-gate field effect transistors, *IEEE Trans. Microwave Theory Tech.*, **22**, 510, 1974.
3.13 Tserng, H. Q., Nelson, S. R. and Macksey, H. M. 2–18 GHz, high-efficiency, medium-power GaAs FET amplifiers, *IEEE MTT Symp. Dig.*, p. 31, IEEE, New York, 1981.
3.14 Watkins, E. T., Ho, T. Q. and Post, B. M. FET LNAs designed for advanced EHF systems, *Microwave Syst. News*, **15**, 70, 1985.
3.15 Matthaei, G. L. Synthesis of Tchebycheff impedance-matching networks, filters, and interstages, *IRE Trans. Circuit Theory*, **3**, 163, 1956.
3.16 Levy, R. Explicit formulas for Chebyshev impedance-matching networks, filters and interstages, *Proc. Inst. Electr. Eng.*, **111** (6), 1099, June 1964.
3.17 Ku, W. H. and Petersen, W. C. Optimum gain–bandwidth limitations of transistor amplifiers as reactively constrained active two-port networks, *IEEE Trans. Circuits Syst.*, **22** (6), 523, 1975.
3.18 Pitzalis, O. and Gilson, R. A. Tables of impedance matching networks with approximate prescribed attenuation versus frequency slopes, *IEEE Trans. Microwave Theory Tech.*, **19**, 381, 1971.
3.19 Chen, W. K. and Satyanarayana, C. General theory of broadband matching, *Proc. Inst. Electr. Eng., Part G*, **129** (3), 96, June 1982.
3.20 Mellor, D. J. and Linvill, J. G. Synthesis of interstage networks of prescribed gain versus frequency slopes, *IEEE Trans. Microwave Theory Tech.*, **23**, 1013, 1975.
3.21 Liu, L. C. and Ku, W. K. Computer-aided synthesis of lumped lossy matching networks for monolithic microwave integrated circuits (MMICs), *IEEE Trans. Microwave Theory Tech.*, **32**, 282, 1984.
3.22 Siddiqui, J. *Microwave J.*, **22**, 57, April 1979.
3.23 Carlin, H. J. A new approach to gain–bandwidth problems, *IEEE Trans. Circuits Syst.*, **23**, 170, 1977.
3.24 Carlin, H. J. and Komiak, J. J. A new method of broad-band equalization applied to microwave amplifiers, *IEEE Trans. Microwave Theory Tech.*, **27**, 93, 1979.
3.25 Carlin, H. J. and Amstutz, P. Broadband matching, *Ann. Telecommun.*, **36** (5–6), 281, 1981.
3.26 Carlin, H. J. and Amstutz, P. On optimum broad-band matching, *IEEE Trans. Circuits Syst.*, **28**, 401, 1981.
3.27 Carlin, H. J. and Yarman, B. S. The double matching problem; analytic and real frequency solutions, *IEEE Trans. Circuits Syst.*, **30**, 15, 1983.
3.28 Bode, H. W. *Network Analysis and Feedback Amplifier Design*, Van Nostrand, New York, 1947.

3.29 Fano, R. M. Theoretical limitations on the broadband matching of arbitrary impedances, *J. Franklin Inst.*, **249**, 57, 1950.

3.30 Youla, D. C. A new theory of broad-band matching, *IEEE Trans. Circuit Theory*, **11**, 30, 1964.

3.31 Kuh, E. S. and Rohrer, R. A. *Theory of Linear Active Networks*, Holden Day, 1967.

3.32 Chen, W. K. *Theory and Design of Broadband Matching Networks*, Pergamon, Oxford, 1976.

3.33 Pauker, V. Amplification équilibre large bande à transistors FET, *Acta Electron.*, **23** (4), 299, 1980.

3.34 Ghausi, M. S. *Principles and Design of Linear Active Circuits*, McGraw-Hill, New York, 1965.

3.35 Chien, T. M. A theory of broadband matching of a frequency-dependent generator and load—Part I: Theory, *J. Franklin Inst.*, **298**, 181, 1974.

3.36 Helton, J. W. *Proc. IEEE Conf. on Circuits and Systems, Phoenix*, p. 628, IEEE, New York, 1977.

3.37 Ha, T. T. *IEEE J. Electron. Circuits Syst.*, **3** (3), 97, May 1979.

3.38 Degout, R., Chambaudu, G. and Obregon, J. *Proc. 8th European Microwave Conf., Paris*, p. 462, Microwave Exhibitions and Publishers, Tunbridge Wells, 1978.

3.39 Basawapatna, G. R. Design and performance of a 2 to 18 GHz medium power GaAs MESFET amplifier, *Proc 8th European Microwave Conf., Paris*, p. 458, Microwave Exhibitions and Publishers, Tunbridge Wells, 1978.

3.40 Young, G. P. and Scanlan, S. O. Matching network design studies for microwave transistor amplifiers, *IEEE Trans. Microwave Theory Tech.*, **29**, 1027, 1981.

3.41 Riddle, A. N. and Trew, R. J. A broad-band amplifier output network design, *IEEE Trans. Microwave Theory Tech.*, **30**, 192, 1982.

3.42 Houng, S. G. Amplifier interstage matching network design, *Microwave J.*, **26,** 103, 1983.

3.43 Houng, S. G. Amplifier interstage matching network design using shunt-open stubs, *Microwave J.*, **26,** 123–126, 1983.

3.44 Apel, T. One-port impedance models prove useful for broadband RF power amplifier design, *Microwave Syst. News*, **14,** 97, 1984.

3.45 California Eastern Laboratories Inc., April 1986.

3.46 Davis, W. A. *Microwave Semiconductor Circuit Design*, Chapter 16, Van Nostrand Reinhold, New York, 1984.

3.47 Niclas, K. B. and Pereira, R. R. Feedback applied to balanced FET amplifiers, *Microwave Syst. News*, **10,** 66, 1980.

3.48 Mellor, D. J. and Pierson, L. G. Calculator-based synthesis routine speeds microwave amplifier design, *IEEE MTT Symp. Dig.*, p. 130, IEEE, New York, 1977.

3.49 Riddle, A. N. and Trew, R. J. Odd order impedance matching networks for low cost microwave circuits, *IEEE MTT Symp. Dig.*, p. 459, IEEE, New York, 1982.

3.50 Obregon, J., Funck, R. and Barvet, S. *IEEE Int. Solid-State Circuits Conf. Digest of Technical Papers*, p. 66, IEEE, New York, 1981.

3.51 Camargo, E., Consoni, D. and Soares, R. A. Reactance compensation matches FET circuits, *Microwaves Radiofreq.*, **24,** 93, June 1985.

3.52 Houng, S. G. A practical approach to amplifier matching, *Microwaves Radiofreq.*, **24,** 103, July 1985.

3.53 Medley, M. W. and Allen, J. L. Broad-band GaAs FET amplifier design using negative-image device models, *IEEE Trans. Microwave Theory Tech.*, **27**, 784, 1979.

3.54 Nilson, S. *et al.* Power amplifier leads MMIC line, *Microwaves Radiofreq.*, **25,** 125–132, December 1986.
3.55 Halladay, R., Jones, M. and Nelson, S. A producible 2 to 20 GHz monolithic power amplifier, *IEEE 1987 Microwave and Millimeter-Wave Monolithic Circuits Symp. Digest of Papers*, pp. 19–21, IEEE, New York, 1987.
3.56 Niclas, K. B. On design and performance of lossy match GaAs MESFET amplifiers, *IEEE Trans. Microwave Theory Tech.*, **30**, 1900, 1982.
3.57 Villar, J. C. and Perez, F. Graphic design of matching and interstage lossy networks for microwave transistor amplifier, *IEEE Trans. Microwave Theory Tech.*, **33,** 210, March 1985.
3.58 Niclas, K. B., Wisler, W. T., Gold, R. B. and Hitchens, W. R. The matched feedback amplifier: ultrawide-band microwave amplification with GaAs MESFETs, *IEEE Trans. Microwave Theory Tech.*, **28**, 285, 1980.
3.59 Niclas, K. GaAs MESFET feedback amplifiers, design considerations and characteristics, *Microwave J.*, **23**, 39 – 48, 1980.
3.60 Perez, F. and Ortega, V. A graphical method for the design of feedback networks for microwave transistor amplifiers: theory and applications, *IEEE Trans. Microwave Theory Tech.*, **29**, 1019, 1981.
3.61 Pavio, A. M. A network modeling and design method for a 2–18 GHz feedback amplifier, *IEEE MTT Symp. Dig.*, p. 162, IEEE, New York, 1982.
3.62 Niclas, K. B. and Wilser, W. T. A 2–12 GHz feedback amplifier on GaAs, *IEEE MTT Symp. Dig.*, p. 356, IEEE, New York, 1981.
3.63 Niclas, K. B. Noise in broad-band GaAs MESFET amplifiers with parallel feedback, *IEEE Trans. Microwave Theory Tech.*, **30**, 63, 1982.
3.64 Ahlgren, D. J. and Ku, W. H. Gain–bandwidth properties of a class of matched feedback amplifiers, *IEEE Trans. Microwave Theory Tech.*, **35**, 361, 1987.
3.65 Petersen, W. C., Decker, D. R., Gupta, A. K., Dully, J and Ch'en, D. R. A monolithic GaAs 0.1 to 10 GHz amplifier, *IEEE MTT Symp. Dig.*, p. 354, IEEE, New York, 1981.
3.66 Jutzi, W. A MESFET distributed amplifier with 2 GHz bandwidth, *Proc. IEEE*, **57** (6), 1195–1196, June 1969.
3.67 Ayasli, Y., Worhans, J., Mozzi, R. and Reynolds, L. Monolithic GaAs travelling-wave amplifiers, *Electron. Lett.*, **12**, 413–414, 1981.
3.68 Pengelly, R. S. Application of feedback techniques to the realization of hybrid and monolithic broadband low-noise and power GaAs FET amplifiers, *Electron. Lett.*, **17**, 798–799, 1981.
3.69 Obregon, J., Funck, R. and Barret, S. A 150 MHz–16 GHz FET amplifier, *Int. Solid-State Circuits Conf.*, pp. 66–67, IEEE, New York, 1981.
3.70 Honjo, K. and Takayama, Y. GaAs ultrabroadband amplifiers for 6 bit/s data rate systems, *IEEE Trans. Microwave Theory Tech.*, **29**, 629–636, 1981.
3.71 Perez, F. and Obregon, J. Low frequency limitations of ultrabroadband matched microwave amplifiers, *Electron. Lett.*, **18**, 31–33, 1982.
3.72 Mamodaly, N., Quentin, P., Dueme, P. and Obregon, J. 100 MHz to 17 GHz dual-gate variable-gain amplifier, *IEEE Trans. Microwave Theory Tech.*, **30**, 918–919, 1982.
3.73 Leikuwitz, X. X. Transistor AC amplifier uses multiple feedback, *Electronics*, 84–85, May 1958.
3.74 Perez, F. and Ortega, V. Diseno de amplificadores encadenables aplicacion a CATV en la banda 40–860 MHz, *Mundo Electron.*, 1980.
3.75 Cheadle, D. L. Cascadable amplifiers, *Tech-Notes (Watkins-Johnson Co.)*, **16** (1), 1979.
3.76 Ulrich, E. Use negative feedback to slash wideband VSWR, *Microwaves*, **17,** 66–70, 1978.

3.77 Perez, F. and Ortega, V. A 0.15–12 GHz matched feedback amplifier using commercially available FETs, *IEEE Trans. Microwave Theory Tech.*, **30**, 1289, 1982.

3.78 Niclas, K. B. and Wilser, W. T. A 2–12 GHz feedback amplifier on GaAs, *IEEE Int. Microwaves Conf.*, pp. 132–134, IEEE, New York, 1981.

3.79 Estreich, D. B. A wideband monolithic IC amplifier, *IEEE Int. Solid-State Circuits Conf.*, pp. 194–195, IEEE, New York, 1982.

3.80 Perez, F., Obregon, J. and LeBrun, M. Conception et limitation aux basses fréquences d'amplificateurs adaptés à très large bande, *Journ. Natl. Microondes, Toulouse*, 1982.

3.81 Pavio, A. M., McCarter, S. D. and Saunier, P. A monolithic multistage 6–18 GHz feedback amplifier, *IEEE Microwave and Millimeter-Wave Monolithic Circuits Symp.*, p. 45, IEEE, New York, 1984.

3.82 Beall, J. M., Nelson, S. R. and Williams, R. E. Design and process sensitivity of a two-stage 6–18 GHz monolithic feedback amplifier, *IEEE Trans. Microwave Theory Tech.*, **33**, 1567, 1985.

3.83 Rigby, P. N., Suffolk, J. R. and Pengely, R. S. Broadband monolithic low-noise feedback amplifiers, *IEEE MTT Symp. Dig.*, p. 41, IEEE, New York, 1983.

3.84 Archer, J., Weidlich, H., Pettenpaul, E., Petz, F. and Juber, J. A GaAs monolithic low-noise broad-band amplifier, *IEEE J. Solid-State Circuits*, **16** (6), 648–652, 1981.

3.85 Beyer, J. B., Prasad, S. N., Becker, R. C., Nordman, J. E. and Hohenwarter, G. K. MESFET distributed amplifier design guidelines, *IEEE Trans. Microwave Theory Tech.*, **32**, 268, 1984.

3.86 Chase, E. M. and Kennan, W. A power distributed amplifier using constant-*R* networks, *IEEE MTT Symp. Dig.*, p. 811, IEEE, New York, 1986.

3.87 Niclas, K. B., Wilser, W. T., Kritzer, T. R. and Pereira, R. R. On theory and performance of solid-state microwave distributed amplifiers, *IEEE Trans. Microwave Theory Tech.*, **31**, 447, 1983.

3.88 Ayasli, Y., Reynolds, L. D., Mozzi, R. L. and Hanes, L. K. 2–20 GHz GaAs travelling-wave power amplifier, *IEEE Trans. Microwave Theory Tech.*, **32**, 220, 1984.

3.89 Pauley, R. G., Asher, P. G., Schellenberg, J. M. and Yamasaki, H. A 2 to 40 GHz monolithic distributed amplifier, *IEEE GaAs IC Symp.*, p. 15, IEEE, New York, 1985.

3.90 Niclas, K. B., Remba, R. D., Pereira, R. R. and Cantos, B. D. The declining drain lengths circuit—a computer derived design concept applied to a 2–26.5 GHz distributed amplifier, *IEEE Trans. Microwave Theory Tech.*, **34**, 427, 1986.

3.91 Riaziat, M., Zubeck, I., Bandy, S. and Zdasiuk, G. Coplanar waveguides used in 2–18 GHz distributed amplifier, *IEEE MTT Symp. Dig.*, p. 337, IEEE, New York, 1986.

3.92 Gamand, P. A complete small size 2 to 30 GHz hybrid distributed amplifier using a novel design technique, *IEEE MTT Symp. Dig.*, p. 343, IEEE, New York, 1986.

3.93 Ayasli, Y., Mozzi, R. L., Vorhaus, J. L., Reynolds, L. D. and Pucel, R. A. A monolithic GaAs 1–13 GHz travelling-wave amplifier, *IEEE Trans. Microwave Theory Tech.*, **30**, 976, 1982.

3.94 Becker, R. C. and Beyer, J. B. On gain–bandwidth product for distributed amplifiers, *IEEE Trans. Microwave Theory Tech.*, **34**, 736, 1986.

3.95 LaRue, R. A., Bandy, S. G. and Zdasiuk, G. A. A 12 dB high-gain monolithic distributed amplifier, *IEEE Trans. Electron Devices*, **33**, 2073, 1986.

3.96 Niclas, K. B. and Pereira, R. R. The matrix amplifier: a high-gain module for multioctave frequency bands, *IEEE Trans. Microwave Theory Tech.*, **35**, 296, 1987.

3.97 Ayasli, Y. Decade bandwidth amplification at microwave frequencies, *Microwave J.*, **27**, 71, 1984.

3.98 Swanson, K. W. Millimeter-wave transistors: the pseudomorific HEMT, *Microwaves Radiofreq.*, **26,** 139, March 1987.

3.99 Pustai, J. Millimeter-wave transistors: the permeable-base transistor, *Microwaves Radiofreq.*, **26,** 173, March 1987.

3.100 Wiltse, J. C. The ascent to mm-waves, *Microwaves Radiofreq.*, **26,** 325, March 1987.

3.101 Lawson, J. L. and Whlenbeck, G. E. *Threshold Signals, Radiation Laboratory Series*, McGraw-Hill, New York, 1950.

3.102 Hendler, A. J. and Lebenbaum, M. T. In Sucher, M. and Fox, J. (eds), *Handbook of Microwave Measurements*, Vol. 3, p. 865, Polytechnic Institute of Brooklyn Press, Brooklyn, 1963.

3.103 IRE standards on methods of measuring noise in linear two ports, 1959, *Proc. IEEE*, **48**, 60, 1960.

3.104 Friis, H. T. *Proc. IRE*, **32**, 419, 1944.

3.105 Engberg, J. Simultaneous input power match and noise optimization using feedback, *Proc. 4th European Microwave Conf.*, p. 385, Microwave Exhibitions and Publishers, Tunbridge Wells, 1974.

3.106 Kiiss, A., MITEQ Inc. Personal communication, April 1987.

3.107 Pospieszalski, M. W. On the measurement of noise parameters of microwave two-ports, *IEEE Trans. Microwave Theory Tech.*, **34**, 456, 1986.

3.108 Soares, R. A., de Los Reyes, E. and Guena, J. *Proc. Inst. Electr. Eng., Part H*, **128**, 146, 1981.

3.109 Strid, E. *Microwave Syst. News*, **11**, 62, 1981.

3.110 Strid, E. *Microwave Syst. News*, **11**, 88, 1981.

3.111 Calandra, E. F., Martines, G. and Sannino, M. Characterization of GaAs FETs in terms of noise, gain, and scattering parameters through a noise parameter test set, *IEEE Trans. Microwave Theory Tech.*, **32**, 231, 1984.

3.112 Martines, G. and Sannino, M. A method for measurement of losses in the noise-matching microwave network while measuring transistor noise parameters, *IEEE Trans. Microwave Theory Tech.*, **35**, 71, 1987.

3.113 Cripps, S. C. and Abbott, D. A. Personal communication, 1976.

3.114 Soares, R. A. and Cripps, S. C. *Proc. 6th European Microwave Conf., Rome*, p. 19, Microwave Exhibitions and Publishers, Tunbridge Wells, 1976.

3.115 Goloubkoff, M. and Leost, J. C. Extension d'un analyseur de réseaux hyperfréquences à la mesure des paramètres de bruit, *VRSI Proc. on Measurements in Telecommunications, Lannion*, pp. 178–181, 1977.

3.116 Abbott, D. A. and Shrumer, H. V. Automatic noise figure measurements with computer control and correction, *Radio Electron. Eng.*, **52** (10), 468–474, 1982.

3.117 Lane, R. Q. Derive noise and gain parameters in 10 seconds, *Microwaves*, **17,** 53–57, August 1978.

3.118 Mitama, M. and Katoh, H. An improved computational method for noise parameter measurement, *IEEE Trans. Microwave Theory Tech.*, **27** (6), 612–615, 1979.

3.119 Fukui, H. Available power gain, noise figure, and noise measure of two-ports and their graphical representations, *IEEE Trans. Circuit Theory*, **13**, 137–142, 1966.

3.120 Eisenberg, J. A. Designing amplifiers for optimum noise figure, *Microwaves*, **12,** 36–44, April 1974.

3.121 Anastassiou, A. and Strutt, M. J. O. Effects of a source lead inductance on the noise figure of a GaAs FET, *Proc. IEEE*, **62**, 406–408, 1974.
3.122 Weinreb, S., Fenstermarcher, D. L. and Harris, R. W. Ultra low-noise 1.2–1.7 GHz cooled GaAs FET amplifiers, *IEEE Trans. Microwave Theory Tech.*, **30**, 849, 1982.
3.123 Nevin, L. and Wong, R. L-band GaAs FET amplifier, *Microwave J.*, **22** (4), 82, 1979.
3.124 Lehmann, R. E. and Heston, D. D. X-band monolithic series feedback LNA, *IEEE Trans. Microwave Theory Tech.*, **33**, 1555, 1985.
3.125 Närhi, T. Use internal feedback to drop amplifier VSWR, *Microwaves Radiofreq.*, **25,** 71, April 1986.
3.126 Leost, J. C. and Lorious, B. Propriété à basse température des transistors GaAs à effet de champ hyperfréquences, *Onde Electr.*, **54** (10), 514–522, 1974.
3.127 Weinreb, S. Low-noise cooled GASFET amplifiers, *IEEE Trans. Microwave Theory Tech.*, **28** (10), 1041, 1980.
3.128 Baechtold, W. Noise behavior of GaAs field-effect transistors with short gate lengths, *IEEE Trans. Electron. Devices*, **19**, 674–680, 1972.
3.129 Beaudin, M. and Demerm, M., Observatoire de Meudon, France. Personal communication, 1982.
3.130 Tomasetti, G., Ambrosini, R. and Sinigaglia, G. An original coaxial structure allows operation of packaged MESFET at top noise and gain performance up to 24 GHz, *Alta Freq.*, **51** (3), 1982.

Chapter 4

Linear Simulation and Computer-aided Engineering

YVES-MICHEL MARTI

The purpose of this chapter is to present clearly the main tools of computer-aided engineering (CAE) for microwaves, with emphasis on monolithic microwave integrated circuits (MMICs).

Microwave technology presents many difficulties. It is of interest to compare two state-of-the-art monolithic circuits: an X band amplifier with less than a dozen transistors, and a 32-bit microprocessor with several hundred thousand transistors. Because of modelling difficulties there has been a reluctance to adopt CAE techniques in the field of microwaves. Only a minority of engineers in this field used simulation software (simulators) before 1980. This situation has changed dramatically in the last few years with the availability of powerful software and cheap computers. The pace of change is accelerating quickly with the introduction of CAE techniques from the domains of very-large-scale integration and computer-aided software engineering.

To some extent the outline of the chapter follows the design process, as CAE techniques are presented in the following sequence: modelling, analysis, optimization, statistical analysis, layout and databases. The basic principles of optimization are presented in Section 4.4. Our aim is to present the subject as clearly as possible without unnecessary detail, since optimization has been comprehensively described in many books. In Section 4.5 we avoid complex statistical methods, as we feel that Monte Carlo analysis and worst-case optimization are largely under-used. We hope that the reader will find Section 4.6 of interest since the subject of automated layout tools, which is treated in this section, has received little attention in previous publications. We do not discuss synthesis techniques as these have been described in Chapter 3.

4.1 The computer-aided engineering approach

Although computers have been used to solve electrical engineering problems since they were first developed, today's explosion in CAE is the result of a combination of necessity and the rapid increase in computer power. A modern personal computer is as powerful as the large mainframes of the 1960s, and a 32-bit work station as powerful as the mainframes of the 1970s can be

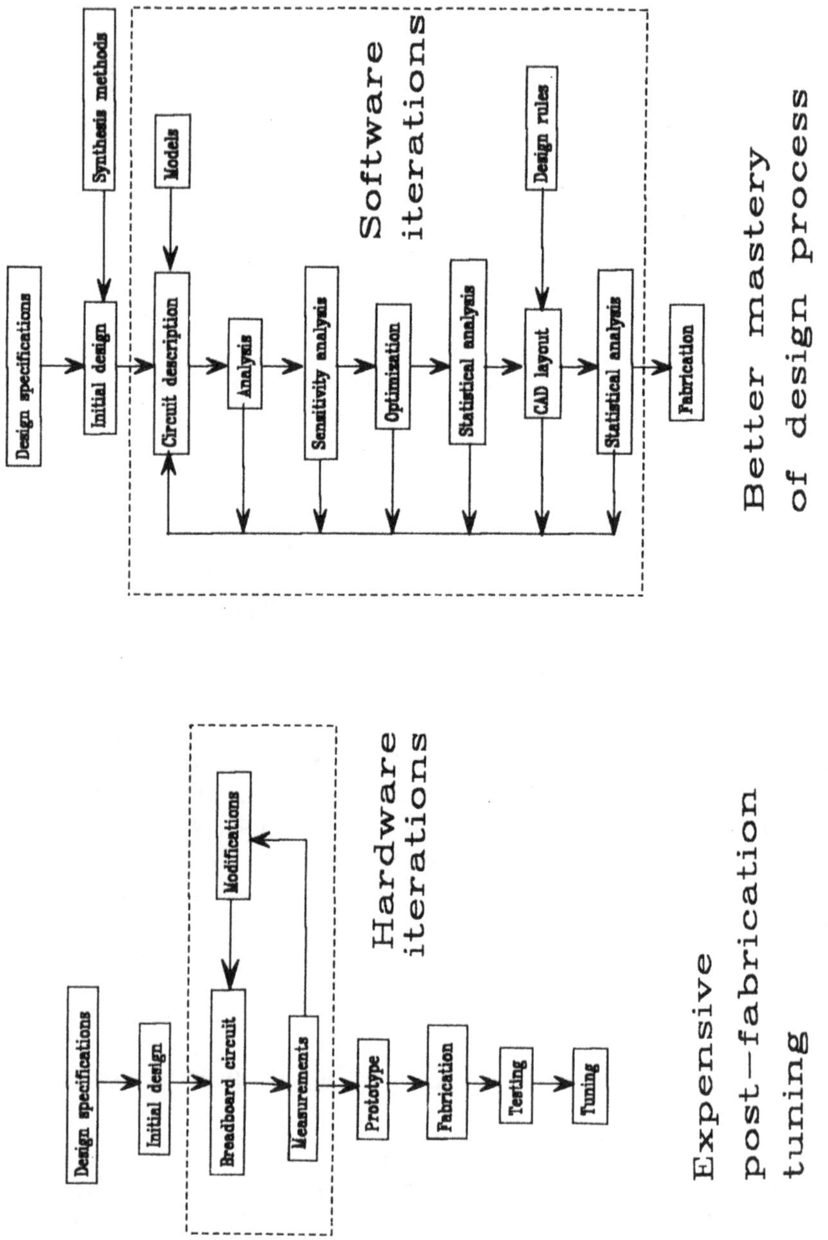

Fig. 4.1 Classical approach versus CAE approach.

obtained for about $20 000. Software tools have also evolved, and far-reaching changes have taken place in design methods (Fig. 4.1) so that design iterations can now be carried out in software rather than hardware.

CAE helps the designer to master the challenge of more complex circuits and higher frequencies, to shorten design cycles, to reduce costs by maximizing yields and to increase the reliability of systems.

4.2 Modelling of circuit elements

4.2.1 Approximation of reality

4.2.1.1 Circuit decomposition

Microstrip is widely used in both hybrid and monolithic microwave integrated circuits. Complex structures such as that shown in Fig. 4.2(a) are not unusual.

The designer needs to understand the electrical behavior of his device. Rigorous methods such as solving the Maxwell equations for the whole circuit are not realistic because of the large amount of computing power required. A more cartesian approach is traditionally used: decompose the problem into simpler elements (Fig. 4.2(b)) which can be solved separately. The decomposition is not unique because the designer can choose between different trade-offs. Logic and physical constraints should dictate the decomposition but this is often not the case: the designer will design a circuit using only elements which he is able to model. Thus in this initial step of circuit decomposition very important choices must be made that will have consequences for the rest of the design process.

4.2.1.2 Element modelling

Let us now consider the elements of the circuit. We may end up with resistive deposited layers, transmission lines, spiral inductors and so on. For example, at low frequencies a real resistor exhibits almost ideal behavior. However, at higher frequencies parasitic effects appear which can be modelled by a series inductance and a shunt capacitance (Fig. 4.3). When the intrinsic resistance R is much higher than the reactive impedance of the inductor we can

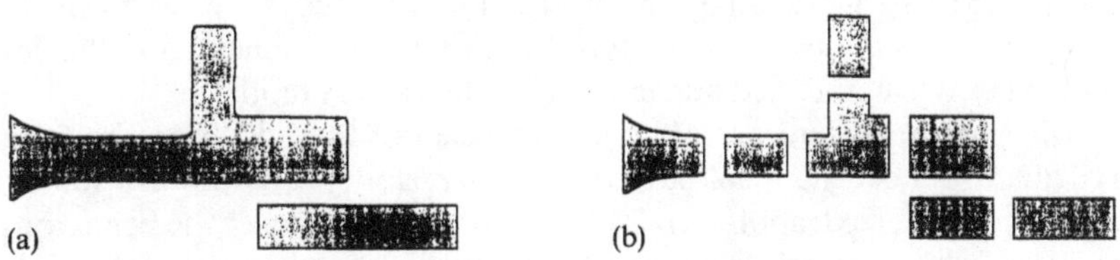

Fig. 4.2 Decomposition of a microstrip structure into elements.

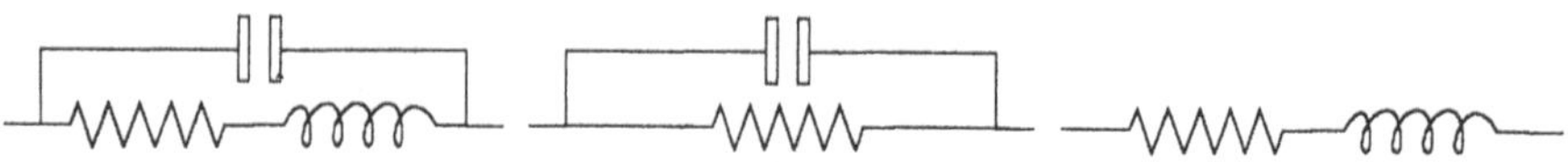

Fig. 4.3 Models of high frequency resistors: (a) full model; (b) simplified model ($R \gg L\omega$); (c) simplified model ($R \ll 1/C\omega$).

simplify the model by neglecting the parasitic inductance, and when the reactive impedance of the capacitor is much higher than the intrinsic resistance R we can neglect the parasitic capacitance. This example shows that the *frequency range* and the *parameter values* should be taken into account, even for a simple model.

4.2.1.3 Importance of models

Trade-offs have to be made in circuit decomposition and element modelling. These trade-offs must be carefully thought out because modelling is a very critical step: it is the link between physical reality and computation. This is the major difficulty for novice CAE users: the first step in the work—modelling—is the most difficult.

4.2.2 Types of model

We are concerned here with ways of translating the electrical behavior of a device into a form understandable by the computer. A computer simulator accepts basically the following types of model (in increasing order of complexity): data files, catalog models, equivalent circuits, interpreted models and user-compiled models. The choice of any of these types of model depends on the device data available to the designer, his understanding of its function and, of course, the capabilities of the simulator.

4.2.2.1 Data files

Measurement results obtained from a network analyzer, or any other suitable test bench, can be stored in a data file. Most simulators accept data files as input in the form of "black-box elements". These elements transform the data so that they can be processed by the simulator. One interesting feature of the TOUCHSTONE simulator [4.1] is its direct connection to the Hewlett-Packard 8510 network analyzer. Every time an analysis is performed the simulator orders the network analyzer to take a measurement, and the device data are then de-embedded and fed into the black-box model.

Measurement data are of course precise but they do have limitations: large quantities of data must be stored in the memory; no optimization of the device is possible; extrapolation is difficult (e.g. prediction of the behavior of a transistor with a larger gate width); extrinsic parasitic elements such as connection inductances cannot be eliminated.

4.2.2.2 Catalog models

All major commercial simulators provide a catalog of models. In fact it is one of the features that users can easily compare. There are four main groups of catalog models:

(1) ideal elements—ideal resistors, inductors, transmission lines etc.;

(2) electrical models—capacitances with a quality factor, transmission lines described by characteristic impedance, electrical length and attenuation etc.;

(3) physical models—these require input data such as physical lengths, dielectric constants, dielectric or magnetic loss tangents etc.;

(4) special-purpose elements—gyrators, gain correctors etc.

It is difficult to estimate the parameters of electrical models. The parameters of physical models are more relevant to the user, and they relate directly to process monitoring by manufacturers. However, they require accurate measurement methods (see Chapter 1).

4.2.2.3 Equivalent circuits

Nearly all designers have some physical understanding of circuits. For example, reflection coefficients that lie on the lower half of a Smith chart indicate some capacitive behavior. By using this kind of reasoning, circuits that are the electrical equivalents of the real circuit can be created (Fig. 4.4) (see Chapter 2). The use of equivalent circuit models has many advantages.

(a) The parameters have a physical meaning: C_{GS} is the capacitance between the gate and the source, g_m is the transconductance etc.

(b) Once the element values are established, the extrinsic parasitic elements such as the connection inductances are known and the behavior of the intrinsic device can be estimated.

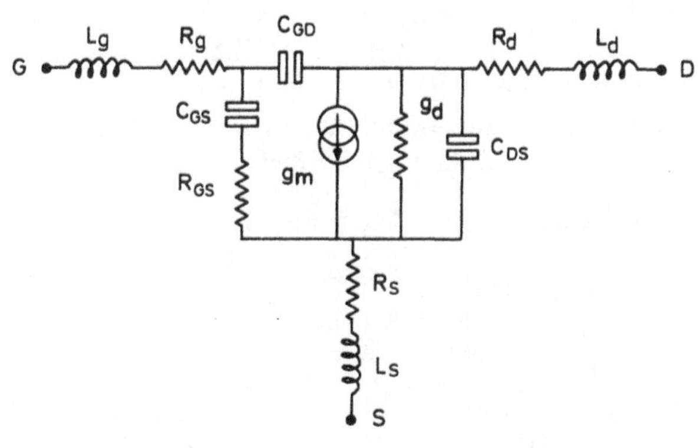

Fig. 4.4 Equivalent circuit of a MOSFET.

(c) Only a few parameters are needed to describe the device over the whole frequency range.

(d) The element values can give an insight into the manufacturing process. For example, the capacitance C_{GD} is a good measure of the quality of the passivation oxide over a MOSFET.

(e) It is possible to interpolate and even extrapolate the measurement results.

(f) The element values of equivalent circuits are often known functions of some physical or geometrical parameter. For example, for a MESFET, C_i and g_m depend linearly on the gate width W, and R_{DS} and R_s depend linearly on $1/W$. With a few sets of measurements for different gate widths it is possible to know the response of a MESFET of any width. An example is shown in Fig. 4.7 below.

The major problem with equivalent circuits is that the element values are not easy to obtain (see Section 4.2.3) and they must be recalculated from new measurement data when the technology changes.

4.2.2.4 Interpreted models

Interpreted models are defined in the description file of the simulator and are interpreted at run time. They have the advantages of being quickly written and easy to debug and maintain, while allowing the user great flexibility in defining personalized models. The following fall into the category of interpreted models: expressions, functions and parameterized circuit blocks.

Model parameters defined by arithmetical expressions Element values can be defined by expressions involving operators such as +, −, *, /. This is illustrated by the example shown in Fig. 4.5, where a circuit can be analyzed for different values of an inductance L_1 with the sum of the inductances remaining constant and equal to 12 nH.

Model parameters defined by complex mathematical functions Simulators such as ESOPE [4.2] allow the user to define functions with control structures such as IF ... THEN ... ELSE, and to call a number of FORTRAN functions such as EXP, SQRT, LOG, COS, SIN etc. In Fig. 4.6 the parasitic capacitance and inductance of a deposited resistive layer are functions of the width W and the intrinsic resistance R.

Parameterized circuit blocks Most simulators allow the user to define a

```
BLK
  IND 1 2 L=L1
  IND 2 3 L=12NH - L1
  SELF22 : 2POR 1 3
END
```

Fig. 4.5 Element values defined by expressions.

```
$FONC
A(W) = IF (W<=30MC) THEN ALOG ( 240/W + W/120 )
                    ELSE 6.3 / ((W/30)+1.4+0.67*ALOG((W/30+1.44))
E(W) = 6.75N + 5.8*SQRT(1+360/W)
Lresi (R,W) = (W*R/170) * A(W)
Cresi (R,W) = (W*R/5.1) * E(W) / A(W)
```

Fig. 4.6 Element values defined by functions.

block of circuit elements, to label it and to call it several times in the rest of the circuit. These blocks are defined once and for all, and they only allow for the repetition of identical cells. A more powerful description is possible with the ESOPE simulator where parameters can be re-initialized when calling the blocks. These blocks are called parameterized blocks because they can be compared with FORTRAN subprograms where parameters are re-initialized. Figure 4.7 shows part of the description file of a MESFET model. The MESFET is modelled as an equivalent circuit where the element values are functions of the normalized gate width *w*.

When all these features are found in a simulator, they allow versatile applications such as distributed FET models, temperature- and bias-dependent FET models, distributed amplifiers, biased subcircuits etc. Interpreted models are difficult to manage because multiple versions exist in different description files and not all of them are up to date. Furthermore, there is not much that the user can do when convergence problems arise.

4.2.2.5 User-compiled models

Software companies that produce commercial simulators cannot always cope with all the needs of their users. Many users have very specific needs, and

```
$FONC
  ci(w)  = ci0  * w
  ri(w)  = ri0  / w
  cdg(w) = cdg0 * w
  gm(w)  = gm0  * w
  cds(w) = cds0 * w
  rds(w) = 1/(gds0*w)
$MODEL
  tec  (g,d,s)    w              ! Model name, ports, calling parameter
  ci1  (g,g1)     ci(w)
  ri1  (g1,s)     ri(w)
  cdg1 (g,d)      cdg(w)
  j11  (d,s,g,s)  gm(w) tau      ! voltage controlled current source
  cds1 (d,s)      cds(w)
  rds1 (d,s)      rds(w)
$CIRC
  tec1 (1,2,1) w=50              ! calling tec model with widths
  tec2 (3,4,1) w=75              ! w=50, w=75, w=100.
  tec3 (5,6,1) w=100
```

Fig. 4.7 MESFET model parameterized by gate width.

developing special models for them would only benefit a small number of clients. These users may find that the modelling capabilities of their simulator are inadequate: a good example of this is if they want to use a powerful scientific subprogram such as matrix inversion. Moreover, some knowledgeable users develop their own models and want to keep them confidential.

Some simulators meet these needs by allowing users to code their own models in a high level language. The basic steps for introducing this kind of model into a simulator are as follows [4.3].

(1) Derive the analytical equations of the model in terms of the Z, Y or S matrix.

(2) Code the model, usually in the same high level language as that used by the simulator (PASCAL, FORTRAN, C etc.).

(3) Compile the subroutine and link it to the simulator.

(4) TEST!!! If necessary, iterate steps (1)–(4).

User-compiled models are very attractive for many reasons: they are fast (two or three times faster than interpreted models); sophisticated external mathematical subprograms can be used (finding the zeros of an equation, finding the eigenvalues of a matrix, inverting a matrix, optimization routines, splines etc.); confidentiality is assured; the management of models is easier as there is only one copy of the model and it is built into the simulator. However, there are some disadvantages: data checking (division by zero, taking the logarithm of negative numbers etc.), testing and documentation must be done by the user; the program must be linked again for each new version of the model; these models can be used in sensitivity analysis and optimization, but for efficient performance it is necessary to derive the differential expressions of the model which can be very difficult and may require special tools such as symbolic equation solvers.

In conclusion, user-compiled models are very efficient but they should be managed by a CAE specialist working closely with the microwave engineers.

4.2.3 Estimation of model parameters

The estimation of model element values is often a challenge to the CAE user. It is not an easy task, as is shown by the caution with which experienced microwave engineers approach constructors' data. Measurement methods abound, and most are specific to a particular model [4.4–4.6]. There are also many numerical methods, but they are beyond the scope of this chapter: conformal mapping in cases presenting some symmetry [4.7], finite-difference methods (ref. 4.8, p. 53), finite-element methods [4.9, 4.10] and image methods using Green functions [4.11, 4.12]. We shall discuss only two methods: optimization and extraction protocols.

4.2.3.1 Fit by optimization

Let us assume that we have a set of S parameters measured over some range of frequencies. We want to set up a computer model of the device and use it to estimate the S parameters at frequencies where we do not have measured data. An optimization method can be used (see Section 4.4) which acts on the model parameters until the model response fits the measurements. All the equivalent circuit parameters are extracted simultaneously. It is a very general method that can be used for any kind of model. This procedure was employed for the equivalent circuit derived in Chapter 2.

Convergence can be very fast when a good set of starting values is used, but this very appealing method has some drawbacks.

(a) Some parameters have little influence on the circuit response. They can vary widely and reach nonphysical values without altering the accuracy of fit.

(b) Model parameters are not always independent. This can lead to values which are nonphysical even though they form the best set of fitting parameters. An annoying side effect of dependent model parameters is the much longer optimization.

(c) Optimization time increases substantially with the number of variables. In practice, current simulators do not handle more than ten variable parameters efficiently.

4.2.3.2 Extraction protocols

In order to overcome the limitations of rigorous optimization, more intelligent methods using "divide and rule" strategies have been proposed. For example, the following physical observations can be made for an FET: R_{DS} and C_{DS} have a strong influence on S_{22}, C_{GS} mainly influences S_{11} etc. Instead of extracting all the equivalent circuit parameters at once, parameters can be extracted in groups related to a particular circuit response. The method presented in Table 4.1 has proved to be very efficient [4.13]. It should be noted that the order of the extraction steps is important.

Table 4.1 Example of extraction protocol

Step	*Elements optimized*	*S parameter fitted*	*Frequency range*
1	R_{DS}, C_{DS}	S_{22}	*All*
2	C_{GS}	S_{11}	*All*
3	C_{DG}, R_s	S_{12}	*All*
4	g_m	S_{21}	*All*
5	R_d, L_d	S_{22}	Upper half-band
6	R_g, R_i, L_g	S_{11}	Upper half-band
7	L_s	S_{12}	Upper half-band
8	τ	S_{21}	Upper half-band

Alternativèly, some parameters can be obtained by d.c. or low frequency MESFET measurements (Chapter 1) and then kept constant. Others are calculated from low frequency (1–2 GHz) r.f. measurements (Chapter 2).

4.2.4 Caution with models

As we have said above, modelling is a critical step and we intend here to give some practical advice.

4.2.4.1 Outliers

Designers sometimes obtain odd points called "outliers" in their data. From a statistical point of view, it is wrong to remove an outlier from a set of experimental points. The correct action is either to repeat the measurement until no outlier appears or to make a large number of measurements and take the average value.

4.2.4.2 Validity domain of models

Many of today's catalog models are based on research performed a few years ago. However, the technology has evolved and some of these models are no longer relevant. A good example is the change from hybrid circuits to GaAs integrated circuits (Fig. 4.8): the physical dimensions of the microstrip transmission lines have changed and the classical physical models no longer hold.

4.2.4.3 Choosing the right model

When fitting a model to experimental data, it is not always possible to choose a model structure on the basis of knowledge of the internal mechanism of the device. A choice must often be made between models with different numbers of parameters or with a variety of structures. Goodness-of-fit tests (χ^2, Kolmogorov) alone are not sufficient. This is because a model with too many parameters will tend to model the measurement noise and not the device itself, so that it will not perform so well with a different data set or in its intended application.

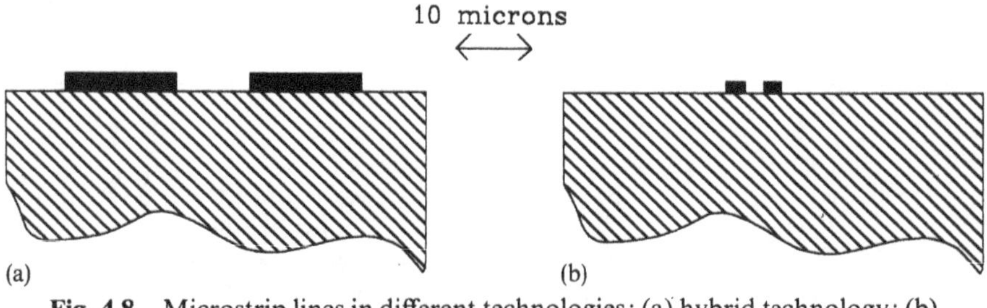

Fig. 4.8 Microstrip lines in different technologies: (a) hybrid technology; (b) monolithic technology.

Many tests are available [4.14]. One of the easiest to use is the final prediction error (FPE) test in which the variance σ^2 of the residual errors is balanced against the number n of free parameters in the model:

$$\text{FPE} = \sigma^2 \frac{N+n+1}{N-n-1}$$

where N is the number of data points. The model with the lowest value of FPE is chosen.

4.3 Linear circuit analysis

In this section we consider the computational methods used to predict the behavior of electrical circuits.

4.3.1 Analysis of cascaded two-ports

A chain of network elements can be analyzed simply by multiplying the individual *ABCD* or *R* matrices to yield the *ABCD* or *R* matrix of the overall network (see Chapter 2):

$$\begin{bmatrix} A & B \\ C & D \end{bmatrix} = \begin{bmatrix} A_1 & B_1 \\ C_1 & D_1 \end{bmatrix} \begin{bmatrix} A_2 & B_2 \\ C_2 & D_2 \end{bmatrix} \begin{bmatrix} A_3 & B_3 \\ C_3 & D_3 \end{bmatrix} \cdots \begin{bmatrix} A_n & B_n \\ C_n & D_n \end{bmatrix}$$

$$[R] = [R_1][R_2][R_3]\ldots[R_n]$$

This was the foundation of the first simulation programs. The R matrix representation was more popular because the R matrices are directly related to the S matrices commonly used to characterize microwave circuits (Chapter 2, Appendix 2.1). These programs were later improved to handle parallel branches by using Y matrix representation and dummy connection components locally, so that simple feedback circuits such as that shown in Fig. 4.9 could be analyzed.

The topologies that could be analyzed this way were limited, and several methods were developed to overcome these restrictions (ref. 4.15, Chapter 11).

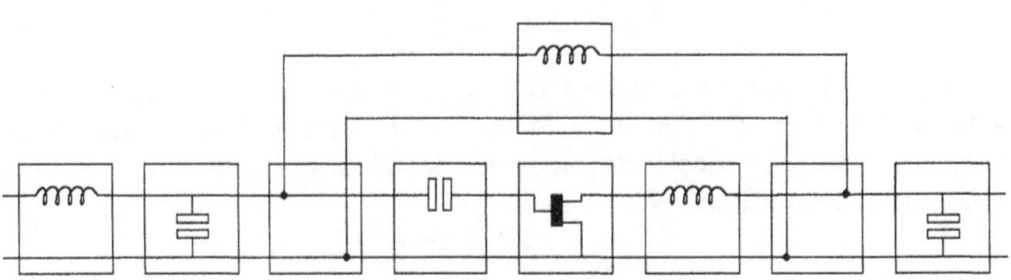

Fig. 4.9 Cascaded network with feedback.

The main advantage of this type of analysis is its speed, as the computation time increases only linearly with the number of circuit components.

4.3.2 Principles of nodal analysis

Arbitrarily connected networks are best analyzed using nodal analysis. The method uses the general nodal admittance matrix of the network which connects the node voltage vector to the source current vector:

$$\boldsymbol{I} = \boldsymbol{Y}\boldsymbol{V}$$

This matrix can be established quite easily, as shown in Fig. 4.10. A datum node is chosen so that the order of the matrix is $N-1$ where N is the number of nodes. The principles used have already been outlined in Chapter 2 but are

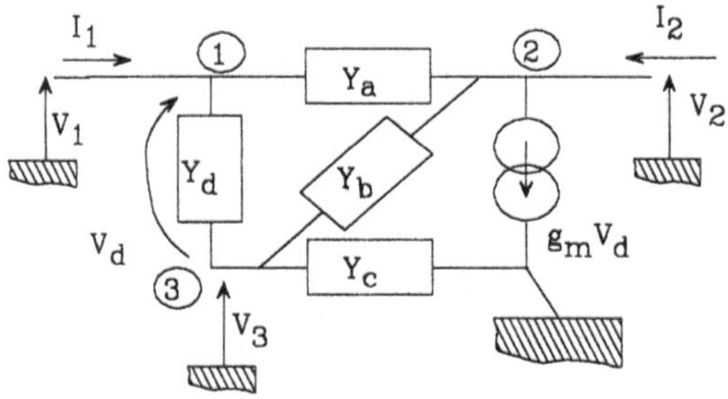

$$\begin{bmatrix} I_1 \\ I_2 - g_m V_1 + g_m V_3 \\ 0 \end{bmatrix} = \begin{bmatrix} Y_a + Y_d & -Y_a & -Y_d \\ -Y_a & Y_a + Y_b & -Y_b \\ -Y_d & -Y_b & Y_b + Y_c + Y_d \end{bmatrix} \begin{bmatrix} V_1 \\ V_2 \\ V_3 \end{bmatrix}$$

Fig. 4.10 Nodal admittance matrix of a simple network.

repeated here for completeness: the diagonal terms Y_{ii} are the sums of all the admittances connected to node i; the nondiagonal terms Y_{ij} are the negative sums of all the admittances connected between nodes i and j. The matrix rows and columns are rearranged so that internal (i) and external (e) nodes are grouped together:

$$\begin{bmatrix} \boldsymbol{I}_e \\ \boldsymbol{I}_i \end{bmatrix} = \begin{bmatrix} \boldsymbol{Y}_{ee} & \boldsymbol{Y}_{ei} \\ \boldsymbol{Y}_{ie} & \boldsymbol{Y}_{ii} \end{bmatrix} \begin{bmatrix} \boldsymbol{V}_e \\ \boldsymbol{V}_i \end{bmatrix}$$

where $\boldsymbol{I}_e$ and $\boldsymbol{V}_e$ are the current and voltage vectors relative to the external nodes, and $\boldsymbol{I}_i$ and $\boldsymbol{V}_i$ are the internal source current and voltage vectors. Since no generators are connected to the internal nodes, we can take $\boldsymbol{I}_i = 0$. Then expanding the equations gives

$$\boldsymbol{V}_i = -\boldsymbol{Y}_{ii}^{-1}\boldsymbol{Y}_{ie}\boldsymbol{V}_e$$

and

$$\boldsymbol{I}_e = (\boldsymbol{Y}_{ee} - \boldsymbol{Y}_{ei}\boldsymbol{Y}_{ii}^{-1}\boldsymbol{Y}_{ie})\boldsymbol{V}_e$$

Therefore the admittance matrix of the network is given by

$$Y = Y_{ee} - Y_{ei} Y_{ii}^{-1} Y_{ie}$$

Its order is N_e, where N_e is the number of external nodes of the network. We can see that in order to solve the problem (i.e. find the admittance matrix Y of the network), the internal nodal admittance matrix Y_{ii} must be inverted. This is a costly process. If a direct inversion technique (Cramer method) is used, the computational time is proportional to $N_i!$, where N_i is the number of internal nodes. More efficient methods such as Gaussian elimination or LU decomposition, which have a computation time proportional to N_i^3, are generally used [4.16].

Conversion routines are used before the analysis if some elements are described by parameters other than Y parameters. Conversion routines are used again after the analysis if the user wants the results expressed in terms of Z or S parameters. A complete set of conversions between Y, Z, $ABCD$, S and R parameters for two-port networks is given in Chapter 2, Appendix 2.1.

Other analysis methods exist, and a good presentation can be found in ref. 4.15.

4.3.3 Parametric analysis

After a designer has carried out a nominal analysis he often needs to know how his circuit will respond to variations in element values or process parameters. If he wants to tune a circuit, he may want to know which elements can be adjusted to improve the performance. If he is faced with a system with a low production yield, he may want to identify unduly sensitive components and estimate maximum acceptable tolerances.

This information can be obtained very simply by performing repeated analyses with various values of the parameter under study. Most simulators let the user define a variable circuit parameter and specify its values. The corresponding analyses are performed automatically and families of curves are obtained (Fig. 4.11(a)). Some simulators such as TOUCHSTONE provide an

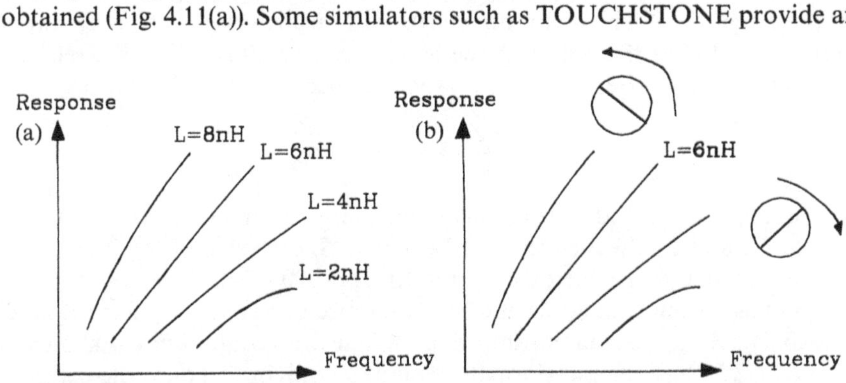

Fig. 4.11 Drawing families of curves: (a) static mode; (b) tuning mode.

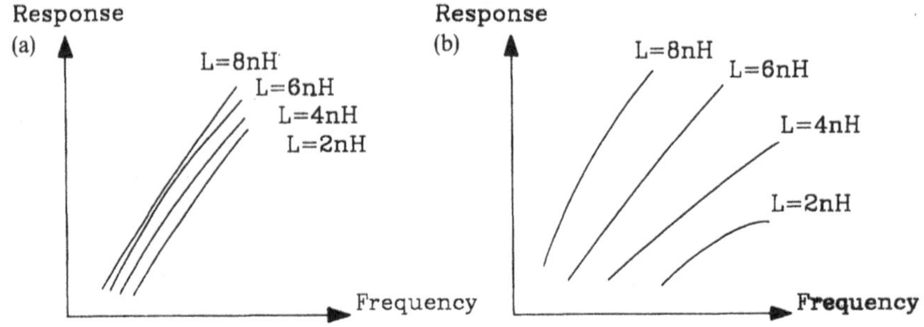

Fig. 4.12 Qualitative sensitivity study with parametric analysis: (a) insensitive circuit; (b) sensitive circuit.

interesting interaction with the user: the curve varies as the user rotates a knob (Fig. 4.11(b)).

Parametric analysis can give a qualitative understanding of sensitivity: in Fig. 4.12(a) the curves are very close and the circuit is rather insensitive to parameter variation, whereas in Fig. 4.12(b) the curves are widely separated and indicate a very sensitive circuit response. This method gives qualitative information about the sensitivity of a circuit parameter to large changes but it is very costly: a complete analysis of the circuit must be performed for each value of the parameter.

4.3.4 Mapping

The mapping feature of the SUPER-COMPACT simulator [4.17] provides a much more efficient method. Let us consider the one-port circuit shown in Fig. 4.13. The variable element of interest can be extracted from the circuit and its terminals considered as external ports, so that the circuit is now studied as a two-port. The scattering matrix of the remaining circuit under a 50 Ω reference is S. The scattering matrix with references of 50 Ω at the input plane and of Z at the output plane is S'. It can be shown that the relation between any parameter S'_{ij} and the varying impedance Z is bilinear, i.e.

$$S'_{ij} = \frac{a+bZ}{c+dZ} \Leftrightarrow S'_{ij} = \alpha + \beta\frac{Z-\delta}{Z+\delta}$$

where a, b, c, d, α, β and δ are complex constants determined by the frequency of interest and by the elements of the matrix S, exclusive of the impedance $Z = R + \mathrm{j}X$ of the variable element (ref. 4.18, pp. 100–104).

Let us assume that we want to study the circuit response for fixed R or X. The loci of Z on the complex impedance plane are straight lines as shown in Fig. 4.14. The bilinear transformation maps a straight line onto the complex reflection plane of the S' reflection parameters as follows: first a conformal

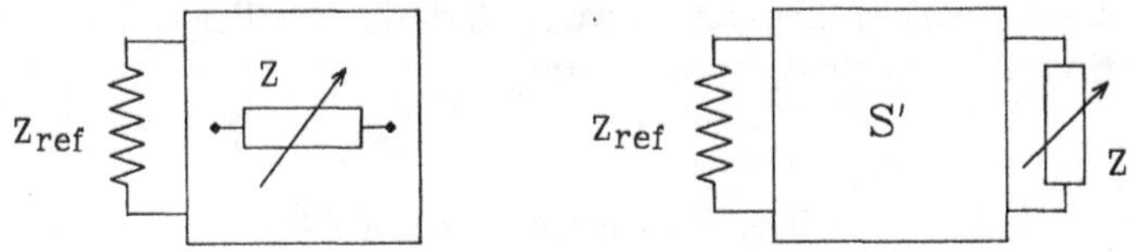

Fig. 4.13 Principles of mapping.

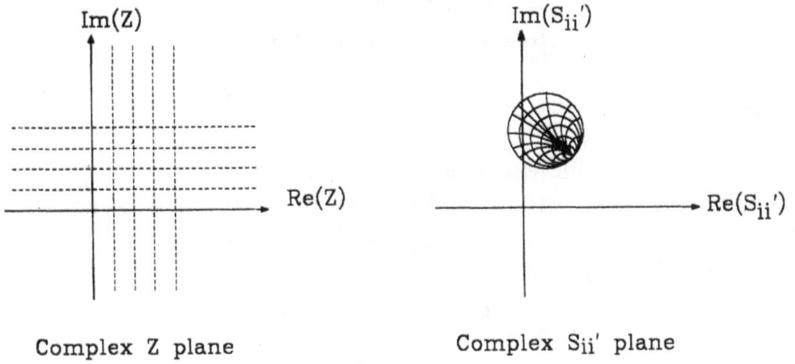

Fig. 4.14 Mapping a variable impedance through a two-port.

transformation that maps straight lines into circles

$$Z \to \frac{Z-\delta}{Z+\delta}$$

then a magnification by $|\beta|$ and a rotation by $\arg(\beta)$

$$Z \to \beta Z$$

and finally a translation by α

$$Z \to \alpha + Z$$

Thus the loci of all the circuit responses for varying Z can be obtained from just one computation of the circuit.

This method is very useful for analyzing feedback amplifiers [4.19]. S'_{11} is zero for R and X values such that the constant-R and constant-X circles both cross the origin of the complex reflection plane. It is also used in Chapter 7 to obtain a negative impedance in the S'_{ii} plane for oscillator design.

4.3.5 Sensitivity analysis

In Section 4.3.3 we saw how parametric analysis gives qualitative information about the sensitivities of a circuit parameter to large changes in the circuit response. In this section we shall describe efficient methods for the computation of small change sensitivities.

4.3.5.1 Definition

The normalized sensitivity of a circuit response R to a circuit parameter X is formally defined as the limit

$$S = \lim_{\Delta X \to 0} \frac{\Delta R/R}{\Delta X/X}$$

There are three main methods for computing sensitivity: the finite-difference method, the incremental network approach and the adjoint network approach.

4.3.5.2 Finite-difference method

In the finite-difference method the circuit parameter X is varied slightly and the change in the circuit response R is calculated. The sensitivity is estimated as follows:

$$S = \frac{R(X+\Delta X)-R(X)}{R(X)} \frac{X}{\Delta X}$$

This method is inaccurate because the difference between two very close real numbers is itself a small real number that is not always accurately represented in the computer. It is also computationally expensive since the network must be analyzed at each frequency with respect to the variations in each parameter.

4.3.5.3 Incremental network method

The basic principle of the incremental network method is as follows. Let us consider the nodal equilibrium relations of a linear network written in the following matrix form:

$$\boldsymbol{I} = \boldsymbol{Y}\boldsymbol{V}$$

By differentiation we obtain

$$\delta \boldsymbol{I} = \boldsymbol{Y}\delta \boldsymbol{V} + \delta \boldsymbol{Y}\boldsymbol{V}$$

The first term on the right-hand side represents the nodal equilibrium relations of the same network Y but with incremental currents and voltages δI and δV. All independent voltage and current sources are cancelled. The second term on the right-hand side gives the perturbation in the variable circuit element Y_{jk} which can be represented by a shunt current source of value $\delta Y_{jk} V_{jk}$. Therefore the incremental circuit obtained by differentiating the first network is obtained as shown in Fig. 4.15. Since the network is linear, we can divide all the incremental currents and voltages by $\delta Y_{jk} V_{jk}$ and represent the perturbation on the variable element by a 1 A current source.

The incremental circuit is analyzed using the same simulator as the

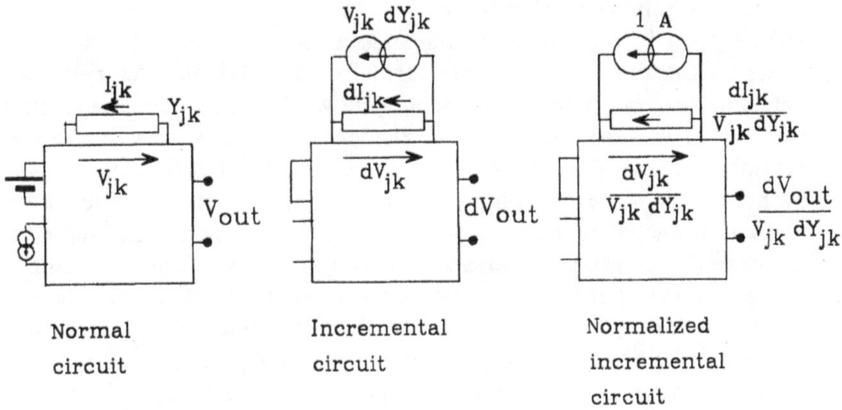

Fig. 4.15 Method of obtaining the incremental network.

normal circuit. Analysis of the normal network gives the branch voltage V_{jk} across the variable element. Analysis of the incremental network gives the output voltage of the network:

$$V'_{\text{out}} = \frac{\delta V_{\text{out}}}{\delta Y_{jk} V_{jk}}$$

Thus the unnormalized sensitivity is given by

$$\frac{\delta V_{\text{out}}}{\delta Y_{jk}} = V_{jk} V'_{\text{out}}$$

This method does not have the accuracy problems of the finite-difference method, but it is no faster; an analysis must still be made for each variable element at each frequency.

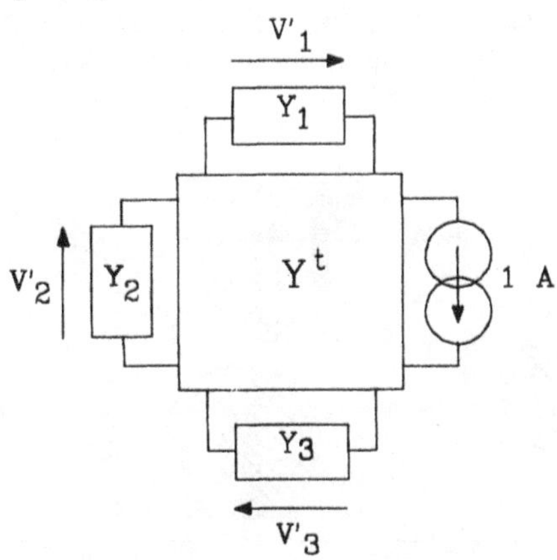

Fig. 4.16 The adjoint network.

4.3.5.4 Adjoint network method

In the preceding method an incremental network must be built and analyzed over the frequency range of interest for each variable circuit element. The only difference between the incremental networks is the 1 A current source applied successively to each variable-element branch. A very interesting observation is made if we analyze an incremental network whose nodal admittance matrix is the transpose of Y and which has a current source of 1 A at the output branch (Fig. 4.16). It can be proved that the voltages observed at the branches of the variable elements in this adjoint network are the same as the voltages observed at the output ports of each of the incremental networks [4.20]. Therefore only two analyses are necessary to obtain all the unnormalized sensitivities of any circuit response to any variable parameter. Analysis of the normal network gives the node voltages V_{jk} and analysis of the adjoint network gives the branch voltages V'_{jk}:

$$\frac{\delta V_{\text{out}}}{\delta Y_{11}} = V_{11} V'_{11} \qquad \frac{\delta V_{\text{out}}}{\delta Y_{jk}} = V_{jk} V'_{jk} \ldots$$

The adjoint network method is both fast and accurate. It must be noted, however, that this technique cannot be adopted with a simulator that does not use a nodal resolution method. A very clear description of all these methods has been given by Brayton and Spence [4.20].

4.3.5.5 Display of sensitivity

The results of sensitivity analyses are usually presented in the form of numerical tables. However, a very clever solution to the problem of representing sensitivities has been proposed [4.20]: circles are drawn at each sensitive element of the circuit and the radii of the circles are equal to the magnitude of the sensitivity (Fig. 4.17). The circles grow and shrink as a knob is rotated through the frequencies of interest. Resonant combinations of elements become evident when two circles suddenly increase in size. This method can be refined by using different colors for positive or negative sensitivities.

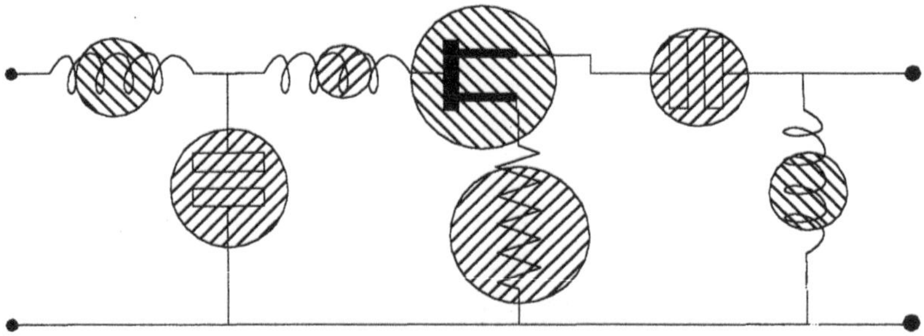

Fig. 4.17 Graphical display of sensitivities.

4.3.5.6 Applications

Sensitivity analysis has many applications.

(a) Model simplification: elements that have little influence on the model response can be neglected.

(b) Yield estimation: a very sensitive element that has large fabrication tolerances is likely to alter the fabrication yield.

(c) Temperature, bias and aging effects: designers try to obtain ultra-stable circuits whose performance will not be degraded much by aging and environmental effects.

(d) Potentially hazardous parasitics: during the layout of the actual circuit the designer may ask himself if it is safe to put two circuit elements close together. It can be helpful to insert a zero-value capacitor between the two elements and to perform a sensitivity analysis on it. If it shows high sensitivity, the capacitive parasitics at this point are very dangerous and the two elements should be widely separated.

4.4 Optimization

An engineer often adjusts circuit elements such as potentiometers, variable capacitors etc. on a test bench. At each step he checks to see whether the circuit response is closer to the desired specifications. An automatic optimization program follows the same iterative pattern.

(1) The network is analyzed and its performance is estimated.

(2) An error function that measures the discrepancy between the circuit responses and specifications is constructed.

(3) The algorithm defines which modifications to the element values will minimize the error function.

(4) The circuit element values are modified.

(5) Steps (1)–(4) are repeated until the error function vanishes or becomes less than some predefined value, or a predefined maximum number of iterations have been performed.

The terms "error function" and "optimization algorithm" are defined in the following sections. We then describe a typical application and give some practical advice on how to manage optimization.

4.4.1 The error function

4.4.1.1 Definition

Let us assume that we want to adapt impedances with a matching network. Matching is achieved when the input reflection coefficient ρ at the input plane is minimum. The following quantity can be used as a measure of the goodness of match:

$$E = \sum_{f=f_1}^{f=f_n} |\rho|^2$$

The summation is performed over the n frequencies of interest, and ρ depends on the topology and component values of the matching network. E is called an error function or an objective function. More complex objective functions with different circuit responses and different frequency bands can be defined, e.g.

$$E = \sum_{f=f_a}^{f=f_b} |\mathrm{MS}_{11}|^2 W_1 + \sum_{f=f_c}^{f=f_d} |\mathrm{DBS}_{21} - 10\mathrm{dB}|^2 W_2$$

where MS_{11} is the magnitude of the reflection coefficient S_{11}, DBS_{21} is the magnitude of S_{21} in decibels and 10 dB is the desired objective for DBS_{21}. W_1 and W_2 are scalar quantities that are used to increase the weighting on the desired partial errors.

The objective functions described above are least-squares functions. Another common type is the minimax objective function which is defined as the maximum partial error between the actual circuit response $R(f)$ and the desired circuit response $O(f)$ over the frequencies of interest. The name minimax is derived from the objective of *min*imizing the *max*imum of the error function:

$$E = \max_{f=f_1}^{f=f_n} |R(f) - O(f)|^2$$

The least-squares and minimax error functions are special cases of a more general type, the least-pth error function:

$$E = \sum_{f=f_1}^{f=f_n} |R(f) - O(f)|^p \qquad \text{with } p \text{ even}$$

For $p \to 2$, the limit of the least-pth error function is the least-squares error function. When $p \to \infty$, the greatest partial error predominates and the limit of the least-pth error function is the minimax error function.

The following comments can be made regarding the different types of error function.

(a) The minima obtained for various values of p in a least-pth error function generally do not coincide.

(b) Although the objective function for $p = 2$ behaves smoothly around the minimum, when p goes to infinity the minimax function exhibits discontinuities in the derivatives. This creates difficulties for the optimization process.

4.4.1.2 Choosing an error function

These different types of objective function do not have the same uses.

(a) The minimax error function tends to give an optimized circuit response with ripples of equal amplitude since it focuses only on the greatest partial error. This explains why the minimax error function is very sensitive to outlier points in attempts to fit a model to noisy measurement data. Minimax error functions are better suited in constrained specifications such as for filters.

(b) The least-squares error function gives the best solution when a model is being fitted to noisy measurement data if the noise is random with a Gaussian distribution (ref. 4.21, pp. 339–342).

4.4.1.3 Weights

Care must be taken in the choice of weights, which is often a problem for the inexperienced user. Some simulators automatically scale the partial errors so that they all have similar orders of magnitude. If this is not the case, then greater weight should be assigned to partial errors that have low magnitudes, e.g. MS_{12} should have a greater weight than MS_{21} in the case of an amplifier. If the error function is defined as follows, where the weights W_i have the same exponent as the actual circuit responses R_i and the desired circuit responses O_i,

$$E = \sum_{f=f_a}^{f=f_b} |(R_1 - O_1)W_1|^p + \sum_{f=f_c}^{f=f_d} |(R_2 - O_2)W_2|^p + \dots$$

then a good choice for the weights W_i is $1/T_i$, where T_i is the maximum tolerance allowed for the response R_i. In this case, if R_i is at the limit of tolerance, the partial error function will have a value of unity. This method is an implicit yield optimization technique since it takes into account the acceptable limits of the circuit responses. It is used in the CADEC simulator [4.22].

4.4.2 Algorithms

The behavior of the error function is unknown except at the operating point (i.e. a set of values for the varying circuit elements) where an analysis has been performed. In the case of a one-dimensional problem, when only one variable can be optimized, an error function may have the form shown in Fig. 4.18. When two circuit variables are optimized at the same time, the problem is two-dimensional and the error function looks like a mountain range. The optimization problem can be stated as follows: how can a blindfolded man lost on a slope of the Himalayas reach the village in the lowest valley?

One solution is to follow the steepest slope downwards. This steepest-descent algorithm was first proposed by Cauchy in 1847 [4.23]. At each point the optimizer computes the derivative of the error function with respect to the optimization variables. It then "takes a step" by moving in the direction of the negative gradient, which minimizes the error function

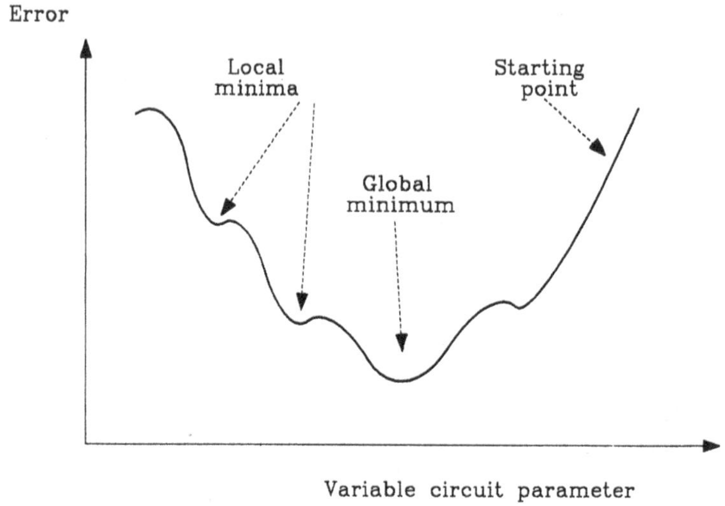

Fig. 4.18 A one-dimensional error function.

$$X_{n+1} = X_n - \alpha \, \text{grad}(E_n)$$

where

$$X_n = [x_n \, y_n \ldots z_n]^t$$

and

$$\text{grad}(E_n) = \left[\frac{\delta E}{\delta x} \frac{\delta E}{\delta y} \cdots \frac{\delta E}{\delta z}\right]^t_{(n)}$$

where the vector X_n is the position in the parameter space and E_n is the value of the error function at step n. The positive scalar α can be adjusted to control the step length. Large step lengths allow for fast progress along steep slopes, and smaller step lengths are necessary to avoid missing nearby minima.

Convergence is assured with the steepest-descent method, but it can lead to a local minimum that is not necessarily the best. A way out can be to use different starting points and see whether a better minimum is found. However, the most convenient method of escaping from local minima is the random method, where the optimizer randomly chooses different sets of values for the circuit elements. The resulting circuit is analyzed and the error function is computed. If the error function is poor, this set of values is abandoned and another trial is performed. If the error function is better, this set of values is kept and optimization continues.

However, random methods are much less efficient than steepest-descent methods in making progress towards a nearby minimum. As shown in Fig. 4.19, the two methods are complementary: the steepest-descent method makes

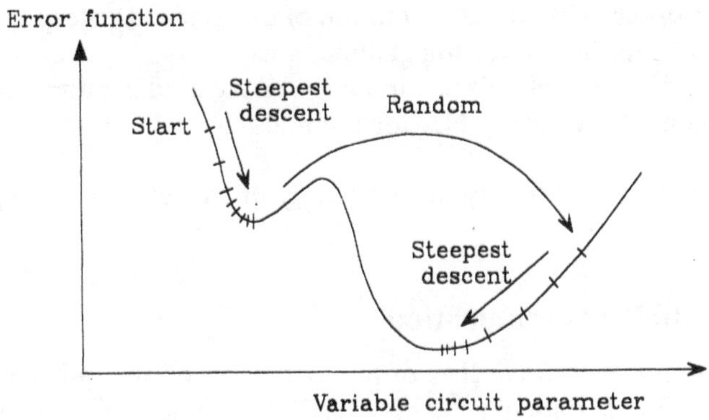

Fig. 4.19 Steepest-descent and global methods of finding the global minimum.

quick progress along a slope; the random method enables the optimizer to avoid local minima.

The steepest-descent and random methods presented here are of course generic. The actual algorithms used in the simulators were the object of much attention in the 1960s. Optimization methods can be classified in three groups according to the information they use about the error function.

(1) Direct methods (no gradient computation): random methods, rotating coordinates, conjugate directions, pattern search methods, simplex or polytope methods etc.

```
BLK                                                      !
  FET 1 2 0  G   = ?.05?    T   = ?4E-12?              ! description
+       CGS= ?1PF?    CDG= ?1PF?   CDS= ?1PF?          ! of FET model
+       RG = ?.5?     RD = ?1?      RI = ?1?           ! to be
+       RS = ?1?      GDS= ?.01?                       ! optimized
+       LS = ?10PH?   LG = ?10PH?  LD = ?10PH?         !
  EQU : 2POR 1 2
END
BLK
  TWO 1 2 Q                                     ! black box (TWO) with label Q
  MES : 2POR 1 2                                ! pointing to measurement data
END
FREQ
  STEP 1GHZ 19GHZ 1GHZ                          ! frequencies for optimization
END
DATA
 Q : POLAR1 FILE= M05GA121                      ! measurement file
END
OPT                                  ! definition of optimization objective :
 EQU S=MES                           ! for circuit EQU, the S-parameters should
END                                  ! equal those of black box MES.
```

Fig. 4.20 SUPER-COMPACT description file for optimization.

(2) First-order methods (computation of the gradient): steepest-descent, conjugate gradients, quasi-Newton methods etc.

(3) Second-order methods (computation of the gradient and the Hessian (matrix of second derivatives)): Newton methods etc.

These methods are discussed extensively in two excellent books [4.24, 4.25].

4.4.3 Example of optimization

The example chosen consists of the extraction of an equivalent circuit from the S parameter data of a GaAs MESFET obtained using an automatic network analyzer. Figure 4.20 shows the description file for optimization where the values of the variable components are enclosed in question marks. The optimization run is shown in Fig. 4.21. The error function is constructed for each frequency using the measured S values. The gradient method is used

```
CMD > opt              ! starting gradient optimization
Gradient, Random or Quit? (G/R/Q) g

     Variables                        Gradients
  ( 1)  0.50000E-01               ( 1) -3.8528
  ( 2)  0.40001E-11               ( 2)  0.15259   ! initial
  ( 3)  1.0000      PF            ( 3) -4.8447    ! values of variables
  ( 4)  1.0000      PF            ( 4)  8.6308    ! and gradients
          ""                             ""
 Err. F.=          13.147

Number of iterations? (x/<0>) 30
 Err. F.=           4.948
 Err. F.=           0.161          ! Fast progression of
 Err. F.=           0.074          ! the gradient algorithm
 Err. F.=           0.005

     Variables                        Gradients
  ( 1)  0.85912E-01             ( 1)  0.30920E-02  ! final values.
  ( 2)  0.35860E-11             ( 2) -0.32783E-02  ! gradients vanish
  ( 3)  1.1504      PF          ( 3) -0.30547E-02  ! close to the minimum.
  ( 4)  0.59337E-01PF           ( 4)  0.18626E-03
          ""                             ""
 Err. F.=           0.004
Gradient termination

Gradient, Random or Quit? (G/R/Q) r
Number of trials? (n/<0>)         400
  Trial        Err. F.           ! random optimization to check for a
  326           0.004            ! better minimum :
  352           0.004            ! none is found.
```

Fig. 4.21 Optimization run with SUPER-COMPACT.

first. When it has converged, the random method is applied to check whether a better minimum can be found.

4.4.4 Managing optimization

The following rules should be borne in mind when microwave circuits are being optimized.

(1) Use conventional matching theory with a Smith chart or synthesis techniques to obtain starting values of the elements for optimization.

(2) Avoid large numbers of frequencies and variables which will result in excessive computer time and cost. It is often possible to select the most sensitive elements by performing a few trial analyses or a sensitivity analysis. During the optimizing process it can be useful to perform a sensitivity analysis on all the circuit parameters to see whether some "sleepy" variables with little influence at the beginning of optimization have "woken up" and become more influential.

(3) Do not over-optimize circuits! Figure 4.22 shows two types of minima. The basin type of minimum is the easiest to find. The ravine type of minimum is more appealing: the value of the error function is much smaller and the circuit should perform better. However, if some component values change, even by a small amount, the circuit response is significantly degraded, whereas the basin type of minimum corresponds to a more insensitive circuit and gives a much higher production yield.

(4) A final analysis with a much finer interval between successive frequencies can be useful to check for undesirable effects on frequencies that were not analyzed during optimization.

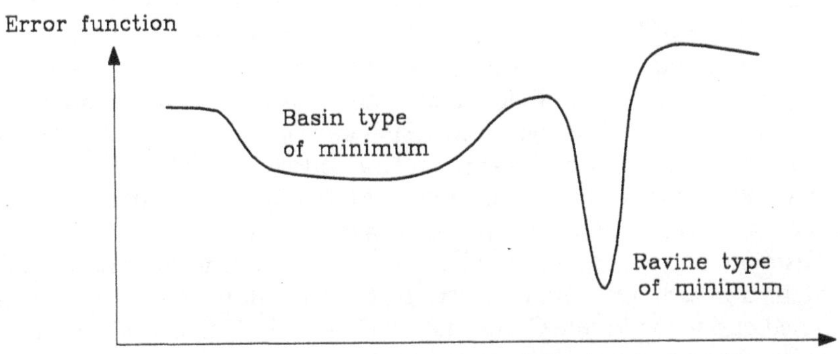

Fig. 4.22 Different types of minima.

4.5 Statistical methods

A recurring engineering problem is the circuit design that functions perfectly in the engineering department as a breadboard or simulation but operates poorly or not at all in early production versions. A major cause is the spread of parameter values for the components used in production.

Classical microwave fabrication techniques allow for some corrections after the circuit is realized: scalpels are used to scratch microstrip stubs, lumped capacitors are connected using small gold wires etc. However, this approach tends to become obsolete because of the cost of labor and test equipment, the small dimensions of higher frequency circuits that make these operations more difficult and the growing importance of monolithic integration techniques where tuning is almost impossible.

The designer should try to forget about tweaking his circuits and start thinking about improving overall fabrication yields. Traditional methods such as parametric analysis and sensitivity analysis will give way to methods such as worst-case analysis, Monte Carlo analysis and other statistical techniques.

4.5.1 Monte Carlo analysis

Monte Carlo analysis is used to predict the manufacturing yield of circuits. The simulator mimics the fabrication process by analyzing circuits whose element values are assigned randomly according to estimated fabrication tolerances and probability laws. The yield is estimated by the ratio of circuits that meet specifications to the total number of circuits analyzed. For a Monte Carlo analysis to take place the user must provide the following information: identification of the circuit components subject to tolerances, the probability distribution laws of the element values and the specifications of acceptable circuit behavior (upper and lower bounds).

Figure 4.23 shows the Monte Carlo analysis of a lowpass filter using SUPER-COMPACT. The probability laws can be defined as uniform or Gaussian within the tolerance range. The outputs of the Monte Carlo analysis are the estimated fabrication yield and histograms that show the probable response distribution. A histogram corresponds to *one* circuit response at *one* operating frequency for variations in *all* toleranced element values. Figure 4.24 shows a typical histogram. If it is necessary to study other responses of the circuit or other frequencies, other histograms must be drawn.

Histograms give quantitative information, but the same information can be obtained in a more qualitative way by drawing all the circuit responses computed during the Monte Carlo analysis (Fig. 4.25). A high local density of curves corresponds to a tall histogram bar.

A Monte Carlo analysis is a useful tool with a dangerous appearance of simplicity. Some precautions must be taken in its use.

```
LAD                                     ! Network description :
  CAP 1 0 C = ? 3PF   10%      ?
  IND 1 2 L = ? 3NH 3.5NH 4NH ?         ! Toleranced elements values are
  CAP 2 0 C = ? 4.5PF 10%      ?        !        between question marks.
  IND 2 3 L = ? 3NH 3.5NH 4NH ?         ! Tolerances : mini nominal maxi.
  CAP 3 0 C = ? 3PF   10%      ?        ! Tolerances :         percentage.
 FILT : 2POR 1 3
END
FREQ                                    ! Frequency definition
  STEP 1GHZ 6GHZ .5GHZ
END
STAT                                    ! Circuit specifications :
 FILT                                   !
 F=1GHZ 2GHZ MS21 = -1DB   GT           ! Ripple less than 1dB in passband
 F=4GHZ 6GHZ MS21 = -30DB  LT           ! Rejection better than -30dB in
END                                     !                          stopband.

   CMD >  stat                          ! Analysis
   STAT > go
   Number of trials ? 200
   Trials     Yield
       25          9
       50         20
       75         32
      100         38
      125         48
      150         60
      175         69
      200         80
   Yield =        80/   200  ( 40.00PCT)
```

Fig. 4.23 Monte Carlo analysis of a lowpass filter.

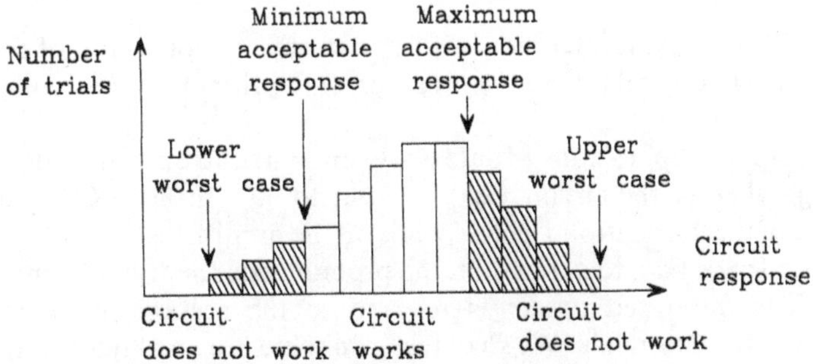

Fig. 4.24 Histogram for Monte Carlo analysis.

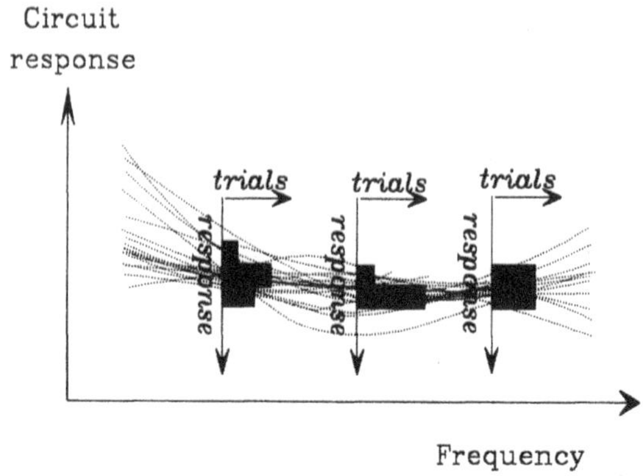

Fig. 4.25 Histograms and circuit responses.

(1) In a Monte Carlo analysis it is assumed that all circuit elements vary independently, and this is almost never the case [4.26, 4.27]. For example, if a mask is misaligned during the fabrication of an integrated circuit, the lengths and widths of transmission lines running in the same direction will vary by the same amount. Some simulators such as ESOPE take this into account by using statistical models of elements: instead of choosing the element values in a model independently, whole sets of element values can be taken from an assortment of computed or experimental data.

(2) The distribution laws governing element values are much more difficult to obtain than the tolerances. In the case of a monolithic circuit the element value can be considered as the outcome of a random variable that is the sum of a large number of independent random variables. Then, according to the central limit theorem, we can assume that the element value probability law is Gaussian. However, in the case of discrete elements the manufacturer obtains Gaussian distributions of element values which are measured and sorted into different tolerance classes and sold at different prices, as shown in Fig. 4.26.

(3) Additionally, it must not be forgotten that the precision of the yield estimation is directly related to the precision of the circuit simulation.

In most real applications Monte Carlo analysis can only indicate general trends, e.g. whether the circuit yield is closer to 20% or 80%. Of course the estimation of the fabrication yield depends on the number of circuits analyzed. We have no space here to expand on the probabilistic theory of Monte Carlo analysis. The interested reader is referred to the sections on confidence intervals in refs 4.21 or 4.28. We shall just give some practical information.

(a) A minimum of 30 analyses is required. It is common practice to

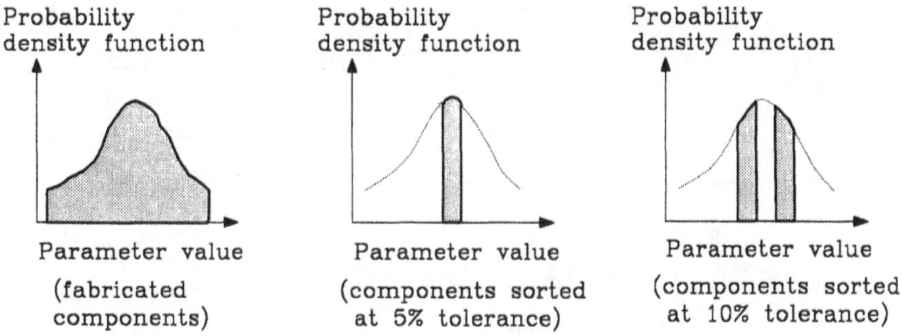

Fig. 4.26 Fabrication distribution laws.

continue circuit analyses until the estimated fabrication yield roughly stabilizes.

(b) The precision of the yield estimation is proportional to the square of the number of analyses performed. In order to double the precision, the number of analyses must be multiplied by four.

(c) The estimation accuracy is independent of the number of toleranced elements in the circuit, the size of the circuit and the number of different circuit responses that define the circuit specifications.

4.5.2 Worst-case analysis

Using Monte Carlo analysis a designer can choose to design a circuit that has less than 100% yield if the economics of the situation warrant it. There are other cases where it is necessary to be sure that *all* the manufactured circuits meet the specifications: low cost circuits where the price depends on the fabrication yield; high reliability circuits needed in military or space applications where the circuit must still meet the specifications even if the components are subject to changes from aging or environmental conditions such as radiation or temperature. The proper CAE tool is then worst-case analysis.

When the designer is concerned with one circuit response at one frequency, the worst cases are given by the limits of a histogram, as shown in Fig. 4.27(a). When it is necessary to know the worst cases over a whole frequency band, they are given by the envelope curves of all the circuit responses during a Monte Carlo analysis, as shown in Fig. 4.27(b) for the ESOPE simulator.

A full Monte Carlo analysis allows a precise estimation of the worst case but it involves too many analyses. Some simulators, such as ESOPE, use the following algorithm to reduce the computation time.

(1) A sensitivity analysis is performed on each of the toleranced circuit

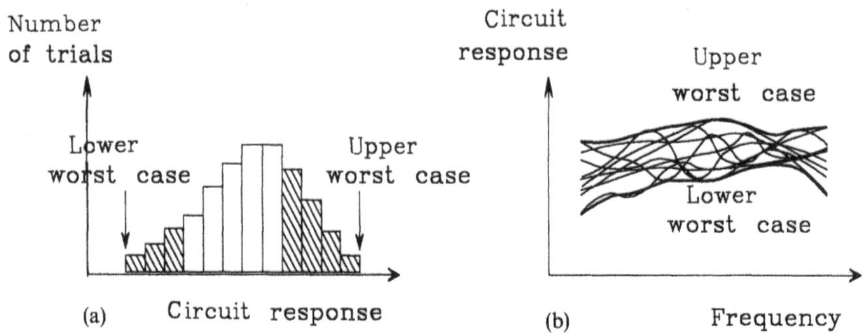

Fig. 4.27 Worst cases as the limits of Monte Carlo analysis: (a) worst cases for one frequency; (b) worst cases for a frequency band.

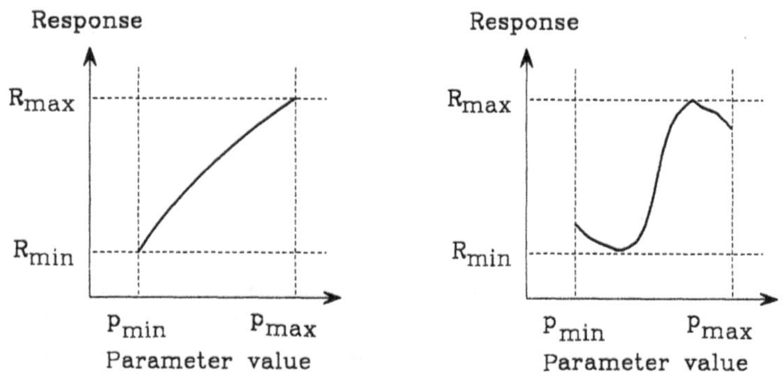

Fig. 4.28 Maximum response deviation in component tolerance intervals.

parameters $p_1, p_2, \ldots, p_n$ at their nominal values.

(2) The upper worst case is obtained by a circuit analysis with all the circuit parameters swung to their tolerance limits in the following manner: $p_i + \Delta p_i$ if the sensitivity of the parameter p_i is positive, and $p_i - \Delta p_i$ if the sensitivity of p_i is negative.

(3) The lower worst case is obtained in the same fashion, with $p_i - \Delta p_i$ for a positive sensitivity and $p_i + \Delta p_i$ for a negative sensitivity.

This algorithm is very efficient because it requires only one sensitivity analysis and two circuit analyses compared with the Monte Carlo technique that requires at least 30 circuit analyses.

This worst-case estimation is good only if all the component tolerances are small because it assumes that the maximum deviation in the circuit response occurs when all components are at their tolerance limits. If one component tolerance is large, the circuit response may not be a monotonic function of this parameter and therefore the maximum response deviation may be obtained for a component value inside the tolerance interval, as shown in Fig. 4.28. This algorithm is not incorrect, but it is rudimentary and must be used with caution.

4.5.3 Worst-case optimization

Many designers would like to optimize the fabrication yield of their circuits. Unfortunately this is too costly, since a full Monte Carlo analysis must be performed at each iteration of the optimization process. It is much more efficient to optimize on the upper and lower worst cases of the circuit responses of interest if a fast algorithm is used (such as the one presented in Section 4.5.2). An example using the ESOPE simulator is shown in Fig. 4.29. The total error function is the sum of the error functions computed for the upper and lower worst cases. The optimization variables are the nominal values of the parameters and the tolerance ranges remain constant.

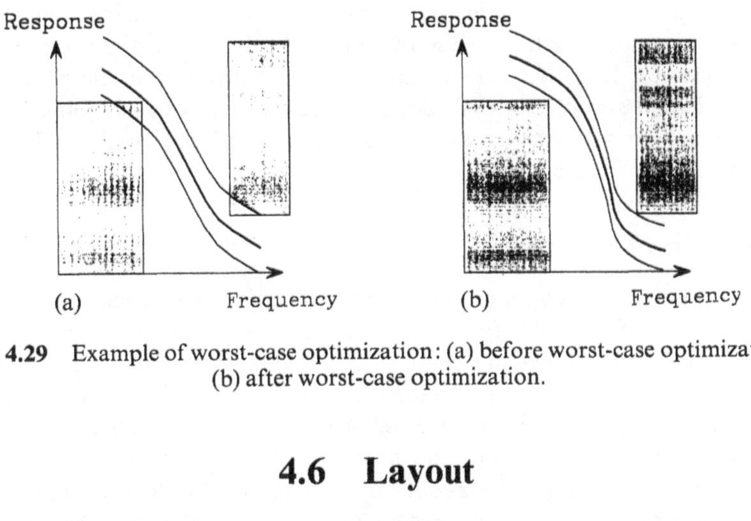

Fig. 4.29 Example of worst-case optimization: (a) before worst-case optimization; (b) after worst-case optimization.

4.6 Layout

Selective chemical processes, which act on some surfaces and leave others intact, are used in the manufacture of today's electronic circuits. Examples include interconnect lines on printed-circuit boards, deposited components such as metal–insulator–metal (MIM) capacitors on hybrid circuits, transistors on monolithic circuits etc. The processes most commonly used are photolithography, electronic masking and direct ion implantation. In all these cases, information must be provided about which zones are to be subjected to chemical action and which are to be preserved. In the case of photolithography

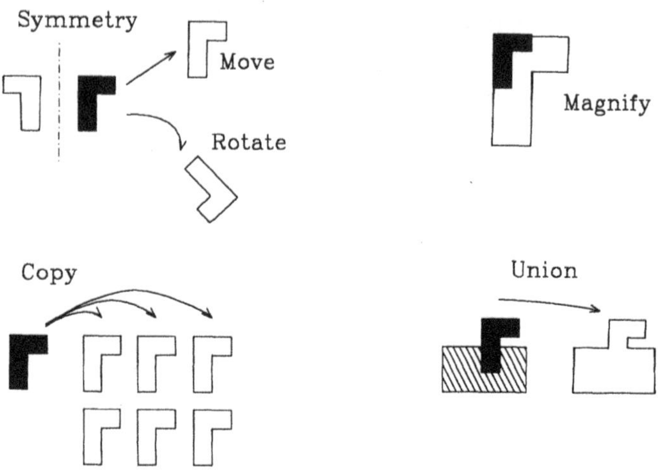

Fig. 4.30 Some graphic operations.

this information is manifested as a glass or plastic mask rendered opaque in certain areas. In the case of electronic masking and direct ion implantation, it is a list of X, Y coordinates in a computer file. The list of coordinates is a virtual mask, by analogy with the real mask of the photolithographic process. In general several successive processes are necessary, and this requires several masks which are also called mask layers or mask levels. The object of layout is the realization of these masks.

4.6.1 Graphic editors

Graphic editor programs allow simple graphic shapes such as rectangles, straight lines, circles etc. to be created and manipulated. More complex patterns can be obtained by using graphic operations, some of which are illustrated in Fig. 4.30. Complex shapes can be generated from several simpler shapes by using the union operation. All these graphic operations can be performed on a single object or on a group of selected objects.

Paper plots of the circuit layout can be generated for the mask fabrication process. Over the years postprocessors have been added in order to automate the manufacturing process further: (i) the plotter pen has been replaced by microknives that directly cut the masking film on the plotter itself [4.29]; (ii) computer files containing the X,Y coordinates of the circuit patterns can be transferred to the mask-making machines. However, these programs still have the following limitations: only one mask level can be drawn at a time; if the same object is repeated several times in the circuit, a modification requires several updates with the corresponding risk of error; no standard interface exists to exchange data between different mask systems.

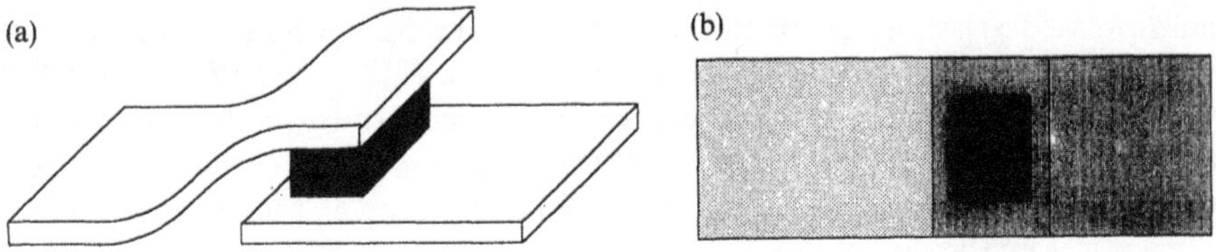

Fig. 4.31 Layers and masks of a MIM capacitor: (a) actual component; (b) mask layers.

4.6.2 Layout systems

Layout systems are programs specifically designed for the layout of integrated circuits. Their power comes from the following features: simultaneous editing of several mask levels, hierarchy, graphic macrocommands, element libraries and standard data interfaces.

4.6.2.1 Editing of several mask levels

The fabrication of a MIM capacitor, for example, requires at least three successive operations (Fig. 4.31(b)): deposition of the first metal, deposition of the insulating layer and deposition of the second metal. Each of these processes requires its own mask layer, and these mask layers must be precisely superposed as in Fig. 4.31(b). By using a layout system it is possible to select the three mask shapes on the different layers and to manipulate them as a single group.

4.6.2.2 Hierarchy

Sometimes many shapes on different mask layers must be moved together and their relative separations preserved. Unfortunately, it is very easy to forget one of the shapes or to take along some undesirable graphic object. It is thus useful to define sets of mask shapes. These sets are called cells and are the basic concept of hierarchical layout systems. A cell can be given a name and can be

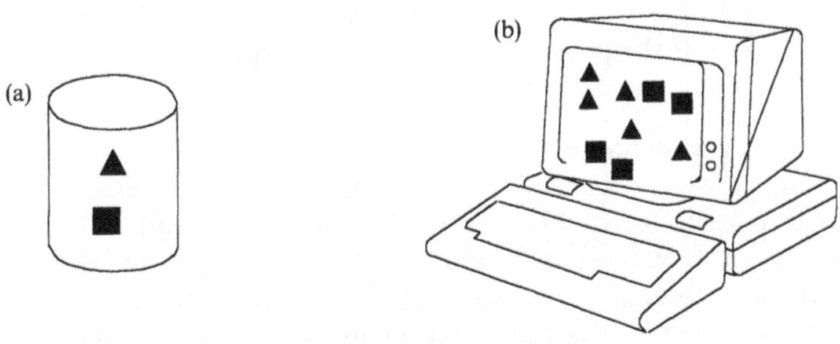

Fig. 4.32 The concept of a graphic cell: (a) cells A and B are stored on disk; (b) instantiations of cells A and B.

manipulated as a single element. It is stored once in the computer memory, and it can be called (instantiated) many times on a drawing, as shown in Fig. 4.32. One obvious advantage is that less memory is used. However, the real benefit accrues when a cell must be updated: the definition of the cell is modified once on disk, and when the drawing on the screen is refreshed all the instances of the cell are automatically updated.

As circuits become more complex, the human mind has to resort to simplifying tools. The most obvious method is subdivision into several hierarchical levels of layout. The circuit is considered as one cell partitioned into several subcells that can themselves be further decomposed. In this way complex circuits can be laid out relatively easily using structured top-down or bottom-up approaches.

One aspect of hierarchy is that when a cell is well defined its full contents need not be shown and a simpler symbolic representation can be used. This reduces dramatically the time required to draw the circuit on the screen.

4.6.2.3 Graphic macrocommands

The user can instruct the program to draw a shape by, for example, interactively typing the following commands:

```
CMD>INIT 10 10              ! Defines initial point
CMS>RECT 20 30 METAL1       ! Draws rectangle to point 20,30 on
                                                         METAL1
```

All the successive commands needed to draw an element can be saved in an executable file (Macrocommand). By just typing the name of the file, the graphic commands are successively executed and the element is drawn.

Some layout systems allow for the creation of very sophisticated executable files with the following features: labels representing variables, control structures such as IF ... THEN ... ELSE and calls to external executable routines. The labels are assigned, and the controlled structures are interpreted at run time. The simple macrocommand of Fig. 4.33 draws square capacitors of variable dimensions. The software used is CHIPGRAPH [4.30].

One interesting consequence of graphic macrocommands is that journal files can be used to record all the commands executed since the beginning of the work session. If the program accidentally aborts, the journal file can be executed, thus reconstructing the drawing.

4.6.2.4 Element libraries

During the layout of a circuit the same elements are often used repeatedly, and thus it can be useful to store them in libraries. For example, a design center that uses several foundries to manufacture monolithic circuits can have one graphic-element library for each foundry. Graphic elements can be stored in two forms: (1) binary files that contain the description of the graphical segments of the cell; (2) editable files which are macrocommands that redraw the cell. Binary files are used for fixed elements such as bonding or test pads.

```
input  'Click lower left corner              :'  -loc x0 y0 w
input  'Value of unit capacitance  in pF/mm2  ?'  capuni
input  'Value of desired capacitor in pF      ?'  cap

assign len ^(((cap/capuni)**0.5)*1E3)  ! computation of capacitor size

init   ^x0         ,^y0       , ^w
rect   ^(x0+len), ^(y0+len), ^w        ! First metalization
add shape   METAL1                     ! on layer METAL1

init   ^x0         ,^y       0, ^w
rect   ^(x0+len), ^(y0+len), ^w        ! Insulator layer
add shape   INSU

init   ^x0         ,^y       0, ^w
rect   ^(x0+len), ^(y0+len), ^w        ! Second metalization
add shape   METAL2                     ! layer METAL2
```

Fig. 4.33 Graphics macrocommand for drawing a capacitor.

Macrocommands are used for parameterized circuit elements that are drawn with a variety of dimensions.

4.6.2.5 Data interfaces

Standard normalizations, such as GDS2, DAVIDMAN etc., exist to describe the mask patterns.

4.6.3 Design-rule checking

Most manufacturing processes impose layout rules for a good production yield.

4.6.3.1 Necessity for design-rule checking

As circuit complexity increases, design rules become more critical: monolithic integrated circuit techniques use as many as 15 mask layers, and the number of design rules to be met increases; as the number of components in the circuit increases, many more verifications must be made.

In a hierarchical graphic editor the complexity of verifications can be reduced somewhat by careful design—by cautiously designing each library graphic cell, and by careful data checking inside the macrocommands. As long as the designer only uses library cells and macrocommands, he can be reasonably sure that the design rules are not violated inside the assembled cells themselves. However, this is no guarantee for design rules that apply between layers of different cells. Furthermore, at the end of the layout step some graphic shapes can be added to complete the design or some cells can be modified in order to improve the design. Therefore an overall automatic verification is necessary in most cases.

4.6.3.2 Structure of a design-rule checking program

Two kinds of information must be provided.

(1) Information concerning the graphic elements in the layout: position, size and layer of each shape. This information is coded in a binary file with a normalized language such as GDS2, DAVIDMAN etc.

(2) Information about the technology rules. This is coded in a description file using a syntax specific to the design-rule checker.

The program then verifies the design rules for each pattern of the layout. The output is an error file that contains the faulty patterns and indicates which rules are violated. This file can be interpreted by the layout system which highlights erroneous shapes. The principles of a design-rule checking program are shown in Fig. 4.34. Examples of design-rule checkers are DRACULA [4.31] and PHOENIX.

Design-rule checkers often offer additional features such as electrical rule check (ERC) which checks for electrical continuity, short-circuits etc., layout versus schematics (LVS) which verifies that the circuit schematics correspond to the actual layout and layout parameter extraction (LPE) which extracts electrical information from the layout (e.g. the overall capacity of a transmission line).

4.7 Data management

4.7.1 Unforseen needs

Experience has shown that data processing in general and computer-

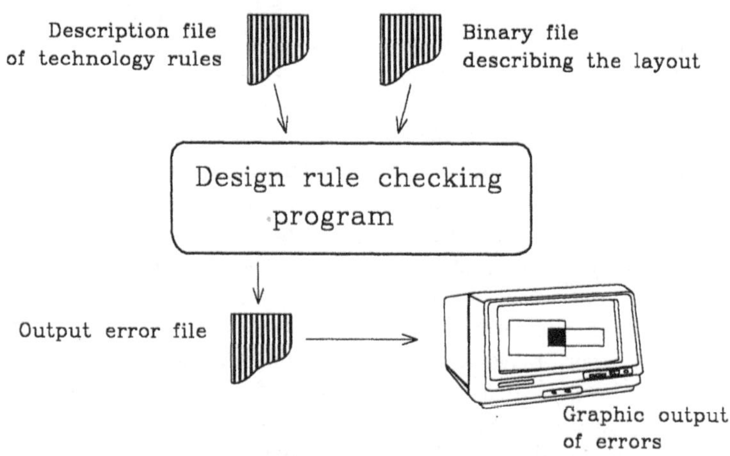

Fig. 4.34 Principles of a design-rule checking program.

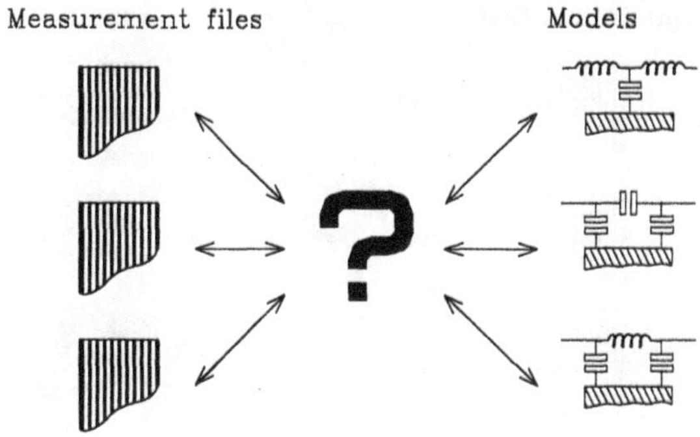

Fig. 4.35 Management of measurements and models.

aided design in particular generate enormous amounts of data. One of the reasons for this is the temptation to keep multiple versions of programs, simulation files, layouts, measurements etc. Each of the CAE steps generates its own data.

Measurements Test circuits are regularly measured in order to keep track of technological developments.

Modelling Several versions of a model under development may coexist, and each of these versions must be checked over many measurement files for verification. It must be possible to find the corresponding model of a given measurement file and vice versa (Fig. 4.35). Furthermore, several designers may work on the same project and they may not all have the same updates of models or measurements.

Simulation and layout A designer often works with different versions of a circuit and is forced to manage different simulation files and layouts (Fig. 4.36). Also, complex designs are often prepared by teams where each designer is responsible for a different part of the system.

4.7.1.1 Data archives

Once a circuit is constructed, its design data must often be maintained for a number of reasons: after-sales service, development of new versions, legal liabilities etc. It may be necessary to recreate the whole design several months or years later, even if the measurement methods, models, technology or design rules have changed. Put simply, with CAE, a large amount of heterogeneous data must not only be stored but also be easily accessible.

All this is usually done manually. However, as teamwork becomes more critical and systems integration proceeds, some automated help becomes indispensable. Data management is one of the least developed fields of CAE, not because data processing tools are rare but, above all, because users do not have a clear idea of their needs.

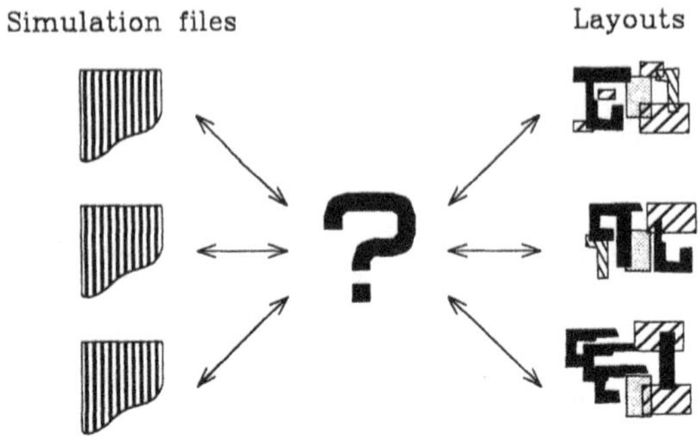

Fig. 4.36 Managing different versions of a circuit.

4.7.2 Solutions from data processing

File management systems were historically the first solution to the data management problem in the CAE field. These systems can retrieve files and sort them according to criteria defined once and for all in the structure of the program. A good example of this kind of system is the BANK command of SUPER-COMPACT which manages measurement files (see Section 4.7.3). However, there are severe limitations:

(1) If another computer, or another operating system, has to be used, an important part of the program must be rewritten.
(2) The program needs serious modifications if data structures change.
(3) The types of queries the user can make on the data cannot be changed unless the program is modified.

Furthermore, every modification of the program for one of the above reasons implies extensive and costly testing.

More flexible systems—database management systems (DBMS)—have been developed in the business field (see Section 4.7.4 and ref. 4.32). From the user point of view, their main interest is that they provide a specialized language for data manipulation. This language is independent of the data, and as long as the syntax is respected the user can make all the queries he wants concerning the data. These systems can easily call and be called by other programs written in high level languages such as FORTRAN, PASCAL, C etc.

4.7.3 Example of a file management system

A data bank containing a few hundred measurement files of commercial transistors is provided with the SUPER-COMPACT software. With the BANK command, these files can be sorted according to several search criteria

```
CMD> BANK
BNK> KFACT > 1
BNK> FREQ > 18E9
BNK> GMAX > 6
BNK> SHOW

Current Transistor Data Bank Search criteria :

Data Bank name...... PUBLIC
Manufacturer........ ALL
Frequency Range..... > 18 GHz
Magnitude of S11.... NOT USED
Magnitude of S22.... NOT USED
Magnitude of S21(dB) NOT USED
Noise data range.... NOT USED
Gmax or MSG......... > 6
K factor............ > 1
FT.................. NOT USED

BNK> SEARCH
 Searching. . .
 PLEJJ PLESSEY SEMI. GAT6 FULL CHIP V = 5v I = 10ma
 End of search.   1 transistor found.

BNK> PRINT PLEJJ

Freq      S11          S21           S12           S22        Gmax     K
(MHz)  Magn< Ang)  (Magn< Ang)  (Magn< Ang)  (Magn< Ang)
 2000   .97< -26    1.82< 153    .030< 74      .79< -18   17.83  .2174
 3000   .95< -34    1.74< 144    .035< 67      .79< -20   16.96  .3968
 4000   .93< -43    1.66< 134    .039< 64      .79< -24   16.29  .4886
  ""     ""          ""           ""            ""         ""      ""

BNK> KFACT ALL
BNK> SEARCH
 Keep transistor data from last search (<Y>/N) ? Y
 New transistor data will be appended to data already found
 Searching. . .

    PLEHH PLESSEY SEMI. GAT6 HALF CHIP V = 5v I = 5ma
    PLEII PLESSEY SEMI. GAT6 HALF CHIP V = 5v I = 5ma
    PLEJJ PLESSEY SEMI. GAT6 HALF CHIP V = 5v I = 10ma

 End of search.   3 transistors found
```

Fig. 4.37 File management system: SUPER-COMPACT.

and then be printed, plotted or used in the simulator.

In Fig. 4.37 we search for a transistor that works at frequencies above 18 GHz with a stability factor greater than unity and a maximum stable gain greater than 6 dB. User input is underlined.

4.7.4 Example of a database management system

We want to manage measurements of self-inductors. The items of interest are as follows.

FILENAME	:	name of the measurement file
LX, LY	:	length and height of the self-inductor
NT	:	number of turns
W	:	metallization width
S	:	spacing between the conductors
IND	:	value in nanohenrys of the intrinsic self-conductance

The very simple data structure is as follows:

FILENAME	LX	LY	NT	W	S	IND

The user describes this structure in a schema file as shown in Fig. 4.38 using the specific syntax of a DBMS such as NOMAD2 [4.33]. This schema file is then interpreted by the DBMS, which creates the data structure and loads the data. The engineer can now use the DBMS query language to retrieve the information he needs, as in Fig. 4.39.

DBMS commands can be put into an executable file, and can be successively executed by just typing the name of the file. As in graphic executable files (or macrocommands), variables can be labelled and control structures added. Some macrocommands can become very sophisticated programs, and complete applications can be developed where the user simply answers questions and does not need to know the syntax of the DBMS query language.

Other interesting examples can be found in ref. 4.34.

4.8 Future developments

In this chapter we have stressed the importance of CAE methods in the design of monolithic circuits. However, CAE is still in its infancy and improvements are likely to appear in the following areas: user interface, applications of artificial intelligence techniques, statistical design methods and hardware.

4.8.1 User interface

Perhaps the most important development in CAE in the last decade has been the move from analysis programs running on large computers—with

```
MASTER self INDEXED BY filename,lx,ly,nt,w,s,ind;
ITEM filename     as a8;
ITEM lx           as 999;
ITEM ly           as 999;
ITEM nt           as 99.9;
ITEM w            as 99;
ITEM s            as 99;
ITEM ind          as 999.99;
```

Fig. 4.38 Schema file describing a data structure.

```
NOMAD2: LIST BY ind BY nt BY filepath WHERE nt>3

! Lists spiral inductors with number of turns greater than 3

  IND    NT  FILEPATH
 ----  ----  --------
 1.05  4.00  self0243
 1.10  4.00  self0225
 2.35  3.75  self0111
 2.42  4.00  self0275
 4.00  4.00  self0254
 5.34  4.00  self0283
 5.85  6.00  self0333
 6.48  5.00  self0314
 7.24  6.00  self0453

NOMAD2: LIST BY ind BY lx BY ly BY filepath
        WHERE (ind>2) AND ((lx<200) OR (ly<200))

! Lists spiral inductors of value greater than 2nH,
! and with one dimension less than 200 microns.

  IND   LX   LY  FILEPATH
 ----  ---  ---  --------
 2.42  220  135  self0275
 4.00  375  135  self0254
 5.85  157  350  self0344
```

Fig. 4.39 Queries in a DBMS.

exclusively textual input and output—to work stations or personal-computer-based systems with greater emphasis on graphical user interfaces which cover a greater part of the design cycle. This trend should become more important with a generalized use of schematics capture, user-defined models, element libraries from foundries and sophisticated graphical representation of results.

Users are acquiring a better understanding of CAE and some are developing their own software which functions as preprocessors or postprocessors of commercial software. A natural demand of these users will be access to commercial software in order to adapt it to their own design context.

4.8.2 Artificial intelligence

Artificial intelligence techniques (LISP and PROLOG programming languages, expert systems etc.) are the object of active research. The consequences will probably be as great as the introduction of high level programming languages such as FORTRAN a few decades ago. We can expect fall-out in the following areas: symbolic analysis (as opposed to today's numerical analysis); testing; "intelligent" user interfaces such as design assistants [4.35].

4.8.3 Statistical design methods

Monte Carlo analysis and worst-case optimization are rather crude ways of designing insensitive circuits, and more sophisticated methods are under investigation [4.36, 4.37]. However, it is likely that acceptance of these techniques will be slow owing to conservative attitudes among users.

4.8.4 Hardware

Work stations based on 16-bit personal computers lack the computing power for complex applications. The real breakthrough will probably come with the availability of cheap 32-bit work stations which will be connected via high speed local area networks with access to specialized processing units such as floating-point accelerators and database machines.

References

4.1 *TOUCHSTONE User Manual*, EESOF, Westlake Village, CA, 1982.

4.2 *ESOPE Manuel de l'Utilisateur*, Version 8, Thom'6 Co., Paris, August 1986.

4.3 Hamilton, S. E. and Morton, D. M. User defined elements expand CAE software capabilities, *Microwave Sci. News & C.T.*, 111–117, April 1986.

4.4 Ward, D. E. and Doganis, K. Optimized extraction of MOS model parameters, *IEEE Trans. Comput.-Aid. Des.*, **1** (4), 163–168, 1982.

4.5 Diamand, X. X., Montoriol, X. X. and Rumelhard, C. Modélisation du FET AsGa à tension drain–source nulle: application à la mesure des éléments extrinsèques, *Journ. Natl. Microondes, Dig.*, 60–61, 1984.

4.6 Laguna, G. A. and Treece, R. K., VLSI modelling and design for radiation environments, *Int. Conf. on Computer-aided-design VLSI in Computers, Processes and Design, ICCD'86*, pp. 380–384, IEEE, New York, 1986.

4.7 Wheeler, H. A. Transmission line properties of parallel wide strips by a conformal mapping approximation, *IEEE Trans. Microwave Theory Tech.*, 280–289, May 1964.

4.8 Aperce, G., Dueme, Ph., Ouvrard, P. L., Marti, Y. M. and Seguy, B. X-band GaAs monolithic compact power amplifier successfully manufactured in one foundry pass thanks to precise modelling, *5e Journ. Natl. Microondes, Nice, 1987.*

4.9 Gardiol, F. *Traité d'Electricité*, Vol. 3, *Electromagnétisme*, Georgi, Lausanne, 1979.

4.10 Segerlind, L. J. *Applied Finite Element Analysis*, Wiley, New York, 1976.

4.11 Truyen, N. N. Characterization of multi-port microwave networks using the finite element method. In *Computational Electromagnetics*, pp. 263–274, Elsevier, Amsterdam, 1986.

4.12 Samanski, P. Calcul des lignes hyperfréquences platés (microstrip, microstrip recouvert, microstrip suspendu, triplaque bifilaire) et de leurs discontinuités, *Rev. Tech. Thomson-CSF*, **5**, 505, 1973.

4.13 Kondoh, H. Accurate FET modelling from measured *S*-parameters, *IEEE Microwave Theory and Techniques Symp. Dig.*, pp. 377–380, IEEE, New York, 1986.

4.14 Freeman, T. G. Selecting the right model to fit data, *Math. Comput. Simulation*, **27**, 137–140, 1985.

4.15 Gupta, K. C., Garg, R. and Chadha, R. *Computer-aided Design of Microwave Circuits*, Artech House, Dedham, MA, 1981.

4.16 Director, S. W. *Circuit Theory: A Computational Approach*, Wiley, New York, 1975.

4.17 *SUPER-COMPACT User Manual*, Version 1.8, Communication Consulting Corporation, Paterson, NJ, November 1984.

4.18 Spence, R. *Linear Active Networks*, Wiley, New York, 1970.

4.19 Rosenbaum, G. Feedback mapping applied to the GOES RF amplifier, *Microwave Sci. News & C.T.*, 145–155, May 1985.

4.20 Brayton, K. and Spence, R. *Sensitivity and Optimization*, Elsevier, Amsterdam, 1980.

4.21 Ventzel, H. *Théorie des Probabilités*, Mir, Moscow, 1973.

4.22 Rohde, U. L. CAD packages improve circuit optimization methods, *Microwave Syst. News & C.T.*, 117–135, May 1985.

4.23 Cauchy, A. L. Methode générale pour la résolution des systèmes d'équations simultanées, *C. R. Acad. Sci., Paris*, **25**, 536–538, 1847.

4.24 Gill, P., Murray, W. and Wright, M. *Practical Optimization*, Academic Press, New York, 1981.

4.25 Minoux, M. *Programmation Mathématique*, Vol. 1, *Théorie et Algorithmes*, Dunod, Paris, 1983.

4.26 Liu, S. and Singhal, K. A statistical model of MOSFETS, *IEEE Int. Conf. on Computer-aided Design*, pp. 78–80, IEEE, New York, 1985.

4.27 Herr, N. and Barnes, J. J. Statistical modelling for circuit simulation of CMOS VLSI, *IEEE Int. Conf. on Computer-aided Design*, IEEE, New York, 1985.

4.28 Miller, I. and Freund, J. E. *Probability and Statistics for Engineers*, Prentice-Hall, Princeton, NJ, 1985.

4.29 Zurcher, J. F. Réalisation de masques pour circuits en lignes microrubans à l'aide d'une calculatrice de table, *Journ. Natl. Microondes Dig.*, 134–135, 1984.

4.30 *CHIPGRAPH User Manual*, Mentor Graphics Corporation, Beaverton, OR, February 1986.

4.31 *DRACULA User Manual*, ECAD Co., Santa Clara, CA, 1982.

4.32 Date, C. J. *An Introduction to Data-base Systems*, Addison-Wesley, Reading, MA, 1981.

4.33 *NOMAD2 Reference Manual*, D&B Computing Services Inc., 1983.

4.34 Faust, M. G. Using a relational database for SPICE results analysis, *IEEE Int. Conf. on Computer-aided Design*, pp. 247–249, IEEE, New York, 1985.
4.35 A new easy way to design ASICs, *Electronics*, 53–57, 16 October 1986.
4.36 Spence, R. Tolerance analysis and design of electronic circuits, *Comput.-Aid. Eng. J.*, 91–99, April 1974.
4.37 Special Issue on Tolerance Analysis and Design, *Proc. Inst. Elect. Eng., Part G*, **129** (4), 1982.

Chapter 5

Nonlinear Analysis

J. BAUCELLS, A. MEDIAVILLA AND A. TAZON

5.1 Introduction

One of the more difficult domains of mathematics and circuit theory is that involving nonlinear functions. Several different definitions of nonlinear systems are possible, but as this book is concerned with the applications of microwave GaAs MESFETs we shall restrict ourselves to the field of circuit theory. In this context a nonlinear system can be described as one in which the outputs vary with respect to the phase and amplitude of one or more vector input signals in a manner which cannot be described by a simple linear function. In order to be able to treat this problem in a manner compatible with computer-aided design techniques, we shall only consider those functions with a unique solution. Effects of the strong hysteresis type are disregarded. The chapter is thus made up of three parts.

Mathematical methods that are currently used in nonlinear analysis are described in the first part. There are essentially four of these: Volterra series, describing functions, time-domain analysis and harmonic balance. These methods are chosen not because they are the only ones in existence but because they are the most frequently used in the microwave field. Each method has its own advantages and disadvantages and is selected according to its relevance to the circuit to be analyzed, the type of result required and the accuracy required of the solution.

(1) Time-domain methods give the most complete results in that they allow an evaluation of the transient as well as the steady state response of the circuit. Depending on the type of circuit being analyzed and the device models being used, however, they may suffer from instabilities (e.g. they may not converge to a unique solution) or prove to be very slow.

(2) Harmonic balance methods use the time domain and frequency domain in the same analysis. These give just the steady state response and are very useful when the excitation is at only one frequency.

(3) Systems which use nonlinear transfer functions (Volterra series and power series) are useful for analyzing weakly nonlinear circuits with several carriers.

(4) Describing-function methods are approximation methods which involve changing from a nonlinear system to an equivalent linear system.

Models of the GaAs MESFET are dealt with in the second part of the chapter. These can essentially be treated in terms of the semiconductor equations or with experimentally determined parameters. The latter technique is the most widely used, once again because of the ease with which it can be integrated into large-signal computer analysis programs in which many iterations are often required. Therefore we shall describe in detail the experimental GaAs MESFET models in use after a brief review of a simple physical model.

The final part of the chapter draws on the first two to demonstrate how a mathematical technique can be used with a transistor model to analyze a given circuit function. The four functions described are large-signal amplifier, oscillator, doubler and mixer operation.

5.2 Volterra series

The relation between the output signal $y(t)$ and the input signal $x(t)$ in a linear system (Fig. 5.1) is given by the theorem of convolution [5.1]:

$$y(t) = \int_{-\infty}^{t} h(t-\tau)\, x(\tau)\, \mathrm{d}\tau = \int_{-\infty}^{t} h(\tau)\, x(t-\tau)\, \mathrm{d}\tau = h(t) * x(t) \tag{5.1}$$

In the frequency domain

$$Y(f) = H(f)\, X(f) \tag{5.2}$$

where

$$X(f) = \int_{-\infty}^{\infty} x(t) \exp(-\mathrm{j}\omega t)\, \mathrm{d}t \tag{5.3}$$

with

$$\omega = 2\pi f$$

and

$$x(t) = \frac{1}{2\pi} \int_{-\infty}^{\infty} X(f) \exp(\mathrm{j}\omega t)\, \mathrm{d}f \tag{5.4}$$

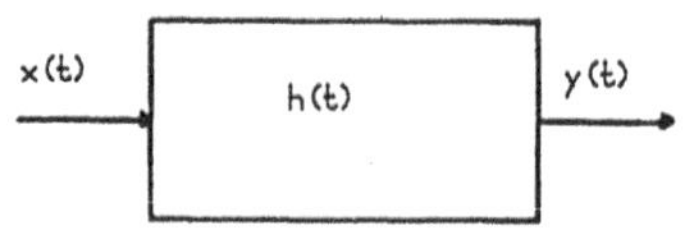

Fig. 5.1 Relationship between input and output signals in a linear system; $h(t)$ is the impulse response from the linear system.

In a nonlinear system the relation between $y(t)$ and $x(t)$ cannot be defined by a single function. It is necessary to use a series of functions, also called functionals. One of these series of functions is the Volterra series, which is an extension of the theorem of convolution from linear (eqn (5.1)) to nonlinear systems (Fig. 5.2).

When the Volterra series is used [5.2],

$$y(t) = \int_{-\infty}^{\infty} h_1(\tau)\, x(t-\tau)\, d\tau + \int_{-\infty}^{\infty} h_2(\tau_1,\tau_2)\, x(t-\tau_1)\, x(t-\tau_2)\, d\tau_1\, d\tau_2 + $$
$$+ \ldots + \int_{-\infty}^{\infty} \cdots \int_{-\infty}^{\infty} h_n(\tau_1,\ldots,\tau_n)\, x(t-\tau_1)\ldots x(t-\tau_n)\, d\tau_1 \ldots d\tau_n$$

or more succinctly

$$y(t) = \sum_{n=1}^{\infty} y_n(t)$$

where

$$y_n(t) = \int_{-\infty}^{\infty} \cdots \int_{-\infty}^{\infty} h_n(\tau_1,\ldots,\tau_n)\, x(t-\tau_1)\ldots x(t-\tau_n)\, d\tau_1 \ldots d\tau_n \qquad (5.5)$$

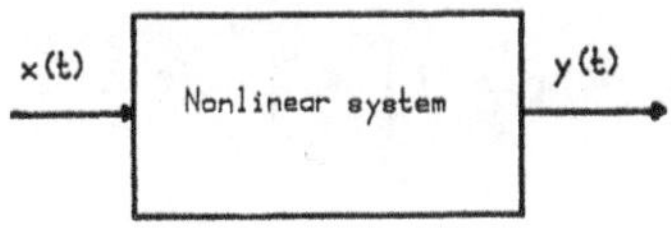

Fig. 5.2 Extension of the theorem of convolution to nonlinear systems.

The nth-order kernel $h_n(\tau_1, \ldots, \tau_n)$ is also called the "nth-order impulse response". For a linear system $h_n(\tau_1, \ldots, \tau_n) = 0$ for all $n \neq 1$.

The nth-order transfer function is defined as

$$H_n(f_1,\ldots,f_n) = \int_{-\infty}^{\infty} \cdots \int_{-\infty}^{\infty} h_n(\tau_1,\ldots,\tau_n) \prod_{i=1}^{n} \exp(-j2\pi f_i \tau_i)\, d\tau_i \qquad (5.6)$$

Substituting in eqn (5.5) we obtain

$$y_n(t) = \int_{-\infty}^{\infty} \cdots \int_{-\infty}^{\infty} H_n(f_1,\ldots,f_n) \prod_{i=1}^{n} X(f_i) \exp(j\omega_i t)\, df_i \qquad (5.7)$$

We now take the Fourier transform of eqn (5.7) in order to obtain the nth-order frequency-domain output signal:

$$Y_n(f) = \int_{-\infty}^{\infty} \cdots \int_{-\infty}^{\infty} H_n(f_1,\ldots,f_n)\, \delta(f-f_1,\ldots,f-f_n) \prod_{i=1}^{n} X(f_i)\, df_i \qquad (5.8)$$

where

$$\int_{-\infty}^{\infty}\cdots\int_{-\infty}^{\infty}\delta(f-f_1,\ldots,f-f_n)\,\mathrm{d}f\ldots\mathrm{d}f_n = 1$$

The complete frequency-domain output signal is given by

$$Y(f) = \sum_{n=1}^{\infty} Y_n(f) \tag{5.9}$$

The principal use of Volterra series is to enable nonlinear systems to be analyzed when the input signal $x(t)$ is the sum of a group of signals of different frequencies:

$$x(t) = \sum_{k=1}^{p} A_k \exp(\mathrm{j}\omega_k t) \tag{5.10}$$

A_k can be a complex number and ω_k is a real number. With this expression for $x(t)$ eqn (5.5) becomes

$$\begin{aligned} y_n(t) &= \int_{-\infty}^{\infty}\cdots\int_{-\infty}^{\infty} h_n(\tau_1,\ldots,\tau_n)\prod_{i=1}^{n}\sum_{k=1}^{p} A_k \exp\{\mathrm{j}\omega_k(t-\tau_i)\}\,\mathrm{d}\tau_i \\ &= \sum_{k_1=1}^{p}\cdots\sum_{k_n=1}^{p}\prod_{i=1}^{n} A_{k_i}\exp(\mathrm{j}\omega_{k_i}t)\int_{-\infty}^{\infty}\cdots\int_{-\infty}^{\infty} h_n(\tau_1,\ldots,\tau_n)\prod_{i=1}^{n}\exp(-\mathrm{j}\omega_{k_i}\tau_i)\,\mathrm{d}\tau_i \end{aligned} \tag{5.11}$$

By substituting (5.6) in (5.11) we obtain

$$y_n(t) = \sum_{k_1=1}^{p}\sum_{k_n=1}^{p}\left(\prod_{i=1}^{n} A_{k_i}\right) H_n(f_{k_1},\ldots,f_{k_n})\exp\{\mathrm{j}(\omega_{k_1}+\ldots+\omega_{k_n})t\} \tag{5.12}$$

For the particular case in which $p = n$ in eqn (5.12) there are $n!$ terms in the frequency $\omega_1+\omega_2+\ldots+\omega_n$. Each term corresponds to a permutation of the argument of $H_n(f_1,\ldots,f_n)$. If $\omega_1, \omega_2, \ldots, \omega_n$ are linearly independent, it is shown in ref. 5.2 that in $y(t)$ there can be no term in the frequency $\omega_1+\omega_2+\ldots+\omega_n$ that is not one of the $n!$ terms of $y_n(t)$.

Then in general any frequency

$$\omega_0 = \omega_{k_1}+\omega_{k_2}+\ldots+\omega_{k_n}$$

of eqn (5.12) can be expressed as

$$\omega_0 = m_1\omega_1+m_2\omega_2+\ldots+m_p\omega_p$$

where

$$\sum_{k=1}^{p} m_k = n$$

and ω_0 is a frequency generated by the nth-order transfer function.

If $H_n(f_{k_1}, \ldots, f_{k_n})$ is symmetrized [5.2] and all the input frequencies are linearly independent, the output from the nonlinear system at the frequency ω_0 is given by

$$y_{\omega_0}(t) = n!\left(\prod_{k=1}^{p} \frac{A_k^{m_k}}{m_k!}\right) H_n(m_1[f_1], \ldots, m_p[f_p]) \exp(j\omega_0 t) \tag{5.13}$$

where

$$\omega_0 = m_1\omega_1 + m_2\omega_2 + \ldots + m_p\omega_p$$

$$\sum_{k=1}^{p} m_k = n$$

$$m_i[f_i] = \underbrace{f_i, \ldots, f_i}_{m_i \text{ times}}$$

A memoryless nonlinear system can be characterized by a polynomial system or a power series [5.3]

$$y(t) = \sum_{n=1}^{\infty} a_n x(t)^n \tag{5.14}$$

and also by a Volterra series where the nth-order impulse response is

$$h_n(\tau_1, \ldots, \tau_n) = a_n \delta(\tau_1) \ldots \delta(\tau_n) \tag{5.15}$$

The advantage of the Volterra series over the power series is that it allows us to analyze nonlinear systems with memory.

5.2.1 Example of obtaining the nonlinear transfer functions of the Volterra series

The equation of the circuit in Fig. 5.3 is

$$i(t) = C\frac{dV}{dt} + G_1 V + G_2 V^2 \tag{5.16}$$

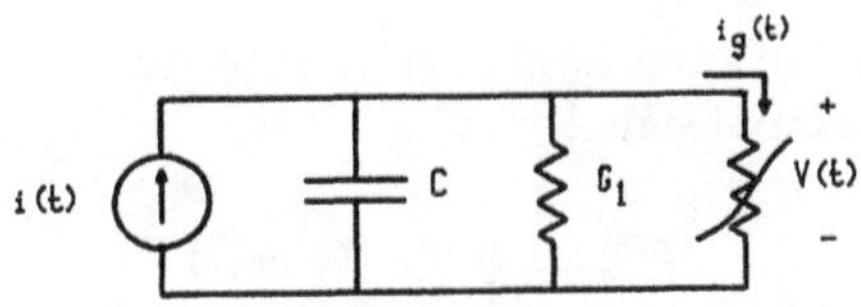

Fig. 5.3 Circuit with a resistive nonlinearity. $i_g(t) = G_2 V^2(t)$.

In this case the output from the system is $V(t)$ and the input is $i(t)$. Suppose that

$$i(t) = \exp(j\omega_1 t) \tag{5.17}$$

From eqn (5.13)

$$V_{\omega_1}(t) = V_1(t) = H_1(f_1)\exp(j\omega_1 t) \tag{5.18}$$

Equation (5.16) at the frequency ω_1 is

$$1 = j\omega CH_1(f_1) + G_1 H_1(f_1) \tag{5.19}$$

Thus

$$H_1(f) = \frac{1}{G_1 + j\omega C} \tag{5.20}$$

If we now consider

$$i(t) = \exp(j\omega_1 t) + \exp(j\omega_2 t)$$

from eqn (5.13)

$$V_{\omega_1}(t) = H_1(f_1)\exp(j\omega_1 t)$$
$$V_{\omega_2}(t) = H_2(f_2)\exp(j\omega_2 t)$$
$$V_{\omega_1+\omega_2}(t) = H_2(f_1,f_2)\exp\{j(\omega_1+\omega_2)t\}$$

Then the component of $V^2(t)$ at the frequency $\omega_1+\omega_2$ is

$$V^2{}_{\omega_1+\omega_2}(t) = 2H_1(f_1)H_1(f_2)\exp\{j(\omega_1+\omega_2)t\}$$

From eqn (5.16) at frequency $\omega_1+\omega_2$ we obtain

$$H_2(f_1,f_2) = -G_2 H_1(f_1)H_1(f_2)H_1(f_1+f_2) \tag{5.21}$$

By continuing this process we can calculate successive nonlinear transfer functions.

5.2.2 Output of a nonlinear system on applying a real signal input

In the preceding sections nonlinear systems were analyzed to obtain the output signal for the input signal

$$x(t) = \sum_{k=1}^{p} A_k \exp(j\omega_k t) \tag{5.22}$$

Using the results already obtained in this chapter, we shall analyze the output signal $y(t)$ for the case when the excitation of the system is a real signal [5.2]:

$$x(t) = \sum_{k=1}^{p} |A_k| \cos(\omega_k t + \phi_k)$$
$$= \sum_{k=1}^{p} \left\{ \frac{A_k}{2} \exp(j\omega_k t) + \frac{A_k^*}{2} \exp(-j\omega_k t) \right\} \tag{5.23}$$

Defining $A_{-k} = A_k^*$ and $\omega_{-k} = -\omega_k$, we can write $x(t)$ as

$$x(t) = \sum_{\substack{k=-p \\ k \neq 0}}^{p} \left\{ \frac{A_k}{2} \exp(j\omega_k t) \right\} \tag{5.24}$$

Any output frequency will be given by

$$\omega_0 = \sum_{\substack{k=-p \\ k \neq 0}}^{p} m_k \omega_k = \sum_{k=1}^{p} (m_k \omega_k - m_{-k} \omega_k) = \sum_{k=1}^{p} (m_k - m_{-k}) \omega_k \tag{5.25}$$

where m_k and m_{-k} are nonnegative whole numbers. If $y_n(t)$ contains the frequency ω_0, then

$$\sum_{\substack{k=-p \\ k \neq 0}}^{p} m_k = n \tag{5.26}$$

The frequencies $\omega_{-p}, \ldots, \omega_p$ are not linearly independent, and so in $y(t)$ at the frequency ω_0 there will be other terms in addition to those in $y_n(t)$. If $\omega_1, \ldots, \omega_p$ are considered to be linearly independent, any ω_0 can be expressed as a linear combination of $\omega_1, \ldots, \omega_p$:

$$\omega_0 = \sum_{k=1}^{p} \alpha_k \omega_k \tag{5.27}$$

Thus the function $y_n(t)$ of a lower order containing terms at the frequency ω_0 will be that in which

$$\sum_{k=1}^{p} |\alpha_k| = n \tag{5.28}$$

with $m_k = \alpha_k$ and $m_{-k} = 0$ for $\alpha_k > 0$, or with $m_k = 0$ and $m_{-k} = -\alpha_k$ for $\alpha_k < 0$. However, the real output signal at the frequency ω_0 will be given by

$$y_{\omega 0}(t) = y_{\omega 0}(t) + y_{-\omega 0}(t) \tag{5.29}$$

5.3 The describing-function method

The object of the describing-function method is to extend the analysis theory for linear systems to nonlinear systems in the frequency domain [5.3].

The method allows us to simplify a nonlinear system and solve it as though it were a linear system. It is a method of approximation.

A describing function of a nonlinear system is an extension of the concept of the transfer function from linear to nonlinear systems; the difference is that the describing function is dependent on the input signal of the system in both form and amplitude [5.3]. The describing-function method is based on the principle of equivalent linearization and the formation of linear operators. The describing functions are quasi-linear operators which describe the characteristics of nonlinear operators from known input signals [5.4]. The nonlinear effect of the system is reflected in the variations in the gain of the describing function with the amplitude of the input signal. Deformation of the signals by nonlinearities is not considered.

Obtaining a describing function associated with a nonlinear system will depend on the form of the describing function, the form of the input signal and the criterion of approximation used. Whatever group of three factors is chosen, the objective is to find an equivalent linear system such that the output signal for the input signal under consideration is as close as possible to the output signal of the nonlinear system for the same input.

The form of the describing function and the criterion considered for evaluating the approximation of the two output signals (from the equivalent linear system and from the nonlinear system) depends on the use to which we wish to put it [5.4]. The criterion of approximation generally used consists in minimizing the mean quadratic error (Fig. 5.4). The forms of the input signal can be classified as follows [5.3]: single sinusoidal signal; multiple sinusoidal signal; random signal.

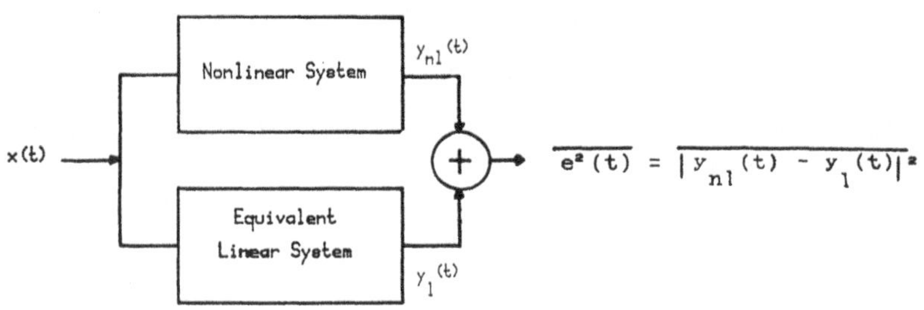

Fig. 5.4 Error determination when applying linearization to a nonlinear circuit.

5.3.1 Methods of obtaining the describing function

The three different methods for obtaining the describing function are distinguished by the form of the describing function and the input signal under consideration [5.4].

(1) In the method proposed by Gelb and Vander Velde [5.5] the input signal $x(t)$ for the system is composed of various random or deterministic signals. The equivalent linear system is made up of linear filters (Fig. 5.5):

$$x(t) = \sum_{n=1}^{N} x_n(t) \tag{5.30}$$

$$y_1(t) = \sum_{n=1}^{N} \int_0^{\infty} h_n(\tau)\, x_n(t-\tau)\, d\tau \tag{5.31}$$

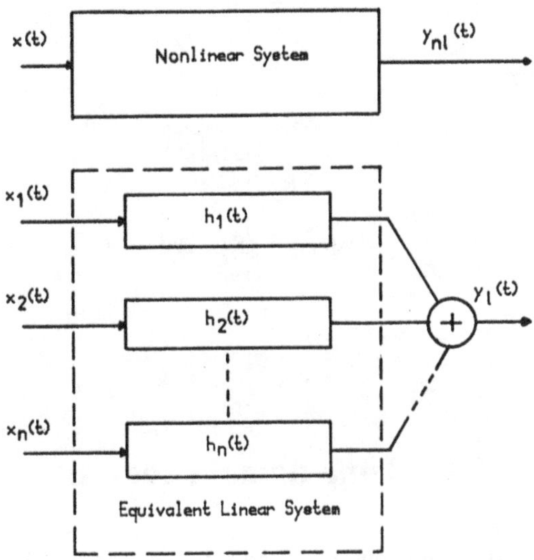

Fig. 5.5 Linearization in the describing-function process.

(2) In a simplification of the above method [5.6, 5.7] the describing function is calculated for particular input signals. The resulting describing function is simplified.

(3) The nonlinear function $f(x)$ is substituted by a linear function $g(x) = \beta x$ [5.3]. The optimum value of β is that which minimizes the mean quadratic error.

Example Let us consider a nonlinear system with the input–output characteristic

$$y_{nl}(t) = a_0 x(t) + a_1 x(t)^2 \qquad a_0 > 0$$

and with

$$x(t) = E\cos(\omega t)$$

The describing function to be found is represented by

$$y_1(t) = \beta x(t)$$

The mean quadratic error is

$$\overline{e^2(t)} = \frac{1}{T}\int_{t_0}^{T+t_0} \{y_1(t) - y_{n1}(t)\}^2 \, dt \tag{5.32}$$

where

$$T = \frac{2\pi}{\omega}$$

Minimizing $\overline{e^2(t)}$ gives

$$\frac{\delta \overline{e^2(t)}}{\delta \beta} = 0 \tag{5.33}$$

$$\int_{t_0}^{T+t_0} \{y_1(t) - y_{n1}(t)\} x(t) \, dt = 0 \tag{5.34}$$

$$\beta = a_0 + \tfrac{3}{4} a_1 E^2 \tag{5.35}$$

5.4 Time-domain analysis

An analysis of a microwave circuit in the time domain consists in establishing the system of equations which characterize it and solving these equations in the time domain. Such a system of equations is obtained by applying Kirchhoff's laws to the circuit to be analyzed. A time-domain analysis presupposes that the relation between voltage and current in each of the elements making up the circuit is defined in the time domain [5.8]. In the case of a nonlinear circuit the system will be composed of nonlinear integro-differential equations.

The solution to the system of equations is found by applying numerical integration methods. Starting from an initial value and using a defined interval Δt, we advance towards a final solution which will be reached once the transient has ended. If the transient is very long, we have to carry out the calculation over many signal periods and it will therefore be slow. This is the main disadvantage of time analyses, and accounts for the term "heavy methods" in connection with them.

The error in the solution depends as much on the integration method used as on the interval Δt chosen. The smaller Δt is, the smaller the error will be, but the number of points to be calculated for each period will be greater, making the calculation slower. Thus a compromise between speed and accuracy has to be made when Δt is chosen.

The system being solved must be carefully considered when choosing which integration method to use. There are many integration methods [5.9–5.12] ranging from the simple trapezoidal method to the sophisticated Gear method [5.12]. The more complicated is the method the more complex are the systems it can solve. However, as it becomes more complicated, a greater amount of information and more calculations are needed, thus reducing the speed of resolution. Therefore the fastest method of analysis for a given system is also the simplest that provides a solution.

There are various time-domain programs: CIRCEC [5.13], SPICE [5.14] (MICROWAVE SPICE [5.15]) and ANAMIC [5.16].

5.4.1 The voltage–current relationship in the time domain of the various elements which can make up a nonlinear system

As stated in the previous section, to conduct an analysis of a nonlinear system in the time domain we need to know the voltage–current relationship of the various elements which make up the system. The elements which can be found in a microwave circuit are (a) resistors, (b) capacitors, (c) inductors and (d) transmission lines.

5.4.1.1 Resistors

The voltage–current relationship of a resistor is given by (Fig. 5.6(a))

$$V = RI \tag{5.36}$$

$$I = GV \tag{5.37}$$

where R is the resistance and G is the conductance. If R is independent of V, and G of I, it is a linear element. However, it is nonlinear if R depends on the current I or if G depends on the voltage V.

5.4.1.2 Capacitors

The voltage–current relationship of a capacitor is (Fig. 5.6(b))

$$i = \frac{\mathrm{d}Q}{\mathrm{d}t} = \frac{\mathrm{d}Q}{\mathrm{d}V}\frac{\mathrm{d}V}{\mathrm{d}t} \tag{5.38}$$

where Q is the charge on the capacitor. The capacitor is linear or nonlinear depending on $\mathrm{d}Q/\mathrm{d}V$: when $\mathrm{d}Q/\mathrm{d}V$ is constant ($\mathrm{d}Q/\mathrm{d}V = C$), the capacitor is linear.

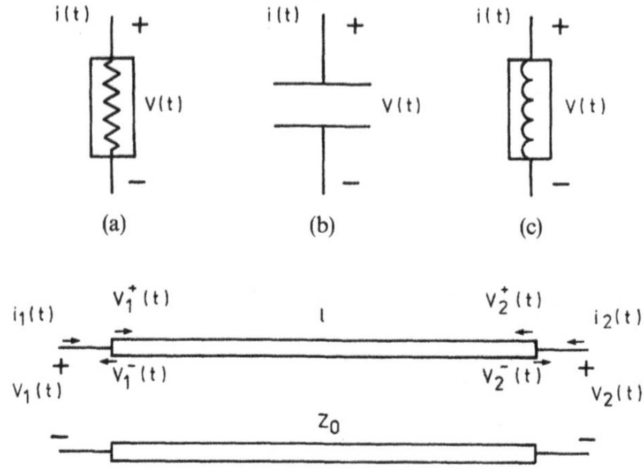

Fig. 5.6 Current–voltage relationships for standard circuit elements.

5.4.1.3 Inductors

The voltage–current relationship of an inductor is given by (Fig. 5.6(c))

$$V = \frac{\mathrm{d}\varphi}{\mathrm{d}t} = \frac{\mathrm{d}\varphi}{\mathrm{d}i}\frac{\mathrm{d}i}{\mathrm{d}t} \tag{5.39}$$

where φ is the magnetic flux across the inductor. As in the case of capacitors, an inductor is linear when $\mathrm{d}\varphi/\mathrm{d}i$ is constant ($\mathrm{d}\varphi/\mathrm{d}i = L$).

5.4.1.4 Transmission lines

An ideal transmission line is a linear two-port with the following voltage and current relations (Fig. 5.6(d)):

$$V_1(t) = V_1^+(t) + V_1^-(t) \tag{5.40}$$

$$V_2(t) = V_2^+(t) + V_2^-(t) \tag{5.41}$$

$$i_1(t) = \frac{1}{Z_0}\{V_1^+(t) - V_1^-(t)\} \tag{5.42}$$

$$i_2(t) = \frac{1}{Z_0}\{V_2^+(t) - V_2^-(t)\} \tag{5.43}$$

where

$$V_1^+(t) = V_2^-(t+T) \qquad V_2^+(t) = V_1^-(t+T) \tag{5.44}$$

$$T = l/c$$

l is the length of the line, c is the speed of propagation along the line and T is the delay on the line.

5.5 The harmonic balance method

The harmonic balance method takes advantage of the fact that all nonlinear systems are composed of nonlinear and linear elements. It breaks down the nonlinear system into two clearly differentiated subsystems: a linear subsystem and a nonlinear subsystem (Fig. 5.7). Both these subsystems can be broken down into further subsystems. The complete system is solved by solving each subsystem and taking into account the links which exist between them.

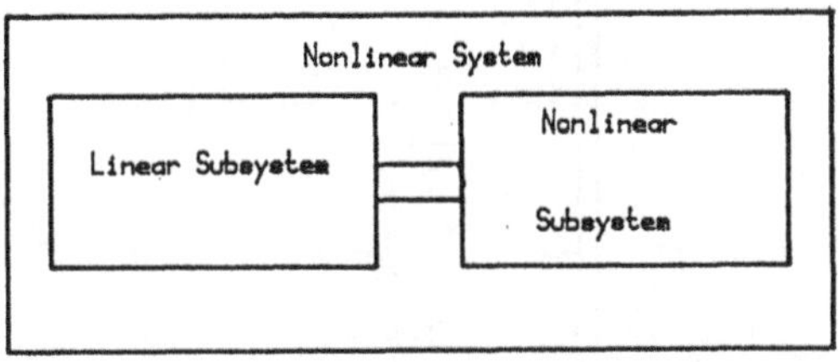

Fig. 5.7 Partitioning of a nonlinear circuit.

The method allows us to adopt the most appropriate method for solving each subsystem, thus making it possible to use both frequency-domain and time-domain methods. Usually, frequency-domain methods are used to solve the linear subsystems and time-domain methods are used for the nonlinear subsystems. The subsystems are related by Fourier series. The use of the frequency-domain method for solving the linear subsystems means that the steady state solution will be obtained for the system. The behavior of the whole system is known when the harmonic components of the signals used to link the various subsystems are known.

The advantage of partitioning the nonlinear system into subsystems is that it allows us to use simpler resolution methods as the complexity of the problem is reduced. Once the system is partitioned, an iterative process is used to obtain a solution. One of the subsystems (usually the linear subsystem) is solved starting from the initial conditions. Solution of this subsystem gives the initial conditions which enable us to solve the other subsystem. Solution of the second subsystem (usually the nonlinear system) gives the initial conditions required to go back and solve the first subsystem once more, and the whole process is repeated. The cycle described is called an iteration (Fig. 5.8).

A solution for the whole system is considered to have been obtained when the solution of one of the principal subsystems in a given iteration does not differ too much from the solution of the same subsystem in the preceding iteration. In order to take this decision, it is necessary to define an error function and its associated convergence criterion.

Whether the harmonic balance method is more efficient than the other

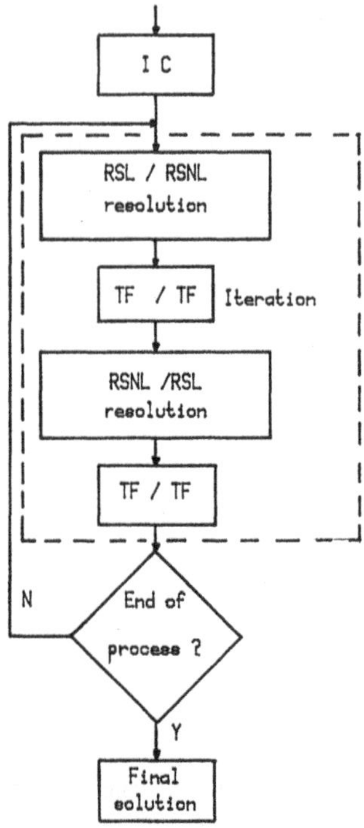

Fig. 5.8 Flowchart of a harmonic balance resolution: RSL, resolution of the linear subsystem; RSNL, resolution of the nonlinear subsystem.

methods depends on a number of factors: the nonlinear system to be solved and its partition into subsystems; the methods used to solve each subsystem; the number of harmonics considered in the analysis; the initial conditions.

5.5.1 The nonlinear system to be solved and its partition into subsystems

A system is partitioned when the resulting subsystems are easier to solve than the whole system. Let us consider the circuit analyzed by Fleri and Cohen [5.17] (Fig. 5.9). Depending on whether the model of the Schottky diode used is as shown in Fig. 5.9(a) or in Fig. 5.9(b), the equations of the circuit are respectively

$$V_p \cos(\omega_p t) = (R_g + R_s) i_g + V_d \tag{5.45}$$

$$V_p \cos(\omega_p t) = (R_g + R_s)\left(i_g + C_j \frac{dV_d}{dt}\right) + V_d \tag{5.46}$$

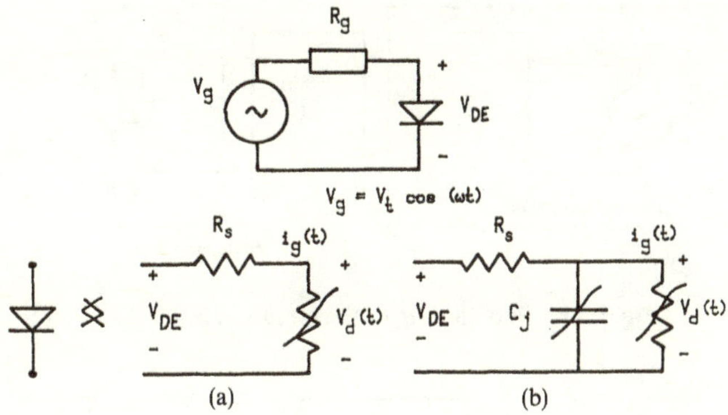

Fig. 5.9 Analysis of a diode: $i_g(t) = I_s\{\exp(\alpha V_d) - 1\}$; $C_j = C_0(1 - V_d/\phi)^{-\gamma}$.

Equations (5.45) and (5.46) can be solved directly using the Newton–Raphson method [5.9] and the Runge–Kutta method [5.10, 5.11] respectively.

Any partition of this circuit into subsystems will result in nonlinear circuits as complicated as the original system. Once the system is broken down, an iterative process is used to solve it, where each iteration is equivalent to solving an equation of type (5.45) or (5.46) according to the diode model used. In this case solution using the harmonic balance method will be much slower than solution using a direct method.

If we go on to a more complicated circuit such as that shown in Fig. 5.10 and take as the diode model either that of Fig. 5.9(a) or that of Fig. 5.9(b), the solution of this system, depending on the linear quadrupole, will be much easier if it is broken down into subsystems than if it is treated as a whole. A useful breakdown is that described in ref. 5.18 (Fig. 5.11).

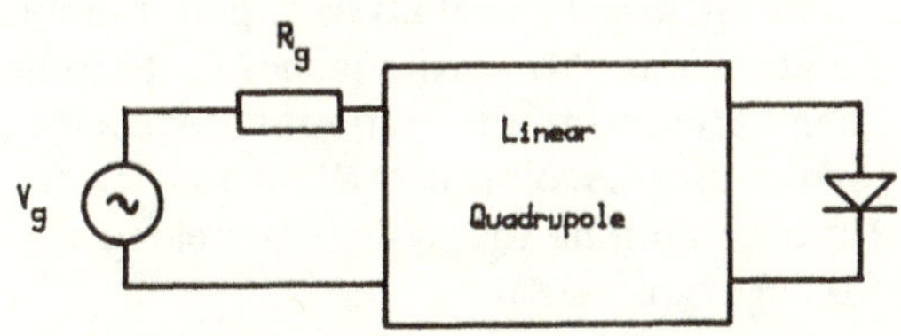

Fig. 5.10 Analysis of a circuit including a diode; $V_g = V_t \cos(\omega t)$.

5.5.2 Methods used in the resolution of each of the subsystems

A nonlinear system which is partitioned into linear and nonlinear subsystems can be illustrated as in Fig. 5.12.

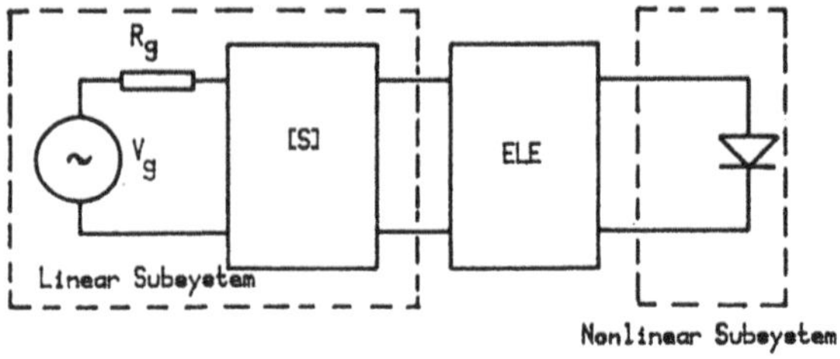

Fig. 5.11 Partitioning of the circuit in Fig. 5.10.

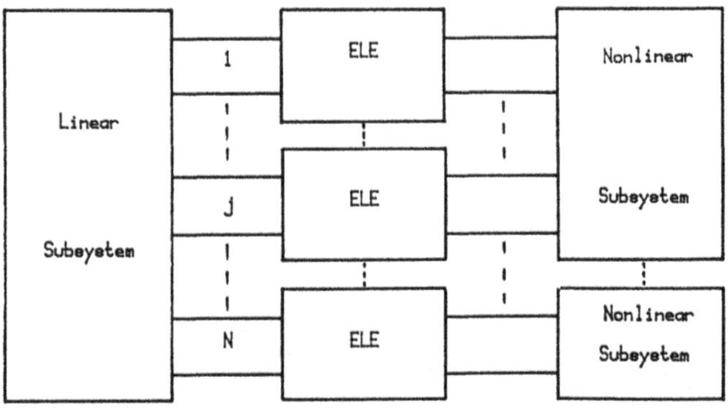

Fig. 5.12 Harmonic balance analysis of a complex nonlinear circuit.

5.5.2.1 Linear subsystem

The linear subsystem forms a multiport in which N of the ports link up with the nonlinear subsystems. If the analysis is performed in the frequency domain and the excitation is at only one frequency f_p, the multiport should be characterized for each harmonic of the excitation frequency. The solution of the linear subsystem becomes the solution of a linear system for each frequency nf_p $(n \geqslant 1)$ (under d.c. conditions the system is treated in a different way because of its special behavior [5.18]).

Although the linear multiport can be characterized using any type of parameter, the most useful are S parameters as they are always present. When S parameters are used, the variables of the system will be the incident and reflected voltages in each of the ports of the multiport. This gives rise to a system of equations such as

$$\boldsymbol{B}_n = \boldsymbol{S}_n \boldsymbol{A}_n \qquad n \geqslant 1 \tag{5.47}$$

where $\boldsymbol{S}_n$ is the S matrix of the linear system at the frequency nf_p. $\boldsymbol{B}_n$ and $\boldsymbol{A}_n$ are the reflected and incident voltage vectors at the frequency nf_p (Fig. 5.13).

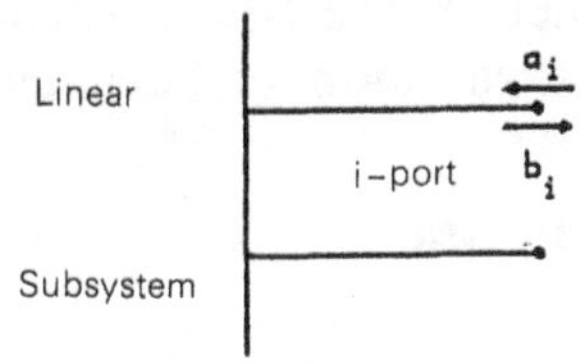

Fig. 5.13 Use of S parameters to analyze linear circuit performance at different harmonics.

5.5.2.2 Link between the subsystems

The variables that link the linear with the nonlinear system are a_i and b_i of ports 1 to N (Fig. 5.12). The relation between the linear and nonlinear systems represented by the ELE block in Fig. 5.12 is given by a Fourier series in each port:

$$x_i(t) = \sum_{k=1}^{M} \mathrm{Re}\{X_k \exp(jk\omega_p t)\} \qquad 1 \leqslant i \leqslant N \tag{5.48}$$

where M is the maximum number of harmonics of $f_p\,(\omega_p)$ considered.

5.5.2.3 Nonlinear subsystems

These subsystems can be solved using either the analysis method in the time domain which is best suited to the system [5.19] (Section 5.4) or methods in the frequency domain with calculations in the time domain [5.20, 5.21].

5.5.3 Number of harmonics considered in the analysis

The representation of any periodic signal in the time domain using Fourier series is given by

$$x(t) = \sum_{k=-\infty}^{\infty} X_k \exp(jk\omega t) = \sum_{k=0}^{\infty} \{a_k \cos(k\omega t) + b_k \sin(k\omega t)\} \tag{5.49}$$

Working with Fourier series using a computer means that the number of harmonics must be limited to M:

$$x(t) = \sum_{k=-M}^{M} X_k \exp(jk\omega t) = \sum_{k=0}^{M} \{a_k \cos(k\omega t) + b_k \sin(k\omega t)\} \tag{5.50}$$

The larger the value of M is, the closer the approximation of the Fourier series will be to the real signal, but the number of unknowns in the system will be greater and so too will be the calculation time. Thus it is necessary for M to have as low a value as possible in order to reduce the resolution time to a minimum. However, if M is too small, the solution of the nonlinear system cannot be taken as valid or it will not exist as the number of harmonics will not be sufficient to represent the waveforms of the circuit [5.18, 5.19].

5.5.4 Convergence criterion

The convergence criterion allows us to stop the iterative process and thus stop the resolution of the nonlinear system. Given a variable X to compare, a good criterion of comparison is

$$\left| \frac{X_{n+1} - X_n}{X_n} \right| < \varepsilon \tag{5.51}$$

where n is the number of iterations. The choice of the variable X will depend on the applications for the circuit under analysis or on what sort of results we wish to obtain for the system [5.18].

If incident and reflected voltages are used as link variables between the linear and nonlinear subsystems and as exact a solution as possible is required for the system under analysis, either incident or reflected voltages will be taken as the X variables for each port of the linear multiport and for each frequency. For reflected voltages eqn (5.51) becomes

$$\left| \frac{(V_{kp})_{n+1}^{-} - (V_{kp})_n^{-}}{(V_{kp})_n^{-}} \right| < \varepsilon \tag{5.52}$$

where n is the number of iterations, k is the number of ports and p is the number of harmonics.

If, however, we are looking for results that will be of practical use, it is much better to use the result we want from the system as a variable. Therefore if a mixer is analyzed the conversion loss or gain will be compared; if a multiplier is analyzed the conversion yield will be used; in the case of an amplifier the gain will be used etc. [5.18]. Although this approach usually means that a few more calculations per iteration must be performed, it enables us to make a better decision when choosing ε in eqn (5.51) and to perform fewer iterations to arrive at a good solution. Once we have decided to which variable it will be applied, the value of ε in eqn (5.51) will depend on the precision needed in the solution. The greater the value of ε is, the less precise the result will be but the quicker a solution will be obtained. The smaller ε is, the greater is the precision, but the calculation time will also be greater and could lead to convergence problems.

5.5.5 Initial conditions

It is best if the initial conditions are as close to the final solution as possible, both to ensure the convergence of the system and to cut down the resolution time.

5.6 Advantages and disadvantages of each method

5.6.1 Volterra series

When a complete knowledge of $y(t)$ is needed, use of the Volterra series is a useful method if the system is weakly nonlinear because it allows us to discard the terms greater than third order. The degree of complexity in the calculation of $H_n(f, \ldots, f_n)$ increases rapidly with increasing n. This can already be seen in the relatively simple example dealt with in Section 5.2.1. The possibility of analyzing nonlinear systems where the input signal is made up of signals at various frequencies makes this method ideal for analyzing the intermodulation products in amplifiers [5.22–5.25]. This is the main area in which Volterra series are used. Although the method has also been used to analyze frequency multipliers and mixers [5.26], there are alternative methods of analysis for these elements which are more effective.

5.6.2 Describing function

The describing function method allows us to analyze systems in which the level of nonlinearity is low. Its use in the analysis of strongly nonlinear systems can lead to very erroneous results. It is fairly difficult to evaluate the errors made using this method, and thus it is more a conception method than an accurate simulation technique. It is quite a useful method for predicting the stability of systems without excitation, but gives little information about their transient characteristics. Applications of describing functions are limited by the difficulty in defining the filters, either because of the large number of components at the input of the nonlinear system or because of the complexity of the nonlinear expression.

5.6.3 Time domain

The only disadvantage with time-domain methods is that all the transients have to be calculated, and for most applications in microwave circuits, where we are only interested in the steady state solution, the transients turn out to be extremely slow. Nevertheless the only limitations on these methods are those imposed by the computer used. They allow a very reliable representation of physical phenomena and are very useful for analyzing nonlinear systems which present nonperiodic or nonstationary solutions.

5.6.4 Harmonic balance

The great advantages of the harmonic balance method are that it allows a better understanding of the behavior of each part of the system, that it allows the application of the best methods for solving each part of the system, thus obtaining a quicker solution, and that as a result it allows us to deal with very complicated problems more easily. The disadvantage is that it is difficult to analyze circuits where different frequencies have to be considered, e.g. amplifier circuits with several carriers.

5.7 GaAs MESFET models

5.7.1 Introduction

Although William Shockley first introduced the operating principle of the field-effect transistor in 1952 [5.27], the successful development of microwave MESFETs only began in 1970. *IEEE Transactions on Microwave Theory and Techniques* [5.28] and *Microwave Systems News* [5.29] both dedicated monographs to studies of MESFET transistors in 1976. The spectacular recent advances of this solid state device are mainly due to the unusual features of GaAs.

(a) The semi-insulating GaAs substrate has an intrinsic resistivity five times greater than that of silicon ($10^7 \Omega\,\mathrm{cm}^{-1}$). This feature overcomes the isolation problems inherent in silicon junction FET or MESFET devices.

(b) The electron mobility in GaAs is about six times greater than that in silicon and the drift velocity is twice that in silicon. This leads to higher frequencies with low noise behavior.

The MESFET chip geometry shown in Fig. 5.14 is an example of the most common structure. In this case the parasitic resistances are reduced by using a relatively thick epitaxial layer and creating a deep recess for the gate. The round bottom of the recess implies a lower depletion length than would result with a flat surface; the effective gate length is less than its actual value. Furthermore, a T-shaped gate has a lower resistance.

An interesting device that merits special study is the dual-gate GaAs MESFET (Fig. 5.15). Its structure is similar to that of the single-gate device

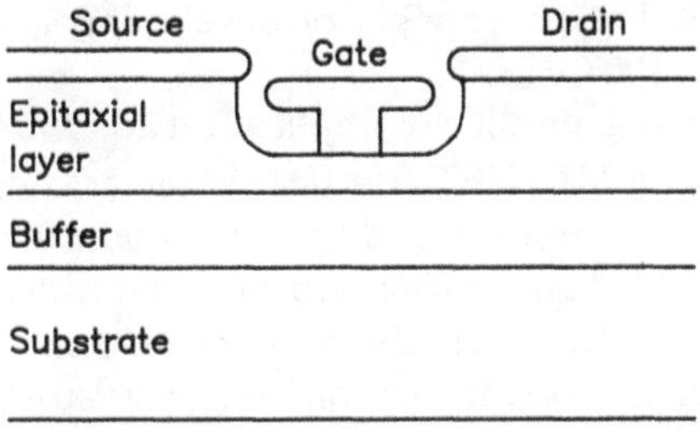

Fig. 5.14 MESFET chip geometry: a recessed gate with a T-shaped cross-section.

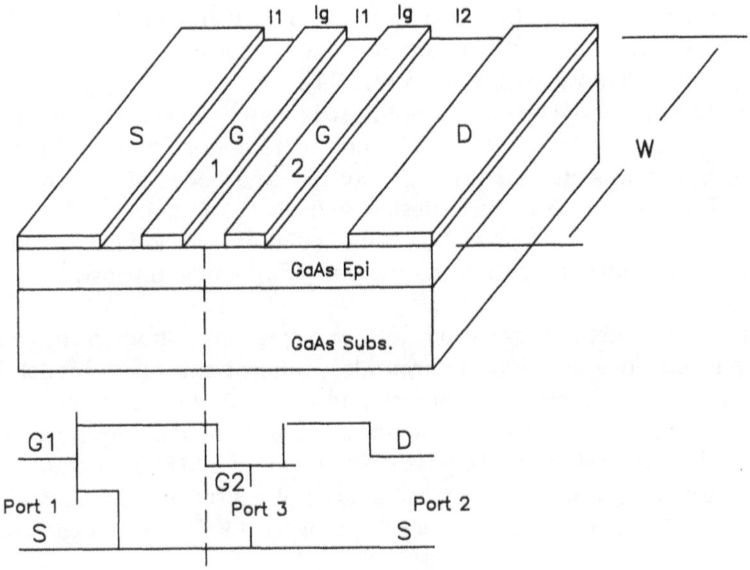

Fig. 5.15 Dual-gate MESFET structure.

except that it includes a second gate between the first gate and drain. The device can be made with low noise and a wide dynamic range at microwave frequencies. The principal advantages [5.30, 5.31] of the dual-gate structure compared with the single-gate MESFET are (1) greater functional capabilities owing to the two independent gates and (2) lower feedback which improves the power gain and stability. The ability to control gain by a second gate bias is an example of the merit of the device because this capability is particularly valuable [5.32] for the input amplifier of a communications receiver, which may have to accommodate a very large range of r.f. input levels.

Currently, GaAs MESFETs are not limited to microwave small-signal low noise applications. Their use extends to large-signal applications such as power amplifiers, oscillators, mixers, multipliers, pulsed circuits etc. Furthermore, existing technology is capable of developing GaAs MESFETs that oscillate at as high as 100 GHz.

Another interesting nonlinear application for GaAs MESFETs is as microwave mixers. The MESFET transfer characteristics follow an approximate square law, and the noise and third-order intermodulation behavior are lower than is the case for bipolar transistors or Schottky diodes with less local oscillator (LO) power. However, the most significant difference between a MESFET mixer and a diode mixer (bipolar transistors have serious limitations in frequency) is that the former is capable of achieving conversion gain. This feature reduces the need for amplifier stages in a conventional receiver. If we use a dual-gate MESFET instead of a single-gate device, one of the gates is used as the LO port and the other as the r.f. port, thereby increasing r.f.–LO isolation. We can also use this dual-gate device as a self-oscillating mixer in low cost commercial applications [5.33, 5.34].

All these features can only be optimized by using an accurate description of device behavior. In fact, the use of power transistors, microwave mixers, monolithic circuits etc. implies an exhaustive knowledge of their nonlinear nature. Today it is not enough to design on the basis of small-signal scattering parameters along with "cut-and-try" manipulations of the final circuit; it is necessary to understand and model the transistor's intrinsic nonlinear behavior.

Two main methods are applicable: experimental characterization using load–pull methods or use of an accurate nonlinear equivalent model. Both methods have their own advantages and disadvantages, but the physical nature is better understood by extracting a nonlinear model. Therefore we first make a qualitative study of the physical behavior of a MESFET transistor in this section. Afterwards we propose analytical expressions and calculation methods for the nonlinear elements in order to fit some experimental measurements.

5.7.2 GaAs MESFET operation

The particular band structure of GaAs implies that the variation in drift velocity with electric field is as shown in Fig. 5.16. This curve can be considered as a combination of two straight lines—one for the lower valley and the other for the upper valley—although a more exact treatment requires the solution of two coupled Boltzmann equations [5.34]. Moreover, there are three main regions in this figure: a low field region ($E < E_{s_1}$) with a constant mobility (constant resistance), an intermediate field region ($E_{s_1} < E < E_{s_2}$), where the drift velocity increases with a constant slope until the electric field achieves a peak value (E_p) and the drift velocity starts to drop (negative differential mobility (Gunn zone)), and a high field region ($E > E_{s_2}$) where the mobility is

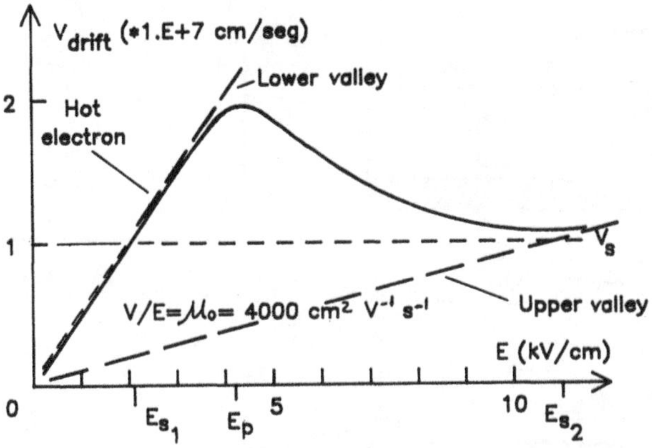

Fig. 5.16 Electron drift velocity versus electric field in GaAs (300 K).

approximately constant and the drift velocity is saturated. An understanding of this behavior is essential for the development of an accurate model.

The existence of a metal–semiconductor contact (gate) between the source and the drain creates a layer in the semiconductor that is completely depleted of free-carrier electrons. This depletion layer acts like an insulating region that modulates the cross-section available for electronic current flow in the active layer. In Fig. 5.17(a) the gate and drain terminals are shorted together and a small voltage is applied at the drain terminal. Under these conditions the depletion region grows and the layer thickness d is smaller than would be the case in the absence of the metal–semiconductor junction. Consequently, the effective resistance between the source and the drain increases. The current I_{DS} flowing from drain to source will be [5.35, 5.36]

$$I_{DS} = wqn(x)\,v(x)\,d(x) \tag{5.53}$$

where w is the gate length, q is the electron charge, $n(x)$ is the density of the conducting electrons, $v(x)$ is the drift velocity, $d(x)$ is the thickness of the conductive layer and x is the coordinate in the direction of electron movement. The electron density n is equal to the donor density N_d when the electric field is below the critical value E_c.

An increase in the drain voltage forces the depletion layer to become wider and the channel cross-section to become narrower. The decrease in the conductive cross-section $d(x)$ must be compensated by an increase in the electric field and the electron velocity v to maintain a constant current through the channel. A further increase in the drain voltage makes the electrons attain the limiting velocity V_s (Figs 5.16 and 5.17(b)). The channel is constricted to the smallest thickness d_0 under the gate. The electric field reaches the critical value E_c and the current starts to saturate.

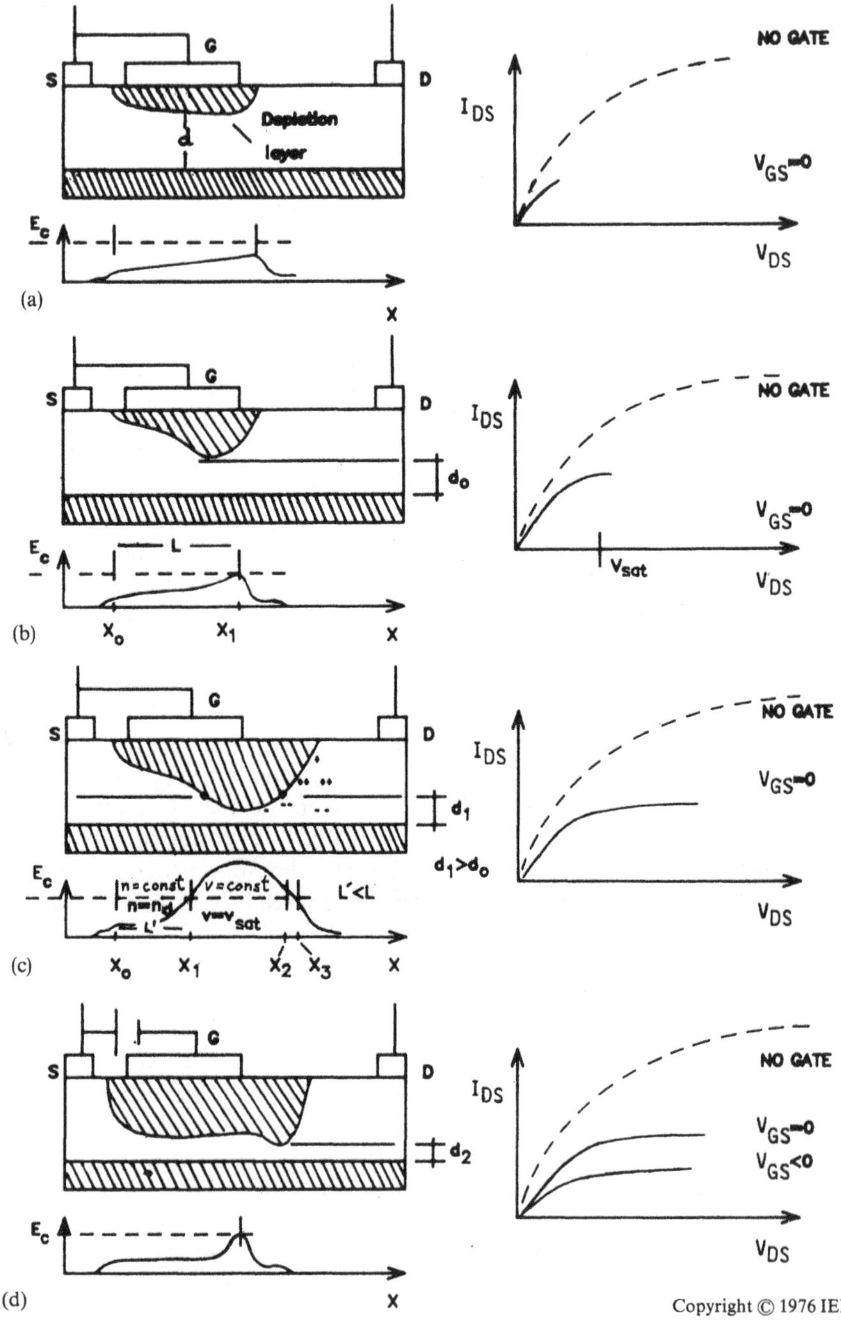

Fig. 5.17 (a), (b), (c) Current controlled by the depletion layer under a Schottky gate shorted to the source; (d) the condition for a negative gate bias [5.36].

When the drain voltage exceeds V_{sat}, the depletion zone grows toward the drain. The point x_1 where the electrons reach V_s moves slightly toward the source (Fig. 5.17(c)) causing a voltage drop at x_1. As a consequence d_1 increases ($d_1 > d_0$) and more current is injected into the velocity-limited region. This results in a positive slope in the I_{DS} curve and a finite drain–source resistance exists beyond saturation [5.37]. This effect is particularly important in microwave MESFETs.

In fact the channel potential grows from x_1 to the drain zone, the depletion layer widens and d becomes narrower ($d < d_1$). When the electron velocity saturates, the change in the channel cross-section must be compensated by a change in the carrier concentration in order to maintain constant current; this implies electron accumulation between x_1 and x_2 ($d < d_1$). At x_2 the cross-section is again d_1 (Fig. 5.17(c)) and the negative space charge changes to positive in order to maintain constant current. The positive space-charge accumulation is mainly due to the partial electron depletion. The electron velocity remains saturated between x_2 and x_3 because of the additional field due to the negative space charge [5.37].

When a negative voltage is applied at the gate port (Fig. 5.17(d)), the depletion-layer thickness increases ($d_2 < d_1$). At small values of the voltage V_{DS} the channel acts as a linear resistor, but this resistance is larger owing to the narrower cross-section available for current flow. An increase in V_{DS} implies that the electric field reaches its critical value for currents lower than is the case for $V_{GS} = 0$ (the actual channel resistance is larger).

Because of the particular properties of GaAs materials (Fig. 5.16), the behavior described above is only valid for GaAs MESFETs working in the low field region. Figure 5.18 shows GaAs MESFET behavior in the saturated region. The narrowest channel cross-section is located under the drain end of the gate. The drift velocity reaches a peak value at x_1 and then decreases until it attains the saturated value V_s under the gate edge. In order to maintain the continuity of the current according to eqn (5.53) electrons accumulate in this region [5.36], i.e. the electron velocity decreases with increasing x. Between x_2 and x_3 the situation is exactly the opposite: the channel cross-section increases and the electrons move faster. The charges in the accumulation and depletion layers are almost equal and most of the drop in drain voltage is across this dipole layer.

There is an additional effect that takes place at the edge of the gate finger close to the drain (Fig. 5.18) in the highest field zone; this is the breakdown effect [5.38, 5.39]. The current I_{DG} is due to the generation of electron–hole pairs and the breakdown behavior is considered [5.35] to be due to the formation of a dipole layer behind the depletion layer near the drain which absorbs part of the voltage drop across the drain and the gate (Fig. 5.19). This effect facilitates field accumulation near the gate edge and an increase in the total potential between the gate and the drain before the onset of breakdown.

The other currents that arise in a MESFET device are the channel current

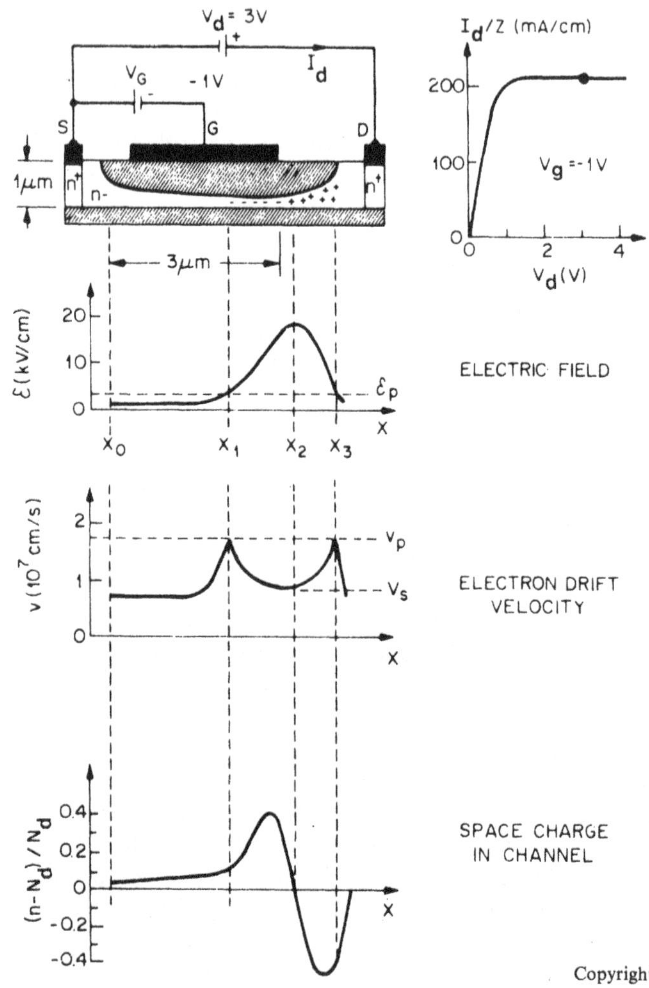

Fig. 5.18 Channel cross-section, electric field, drift velocity and space-charge distribution in the channel of a GaAs MESFET operated in the saturated-current region [5.36].

I_{DS} between the drain and the source and the rectified current I_{GS} between the gate and the source, although the latter exists only when the MESFET is operating in the large-signal regime.

So far a necessary qualitative description of MESFET behavior has been developed, but it is inadequate for design purposes. A quantitative nonlinear model is required, i.e. a model which relates MESFET material characteristics and dimensions to the parameters of a lumped equivalent circuit. In order to make realistic use of a given transistor it is important to be able to determine

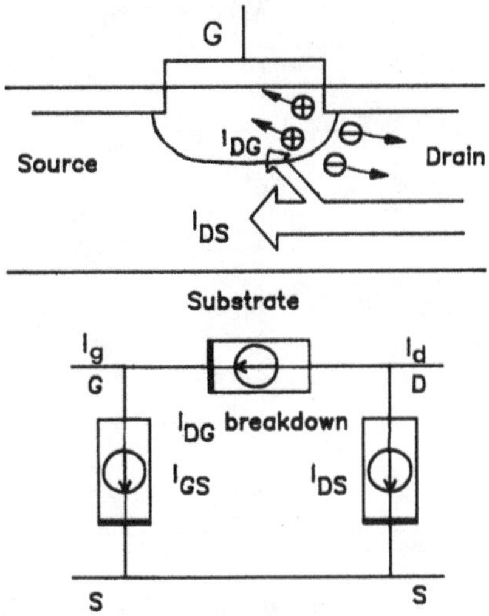

Fig. 5.19 Breakdown effect.

all the parameters of the circuit, particularly their nonlinearities, from the model. This problem will be treated in the following sections.

5.7.3 Physical modelling

An interesting approach to modelling a MESFET has been described by Madjar and Rosenbaum [5.40], and this model has also been used for dual-gate MESFET operation [5.41]. Figure 5.20 shows the intrinsic model proposed together with the rest of the linear elements. The diodes represent the Schottky gate–source and gate–drain junctions, which are assumed to be ideal [5.40]. R_{SD} and R_{SS} represent the bulk resistance of the epitaxial layer, and R_{sub} represents the substrate leakages. Madjar and Rosenbaum used approximate analytical solutions to the semiconductor differential equations which depend upon the device geometry and material parameters. The model tries to express the total terminal currents (I_d, I_g, I_s) as explicit functions of the terminal voltages (V_{GS}, V_{DS}) and their time derivatives. Obviously, these expressions can be used with conventional time-domain circuit analysis programs.

The transport properties in a MESFET device are governed by the semiconductor equations [5.42, 5.43]

$$\nabla^2\psi = -\frac{e(N_d - n)}{\varepsilon} \tag{5.54}$$

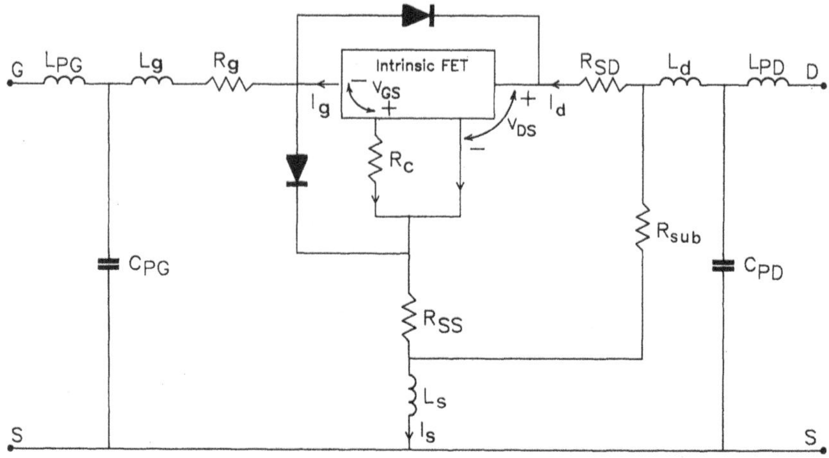

Fig. 5.20 Madjar's large-signal MESFET model [5.40].

$$q\frac{\partial n}{\partial t} = \nabla \boldsymbol{J} \tag{5.55}$$

$$\boldsymbol{J} = -en\bar{\mathbf{v}} + eD\nabla n \tag{5.56}$$

$$\boldsymbol{J}_t = \boldsymbol{J} + \varepsilon\frac{\partial \boldsymbol{E}}{\partial t} \tag{5.57}$$

where N_d is the donor concentration, e is the electron charge, ε is the dielectric permittivity, $\boldsymbol{J}$ is the conduction current density, $\boldsymbol{E}$ is the electric field, $\mathbf{v}$ is the average carrier velocity, and n is the carrier density. The system unknowns are the instantaneous carrier distribution n, the electric potential and the total current density $\boldsymbol{J}_t$. The solution of these equations needs to take into account the geometrical boundary conditions, the externally applied voltages and the velocity–electric field relation for the carriers.

Madjar and Rosenbaum [5.40] approximated the $v(E)$ curve by a piecewise linear characteristic (Fig. 5.16) whose equation is [5.42]

$$v(E) = \mu E \qquad E \leqslant E_{p_1} \tag{5.58a}$$

$$v(E) = V_s \qquad E > E_{p_1} \tag{5.58b}$$

The complete equation, including the negative mobility zone, has been given by Yamaguchi *et al.* [5.43].

If we could find an analytical expression for the carrier density n, it would be easy to deduce the current density and the electric field. For this purpose the active region under the gate can be subdivided into three regions (Fig. 5.21): a depletion region under the gate barrier ($n = 0$), a free-channel region ($n = N_d$) and a transition region in which n varies from 0 to N_d along the y coordinate.

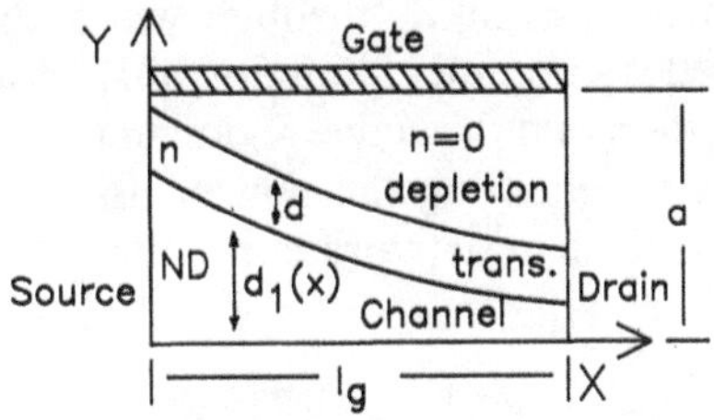

Fig. 5.21 Subdivisions of the active region under the gate [5.40].

The height d of the transition zone is taken as six times the Debye length. The channel height d_1 varies strongly with V_{GS} but depends weakly on V_{DS}. The transition region can be described by [5.42]

$$n(x, y) = N_d\{1-\alpha(x-\Gamma)\}\frac{1+\cos\{\pi(y-d_1)/d\}}{2} \tag{5.59}$$

where α and Γ are derived from the boundary conditions. Once the carrier distribution is known, an expression for the electric field can be derived by solving the Laplace and Poisson equations [5.40, 5.42]. The currents can then be derived numerically, although Yamaguchi and Kodera [5.42] used an analytical approach.

The displacement currents are computed from

$$\boldsymbol{J}_{\text{dis}} = \varepsilon\frac{\partial \boldsymbol{E}}{\partial t} \tag{5.60}$$

by integrating over the appropriate electrode boundaries. Since the electric field is a function of V_{GS} and V_{DS}, the displacement current depends on the time derivatives of these voltages:

$$I_g = GV_{GS}\frac{dV_{GS}}{dt}+GV_{DS}\frac{dV_{DS}}{dt} \tag{5.61}$$

$$I_d = I_{con}+DV_{GS}\frac{dV_{GS}}{dt}+DV_{DS}\frac{dV_{DS}}{dt} \tag{5.62}$$

$$I_s = I_{con}+SV_{GS}\frac{dV_{GS}}{dt}+SV_{DS}\frac{dV_{DS}}{dt} \tag{5.63}$$

where I_{con} is the conduction current. The other parameters are nonlinear capacitive coefficients which, together with I_{con}, depend on the instantaneous values of V_{GS} and V_{DS}. From the current continuity condition, the coefficients are related by

$$SV_{GS} = DV_{GS}-GV_{GS} \tag{5.64}$$

$$SV_{DS} = DV_{DS}-GV_{DS} \tag{5.65}$$

The effective charge resistor R_c through which the depletion region is charged and discharged is a nonlinear function of V_{GS} and V_{DS}. The calculation is based on the assumption that the active region under the gate is a distributed RC transmission line. By equating the power dissipated in this distributed network to that in a equivalent network containing R_c and the total channel capacitance we can write

$$R_c = \int_0^{l_g} \left\{1 - \frac{C(x)}{C_t}\right\}^2 R(x)\,dx \tag{5.66}$$

where C_t is the total channel capacitance, $C(x)$ is the channel capacitance between 0 and x, and $R(x)$ is the channel resistance per unit length.

Since the relationship between the electron velocity and the electric field is known, the transit time is computed from the electric field distribution. The other parameters defining the model have already been treated.

5.7.4 Large-signal model of a GaAs MESFET

Figure 5.22(a) shows the lumped-element equivalent circuit given by Liechti [5.36]. This model to a first approximation follows the physical behavior of a classical MESFET structure (Fig. 5.22(b)).

An exhaustive study of the work related to the problem of modelling MESFET transistors [5.15, 5.31, 5.38, 5.44–5.52] shows that each researcher develops his own model by including or removing some elements in order to improve the simulation of a particular characteristic (physical semiconductor properties or experimental measurements). The reduction of all these models to a supermodel would make the final result so complicated as to be useless. Moreover, the use of a sophisticated model does not allow the physical phenomena to be completely understood.

In this section we use the circuit model shown in Fig. 5.23. Apart from the different alternatives for the analytical expressions for the nonlinear elements, this model is basically the same as that proposed by Materka and Kacprzak [5.44] and Obregon [5.49]. It is composed of 17 intrinsic and extrinsic elements that depend on external voltages. This dependence is related to the nature of the physical phenomenon to be modelled. Following the notation given in Chapter 1, R_g, R_d, R_s, L_g, L_d, L_s, L_{PG}, L_{PD}, C_{PGS} and C_{PDS} are the extrinsic elements. Some workers [5.27] classify R_g, R_d and R_s as intrinsic elements, but because of their small bias dependence we shall classify them as extrinsic.

The intrinsic part of this model can be divided into three main blocks.

(a) The first block models the internal drain current. C_{DS} is the drain-to-source capacitance due to the substrate (Fig. 5.22) and the generator I_{DS} represents the channel current flowing in the drain-to-source direction. This

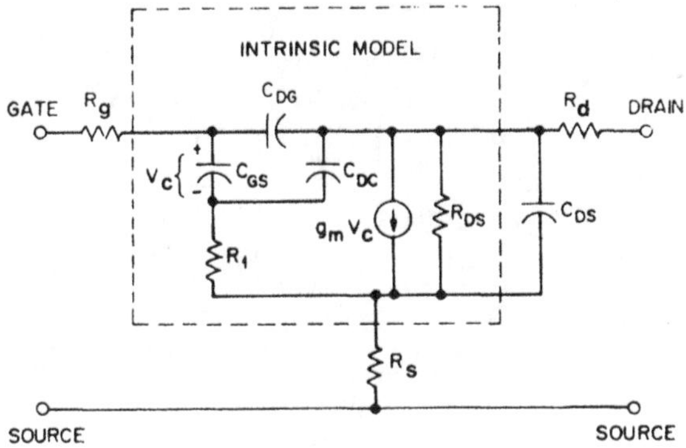

(a)

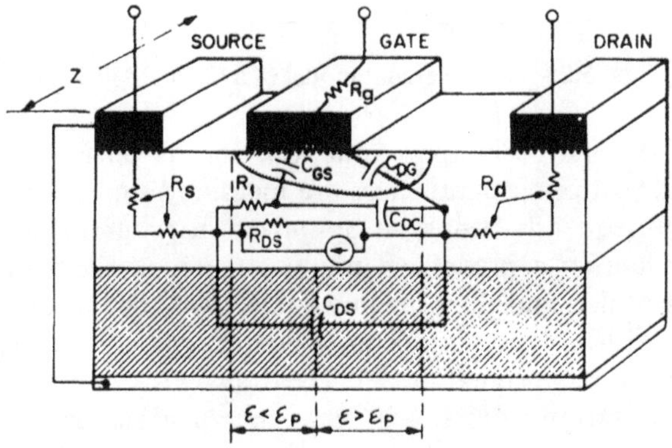

(b)

Fig. 5.22 (a) Equivalent circuit of a MESFET; (b) physical origin of the circuit elements [5.36].

current generator will depend mainly on the internal gate and drain voltages V_{gi} and V_{di}.

(b) The second block corresponds to the gate forward current I_f and represents a Schottky junction [5.53]. I_{GS} is the gate–channel current, C_{GS} is the gate–source equivalent capacitance and R_i takes into account the ohmic part of the depletion layer under the gate port toward the drain.

(c) Finally, the third block models the reverse current I_r, where I_{DG} is the breakdown current between the drain and the gate and C_{GD} is the parasitic gate–drain capacitance.

Some workers [5.15, 5.46, 5.47, 5.54] have deduced the MESFET parameters of the large-signal model by starting from a linear model which

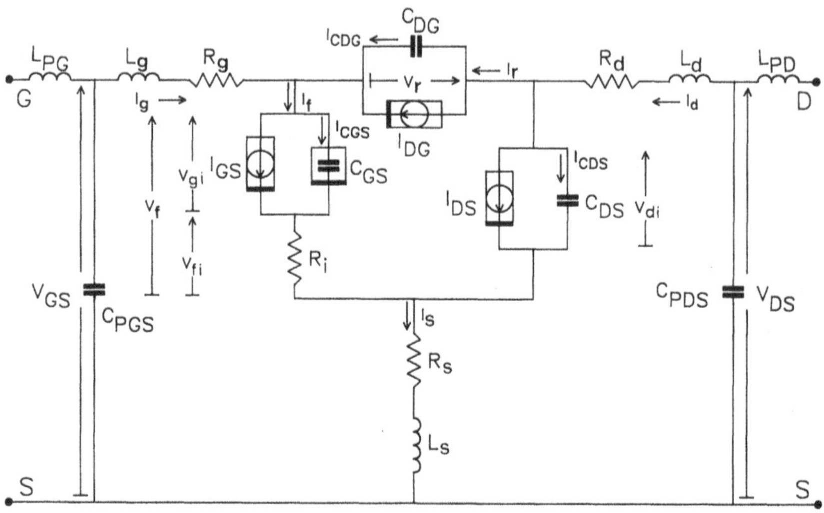

Fig. 5.23 Large-signal model of a GaAs MESFET.

they consider as a first-order approximation. For this purpose a large number of small-signal scattering parameters are measured under a number of bias conditions. Subsequently, using an optimization program, we can fit these experimental data to a predefined nonlinear model. Elements which vary slightly are identified and their mean value is used. The results obtained are used together with polynomial forms which approximate these bias dependences to derive a large-signal model. The drawback of this approach is that the model optimization requires a large number of measurements. In addition, it is impossible to derive all the element values unambiguously (Chapter 2). It is thus necessary to make d.c. and low frequency measurements to fix some model elements and low frequency S parameter measurements to fix others [5.55] before optimizing the microwave results.

An alternative approach to the large-signal characterization of the microwave GaAs MESFET is to consider that the large-signal device properties are governed primarily by the transistor d.c. characteristics [5.38, 5.44, 5.45, 5.48, 5.49]. In any case, extrapolation of these results to predict dynamic MESFET behavior implicitly invokes the quasi-static assumption. This means that the charge in the MESFET channel is defined uniquely from the instantaneous terminal quantities. The rates of change of the depletion layer and the carrier transit time impose limitations upon the maximum frequencies for which the assumption remains valid.

Consequently, in this section we shall attempt to give some analytical expressions and methods for a complete development of the nonlinear model shown in Fig. 5.23. We shall treat only the four nonlinearities I_{DS}, I_{DG}, I_{GS} and C_{GS} because they are the most important in a GaAs MESFET.

5.7.4.1 I_{DS} source

Many analytical solutions have been developed to model the drain current I_{DS}. All these equations attempt to reproduce the d.c. characteristics as exactly as possible.

Gopinath and Rankin [5.45] used the model shown in Fig. 5.24. They deduced the parameter values from small-signal scattering measurements and d.c. MESFET characteristics. Figure 5.25 shows I_{DS} versus V_{DS} with V_{GS} as parameter. These curves can be fitted by the following analytical function:

$$I_{DS} = I_{DSS} F_g(V_{GS}) F_d(V_{GS}, V_{DS}) \tag{5.67}$$

where

$$F_g(V_{GS}) = \left(1 - \frac{V_{GS}}{V_p}\right)^2 \tag{5.68}$$

$$F_d(V_{DS}) = 1 + \frac{V_{DS}}{I_{DSS}/R_{DSO}} \qquad V_{DS} > V_{tan} \tag{5.69a}$$

$$F_d(V_{GS}, V_{DS}) = 1 + \frac{V_{DS}}{I_{DSS}/R_{DSO}} - \left(V_{tan} - \frac{V_{DS}}{V_{tan}}\right)^3 \qquad V_{DS} < V_{tan} \tag{5.69b}$$

V_{tan} follows a parabola (broken curve) that starts at the origin and intersects the point (V_{dt}, I_{dt}) defined by

$$I_{dt} = I_{DSS} + \frac{V_{dt}}{R_{DSO}} \tag{5.70}$$

$$V_{tan} = V_0\left\{1 + \left(1 + 2\frac{R_{DSO} I_{DSS}}{V_0}\right)^{1/2}\right\} \tag{5.71}$$

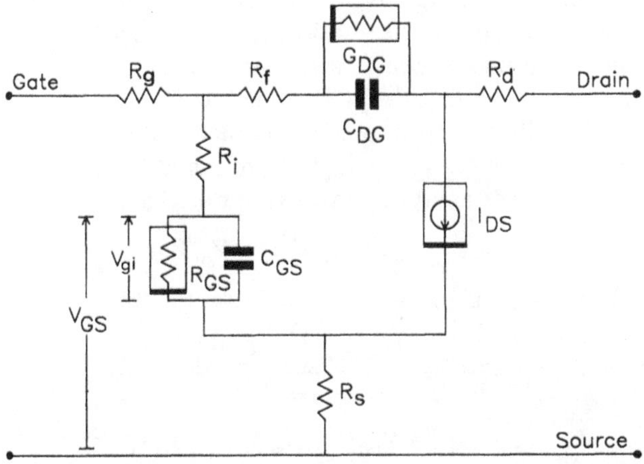

Fig. 5.24 Gopinath's equivalent circuit [5.45].

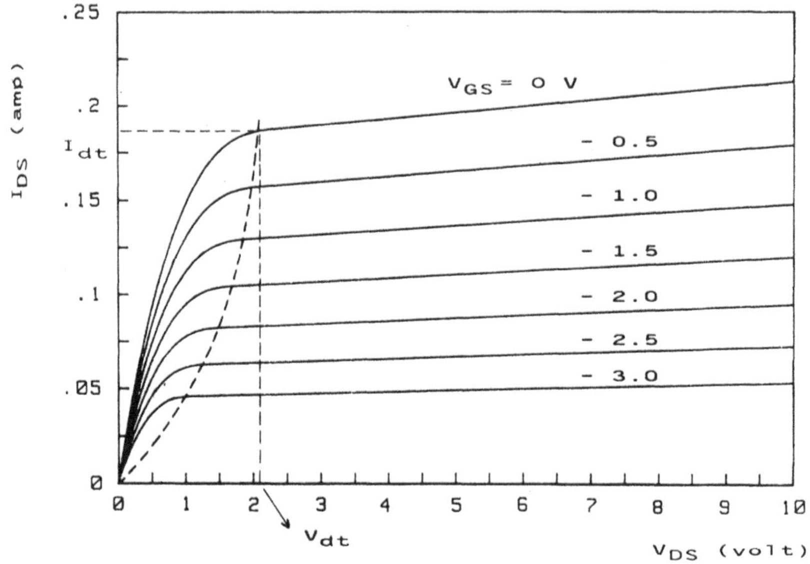

Fig. 5.25 I_{DS}–V_{DS} characteristics [5.45].

where

$$V_0 = \left(1 - \frac{V_{GS}}{V_p}\right)^2 \frac{V_{dt}^{\,2}}{2(I_{DSS} R_{DSO} + V_{dt})} \tag{5.72}$$

V_p is the gate pinch-off voltage, R_{DSO} is the output conductance at $V_{GS} = 0$ and I_{DSS} is the drain–source saturation current.

It should be noted that there are two equations for F_d. The first is valid for $V_{DS} < V_{tan}$ where the d.c. characteristics are strongly nonlinear with respect to V_{DS}, while the second holds for the quasi-linear zone where V_{DS} is greater than V_{tan}. Furthermore, the independent control variables are V_{GS} and V_{DS} (Fig. 5.24) rather than the internal voltage drops V_{gi} and V_{di} defined in Fig. 5.23.

Tajima *et al.* [5.48] have proposed a general equation for I_{DS} with V_{gi} and V_{di} as independent variables. Obregon [5.49] has modified and completed this interesting approach. The analytical expressions used are

$$I_{DS} = I_{DSS} F_g(V_{gi}, V_{di}) F_d(V_{gi}, V_{di}) \tag{5.73}$$

where the functions are defined as

$$F_g = \frac{V_{g\,in} - \{1 - \exp(-M_{ss} V_{g\,in})\}/M_{ss}}{K_{ss}} \tag{5.74}$$

$$F_d = 1 - \exp\{-V_{d\,in}(1 + A_{ss} V_{d\,in} + B_{ss} V_{d\,in}^{\,2})\} \tag{5.75}$$

$$K_{ss} = 1 - \frac{1 - \exp(-M_{ss})}{M_{ss}}$$

$$V_{g\,in} = 1 + \frac{V_{gi} - V_{phi}}{V_p}$$

$$V_{d\,in} = \frac{V_{di}/V_{di\,ss}}{1 + W_{ss}\,V_{gi}/V_p} \tag{5.76}$$

$$V_p = V_{po} + P_{ss}\,V_{di} + V_{phi}$$

M_{ss}, A_{ss}, B_{ss}, P_{ss}, W_{ss}, $V_{di\,ss}$ and V_{phi} are weighting factors.

Materka and Kacprzak [5.44] have published another interesting model for I_{DS}. The main equation is identical with eqn (5.73) but F_g and F_d are given by two simple expressions:

$$I_{DS} = I_{DSS}\left(1 - \frac{V_{gi}}{V_p}\right)^2 \tanh\left(\frac{\alpha V_{di}}{V_{gi} - V_p}\right) \tag{5.77}$$

$$V_p = V_{po} + \gamma V_{di}$$

where α and γ are weighting parameters. The hyperbolic tangent function is predominant for small values of V_{di}, whereas the expression $(1 - V_{gi}/V_p)^2$ is predominant in the zone where I_{DS} varies more slowly. This is logical because the hyperbolic tangent approaches unity with increasing V_{di}.

A further approach that merits attention [5.56] has been implemented in the well-known circuit-simulation program SPICE [5.57] by following the model developed by Curtice [5.58]:

$$I_{DS} = \beta(V_{GS} - V_t)^2 (1 + \Gamma V_{DS}) \tanh(\alpha V_{DS}) \tag{5.78}$$

where V_t is the threshold voltage, β is the transconductance parameter, Γ is the channel-length modulation parameter and α is the hyperbolic tangent function parameter. The use of a hyperbolic tangent function greatly improves the usefulness of the equation below saturation.

In all these cases the instantaneous dynamic current is given by

$$I_{DS}(t) = F\{V_{gi}(t - \tau), V_{di}(t)\} \tag{5.79}$$

where τ is the transit time which strictly is nonlinear but is considered as a linear part of the model.

5.7.4.2 Schottky junction

Two parallel nonlinear elements—the source I_{GS} and the capacitance C_{GS}—are also present in Fig. 5.23. These elements, together with the resistors R_g, R_i and R_s, represent a real Schottky junction [5.44, 5.45, 5.53, 5.54].

The nonlinear equation for the gate–source capacitance is given by

$$C_{GS} = C_{GSO}\left(1 - \frac{V_{gi}}{V_{bi}}\right)^{-\gamma} \qquad V_{gi} < kV_{bi} \tag{5.80a}$$

where C_{GSO} is the gate–source capacitance at $V_{gi} = 0$, V_{bi} is the built-in voltage, and γ and k are weighting parameters for a given transistor. Equation (5.80a) increases very rapidly for $V_{gi} > kV_{bi}$, which presents a difficult problem in computer simulation. In order to avoid this inconvenience, the C_{GS} curve can be approximated by a constant which maintains the continuity of the first derivative:

$$C_{GS} = C_{GSo}(1-k)^{-\gamma-1}\left\{\gamma\frac{V_{gi}}{V_{bi}}+1-k(1-\gamma)\right\} \qquad V_{gi} > kV_{bi} \qquad (5.80b)$$

The classical equation for a junction diode can be used to represent the behavior of the current I_{GS}:

$$I_{GS} = I_{ns}\{\exp(\alpha_s V_{gi})-1\} \qquad (5.81)$$

where I_{ns} is the saturation current and α_s is the exponential factor. Equation (5.81) exhibits a similar problem to that of the equation for C_{GS}: the term $\exp(\alpha_s V_{gi})$ may cause computer convergence problems. As in the previous case, one way of eliminating possible failures is to approximate I_{GS} by a parabola beyond a certain limiting value M for the product $\alpha_s V_{gi}$, i.e.

$$P(V_{gi}) = \exp(M)\left\{1+(\alpha_s V_{gi}-M)+\frac{(\alpha_s V_{gi}-M)^2}{2}\right\} \qquad \alpha_s V_{gi} > M \qquad (5.82a)$$

$$P(V_{gi}) = \exp(\alpha_s V_{gi}) \qquad \alpha_s V_{gi} \leqslant M \qquad (5.82b)$$

and consequently eqn (5.81) can be rewritten as

$$I_{GS} = I_{ns}\{P(V_{gi})-1\} \qquad (5.83)$$

5.7.4.3 Breakdown effect

The physical nature of the breakdown effect was discussed in Section 5.6.2 and here again there are many models. Figure 5.26(a) shows I_{DG} versus V_{DS} with V_{GS} as parameter [5.38], while Fig. 5.26(b) shows I_{DG} versus the reverse voltage V_{DG}. The latter curves indicate that the breakdown current depends exclusively on the voltage drop between the gate and the drain when the MESFET is in pinch-off.

Tajima and Miller [5.38] simulate the current I_{DG} by a pair of equations depending on the breakdown voltage V_b and a series resistance R_b:

$$I_{DG} = 0 \qquad \text{for } V_r < V_b \qquad (5.84a)$$

$$I_{DG} = \frac{V_r - V_b}{R_b} \qquad \text{for } V_r \geqslant V_b \qquad (5.84b)$$

These parameters are also linear functions of the current I_{DS}:

$$V_b = V_{bo} + R_1 I_{DS} \qquad (5.85a)$$

$$R_b = R_{bo} + R_2\frac{I_{DS}}{I_{DSS}} \qquad (5.85b)$$

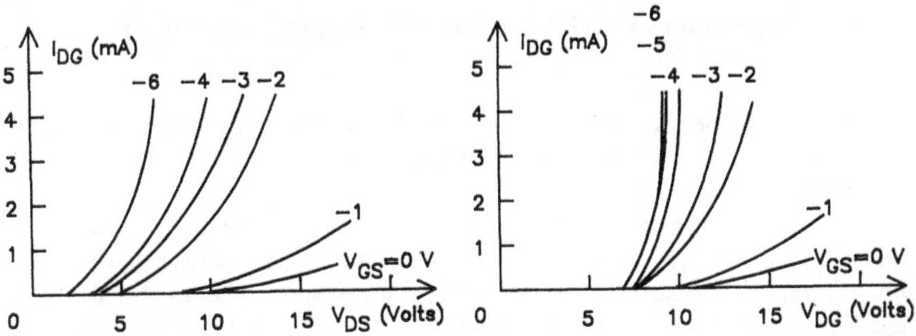

(a) (b)

Fig. 5.26 Breakdown current as a function of (a) the drain–source voltage and (b) the drain–gate voltage [5.38].

V_{b0}, R_{b0}, R_1, R_2 and I_{DSS} are parameters that must be fitted to experimental measurements.

Obregon [5.49] used an exponential function to describe this effect:

$$I_{DG} = I_{nsr}\{1 + I_{nsr0} V_{di}{}^{\alpha_{sr0}}\}^{\alpha_{sr}} \tag{5.86a}$$

$$\alpha_{sr} = A_{sr} - B_{sr} V_{gi} \tag{5.86b}$$

Equations (5.86) are a direct function of the independent internal voltages V_{gi} and V_{di} (Fig. 5.23) and use five fitting parameters I_{nsr}, I_{nsr0}, α_{sr0}, A_{sr} and B_{sr}.

Materka and Kacprzak [5.44] utilized a diode characteristic approximation:

$$I_{DG} = I_{nsr}\{\exp(\alpha_{sr} V_r) - 1\} \tag{5.87}$$

The fundamental difference lies in the fact that the above equation does not contain the parameters generally used in defining a junction diode. This expression suggests a treatment analogous to that applied to eqn (5.72), but using M_r instead of M:

$$P_r(V_r) = \exp(M_r)\frac{1 + (\alpha_{sr} V_r - M_r) + (\alpha_{sr} V_r - M_r)^2}{2} \qquad \alpha_{sr} V_r > M_r \tag{5.88a}$$

$$P_r(V_r) = \exp(\alpha_{sr} V_r) \qquad \alpha_{sr} V_r \leqslant M_r \tag{5.88b}$$

Then eqn (5.87) becomes

$$I_{DG} = I_{nsr}\{P_r(V_r) - 1\} \tag{5.89}$$

The advantage of this expression over eqns (5.84) and (5.86) lies in the fact that only two parameters are introduced, but the independent variable is V_r instead of the internal voltages V_{gi} and V_{di} as in eqn (5.86). In any case we must carefully verify the usefulness of the equation over the range of utilization. In fact, as the derivation of a model can be very complicated, it is not possible to know *a priori* whether the equations will be monotonic. Some V_{GS}, V_{DS} pairs can give surprising results.

5.7.5 Experimental determination of the model parameters

Although the physical meaning of MESFET parameters is treated extensively in Chapter 1, we shall reconsider some aspects in this section so as to achieve a self-contained chapter.

As previously noted, there are some methods where the derivation of MESFET parameters is exclusively based on small-signal scattering measurements (quasi-static approach). However, the existence of parasitic elements and thermal problems associated with particular bias points means that the modelling of a given transistor must be performed in several steps, each of which is only used for the derivation of some of the elements.

(a) D.c. and pulsed measurements: the equivalent circuit consists only of resistive elements and voltage-controlled current sources.

(b) Scattering measurements at low frequencies (below 2 GHz): in this case we can simplify the model by eliminating the parasitic reactances.

(c) Scattering measurements at microwave frequencies: these measurements will serve to determine the parasitic elements.

Pulsed measurements are used instead of d.c. measurements to minimize the unwanted thermal effects associated with the modelling of the intrinsic current sources of a transistor. Figure 5.27 shows an experimental set-up [5.49] used for pulsed measurements on a given device. The absorbing loads avoid high frequency oscillations, and the signal applied at the two terminals consists of synchronous rectangular pulses superposed on d.c. voltages. As we shall need large drain voltage excursions, the pulse at the drain is amplified beforehand and filtered by an *RC* network in order to avoid overshoot phenomena capable of destroying the transistor. The same precaution is used for the gate. The pulse frequency is between 10 and 100 kHz, and the pulse width is maintained about two orders of magnitude lower than the period. Pulses are superposed on the operating d.c. voltages in order to keep the best thermal agreement between measured and posterior operation.

In the low frequency range (less than 2 GHz) the parasitic reactances and bonding inductances (L_{PG}, L_{PD}, C_{PGS}, C_{PDS}, L_g,L_d, L_s) can be eliminated from the model. Moreover, if we use a range of bias points where the breakdown and Schottky currents are negligible, the model is substantially simplified. The measurement of small-signal scattering parameters at different bias points will allow us to determine [5.55] the evolution of the nonlinear capacitances C_{GS} and C_{DG} (normally C_{DG} can be approximated by a linear capacitance by using the mean value over the entire range of bias points). Finally, small-signal scattering measurements over a broad range of microwave frequencies are used to calculate the parasitic elements that complete the model.

Therefore in this section we shall treat the general problem of the

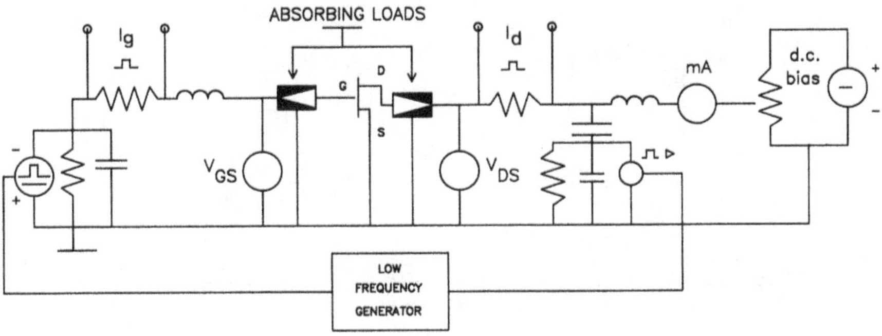

Fig. 5.27 Experimental set-up for pulsed measurements [5.49].

experimental determination of model parameters by using the stepwise procedure outlined above.

5.7.5.1 Calculation of R_d and R_s

In order to measure the source resistance R_s, a d.c. current source is attached to the gate–source junction with the drain terminal open. The test bench is shown in Fig. 5.28(a). Starting from the measurement of V_{DS} we can write [5.44, 5.54]

$$R_s = \frac{\Delta V_{DS}}{\Delta I_g} \qquad \text{floating drain} \tag{5.90}$$

where ΔV_{DS} and ΔI_g are the incremental drain–source voltage and gate current respectively. The main problem in this case is the strong dependence of R_s on the gate current I_g. The problem can be avoided using the average value obtained over a wide range of bias currents.

The resistance R_d is found in a similar manner by forward biasing the gate–drain junction and measuring the voltage drop at the floating source (Fig. 5.28(b)):

$$R_d = \frac{\Delta V_{DS}}{\Delta I_g} \qquad \text{floating source} \tag{5.91}$$

The measuring problem is the same as that for R_s.

5.7.5.2 Schottky junction—R_g, R_i, α_s, I_{ns}

The process performed above is approximate not only because the source resistance R_s depends strongly on the gate current but also because the dependence varies when a drain bias is applied. Kurita and Morita [5.53] demonstrated this dependence and gave an interesting method for the overall determination of a real Schottky junction. For this purpose, they used the equivalent unilateral model (Fig. 5.29(a)) where the dependences of I_{DS} on V_{gi}

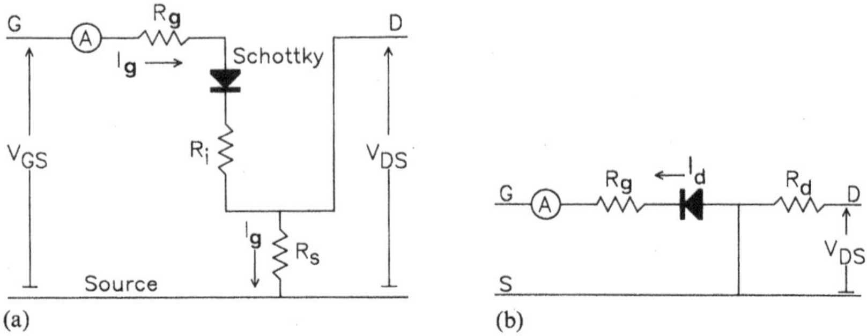

Fig. 5.28 Measurement circuits for (a) R_s and (b) R_d.

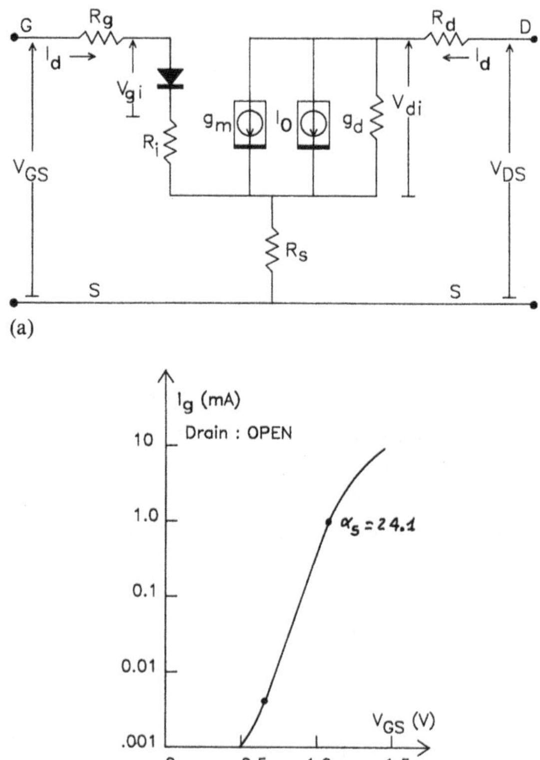

Fig. 5.29 (a) Unilateral MESFET model; (b) experimental logarithmic I_g–V_{GS} characteristic with nonbiased drain circuit [5.53].

and V_{di} are separated. In the unilateral approach it is assumed that we are far from the breakdown effect (where the I_{DS} current is nearly constant with V_{DS})

and can thus neglect this dependence. With the drain terminal open, the voltage drop across the diode is

$$V_{\mathrm{diode}} = V_{\mathrm{GS}} = I_g R + \frac{1}{\alpha_s} L_n \frac{I_g}{I_{ns}} \tag{5.92}$$

where $R = R_g + R_i + R_s$. We can construct an experimental curve (Fig. 5.29(b)) and for small values of I_g ($I_g \approx 100\,\mu\mathrm{A}$) we are on the linear part of the logarithmic curve. Therefore we have

$$I_g R \ll \frac{1}{\alpha_s} L_n \frac{I_g}{I_{ns}} \tag{5.93}$$

We can obtain $1/\alpha_s$ and the saturation current I_{ns} directly from the slope of this linear region. The intrinsic parameters (independent of the operating point) are thus determined.

We can derive R_g by using eqn (5.92) at points where the curve is not already linear and using eqn (5.90) for R_s [5.54]:

$$R_g = \frac{R - R_s}{3} \tag{5.94}$$

When I_d is saturated with respect to the drain voltage V_{DS}, we can write, with reference to Fig. 5.29(a),

$$V_{\mathrm{GS}} = V_{gi}(I_g) + I_g R' + \{g_m V_{gi}(I_g) + I_0\} R_s' \tag{5.95}$$

$$I_d = I_0 + g_m V_{gi}(I_g) \tag{5.96}$$

where R' and R_s' represent R and R_s when the drain terminal is biased. I_0 is the d.c. bias current flowing from drain to source and is included in order to describe the variation in the bias voltage in the gate circuit [5.53]. At this point we can construct the experimental curves shown in Fig. 5.30. At the points where I_g is so small that $I_g R'$ can be neglected (the linear zone of Fig. 5.30(a)), we obtain from eqn (5.96)

$$V_{\mathrm{GS}} = \left(\frac{1}{g_m} + R_s'\right) I_d - \frac{I_0}{g_m} \tag{5.97}$$

and by using Fig. 5.30(b) it is easy to derive the terms $1/g_m + R_s'$ and I_0/g_m. By combining Figs 5.29(b) and 5.30, the parameters g_m, I_0, R_s and R' are rigorously defined. Since we know R_s and the terms $1/g_m + R_s'$ and I_0/g_m in eqn (5.97), we can derive the curve $V_{gi}(I_g)$. If we take any value of I_g on the linear part of Fig. 5.30(a) and use eqn (5.96), R' is known and R_g' is obtained from

$$R_g' = \frac{R' - R_s'}{3} \tag{5.98}$$

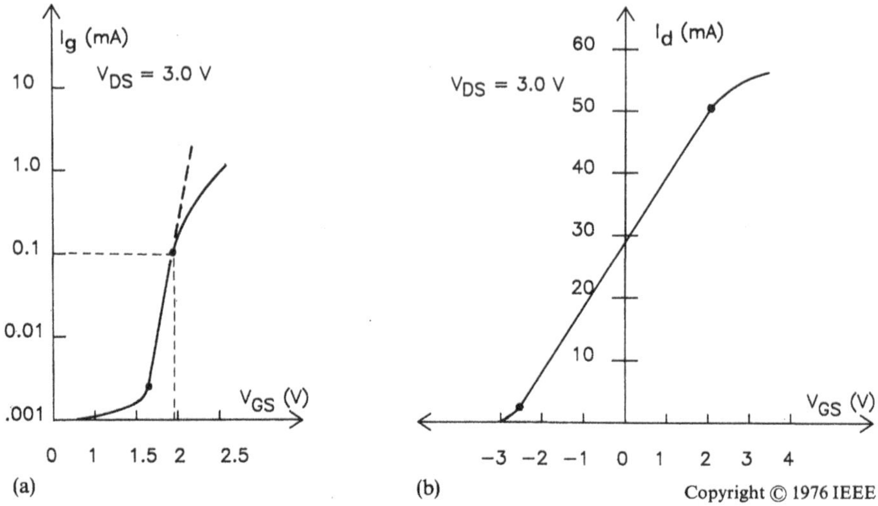

Fig. 5.30 Experimental (a) I_g–V_{GS} and (b) I_d–V_{GS} characteristics with biased drain circuit [5.53].

Since $R_i = R' - R_g' - R_s'$ we have now obtained all the parameters.

Kurita and Morita [5.53] give the following values:

$I_{ns} = 1.14\,\text{pA}$	$A_s = 24.1$	$R = 18.5\,\Omega$
$I_g = 100\,\mu\text{A}$	$R_s = 21.9\,\Omega$	$R' = 60\,\Omega$
$I_0 = 39.5\,\text{mA}$	$g_m = 14.1\,\text{mmho}$	

It is interesting to note the marked difference between R calculated under floating-drain conditions and R' calculated with a bias voltage V_{DS}.

5.7.5.3 Breakdown parameters

In dealing with the breakdown effect the number of parameters to be considered depends upon the choice of model. We can intuitively see that when the Schottky junction is inversely biased the current I_{GS} is almost zero. A measure of the gate current will be a measure of the current I_{DG}, and consequently, as the resistances R_g and R_d are already known, we can derive the reverse voltage V_r. Therefore the breakdown current can be derived as a function of V_r (Fig. 5.26(b)) or V_{DS} (Fig. 5.26(a)). In any case, if we are able to measure the current I_{DG} experimentally (using pulse methods because of the thermal problems), a suitable fitting program will allow us to obtain the desired parameters (Fig. 5.23).

5.7.5.4 Parameters of I_{DS}

When we want to calculate the formal parameters defining the I_{DS} source without prior knowledge of the I_{DG} source, we must measure the current I_d for

V_{DS}, V_{GS} pairs which do not cause breakdown conflicts. As above, pulsed measurements together with a suitable optimization program will allow us to fit the data to one of the analytical equations describing this source. Conversely, if we know the behavior of I_{DG}, measurement of I_d can go up to the breakdown zones. Figure 5.31 shows the drain characteristics obtained by Materka and Kacprzak [5.44] where the breakdown effects are included.

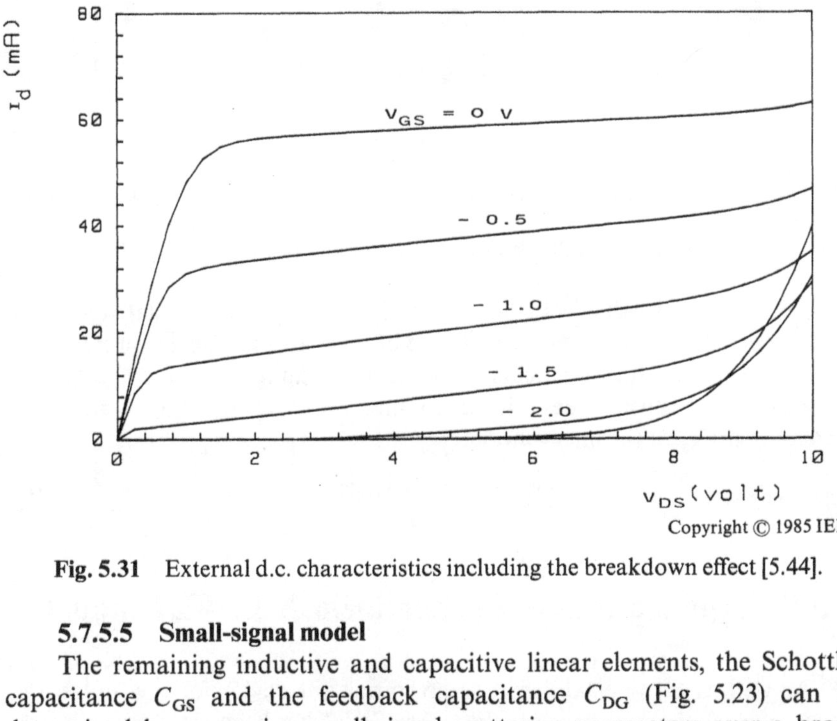

Fig. 5.31 External d.c. characteristics including the breakdown effect [5.44].

5.7.5.5 Small-signal model

The remaining inductive and capacitive linear elements, the Schottky capacitance C_{GS} and the feedback capacitance C_{DG} (Fig. 5.23) can be determined by measuring small-signal scattering parameters over a broad range of frequencies provided that an optimization program will find the appropriate values to fit the measured data over the entire frequency range. Consequently, the small-signal equivalent model is that of Fig. 5.32. This model is only valid for normal small-signal operation, i.e. with the biasing voltages far from breakdown and the Schottky junction inversely biased.

g_m and g_d are the incremental transconductance and drain output conductance for a given bias condition, i.e.

$$g_m = \left(\frac{\partial I_{DS}}{\partial V_{gi}}\right)_o \tag{5.99a}$$

$$g_d = \left(\frac{\partial I_{DS}}{\partial V_{di}}\right)_o \tag{5.99b}$$

where the subscript o indicates derivatives at the operating point (V_{GSo}, V_{DSo}).

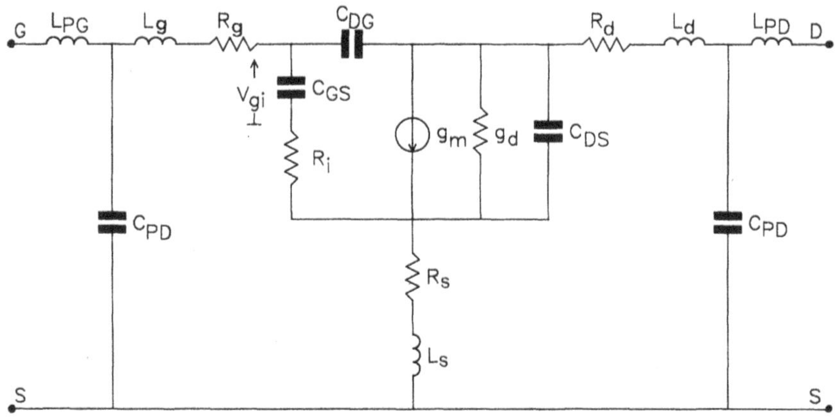

Fig. 5.32 Small-signal equivalent circuit of a MESFET.

These incremental values can be obtained either by small-signal scattering measurements in the low frequency range or from an analytical expression for I_{DS}. The same holds for the calculation of the capacitances C_{GS} and C_{DG} [5.55]. Obviously the transit time τ is taken into account by using a modified expression for g_m in the linear model:

$$g_m = g_m \exp(-j\omega\tau) \tag{5.100}$$

5.8 Applications of the nonlinear MESFET model

As well as being established as an important device for low noise and power microwave amplifiers, the GaAs MESFET has been used in oscillators, frequency multipliers, mixers etc. Large-signal analysis, using an accurate model, is applicable to all these functions. The aim of this philosophy is to predict, rather than measure, the power gain amplification, harmonic distortion, frequency mixing, intermodulation (as a special case of the mixing process), quality of oscillation etc. In this section we give some working methods for using nonlinear models together with large-signal algorithms to derive some practical methods of obtaining accurate predesigns.

5.8.1 Amplifier and oscillator operation

Although continuing efforts to improve GaAs MESFET performance have predominantly been focused on amplifiers, this device is also an attractive candidate for use in efficient microwave oscillators [5.59–5.66]. From the point of view of nonlinear device operation, the two types of circuits are quite similar. Thus, except for a few secondary aspects, device technology and characterization techniques developed for amplifier purposes also apply to GaAs MESFET oscillators.

The general problem of microwave MESFET amplifier (oscillator) design is to derive a systematic procedure for defining the optimum input power, bias conditions and terminating (feedback) impedances that correspond to maximum efficiency, output power, stability etc. In any case, a nonlinear MESFET model is a very useful tool because it enables circuit analysis and/or optimization to be achieved over the entire range of bias voltages and frequencies, thereby avoiding the need for large-signal measurements.

For simplicity consider the amplifier circuit shown in Fig. 5.33, where the MESFET has been modelled by one of the nonlinear models developed above and the input–output lossless networks are the unknowns. Suppose that we want to predict the potential power gain characteristics at the fundamental in order to develop the best amplifier. For a given value of input power to the transistor (maintaining $\mathrm{Re}\{Z_{\mathrm{TI}}(\omega_0)\} > 0$ for stability),

$$P_{\mathrm{i}}(\omega_0) = \tfrac{1}{2}\mathrm{Re}(V_{\mathrm{GS}\,1} I_{\mathrm{g}\,1}) \qquad Z_{\mathrm{TI}}(\omega_0) = \frac{V_{\mathrm{GS}\,1}}{I_{\mathrm{g}\,1}} \tag{5.101}$$

where $V_{\mathrm{GS}\,1}$ and $I_{\mathrm{g}\,1}$ are the phasors representing the fundamental components of the temporal gate voltage and current respectively, we can search for a maximum for the output power at the same frequency (maintaining $\mathrm{Re}\{Z_{\mathrm{TO}}(\omega_0)\} < 0$),

$$P_{\mathrm{o}}(\omega_0) = \tfrac{1}{2}\mathrm{Re}(V_{\mathrm{DS}\,1} I_{\mathrm{d}\,1}) \qquad Z_{\mathrm{TO}}(\omega_0) = \frac{V_{\mathrm{DS}\,1}}{I_{\mathrm{d}\,1}} \tag{5.102}$$

by using a standard optimization procedure together with a large-signal analysis tool (harmonic balance, Volterra series etc.) for the active device.

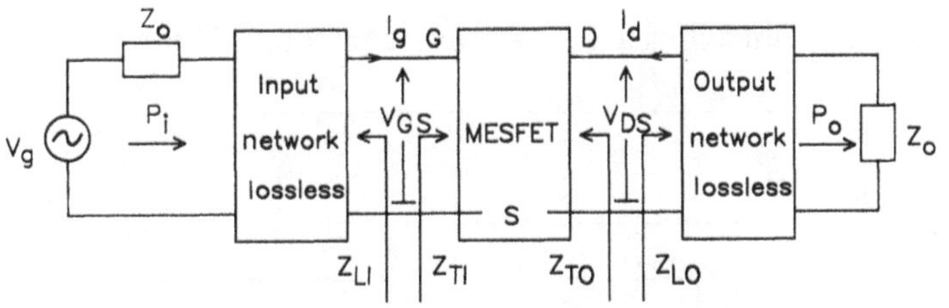

Fig. 5.33 Simple amplifier configuration.

Figure 5.34 shows the power gain characteristics for a standard medium power transistor under various bias conditions. Consequently, at a chosen point on these curves, the synthesis of the input–output linear circuits is evident because

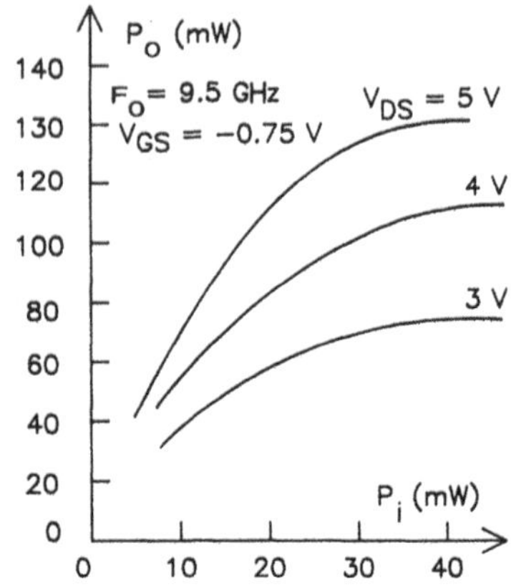

Fig. 5.34 Power gain characteristics.

the optimized voltage and current waveforms at the transistor ports are known:

$$Z_{LI}(\omega_0) = -Z_{TI}(\omega_0) \qquad Z_{TO}(\omega_0) = -Z_{LO}(\omega_0) \tag{5.103}$$

The same philosophy can be used to minimize, at the same time for example, the second-harmonic content of the output power.

The oscillator behavior can be thought of (Fig. 5.35) as a feedback amplifier operation, and the synthesis of the load circuit for oscillator purposes can be treated as a three-port lossless coupling network that yields a simple stable state of oscillation and maximum r.f. power to the load. This can be accomplished by maximizing the added power for the active device at the fundamental frequency (output power in oscillation)

$$\begin{aligned} P_{add}(\omega_0) &= P_o(\omega_0) - P_i(\omega_0) \\ &= \tfrac{1}{2}\mathrm{Re}(V_{DS1} I_{d1}) - \tfrac{1}{2}\mathrm{Re}(V_{GS1} I_{g1}) \end{aligned} \tag{5.104}$$

and if necessary controlling a low harmonic content. As in the amplifier case, a

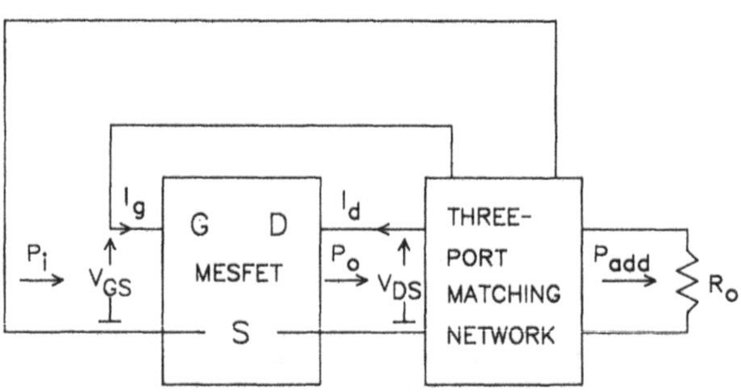

Fig. 5.35 General oscillator configuration.

knowledge of the terminal voltage and current waveforms allows us to obtain the load circuit easily. An obvious precaution is to choose coupling network configurations with minimum complexity in order to avoid unnecessary resonances.

Figure 5.36(b) shows the oscillation frequency and output power characteristics for the simple GaAs MESFET oscillator circuit shown in Fig. 5.36(a). The load circuit was optimized using the techniques developed above. The strong influence of the load resistance R_o on the output power is obvious, i.e. the small-signal approach is no longer valid.

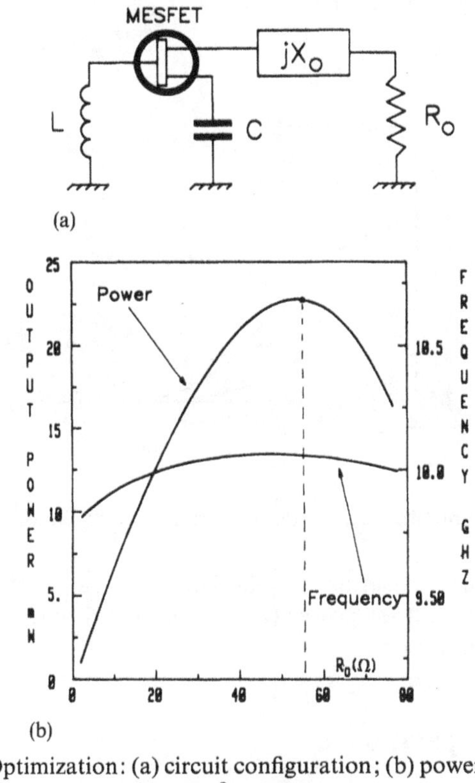

Fig. 5.36 Optimization: (a) circuit configuration; (b) power and frequency performance.

The nonlinear MESFET model can also be used to study the transient behavior and degree of stability of a given oscillator. If we treat the problem as an autonomous nonlinear circuit with a complex time-dependent waveform $s(t) = \sigma(t) + j\omega(t)$ (quasi-static approach), we are able to obtain the time dependences $\sigma(t)$ and $\omega(t)$ of the oscillation growth factor and the frequency respectively as functions of bias conditions or embedding circuit parameters. Figure 5.37 shows the temporal quasi-static simulation of the real and imaginary parts of the complex oscillation for various values of the load resistance R_o (Fig. 5.36(a)). From these curves we can see that a large drain resistance implies a large growth time, i.e. a higher resistance to random

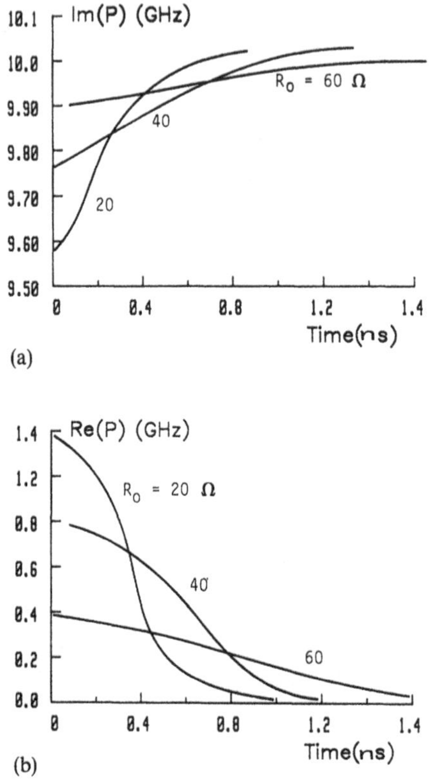

Fig. 5.37 Temporal quasi-static simulation: (a) growth factor; (b) frequency.

departures from the operating point. This particular effect includes a compromise between output power (Fig. 5.36(b)) and quality of oscillation, and ensures oscillation start-up.

5.8.2 Doubler operation

Because of the increased utilization of the millimeter-wave frequency range much effort is being directed toward the exploration of alternatives to generating r.f. power at these frequencies, e.g. using available MESFET transistors as frequency multipliers. Frequency division is also essential in some microwave systems (e.g. phase-locked loops) and frequency-modulated communications systems. GaAs MESFETs operated as frequency multipliers [5.67, 5.68] and dividers [5.69–5.71] are very attractive because they provide a conversion gain over a broad frequency band with isolation between the input and output ports.

There are two main classes of GaAs MESFET doubler: second-harmonic amplification and self-oscillating frequency doublers [5.21, 5.45, 5.72, 5.73]. Figure 5.38 shows a second-harmonic amplifier which, for a frequency generator f_0, will dissipate second-harmonic ($2f_0$) power in the load. One of the aspects of these GaAs MESFET doublers is the importance of parasitic feedback within the transistor, and an extension of this effect is to incorporate an additional external feedback as indicated in Fig. 5.38. The impedance-matching conditions and feedback options involve gain–bandwidth considerations analogous to those in amplifier design except that the input–output frequencies are different. Conversely, in self-oscillating frequency-doubler operation the transistor works as a fundamental frequency oscillator and second-harmonic generator (same circuit as that of Fig. 5.35); an appropriate external fundamental frequency feedback is therefore added.

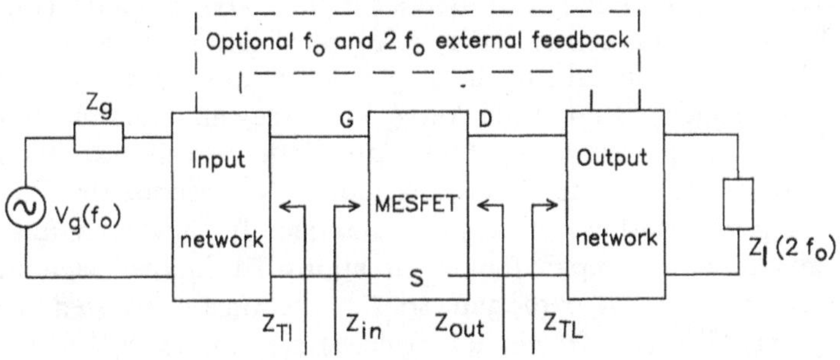

Fig. 5.38 Second-harmonic amplifier circuit.

In contrast with the operation of power amplifiers and oscillators, which can be visualized as an essentially linear circuit problem perturbed by nonlinear effects, frequency multiplication relies specifically on nonlinear device behavior as the basic operating philosophy. Therefore efficient frequency doubling requires circuit elements with strong nonlinear characteristics, giving as much asymmetrical waveform distortion as possible while minimizing parasitic losses.

An effective way of employing a GaAs MESFET as a frequency doubler is to use it as a conventional half-wave rectifier. This can be accomplished by operating the transistor in the vicinity of either forward conduction or pinch-off [5.74]. In fact the two alternatives are almost equivalent regarding the efficiency of conversion from the fundamental to the second harmonic frequency, but the operation near forward conduction involves a higher d.c. drain current and consequently a lower d.c.–r.f. efficiency as well as a higher risk of device failure due to large gate current peaks.

When the MESFET is biased just below the forward conduction point of the gate–source junction and the negative maximum does not reach pinch-off, Fourier analysis of the half-wave-rectified waveform gives a fundamental voltage component magnitude of $V/2$ and a second-harmonic voltage magnitude of $2V/3\pi$. If, as a first approximation, we neglect the nonlinearity of the transfer characteristics, the second-harmonic output power level is about 7.4 dB below that of the fundamental component [5.45]. When the device is biased at the pinch-off voltage, the MESFET is turned on only during the positive half-cycles of the drive waveform. The output will again be a half-wave-rectified waveform, if it is assumed that the gate voltage does not rise to the forward conduction point. If, once again, we neglect the nonlinear transfer characteristics, a similar value of the second-harmonic level will be obtained.

Some workers have employed time-domain analysis by numerical integration of appropriate model differential equations [5.45], but recent methods use the harmonic balance method [5.21, 5.72] for a complete description of this nonlinear process.

Gopinath and Rankin [5.45] used a simple analytical model (Fig. 5.22) with MESFET characteristics (Fig. 5.23) to determine the relationship between the conversion gain of the single-gate frequency doubler and the fundamental frequency gain of the device. First, they analyzed the unilateral equivalent circuit approximation. Omission of the gate–drain branch is an acceptable risk since the gate–drain capacitance C_{GD} becomes small over the relevant range of biasing conditions. Subsequently they used full-model simulation in order to compare the two situations. The input capacitance C_{GS} and the series resistance R_i of the gate–source junction are analyzed as a lossy varactor diode. The transfer characteristic used is eqn (5.54) and the load circuit comprises a resistance R_l in shunt with an r.f. choke (Fig. 5.39).

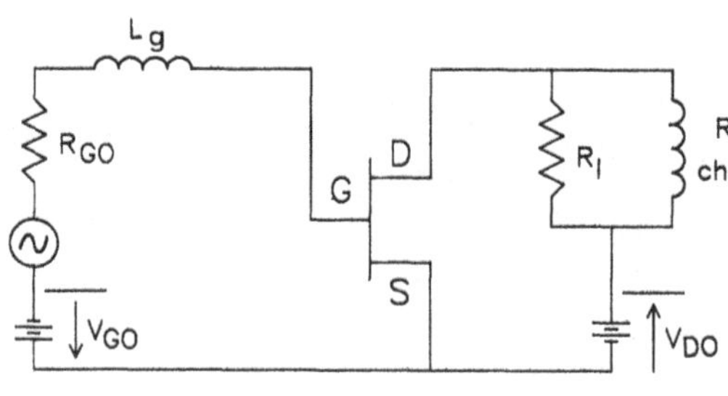

Fig. 5.39 Gopinath's doubler circuit [5.45].

Therefore we can write

$$V_{DS}(t) = V_{DO} - \{I_{DS}(t) - I_{DO}\}\, R_l \tag{5.105}$$

where V_{DO} is the drain supply voltage giving rise to an average current I_{DO}. Assuming a sinusoidal excitation $V_{GS} = V_0 + V_1 \sin(\omega t)$ and substituting in eqns (5.67) and (5.105), we have

$$I_{DS} = \left\{1 - \frac{V_0 + V_1 \sin(\omega t)}{V_p}\right\}^2 \frac{(V_{DO} + I_{DO}R_l)/R_{DSO} + I_{DSS}}{1 + [1 - \{V_0 + V_1 \sin(\omega t)\}/V_p]^2 (R_l/R_{DSO})} \tag{5.106}$$

This expression enables us to determine the conversion gain and the fundamental gain easily. It should be noted that the transit time τ is approximated by a constant value although in reality it is a nonlinear function of the gate voltage V_{GS}. This becomes important at input or output frequencies approaching $f_t = 1/\tau$.

Figures 5.40 and 5.41 show some results for the unilateral analysis and the full-model analysis respectively. The device used in this work was a medium power packaged MESFET (MSC 88001).

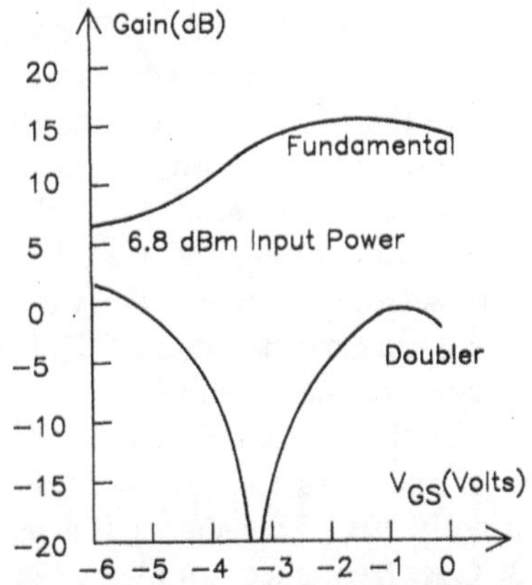

Fig. 5.40 Gain characteristics of a 4 GHz MESFET doubler versus gate bias (V_{DS} = 5 V) (unilateral approach) [5.45].

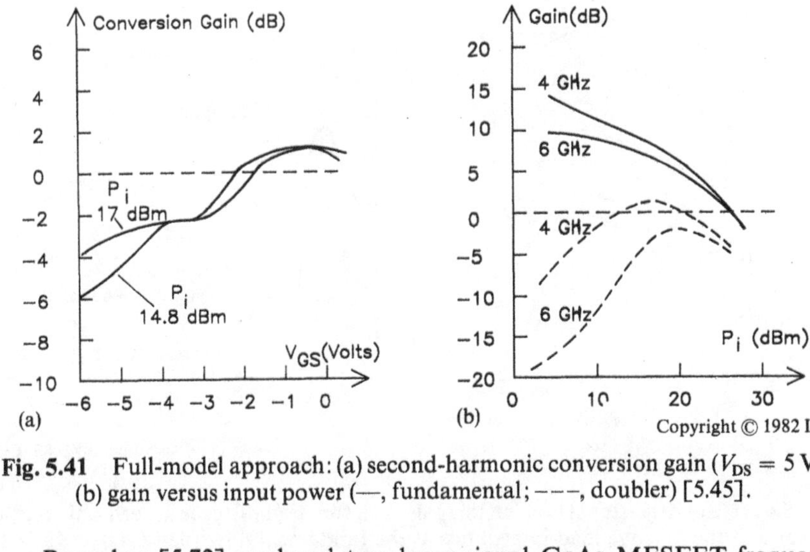

Fig. 5.41 Full-model approach: (a) second-harmonic conversion gain (V_{DS} = 5 V); (b) gain versus input power (—, fundamental; – – –, doubler) [5.45].

Rauscher [5.72] analyzed two large-signal GaAs MESFET frequency doublers: a second-harmonic amplification frequency doubler and a self-oscillating frequency doubler. He took into account only those nonlinearities which participate significantly in the frequency-doubling process. Thus only the drain current I_{GS}, which describes the intrinsic dynamic I–V characteristics of the device [5.75, 5.76], needs to be treated as a full nonlinear element. The nonlinearities associated with C_{GS}, R_i and C_{DG} are represented by linearized large-signal time-averaged values for their respective voltage-dependent values.

The basic frequency-doubler configuration of Fig. 5.38 was analyzed using the harmonic balance technique and the solution was obtained by minimizing a least-squares error function. The network matching the gate–source port of the device to the generator was analyzed at the fundamental frequency, and the network matching the drain–source port to the external load was analyzed at the second harmonic frequency. Rauscher [5.72] used a 0.5 μm, 250 μm transistor at a d.c. bias $V_{DS} = 3$ V and $V_{GS} = -1.2$ V with a pinch-off voltage of -1.5 V. The gate bias voltage corresponds to the point where the second derivative of the drain current versus the gate voltage is a maximum. Figure 5.42(a) shows the large-signal conversion gain as a function of the second-harmonic output power and the fundamental frequency reactive load impedance $Z_{TL}(f_0)$, and Fig. 5.42(b) shows the importance of the parasitic feedback at the second harmonic ($Z_{TI}(2f_0) = jX_{TI}(2f_0)$). Finally Fig. 5.43 shows the measured and predicted performances of a 15–30 GHz frequency doubler.

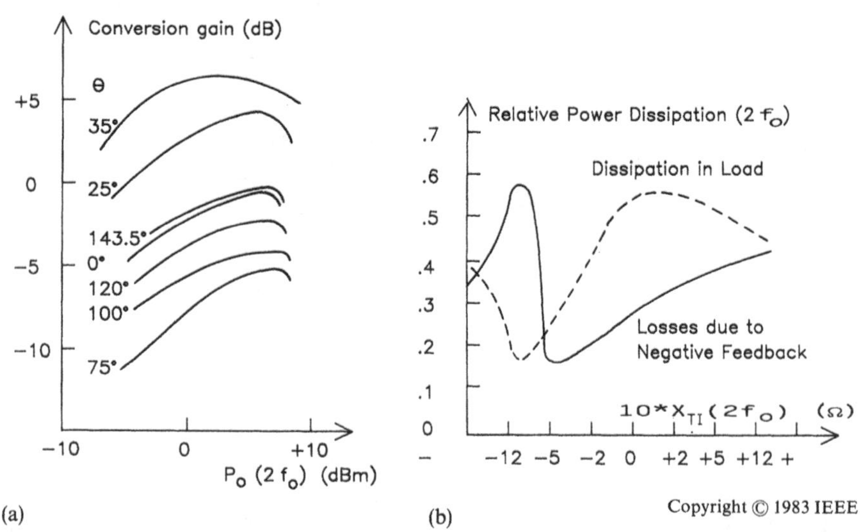

Fig. 5.42 (a) Large-signal conversion gain as a function of second-harmonic output power and the reactive load impedance at the fundamental frequency ($f_0 = 15$ GHz, $Z_l(f_0) = j50 \tan\theta$, $Z_{TI}(2f_0) = 0$); (b) relative second-harmonic power dissipation versus second-harmonic feedback [5.21].

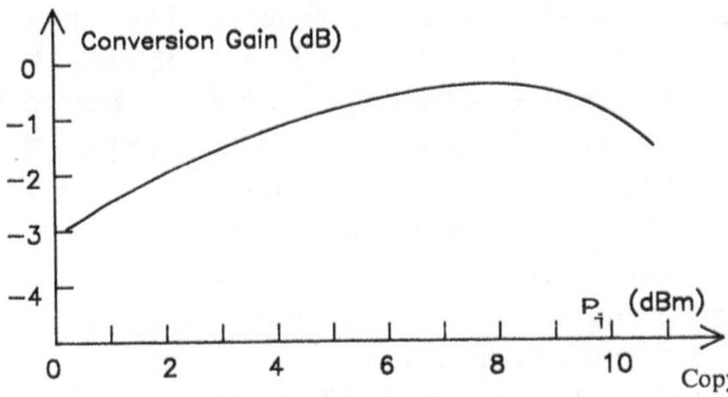

Fig. 5.43 Measured and predicted performances of a 15–30 GHz frequency doubler (f_0 = 15 GHz) [5.21].

The iteration of the device circuit at the fundamental frequency and at the second harmonic can be treated as two independent phenomena [5.72]. This property has been used in the design of self-oscillating doublers, and the fundamental frequency oscillation problem is treated separately from the optimization of the second-harmonic output power. The results obtained by Rauscher for a self-oscillating doubler device are summarized in Fig. 5.44 together with the predictions for the bias point.

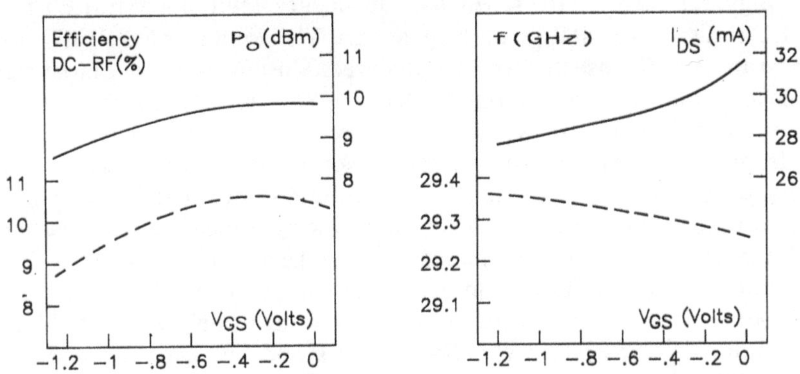

(a) (b)

Fig. 5.44 (a) Second-harmonic output power and d.c.–r.f. efficiency versus gate bias voltage; (b) d.c. drain–source current and output frequency versus gate bias voltage. V_{DS} = 3 V.

5.8.3 Mixer operation

One of the most interesting applications of a nonlinear MESFET model is the design or optimization and characterization of microwave mixers. The majority of papers on this theme are experimental. The reason for the relative lack of interest in these components is that the noise figure of MESFET mixers reported in the technical literature has on the whole been disappointing in comparison with the noise behavior of MESFET amplifiers. It is possible, however, to design MESFET mixers with comparable noise performance and conversion gain and lower LO effects than those of a diode mixer. In this sense, the interest in nonlinear modelling of the MESFET is the derivation of closed-form equations for the conversion gain and load impedances in order to minimize or optimize specific effects.

Many attempts have been made to analyze MESFET mixers from a nonlinear model point of view [5.77–5.84]. Some of them assume simplified models for the transistor, while others are restricted to computer approximations (Volterra series, harmonic balance and time domain) which ignore some effects of higher-order mixing products or LO harmonics. In any case the aim of large-signal analysis is to obtain the voltage waveforms for the independent control variables in the model, so that the small-signal resistance, capacitance or transconductance waveforms and subsequently the input and output impedances and conversion gain can easily be found. In fact this means that the r.f. signal is applied to a small-signal equivalent circuit, some of whose elements are time varying under large-signal LO excitation.

Figure 5.45 shows the schematic equivalent circuit of a MESFET mixer for LO sinusoidal excitation at the gate port. The nonlinear MESFET model can be any one of those that we have analyzed. The purpose of this large-signal analysis (Volterra series, harmonic balance etc.) is to determine the time variation under LO excitation of the nonlinear elements that define our model: gate–source capacitance C_{GS}, gate–source conductance G_{GS}, transconductance g_m, gate–drain capacitance C_{DG}, drain–source conductance g_d etc. The accuracy of this analysis will depend on the agreement between the model and the transistor and the number of LO harmonics considered in the computer simulation. If we apply a very low r.f. signal level $V_{rf}\cos(\omega_s t)$ to the gate port and try to extract the i.f. signal from the drain port we will have a conventional MESFET mixer. The currents and voltages in the pumped MESFET will now have components at frequencies $\omega_n = \omega_i + n\omega_0$ ($n = -\infty, \ldots, -1, 0, 1, \ldots, \infty$), where $\omega_i = \omega_0 - \omega_s$ (usually the i.f.). The analysis of such a circuit is very easy if we know the time variations of the nonlinear elements under LO excitation (small-signal analysis).

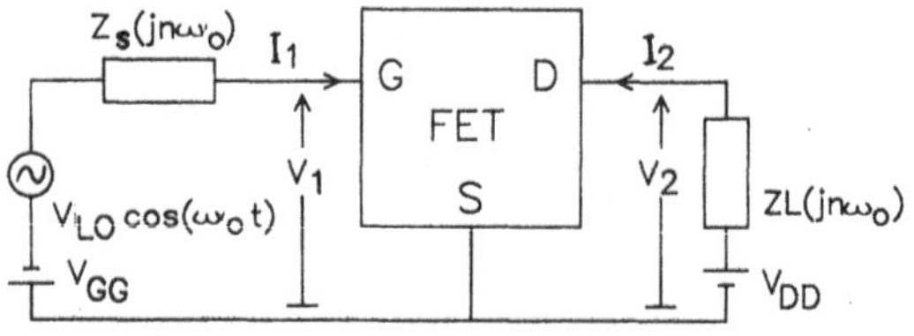

Fig. 5.45 Schematic equivalent circuit of a MESFET mixer under LO.

For example, the Fourier series representation of a time-varying conductance $G(t)$ with fundamental frequency ω_0 is

$$G(t) = \sum_{n=-\infty}^{\infty} G_n \exp(jn\omega_0 t) \qquad G_n \text{ complex} \tag{5.107}$$

The current response $I(t)$ of this temporal conductance to a voltage signal $V(t)$ with components at the frequencies $\omega_n = \omega_i + n\omega_0$ can be written as

$$I(t) = G(t)\, V(t) \tag{5.108}$$

$$\sum_{k=-\infty}^{\infty} I_k \exp\{j(\omega_i + k\omega_0)t\} = \sum_{m=-\infty}^{\infty} \sum_{n=-\infty}^{\infty} G_m V_n \exp[j\{\omega_i + (n+m)\omega_0\}t]$$

where

$$I(t) = \sum_{n=-\infty}^{\infty} I_n \exp\{j(\omega_i + n\omega_0)t\}$$

$$V(t) = \sum_{n=-\infty}^{\infty} V_n \exp\{j(\omega_i + n\omega_0)t\}$$

It is obvious that the same frequencies are present on each side of the equation, and so we can expand in matrix form:

$$\begin{bmatrix} I_{-N} \\ I_{-N+1} \\ \vdots \\ I_0 \\ \vdots \\ I_{N-1} \\ I_N \end{bmatrix} = \begin{bmatrix} G_0 & G_{-1} & G_{-2} & \cdots & G_N & \cdots & G_{-2N+1} & G_{-2N} \\ G_1 & G_0 & G_{-1} & \cdots & G_{-N+1} & \cdots & G_{-2N+2} & G_{-2N+1} \\ \vdots & \vdots & \vdots & & \vdots & & \vdots & \vdots \\ G_N & G_{N-1} & G_{N-2} & \cdots & G_0 & \cdots & G_{-N+1} & G_{-N} \\ \vdots & \vdots & \vdots & & \vdots & & \vdots & \vdots \\ G_{2N-1} & G_{2N-2} & G_{2N-3} & \cdots & G_{N-1} & \cdots & G_0 & G_{-1} \\ G_{2N} & G_{2N-1} & G_{2N-2} & \cdots & G_N & \cdots & G_1 & G_0 \end{bmatrix} \begin{bmatrix} V_{-N} \\ V_{-N+1} \\ \vdots \\ V_0 \\ \vdots \\ V_{N-1} \\ V_N \end{bmatrix}$$

or

$$\boldsymbol{I} = \boldsymbol{G}\boldsymbol{V} \tag{5.109}$$

where the elements of this conversion matrix are simply the Fourier series elements (up to truncation order N) of the time-varying conductance under LO excitation. The same applies to the other nonlinear elements in the nonlinear equivalent circuit of the transistor.

Since sinusoids at different frequencies are linearly independent, we can easily determine [5.82], using Kirchhoff's laws, the input–output impedances and conversion gain for each mixing product. In effect, a transfer admittance matrix that relates the excitation vector $\boldsymbol{V}_{\mathrm{LO}}$ to the output current vector $\boldsymbol{I}_2$ can be derived:

$$\boldsymbol{I}_2 = \boldsymbol{Y}_{\mathrm{t}} \boldsymbol{V}_{\mathrm{LO}} \tag{5.110}$$

Consequently the conversion gain between the input frequency ω_m and the output frequency ω_n is

$$G_c(\omega_n, \omega_m) = 4|Y_t(\omega_n, \omega_m)|^2 \operatorname{Re}\{Z_s(\omega_m)\} \operatorname{Re}\{Z_l(\omega_n)\} \tag{5.111}$$

The same approach can be used to derive the input and output impedance matrices:

$$\boldsymbol{V}_1 = \boldsymbol{Z}_{\text{in}} \boldsymbol{I}_1 \qquad \boldsymbol{V}_2 = \boldsymbol{Z}_{\text{out}} \boldsymbol{I}_2 \tag{5.112}$$

The advantage of this closed-form formulation is that the optimum impedances and LO power for maximum conversion gain to the intermediate frequency, image rejection etc. can be predicted. In fact the final problem is a linear synthesis problem for the input and output circuits.

Figure 5.46(a) shows a schematic diagram [5.82] of a MESFET mixer with r.f. and/or LO at the gate and i.f. at the drain, and Fig. 5.46(b) shows the conversion gain for two LO levels. It can be seen that the conversion gain versus LO power for a given device can eventually be optimized (computer optimization).

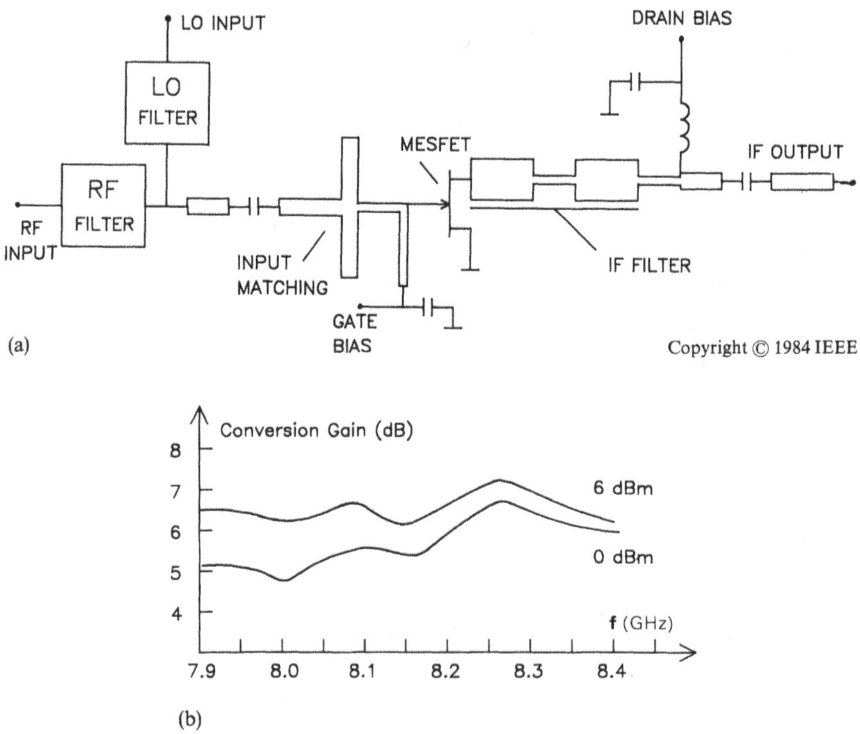

Fig. 5.46 (a) Microstrip MESFET mixer [5.82]; (b) conversion gain characteristics.

References

5.1 Obregon, J. Non-linear analysis and design of microwave circuits, *European Microwave Conf., Paris*, pp.1089–1093, Microwave Exhibitions and Publishers, Tunbridge Wells, 1985.

5.2 Ha, T. T. *Solid-state Microwave Amplifier Design*, Wiley, New York, 1981.

5.3 Loembe, A. *Linéarisation des Amplificateurs de Puissance à Transistors à Effet de Champ GaAs en Hyperfréquences*, ENST, Paris, 1986.

5.4 Correa de Albuquerque, M. Contribution à l'analyse et à la conception optimale des circuits non linéaires microondes, *Thèse*, Université de Limoges, 1986.

5.5 Gelb, A. and Vander Velde, W. E. *Multiple-input Describing Functions and Non-linear System Design*, McGraw-Hill, New York, 1968.

5.6 Blaquière, A. *Analyse des Systèmes Non-linéaires*, Institut National des Sciences et Techniques Nucléaires, Saclay, 1966.

5.7 Ogata, K. *Ingenieria de Control Moderna*, Prentice-Hall, Englewood Cliffs, NJ, 1974.

5.8 Sobhy, M. I. and Jatrzebski, A. K. Direct integration methods of non-linear microwave circuits, *European Microwave Conf., Paris*, pp.1110–1118, Microwave Exhibitions and Publishers, Tunbridge Wells, 1985.

5.9 Angot, A. *Complements de Mathèmatiques à l'Usage des Ingenieurs de l'Electronique et des Télécommunications*, Masson, Paris, 1972.

5.10 Chua, L. O. and Pen-Min-Lin, *Computer Aided Analysis of Electronic Circuits: Algorithms and Computational Techniques*, Prentice-Hall, Englewood Cliffs, NJ, 1975.

5.11 Abramowitz, M. and Stegun, I. A. *Handbook of Mathematical Functions with Formulas, Graphs and Mathematical Tables*, Dover Publications, New York, 1972.

5.12 Gear, C. W. Simultaneous numerical solution of differential algebraic equations, *IEEE Trans. Circuit Theory*, **18** (1), 89–95, 1971.

5.13 *CIRCEC: A Time Domain Program Analysis for Non-linear Circuits*, Thomson Information Service, Paris.

5.14 Pederson, D. O. A historical review of circuit simulation, *IEEE Trans. Circuits Syst.*, **31** (1), 103–111, 1984.

5.15 Eesof Inc. Non-linear circuit analysis comes to microwave, *Microwave J.*, **28,** 153–156, August 1985.

5.16 Jastrzebski, A. K. and Sobhy, M. I. Analysis of non-linear microwave circuits using state-space approach, *Proc. IEEE Int. Symp. on Circuits and Systems, Montreal, 1984*, Vol. 2, pp.1119–1122, IEEE, New York, 1984.

5.17 Fleri, D. A. and Cohen, L. D. Non-linear analysis of the Schottky mixer diode, *IEEE Trans. Microwave Theory Tech.*, **21** (1), 39–43, 1973.

5.18 Baucells Boix, J. and Perichon, A. CAO des mélangeurs à diode Schottky, *Note Technique NT/LAB/MER 249*, Centre National d'Etudes des Télécommunications, 1986.

5.19 Filicori, F., Monaco, V. A. and Naldi, C. Simulation and design of microwave class-C amplifiers through harmonic analysis, *IEEE Trans. Microwave Theory Tech.*, **27** (12), 1043–1051, 1979.

5.20 Camacho-Peñalosa, C. *Nueva Técnica de Analisis no Lineal en el Dominio de la Frecuencia de Transistores MESFET de Microondas*, UPM, Madrid, 1982.

5.21 Camacho-Peñalosa, C. Numerical steady-state analysis of nonlinear microwave circuits with periodic excitation, *IEEE Trans. Microwave Theory Tech.*, **31** (9), 724–730, 1983.

5.22 Hu, Y., Mollier, J. C. and Obregon, J. A new method of third-order intermodulation reduction in non-linear microwaves systems, *IEEE Trans. Microwave Theory Tech.*, **34** (2), 245–250, 1986.

5.23 Lambranianou, G. M. and Aitchison, C. S. Optimization of third-order intermodulation product and output power from an X-band MESFET amplifier using Volterra series analysis, *IEEE Trans. Microwave Theory Tech.*, **33** (12), 1395–1403, 1985.

5.24 Minasian, R. A. Intermodulation distortion analysis of MESFET amplifiers using the Volterra series representation, *IEEE Trans. Microwave Theory Tech.*, **28** (1), 1–8, 1980.

5.25 Javed, A., Goud, P. A. and Syrett, B. A. Analysis of microwave feedforward amplifier using Volterra series representation, *IEEE Trans. Commun.*, **25,** 355–360, 1977.

5.26 Minasian, R. A. Volterra series analysis of MESFET mixers, *Int. J. Electron.*, **50** (3), 215–219, 1981.

5.27 Shockley, W. A unipolar field effect transistor, *Proc. IRE*, **40**, 1365, 1952.

5.28 Special Issue on Microwave Field Effect Transistors, *IEEE Trans. Microwave Theory Tech.*, **24** (6), 1976.

5.29 The FET revolution, *Microwave Syst. News*, **6** (2), 1976.

5.30 Ogawa, M., Ohata, K., Furutsuka, T. and Kawamura, N. Submicron single-gate and dual-gate GaAs MESFETs with improved low noise and high gain performance, *IEEE Trans. Microwave Theory Tech.*, **24** (6), 300–305, 1976.

5.31 Asai, S., Murai, F. and Kodera, H. GaAs dual-gate Schottky-barrier FETs for microwave frequencies, *IEEE Trans. Electron. Devices*, **22** (10), 897–904, 1975.

5.32 Liechti, C. Characteristics of dual-gate GaAs MESFETs, *European Microwave Conf., Montreux*, pp.87–91, Microwave Exhibitions and Publishers, Tunbridge Wells, 1974.

Liechti, C. Performance of dual-gate GaAs MESFETs as gain-controlled low-noise amplifiers and high-speed modulators, *IEEE Trans. Microwave Theory Tech.*, **23** (6), 461–469, 1975.

5.33 Oxner, E. *FETs in Balanced Mixers*, Siliconix Inc., Santa Clara, CA, 1979.

5.34 Dascalu, D. *Electronic Processes in Unipolar Solid-state Devices*, Abacus, Tunbridge Wells, 1977.

5.35 Mead, C. A. Schottky-barrier gate field effect transistor. *Proc. IEEE*, **54**, 367–380, 1966.

5.36 Liechti, C. Microwave field-effect transistors—1976, *IEEE Trans. Microwave Theory Tech.*, **24** (6), 279–300, 1976.

5.37 Lehovec, K. and Miller, R. Field distribution in junction field-effect transistors at large drain voltages, *IEEE Trans. Electron. Devices*, **22** (5), 273–281, 1975.

5.38 Tajima, Y. and Miller, P. D. Design of broad-band power GaAs FET amplifiers, *IEEE Trans. Microwave Theory Tech.*, **32** (3), 261–267, 1984.

5.39 David, J. P. R., Sitch, J. E. and Stern, M. S. Gate–drain avalanche breakdown in GaAs power MESFETs, *IEEE Trans. Electron. Devices*, **29** (10), 1548–1552, 1982.

5.40 Madjar, A. and Rosenbaum, F. J. A large-signal model for the GaAs MESFET, *IEEE Trans. Microwave Theory Tech.*, **29** (8), 781–788, 1981.

5.41 Madjar, A. and Dreifuss, J. Large-signal microwave performance prediction of dual-gate GaAs MESFET using an efficient and accurate model, *IEEE Trans. Microwave Theory Tech.*, **33** (7), 639–642, 1985.

5.42 Yamaguchi, K. and Kodera, H. Drain conductance of junction gate FETs in the hot electron range, *IEEE Trans. Electron. Devices*, **23** (6), 543–553, 1976.

5.43 Yamaguchi, K., Asai, S. and Kodera, H. Two-dimensional numerical analysis of stability criteria of GaAs FETs, *IEEE Trans. Electron. Devices*, **23** (12), 1283–1290, 1976.

5.44 Materka, A. and Kacprzak, T. Computer calculation of large-signal GaAs FET amplifier characteristics, *IEEE Trans. Microwave Theory Tech.*, **33** (2), 129–135, 1985.
5.45 Gopinath, A. and Rankin, J. B. Single-gate MESFET frequency doublers, *IEEE Trans. Microwave Theory Tech.*, **30** (6), 869–875, 1982.
5.46 Willing, H. A., Rauscher, C. and Santis, P. A technique for predicting large-signal performance of a GaAs-MESFET, *IEEE Trans. Microwave Theory Tech.*, **26** (12), 1017–1023, 1978.
5.47 Rauscher, C. and Willing, H. A. Simulation of nonlinear microwave FET performance using a quasi-static model, *IEEE Trans. Microwave Theory Tech.*, **27** (10), 834–840, 1979.
5.48 Tajima, Y., Wroma, B. and Mishima, K. GaAs FET large-signal model and its application to circuit designs, *IEEE Trans. Electron. Devices*, **28** (2), 171–175, 1981.
5.49 Obregon, J. Small signal and non-linear microwave active devices, *Course–Colloquium on Millimeter Waves, Universidad de Santander, July 1986.*
5.50 Dufond, P. and Derewonko, H. Power GaAs FET characterization, *European Microwave Conf., Rome*, pp. 591–595, Microwave Exhibitions and Publishers, Tunbridge Wells, 1976.
5.51 Vendelin, G. D. Feedback effects in the GaAs-MESFET model, *IEEE Trans. Microwave Theory Tech.*, **24** (6), 383–385, 1976.
5.52 Minasian, R. A. Modelling the MESFET output nonlinearity, *Electron. Lett.*, **15** (17), 515–516, 1979.
5.53 Kurita, O. and Morita, K. Microwave MESFET mixer, *IEEE Trans. Microwave Theory Tech.*, **24** (6), 361–366, 1976.
5.54 Tsironis, C. and Meierer, R. Microwave wide-band model of gate MESFETs, *IEEE Trans. Microwave Theory Tech.*, **30** (3), 243–251, 1982.
5.55 Soares, R., Obregon, J. and Graffeuil, J. *Applications des Transistors à Effet de Champ en Arseniure de Gallium*, Eyrolles, Paris, 1984.
5.56 Sussman-Fort, S. E., Narasimhan, S. and Mayaram, K. A complete GaAs MESFET computer model for SPICE, *IEEE Trans. Microwave Theory Tech.*, **32** (4), 471–473, 1984.
5.57 Nagel, L. W. SPICE 2: A computer program to simulate semiconductor circuits, *Memo. ERL-M520*, Electronics Research Laboratory, University of California, Berkeley, CA, 1975.
5.58 Curtice, W. R. A MESFET model for use in the design of GaAs integrated circuits, *IEEE Trans. Microwave Theory Tech.*, **28** (5), 448–456, 1980.
5.59 Abe, H. A GaAs MESFET self-bias mode oscillator, *IEEE Trans. Microwave Theory Tech.*, **34** (1), 167–172, 1986.
5.60 Siweris, H. J. and Schiek, B. Analysis of noise upconversion in microwave FET oscillators, *IEEE Trans. Microwave Theory Tech.*, **33** (3), 233–242, 1985.
5.61 Ehlers, E. R. An empirical design technique for microwave oscillators, *IEEE Trans. Microwave Theory Tech.*, **32** (5), 556–559, 1984.
5.62 Basawapatna, G. R. and Stancliff, R. B. A unified approach to the design of wide band microwave solid state oscillators, *IEEE Trans. Microwave Theory Tech.*, **27** (5), 379–385, 1979.
5.63 Mediavilla, A., Tazon, A. and Garcia, J. L. CAD for improved performance prediction of microwave FET oscillators, *Proc. Mikrowellentechnologie und Optoelektronik, Fachmesse und Konferenz für Höchstfrequenztechnologie, Wiesbaden, 1986.*
5.64 Johnson, K. Large signal GaAs MESFET oscillator design, *IEEE Trans. Microwave Theory Tech.*, **27** (3), 217–227, 1979.
5.65 Mitsui, Y., Nakatani, M. and Mitsui, S. Design of GaAs MESFET oscillator using large signal *S*-parameters, *IEEE Trans. Microwave Theory Tech.*, **25** (12),

981–984, 1977.

5.66 Rauscher, C. Large signal technique for designing single-frequency and voltage-controlled GaAs FET oscillators, *IEEE Trans. Microwave Theory Tech.*, **29** (4), 293–304, 1981.

5.67 Allamando, E., Radhy, N. E. and Constant, E. Broad-band high-order frequency multipliers by using dual-gate MESFETs, *European Microwave Conf., Paris*, pp. 617–662, Microwave Exhibitions and Publishers, Tunbridge Wells, 1985.

5.68 Cuhaci, M. and Stubbs, M. G. A self-biased FET multiplier, *European Microwave Conf., Liège*, pp. 280–283, Microwave Exhibitions and Publishers, Tunbridge Wells, 1984.

5.69 Jacob, A. Transfer properties of FET frequency dividers, *European Microwave Conf., Liège*, pp. 284–289, Microwave Exhibitions and Publishers, Tunbridge Wells, 1984.

5.70 Rauscher, C. Regenerative frequency division with a GaAs FET, *IEEE Trans. Microwave Theory Tech.*, **32** (11), 462–472, 1984.

5.71 Honjo, K. and Madihian, M. Novel design approach for X-band GaAs monolithic analog 1/4 frequency divider, *IEEE Trans. Microwave Theory Tech.*, **34** (4), 436–441, 1986.

5.72 Rauscher, C. High-frequency doubler operation of GaAs field-effect transistors, *IEEE Trans. Microwave Theory Tech.*, **31** (6), 462–472, 1983.

5.73 Soares, R., Castelleto, J. P., Legaud, P. and Armand, M. Performance of an InP MISFET X-band oscillator and K-band oscillator–doubler, *European Microwave Conf., Liège*, pp. 263–267, Microwave Exhibitions and Publishers, Tunbridge Wells, 1984.

5.74 Camargo, E., Soares, R., Perichon, R. A. and Goloubkoff, M. Sources of non-linearity in GaAs MESFET frequency multipliers, *1983 IEEE Microwave Theory and Techniques Symp. Dig.*, pp.343–345, IEEE, New York, 1983.

5.75 Willing, H. A., Rauscher, C. and de Santis, P. A technique for predicting large-signal performance of a GaAs MESFET, *IEEE Trans. Microwave Theory Tech.*, **26** (12), 1017–1023, 1978.

5.76 Rauscher, C. and Willing, H. A. Simulation of nonlinear microwave FET performance using a quasi-static model, *IEEE Trans. Microwave Theory Tech.*, **27** (10), 834–840, 1979.

5.77 Pucel, R. A., Massé, D. and Bera, R. Performance of GaAs MESFET mixers at X-band, *IEEE Trans. Microwave Theory Tech.*, **24** (6), 351–360, 1976.

5.78 Harrop, P. GaAs FET mixers: theory and applications, *Acta Electron.*, **23** (4), 291–297, 1980.

5.79 Kurita, O. and Morita, K. Microwave MESFET mixer, *IEEE Trans. Microwave Theory Tech.*, **24** (6), 361–366, 1976.

5.80 El Sayed, O. L. Performance analysis of dual-gate MESFET mixers, *European Microwave Conf., Helsinki*, pp. 457–462, Microwave Exhibitions and Publishers, Tunbridge Wells, 1982.

5.81 Hirota, T. and Ogawa, H. A novel K-band balanced FET upconverter, *IEEE Trans. Microwave Theory Tech.*, **32** (7), 679–683, 1984.

5.82 Maas, S. A. Theory and analysis of GaAs MESFET mixers, *IEEE Trans. Microwave Theory Tech.*, **32** (10), 1402–1406, 1984.

5.83 Dreifuss, J., Madjar, A., Engel, B. and Bar-Lev, A. Application of the harmonic balance method to analysis of MESFET oscillators and dual gate MESFET mixers, *European Microwave Conf., Paris*, pp. 521–526, Microwave Exhibitions and Publishers, Tunbridge Wells, 1985.

5.84 Tang, O. S. A. and Aitchison, C. S. A very wide-band microwave MESFET mixer using the distributed mixing principle, *IEEE Trans. Microwave Theory Tech.*, **33** (12), 1470–1478, 1985.

Chapter 6

Power GaAs MESFET Amplifiers

ROBERT SOARES

6.1 Introduction

The design of narrowband and wideband amplifiers using power GaAs MESFETs requires consideration of several aspects of the device behavior which need not be taken into account for small-signal design. Essentially these relate to the nonlinearity of GaAs MESFETs under large-signal operating conditions and to the fact that the power transistor is realized by connecting several elementary cells in parallel. The former point underlines the necessity of characterizing the device adequately for large-signal operation. Therefore sections on the large-signal measurement and modelling of power GaAs MESFETs are included in this chapter. The parallel connection of elementary structures leads to considerations of impedance matching, even power distribution and combining techniques, and thermal resistance effects. These aspects are all treated in the section on amplifier design.

First, however, we explain why the power GaAs MESFET is different from its small-signal counterpart.

6.2 Device considerations

6.2.1 Nonlinearity in class A operation

The major difference between devices used for low level amplification and those used for high power operation is in linearity. This difference is best appreciated by treating the amplifier thermodynamically. Amplifiers are only a special form of power converter drawing power from a d.c. source to add r.f. power to the input signal. Such a power converter must obey the fundamental laws of thermodynamics [6.1].

Consider the two-port amplifier of Fig. 6.1. The total power into the amplifier must equal the total power out:

$$P_{\rm i}+P_{\rm dc} = P_{\rm o}+P_{\rm d} \tag{6.1}$$

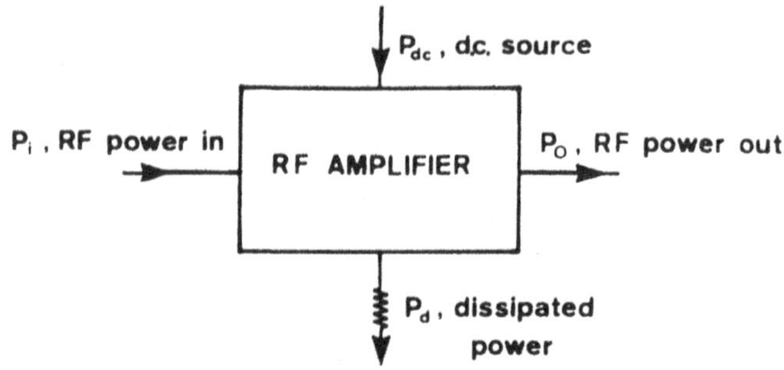

Fig. 6.1 Thermodynamic representation of an r.f. amplifier.

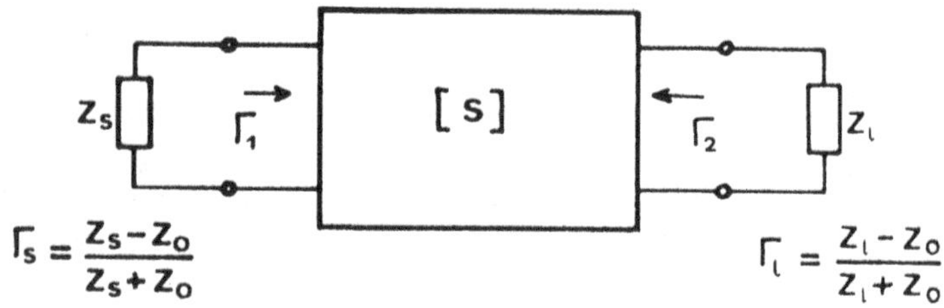

Fig. 6.2 Two-port amplifier.

If the r.f. power gain is defined as

$$G = \frac{P_o}{P_i} \tag{6.2}$$

then

$$P_d = P_{dc} - (G-1)P_i \tag{6.3}$$

Now if the r.f. power gain G were to remain constant and greater than unity and the d.c. power supplied were to remain fixed, at some level of input the dissipated power would become negative. As this is clearly impossible and as maximum d.c. power is limited by device considerations (Chapter 1), the power gain cannot remain constant and we obtain the phenomenon of gain "saturation". The r.f. power gain diminishes with signal drive above a linear operating region. It is important to note from eqn (6.3) that the dissipated power output from the amplifier will diminish with increasing r.f. drive in the amplifier's linear operating region, i.e. under the same bias conditions a power amplifier will run cooler than a small-signal amplifier using the same transistor. This should be taken into account when a power GaAs MESFET is biased for small-signal measurements.

In the two-port amplifier of Fig. 6.2 the active device is inserted between two matching circuits. In terms of the S matrix of the transistor and the source and load impedances Γ_s and Γ_l the amplifier transducer gain is given by (Chapter 2)

$$G_t = |S_{21}|^2 \frac{1-|\Gamma_s|^2}{|1-\Gamma_1\Gamma_s|^2} \frac{1-|\Gamma_l|^2}{|1-S_{22}\Gamma_l|^2} \tag{6.4}$$

where

$$\Gamma_1 = S_{11} + \frac{S_{12}S_{21}\Gamma_l}{1-S_{22}\Gamma_l} \tag{6.5}$$

Under optimum match conditions ($\Gamma_s^* = \Gamma_1, \Gamma_l^* = S_{22}$)

$$G_t = |S_{21}|^2 \frac{1}{1-|\Gamma_1|^2} \frac{1}{1-|S_{22}|^2} \tag{6.6}$$

It can be seen from eqn (6.4) that, for a given passive embedding system, the gain of the amplifier will remain independent of the input signal as long as the transistor S parameters remain constant. As shown by thermodynamic considerations, the gain G_t of the amplifier cannot remain constant for an indefinitely rising input signal level; the S parameters of the device must therefore necessarily change with the onset of saturation. The decrease in G_t may be caused by a fall in S_{21}, or by an increase in $|1-\Gamma_1\Gamma_s|$ and/or $|1-S_{22}\Gamma_l|$ or by a combination of all three. The relative importance of the variations in the various two-port parameters with input signal drive level will be discussed in the section on measurement techniques and nonlinear modelling. An immediate observation can be made, however. If the fall-off in gain with the onset of saturation is due to a parameter other than $|S_{21}|$, it can be noted immediately from eqn (6.6) that a device matched for optimum gain at low signal levels will become mismatched. Thus in order to use a power device to its full potential it is necessary to know either its optimum terminating source and load impedances or the two-port S matrix corresponding to the signal drive conditions desired. Various methods of measuring these parameters are discussed in Section 6.3.

6.2.2 Dynamic load line and physical limitations

6.2.2.1 Idealized conditions

We define the power-added efficiency η_{add} as

$$\eta_{add} = \frac{P_o - P_i}{P_{dc}} \tag{6.7}$$

This establishes the efficiency of the device as a power converter changing d.c. energy into microwave r.f. energy. Equation (6.7) can be rewritten, using eqn (6.1), as

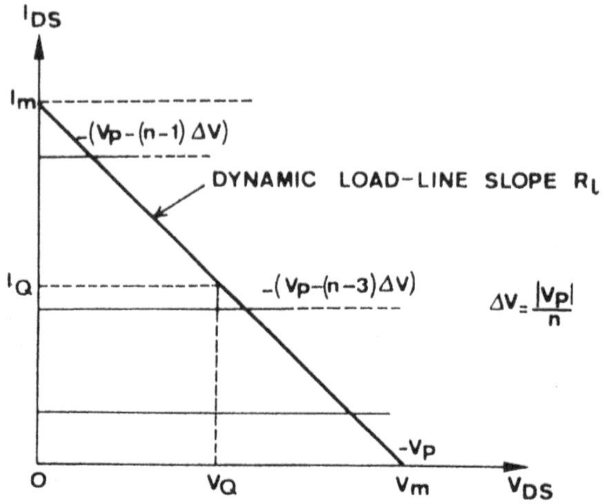

Fig. 6.3 *I–V* characteristics for an idealized active device.

$$\eta_{\text{add}} = 1 - \frac{P_{\text{d}}}{P_{\text{dc}}} \tag{6.8}$$

which indicates the importance of dissipated power with respect to the power-added efficiency. As the dissipated power decreases with the r.f. drive, the power-added efficiency under large-signal operation will be substantially higher than that under small-signal conditions for the same d.c. power.

In class A operation the maximum power-added efficiency is 50%. This is for an ideal voltage or current gain device that is linear in the first quadrant of the *V–I* plane (Fig. 6.3) with idealized characteristics extending to infinity on the positive voltage and current axes. The bias point for the device is set at (V_Q, I_Q) and the dynamic load line, corresponding to a load resistance R_l, is set at 45°. Under these conditions the peak-to-peak signal voltage and signal current swings become $2V_Q$ and $2I_Q$ respectively. The maximum r.f. output power P_{om} is now

$$P_{\text{om}} = \frac{V_{\text{m}} I_{\text{m}}}{8} = \frac{V_Q I_Q}{2} = \frac{P_{\text{dc}}}{2} \tag{6.9}$$

This, of course, corresponds to the area of the triangle $I_Q O V_Q$ and thus the maximum power-added efficiency is, from eqn (6.7),

$$\begin{aligned}(\eta_{\text{add}})_{\text{m}} &= \frac{P_{\text{om}} - P_{\text{im}}}{P_{\text{dc}}} \\ &= \frac{1}{2}\left(1 - \frac{1}{G}\right)\end{aligned} \tag{6.10}$$

Thus the maximum power efficiency of 50% is only attainable for a device with infinite r.f. power gain. Therefore in order to maximize the power-added efficiency of a device for a given output power, it is necessary to set the dynamic load resistance to $R_l = V_Q/I_Q$ while maximizing the r.f. power gain.

6.2.2.2 Main physical limitations

In a practical GaAs MESFET there are four fundamental effects which prevent the idealized load line from being obtained: the drain–source resistance, the maximum channel current I_f, the open-channel avalanche breakdown voltage and the drain–source breakdown voltage. Figure 6.4(a) shows the I_{DS}–V_{DS} characteristics of a typical modern MESFET with a gate width of 1.6 mm (data from manufacturers' data sheets on power GaAs

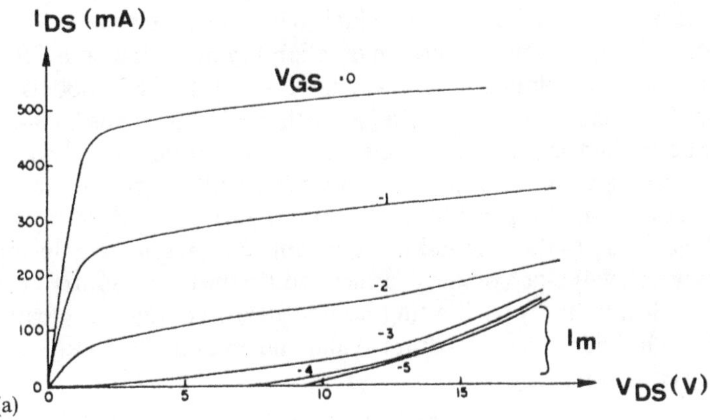

(a)

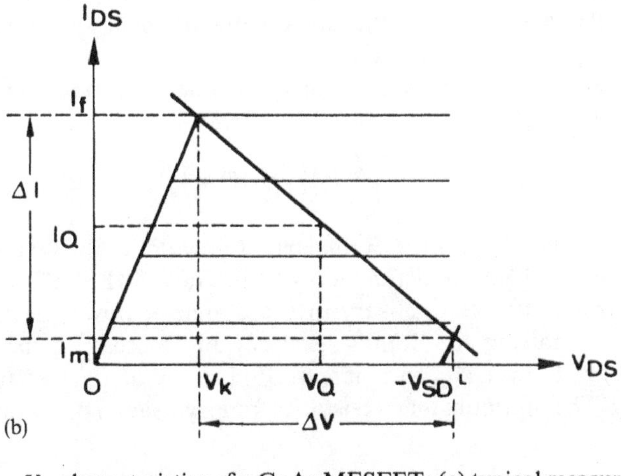

(b)

Fig. 6.4 I_{DS}–V_{DS} characteristics of a GaAs MESFET: (a) typical measured curves; (b) schematic diagram used in calculations.

MESFETs and internally matched devices). The drain–source resistance is minimized by the use of n^+ doping layers under the drain and source contacts. The maximum channel current is limited by the epilayer thickness a and the channel doping density N_d. There is also an important current-limiting effect due to the generation of heat in the channel under high power d.c. bias. This problem is particularly important because GaAs has poor thermal conductivity. Practical solutions for minimizing this effect are described in Section 6.2.3, and power dissipation methods are discussed in Section 6.4.4. The use of recessed-gate structures has increased the open-channel burn-out voltage to over 20 V, so that it no longer sets a fundamental limit on output power, and modern recessed-gate devices are specified for operation at a V_{DS} of 10 V *under large-signal operating conditions* (see the discussion of biasing precautions in Chapter 2). The maximum voltage swing is now limited by the gate–drain avalanche breakdown [6.2–6.4] in which the surface plays a very important role [6.4, 6.5] (see also the discussion of reliability phenomena in Chapter 10). This effect is thus included in large-signal GaAs MESFET models [6.6–6.8] (Chapter 5). Even though r.f. voltages with peaks in excess of $V_{SD}{}^L$ can in practice be applied as avalanche build-up is not instantaneous [6.4], this delay effect has not as yet been quantified and it is best to be prudent and use the schematic diagram of Fig. 6.4(b) for maximum output power calculations.

We define V_p as the external voltage applied between the gate and source contacts which will pinch off the channel (reduce the drain–source current to a predefined lower limit). A limiting source–drain voltage $V_{SD}{}^L$ can now be defined and linked to the gate–drain avalanche breakdown voltage V_{GD} by the following equation [6.2]:

$$V_{SD}{}^L = V_{GD} - V_p$$

Beyond this point excess current which is not modulated by the r.f. signal and therefore will not contribute to the fundamental frequency output power of the device will appear in the drain circuit.

From Fig. 6.4(b), the maximum quasi-linear output power from the transistor is (for $I_m/I_f \ll 1$)

$$P_{om} = \frac{I_f}{8}\{(V_{GD} - V_p) - V_k\} \tag{6.11}$$

The parameter I_f, which is the drain current flow under the forward gate bias offset voltage V_{bi} [6.9], is a function of the GaAs MESFET gate width w. Therefore trade-offs are necessary in the design of power devices since an increase in I_f (raising the doping level N_d and increasing the active-layer thickness a) will lower V_{GD} and increase V_p. In general, devices are optimized empirically. The optimum load resistance for large-signal operation is given in Fig. 6.4(b):

$$R_{l\,opt} = \frac{V_{GD} - V_p - V_k}{I_f} \tag{6.12}$$

6.2.2.3 Optimum bias conditions

The optimum gate voltage and drain current bias points are thus established statically as $(V_{GD}-V_p-V_k)/2$ and $I_f/2$, and the drain-to-source voltage should not exceed 9 V. A variation in I_{DS} linked either to an excursion of the gate signal into the gate rectifying zone or to the gate–drain avalanche breakdown will appear under large-signal dynamic operation. For best amplifier linearity under maximum r.f. swing, it has been shown experimentally that the gate voltage should either be maintained constant at its static value [6.10] or be allowed to vary slightly so as to maintain a zero average gate current. By keeping the device within this safe operating area we also extend its lifetime and reduce the effects of long-term r.f. power output decay. The price paid is lower short-term output power and power-added efficiency, and we can only reasonably expect an output power of 0.3–0.4 W mm^{-1} of gate width at the 1 dB gain compression point of the transistor.

6.2.3 Cell-combining considerations

Power GaAs MESFETs are made by connecting elementary transistor cells of optimized gate width, gate length and power output in parallel. However, when devices are connected in parallel they do not always perform as expected. Four distinct problems are involved with scaling up.

(a) An electrical problem arises involving the feeding of r.f. power to the individual cells in phase, so that all cells will operate in an identical manner. This also requires homogeneity in doping and active-layer thickness across the power transistor structure.

(b) There is an r.f. power match problem, since r.f. terminal impedances diminish with gate width.

(c) A thermal problem is associated with the dissipation of a large amount of heat in a small transistor chip volume.

(d) A parasitic problem concerns the physical interconnection of the source pads (dielectric crossover, air bridge, flip-chip, via hole or wire bonding) and their connection to an r.f. ground. This is a major problem because of the effect of the source lead inductance on the gain.

The thermal problem encountered in high power GaAs MESFET structures is dealt with very fully in ref. 6.11. Essentially, a compromise has to be made between minimizing the electrical phase difference along the gate and drain busbars by using compact structures and reducing the heat build-up by using a more dispersed structure with wider separation between the cells. Heat is also evacuated out of the transistor by using thinned substrates (typically 50 μm) mounted on plated heat sinks with via holes in the source pads.

Power combining at chip and amplifier level is treated in Section 6.4, as is impedance matching. However, a few introductory device-related concepts are presented here.

An equivalent circuit representation of a single-cell GaAs MESFET transistor is given in Fig. 6.5. If n such cells are connected in parallel to give a power transistor, the input impedance will consist of R_i/n in series with nC_{GS} and an output conductance ng_{do} in parallel with nC_{DS}. The gate-to-drain feedback capacitance and the ohmic contact parasitics R_s and R_d as well as the gate metal resistance R_g will also scale with n. However, when the cells are combined the source lead inductance L_s is not reduced proportionally as cells are added, even if flip-chip mounting is used, because of mutual inductance effects. This has a direct bearing on power gain performance, since the small-signal maximum available gain (MAG) is [6.12]

$$\mathrm{MAG} = \left(\frac{f_T}{f}\right)^2 \frac{1}{4g_{do}(R_g+R_i+R_s+\pi f_T L_s)+4\pi f_T C_{GD}(2R_g+R_i+R_s+2\pi f_T L_s)}$$

where $f_T \approx g_{mo}/2\pi C_{GS}$ is the frequency at which the current gain is unity and f is the operating frequency.

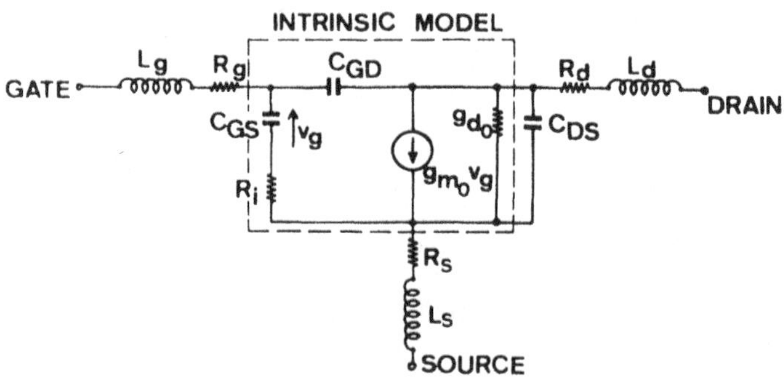

Fig. 6.5 GaAs MESFET equivalent circuit used in the scaling-up exercise.

Thus unless special care is taken, such as using periodic via holes in the source pad, the overall gain will degrade as cells are combined and the series inductance of the center cells increases. Another source of gain degradation derives from the reduction of input resistance with gate width. For example, for a device with a nominal gate length of 1 μm, the input resistance $R_i + R_g + R_s$ decreases from 4.5 Ω to 0.8 Ω as the gate width is increased from 1200 μm to 11 200 μm. Beyond a certain gate width, matching losses increase disproportionately, and it is preferable to prematch medium power transistor cells, particularly at the gate terminal, before combining them. In Section 6.4.2 we describe simple L-section and π-section lumped-element design techniques commonly used in the design of prematching circuits. In Section 6.4.3 we study power dividing–combining structures used to connect several prematched power cells in parallel. Most manufacturers limit the maximum gate width on a single chip to about 8 mm, which corresponds at C band to an output power of about 3 W at a gain compression of 1 dB. In order to limit matching losses due to high voltage standing-wave ratios (VSWRs) in the microstrip lines between the chip and the matching circuit, manufacturers now commonly offer packaged devices using one or several prematched chips. These power GaAs MESFETs have a 50 Ω bandwidth of 1 GHz centered on the various communications bands. (These data are obtained from manufacturers' data sheets on power GaAs MESFETs and internally matched devices, and from application notes by NEC, Toshiba, MSC, Fujitsu etc.)

Figure 6.6 shows a typical medium power transistor. In this case the source-to-ground lead inductance is minimized by butting the device against a metal ledge and using a large number of very short connections. Figure 6.6(b)

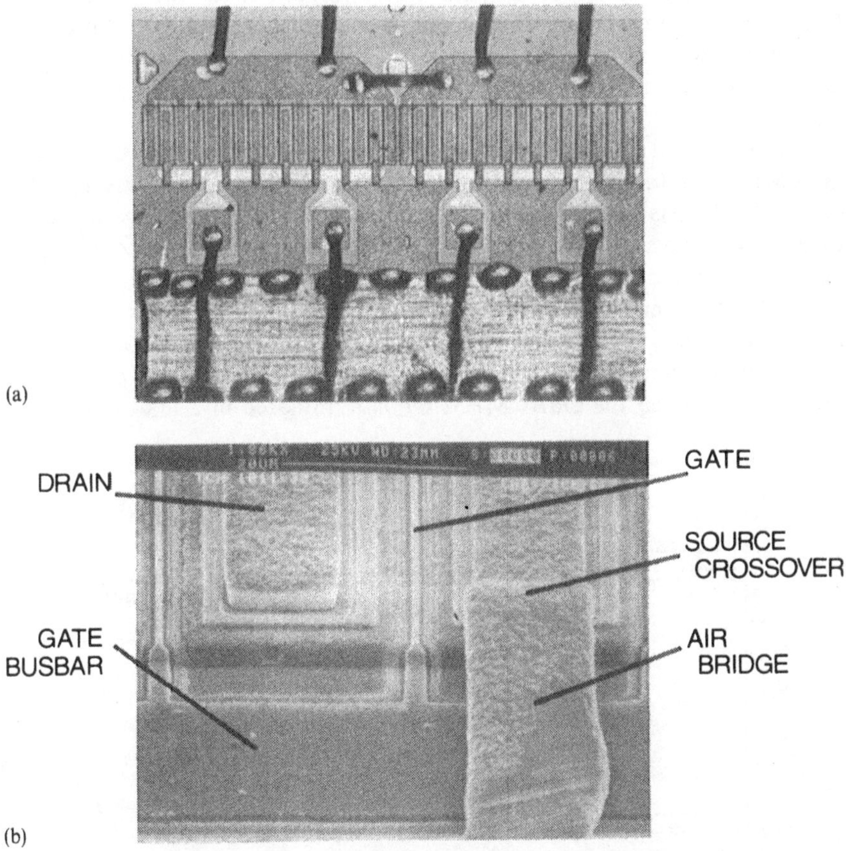

Fig. 6.6 Microphotographs of a medium power GaAs MESFET: (a) chip; (b) detail showing the air bridge.

shows an air bridge used to connect individual transistor source pads to a common busbar. The bridge hops over the gate busbar.

6.3 Large-signal characterization and modelling

As already noted, under large-signal conditions the S parameters of any two-port amplifying device must alter in order to comply with the laws of thermodynamics. In the case of a GaAs MESFET amplifier this is confirmed by experiment [6.13].

Figure 6.7 shows gain against output power performance of a GaAs MESFET with various terminating loads. High gain but low power output saturation is obtained under small-signal conjugate matching conditions. Other load terminations improve the power saturation performance considerably at the expense of power gain. The optimum loads for an amplifier matched first to maximum small-signal gain and then to maximum output power are shown in Fig. 6.8. It can be seen that the large-signal behavior of the GaAs MESFET is affected more by the output load Γ_l than by the source load Γ_s.

Conventional small-signal S parameters are of little use when large-signal GaAs MESFET amplifiers are being designed, and it becomes necessary either to measure the load impedances which correspond to a desired gain–output power performance at a given frequency or to resort to large-signal two-port measurements. However, there are certain short cuts which can be applied when amplifier performance specifications are not too rigorous. These are based on the following considerations.

Observation of the amplifier gain–output power curves of Fig. 6.7 indicates that when the GaAs MESFET is terminated in a load impedance

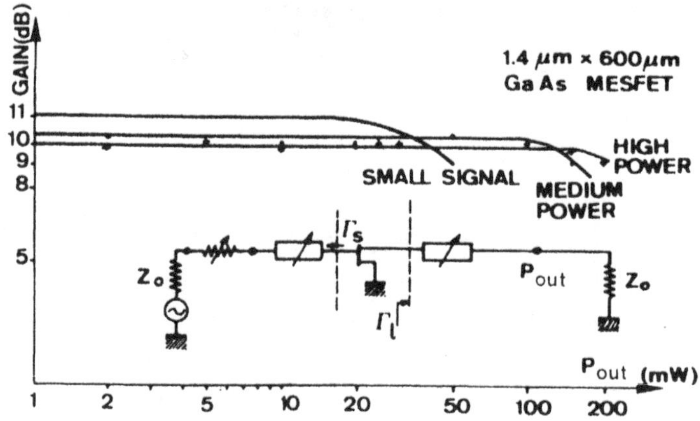

Fig. 6.7 Variation in gain with signal drive and terminating loads Γ_s and Γ_l.

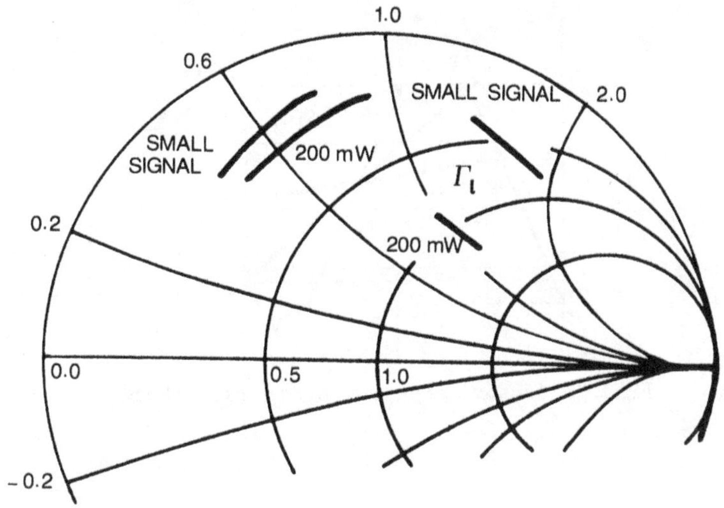

Fig. 6.8 Variation in the optimum loads Γ_s and Γ_l with signal drive.

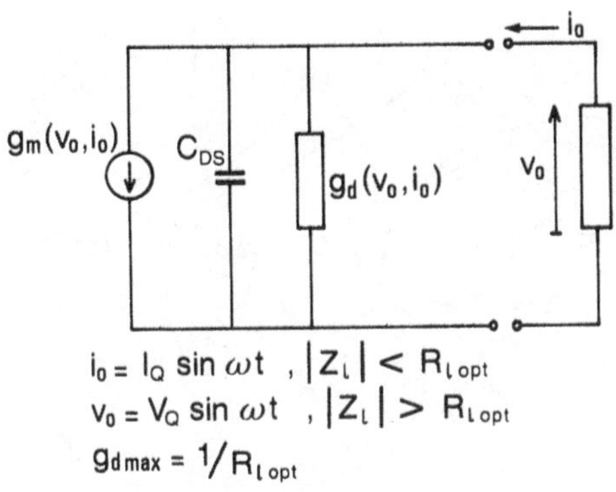

Fig. 6.9 Norton equivalent circuit representation of the output large-signal behavior of a GaAs MESFET.

which is conducive to a larger output power than is the case with the small-signal optimum termination it operates in a quasi-linear manner up to a higher gain compression level. Since large-signal output power depends essentially on the output load (see Fig. 6.8), we can represent the output power performance of the transistor by the simple Norton equivalent circuit of Fig. 6.9. Whereas rigorously the output current generator and the associated output admittance are functions of both input and output signal level, in the quasi-linear operating region we can loosely associate both elements with the output load (i.e. the output power is "load pulled"). The optimum output power P_{om} (eqn

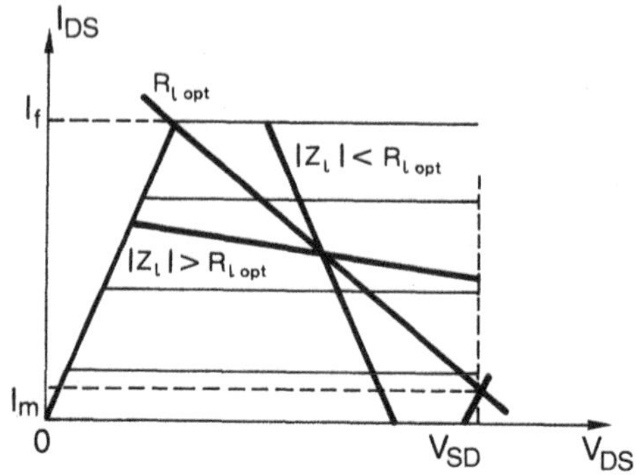

Fig. 6.10 Limitations imposed by various load lines.

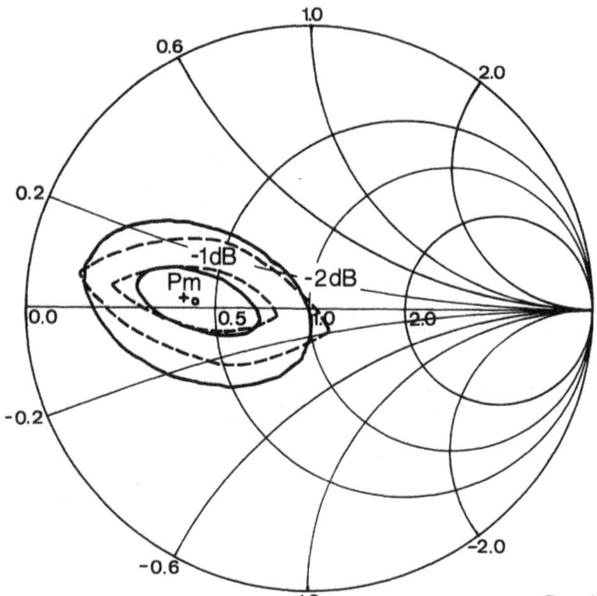

Fig. 6.11 Comparison of experimental (+, ——) and predicted (○, – – –) results using the simplified load-line theory [6.14].

(6.11)) is obtained for a load impedance with real part equal to $R_{1\text{opt}}$. Cripps [6.14] has argued that in order to obtain constant-output-power contours less than P_{om} it is sufficient to consider two cases (Fig. 6.10): $|Z_1| < R_{1\text{opt}}$ when the output power is current limited and $|Z_1| > R_{1\text{opt}}$ when the output power is voltage limited. We obtain the following relations for the two cases:

$$\frac{P_o}{P_{om}} = \frac{R_l}{R_{l\,opt}}$$
$$X_l^2 \leqslant R_{l\,opt}^2 - R_l^2 \qquad (6.13)$$

for $|Z_l| < R_{l\,opt}$, and

$$\frac{P_o}{P_{om}} = \frac{G_l}{G_{l\,opt}}$$
$$B_l^2 \leqslant G_{l\,opt}^2 - G_l^2 \qquad (6.14)$$

for $|Z_l| > R_{l\,opt}$. The familiar ellipsoidal contours (Fig. 6.11) are obtained by following constant-resistance contours on the Smith chart for eqn (6.13) and constant-admittance contours for eqn (6.14), within the bounds fixed for X_l and B_l respectively, and adding in the bond-wire parasitics.

This technique, which is rapid and simple, is only partially valid since the output signal of a MESFET operating under large-signal conditions is *not* a sinusoid irrespective of the load impedances. In addition, the locus of the dynamic load line (ratio of the instantaneous output voltage to the output current) is not linear but elliptical [6.15]. For *accurate* power amplifier design over large frequency bandwidths, it is thus better either to perform large-signal measurements at all important design frequencies or to use more accurate nonlinear transistor models. Such models are treated in detail in Chapter 5, but for completeness a simple example of their use is given in Section 6.3.2. We begin, however, with a review of standard large-signal and quasi-large-signal measurement methods as summarized in Table 6.1.

Table 6.1 Standard large-signal measurement methods

Large-signal parameter	*Measurement technique*	*Section*	*Reference*
S_{22}	Small-signal S_{22} with $V_{DS} < 2$ V	6.3.1.2	[6.16]
Gain/P_o performance, and associated Γ_l contours (Γ_s fixed)	(a) Active load-pull (b) Semi-automatic method (c) Output tuner Y parameter method (Γ_l only)	6.3.1.3 6.3.1.2 6.3.1.2	[6.15, 6.17] [6.18] [6.19]
Gain/P_o performance and associated terminal impedances (Γ_s, Γ_l)	(a) Classical load-pull (b) Peeling method	6.3.1.1 6.3.1.2	[6.15, 6.18, 6.20] [6.21, 6.22]
Large-signal S parameters (50 Ω)	Large-signal S_{11}, S_{22}, S_{21}, S_{12}	6.3.1.4	[6.23–6.26]
Large-signal S parameters for specified (Γ_s, Γ_l) or gain/P_o	Large-signal S_{11}, S_{22}, S_{21}, S_{12} under amplifier operating conditions	6.3.1.5	[6.21, 6.27]

6.3.1 Large-signal measurements

6.3.1.1 Basic load-pull method

Load-pull is the only practical method of measuring those terminal impedances which, when applied to a device, enable it to have particular linear or nonlinear characteristics. Classically, the device is tuned with removable passive tuners until the required performance characteristics are obtained. The tuners are then removed, and the impedances presented to the device are measured on a slotted-line or network analyzer. This procedure is repeated at each frequency point of interest. A circuit which presents the measured impedances to the device can now be synthesized using normal synthesis techniques. This time-honored technique has been widely used in the design of oscillators, mixers and multipliers, as well as class A, B, C and pulsed-mode amplifiers.

A practical load-pull test bench for measuring the output large-signal load corresponding to a power GaAs MESFET performance is shown in Fig. 6.12. At a fixed frequency and for a given input power, measured on P_1, the input tuner T_1 is adjusted for optimum return loss as measured on power meter P_3 (typically, 25 dB return loss). With the single-pole double-throw switch connected to branch 1, the output of transistor T_2 is tuned for the output power performance desired. The load applied to the output port of the transistor can be read directly if the electrical lengths of branches 1 and 2 are adjusted to be identical and T_2 is switched to branch 2. It corresponds to a particular output power–gain performance for a given device at a given frequency.

In order to obtain the power out referred to the device plane a correction must be made to P_2 for tuner matching losses. This can be done experimentally by removing tuner T_2, connecting another tuner to it and tuning for the minimum insertion loss which is then measured. Alternatively, if the load reflection coefficient Γ_2 and the line losses of branch 1 are known, we can use the correction formula given in Chapter 3:

$$P_{2\,\mathrm{cor}}(\mathrm{dB}) = P_2(\mathrm{dB}) + 10\log_{10}\left\{\frac{A(1-\Gamma_2{}^2)}{1-A^2\Gamma_2{}^2}\right\} \tag{6.15}$$

where A is the insertion loss of branch 1 terminated in 50 Ω.

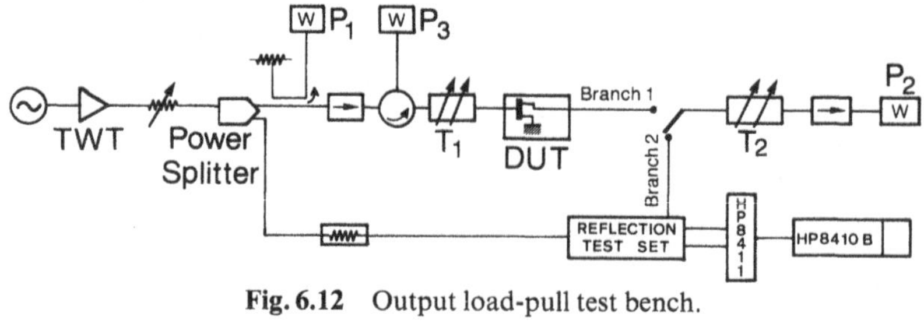

Fig. 6.12 Output load-pull test bench.

If it is necessary to measure the input impedance and the output load contours corresponding to various power levels for a fixed input match, a two-port load-pull test set-up is required (Fig. 6.13) [6.15]. The dual directional couplers are associated with power splitters so that the reflection coefficients of the input and output tuners are measured simultaneously with net power into and out of the GaAs MESFET. Thus the gain and output power of the device corresponding to various tuner loads are obtained, and the output power contours are easy to trace.

It should be noted that, whereas the output reflectometer reads the tuner impedance directly, the input reflectometer measures the reciprocal of the input tuner reflection coefficient. The measurement principle is summarized in

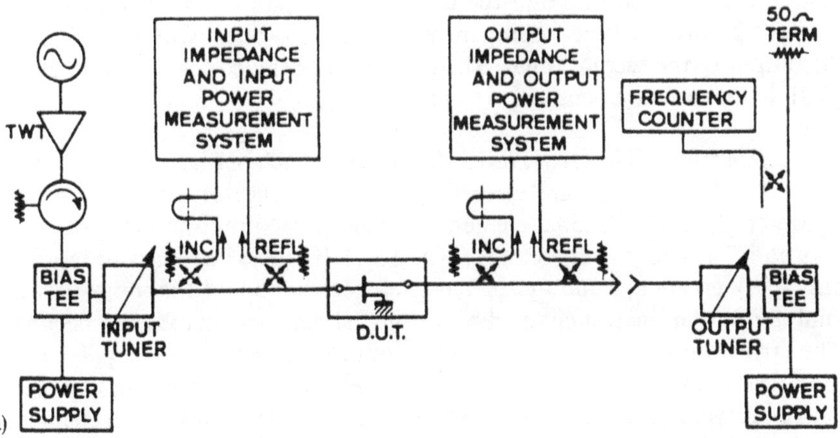

(a)

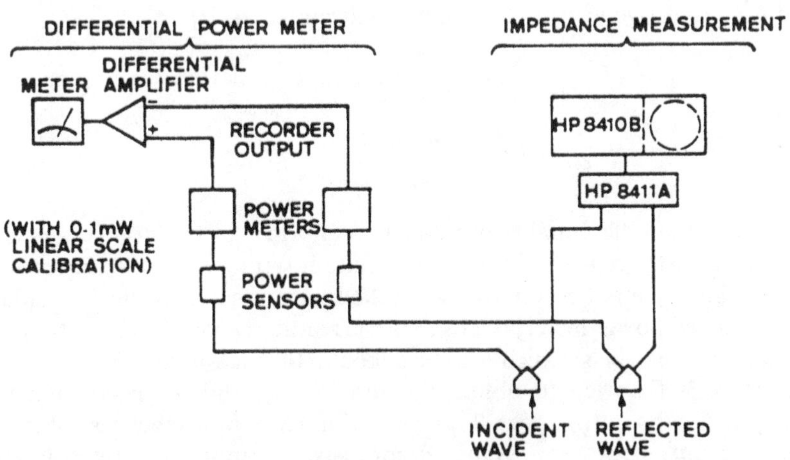

(b)

Fig. 6.13 (a) Two-port load-pull measurement system; (b) details of impedance and power measurement (the input and output systems are identical).

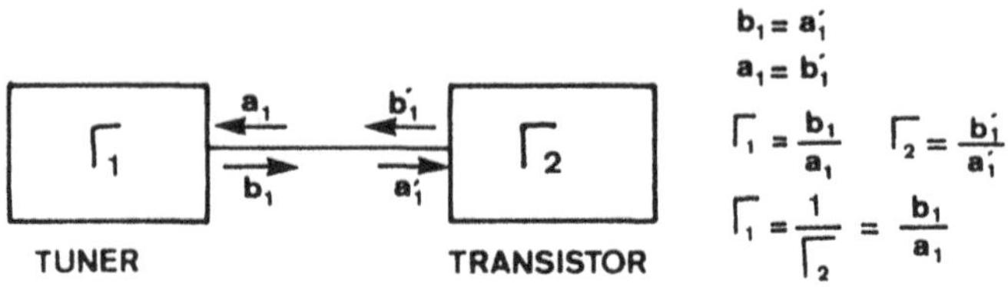

Fig. 6.14 Principle of the direct measurement of the tuner reflection coefficient.

Fig. 6.14. In order to reduce error, great care must be taken to minimize the VSWR of all components used in the test bench including the test fixture, the dual-directional couplers and the tuners. The maximum VSWR should not exceed 1.2, with the tuners being measured back-to-back with a 50 Ω load on the output of the second tuner. The directivity of the couplers should exceed 25 dB in the measurement frequency range.

Calibration The error models for the test bench of Fig. 6.13 are shown in Fig. 6.15 [6.20]. These are essentially standard reflectometer error models (Chapter 2) with additional eleménts added for the input and output tuners (Γ_{1T} and Γ_{2T}) and power meters (e_{16} and e_{26}). It is also necessary to represent directivity errors e_{15} and e_{25} in the reference channel of the input network analyzer and in the test channel of the output network analyzer respectively. The error terms e_{11}, e_{13}, $e_{12}e_{14}$ on the input port and e_{21}, e_{23}, $e_{22}e_{24}$ at the output port are evaluated using one of the techniques described in Chapter 2. This enables us to calculate the corrected transistor input reflection coefficient $\Gamma_{\text{in cor}}$ and load reflection coefficient $\Gamma_{\text{l cor}}$ in terms of the measured values ($\Gamma_{\text{in m}}$ and $\Gamma_{\text{l m}}$ respectively) using the following formulae [6.20]:

$$\Gamma_{\text{in cor}} = \frac{\Gamma_{\text{in m}} - e_{11}}{e_{12}e_{14} - e_{11}e_{13} + e_{13}\Gamma_{\text{in m}}}$$
$$\Gamma_{\text{l cor}} = e_{23} + \frac{e_{22}e_{24}\Gamma_{\text{l m}}}{1 - e_{21}\Gamma_{\text{l m}}} \tag{6.16}$$

The input differential power meter is calibrated at two points: the first is for the transistor replaced by a short circuit for setting zero, and the second is for an arbitrary input power level when the device under test (DUT) is replaced by a matched power meter; the ratio of the readings gives the correction term $|e_{12}/e_{16}|^2$ (in decibels) which is to be added to the measured power. A similar procedure is followed to obtain the zero setting and the correction term $|e_{24}e_{26}|^2$ for the output port. This correction term (in decibels) is subtracted from the output power reading. Rigorously, the input and output power readings should be corrected for the VSWRs of the input and output couplers respectively. The exact formulae are given in ref. 6.20.

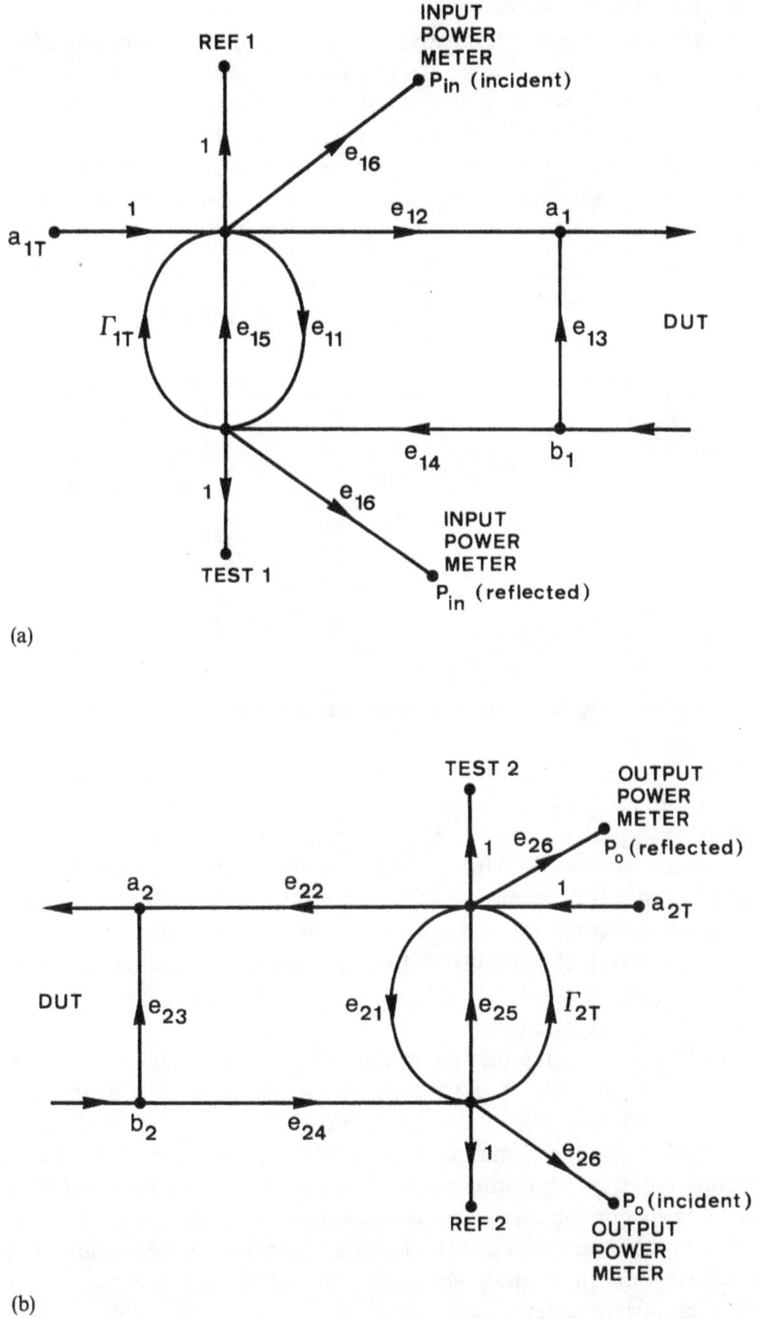

Fig. 6.15 Signal flow graph error model of the load-pull bench of Fig. 6.13 (after ref. 6.20): (a) input port; (b) output port.

6.3.1.2 Peeling method

Sometimes only the load and source impedances corresponding to a given performance are required. In this case the peeling method can be used and gives extremely accurate results if the test fixture is properly characterized [6.21, 6.22]. Essentially, the technique is to match the transistor under large-signal conditions to give a large-signal performance and then to reduce the signal level to small-signal proportions in order to carry out two-port peeling (Chapter 2).

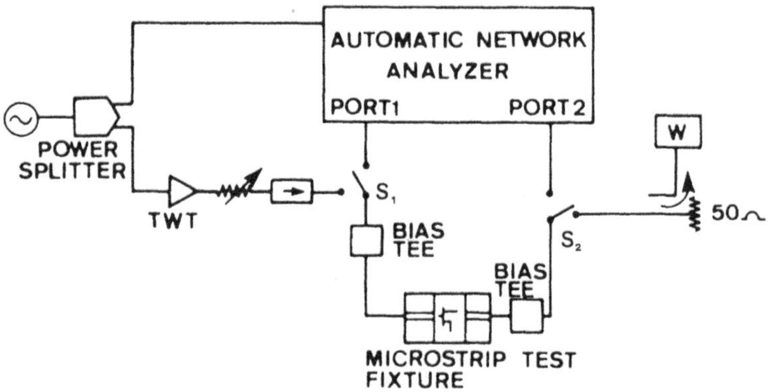

Fig. 6.16 Set-up for large-signal peeling.

The set-up for doing this is shown in Fig. 6.16. It is a combined large-signal–automatic S parameter test bench. The switches S_1 and S_2 are initially set for large-signal operation. The transistor mounted in a microstrip test fixture, which includes 50 Ω microstrip lines just over half a wavelength long at the input and output terminals of the transistor, is disk tuned to the output power–gain performance requirement. The switches are now thrown to connect the matched "amplifier" to the automatic network analyzer. The peeling sequence described in Chapter 2 is then applied in order to obtain source and load impedances.

Another characterization technique which relies on large-signal matching followed by small-signal S parameter measurement is that proposed in ref. 6.19. The method is based on the principle that the small-signal output conductance of an input matched transistor is a function of d.c. bias and is independent of the output tuner load. The transistor is r.f. matched to give the desired large-signal performance using output and input tuners. The two-port tuner load matrix can then be calculated by successively measuring the output impedance of the input matched device at various d.c. bias points with and without the output tuner.

A simple and rapid technique of simulating large-signal S_{22} is to measure the small-signal S_{22} under a reduced drain–source bias voltage [6.16]. In ref.

6.16 it is stated that optimum matching of a 1 μm × 2400 μm transistor to a small-signal S_{22} measured at a $V_{DV} = 2$ V gave a large-signal 1 dB gain compression output power of 30 dBm at $V_{DS} = 8$ V, which is about the optimum performance expected for the device.

6.3.1.3 Active load-pull method

It is difficult to obtain high VSWRs in passive tuners owing to losses in the tuners themselves and in the transmission line between the tuner and the active device. For example, even if a tuner were capable of generating a reflection coefficient of unity, a loss of 1 dB in the transmission line between the tuner and the device would reduce the maximum reflection coefficient obtainable at the device plane to 0.79 (eqn (6.15)).

An elegant means of overcoming this problem [6.17] is to generate the reflected wave back onto the transistor output by means of an external coherent source instead of a passive tuner. The magnitude of the reflected wave can be made superior to that of the incident wave in order to compensate for losses between the "tuner" and the device. A simple means of realizing such an active tuner is to replace the output tuner used in Fig. 6.13(a) by the set-up shown in Fig. 6.17 [6.15]. By suitably varying the phase shifter and the variable attenuator it is now possible to cover the whole of the Smith chart. Another advantage of this technique is that tuning is simplified: constant Γ_1 contours can be generated on the output load plane by simply varying the phase shifter. Calibration procedures are identical with those used for the passive tuner. An example of output load impedance contours for constant-power outputs for a 1.2 μm × 1600 μm GaAs MESFET is given in Fig. 6.18.

6.3.1.4 Load-pull intermodulation measurements

Linearity is an important consideration in power amplifiers used in multichannel links. If the transfer response of the transistor is represented by a polynomial (see Chapter 5), the odd-order elements lead to the creation of intermodulation products that fall within the amplifier passband, particularly for closely spaced carrier frequencies. It is shown in Chapter 5 that even-order elements also play a role in the generation of intermodulation products by a mixing process. The intermodulation products are specified with respect to the output power of the amplifier and are expressed as a carrier-to-intermodulation (C/I) ratio (in decibels). As transistor linearity is directly

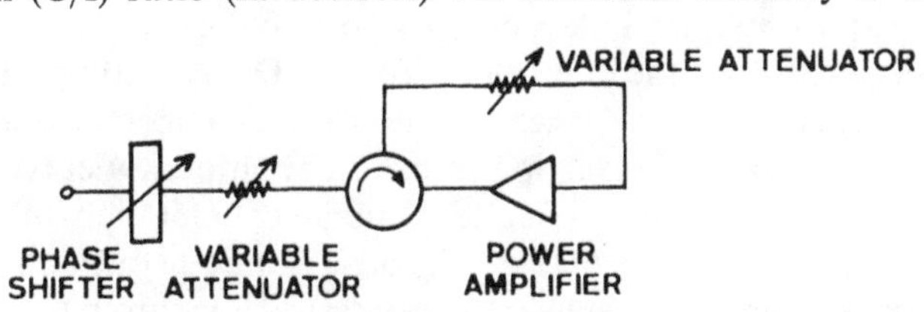

Fig. 6.17 Active tuner.

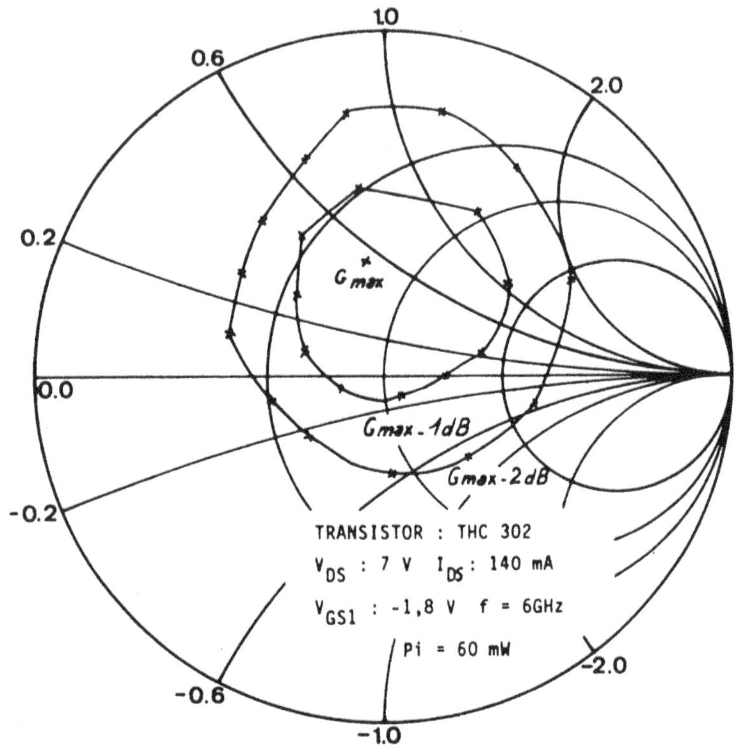

Fig. 6.18 Output power–load contours for a GaAs MESFET amplifier.

related to the degree of gain compression, the most common approach used to obtain a C/I ratio at a given output power is to design the amplifier to a higher linear power (1 dB gain compression) and "back-off" into a lower power and, normally, more linear operating region. This "brute force" approach necessitates the use of transistors which are overdimensioned with respect to the operating output power. A better method is to measure the intermodulation product as a function of the output power load-pull using a bench of the type shown in Fig. 6.19 [6.28]. In this set-up two signals, separated by typically 5 MHz, are amplified independently to avoid generating intermodulation products and fed via the input tuner into the DUT. A computer-driven output tuner is then used to generate the various constant-output-power and constant-C/I contours given in Fig. 6.20. The output tuner is either precharacterized or measured using a dual-directional coupler–network analyzer set-up of the type shown in Fig. 6.12 or Fig. 6.13. A unique output impedance which allows maximum output power to be obtained for a given C/I requirement can now be defined for a given frequency, input power level and bias point.

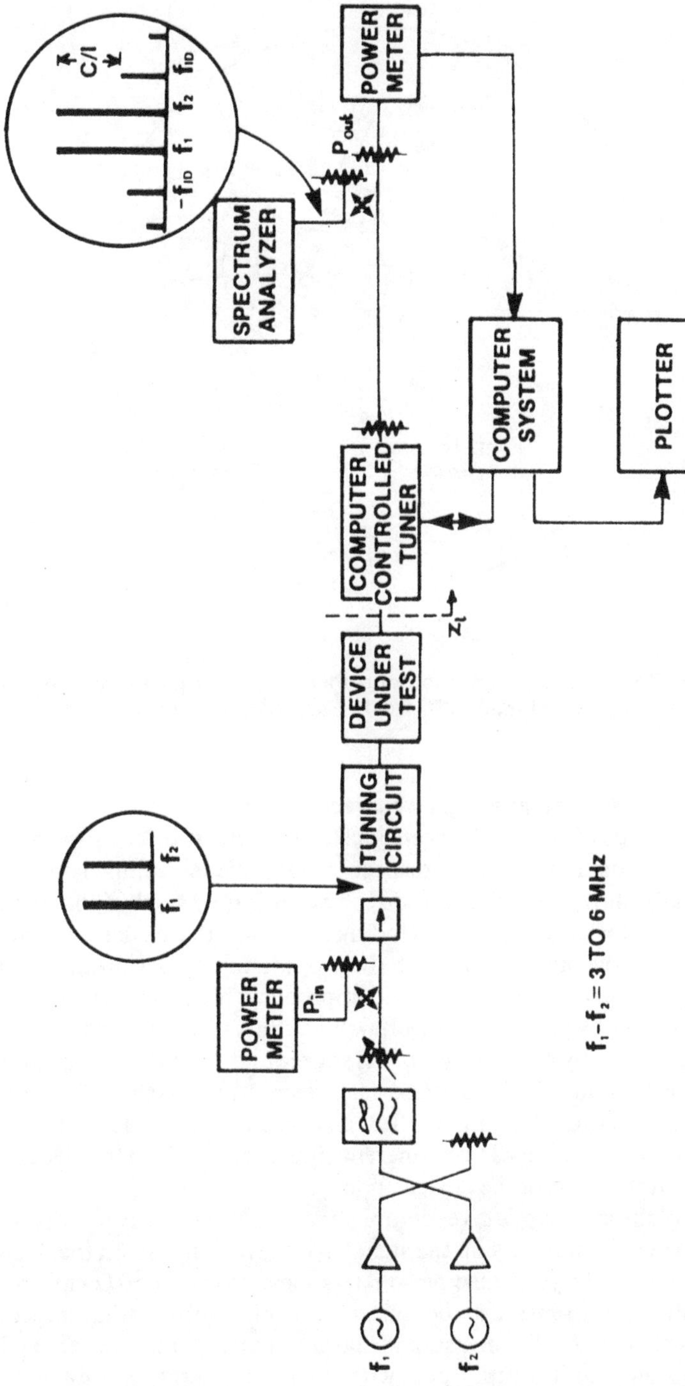

Fig. 6.19 Intermodulation and output power load-pull test bench [6.28].

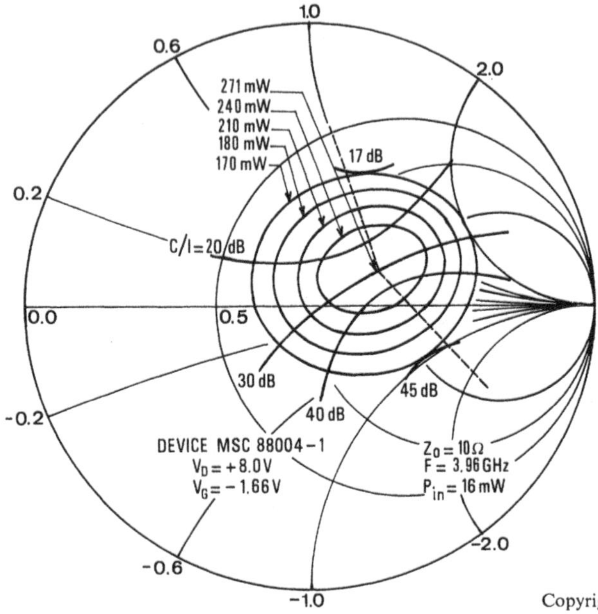

Fig. 6.20 Output load impedance contours for constant output power and constant C/I ratio for an MSC 88004 GaAs MESFET [6.28].

6.3.1.5 Two-port large-signal characterization

The first type of two-port large-signal characterization consists in making S parameter measurements with respect to 50 Ω reference impedances. There are essentially three methods: the direct extension of small-signal measurement methods to large signals, including reversing the device and injecting high power into the output port (with the input terminated in 50 Ω) to measure S_{22} and S_{12} [6.23, 6.24]; a more refined approach which consists in measuring the S parameters by small-signal techniques whilst ensuring that the input of the device is driven by a large-signal 50 Ω source and the output is connected to a passive 50 Ω load [6.25]; a quasi-large-signal approach [6.26] whereby large-signal S parameters in a 50 Ω environment are measured by the simultaneous application of two coherent signals at the same frequency to the input and output ports of the device under test.

None of these methods gives complete satisfaction, however. Large-signal parameters depend not only on the signal drive level but also on the transistor port terminations [6.18], since parameters measured in a 50 Ω environment differ significantly from those measured with other terminating impedances [6.29]. It is also illogical to measure the large-signal S_{22} and S_{12} of a device by injecting power into its output port with the input passively terminated. For these parameters to be meaningful, they should be measured with the transistor under the input drive condition (and with the load terminations) likely to be encountered in practice. Of course, this necessarily complicates the measurement procedure, but the additional information provided makes the two-port method worthwhile. The results obtained can be directly applied to large-signal amplifier design or used to establish the manner in which certain nonlinear elements in the electrical equivalent circuit of the device vary with the signal drive level [6.30]. A large-signal equivalent circuit for the GaAs MESFET which is independent of both frequency and terminating impedances can be obtained by associating the behavior of these nonlinear elements with their terminal voltages or currents.

6.3.1.6 Large-signal characterization between arbitrary loads

The measurement procedure adopted [6.27, 6.31] is to match the GaAs MESFET to a desired output power–gain condition and then to measure the two-port parameters of the resulting class A–B amplifier at a fixed frequency for successive drive levels. The transistor large-signal parameters corresponding to each drive level are obtained by abstracting the load terminations via the small-signal measurements which are performed on the same test bench without removing the amplifier (Fig. 6.21).

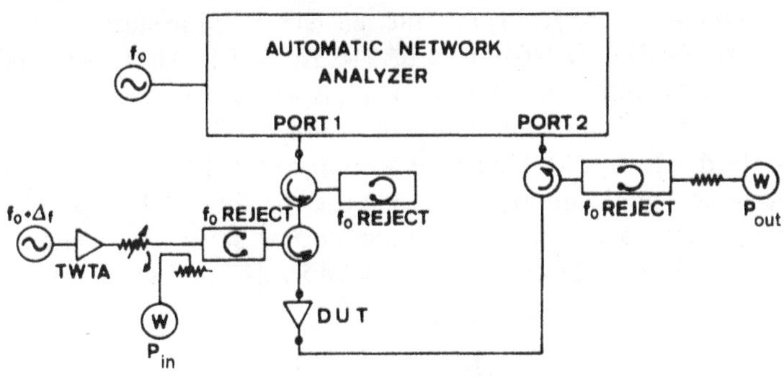

Fig. 6.21 Two-port large-signal test bench.

The large-signal two-port parameters of the amplifier are obtained under large-signal drive by using a small-signal network analyzer with its signal source offset by 10 MHz with respect to the large-signal generator. Frequency separation of the signals is effected by the use of high Q cavity filters and high isolation circulators to channel each signal along the appropriate path. Large-signal drive to the amplifier is now turned off, and the input and output network matrices are determined on the same bench using the peeling method. Finally, transistor large-signal parameters are recovered from amplifier large-signal measurements by de-embedding (Chapter 2).

This procedure can be repeated for any desired source and load-terminating impedances and for any number of frequencies.

6.3.2 Large-signal GaAs MESFET models

The development of large-signal simple models is of obvious interest in the design of nonlinear components using active devices such as GaAs MESFETs. Although significant advances have been made in this domain in recent years, universal turnkey nonlinear analysis programs do not yet exist. However, large-signal models and nonlinear analysis techniques *have* reached the point where a careful designer, who has a difficult specification to meet, can find a large number of tools available in the literature [6.32–6.40]. Essentially the designer has to select an appropriate large-signal model and then apply a suitable analysis technique. As the subject is treated thoroughly in Chapter 5, we shall only give a simple example of a practical application in this section.

6.3.2.1 Large-signal model

There are essentially three types of model in use: physical models based on a solution of transport equations in the semiconductor [6.32–6.34], quasi-static models using small-signal S parameter measurements made over a wide range of bias conditions [6.38] and models based on the static characteristics of the GaAs MESFET [6.6–6.8, 6.35–6.37, 6.39]. All these models are comprehensively treated in Chapter 5. Thus we shall only consider a simple case: the large-signal characterization of a 1 μm × 600 μm GaAs MESFET and its use in the design of a 6 GHz power amplifier [6.35].

The large-signal equivalent circuit used is that of Fig. 6.22. The four nonlinear elements selected are the output current source I_{DS}, a current source I_{DG} to simulate drain–gate avalanche breakdown, a nonlinear capacitance C_{GS}

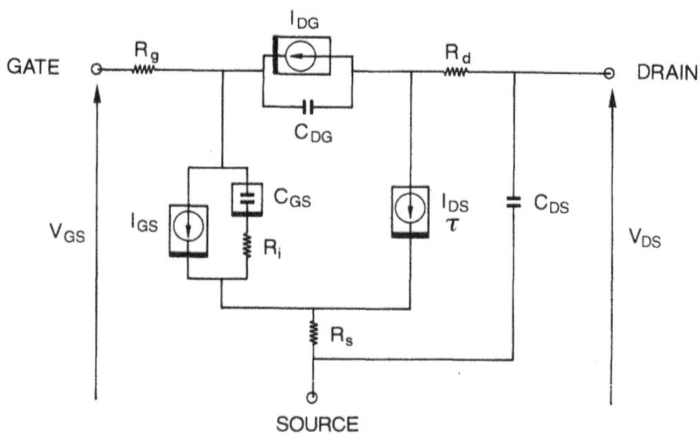

Fig. 6.22 Nonlinear equivalent circuit of a GaAs MESFET.

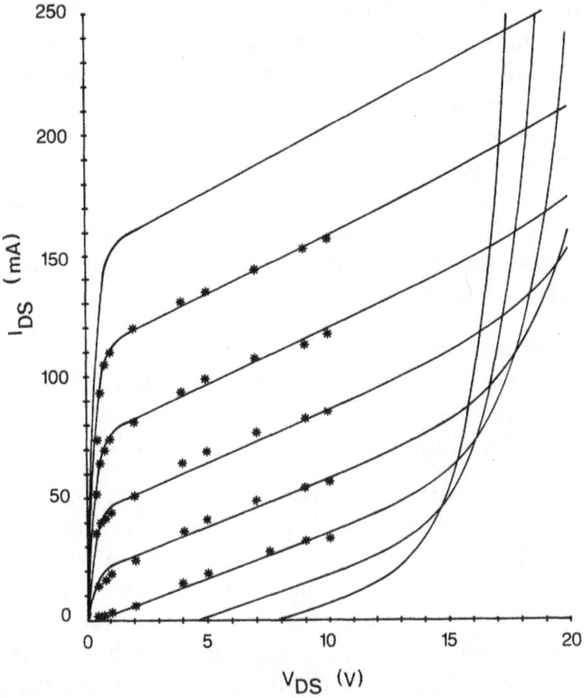

Fig. 6.23 Measured and simulated pulsed measurements of I_{DS} versus V_{DS}.

simulating the variation in the gate depletion region with signal level and a current source I_{GS} representing gate behavior in the forward conduction region. The current source is simulated by fitting a polynomial function of V_{GS} and V_{DS} to a static I_{DS}–V_{DS} characteristic (Fig. 6.23) obtained using the d.c. pulse measurement bench shown in Fig. 5.27. Two important aspects of this type of model should be emphasized.

(1) The use of d.c. pulsed measurements to explore the static characteristics of the GaAs MESFET eliminates thermal distortion effects under conditions of large drain current and high drain voltage. It also protects the device from burnout.

(2) From these measurements it is necessary to represent accurately the dynamic output current and voltage waveforms at a given signal drive level and d.c. bias point V_Q, I_Q. This is done by representing the instantaneous output current $I_{DS}(t)$ as a polynomial, and thus closed-form analytical, function of the gate voltage $V_{GS}(t)$ and the drain voltage $V_{DS}(t)$. The polynomial fit to the measured d.c. characteristics is of major importance if the model is to represent large-signal transistor behavior effectively. This is why the whole of Section 5.6.5.1 in Chapter 5 is devoted to this subject.

A polynomial representation of the transistor V_{DS}–I_{DS} breakdown as measured on the pulse bench is also used to represent the current source I_{DG}. The Schottky function (Fig. 6.22) is represented by two nonlinear parameters C_{GS} and I_{GS}. C_{GS} is obtained by low frequency capacitance bridge measurements made at various gate voltages V_{GS} to which a polynomial function is fitted, and I_{GS} is represented by a mathematical function (Chapter 5, Section 5.6.5.2). The other elements of the equivalent circuit are obtained by d.c. measurement (R_g, R_d and R_s) and small-signal S parameter fitting to measurements performed over a very wide frequency range at the intended I_{DS} bias point (normally $I_{DSS}/2$) and $V_{DS} \approx 5$ V.

6.3.2.2 Nonlinear analysis

Having obtained a suitable model for the large-signal behavior of the GaAs MESFET, we must now select an analysis method. In the case of nonlinear amplifiers, we have in general two major goals.

(1) We desire maximum linearity, i.e. the minimization of intermodulation products, amplitude-to-phase (AM–PM) distortion and amplitude-to-amplitude (AM–AM) distortion. The methods used to analyze AM–PM and AM–AM distortion can be of the describing-function type, the harmonic balance type or the time-domain type, and intermodulation distortion is analyzed using a Volterra series approach, a harmonic balance technique or a time-domain technique. Harmonic products are also accessible using any of the last three methods. In general, however, the elements of Fig. 6.23 are too nonlinear for a Volterra series analysis and time-domain or harmonic balance techniques are preferred for amplifier distortion analysis.

(2) The second goal is optimum output power and maximum power-added efficiency. Time-domain analysis programs such as ASTEC III and MICROWAVE SPICE are commercially available and have been used successfully to optimize these large-signal amplifier performance parameters. However, harmonic balance methods are better suited to this problem and can be made interactive.

In the example presented here the in-house time-domain program CIRCEC was employed. A major disadvantage of time-domain analysis is that a steady state solution, which is the only one of interest in amplifier design, often requires several cycles and thus a long computation time. The main culprits are the external linear circuit elements, and it may be necessary to restrict certain element values in order to limit execution time. The time-domain program may also run into numerical instabilities and thus demand a different type of description for the nonlinear element (Section 5.6.5.2). Despite these problems, the device analyzed gave the results shown in Fig. 6.24. Five transistors taken from a single slice were measured under the same optimized load and remarkably good agreement with theory was obtained (Figs 6.24(a) and 6.24(b)). The program also enabled the harmonic second- and third-order

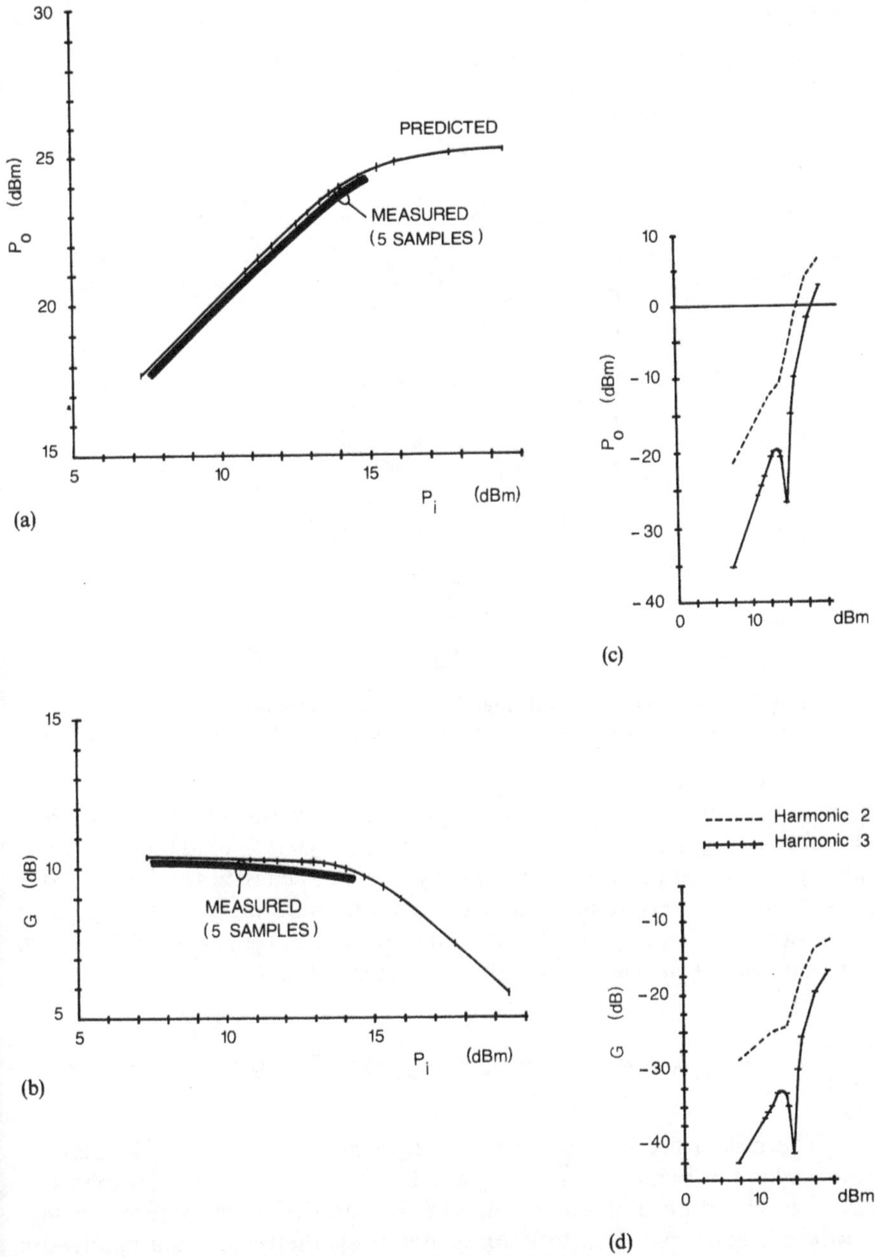

Fig. 6.24 Calculated and measured large-signal performance for a medium power GaAs MESFET: (a) output power versus input power; (b) gain versus input power; (c) harmonic power versus fundamental signal input power; (d) harmonic product gain versus fundamental signal input power.

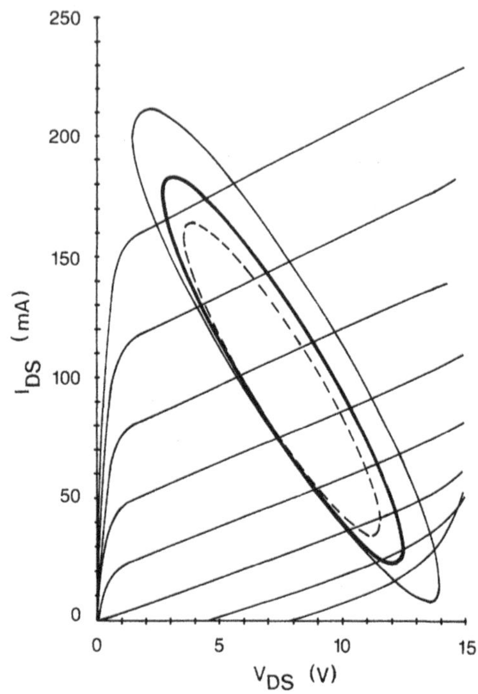

Fig. 6.25 Simulated dynamic load line for various signal drive levels: – – –, $P_i = 8.5$ dBm; ——, $P_i = 10.5$ dBm; ——, $P_i = 16.5$ dBm.

products to be monitored via the Fourier analysis of output current waveforms (Figs 6.24(c) and 6.24(d)). Finally the dynamic load line at various input power levels was traced onto the I_{DS}–V_{DS} characteristics of the transistor (Fig. 6.25). All measurements and calculations were made at 6 GHz for a drain–source voltage V_{DS} of 7 V and a gate–source voltage V_{GS} of -0.7 V which gave a quiescent current value slightly superior to $I_{DSS}/2$.

6.4 Power amplifier design

The design procedure for amplifiers using GaAs power MESFETs is reviewed in this section. General criteria for the selection of devices to meet a given specification and a detailed step-by-step design procedure are first outlined. Next, some of the techniques used to synthesize medium-bandwidth power amplifier matching networks are reviewed and illustrated by specific design examples. In order to provide high output power, matched transistors must be combined at either chip level or circuit level, and various power-combining schemes are described. Finally, with the high output powers now available from modern devices, the problem of heat sinking and power dissipation must be carefully considered.

6.4.1 Amplifier design procedure

Power GaAs MESFET amplifier design can be broken down into several steps.

(1) The first and most important step is to define the design goals: power output, linearity, r.f.–d.c. efficiency, gain and bandwidth. For example, the appropriate device gate width is 3 mm per watt of output power at 1 dB gain compression and 2.2 mm per watt of output power at saturation. The r.f.–d.c. efficiency varies from under 10% for high gain driver stages providing output power under 500 mW to over 33% for 8 W devices. Thus for very high output power amplifiers particular attention must be paid to mechanical layout and adequate provision for cooling must be made at the outset. Packaged devices, internally matched or not, should be used wherever possible [6.41–6.43]. Most device manufacturers now provide high power transistors in prematched packages covering the major telecommunication bandwidths up to 14–14.5 GHz. If chip devices are used it is always advisable to connect output stages in parallel in order to ease input circuit matching problems inherent in transistors with very wide gate widths. Carefully designed combining–matching circuits adapted to the devices being used enable matching losses and r.f. power wastage to be minimized. In all cases transistors should be selected to ensure that, when all combining and matching losses are taken into account, the output stage is the first to compress. Different mounting configurations are shown in Fig. 6.26.

(2) Bias points are selected for linearity, output power or high gain for each stage of the amplifier. Bias supply design, as detailed in Chapter 2, should now follow: self-bias should be used for high gain devices with low to medium output power, whereas for high output power devices a dual-polarity power supply with a sequencing circuit should be employed to ensure that the gate is always negatively biased before the drain is positively biased. Protection of the gate under r.f. large-signal swings into forward conduction is ensured by a series resistance which limits the gate current to about 1.2 mA per millimeter of gate width. Clamping will occur above this level and r.f. voltage swing into forward conduction is thereby limited. An important precaution to take in the design of the high drain current stage is to ensure that the physical dimensions (strip width and thickness) of the highest impedance microstrip lines and bond-wires used in the d.c. drain current path are capable of supplying the transistor without overheating. In the case of an 8 W device the drain current is typically 2.5 A, and this requirement severely limits the maximum impedance of the d.c. bias and matching-circuit elements.

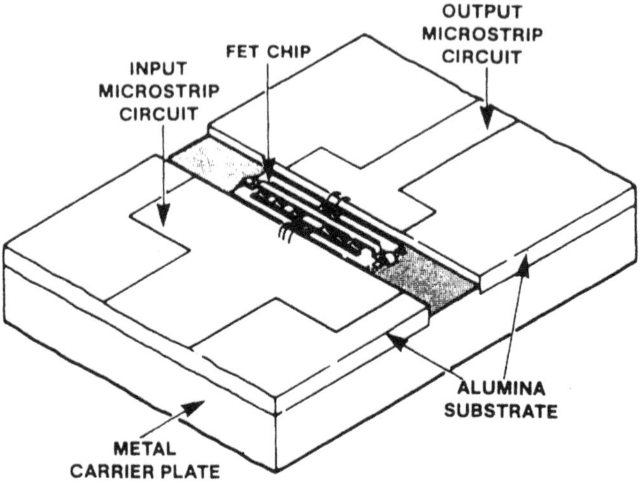

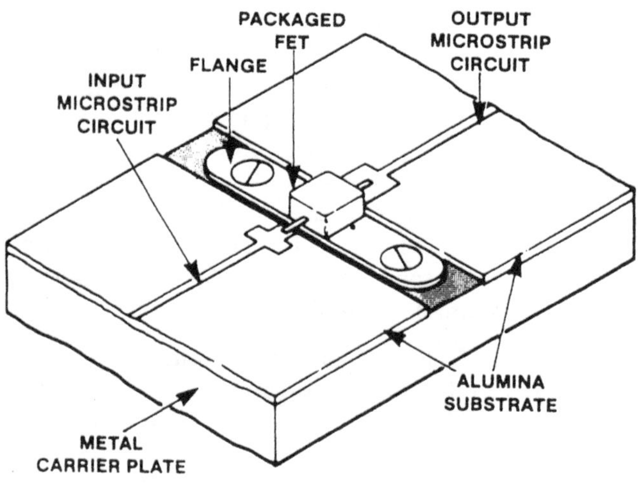

Fig. 6.26 Different power GaAs MESFET mounting configurations.

(3) Each device is characterized under large-signal operation using load-pull measurements or other impedance characterization technique outlined in Section 6.3. Optimum source and load impedances giving the required performance at several frequency points within the band are obtained. Owing to the difficulty of accurately measuring the *small-signal* parameters of transistors with large gate widths, these measurements are in general made on GaAs MESFETs with a maximum gate width of 2 mm using an automatic

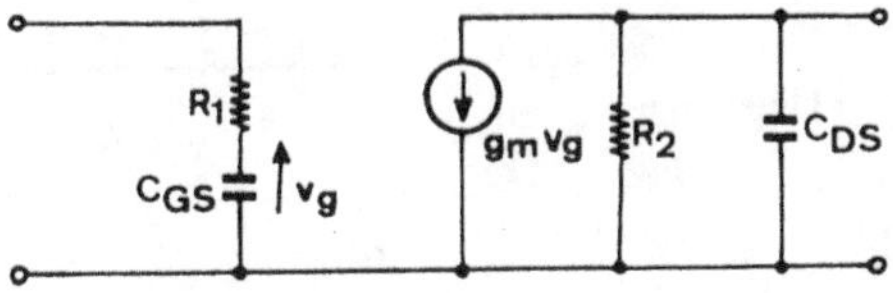

Fig. 6.27 Large-scale circuit used in the scaling-up exercise: $g_m = 80\,\text{mS}\,\text{mm}^{-1}$; $C_{GS} = 1.2\,\text{pF}\,\text{mm}^{-1}$; $C_{DS} = 1.2\,\text{pF}\,\text{mm}^{-1}$; $R_1{}^{-1} = (R_g + R_i + R_s)^{-1} = 250\,\text{mS}\,\text{mm}^{-1}$; $g_2 = R_2{}^{-1} = 15\,\text{mS}\,\text{mm}^{-1}$. The values given for g_m and R_2 are for optimal output power. Both elements are highly nonlinear and values should be checked experimentally.

network analyzer over a wide frequency band (e.g. 2–18 GHz). As described in Chapter 2, an optimization program is then used to obtain an equivalent circuit representation of transistor performance. In a similar manner large-signal impedance measurements are also made on cells of gate width less than 2 mm. Small- and large-signal element values for the equivalent circuit are now obtained by scaling up to the desired gate width [6.44]. We can thus directly relate equivalent circuit element values to gate width for devices fabricated using the same technology. An example of a simplified X band power GaAs MESFET equivalent circuit, ideal for matching-circuit design, is given in Fig. 6.27 with element values referred to a gate width of 1 mm. This equivalent circuit relates to a power device with a nominal gate length of 1 μm and finger widths of 100–150 μm.

(4) Input and output matching circuits are now synthesized and optimized using the same design procedure as in small-signal amplifier design. Wideband amplifier design has been described in detail in Chapter 3; narrowband and medium-band design is described in Section 6.4.2 and optimization techniques are given in Chapter 4. Bias tees and decoupling capacitors can be included in the final circuit realization, as can elements to ensure that the GaAs MESFET is always resistively loaded at out-of-band frequencies [6.45] (Fig. 6.28).

(5) Each stage is realized in microstrip form, and power-combining dividing circuits are realized as described in Section 6.4.3. The individual stages are tested and "tweaked" to give the desired performance. Sensitivity analysis and "reverse design" procedures can be used to facilitate this part of the process by indicating which circuit elements will have the most effect on overall performance.

(6) Finally the individual stages are cascaded with the appropriate combining networks to give the overall performance. At this stage careful thought must be given to heat-sinking requirements, and some simple procedures are described in Section 6.4.4.

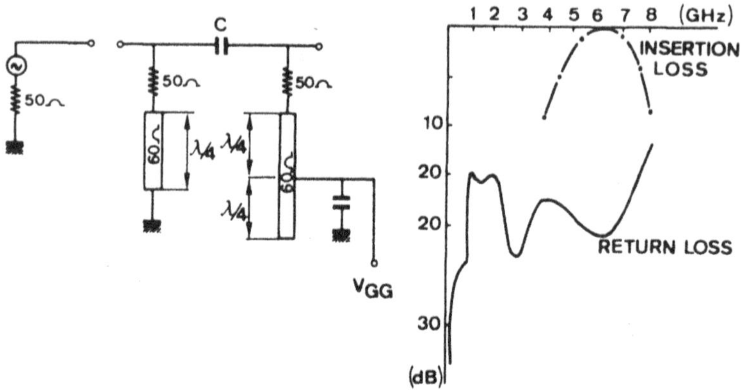

Fig. 6.28 Out-of-band damping circuit.

6.4.2 Design of power amplifier matching circuits

In order to synthesize a practical amplifier circuit we can either use an *RC* representation of the impedance and then one of several synthesis techniques or trace the points on a Smith chart and make use of its various properties to arrive at an acceptable circuit. Smith chart design techniques are documented in refs 6.46 and 6.47, and wideband synthesis techniques have been treated in Chapter 3, Section 3.2. In this section we shall therefore consider practical matching techniques applicable to narrowband and medium-band amplifier designs (maximum bandwidth 50%). However, the power GaAs MESFET is first characterized at radio frequency to give the performance required.

In synthesizing matching circuits for large-signal operation it is necessary to design not only to a gain–bandwidth specification but also to an output power–bandwidth performance. As already seen (Section 6.3), the optimal load impedance corresponding to a particular gain–output power performance can be approximated by a resistance in parallel with a capacitor. It is important to try to synthesize an output circuit with a transformed conductance close to this transistor large-signal conductance. This necessitates the use of a larger number of matching sections than for the design of a small-signal amplifier of the same bandwidth. In the case of amplifiers with a wide frequency band, gain flatness should be dealt with by including mismatch either in the input circuit or in some noncompressing preliminary high gain stage. The case of an amplifier designed to give small-signal gain flatness and running into uneven gain compression problems at large signals is shown in Fig. 6.29 [6.6]. It should be noted that a load resistance lower than $R_{1\,\mathrm{opt}}$ leads to current-limited compression as evidenced by the forward gate current I_{GS}, whereas a load

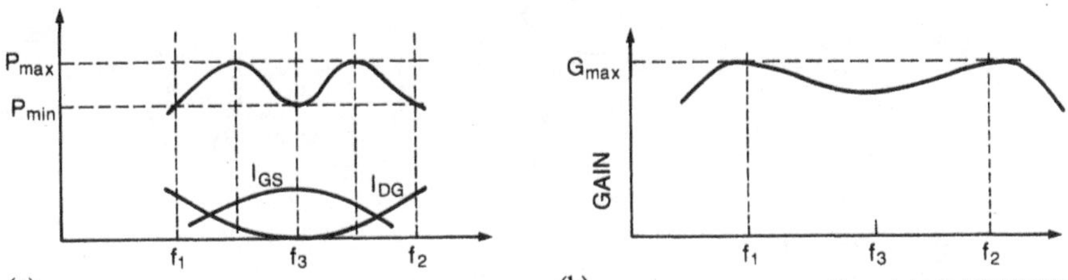

Fig. 6.29 (a) Saturation power and (b) small-signal gain of a typical power amplifier [6.6].

resistance higher than $R_{1\text{opt}}$ leads to voltage-limited compression and a drain–gate breakdown current I_{DG}. The maximum power output occurs when I_{GS} and I_{DG} are equal and opposite; gain compression is a maximum when one or other predominates. It makes good design sense to synthesize matching circuits with bandwidths exceeding that specified.

We shall describe two design techniques that are widely used in the design of medium-bandwidth power amplifiers: matching networks using lumped elements, and quarter-wavelength transmission line matching.

6.4.2.1 L matching networks

In order to transform a complex impedance to any other complex impedance value at a single frequency, a lossless matching network of at least two reactive elements—one in parallel across the impedance to be matched and the other in series with the resulting network—must be used (Fig. 6.30) [6.48]. Alternatively, a single length of transmission line of characteristic impedance and length determined as shown in ref. 6.47 can be used.

For an L network the shunt reactance is placed in parallel with the larger of the two resistances, thereby transforming Z_0 to equal the real part of the transistor input impedance R_1. The second reactive element then compensates for the reactive part $-j/\omega_0 C_{GS}$ of the transistor impedance and also cancels out the reactive component of the transformed value of Z_0:

$$X_2 = \pm \frac{nR_1}{(n-1)^{1/2}} \tag{6.17}$$

$$X_1 = R_1\{\mp(n-1)^{1/2} + Q\} \tag{6.18}$$

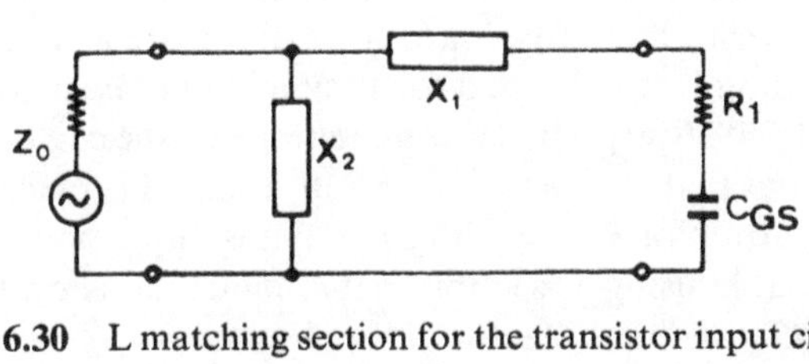

Fig. 6.30 L matching section for the transistor input circuit.

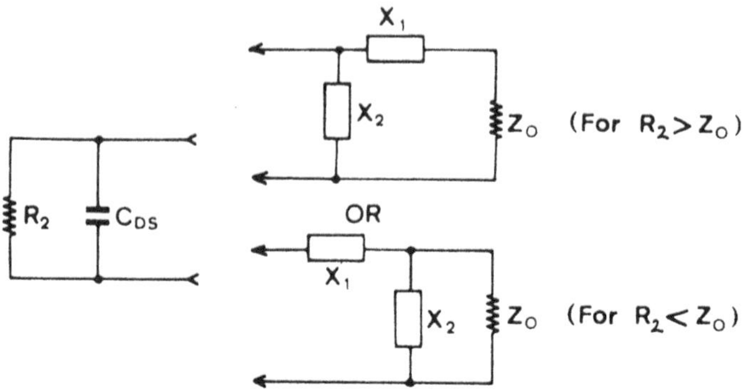

Fig. 6.31 Power GaAs MESFET output-matching schemes.

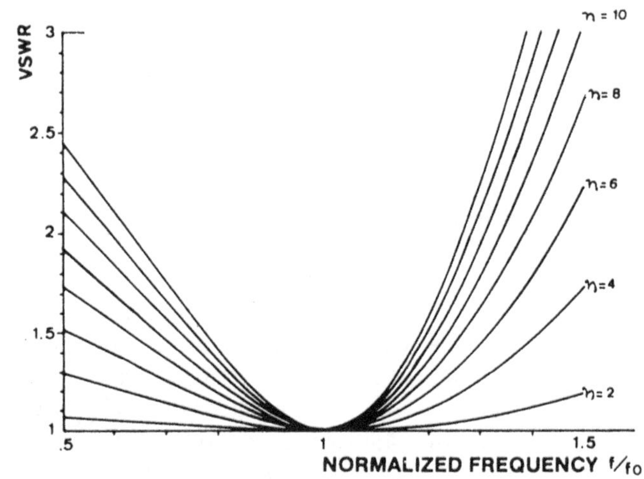

Fig. 6.32 Bandwidth for various L-section transformation ratios.

where $n = Z_0/R_1$ is the transformation ratio and $Q = (\omega_0 C_{GS} R_1)^{-1}$ is the quality factor of the transistor. It should be noted that X_2 can be either positive or negative (inductive or capacitive), but that X_2 and X_1 must be of opposite sign. In general, for GaAs MESFET matching, a bond wire can be used to provide a series inductive load: X_2 is therefore a shunt capacitance. This type of matching is widely used for prematching GaAs MESFETs with large gate widths [6.49, 6.50]. The same matching scheme can be applied to the output circuit except that in this case the shunt element is placed immediately at the transistor output for $R_2 > Z_0$ (Fig. 6.31). It is important to note that the bandwidth obtainable using L-section matching depends on the transformation ratio $n = R/Z_0$ and the Q factor of the circuit.

The bandwidth for a given VSWR and various values of n is given in Fig. 6.32. It can be seen that a 40% bandwidth (VSWR = 1.2) can be achieved with ideal lumped elements for $n = 8$. For a 2 W GaAs power MESFET the transformation ratio n is greater than 30 [6.51]. It is therefore necessary to incorporate a second transforming network (which may be a quarter-wave section of line or a second L section) in order to obtain a match over a reasonable bandwidth. This is best appreciated by considering the following design example.

6.4.2.2 Design of internally matched modules

Consider the design of a 2.7 W amplifier at 1 dB gain compression using lumped elements over the frequency band 5.5–6.5 GHz.

To limit losses due to the matching out of too high an input VSWR, it is decided to use two prematched cells in parallel with a Wilkinson combiner–divider circuit. Each cell is a 1.2 μm × 4800 μm device (24 fingers). This transformation must take place over a frequency bandwidth of 17%. There are two possibilities: cascade two L-section networks, or use a single L section followed by a quarter-wave transmission line transformer. The latter approach was adopted because the physical layout fits in well with the Wilkinson splitter. The design strategy used is summarized in Fig. 6.33. From Fig. 6.32 we select $n = 12$ to give a VSWR < 1.2 across the band. From eqn (6.18)

$$Q = \frac{1}{\omega_0 C_{GS} R_1}$$

$$X_1 = 1.6\{(11)^{1/2} + Q\} = 9.9\,\Omega$$

This is obtained at 6 GHz by using a series inductance of 0.26 nH. From eqn (6.17)

$$X_2 = 19.2/(11)^{1/2} = 5.8\,\Omega$$

This is realized at 6 GHz by using a shunt capacitance of 4.5 pF. The series inductance is realized by connecting two 25 μm bond wires (typically 1 nH mm^{-1} per bond wire) in parallel; the shunt capacitor is a high dielectric ceramic chip (0.63 mm × 0.63 mm). The second transformation is $n = 5.2$ which requires a quarter-wave line of characteristic impedance 44 Ω.

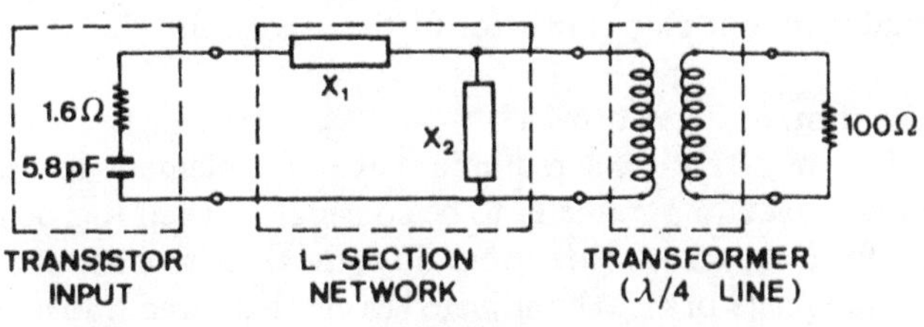

Fig. 6.33 Input design for the internally matched module (total impedance transformation $n = 100/1.6 = 62.5$).

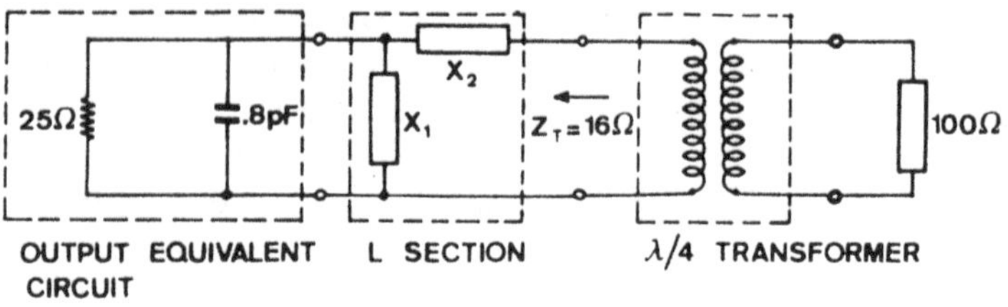

Fig. 6.34 Output design for the internally matched module.

The matching scheme adopted for the output (Fig. 6.34) is dictated by the physical layout. A quarter-wave length of 40 Ω transmission line is used to match 100 Ω down to 16 Ω. In this case the relevant formulae are

$$X_2 = \pm\frac{R_2(n-1)^{1/2}}{n} \tag{6.19}$$

$$X_1 = -X_{\text{DS}} \pm \frac{R_2}{(n-1)^{1/2}} \tag{6.20}$$

where

$$X_{\text{DS}} = (\omega_0 C_{\text{DS}})^{-1}$$

In this case ($n = 1.6$) the relative values of n, R_2 and C_{DS} are such that the series equivalent resistance of the parallel output circuit is 16 Ω, i.e. $X_1 = 0$—no shunt element is required. The series element X_2 is given by

$$X_2 = \frac{25 \times (0.6)^{1/2}}{1.6} = 12\,\Omega$$

This can be obtained by simply extending the series 40 Ω transmission line by 0.04λ.

The complete circuit is realized on 0.4 mm alumina mounted in an Ampac case (Fig. 6.35). The resistive Nichrome film inserted between the parallel branches to ensure good isolation between the matching circuits should be noted. This is often necessary to avoid spurious oscillation caused by the parasitic loading of one chip transistor by the other chip.

6.4.2.3 T matching networks

The addition of an extra redundant reactive element to the L-section network allows an extra parameter to be considered when R is being matched to Z_0. Apart from the transformation ratio n, the three reactive elements of the lossless matching network can be adjusted to give a desired transmission phase [6.48], to allow a measure of control over the VSWR bandwidth [6.52] or to permit more practical realizable element values to be selected. As T matching

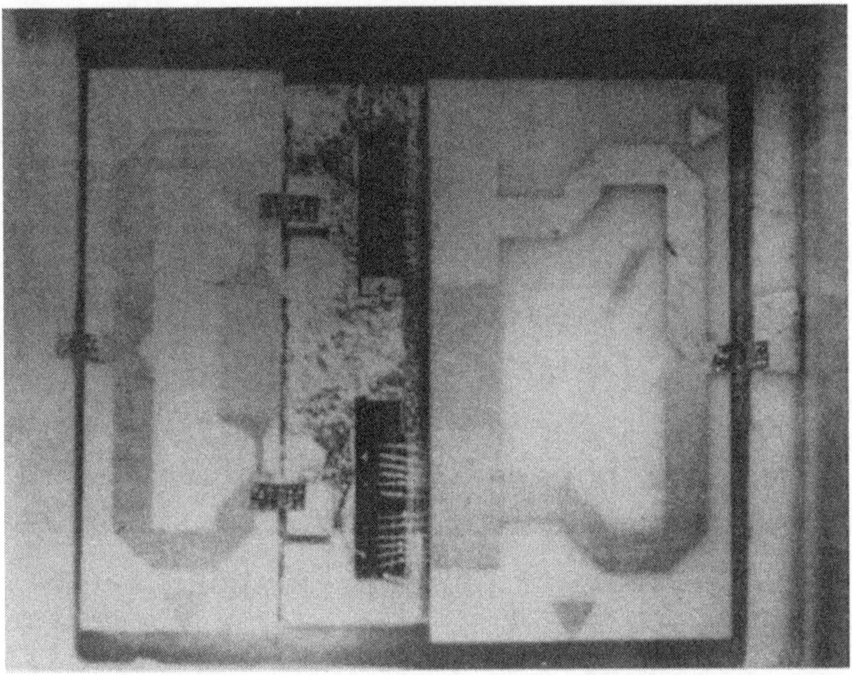

Fig. 6.35 Photograph of an amplifier with L-section prematching.

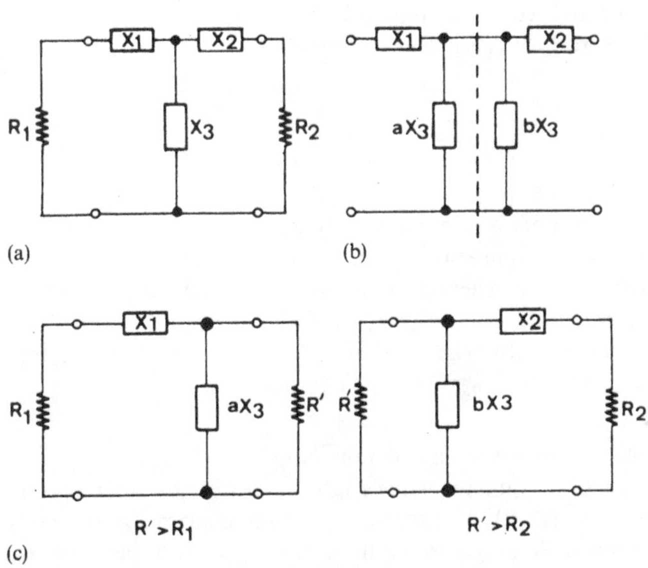

Fig. 6.36 Analysis of a T-section network.

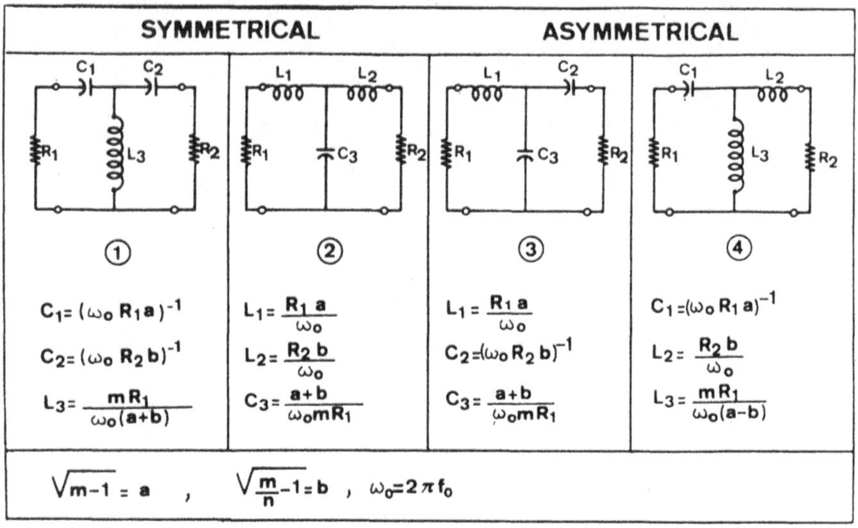

Fig. 6.37 T-section network configurations.

networks are treated thoroughly in ref. 6.52, only a brief outline will be presented here.

Essentially a T-section network can be split into two back-to-back L-section networks (Fig. 6.36). This allows element X_3 of the T matching network to be divided into two parallel elements aX_3 and bX_3. A parameter R' can now be introduced (Fig. 6.36(c)) such that

$$R' > R_1 \qquad R' > R_2 \tag{6.21}$$

$$\frac{ab}{a+b} = 1 \tag{6.22}$$

The different signs of a and b that satisfy eqn (6.22) lead to the definition of four different circuit configurations (Fig. 6.37). A bandwidth parameter m defined as R'/R_1 can now be selected, following the bandwidth requirements of the circuit, independently of the transformation ratio n provided that $m > n$. A group of four π networks can also be obtained by applying a Y-to-Δ transformation to the T networks of Fig. 6.37.

6.4.2.4 Quarter-wavelength matching

The use of a quarter-wavelength transmission line of characteristic impedance $Z_1 = (Z_0 R)^{1/2}$ to match a real impedance Z_0 to a resistance R is well established. A single section of this type can be used to obtain a transformation ratio $n = 4$ over a 60% bandwidth (VSWR = 2). Alternatively,

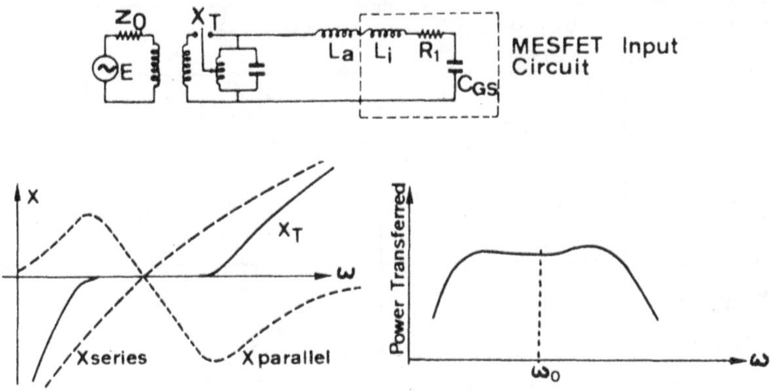

Fig. 6.38 Reactive compensation of a GaAs MESFET input impedance [6.53].

two quarter-wavelength sections can be cascaded in order to allow either a larger n or a lower VSWR over the same bandwidth.

For complex impedance matching, however, requirements are more stringent: not only must the resistive parts of the device impedance and synthesized load be equal, but their reactive parts must also be equal and opposite. An attractive method of achieving these conditions over bandwidths of up to 20% in the case of the GaAs MESFET input circuit is to bring the input impedance to series resonance at the center frequency and then to compensate the reactance slope $\mathrm{d}X/\mathrm{d}\omega$ at this frequency by the negative reactance slope $-\mathrm{d}X/\mathrm{d}\omega$ of an antiresonant circuit [6.53] (Fig. 6.38). Thus two requirements must be imposed on the transforming circuit.

(a) The circuit must be capable of realizing the transformation ratio n required to transform the real part of the device impedance to the real part of the load. In the case of power GaAs MESFETs, it is sometimes necessary to carry out prematching to enable this condition to be met.

(b) The inverse of the source impedance susceptance slope parameter β must be equal and opposite to the reactance slope parameter χ of the GaAs MESFET input impedance, where

$$\beta = \frac{\omega_0}{2} \frac{\mathrm{d}B}{\mathrm{d}\omega}\bigg|_{\omega_0}$$

$$\chi = \frac{\omega_0}{2} \frac{\mathrm{d}X}{\mathrm{d}\omega}\bigg|_{\omega_0}$$

The same technique can also be applied to the output circuit, replacing each circuit by its dual. The various matching schemes possible for each case are summarized in Fig. 6.39 [6.54]. Essentially, each of these matching schemes, which use distributed elements, requires two parameters (character-

CIRCUIT TYPE	MATCHING NETWORK	DESIGN EQUATION
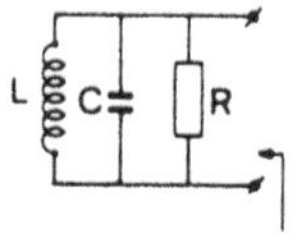 Q $=\omega_0 RC$	(1) 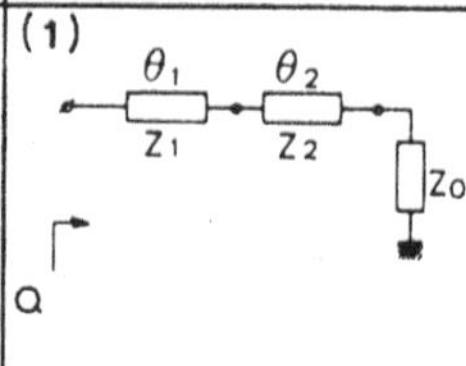	$\theta_1 = 90^\circ \quad \theta_2 = 180^\circ$ $Z_1 = (RZ_0)^{1/2}$ $T = \frac{2Q}{\pi} - \frac{1}{2}\left(\frac{Z_0}{Z_1} - \frac{Z_1}{Z_0}\right)$ $Z_2 = -\frac{TZ_0}{2} + \left\{\left(\frac{TZ_0}{2}\right)^2 + Z_0^2\right\}^{1/2}$
	(2) 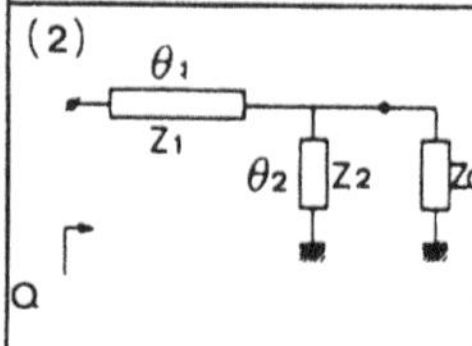	$Z_1 = (Z_0 R)^{1/2}$ $Z_2 = \frac{\pi Z_1 Z_0^2}{\pi Z_1^2 - \pi Z_0^2 + 4Z_1 Z_0 Q}$ $\theta_1 = \theta_2 = 90^\circ$
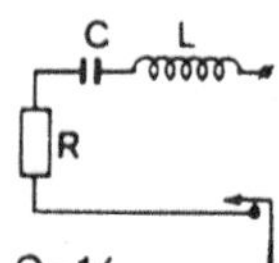 Q $= 1/\omega_0 RC$	(3) 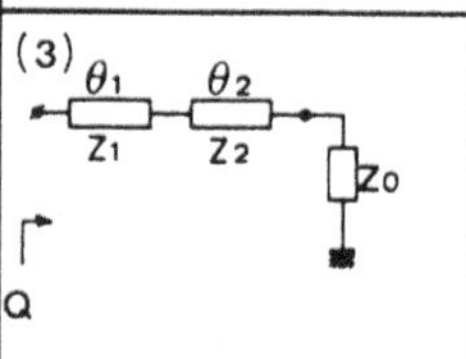	$\theta_1 = \theta_2 = 90^\circ$ $Z_1 = AZ_2$ $Z_2 = BZ_0$ $A = (R/Z_0)^{1/2}$ $B = \frac{A}{1+A}\left[-\frac{2}{\pi}Q \pm \left\{\left(\frac{2Q}{\pi}\right)^2 + \frac{(1+A)^2}{A}\right\}^{1/2}\right]$ $A > B$
	(4) 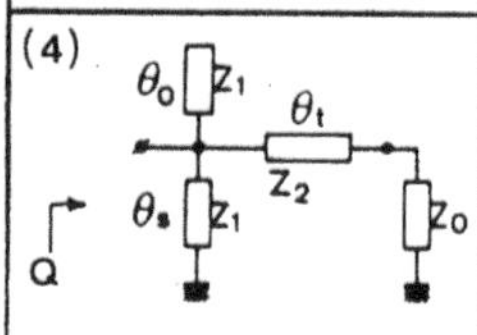	$\theta_t = 90^\circ \quad \theta_s = 30^\circ$ $\theta_o = 60^\circ$ $Z_2 = (Z_0 R)^{1/2}$ $Z_1 = 4\frac{Z_2^2}{Z_0}\left\{\frac{4Q}{\pi} + \left(\frac{Z_0}{Z_2} - \frac{Z_2}{Z_0}\right)\right\}^{-1}$

Fig. 6.39 Quarter-wavelength matching-circuit configurations.

istic impedances of line 1 and line 2) to be calculated in order to satisfy the two requirements of impedance transformation and slope matching. All the necessary equations are provided in Fig. 6.39. It should be noted that in each case the Q of the transistor is equal to the Q of the load. The use of Fig. 6.39 is best illustrated by an example.

6.4.2.5 Design of X band matching circuits

Consider the design of a 500 mW amplifier over 10.7–11.7 GHz. A balanced amplifier configuration using two MSC 88102 transistors allows the specification to be met comfortably. It should be noted that, although each transistor typically gives 500 mW at 1 dB gain compression at 12 GHz, in order to allow for production spread and circuit loss due to matching over the 10% bandwidth an output power of 300 mW is a more realistic target. The input equivalent circuit of the 88102 CHIP PAC transistor is given in Fig. 6.40.

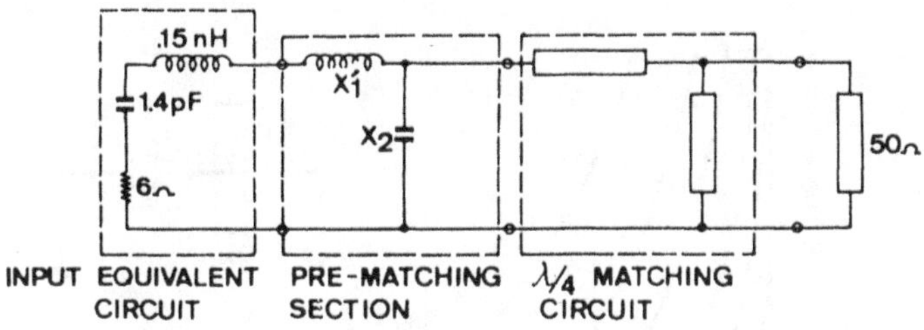

Fig. 6.40 Input circuit matching of a power GaAs MESFET.

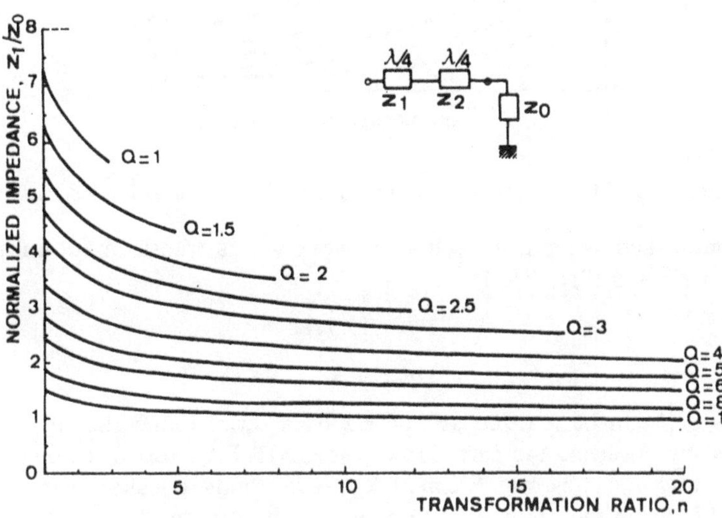

Fig. 6.41 Design curves for cascaded quarter-wavelength sections.

Referring to Fig. 6.39, it can be seen that two matching schemes are available: the two cascaded series quarter-wavelength lines, or the shunt stubs followed by a series quarter-wavelength line. The design curves for these two cases are given in Figs 6.41 and 6.42 for $n = Z_0/R > 1$. In this case

$$n = 50/6 = 8.3$$

$$Q = (\omega_0 C_{GS} R_1)^{-1}$$

The curves obtained from the formulae show that solution (3) is not feasible, and that in solution (4) the shunt stubs require a line of characteristic impedance 11 Ω which is not feasible in microstrip on an alumina substrate 0.635 mm thick. Prematching of the transistor input is thus necessary, and can

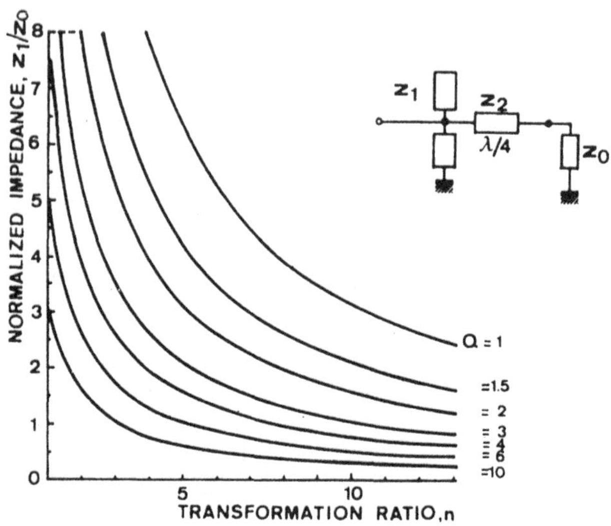

Fig. 6.42 Design curves for shunt stubs–series line-matching sections.

be accomplished using an L-section network with a transformation ratio of 3.7. From eqns (6.17) and (6.18)

$$X_2 = 13.5\,\Omega$$

$$X_1' = -\omega_0 L + X_1 = 9.9\,\Omega$$

Element X_1' can be realized at 11 GHz by a series inductance of 0.14 nH obtained by offsetting the gate lead of the CHIP PAC transistor by 0.6 mm from the alumina substrate. Element X_2 corresponds to a shunt capacitance obtainable using two parallel 50 Ω open-circuit stubs, each 0.17λ long. The effect of the L section is to transform the series resonant circuit of the transistor input into a shunt resonant circuit (Fig. 6.43) where both R_p and L_p are functions of frequency. However, over a 10% bandwidth they can be treated as constant. Matching network (2) is selected from Fig. 6.39, and the element values are calculated according to the formulae

$$Z_1 = (Z_0 R_p)^{1/2} = 34\,\Omega$$

$$Z_2 = \frac{\pi Z_1 Z_0^2}{\pi Z_1^2 - \pi Z_0^2 + 4 Z_1 Z_0 Q} = 42\,\Omega$$

The resulting circuit has a VSWR < 1.6 from 10.6–11.8 GHz before optimization.

The output circuit can be synthesized using the same type of network. In this case (Fig. 6.44) a large-signal VSWR of better than 1.2:1 is obtained over the same bandwidth, which is a significantly better result. The frequency

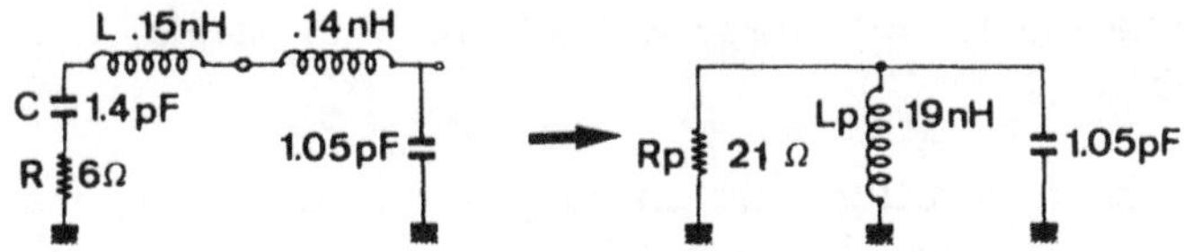

Fig. 6.43 Effect of prematching.

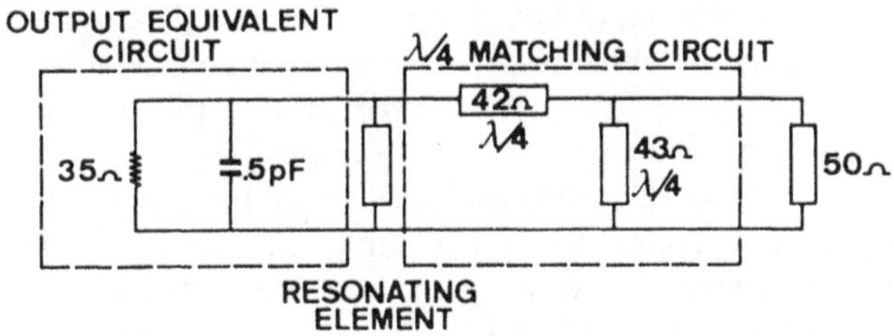

Fig. 6.44 MSC 88102 output matching circuit.

limitations of the prematching at the transistor input are the main cause of the poorer input circuit performance.

6.4.3 Power combining

It is often preferable, and for very high output power essential, to combine medium power GaAs MESFETs to achieve high power rather than to use a single oversize device. This is particularly true at frequencies above X band where problems of phase imbalance, unequal power sharing between devices, heat accumulation at the center of very large cells and transverse propagating modes become important. Thus the study of low loss combining structures capable of dealing with large numbers of devices has become increasingly important [6.55–6.63].

Four basic considerations apply to the parallel connection of identical power transistors on a single hybrid substrate. There should be

(i) an equal power split with identical transmission phase between the common input port and the common output port independent of the physical position of the various cells (transmission loss should obviously be kept to a minimum in order to retain optimum power-added efficiency),

(ii) the best possible r.f. power match between the common port and each device port over the operating bandwidth,

(iii) effective isolation between device ports at all frequencies to minimize intradevice interaction and

(iv) a sufficiently dispersed structure to avoid excessive heat accumulation.

Combining structures are generally separated into two major groups: progressive power combiners in which successively increasing numbers of devices are combined, and N-way structures in which power is fed to and recovered from N transistors in a single step. Progressive combining–dividing structures are further divided into corporate or tree structures and serial or chain structures. In general, although both types of hybrids can be used to connect either transistor chips or amplifiers in parallel, N-way hybrids are often preferred for chips and progressive hybrids for amplifiers [6.59, 6.60].

A serious problem often encountered in connecting transistors in parallel is the failure of one or more devices and its effect on the overall output power. Ideally the hybrid design chosen should exhibit "graceful degradation", i.e. the output power level should decrease as a function of the number of failed devices and not catastrophically fall to zero. It is intuitively obvious that progressive structures and perfectly symmetrical N-way hybrids exhibit the best graceful degradation: the loss in output power is independent of the position of the device in the structure, provided that intradevice isolation is perfect and that the isolation resistors are capable of handling the reflected mismatch power. This complex problem is well treated in refs 6.56 and 6.62. Combining hybrids are discussed in more detail in refs 6.55–6.63.

In this section the generic types of combining hybrids are summarized and a few examples of applications are given. It should be noted that in general identical hybrid structures are used at input and output for any particular design.

6.4.3.1 Basic three-port combining structures

The main circuit elements used in the realization of progressive combining hybrids are three-port reciprocal devices in which the output ports are either in phase or in quadrature phase.

Consider first the quadrature-phase coupler used as a combiner–divider. The S parameter matrix for an ideal reciprocal lossless coupler, be it a branch-arm hybrid or a coupled-line device, can be written as follows:

$$\boldsymbol{S} = \begin{bmatrix} 0 & 0 & \alpha & j\beta \\ 0 & 0 & j\beta & \alpha \\ \alpha & j\beta & 0 & 0 \\ j\beta & \alpha & 0 & 0 \end{bmatrix} \tag{6.23}$$

where $\alpha^2 + \beta^2 = 1$. It is assumed that the network is perfectly matched at all ports with perfect isolation between ports 1 and 2 and between ports 3 and 4. Placing a matched termination on port 4 and equalizing the coupling between ports 3 and 1 and between ports 3 and 2 enables us to rewrite eqn (6.23) for a

3 dB coupler as follows:

$$\boldsymbol{S} = \begin{bmatrix} 0 & 0 & \alpha \\ 0 & 0 & j\alpha \\ \alpha & j\alpha & 0 \end{bmatrix} \tag{6.24}$$

If a network is now made up of a quadrature coupler at the input port A (Fig. 6.45) with identical amplifiers connected to ports B and C followed by a reversed but identical quadrature coupler with inputs D and E and output F, the resulting structure satisfies the combiner–divider requirements outlined in Section 6.4.3. The transmission phase between ports B and D and between ports C and E should be equal to ensure that individual stages are not driven at differential angles which depart significantly from 90°. Overall output power is now simply the sum of the powers out of amplifier ports D′ and E′, as a simple multiplication of the R matrices of the three components will show. In addition identical mismatch at amplifier ports B′ and C′ will cancel at port A, and identical mismatch at ports D′ and E′ will cancel at output port F.

At wide bandwidths it is often necessary to use Lange couplers with complex bond-wire interconnections. To limit the use of such fragile labor-intensive couplers the split-tee or Wilkinson combiner is often preferred. The S parameter matrix for the Wilkinson coupler is

$$\boldsymbol{S} = \begin{bmatrix} 0 & \alpha & \beta \\ \alpha & 0 & 0 \\ \beta & 0 & 0 \end{bmatrix} \tag{6.25}$$

where $\alpha^2 + \beta^2 = 1$, and it is assumed that the coupler is perfectly matched and that there is perfect isolation between the output ports 2 and 3. The ratio of the power split between the ports is

$$c^2 = \frac{\text{power at port 2}}{\text{power at port 3}} = \frac{\alpha^2}{\beta^2}$$

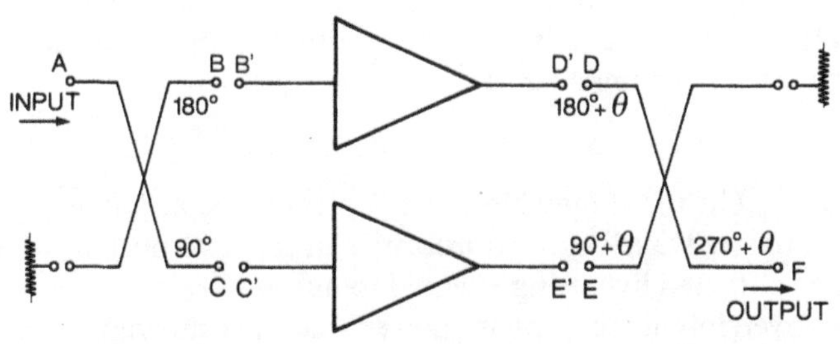

Fig. 6.45 Power combining using quadrature hybrids.

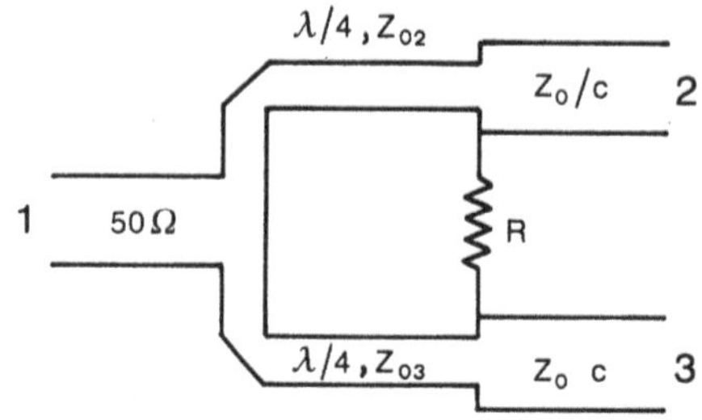

(a)

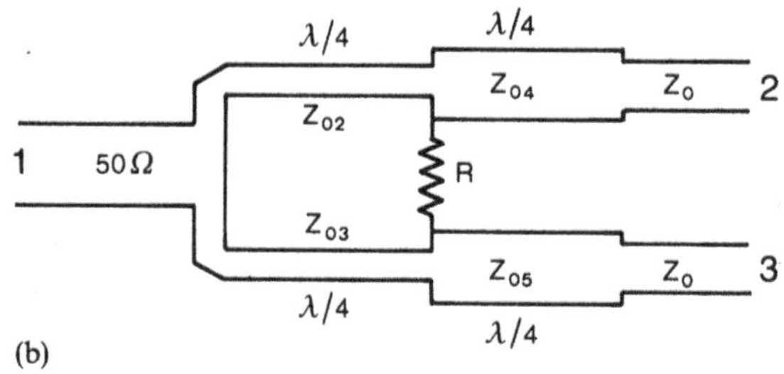

(b)

Fig. 6.46 Uneven power split using split-tee hybrids: (a) for non-50 Ω terminations on ports 2 and 3; (b) for 50 Ω terminations on ports 2 and 3.

The ratio of the characteristic impedances of the lines for the uneven power split is c^2 and is given by (Fig. 6.46) [6.64]

$$\begin{aligned} Z_{02} &= Z_0\left(\frac{1+c^2}{c^3}\right)^{1/2} \qquad c^2<1 \\ Z_{03} &= Z_{02}c^2 \end{aligned} \tag{6.26}$$

The value of the isolation resistance R is

$$R = Z_0\frac{1+c^2}{c} \tag{6.27}$$

If we wish to match to an impedance Z_0 at ports 2 and 3 rather than Z_0/c and Z_0c respectively, it is necessary to add another quarter-wavelength transformer of characteristic impedance

$$Z_{04} = Z_0/c^{1/2} \qquad Z_{05} = Z_0c^{1/2} \tag{6.28}$$

to each branch. The input match of the coupler can be improved by adding a Chebyshev quarter-wavelength transformer in front of port 1 [6.64]. Equal power split is obtained by setting c^2 equal to unity.

Another version of the in-phase power splitter is shown in Fig. 6.47. This coupler is made up of a central line edge coupling to two symmetrically disposed lines on each side of it via quarter-wavelength open-circuit coupled-

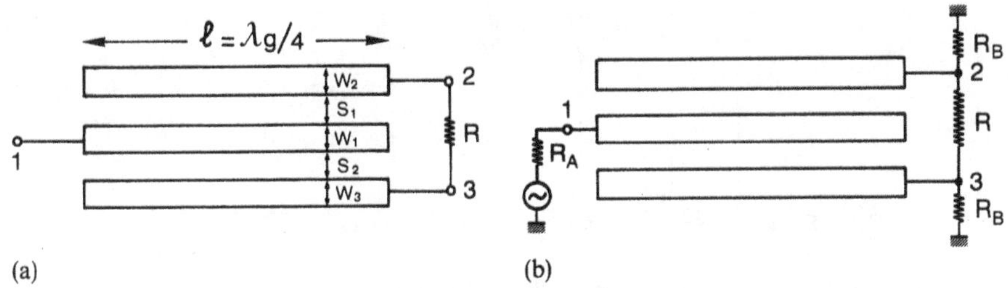

Fig. 6.47 Schematic diagram of a coupled-line power splitter [6.65].

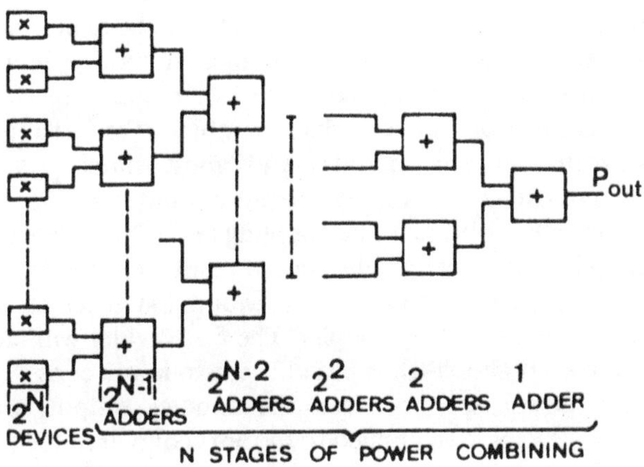

Fig. 6.48 Corporate structure for power combining.

line sections [6.65]. The coupler can be designed such that it simultaneously feeds power into low resistance devices and matches them to 50 Ω.

6.4.3.2 Corporate and serial combining structures

A corporate structure (or tree) for combining (or dividing) power is shown in Fig. 6.48. The principal elements used to realize a tree are the quadrature and in-phase couplers described in Section 6.4.3.1. The number of devices combined in this structure is binary, i.e. it must equal 2^N where N is a positive integer. When combining amplifiers it is customary to use symmetrical trees made up throughout of either in-phase or quadrature couplers at the input and the output. In some cases, however, a mix of couplers is used. Thus a 4 W amplifier for the 6–12 GHz band has been realized using Wilkinson couplers for the outer combining network and Lange couplers for the inner combining network.

A chain or serial combiner is shown in Fig. 6.49. Here each successive stage of an N-stage combiner adds $1/N$ of the power delivered to the output.

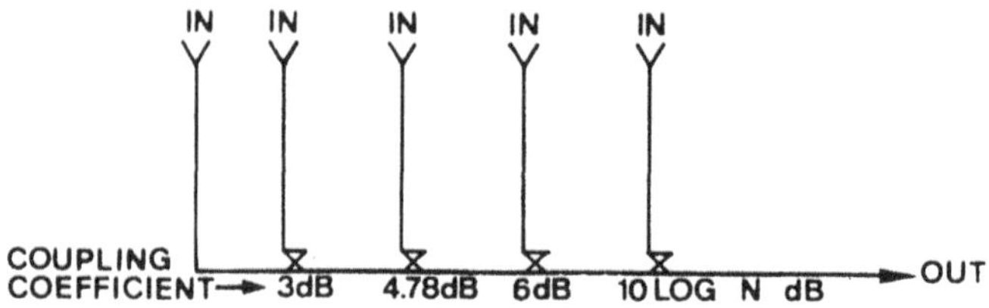

Fig. 6.49 Chain combiner.

The great advantage of this type of coupler is that amplifier stages do not need to be combined in binary units. Also, the structure is extremely flexible: to add a port to an existing N-way combiner a coupler with a coupling coefficient of $1/(N+1)$ is added to the string, and the other coupling coefficients are unchanged. This form of coupler, which combines the advantages of good isolation between amplifier modules and good match, was difficult to implement on microstrip until recently because some coupling factor requirements were too weak for branch-line coupling and others were too strong for proximity coupling. In the travelling-wave combiner these problems have been cleverly resolved by the use of a distributed split-tee structure [6.58]. Consider the case of a four-way coupler. The first divider will take 6 dB from the input power P_i. It should thus be sufficient to apply eqns (6.26)–(6.28) to obtain the $\frac{1}{4}$:$\frac{3}{4}$ power split between the coupler output branches. The remaining power, $\frac{3}{4}P_i$, is then applied to the next coupler in the chain, and the same design procedure is followed for a $\frac{1}{3}$:$\frac{2}{3}$ power split. The remaining power is then fed into the final divider where an equal power split completes the four-way equal-power division $P_i/4$ of the chain. In practice additional quarter-wavelength sections and further resistances can be added to enlarge bandwidth, reduce VSWR at the input and output ports and increase intradevice isolation. The combiner shown in Fig. 6.50 is used to connect three devices in parallel in the first stage and four in the second stage to give an amplifier with power 1.5 W and 9 dB gain in the frequency range 6–12 GHz. Once again identical but reversed structures are used at amplifier input and output ports to meet the identical transmission phase shift criterion (see p. 329).

In conclusion, both chain and tree structures are used to combine small numbers of amplifiers with high combining efficiency (greater than 90% in the case of the four-way travelling-wave structure [6.58]) over bandwidths of an octave or more.

6.4.3.3 *N*-way combiners

Essentially three types of N-way combiners—Wilkinson, radial and planar—are used for combining large numbers of amplifiers [6.60–6.62] and devices [6.63]. The great advantage of N-way combiners is that they allow large numbers of devices ($N \geqslant 6$) to be connected in parallel with minimal loss.

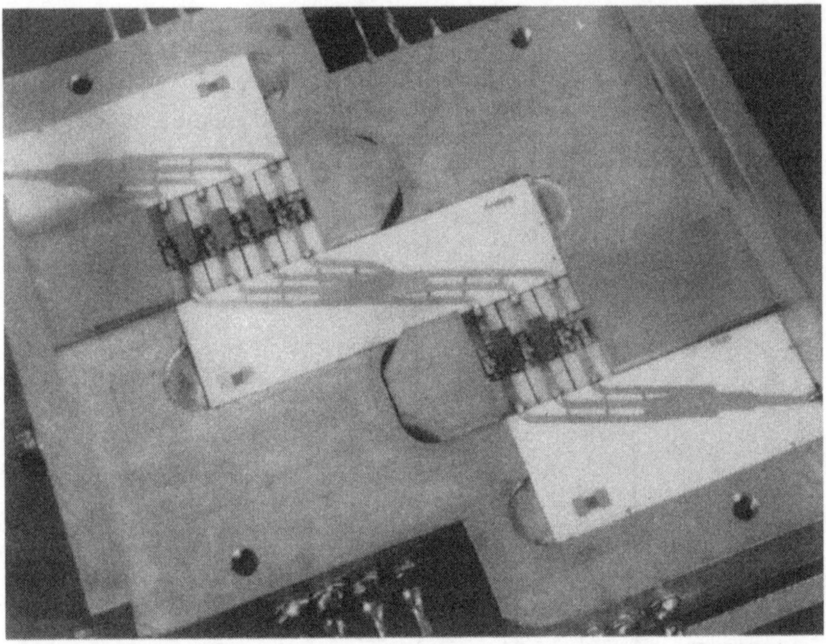

Fig. 6.50 Photograph of a two-stage amplifier using a travelling-wave combiner.

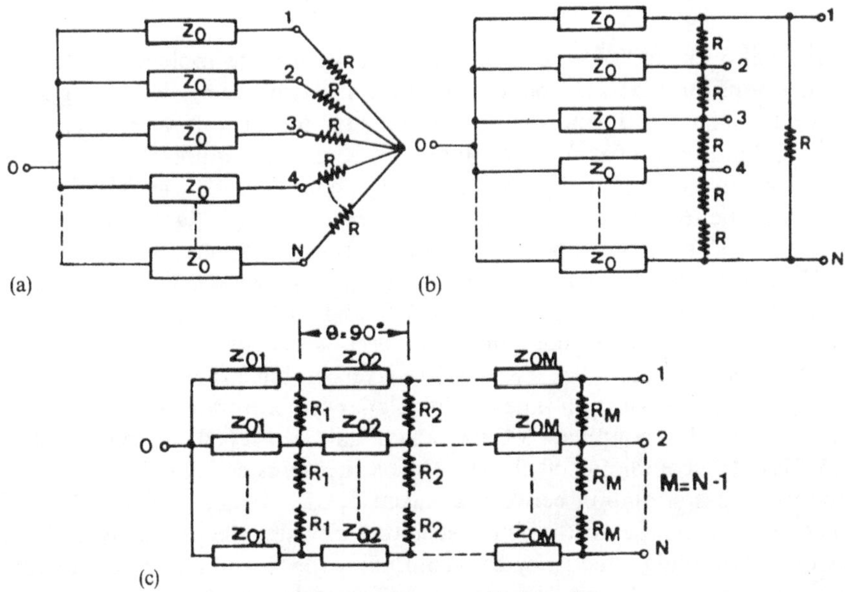

Fig. 6.51 *N*-way combiners.

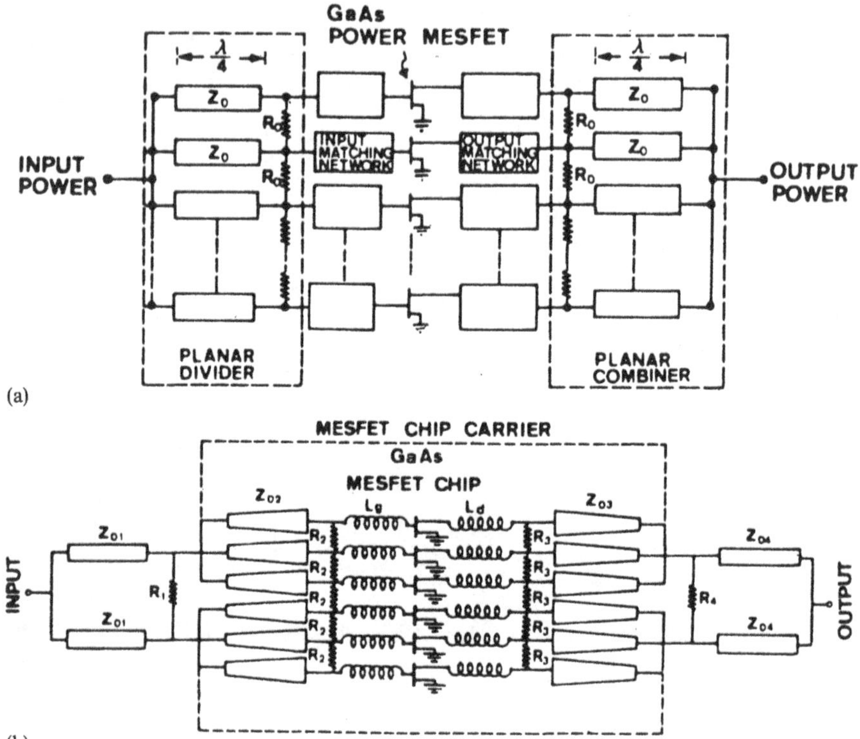

Fig. 6.52 Amplifiers realized using planar N-way combiners.

Wilkinson and radial combining networks must be implemented in three dimensions for $N > 2$. The planar structure, as its name implies, is designed to operate on microstrip and was first presented in the form shown in Fig. 6.51(c) [6.61]. In the case of an N-way combiner this version requires that each of the N segments is composed of $N-1$ homogeneous quarter-wavelength transmission line sections and that each of the line sections is flanked by resistive networks. Fortunately this type of combiner can be simplified as shown in Fig. 6.52(a) or Fig. 6.52(b) [6.62, 6.63] while maintaining good isolation properties and a combining efficiency of between 78% and 90% for a six-cell device.

While the planar combiner has been successfully used to connect amplifier stages in parallel, it is as a chip-level combiner that it shows the most promise. The two-section tapered-line planar structure shown in Fig. 6.52(b) has been used to combine six cells, each of gate width 600 μm, to give 2 W at 15 GHz. The combining efficiency of the structure is estimated at 90% while the worst-case isolation between adjacent ports is better than 16 dB [6.63]. The use of this type of combining structure obviates the problems discussed with regard to high frequency chip combining (phase imbalance and unequal power sharing) and can be used to good effect up to frequencies of 20 GHz.

6.4.4 Power dissipation requirements

A major problem with high power semiconductor devices is posed by the need to evacuate the heat generated within the device while maintaining the active region at a reasonable temperature. In GaAs MESFETs a maximum channel temperature of 150 °C is deemed acceptable for most devices. It is worth noting, however, that reducing the channel temperature can greatly increase the device lifetime as thermally activated degradation mechanisms will be increasingly inhibited. For instance, reducing the channel temperature from 120 to 100 °C yields an improvement of a factor of 4.9 in median life for a process with an activation energy of 1 eV [6.11].

As indicated in Section 6.2.1, as the r.f. drive to a power GaAs MESFET amplifier is increased (under class A–B operation), the r.f.–d.c. efficiency increases and the power dissipated tends to peak (Fig. 6.53).

In order to enhance the evacuation of heat from the channel of a power GaAs MESFET chip, i.e. to reduce the thermal resistance of the chip, the thermal design techniques described in Section 6.2.3, and in greater detail in ref. 6.11, are applied. The two main techniques are flip-chip mounting and the thinning of the GaAs MESFET substrate to 50 μm before plating the rear face with 50 μm or more of gold. Periodic holes (vias) in the transistor pattern, gold plated to permit mechanical connection through the GaAs MESFET substrate between the source pads and the plated gold on the obverse side of the chip, are also associated with the latter technique to reduce the device thermal resistance further. The vias also serve to reduce source inductance.

The performances of some high power 5.9–6.4 GHz internally matched GaAs MESFETs are given in Table 6.2. It is seen that the highest power devices are currently capable of furnishing 18 W at the 1 dB gain compression point while dissipating typically 50 W. The performances given in the table are for a maximum specified case temperature. Some manufacturers, however,

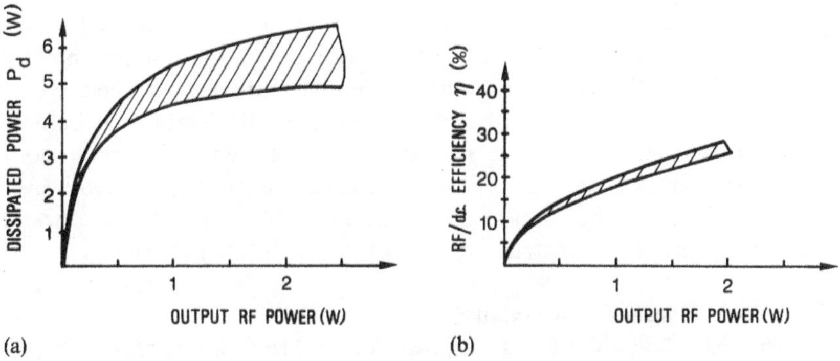

(a) (b)

Fig. 6.53 Class A–B output power amplifier characteristics.

Table 6.2 Typical power GaAs MESFET performances

Manufacturer	*Transistor reference*	P_1[b] *at 6 GHz* (dBm)	R_{TH} *(chip/case)* (°C W^{-1})	*Power dissipation* (W)	*Maximum flange temperature*[a] (°C)
Fujitsu	FLM 5964–4C	36	5	11	⩽ 75
	FLM 5964–8D	39	3	25	⩽ 75
	FLM 5964–14	42.5	1.6	50	⩽ 70
MSC	MSM 5964–5	37	3.5	18	⩽ 75
	MSM 5964–10	40	1.8	35	⩽ 75
Toshiba	S 8846	42	1.6	60	⩽ 60

[a] Ambient temperature, 50 °C.
[b] P_1 represents output power at 1 dB gain compression.

particularly for transistor chips, do not specify the temperature conditions under which the device gives the reported performances. In such cases it is advisable to consider that the gain and output power performance have been obtained with the transistor connected to an infinite heat sink maintained at 17 °C. In real amplifier design such conditions are unattainable, and a certain derating of the output power and gain performance is inevitable. In order to minimize this deterioration in r.f. performance and to prolong the device lifetime, careful attention must be paid to the thermal layout of the transistor and the heat-sink and cooling requirements. An example of a thermal design which has been well thought out is shown in Fig. 6.54 [6.66]. Three internally matched 6 GHz 8 W power GaAs MESFETs are mounted so as to ensure that the central device does not receive the heat dissipated by the two lateral transistors. This is particularly important in this case where each device dissipates 25 W. Temperature measurements confirm that there is no heat accumulation at the center and that the case temperatures of all three devices are 52 °C. The devices are interconnected via two three-way distributed split-tee networks.

Essentially three methods of cooling are possible: natural convection, forced convection and liquid cooling. In each case it is necessary to establish a "balance sheet" in which the power to be dissipated, the maximum channel temperature and the maximum ambient temperature are specified. Next the thermal resistances of the chip and the case must be established before the thermal resistance of the heat sink can be calculated. Finally the cooling method to be used is decided by considering the physical space available, the energy available for the ventilator and the availability of a source of water or other cooling fluid. It is instructive to consider each case in turn.

6.4.4.1 Natural convection

The case of an 8 W r.f. output power GaAs MESFET operating at 6 GHz is illustrated in Fig. 6.55. The thermal resistance of this device is 3 °C W^{-1}

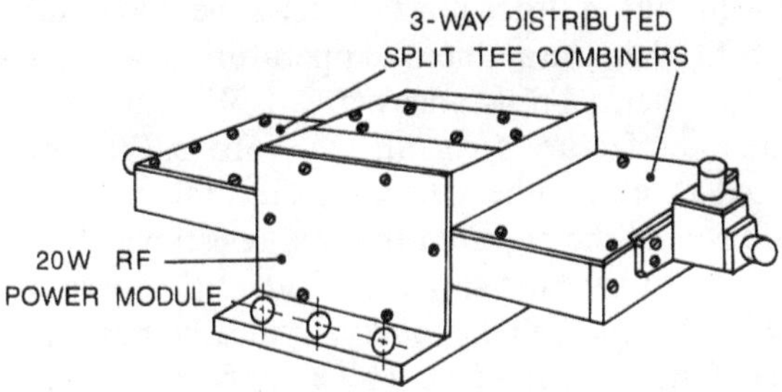

(a)

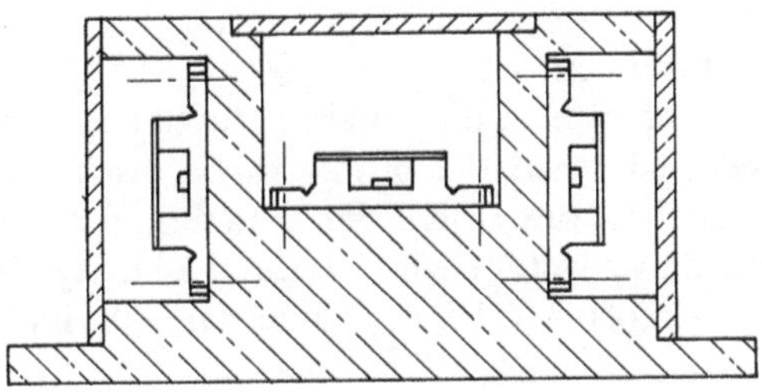

(b)

Fig. 6.54 Mounting configuration to optimize heat transfer [6.66].

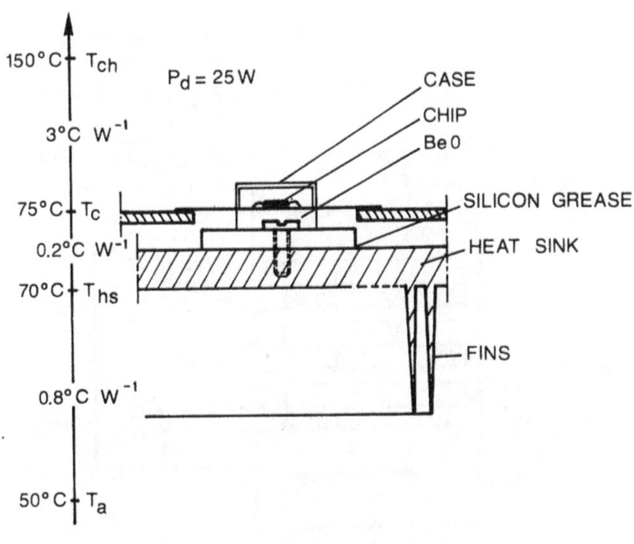

Fig. 6.55 Cooling by natural convection.

(Table 6.2). It is seen that in order not to exceed the maximum specified case temperature of 75 °C when the ambient temperature is 50 °C, it is necessary to use a heat sink with a thermal resistance of 0.8 °C W^{-1}. By examining the data sheets of heat-sink manufacturers (e.g. SEEM, Vanves, France) it is possible to obtain an anodized aluminum heat sink with a total surface area (including fins) sufficient to provide the required thermal resistance. It should be noted that the values apply for the heat sink mounted vertically to enhance natural convection. The heat sink will be bulky, typically having dimensions of 10 cm × 8.2 cm × 6.9 cm including fins. This may be physically too large and thus incompatible with the compactness required of GaAs MESFET amplifiers.

6.4.4.2 Forced convection

The physical size of the heat sink can be reduced if a fan is used. In general this solution is selected when the power to be dissipated exceeds 50 W. Depending on the air flow rate (Fig. 6.56), the thermal resistance of the heat sink can be multiplied by a factor *a* which is less than unity. Thus in the case of a heat-sink requirement of 0.8 °C W^{-1}, a value of $a = 0.4$ is obtained for an air

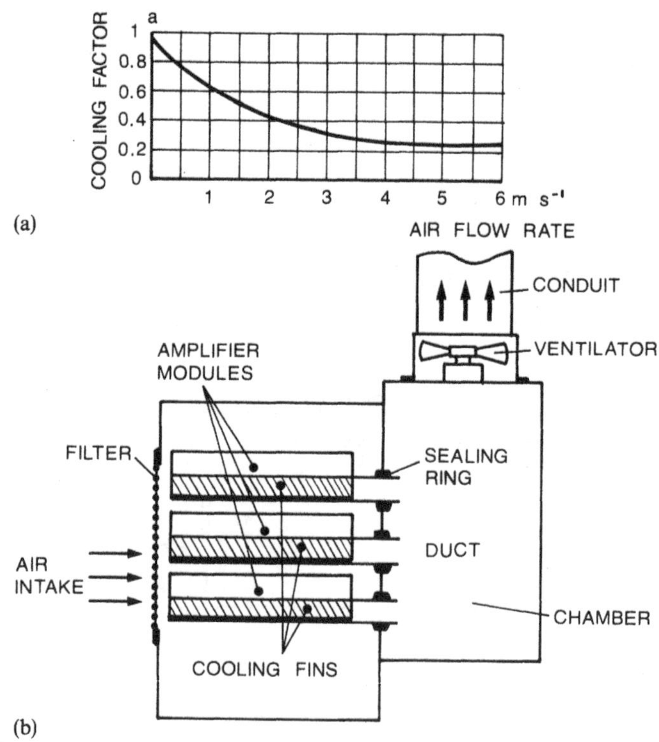

Fig. 6.56 Cooling by forced convection.

flow rate of $2\,\mathrm{m\,s^{-1}}$. This permits the selection of a heat sink with a thermal resistance of $2\,^{\circ}\mathrm{C\,W^{-1}}$, i.e. dimensions of 9 cm × 4.75 cm × 1.9 cm, which gives a volume reduction factor of 7. In practice such an approach becomes worthwhile only when there is a requirement to remove a large amount of heat from a confined volume.

6.4.4.3 Liquid cooling

Liquid cooling systems are in practice only used when it is necessary to remove hundreds of watts from a compact volume. The physical principle involved is the fact that the calorific capacity of water is 4000 times greater than that of air. This permits the evacuation of several kilowatts of heat with a flow rate of a few liters per minute through pipes of small cross-section [6.67]. If an open-loop cooling system is used there is a requirement for a source of cold water.

A more promising method of heat transfer is that offered by the heat pipe. This is not strictly a cooling technique but rather a method of transferring heat from an area where physical compactness precludes the use of heat sinks to an area where an adequate cooling system can be installed. The principle of the heat pipe is shown in Fig. 6.57. Heat is transferred from the area to be cooled to

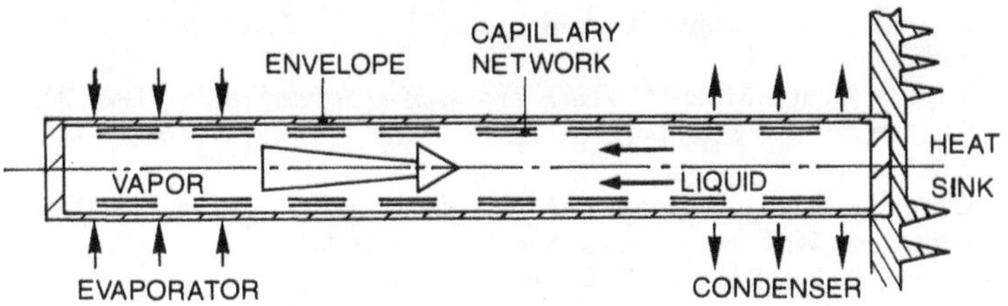

Fig. 6.57 Heat-pipe cooling.

the area of dissipation by causing a change of state of the fluid in the pipe from the liquid to the vapor phase. Thus calories are absorbed by the latent heat of vaporization at the interface to be cooled and transferred by various physical forces (pressure, convection currents) to the cooled region where liquefaction takes place and the heat is released. The liquid is then moved back to the area to be cooled by a combination of capillary and gravitational forces (DATE, La Motte d'Aveillans, France). Heat pipes have been successfully used to help cool a compact 4.5 GHz 40 W GaAs MESFET amplifier which dissipated 300 W [6.68].

Acknowledgments

The author is indebted to E. Camargo and A. Bert for their contributions to this chapter. He also wishes to thank J. Graffeuil, J. Obregon and R. Wroblewski for their helpful suggestions and J. P. Le Rolland, B. Mathieu, J. Guilcher and M. Bouguennec for doing the artwork. Finally the permission of the directors of the Centre National d'Etudes des Télécommunications to publish this chapter is gratefully acknowledged.

References

6.1 Perlow, S. M. *RCA Rev.*, **37**, 234, 1976.
6.2 Di Lorenzo, J. and Wisseman, W. *IEEE Trans. Microwave Theory Tech.*, **27**, 367, 1979.
6.3 Wroblewski, R., Salmer, G. and Crosnier, Y. *IEEE Trans. Electron Devices*, **30**, 154, 1983.
6.4 David, J., Sitch, J. and Stern, M. *IEEE Trans. Electron Devices*, **29**, 1548, 1982.
6.5 Dumas, J., Lecrosnier, D. and Bresse, J. *IEEE Electron Devices Lett.*, **6**, 192, 1985.
6.6 Tajima, Y. and Miller, P. *IEEE Trans. Microwave Theory Tech.*, **32**, 261, 1984.
6.7 Curtice, W. and Ettenberg, M. *IEEE Trans. Microwave Theory Tech.*, **33**, 1383, 1985.
6.8 Obregon, J. *Small Signal and Non-linear Microwave Active Devices, Course Colloq. on Millimeter Waves, Santander, July 1986.*
6.9 Fukui, H. *Bell Syst. Tech. J.*, **57**, 771, 1978.
6.10 Cheng, T. and Shurmer, H. *Electron. Lett.*, **20**, 788–790, 1984.
6.11 Di Lorenzo, J. V. and Khandelwal, D. D. *GaAs FET Principles and Technology*, Artech House, Dedham, MA, 1982.
6.12 Fukuta, M., Suyama, K., Suzuki, H. and Ishikawa, H. *IEEE Trans. Electron Devices*, **23**, 388, 1976.
6.13 Hornbuckle, D. P. and Kuhlman, L. J. Jr. *IEEE Trans. Microwave Theory Tech.*, **24**, 338, 1976.
6.14 Cripps, S. *IEEE MTT-S Int. Microwave Symp. Dig.*, pp. 221–223, IEEE, New York, 1983.
6.15 Poulin, D. *Microwaves*, **18**, 61, 1980.
6.16 Nakatani, M., Kadowaki, Y. and Ishii, T. *IEEE Trans. Microwave Theory Tech.*, **27**, 1066, 1979.
6.17 Takayama, Y. *IEEE MTT-S Int. Microwave Symp. Dig.*, pp. 218–220, IEEE, New York, 1976.
6.18 Cusack, J. M., Perlow, S. M. and Perlman, B. S. *IEEE Trans. Microwave Theory Tech.*, **22**, 1146, 1974.
6.19 Abe, H. and Aono, Y. *IEEE Trans. Microwave Theory Tech.*, **27**, 394, 1979.
6.20 Tucker, R. and Bradley, P. *IEEE Trans. Microwave Theory Tech.*, **32**, 296, 1984.
6.21 Soares, R. A. *Proc. 7th European Microwaves Conf., Copenhagen*, pp. 113–117, Microwave Exhibitions and Publishers, Tunbridge Wells, 1977.

6.22 Parisot, M., Binet, M. and Rabier, A. *Acta Electron.*, **23**, 137, 1980.
6.23 Chaffin, R. J. and Leighton, W. H. *IEEE G-MTT Int. Microwave Symp. Dig.*, pp. 155–157, IEEE, New York, 1973.
6.24 Johnson, K. M. *IEEE Trans. Microwave Theory Tech.*, **27**, 217, 1979.
6.25 Mitsui, Y., Nakatani, M. and Mitsui, S. *IEEE Trans. Microwave Theory Tech.*, **25**, 981, 1977.
6.26 Mazumder, S. and Van Der Puije, P. D. *IEEE J. Solid-State Circuits*, **12**, 576, 1977.
6.27 Soares, R., Goudelis, M. and de los Reyes Davo, E. *Proc. 10th European Microwaves Conf., Warsaw*, pp. 635–639, Microwave Exhibitions and Publishers, Tunbridge Wells, 1980.
6.28 Sechi, F. *IEEE Trans. Microwave Theory Tech.*, **28**, 1157, 1980.
6.29 Goudelis, M., Personal communication, 1980.
6.30 Soares, R. A., Goudelis, M., de los Reyes Davo, E. and Loriou, B. *IEEE MTT-S Int. Microwave Symp. Dig.*, pp. 63–65, IEEE, New York, 1982.
6.31 Soares, R. A. and Mainguet, B. Unpublished data, 1982.
6.32 Madjar, A. and Rosenbaum, F. J. *IEEE MTT-S Int. Microwave Symp. Dig.*, pp. 399–401, IEEE, New York, 1979.
6.33 Azizi, C., Kandem, J., Graffeuil, J. and Rossel, P. *6th European Solid-State Device Research Conf., Grenoble, September 1980.*
6.34 Kandem, J., Maimouni, R., Azizi, C., Graffeuil, J. and Rossel, P. *Rev. Phys. Appl.*, **18**, 213, 1983.
6.35 Telliez, I., Carnez, B., Rumelhard, C. and Gaynel, M. *Proc. 5e Journées Nationales de Microondes, Nice, June 1987*, p. 230.
6.36 Curtice, W. and Ettenberg, M. *RCA Rev.*, **46**, 321–340, 1985.
6.37 Willing, H. A., Rauscher, C. and de Santis, P. *IEEE Trans. Microwave Theory Tech.*, **26**, 1017, 1978.
6.38 Rauscher, C. and Willing, H. A. *IEEE Trans. Microwave Theory Tech.*, **27**, 834, 1979.
6.39 Tajima, Y., Wrona, B. and Mishima, K. *IEEE Trans. Electron Devices*, **28**, 171, 1981.
6.40 Rauscher, C. and Willing, H. A. *IEEE Trans. Microwave Theory Tech.*, **28**, 1054, 1980.
6.41 Ho, P. T., Pham, C. M. and Mencik, R. L. *IEEE MTT-S Int. Microwave Symp. Dig.*, pp. 128–130, IEEE, New York, 1979.
6.42 Soulignac, P. and Henry, J. P. Oral communication, *Journées Nationales de Microondes, Toulouse, June 1982.*
6.43 Siddiqui, J. *Int. Solid-State Circuits Conf., Dig. Tech. Pap.*, pp. 116–117, IEEE, New York, 1979.
6.44 Bert, A. Personal communication, 1982.
6.45 Arai, Y. and Tokumitse, Y. *Microwave Syst. News*, **9**, 58, October 1979.
6.46 Henming, H. B. *Microwave J.*, **20**, 112, June 1977.
6.47 Day, P. I. *IEEE Trans. Microwave Theory Tech.*, **23**, 772, 1975.
6.48 Everitt, W. L. and Anner, G. E. *Communication Engineering*, McGraw-Hill, New York, 1956.
6.49 Yamamura, S., Shigaki, M. Hidaka, N., Ishikawa, H., Takase, M. and Fukuta, M. *IEEE MTT-S Int. Microwave Symp. Dig.*, pp. 123–125, IEEE, New York, 1981.
6.50 Honjo, K., Takayama, Y. and Higashisaka, A. *IEEE Trans. Microwave Theory Tech.*, **27**, 3, 1979.
6.51 Bessoneau, G., Carnez, B. and Derewenko, H. Personal communication, 1981.
6.52 Chakmanian, J. S. *Microwaves*, **19**, 87, 1981.

6.53 Obregon, J. Personal communication, 1980.
6.54 Camargo, E., Consoni, D. and Soares, R. *Microwaves RF*, **24**, 93, 1985.
6.55 Russell, K. J. *IEEE Trans. Microwave Theory Tech.*, **27**, 472, 1979.
6.56 Ha, T. T. *Proc. Inst. Electr. Eng., Part G*, **127**, 148, 1980.
6.57 Morse, A. W. *Microwaves*, **16**, 70, 1978.
6.58 Bert, A. G. and Kaminsky, D. *IEEE Trans. Microwave Theory Tech.*, **28**, 1468, 1980.
6.59 McCarter, S. D. and Pavio, A. M. *IEEE MTT-S Int. Microwave Symp. Dig.*, pp. 159–161, IEEE, New York, 1982.
6.60 Cohn, M., Geller, B. D. and Schellenberg, J. M. *IEEE MTT-S Int. Microwave Symp. Dig.*, pp. 292–297, IEEE, New York, 1979.
6.61 Nagai, N., Mackawa, E. and Ono, K. *IEEE Trans. Microwave Theory Tech.*, **25**, 1008, 1977.
6.62 Galami, Z., Lampen, J. L. and Temple, S. J. *IEEE Trans. Microwave Theory Tech.*, **29**, 642, 1981.
6.63 Schellenberg, J. M. and Yamasaki, H. *Microwave J.*, **25**, 51, 1982.
6.64 Parad, L. and Moynihan, R. *IEEE Trans. Microwave Theory Tech.*, **13,** 91–95, 1965.
6.65 Soares, R. and Boulouard, A. *French Patent Application No. 86/09375*, 1986.
6.66 Marchand, M. Personal communication, 1987.
6.67 Suray, P., *Rev. Tech. Thomson-CSF*, **13**, 693, 1981.
6.68 Fukuden, N., Ogata, F., Hayakawa, M., Sugawara, H., Takagi, M. and Arai, Y. *IEEE MTT-S Int. Microwave Symp. Dig.*, pp. 66–68, IEEE, New York, 1982.

Chapter 7

Oscillators

A. P. S. KHANNA

7.1 Introduction

Oscillators represent the basic microwave energy source for all microwave systems such as radars, communications, navigation or electronic warfare. A typical microwave oscillator essentially consists of an active device (a diode or a transistor) and a passive frequency-determining resonant element such as a microstrip line, a surface acoustic wave (SAW) cavity or a dielectric resonator for a fixed-frequency oscillator and a varactor or an yttrium iron garnet (YIG) sphere for a tunable oscillator. With the rapid advance of microwave technology, there has been an increasing need for better oscillator performance. The emphasis has been on low noise, small size, low cost, high efficiency, high temperature stability and reliability for all oscillators and additionally, on wider bandwidths, better tuning linearity and reduced settling time for the tunable oscillators.

The origin of solid state microwave oscillators using Gunn and impact avalanche and transit time (IMPATT) diodes dates back to the late 1960s. Prior to that microwave sources were massive klystron or magnetron tubes requiring large power supplies. Solid state oscillators have come a long way in less than two decades. The extension of bipolar transistor oscillators to microwave frequencies and the development of GaAs MESFET devices in the early 1970s has made available highly cost-effective, miniature, reliable and low noise sources for use up to the millimeter-wave frequency range.

The basic active elements which can be used for microwave solid state oscillators are the Gunn diode, the IMPATT diode and transistors. Although the Gunn oscillator has the advantage of low frequency modulation (FM) noise, the IMPATT oscillator has a higher efficiency and is a higher power device than the Gunn device. Transistor oscillators are low noise as well as high efficiency sources. In addition, they do not have the problems of threshold current, heat sinking and tendency to lock at spurious frequencies which are associated with Gunn oscillators. Gunn and IMPATT diodes are inherently negative-resistance devices which means that the application of the required d.c. bias is sufficient to produce a negative resistance. The design of the oscillator is thus simplified to the design of the output matching circuit to deliver the desired power output. Application of d.c. bias to a bipolar or field-effect transistor, however, requires a suitable series or parallel feedback to induce the negative resistance. The frequency range over which the negative resistance is present in the diodes is determined by the physical mechanisms in the device, whereas in the case of transistors this frequency range is also influenced by the feedback topology chosen for the circuit. At present transistors are at a disadvantage with respect to the maximum oscillation frequency. Whereas a Gunn or IMPATT oscillator is capable of delivering power of 100 mW up to 100 GHz, the transistor oscillators are limited to 10 mW at 40 GHz.

Transistor oscillators can be realized using either bipolar or GaAs MESFET devices. Bipolar oscillators have maximum oscillation frequencies lower than those of GaAs MESFET oscillators but the latter are noisier. GaAs MESFET oscillators have been reported to reach frequencies of up to 100 GHz whereas oscillators using bipolar transistors have reached 20 GHz. Typically a bipolar oscillator has 6–10 dB less FM noise very close to the carrier than is obtained with the GaAs MESFET oscillator. The oscillator design approach presented in this chapter is applicable to oscillators using either type of transistor, although the examples given are of GaAs MESFETs only.

The increased capability of modern technology to utilize the pure microwave signal and to measure its characteristics precisely have enforced the development of extremely stable signal sources with minimum FM and amplitude modulation (AM) noise and maximum long-term and short-term frequency stability with time and temperature. If a device is being applied as a local oscillator in a communications system or radar, the frequency stability is often one of the important factors in determining the intermediate frequency bandwidth of the receiver and for bandwidth allocations for communications systems. A narrower intermediate frequency bandwidth improves the receiver performance by improving its noise figure. In Doppler radar the short-term stability of the signal is of utmost importance for the positive detection of moving targets. Low oscillator noise in a transmitter source ensures a high signal-to-noise ratio.

Quartz crystal oscillators provide extremely stable sources but their operation is limited to a few hundred megahertz. At microwave frequencies, stable sources have commonly been realized using frequency multiplication of a quartz crystal oscillator by a factor n. This method increases the FM noise power by a factor of n and has very low efficiency in addition to being very complex and expensive. Stable signals have also been generated in the past using bulky metal high Q cavities in passive cavity stabilization systems or in complicated frequency-discriminator systems.

Owing to their high Q, small size and excellent integrability in microwave integrated circuits (MICs), dielectric resonators can be used directly as frequency-determining elements for realizing stable MIC transistor oscillators. With the advent of recent temperature-stable materials for dielectric resonators, transistor dielectric resonator oscillators are fast becoming an automatic choice for a vast number of applications of fixed-frequency oscillators.

Frequency-tunable oscillators find applications in instrumentation, electronic warfare systems, communications systems and phase-locked oscillator schemes. YIG-tuned oscillators have the widest tuning bandwidths with a tuning capability of 2–20 GHz with a single sphere and a single transistor [7.1]. However, these oscillators require a long time to tune and are bulky and much less efficient than varactor-controlled oscillators (VCOs), although they have much better phase noise and tuning linearity. VCOs can offer a bandwidth of up to 1.5 octaves [7.2] and are commonly used in applications where size and efficiency are the main considerations.

7.2 Concept of negative resistance

The concept of negative resistance is as old as that of oscillators themselves. In contrast with positive resistance, negative resistance is described as a source of electrical energy. A negative resistance implies that the device is active, but an active device does not necessarily imply a negative resistance. A two-terminal device like a Gunn or IMPATT diode inherently has negative resistance, whereas a three-terminal device like a transistor often requires an appropriate impedance to be connected to one or more terminals to create a negative resistance. An important characteristic of negative resistance is that it is, in general, a nonlinear function of the r.f. current flowing through it. When a load R_l is connected to a negative resistance R_n ($R_l < R_n$), an r.f. current starts flowing through the circuit at a frequency at which the imaginary parts of both impedances cancel each other. This r.f. current causes the value of the negative resistance to change until the oscillation condition $-R_n = R_l$ is satisfied. Although the concept of negative resistance plays an important role in our understanding of the oscillation mechanism, it is not very practical to deal with at microwave frequencies. It is not possible to make any significant measurements at these frequencies in terms of R, L and C. In addition, the use of negative resistance for oscillator analysis can be misleading unless negative conductance analysis is used when appropriate [7.3]. The most common approach is to use reflection or transmission coefficients, i.e. S parameters. However, negative resistance or conductance at any terminal can be converted into reflection coefficients by using the following relations:

$$r_n = \frac{R_n}{Z_0} = \frac{1-|\Gamma|^2}{1-2|\Gamma|\cos\theta+|\Gamma|^2} \tag{7.1}$$

$$g_n = \frac{G_n}{Y_0} = \frac{1-|\Gamma|^2}{1+2|\Gamma|\cos\theta+|\Gamma|^2} \tag{7.2}$$

The extended Smith chart shown in Fig. 7.1 shows the negative resistance region, i.e. the area outside the normal Smith chart. This chart can be used to convert a negative impedance to a reflection coefficient and vice versa.

In view of the fact that the S parameter approach is the only practical method at microwave frequencies, S parameters will generally be used for oscillator analysis and design in the remainder of this chapter.

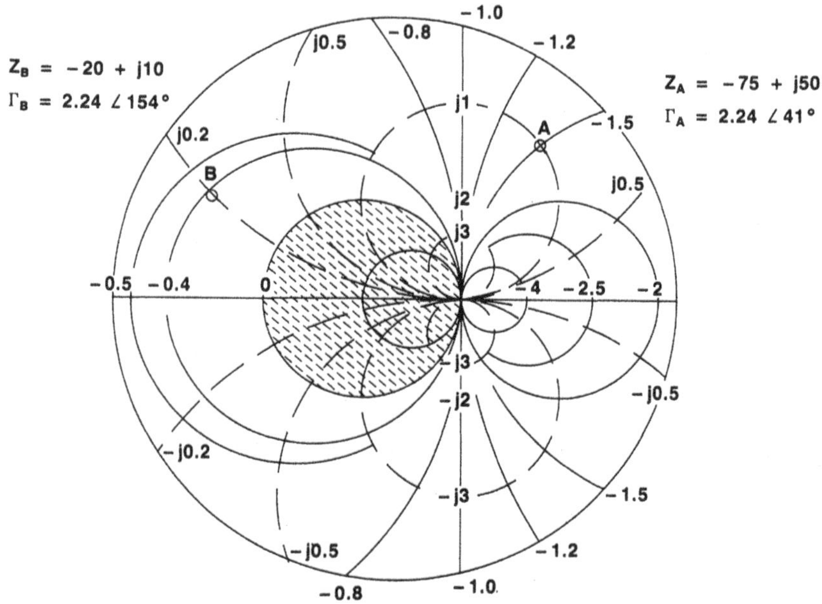

Fig. 7.1 Extended Smith chart.

7.3 Three-port S parameter characterization of transistors

The transistor, which is a three-port device, is generally characterized by its two-port S parameters with one of its ports grounded. The resulting three configurations in the case of a GaAs MESFET are shown in Fig. 7.2. Each configuration has its own advantages (e.g. the common-source configuration is used more often for amplifiers, the common-gate configuration for wideband oscillators and the common-drain configuration for medium power oscillators). Three-port S parameters, although introduced some time ago [7.4], have not often been used owing to the complexity of the analysis involved. The availability of desk-top computers and computer-aided design (CAD) has now made their use practical. The use of three-port S parameters

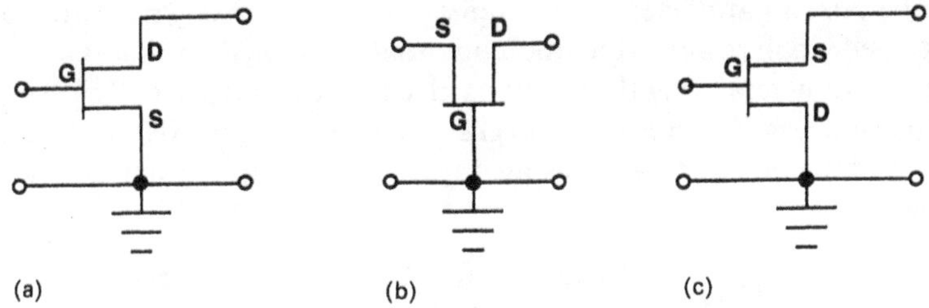

Fig. 7.2 Three transistor configurations: (a) common source; (b) common gate; (c) common drain.

eliminates the otherwise necessary conversion to and from Z and Y parameters to analyze series and parallel feedback effects as shown later in this chapter. The sums of the rows and columns of the three-port indefinite S matrix of the transistor are equal to unity, which helps in the determination and eventual correction of the systematic errors in the measurement or analysis. The use of the three-port S parameters of the transistor is sometimes essential in oscillator design [7.5].

The transistor is shown as a three-port device in Fig. 7.3. The S matrix of the incident and reflected waves is given by (see Chapter 2)

$$\begin{bmatrix} b_1 \\ b_2 \\ b_3 \end{bmatrix} = \begin{bmatrix} S_{11} & S_{12} & S_{13} \\ S_{21} & S_{22} & S_{23} \\ S_{31} & S_{32} & S_{33} \end{bmatrix} \begin{bmatrix} a_1 \\ a_2 \\ a_3 \end{bmatrix} \tag{7.3}$$

Ports 1, 2 and 3 represent the gate, drain and source respectively. This indefinite three-port S matrix satisfies the following conditions:

$$\sum_{j=1}^{3} S_{ij} = 1 \qquad \text{for } i = 1, 2, 3 \tag{7.4}$$

$$\sum_{i=1}^{3} S_{ij} = 1 \qquad \text{for } j = 1, 2, 3 \tag{7.5}$$

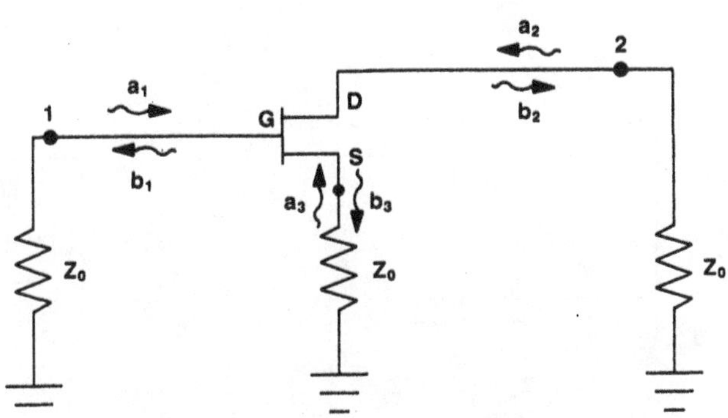

Fig. 7.3 The transistor as a three-port device.

The three-port S parameters of the transistor can be directly measured or can be obtained analytically from the commonly supplied two-port common-source S parameters using the relations given in Appendix 7.1 [7.6]. The two-port S parameters of a typical 0.5 µm GaAs MESFET at 10 GHz and the three-port S parameters calculated using these relations are given below as an example:

$$[S]_{2p} = \begin{bmatrix} 0.73\angle -102^\circ & 0.1\angle 42^\circ \\ 2.23\angle 96^\circ & 0.54\angle -49^\circ \end{bmatrix}$$

$$[S]_{3p} = \begin{bmatrix} 0.85\angle -56.6^\circ & 0.29\angle 66.4^\circ & 0.61\angle 47.3^\circ \\ 1.26\angle 104.7^\circ & 0.83\angle -37.1^\circ & 0.97\angle -47.7^\circ \\ 0.99\angle -30.9^\circ & 0.32\angle 47.2^\circ & 0.28\angle 104.2^\circ \end{bmatrix}$$

7.4 Oscillation and stability conditions [7.7]

Any oscillator can be represented in an arbitrary plane on the output line by a nonlinear impedance Z_{nl}, which is dependent on the current I_0 and has a negative real part, in series with a load impedance Z_l (Fig. 7.4). We assume that the circuit has a sufficiently high Q factor to suppress the harmonic currents. If a current

$$i(t) = I_0 \cos(\omega_0 t) \tag{7.6}$$

flows in the circuit, we can apply the Kirchhoff voltage law and write in the plane PP′

$$\{Z_{nl}(I_0, \omega_0) + Z_c(\omega_0)\} I_0 = 0 \tag{7.7}$$

Let

$$Z_{nl} + Z_c = Z_t = R_t + jX_t \tag{7.8}$$

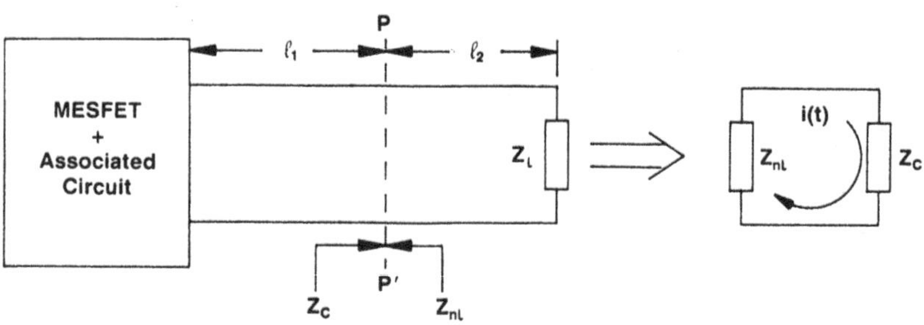

Fig. 7.4 Nonlinear microwave oscillator.

Since I_0 is not equal to zero, eqn (7.7) is satisfied by

$$R_t(I_0, \omega_0) = 0 \tag{7.9}$$

$$X_t(I_0, \omega_0) = 0 \tag{7.10}$$

Since $\mathrm{Re}(Z_l)$ is greater than zero, eqn (7.9) implies that $\mathrm{Re}(Z_{nl})$ is less than zero. Hence the device needs to present a negative resistance in order to be able to oscillate. The frequency of the oscillations is determined by eqn (7.10), i.e. by the requirement that the load reactance is equal and opposite to the device reactance.

Oscillators can also be represented by a nonlinear admittance Y_{nl}. In this case the oscillation conditions are determined using the Kirchhoff current law:

$$\{-Y_{nl}(I_0, \omega_0) - Y_c(\omega_0)\} V_d = 0 \tag{7.11}$$

Let

$$Y_{nl} + Y_c = Y_t = G_t + jB_t \tag{7.12}$$

Since V_d is not equal to zero, eqn (7.2) is satisfied by

$$G_t(I_0, \omega_0) = 0$$

$$B_t(I_0, \omega_0) = 0$$

To ensure that oscillation will start when d.c. bias is applied to the device, it is necessary for the instantaneous r.f. voltage of the device to be superior to that across the load. Thus

$$|v_d| > |v_c|$$

which leads to the necessary oscillation start-up conditions

$$|R_{nl}(I_0, \omega_0)|_{t=0} > |R_c(\omega_0)|$$

$$|G_{nl}(I_0, \omega_0)|_{t=0} < |G_c(\omega_0)|$$

From the above equations it can be seen that at start up the negative resistance (Fig. 7.5(a)) is greater than the load resistance but the negative admittance (Fig. 7.5(b)) is less than the load admittance.

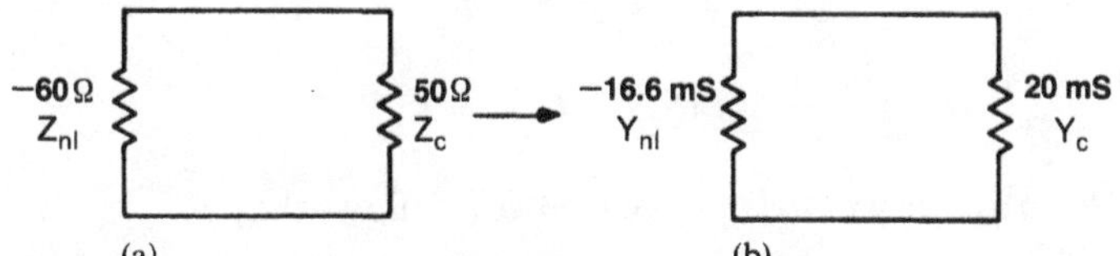

Fig. 7.5

At microwave frequencies it is more convenient to express eqns (7.9)–(7.12) in terms of the corresponding reflection coefficients Γ_{nl} and Γ_c as

$$|\Gamma_{nl}|\,|\Gamma_c| = 1 \tag{7.13}$$

$$\angle\Gamma_{nl} + \angle\Gamma_c = 2\pi n \qquad n = 0, 1, 2, \ldots \tag{7.14}$$

Equation (7.13) implies that the device reflection coefficient $|\Gamma_{nl}|$ should be greater than unity since $|\Gamma_c|$ is unity or less.

An oscillator can be considered as a combination of an active multiport and a passive multiport (the embedding circuit), as shown in Fig. 7.6. If the active device and the embedding circuit are characterized by their scattering matrices, we have for the active device [7.8]

$$\boldsymbol{b} = \boldsymbol{S}\boldsymbol{a} \tag{7.15}$$

and for the embedding circuit

$$\boldsymbol{b}' = \boldsymbol{S}'\boldsymbol{a}' \tag{7.16}$$

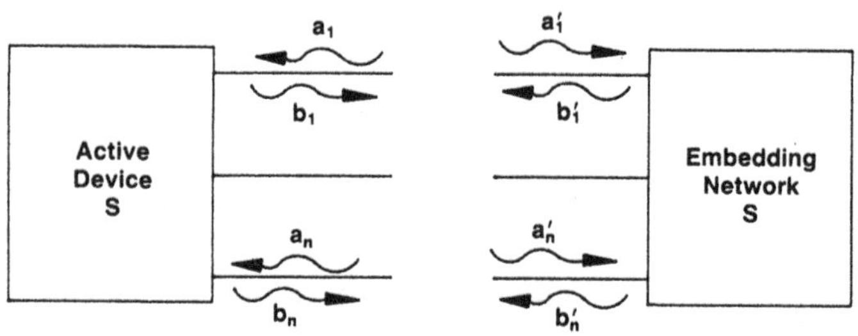

Fig. 7.6 Generalized oscillator configuration.

When the active device and the embedding network are connected we have, for oscillation conditions,

$$\boldsymbol{b}' = \boldsymbol{a} \tag{7.17}$$

$$\boldsymbol{b} = \boldsymbol{a}' \tag{7.18}$$

From eqns (7.15)–(7.18) we can write

$$\boldsymbol{b} = \boldsymbol{S}\boldsymbol{S}'\boldsymbol{a}'$$

or

$$\boldsymbol{a}' = \boldsymbol{S}\boldsymbol{S}'\boldsymbol{a}' \tag{7.19}$$

or

$$(\boldsymbol{S}\boldsymbol{S}' - \boldsymbol{U})\boldsymbol{a}' = 0 \tag{7.20}$$

where $\boldsymbol{U}$ is an identity matrix. Since $\boldsymbol{a}' \neq 0$, it follows that

$$\boldsymbol{M} = \boldsymbol{S}\boldsymbol{S}' - \boldsymbol{U} \tag{7.21}$$

is a singular matrix or

$$\det \boldsymbol{M} = 0 \tag{7.22}$$

which represents the generalized large-signal steady state oscillation condition for an n-port oscillator.

In fact since the scattering matrix of the active device is defined at small-signal level, the n-port oscillation condition for small signals can be represented by

$$|\det(\boldsymbol{S}\boldsymbol{S}' - \boldsymbol{U})| > 0 \tag{7.23}$$

and

$$\arg \det(\boldsymbol{S}\boldsymbol{S}' - \boldsymbol{U}) = 0 \tag{7.24}$$

The oscillations can start as soon as these relations are satisfied and go on building up until the device nonlinearities cause the steady state to be reached. For example, consider an active two-port loaded by two passive impedances as shown in Fig. 7.7. The active device is described by the scattering matrix

$$\boldsymbol{S} = \begin{bmatrix} S_{11} & S_{12} \\ S_{21} & S_{22} \end{bmatrix} \tag{7.25}$$

and the embedding circuit is described by

$$\boldsymbol{S}' = \begin{bmatrix} \Gamma_1 & 0 \\ 0 & \Gamma_2 \end{bmatrix} \tag{7.26}$$

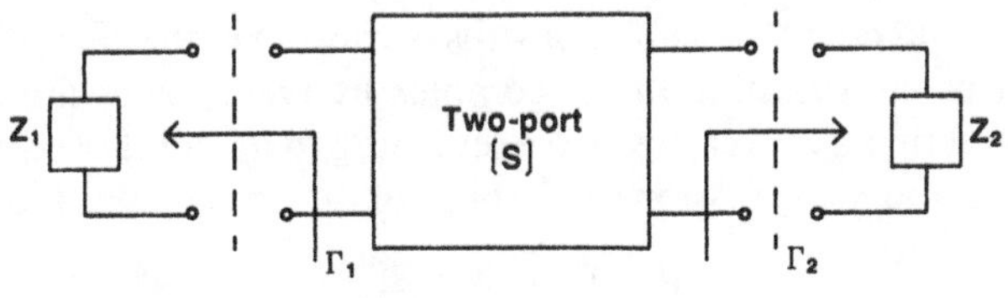

Fig. 7.7 Two-port loaded with two impedances.

From eqn (7.22) the oscillation condition is

$$\det \boldsymbol{M} = \det \begin{bmatrix} S_{11}\Gamma_1 - 1 & S_{12}\Gamma_2 \\ S_{21}\Gamma_1 & S_{22}\Gamma_2 - 1 \end{bmatrix} = 0 \tag{7.27}$$

which gives

$$(S_{11}\Gamma_1 - 1)(S_{22}\Gamma_2 - 1) - S_{12}S_{21}\Gamma_1\Gamma_2 = 0 \tag{7.28}$$

From the above we obtain the following two well-known conditions which are simultaneously satisfied for oscillation [7.9]:

$$S_{11}+\frac{S_{12}S_{21}\Gamma_2}{1-S_{22}\Gamma_2}=\frac{1}{\Gamma_1} \tag{7.29}$$

$$S_{22}+\frac{S_{12}S_{21}\Gamma_1}{1-S_{11}\Gamma_1}=\frac{1}{\Gamma_2} \tag{7.30}$$

The oscillations are considered to be stable if any perturbation in the r.f. voltage or the r.f. current of the oscillator circuit at any instant decays, bringing the oscillator back to its point of equilibrium. Oscillator stability is analyzed [7.10] quasi-statically by applying a small perturbation to the amplitude I_0 (Fig. 7.4). The impedance Z_t defined in eqn (7.8), which is a function of I_0 and the complex frequency $p = \alpha + j\omega$, is developed in a Taylor series about I_0, ω_0. Since the perturbed current is nonzero, i.e. the oscillations continue to exist after the perturbation, we should have in the plane PP′

$$Z_t(I_0,\omega_0)+\frac{\partial Z_t}{\partial p}\delta p+\frac{\partial Z_t}{\partial I_0}\delta I_0=0 \tag{7.31}$$

Since $Z_t(I_0,\omega_0) = 0$ we obtain

$$\delta p = -j\frac{(\partial Z_t/\partial I_0)(\partial Z_t{}^*/\partial\omega)}{|\partial Z_t/\partial\omega|^2}\delta I_0 \tag{7.32}$$

where δp can be decomposed into its real and imaginary parts as follows:

$$\delta p = \alpha + j\delta\omega$$

The oscillator will be stable if, for a positive variation in the current amplitude, the real part of the variation in the complex frequency is negative, i.e. if α is negative, indicating a decreasing wave returning to its point of equilibrium I_0.

From the above expression for δp this condition is realized for

$$\frac{\partial R_t}{\partial I_0}\frac{\partial X_t}{\partial\omega}-\frac{\partial X_t}{\partial I_0}\frac{\partial R_t}{\partial\omega}>0 \tag{7.33}$$

This relation represents the stability condition for an oscillator around an amplitude I_0 and an angular frequency ω_0. Moreover, from eqn (7.32), the imaginary part $\delta\omega$ vanishes if the condition

$$\frac{\partial R_t}{\partial I_0}\frac{\partial R_t}{\partial\omega}+\frac{\partial X_t}{\partial I_0}\frac{\partial X_t}{\partial\omega}=0 \tag{7.34}$$

is fulfilled. This indicates that a variation δI_0 in the amplitude will not produce a variation in the oscillator's real angular frequency ω_0. It can also be deduced from eqn (7.34) that, for maximum stability, the device impedance $Z_{nl}(I_0)$ and the load impedance $Z_c(\omega)$ should intersect at right angles at the oscillation equilibrium point (I_0, ω_0).

7.5 Fixed-frequency oscillators

Oscillators can be subdivided into two main categories: fixed frequency and frequency tunable. Fixed-frequency oscillators are generally used as local oscillators in communications and radar systems, and tunable and wideband tunable oscillators are used in synthesizers, instrumentation and electronic warfare systems. An oscillator in its most general form can be represented as a series or a parallel circuit as shown in Fig. 7.8. The oscillators can be designed so that any one of the immittances shown can contain the output resistive load while the other two are normally reactive. In a fixed-frequency oscillator all the immittances are fixed, whereas in the frequency-tunable oscillator one or more of the immittances are made tunable by using varactor diodes or YIG resonators. Dielectric resonators are now generally used to realize low noise and temperature-stable fixed-frequency oscillators which are known as transistor dielectric resonator oscillators (TDROs).

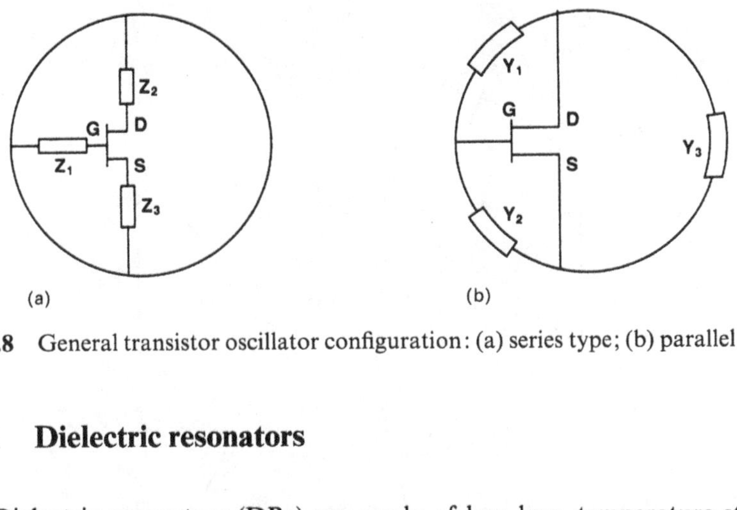

Fig. 7.8 General transistor oscillator configuration: (a) series type; (b) parallel type.

7.5.1 Dielectric resonators

Dielectric resonators (DRs) are made of low loss, temperature-stable, high permittivity and high Q ceramic materials in a regular geometrical form. They resonate in various modes at frequencies determined by the dimensions and shielding conditions. Because of their small size, low price and excellent integrability in MICs, their application in active and passive microwave circuits has been rapidly increasing [7.11].

The dimensions of a DR are much smaller (by a factor of approximately $1/\varepsilon_r^{1/2}$) than those of an empty metallic cavity resonating at the same frequency. If ε_r is high the electric and magnetic fields are confined in and near the DR, and radiation losses are small. The unloaded quality factor Q_u is limited by the losses in the DR. To a first approximation, a DR is a dual of a metallic cavity: the metallic boundary representing an electric short circuit in a cavity is replaced by the dielectric boundary representing a magnetic short circuit. However, the radiation losses of DRs with the most frequently used permittivities are generally much greater than the energy losses in the metallic cavities, which makes suitable shielding of the DR a necessity.

DRs are usually solid cylinders, but tubular, spherical and parallelepiped shapes are also found. A resonant mode frequently used in cylindrical resonators is denoted $TE_{01\delta}$. The magnetic field lines are contained in the meridian plane and the electric field lines form concentric circles around the z axis as shown in Fig. 7.9. This mode appears as a magnetic dipole to a distant observer, and for this reason it is sometimes referred to as a "magnetic dipole mode". When the relative dielectric constant is around 40, more than 95% of the stored electrical energy and more than 60% of the stored magnetic energy of the $TE_{01\delta}$ mode are located within the cylinder. The remaining energy is distributed in the air around the resonator and decays rapidly with distance away from the resonator surface.

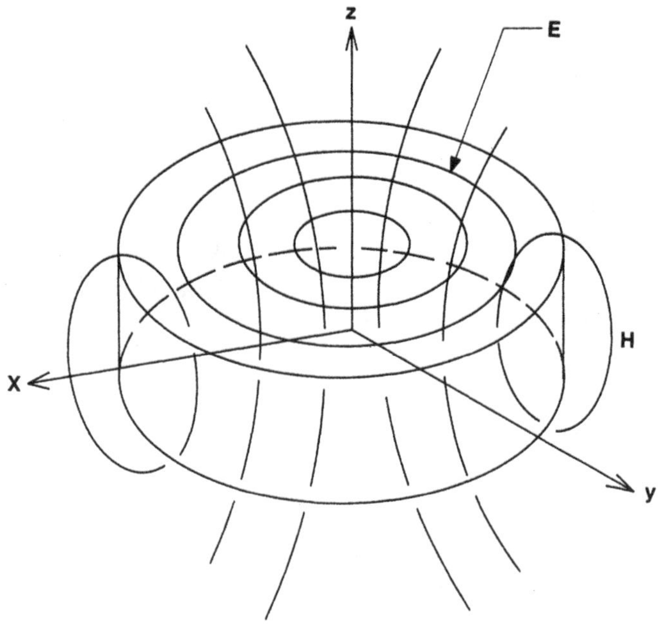

Fig. 7.9 Field lines of the resonant mode $TE_{01\delta}$ in an isolated dielectric resonator.

7.5.1.1 Material

The important properties of the ceramic material to be used for DRs are as follows:

(1) its Q factor which is approximately equal to the inverse of the loss tangent;

(2) the temperature coefficient of the resonant frequency T_f which includes the combined effects of the temperature coefficient of the dielectric constant and the thermal expansion of the dielectric;

(3) its dielectric constant ε_r.

The values of Q, T_f and ε_r required for various applications differ, and in general a suitable combination can be achieved by choosing an appropriate material and composition. Until a few years ago, the lack of suitable materials with acceptable values of Q, T_f and ε_r severely limited DR applications. Materials such as TiO_2 (rutile phase), which has $Q \approx 10000$ at 4 GHz and $\varepsilon_r \approx 100$, were most frequently used for exploratory work. However, TiO_2 has a T_f value of 400 ppm K^{-1} which makes it impractical for most applications.

A number of material compositions have been explored in attempts to develop suitable dielectric materials. These include ceramic mixtures containing TiO_2 [7.12], various titanates and zirconates [7.13, 7.14], glass–ceramic systems [7.15] and alumina-based ceramics. Temperature-stable DRs have been under development for less than a decade. Several ceramic compositions have been developed which offer excellent dielectric properties. The major properties of various materials which have been developed commercially are compared in Table 7.1.

It has not yet been established whether any of the dielectric compositions shown in Table 7.1 have overall superiority since many factors such as ease of ceramic processing and ability to hold tolerances on the dielectric properties must be considered. The performance limitations, if any, of the materials with lower dielectric constants remain to be determined since in most component work reported so far DRs with ε_r in the range 37–100 have been used. The performance of these materials is likely to be more sensitive to shielding owing to an increase in fields outside the resonator.

The temperature coefficient T_f of the resonator can be controlled in some materials, by modifying the composition, to be anywhere between +9 and −9 ppm K^{-1}. Circuit effects also shift T_f by a few parts per million depending on the circuit configuration. An initial resonator T_f of +1 to +4 ppm K^{-1} often results in effective temperature compensation in transistor oscillators. Thus the limit to achievable temperature compensation results from ceramic tolerances and the usual need for frequency tuning that contributes to temperature effects.

The quality factor Q of the DR decreases with increasing frequency. Typically the product f_0Q_0, where f_0 is in gigahertz, is constant for a given material. Some degradation of Q is usually incurred in component applications. Losses due to housing walls, dielectrics, adhesives used to support the resonators and other effects typically reduce Q by 10%–20% as described later. Since variations in the Qs of different materials are significant (Table 7.1), it is

Table 7.1 Dielectric resonator materials

Composition	*Dielectric constant*	*Q*	*Temperature coefficient of frequency* (ppm)	*Frequency range* (GHz)	*Manufacturer*
$Ba_2Ti_9O_{20}$	40	10 000 at 4 GHz	+2	1–100	Bell Labs
$(Zr–Sn)TiO_4$	38	10 000 at 4 GHz	−4 to +10[a]	1–100	Trans Tech Thomson-CSF Murata
$Ba(0.33Zn0.67Ta)O_2$	30	10 000 at 10 GHz	0–10[a]	4–100	Murata
$Ba(0.33Mg0.67Ta)O_2$	25	25 000 at 10 GHz	4	4–100	Sumimoto
$BaO–PbO–Nd_2O_3–TiO_2$	88	5000 at 1 GHz	0–6[a]	<4	Murata–Trans Tech
Al_2O_3	11	50 000 at 10 GHz	0–6[a]	>18?	NTK–Trans Tech

[a] Adjustable with composition.

possible to select a material for a given application. A material with higher Q is preferred for lower noise oscillations and for sharply tuned filters, and a material with lower Q is preferred for frequency-tunable wider-band components.

7.5.1.2 Resonant frequency

The resonant frequency of a resonator is determined by its dimensions and surroundings. Although the geometrical form of a DR is extremely simple, an exact solution of the Maxwell equations is much more difficult than for the hollow metallic cavity. For this reason, the exact resonant frequency of a given resonant mode, such as $TE_{01\delta}$, can only be computed by rigorous numerical procedures. A number of theoretical treatments, which have an accuracy of $\pm 1\%$ for a given configuration, are available in the literature. Unfortunately these methods call for the use of high level computers. Kajfez and Guillon [7.16] have presented an approximate solution of the explicit equations for both the isolated case and the more commonly used MIC configuration. These equations, which are given below, can easily be programmed on a desk-top computer and give accuracies of better than $\pm 2\%$.

Isolated dielectric resonator The resonant frequency is given by

$$f_r(\text{GHz}) = \frac{34}{a\varepsilon_r}\left(\frac{a}{H}+3.45\right) \tag{7.35}$$

where a (in millimeters) is the radius of the resonator and H is its height. This relation is accurate to about 2% in the range

$$0.5 < a/H < 2 \qquad 30 < \varepsilon_r < 50 \tag{7.36}$$

Dielectric resonator in the MIC configuration The inset in Fig. 7.10 represents a DR in an MIC configuration. The various steps used to determine the size of the DR for a given frequency are as follows.

(i) The diameter $D = 2a$ of the DR is selected to be

$$\frac{5.4}{k_0\varepsilon_s{}^{1/2}} > 2a > \frac{5.4}{k_0\varepsilon_r{}^{1/2}} \tag{7.37}$$

where ε_s and ε_r are the relative permittivities of the substrate and the DR respectively, and k_0 is the free-space propagation constant.

(ii) Calculate k' from

$$k' = \frac{2.405}{a} + \frac{Y_0}{2.405a(1+2.43/Y_0+0.291Y_0)} \tag{7.38}$$

where

$$Y_0 = \{(k_0a)^2(\varepsilon_r-1)-2.405^2\}^{1/2} \tag{7.39}$$

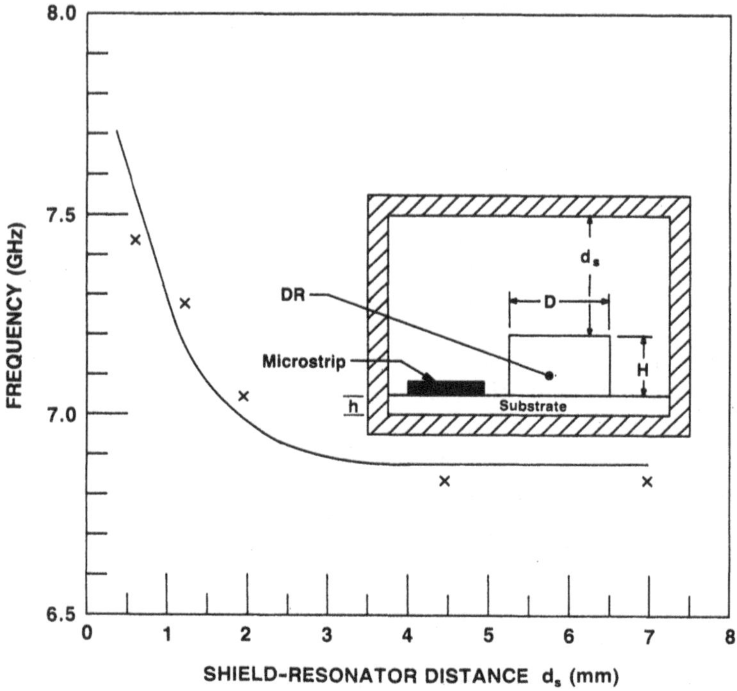

Fig. 7.10 Resonant frequency as a function of cover height: ×, measured; —, calculated (DR, dielectric resonator; $D = 8.5$ mm, $H = 3.7$ mm, $\varepsilon_r = 37.0$; $h = 0.635$ mm; $\varepsilon_s = 9.9$).

(iii) Calculate the propagation constant β for the $TE_{01\delta}$ mode from

$$\beta = (k_0{}^2\varepsilon_r - k'^2)^{1/2} \tag{7.40}$$

(iv) Evaluate the attenuation constants α_1 and α_2:

$$\begin{aligned} \alpha_1 &= (k'^2 - k_0{}^2\varepsilon_s)^{1/2} \\ \alpha_2 &= (k'^2 - k_0{}^2)^{1/2} \end{aligned} \tag{7.41}$$

(v) Find the resonator height H from

$$H = \frac{1}{\beta}\left[\tan^{-1}\left\{\frac{\alpha_1}{\beta}\coth(\alpha_1 h)\right\} + \tan^{-1}\left\{\frac{\alpha_2}{\beta}\coth(\alpha_2 d_s)\right\}\right] \tag{7.42}$$

These relations can also be used to determine the frequency of a DR with known dimensions.

Figure 7.10 shows the variation in the resonant frequency with the cover height. The experimental points show that this method is accurate to within $\pm 2\%$.

7.5.1.3 Coupling of a dielectric resonator in the microwave integrated circuit configuration

DRs are used in a number of different configurations depending upon the application. In order to use DRs effectively in microwave circuits, it is necessary to have an accurate knowledge of the coupling between the resonator and different transmission lines. The $TE_{01\delta}$ mode of the cylindrical resonator can easily be coupled to microstrip line, fin line, magnetic loop, and metallic and dielectric waveguides [7.16]. In this section we shall discuss the most commonly used configuration of the DR, i.e. $TE_{01\delta}$ mode coupling with a microstrip line.

Figure 7.11 shows magnetic coupling between a DR and microstrip. The resonator is placed on the upper surface of a microstrip substrate. The lateral distance between the resonator and the microstrip conductor is the parameter which primarily determines the coupling between the resonator and the microstrip transmission line. The metallic shielding, which is required to minimize the radiation losses (and hence to increase Q), also affects the resonant frequency of the $TE_{01\delta}$ mode. The reason for the modification of the resonant frequency can be explained by cavity perturbation theory: when a metal wall of a resonant cavity is moved inwards, the resonant frequency will decrease if the stored energy of the displaced field is predominantly electric; however, if the stored energy close to the metal wall is primarily magnetic, as in the case of the shielded $TE_{01\delta}$ DR considered here, the resonant frequency will increase when the wall moves inwards.

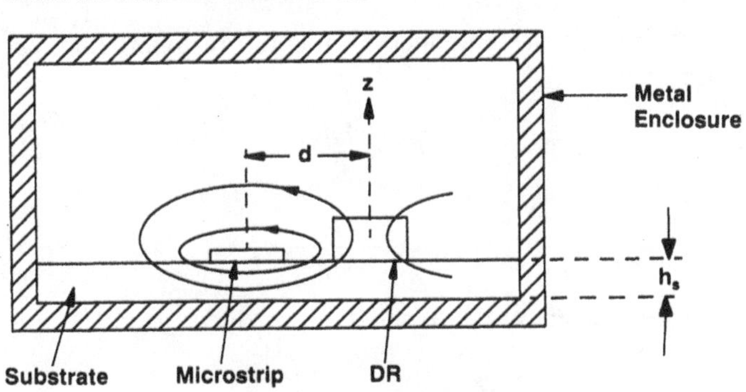

Fig. 7.11 Coupling between a microstrip line and a dielectric resonator.

The $TE_{01\delta}$ mode in a DR can be approximated by a magnetic dipole of moment M. The coupling between the line and the resonator is accomplished by orienting the magnetic moment of the resonator perpendicular to the microstrip plane so that the magnetic lines of the resonator link with those of the microstrip line as shown in Fig. 7.11. The DR located adjacent to the microstrip line operates like a reaction cavity which reflects the r.f. energy at the resonant frequency [7.17]. The equivalent circuit of the resonator coupled to a microstrip line is shown in Fig. 7.12. In this figure L_r, C_r and R_r are the equivalent parameters of the DR, L_1, C_1 and R_1 are the equivalent parameters of the microstrip line and L_m characterizes the magnetic coupling. The transformed resonator impedance Z in series with the transmission line can easily be determined to be

$$Z = j\omega L_1 + \frac{\omega^2 L_m^{\ 2}}{R_r + j\omega(L_r - 1/\omega^2 C_r)} \tag{7.43}$$

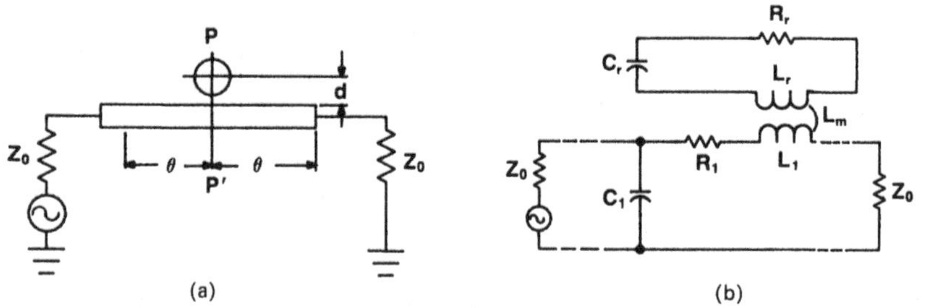

Fig. 7.12 Equivalent circuit of a dielectric resonator coupled with a line.

Around the center frequency, ωL_1 can be neglected and Z becomes

$$Z = \omega Q_u \frac{L_m^{\ 2}}{L_r} \frac{1}{1 + jX} \tag{7.44}$$

where $X = 2Q_u \Delta\omega/\omega$; the unloaded Q and the resonant frequency of the resonator are given by

$$Q_u = \frac{\omega_0 L_r}{R_r} \tag{7.45}$$

$$\omega_0 = \frac{1}{(L_r C_r)^{1/2}} \tag{7.46}$$

At the resonance frequency, $X = 0$ and

$$Z = R = \omega_0 Q_u \frac{L_m^{\ 2}}{L_r} \tag{7.47}$$

Equation (7.47) indicates that the circuit shown in Fig. 7.12 can be represented by the simple parallel tuned circuit shown in Fig. 7.13, where L, R and C satisfy the following equations:

$$L = \frac{L_m^{\ 2}}{L_r} \tag{7.48}$$

$$C = \frac{L_r}{\omega_0^{\ 2} L_m^{\ 2}} \tag{7.49}$$

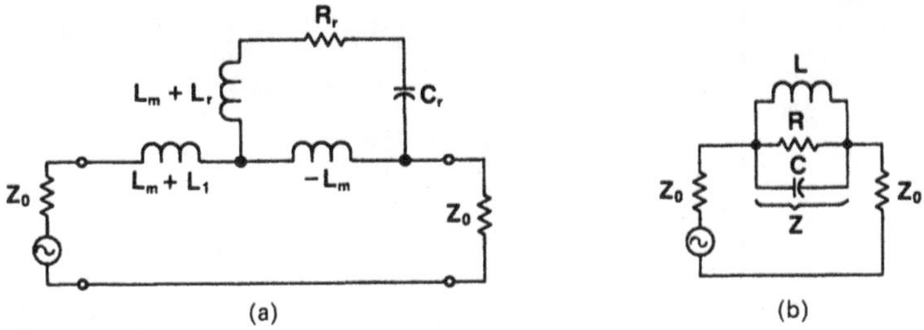

Fig. 7.13 (a) Simplified equivalent circuit; (b) final equivalent circuit of a dielectric resonator coupled with a microstrip line.

$$R = \omega_0 Q_u \frac{L_m{}^2}{L_r} \tag{7.50}$$

The coupling coefficient β at the resonant frequency ω_0 is defined by

$$\beta = \frac{R}{R_{ext}} = \frac{R}{2Z_0} = \frac{\omega_0 Q_u}{2Z_0} \frac{L_m{}^2}{L_r} \tag{7.51}$$

If the reflection and transmission coefficients of the resonance frequency of the DR coupled to the microstrip are defined as S_{110} and S_{210} respectively, β can be shown to be given by [7.18]

$$\beta = \frac{S_{110}}{1 - S_{110}} = \frac{1 - S_{210}}{S_{210}} = \frac{S_{110}}{S_{210}} \tag{7.52}$$

This relation can be used to determine the coupling coefficient from the directly measurable reflection and transmission coefficients. The value of β can also be accurately calculated from a knowledge of the circuit configuration. The quantity $L_m{}^2/L_r$ in eqn (7.51) is a strong function of the distance between the resonator and the microstrip line for given shielding conditions, substrate thickness and substrate permittivity. The analysis of β involves the use of well-known electromagnetic concepts and finite-element techniques.

The relation between different quality factors is well known and is given by

$$Q_u = Q_L(1 + \beta) = Q_e \beta \tag{7.53}$$

The external quality factor $Q_e = Q_u/\beta$ is generally used to characterize the coupling. Figure 7.14 shows an example of the variation in Q_e with the distance between the DR and the line.

The S parameters of the DR coupled to a microstrip with the lengths of transmission lines on input and output, as shown in Fig. 7.12, can be

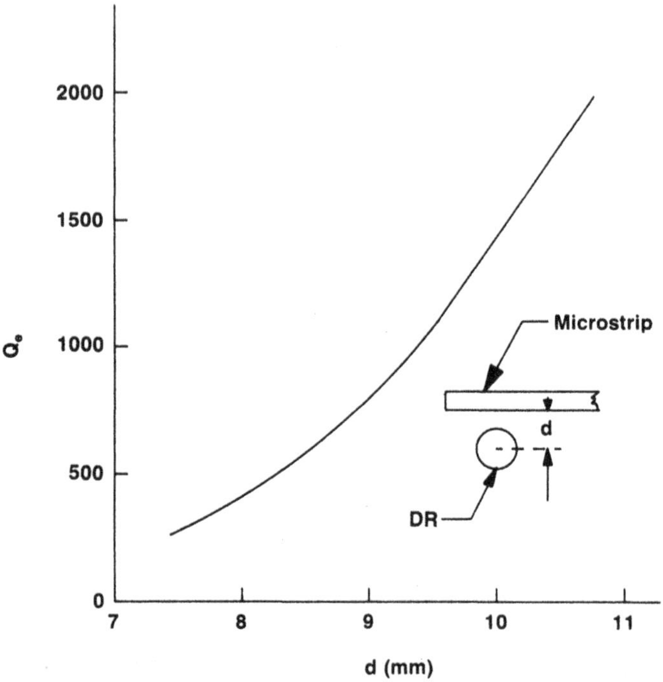

Fig. 7.14 External Q factor as a function of the distance between the line and the resonator: for the resonator, $D = 14$ mm, $H = 7.5$ mm and $\varepsilon_r = 35$; for the substrate, $H = 0.8$ mm and $\varepsilon_s = 2.54$.

determined from the relations presented above and are given by [7.16]

$$S = \begin{bmatrix} \dfrac{\beta}{\beta + 1 + 2jQ_u\Delta\omega/\omega_0}\exp(-2j\theta) & \dfrac{1 + 2jQ_u\Delta\omega/\omega_0}{\beta + 1 + 2jQ_u\Delta\omega/\omega_0}\exp(-2j\theta) \\ \dfrac{1 + 2jQ_u\Delta\omega/\omega_0}{\beta + 1 + 2jQ_u\Delta\omega/\omega_0}\exp(-2j\theta) & \dfrac{\beta}{\beta + 1 + 2jQ_u\Delta\omega/\omega_0}\exp(-2j\theta) \end{bmatrix} \tag{7.54}$$

where 2θ is the electrical line length between the input and output planes.

7.5.1.4 Spurious modes

As in the case of waveguide cavities, there are many possible resonant modes which can be excited in DRs. These modes can be divided into three families: transverse electric (TE), transverse magnetic (TM) and hybrid electromagnetic (HEM) modes. Each of the three families has a large number of individual modes so that the problem of choosing which mode is best suited for a particular application arises. The $TE_{01\delta}$ is the principal or main mode traditionally used, but for certain applications, such as dual-mode filters, the $HEM_{11\delta}$ mode has definite advantages.

The resonant frequencies of the principal mode and the spurious modes are determined by the physical dimensions and the dielectric constant of the resonator for fixed shielding conditions. The resonant frequency of some spurious modes like $TM_{01\delta}$ and $HE_{11\delta}$ may be close to the resonant frequency of the principal $TE_{01\delta}$ mode. Figure 7.15 shows a mode chart of the DR in the shielded configuration indicated [7.19]. The aspect ratio D/L of the DR can be used as a design parameter to place the resonance of spurious modes outside the operating frequency band of the principal $TE_{01\delta}$ mode, thus minimizing undesirable interference. As is clear from Fig. 7.15, an aspect ratio between 2 and 3 results in the best separation of the spurious modes.

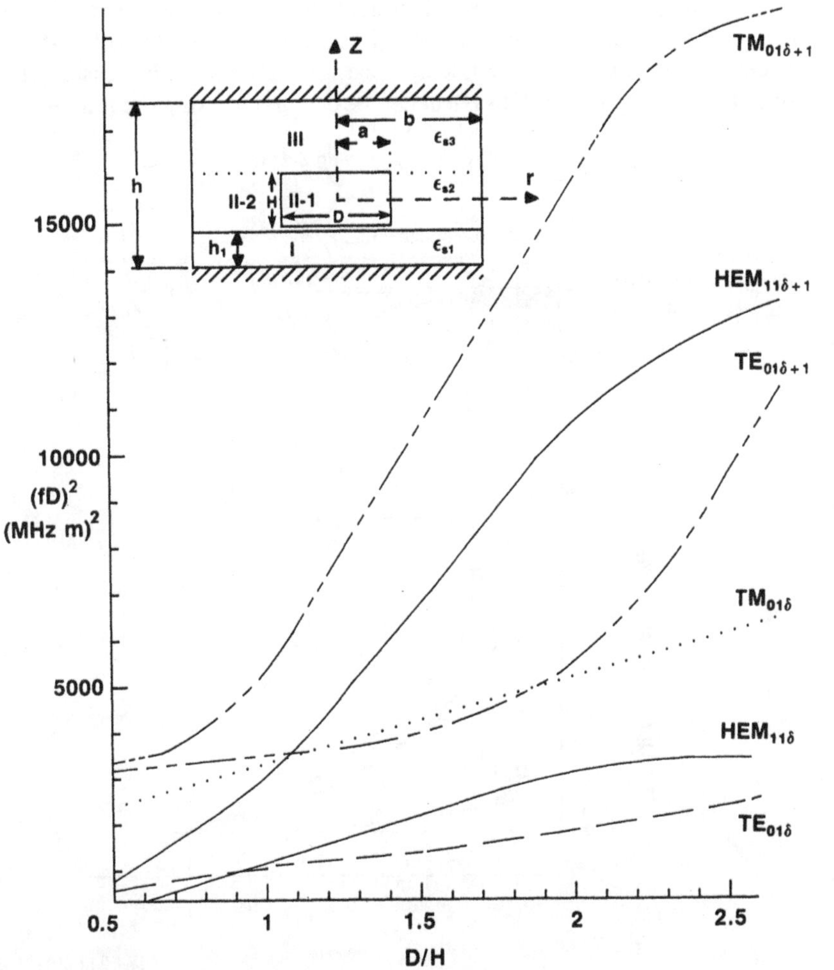

Fig. 7.15 Mode chart of a dielectric resonator in a cavity: $\varepsilon_r = 37.2$; $\varepsilon_{s1} = \varepsilon_{s2} = \varepsilon_{s3} = 1$; $b = 16$ mm; $h_1 = 1$ mm; $H = 21$ mm.

$TM_{01\delta}$ and $HE_{mn\delta}$ modes are far more sensitive than the principal mode to frequency-tuning screws. For example, the same amount of tuning which will tune the principal mode of a 6 GHz DR by 25 MHz can move the resonant frequency of the $TM_{01\delta}$ or $HE_{01\delta}$ mode in the opposite direction across the entire 500 MHz radio band. Frequency tuning must be limited to avoid a significant reduction in mode separation which can bring the spurious responses into the frequency band of interest.

7.5.1.5 Frequency tuning

Most of the applications of the DR demand frequency tunability over a narrow band. The resonant frequency of the DR can be tuned with the help of a tuning screw placed in the metallic shield directly above the resonator as shown in Fig. 7.16. Since the resonant frequency is sensitive to the distance between the resonator and the shield, as explained earlier, it increases with the depth of the tuning screw. This effect can be calculated using the finite-element

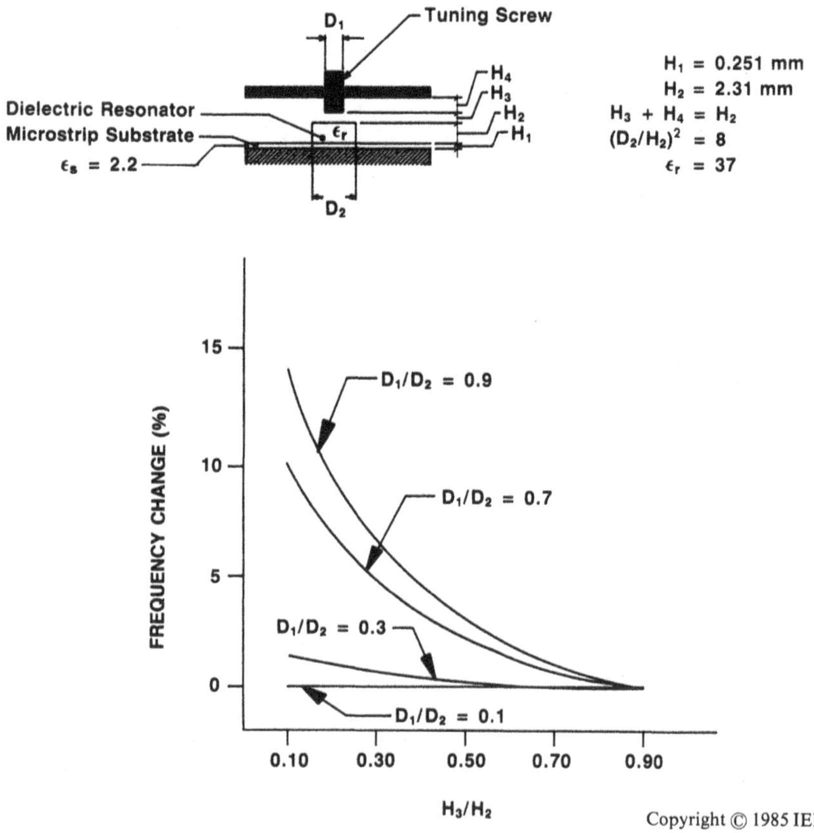

Fig. 7.16 Frequency change of a dielectric resonator for various tuning-screw diameters (after ref. 7.20).

method [7.20]. Figure 7.16 also shows the possible frequency tuning for several tuning-screw diameters. The curves presented in this figure enable an optimum screw diameter to be chosen to obtain the desired tuning margin.

7.5.2 Design of oscillators

The DR can be used to realize one or more of the immittances shown in Fig. 7.8. The commonly used stable TDROs can be divided into two types: one using the DR as a series feedback element and the other using it as a parallel feedback element. We shall discuss the analysis and design of both of these types (ref. 7.16, Chapter 10).

Figure 7.17 illustrates various configurations using the DR as a series feedback element. The configurations shown in Figs 7.17(a)–7.17(c) use the resonator at one terminal pair, whereas the configuration shown in Fig. 7.17(d) uses it as a series feedback element at two terminal pairs of the transistors. As an example we shall now discuss the step-by-step procedure for the realization of the configuration shown in Fig. 7.17(a) using the three-port S parameters of a typical 0.5 μm GaAs MESFET. Comments on the design of other configurations are made where necessary.

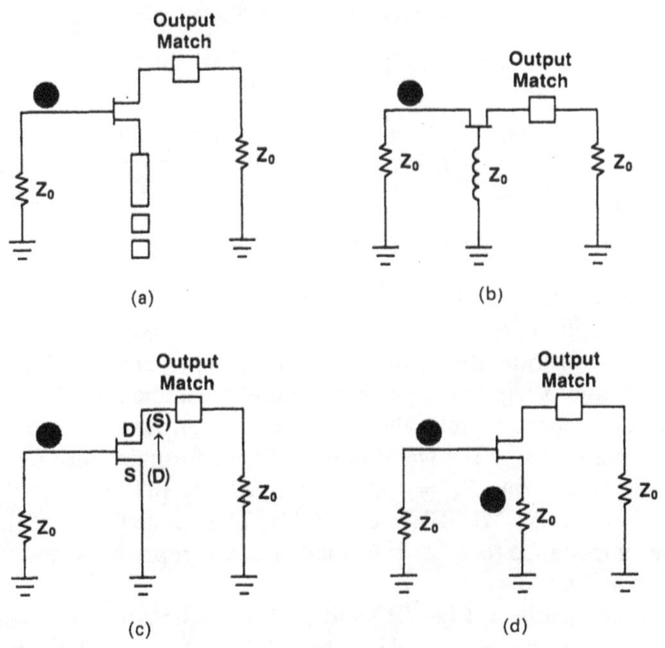

Fig. 7.17 Configurations for series feedback TDROs.

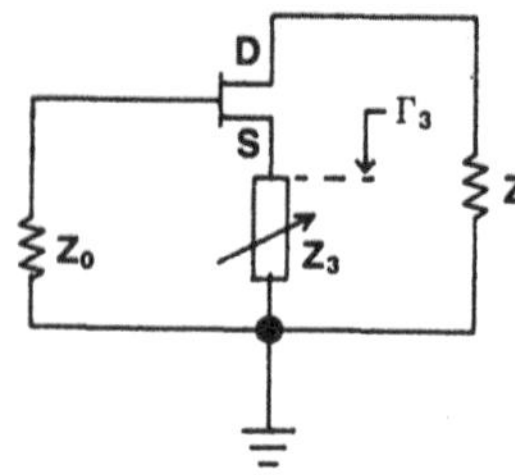

Fig. 7.18 Determination of Z_3.

As a first step in the design procedure we have to determine the impedance Z_3 in Fig. 7.18. When Z_3 is connected to the source terminal, the reflection coefficient Γ_3 becomes

$$\Gamma_3 = \frac{Z_3 - Z_0}{Z_3 + Z_0} \tag{7.55}$$

Substituting in eqn (7.3) we obtain

$$b_1 = S_{11}a_1 + S_{12}a_2 + S_{13}b_3\Gamma_3 \tag{7.56}$$

$$b_2 = S_{21}a_1 + S_{22}a_2 + S_{23}b_3\Gamma_3 \tag{7.57}$$

$$b_3 = S_{31}a_1 + S_{32}a_2 + S_{33}b_3\Gamma_3 \tag{7.58}$$

When b_3 is eliminated from eqns (7.56)–(7.58), the reduced two-port S matrix is given by

$$S_T = \begin{bmatrix} S_{11} + \dfrac{S_{31}S_{13}\Gamma_3}{1 - S_{33}\Gamma_3} & S_{12} + \dfrac{S_{13}S_{32}\Gamma_3}{1 - S_{33}\Gamma_3} \\ S_{21} + \dfrac{S_{31}S_{23}\Gamma_3}{1 - S_{33}\Gamma_3} & S_{22} + \dfrac{S_{23}S_{32}\Gamma_3}{1 - S_{33}\Gamma_3} \end{bmatrix} \tag{7.59}$$

The aim is to determine the value of the series feedback impedance Z_3 for which the modulus of S_{11T} and S_{22T} will be greater than unity and hence create instability in the transistor. For the example considered in Fig. 7.17(a) the open-circuit microstrip line represents a purely reactive impedance. The $|\Gamma_3| = 1$ plane determines a circle when mapped in the input and output reflection coefficient planes using the well-known bilinear formulation [7.21]. Figures 7.19 and 7.20 show the $|\Gamma_3| = 1$ plane mapped into the S_{11T} and S_{22T} planes for the 0.5 μm GaAs MESFET at 10 GHz. The shaded area represents the inductive impedance and the unshaded area represents the capacitive impedance in the source.

It can be seen from Figs 7.19 and 7.20 that a negative reactance greater than -30j Ω is required to make both S_{11T} and S_{22T} greater than unity. A value of -159j Ω is selected which can be realized by either an open-circuit

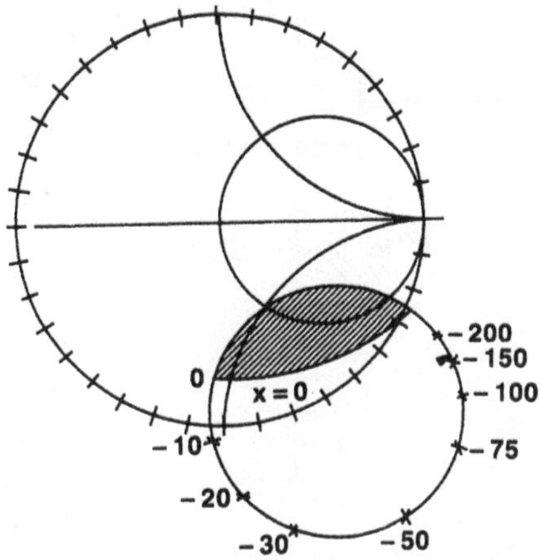

Fig. 7.19 Mapping of $|\Gamma_3| = 1$ in S_{11T}.

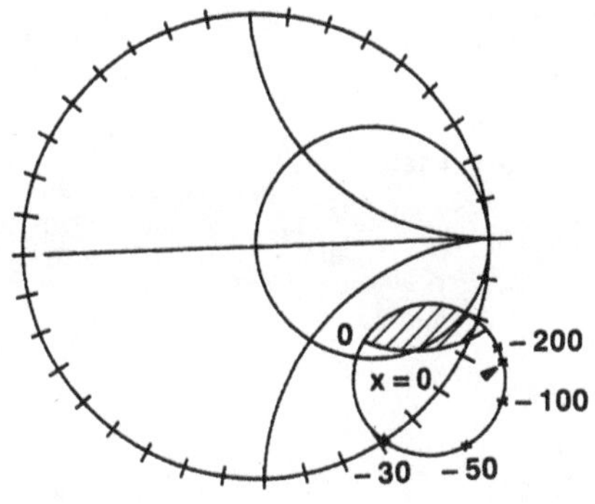

Fig. 7.20 Mapping of $|\Gamma_3| = 1$ in S_{22T}.

stub of length $0.048\lambda_g$ or a 0.1 pF capacitor. The reduced two-port S matrix is now given by

$$\boldsymbol{S}_T = \begin{bmatrix} 1.34\angle -33.5^\circ & 0.49\angle 70.4^\circ \\ 0.49\angle 156^\circ & 1.16\angle -31.9^\circ \end{bmatrix} \tag{7.60}$$

A DR coupled to a microstrip line can also be used as impedance Z_3 in Fig. 7.18 as shown in Fig. 7.17(d). In this case the reflection coefficient Γ_3 is a function of the coupling coefficient β_3 and the distance θ_3 between the transistor plane and the resonator plane as discussed in Section 7.5.1.3. By using the matrix coefficients from eqn (7.59), the Γ_3 plane in terms of β_3 and θ_3 can be mapped into any S parameter of the reduced two-port. Figure 7.21 shows, for example, mapping of the DR reflection coefficient plane into all four S parameters of an X band MESFET at 8 GHz [7.22]. The nonconcentric circles shown in Fig. 7.21 are those for the constant coupling coefficient β_3

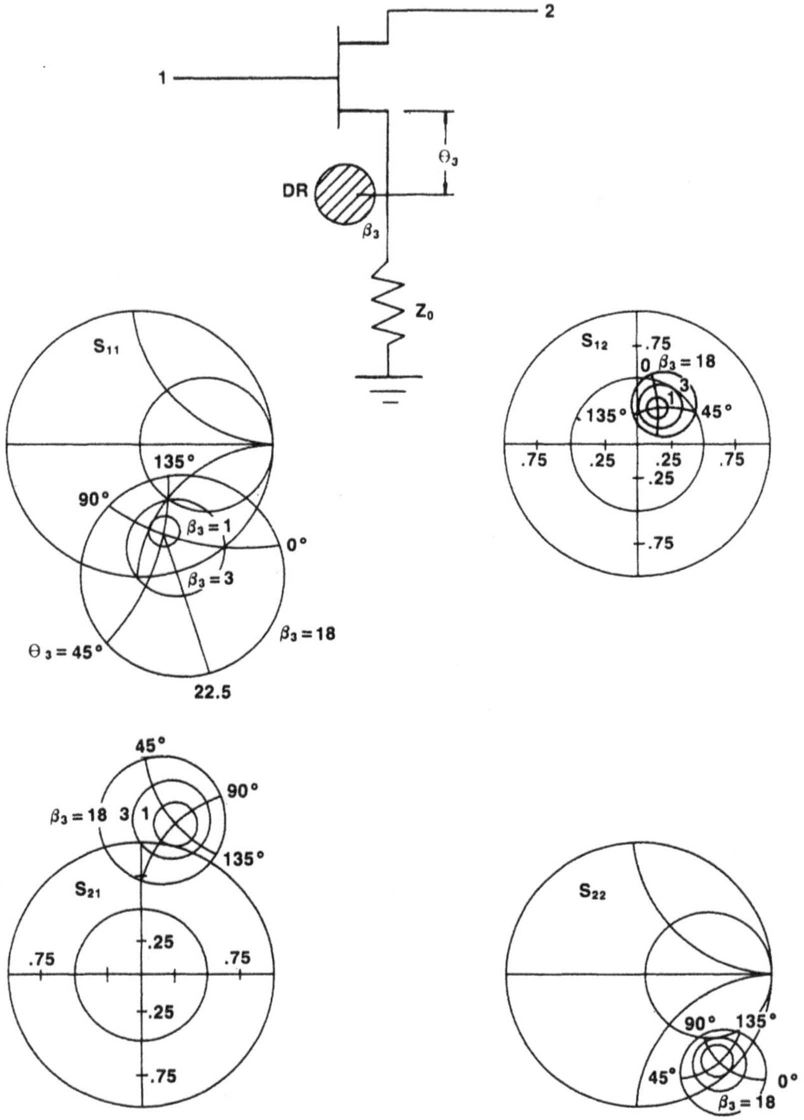

Fig. 7.21 A dielectric resonator as a series feedback element in the source of a MESFET oscillator.

(proportional to $|\Gamma_3|$), and the radial arcs are those for the constant transmission electrical line length θ_3 (proportional to arg Γ_3). The relation between Γ_3, β_3, θ_3 and frequency is given by

$$\Gamma_3 = \frac{\beta_3}{\{(1+\beta_3)^2+\Delta^2\}^{1/2}} \exp\left[-2j\left\{\theta_3+\tan^{-1}\left(\frac{\Delta}{1+\beta_3}\right)\right\}\right] \tag{7.61}$$

where $\Delta = (f-f_0)/f_0$. Figure 7.21 can be used to determine where the DR should be located in order to create the desired instability in the transistor.

Continuing the TDRO design example (Fig. 7.17(a)), we have already determined the impedance Z_3 and the resulting two-port S parameters are given by eqn (7.60). In the second step of the procedure we shall use the relation

$$\Gamma_d = S_{22T} + \frac{S_{12T}S_{21T}\Gamma_1}{1 - S_{11T}\Gamma_1} \tag{7.62}$$

to determine the value of the reflection coefficient Γ_1 which maximizes the reflection coefficient Γ_d at the drain port (Fig. 7.22). In this case the reflection coefficient Γ_1 is realized by a DR coupled to microstrip line and is characterized by the coupling coefficient β_1 and the distance θ_1 between the transistor and the resonator plane.

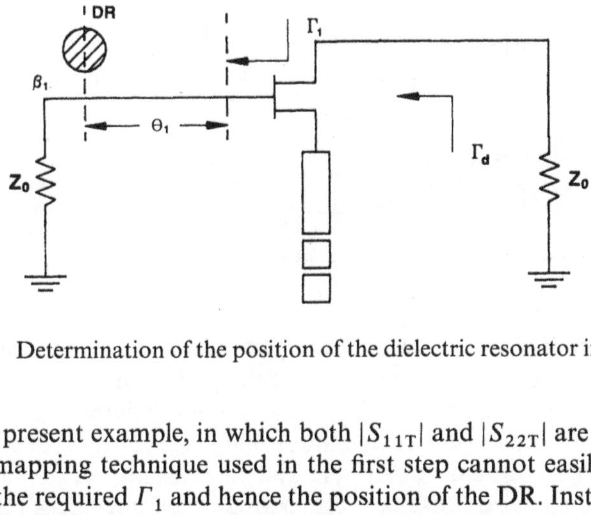

Fig. 7.22 Determination of the position of the dielectric resonator in the gate.

In the present example, in which both $|S_{11T}|$ and $|S_{22T}|$ are greater than unity, the mapping technique used in the first step cannot easily be used to determine the required Γ_1 and hence the position of the DR. Instead, we shall use the constant-reflection-coefficient circles approach, in which the locus of the magnitude $|\Gamma_d|$ of the constant reflection coefficient is drawn on the reflection coefficient plane Γ_1. From eqn (7.62) the radius R and center Ω of the constant-reflection-coefficient circles can be determined as follows:

$$R = \frac{|\Gamma_d S_{12T} S_{21T}|}{|\Gamma_d|^2 |S_{11T}|^2 - |\Delta|^2} \tag{7.63}$$

$$\Omega = \frac{S^*_{11T}(|\Gamma_d|^2 - |S_{22T}|^2)^2 + S_{22T}S^*_{12T}S^*_{21T}}{|\Gamma_d|^2 |S_{11T}|^2 - |\Delta|^2} \tag{7.64}$$

where

$$\Delta = S_{11T}S_{22T} - S_{12T}S_{21T} \tag{7.65}$$

Figure 7.23 shows various circles of constant $|\Gamma_d|$ on the $\Gamma_1(\beta_1, \theta_1)$ plane for the example under consideration. The values of β_1 and θ_1 can now be determined for a high value of $|\Gamma_d|$ ($|\Gamma_d| \gg 1$).

In the first and second steps small-signal S parameters were used to determine the impedances Z_1 and Z_3 to be connected at the gate and source ports respectively in order to achieve a high value of the reflection coefficient at the drain port. A number of approaches can be used to determine the large-signal load impedance Z_2 in Fig. 7.17(a). The two most frequently used, in which large-signal operation is assumed, are the device-line approach and the load-pull approach.

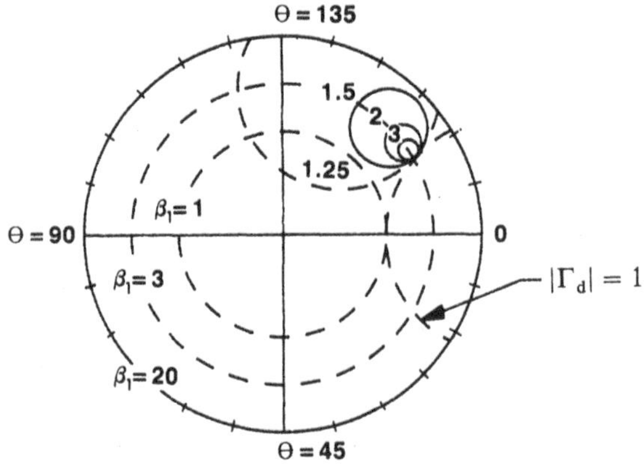

Fig. 7.23 Constant $|\Gamma_d|$ circles in the dielectric resonator reflection coefficient Γ_1 (β_1, θ_1) plane.

The device-line approach [7.23] is based on measurement of the inverse reflection coefficient of the one-port and the power added for varying input power levels. The load impedance for maximum oscillator output power can be determined from the device line thus obtained. This approach, however, is limited by the requirement that the source resistance is greater than the modulus of the device resistance. Otherwise, oscillation takes place and the device line cannot be measured.

The test set-up for measurement using the load-pull approach [7.24] is shown in Fig. 7.24. The oscillator acts as the r.f. power source for the system. The drain port of the transistor circuit (with optimized Z_1 and Z_3 connected) is attached to the load-pull measurement system at the input port of the reflectometer through a 50 Ω line and direct current is applied. The impedance displayed on the polar display will be the impedance presented to the output of the oscillator device. By using the output tuner, contours of constant output power can then be drawn on a Smith chart using the power readings from the output power meter and an x–y recorder connected to the polar display. Typical load-pull data are shown in Fig. 7.25. This load impedance chart can be used to design the output circuit for the transistor.

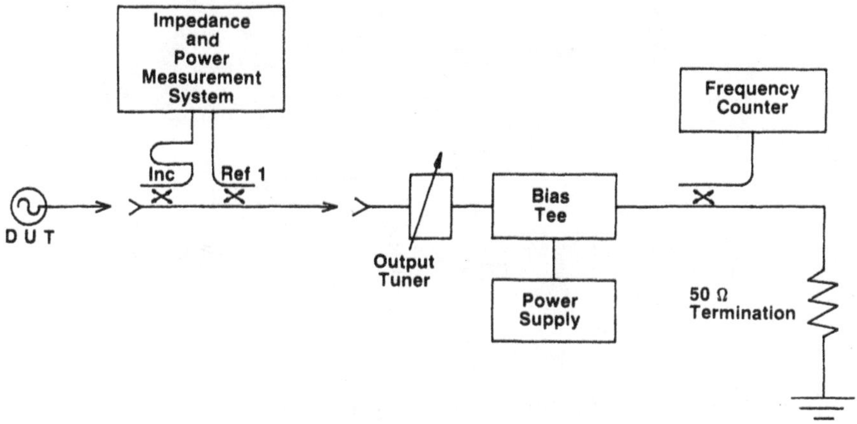

Fig. 7.24 Load-pull set-up for a one-port oscillator: DUT, device under test.

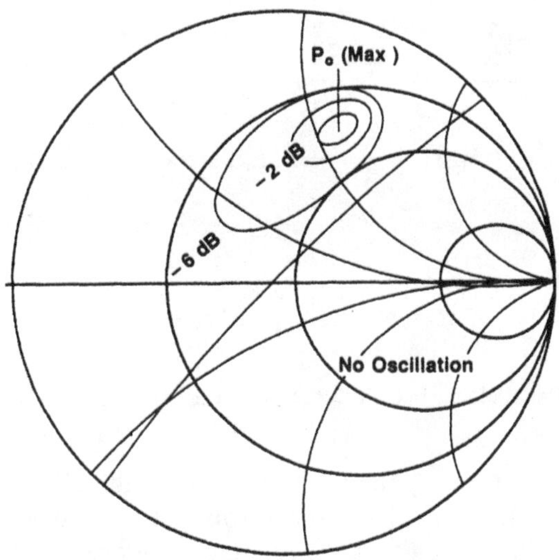

Fig. 7.25 Typical load-pull data (after ref. 7.24).

An alternative way of realizing a stable oscillator is to use the DR simultaneously coupled to two microstrip lines as a parallel feedback element for a transistor. In this case the transistor can be used as a two-port or a three-port device as shown in Figs 7.26 and 7.27 respectively.

In Fig. 7.26 the transistor is treated as a two-port [7.25]. In this case the input and output matching circuits for a common-source transistor are designed for maximum transducer gain at oscillation frequency f_0. Highly selective positive feedback between the input and the output can be used to create stable oscillations. This is achieved by feeding back part of the output

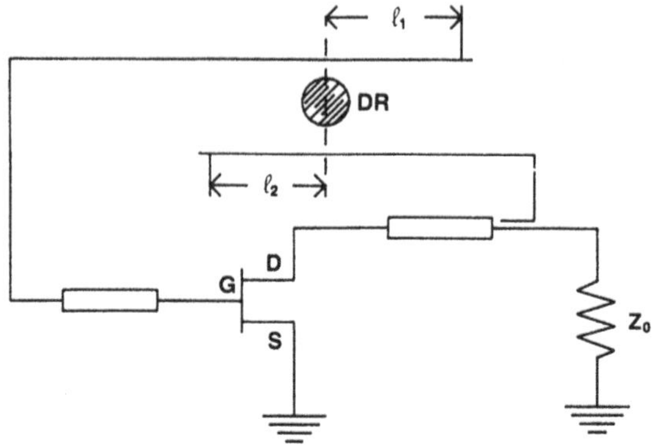

Fig. 7.26 Parallel feedback TDRO using the transistor as a two-port.

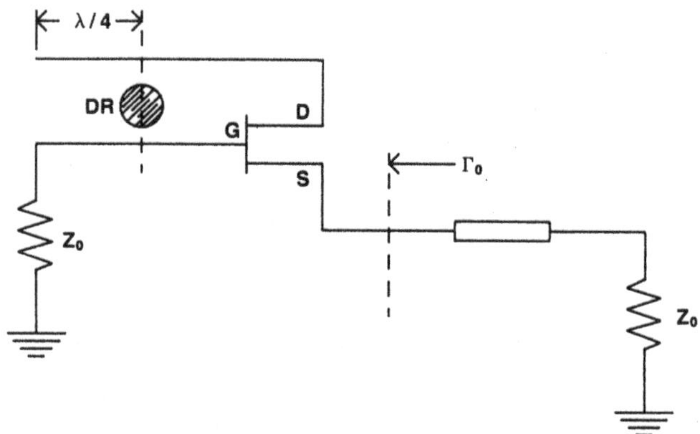

Fig. 7.27 Parallel feedback TDRO using the transistor as a three-port.

signal into the input through the DR transmission filter. The lengths l_1 and l_2 are adjusted to achieve a phase shift around the loop, consisting of the amplifier and feedback circuit, that is equal to an integer multiple of 2π rad at f_0, i.e.

$$\phi_A + \phi_R + \phi_C = 2\pi k \qquad k = 0, 1, 2 \tag{7.66}$$

where ϕ_A, ϕ_R and ϕ_C are respectively the insertion phases of the amplifier, the resonator and the remaining part of the feedback circuit at f_0. The other condition for oscillation to take place is that the open-loop small-signal gain must exceed unity at f_0, i.e.

$$G_A - L_R - L_C > 0\,\text{dB} \tag{7.67}$$

where G_A, L_R and L_C are respectively the amplifier gain, the resonator filter loss and the loss in the other feedback components (in decibels). The relations needed to determine the resonator insertion phase ϕ_R and the insertion loss L_R can be determined from

$$\phi_R = \tan^{-1}\left(\frac{-j2Q_u\delta}{1+\beta_1+\beta_2}\right) \tag{7.68}$$

and

$$L_R = 10\log\left\{\frac{(1+\beta_1+\beta_2)^2}{4\beta_1\beta_2}\right\} \tag{7.69}$$

The extent of the inequality given in (7.67), and hence the amplifier gain compression in the steady state oscillation condition, affects the output power as well as the FM noise performance of the oscillator. Excessive gain compression can adversely affect the oscillator noise owing to an increased amplifier noise figure and AM–PM conversion.

Figure 7.27 shows another configuration for a parallel feedback DRO [7.5, 7.26]. In this case the DR transmission filter is coupled between the two terminals of the transistor and the output is taken from the third terminal. This configuration can be analyzed as a two-port containing a DR as a parallel feedback network connected to a three-port device, the transistor [7.5]. The feedback parameters, i.e. the position and coupling of the DR to the microstrip lines, can be determined to maximize the reflection coefficient (greater than unity) at the output terminal. The output matching circuit can now be designed using the load-pull approach described above.

The series and parallel feedback approaches described above are frequently used for designing TDROs. The series feedback type has the advantage of ease of DR alignment owing to its coupling to microstrip on one side only compared with the parallel feedback type in which the resonator is simultaneously coupled to both sides. In addition, in the latter case the two coupling coefficients cannot be adjusted independently while aligning the oscillator. In the parallel feedback configuration the use of a high gain amplifier in the feedback loop can allow significant decoupling of the resonator from the microstrip lines, resulting in a higher loaded Q factor value for the DR and hence an oscillator with a lower phase noise [7.27].

7.5.3 Temperature stability of transistor dielectric resonator oscillators

Microwave sources with high temperature stability have been realized previously with Invar cavities or by phase locking the source to a frequency-multiplied v.h.f. crystal oscillator in a constant-temperature oven. These approaches are cumbersome and expensive. With the advent of high performance DRs, TDROs now offer a compact, elegant and inexpensive way of achieving high stabilities.

A free-running transistor oscillator is known to have a negative temperature coefficient. A DR with a positive temperature coefficient is thus required to compensate the frequency drift with temperature. At present the temperature coefficient of the DR can be controlled by varying its composition. DRs with T_f values between +9 and −9 ppm K^{-1} are commercially available. The temperature stability of a TDRO can be determined analytically in terms of the coupling coefficient β, the Q factor and the rate of change of the transistor reflection phase with temperature [7.28].

The active device shown in Fig. 7.28, including the circuitry, can be presented at its gate port as a temperature- and power-dependent reflection coefficient:

$$\Gamma_g(T,P) = |\Gamma_g(T,P)| \exp\{j\phi_g(T,P)\} \tag{7.70}$$

with

$$|\Gamma_g(T,P)| > 1 \tag{7.71}$$

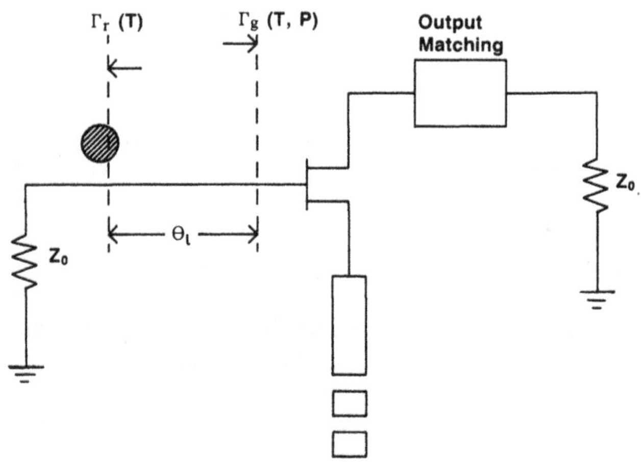

Fig. 7.28 Circuit used for analysis of temperature variation.

The stabilization circuit, which contains a 50 Ω microstrip line loaded by a 50 Ω damping resistor and coupled with the DRs, can be represented at the resonator plane as a temperature-dependent reflection factor:

$$\Gamma_r(T) = \frac{\beta}{1+\beta+j2Q_u\delta(T)} \tag{7.72}$$

where

$$\delta(T) = \frac{f-f_0(T)}{f_0(T)} \tag{7.73}$$

and β is the coupling coefficient between the resonator and the microstrip line. Applying the oscillation condition in the plane of the gate port, we obtain

$$|\Gamma_r(T)|\exp(-j2\theta_1) = \frac{1}{|\Gamma_g(T,P)|}\exp\{-j\phi_g(T,P)\} \tag{7.74}$$

where θ_1 is the electrical length of microstrip transmission line inserted to adjust the phase for the oscillation condition. When this equation is split into real and imaginary parts, the following relationships are obtained for oscillation frequency and power:

$$(\beta+1)\tan\{\phi_g(T,P)-2\theta_1\} = \frac{f}{f_0(T)}-1 \tag{7.75}$$

$$|\Gamma_g(T,P)| = \frac{1+\beta}{\beta} \tag{7.76}$$

$\Gamma_g(T,P)$ is known to be a monotonic decreasing function of temperature and power. A decrease in Γ_g at higher temperatures will be compensated automatically by a decrease in the output power P, thus keeping $|\Gamma_g|$ constant.

To determine the frequency stability with temperature eqn (7.75) is differentiated with respect to temperature and after some approximations this gives

$$\frac{\mathrm{d}f}{f\mathrm{d}T} \approx \frac{\mathrm{d}f_0}{f_0\mathrm{d}T} + \frac{\beta+1}{2Q_u}\frac{\partial\phi_g}{\partial T} \tag{7.77}$$

where $\partial\phi_g/\partial T$ represents the phase drift of the active circuit with temperature. Equation (7.77) can be used to help design temperature-stable DROs. $\partial\phi_g/\partial T$ is known to decrease linearly with temperature, thus necessitating a positive resonator temperature coefficient $\mathrm{d}f_0/f_0\mathrm{d}T$ which varies linearly in the desired temperature range. Equation (7.77) can be rewritten as

$$\tau_f \approx \tau_{f0} + \frac{1}{2Q_l}\frac{\partial\phi_g}{\partial T} \tag{7.78}$$

where τ_f and τ_{f0} are the temperature coefficients of the oscillator and resonator respectively and Q_l is the loaded Q of the resonator. For temperature-stable operation τ_f must be zero, which leads to the requirement

$$\tau_{f0} = -\frac{1}{2Q_l}\frac{\partial\phi_g}{\partial T} \tag{7.79}$$

This relation shows us how to make the oscillation frequency independent of temperature. For a given device, the value of $\partial\phi_g/\partial T$ is fixed but Q_l can be selected by adjusting the coupling coefficient between the DR and the microstrip line. Various combinations of Q_l and τ_{f0} can be tried for optimizing other properties of the DRO such as tunability or power output. When this approach and a suitable DR material have been used, temperature stabilities of 0.1 ppm K^{-1} have been reported at temperatures between 253 and 353 K.

A digital compensation technique can also be used to realize TDROs with high temperature stability [7.29]. In this case a temperature sensor is mounted in the oscillator to detect the temperature changes. These data are digitized and fed to an erasable programmable read-only memory (EPROM) preprogrammed with the temperature characteristics of the oscillator. The look-up table in the ROM provides the necessary digital temperature correction word which is converted to an analog signal (with appropriate synchronization between analog–digital and digital–analog conversion) and applied to the TDRO for frequency correction as shown in the block diagram in Fig. 7.29. The correction signal can be applied to the varactor in the case of a varactor-tuned TDRO [7.29] or can be used to control the phase shift in the feedback loop in the case of a parallel feedback TDRO. A frequency drift of better than ±20 ppm over the temperature range 218–358 K has been achieved using this digital compensation technique.

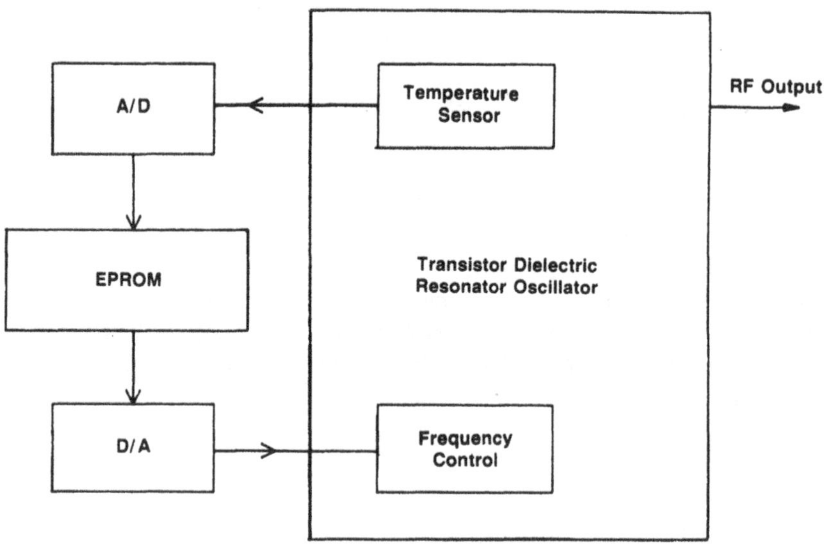

Fig. 7.29 Temperature stabilization of a TDRO using digital techniques.

The temperature stability of a TDRO can also be enhanced significantly by inserting the oscillator package in a temperature-stabilized oven. The package temperature can be maintained within 5 K by using a heater element, a quick-response thermistor and an associated control circuit. The oscillator package itself is isolated from the external environment by using an outer package. The two packages are thermally connected through stand-offs which are determined by taking into account the d.c. dissipation in the oscillator and the desired temperature of the package. The package temperature is generally controlled at 5–10 K above the maximum temperature specified for the DRO. A total frequency stability of better than ±5 ppm can be obtained in the temperature range 233–343 K. This device is suitable for use as a local oscillator in radio communication and test equipment in place of certain phase-locked oscillators.

7.5.4 Tuning of transistor dielectric resonator oscillators

TDROs are basically fixed-frequency oscillators in which the frequency is determined by the permittivity of the resonator material, the dimensions of the resonator and the shielding conditions, as discussed in Section 7.5.1. However, the oscillation frequency can be tuned over a narrow frequency range using different approaches depending on requirements. The frequency tuning of the TDRO can be accomplished mechanically or electrically, as discussed below.

Figure 7.30 shows the configuration used for mechanical frequency tuning of the TDRO. Use is made of the fact that the resonant frequency of the DR is highly sensitive to the shielding, i.e. to the proximity of the ground plane. A tuning screw is inserted from the top cover of the package, immediately above the DR. Increasing the tuning-screw depth d increases the resonant frequency of the DR in the most frequently used $TE_{01\delta}$ mode. Care should be taken to keep the distance h between the resonator and the tuning screw at at least half the resonator height in order not to degrade the Q factor of the DR. A mechanical frequency tuning range of ±1% can be obtained in a TDRO without noticeably affecting the FM noise, output power and phase noise characteristics.

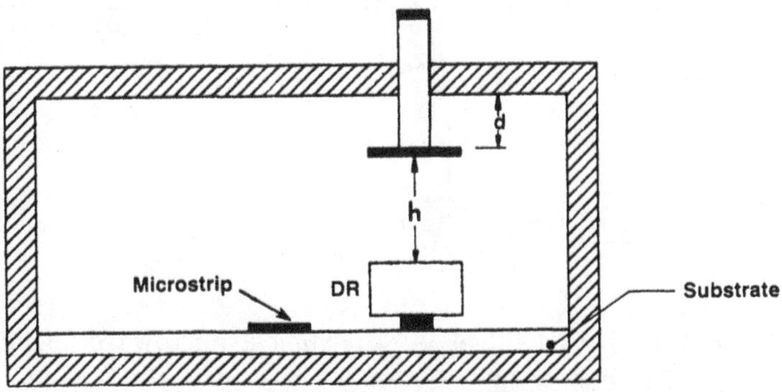

Fig. 7.30 Mechanical tuning of a dielectric resonator.

The frequency of the TDRO can be tuned electrically by a number of different approaches such as varactor tuning, ferrite tuning, bias tuning and optical tuning. Electrical tuning can be effected over a very small bandwidth without significantly affecting the oscillator performance. This fast method can be used for various applications such as digital temperature compensation, low deviation frequency-modulated sources, phase locking etc. A brief description of the various approaches is given below.

In the varactor-tuned TDRO, a varactor associated with a microstrip line is made to resonate around the DR frequency. This resonant circuit is electromagnetically coupled to the DR, so that a pair of mutually coupled resonant circuits is formed. The resonant frequency of the DR, which is coupled to the varactor–microstrip on one side and a 50 Ω microstrip line on the other, can now be tuned by varying the varactor capacitance with the bias voltage. Figure 7.31 shows a typical configuration for coupling the varactor and the DR. Closer coupling between the DR and the varactor will result in a larger frequency control at the cost of a reduced Q factor and consequently FM noise. Figure 7.32 shows the effect of varactor tuning on the FM noise of a low noise TDRO for varactor coupling adjusted for 0.1% frequency control at 11 GHz [7.30]. By using an appropriate varactor diode and a tighter coupling configuration, an electronic frequency control of greater than 1% can be achieved [7.30]. A tuning bandwidth of 0.75% has been reported for another configuration with two varactors on a quartz spacer placed directly above the resonator [7.31].

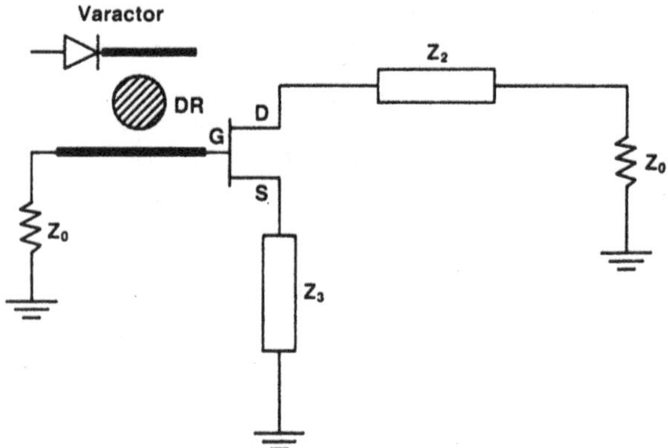

Fig. 7.31 Varactor tuning of a dielectric resonator.

The DR can also be tuned by attaching a microwave ferrite to the resonator and applying a magnetic field to it. The magnetic field controls the magnetic properties of the ferrite and hence the field distributions in and around the DR resulting in a shift in the resonant frequency. Tuning bandwidths of the order of 0.5% [7.32] and 1% [7.31] have been reported. However, this method cannot be used effectively in practice owing to the slow tuning speed, the large size and bulk of the electromagnet required to provide

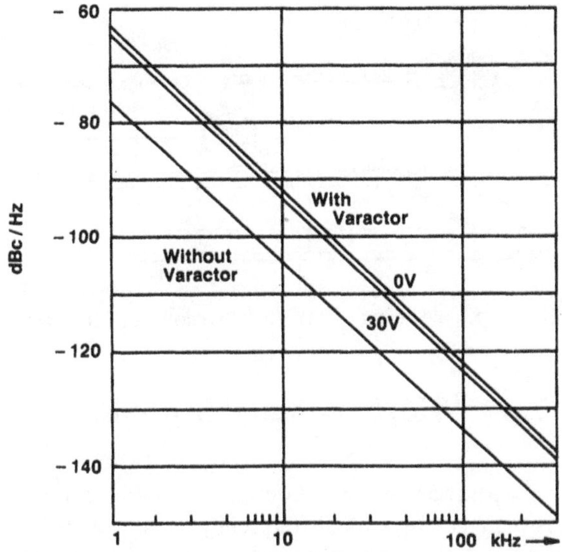

Fig. 7.32 Effect of varactor tuning on FM noise in a dielectric resonator oscillator (after ref. 7.2).

the variable magnetic field and the excessive power consumption of the driving circuit.

Another method of electronic tuning is to use the transistor bias voltage. The frequency of any oscillator is known to be sensitive to the bias voltage, described generally by the pushing figure. Unfortunately the change in bias voltage also affects the output power, thus making it difficult to use for FM purposes. However, the bias circuit can be designed in such a way as to minimize the variation in output power with bias voltage. A frequency-tuning range of 4.5 MHz at 10 GHz with a variation of less than 1 dB in the oscillator power output has been reported using such a technique [7.33].

TDROs have an inherently high Q factor and hence a low pushing figure. This limits the bias-tuned frequency range which can only be increased by intentionally reducing the oscillator Q at the cost of degradation of other oscillator characteristics such as FM noise, temperature stability etc.

The optical control of microwave devices and subsystems is a rapidly growing field of research. The resonant frequency of the DR used in the TDRO can be optically modulated and tuned as shown in Fig. 7.33. A photosensitive material such as high resistivity silicon is placed directly on the dielectric resonator. Light from a laser or a light-emitting diode (LED) is passed through an optical fiber to illuminate the photosensitive material, changing its conductivity and perturbing the electromagnetic field in and around the resonator. This perturbation results in a shift in the center frequency of the TDRO. Tuning bandwidths of better than 0.1% have been achieved at X band using this technique [7.34, 7.35].

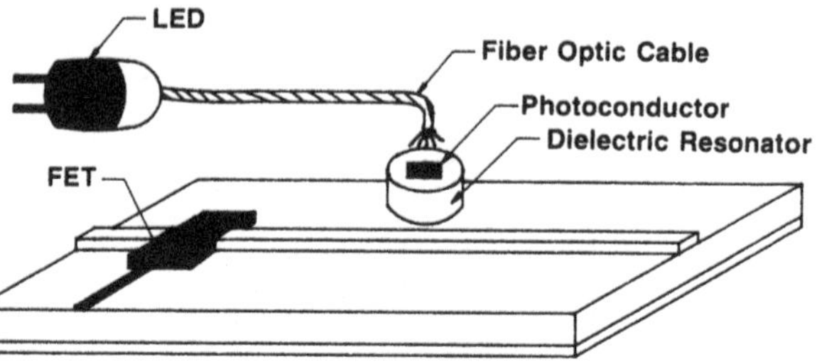

Fig. 7.33 Optical tuning of a dielectric resonator (after ref. 7.35).

7.6 Wideband tunable oscillators

The analysis and design of wideband tunable oscillators are more complex than is the case for a fixed-frequency oscillator. It is important that the active element, e.g. the transistor, can generate a negative resistance or a reflection coefficient greater than unity over the desired bandwidth with the help of suitable series or parallel feedback. The frequency-determining element, e.g. the varactor diode or YIG sphere, should be able to tune with voltage or current such that the oscillation conditions are satisfied over the desired band [7.36]. An analysis using small-signal measured S parameters is sufficient to realize the negative resistance over the desired band. However, in order to calculate and optimize the power output it is necessary to use one of the nonlinear techniques, e.g. the nonlinear model, the device-line approach or the load-pull method.

In the generalized oscillator configuration shown in Fig. 7.8 the load impedance should be the only resistive element in order to obtain a maximum negative resistance in the circuit. In other words the impedances connected to terminals other than the output terminals should be low loss reactive circuits. To realize an electronically tunable oscillator it is also necessary that one of these reactances should be electronically tunable. Although any of the transistor terminals can be used to connect the output load, the MESFET drain and the bipolar collector are generally used. The electronic tuning element can be placed in either of the other two terminals. The tunable oscillator can now be represented as shown in Fig. 7.34.

The first step in the oscillator design consists in determining $Z_2(\omega)$, $X_3(\omega)$ and R_2 so that the tuning element looks into an impedance with a negative real part in the desired frequency band. In wideband tunable oscillators the emphasis is more on maximum frequency band coverage than on the power output. This calls for a design approach different from that used for fixed-frequency oscillators. Equation (7.59) has already given the two-port S matrix in terms of the three-port S parameters. When a load (with a reflection

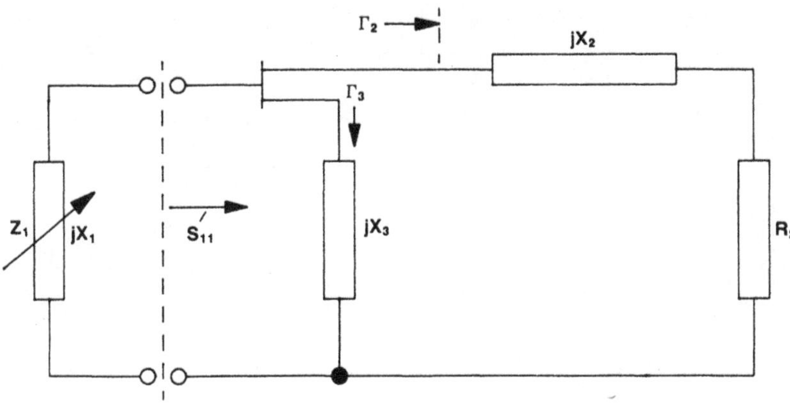

Fig. 7.34 Equivalent circuit representation of a tunable oscillator.

coefficient of Γ_2) is connected to the output terminal, the two-port can be converted to a one-port by using the expression

$$S'_{11} = S_{11\mathrm{T}} + \frac{S_{12\mathrm{T}}S_{21\mathrm{T}}}{1/\Gamma_2 - S_{22\mathrm{T}}} \tag{7.80}$$

where the $S_{ij\mathrm{T}}$ are the two-port S parameters given by eqn (7.59). The above relation can be represented in terms of the three-port S parameters Γ_2 and Γ_3:

$$S'_{11} = S_{11} + v + \\ + \Gamma_2 \frac{S_{12}S_{21}(1 - S_{33}\Gamma_3) + S_{12}S_{31}S_{23}\Gamma_3 + S_{21}S_{13}S_{32}\Gamma_3 + vS_{23}S_{32}\Gamma_3}{(1 - S_{33}\Gamma_3)(1 - S_{22}\Gamma_2) - S_{23}S_{32}\Gamma_2\Gamma_3} \tag{7.81}$$

where

$$v = \frac{S_{31}S_{13}\Gamma_3}{1 - S_{33}\Gamma_3} \tag{7.82}$$

This relation can be used to analyze the negative resistance in a given device for any combination of Γ_2 and Γ_3. The same relation can also be used to optimize the impedances Γ_2 and Γ_3 for negative-resistance bandwidth. Figure 7.35 shows the reflection gain of a typical 0.5 μm GaAs MESFET optimized over 6–18 GHz. YIG spheres and varactors are commonly used (Fig. 7.34) to realize Z_1 in wideband oscillators. We shall now discuss these oscillators before comparing their performances.

7.6.1 YIG-tuned oscillators

YIG-tuned oscillators are frequently used as wideband signal sources in

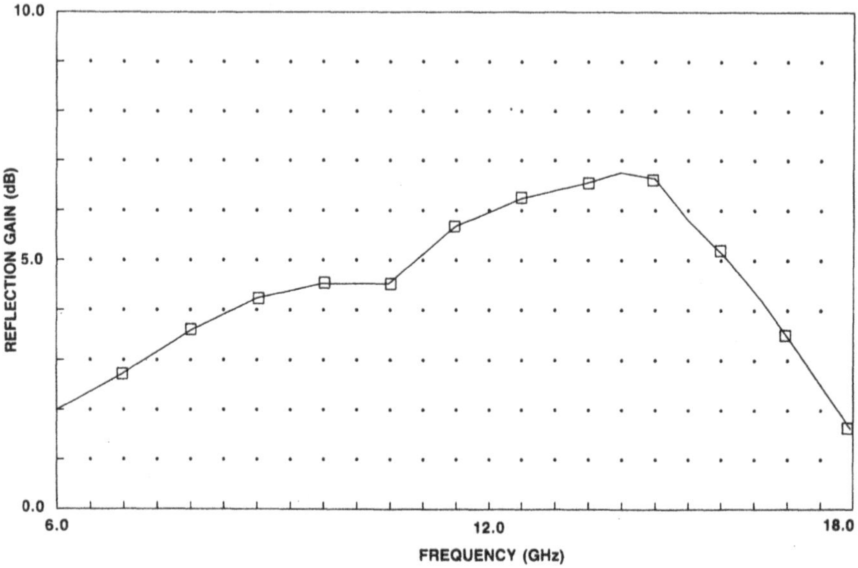

Fig. 7.35 Reflection gain of a GaAs MESFET with a gate length of 0.5 μm.

sweep generators, spectrum analyzers and electronic warfare applications. With the advent of GaAs MESFETS and microwave bipolar transistors, transistor YIG-tuned oscillators have been replacing Gunn YIG-tuned oscillators at frequencies up to 26.5 GHz. The main advantages offered by transistor oscillators over Gunn oscillators are better efficiency and operation at lower oscillation frequencies. Whereas Gunn devices do not oscillate below about 5 GHz, transistor YIG-tuned oscillators are realizable for this range.

7.6.1.1 YIG resonators

YIG resonators are high Q ferrite resonators which can be tuned over a wide band by varying the biasing d.c. magnetic field. Their high performance and convenient size for applications in MICs make them an excellent choice in a large number of commercial and military applications such as filters, multipliers, discriminators, limiters, oscillators etc. A YIG resonator makes use of ferrimagnetic resonance which, depending on the material composition, size and applied field, can be achieved from 500 MHz to 50 GHz [7.37].

Single-crystal YIG and gallium-doped YIG are part of a family of ferrites that resonate at microwave frequencies when immersed in a magnetic field. This resonance is directly proportional to the applied magnetic field and therefore linear tuning is achieved by changing the magnetic field with an electric current. The resonator consists of a YIG sphere, an electromagnet and a coupling loop. A typical YIG resonator in a MIC configuration is shown in Fig. 7.36. YIG tuning magnets are electromagnets with a single air gap. The

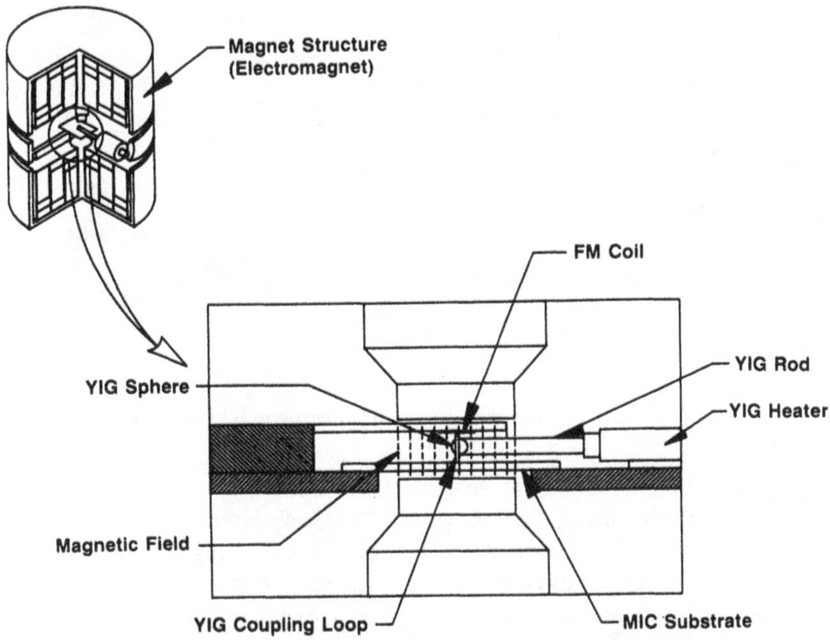

Fig. 7.36 Elements of a YIG oscillator (after ref. 7.37).

d.c. current through the two series-connected main coils provides the required d.c. magnetic field across the gap. The YIG sphere, FM coil and MIC substrate are placed between the poles of the electromagnet.

In theory an electronically tunable microwave resonant element can be formed from any ferrimagnetic material operated at resonance. YIG is preferred over other ferrite materials for the following reasons: lower losses; lower anisotropy (directional dependence of magnetic properties leading to preferred directions of magnetization); no magnetic disorders in the crystal structure; more reasonable value of the saturation magnetization $4\pi M_s$ (magnetic moment per unit volume). The most practical geometry for a YIG resonator is a sphere because it is easily oriented in the magnetic field and the resonant frequency is not strongly dependent on its orientation. The sphere is also easy to prepare with precision.

Resonant frequency The basic ferrimagnetic resonance phenomenon can be explained in terms of spinning electrons which create a new magnetic moment in each molecule of a YIG crystal (Fig. 7.37(a)). Application of an external magnetic biasing field causes these magnetic dipoles to align themselves in the direction of the field, thus producing a strong net magnetization M [7.38]. Any magnetic force at right angles to the field results in precession of the dipoles around the biasing field at a rate which depends on

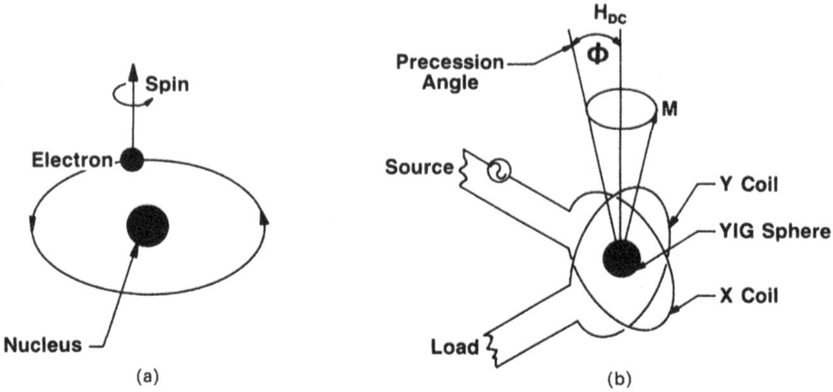

Fig. 7.37 Spin motion and loop coupling in a YIG resonator.

the strength of the magnetic biasing field and to some extent on the fundamental properties of the material [7.39]. Such lateral forces may result from r.f. magnetic fields orthogonal to the biasing magnetic field. If the frequency of these r.f. fields coincides with the natural precessional frequency of the material, a strong interaction occurs resulting in resonance phenomena.

For spherical resonators, this resonant frequency is given by

$$f_0 = \gamma(H_0 \pm H_a) \tag{7.83}$$

where H_0 is the applied biasing field measured in amperes per meter, H_a is an internal field within the crystal which is known as the anisotropy field and γ is the charge-to-mass ratio of the electron which has a value of 28 GHz T^{-1} (note that $\mu_0 = 4\pi \times 10^{-7}$ H m^{-1} is included in this value of γ). The temperature dependence of the anisotropy field can be eliminated by appropriate orientation of the crystal axis of the spherical resonator. R.f. coupling to the YIG is achieved by concentrating the r.f. magnetic field in the vicinity of the sphere with a coupling loop. Orthogonality between the biasing field and the r.f. magnetic field has to be maintained for proper coupling (Fig. 7.37(b)).

Frequency of operation and quality factor The upper frequency of YIG resonance is limited by the available magnetic field and the lower frequency limit of operation is directly proportional to the value of the saturation magnetization $4\pi M_s$, which is a measure of the net density of precessing electron spins in the material. The externally applied tuning field must be sufficient to produce alignment of all the magnetic dipoles within the crystal. A pure YIG material has a $4\pi M_s$ value of 1.78×10^{-1} T at room temperature. YIG with substantially lower saturation magnetization values (typically 2.5×10^{-2} T) can be grown by doping the crystal with gallium. However, doping increases YIG resonator losses which are expressed by the linewidth parameter ΔH.

The linewidth (in amperes per meter) is a direct measure of the unloaded Q_u of the YIG resonator and can be compared with the 3 dB bandwidth of an unloaded cavity. ΔH is related to Q_u and M_s by the following relation [7.40]:

$$Q_u = \frac{H_0 - 4\pi M_s/3}{\Delta H} \tag{7.84}$$

where H_0 is the d.c. biasing field strength at resonance. Typical values of linewidths are less than $40\ \mathrm{A\ m^{-1}}$ for pure YIG and $120\ \mathrm{A\ m^{-1}}$ for doped 4×10^{-2} T YIG. Using eqn (7.83) we have

$$Q_u = \frac{f_0 - 4\pi M_s \gamma/3}{\gamma \Delta H} \tag{7.85}$$

This expression demonstrates an interesting phenomenon in YIG, i.e. Q_u for the resonator decreases with decreasing frequency. The cut-off frequency f_c where $Q_u = 0$ is given by

$$f_c = \frac{4\pi M_s \gamma}{3} \tag{7.86}$$

For pure YIG $f_c = 1670$ MHz.

Figure 7.38 shows typical measured values of $1/Q_u$ for a YIG resonator in the frequency range 6–12 GHz [7.41].

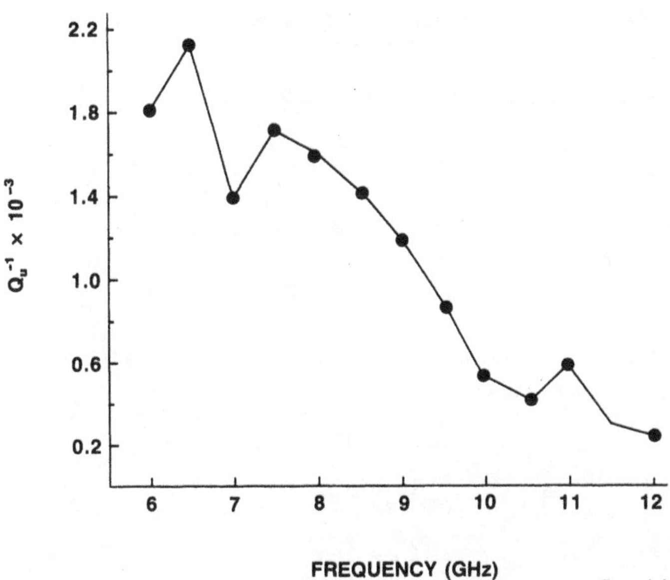

Fig. 7.38 Inverse unloaded Q for a YIG resonator (after ref. 7.41).

The equivalent circuit In order to make the discussion of the crystal properties of YIG more meaningful, it is worth examining the equivalent circuit of the YIG and its relation to the functioning of a microwave resonator. Figure 7.39(a) illustrates the important parameters of a loop-coupled YIG and Fig. 7.39(b) shows its equivalent circuit. In Fig. 7.39(a) the YIG is located at the center of a wire loop. The loop angle can be a full 360° or, as suggested in Fig. 3.39(a), less. It can also be greater than 360° in cases where particularly close coupling is required. The parallel resonant circuit is induced in series with the loop impedance by the coupling of the YIG; L_l and R_l are the inductance and resistance respectively of the coupling loop.

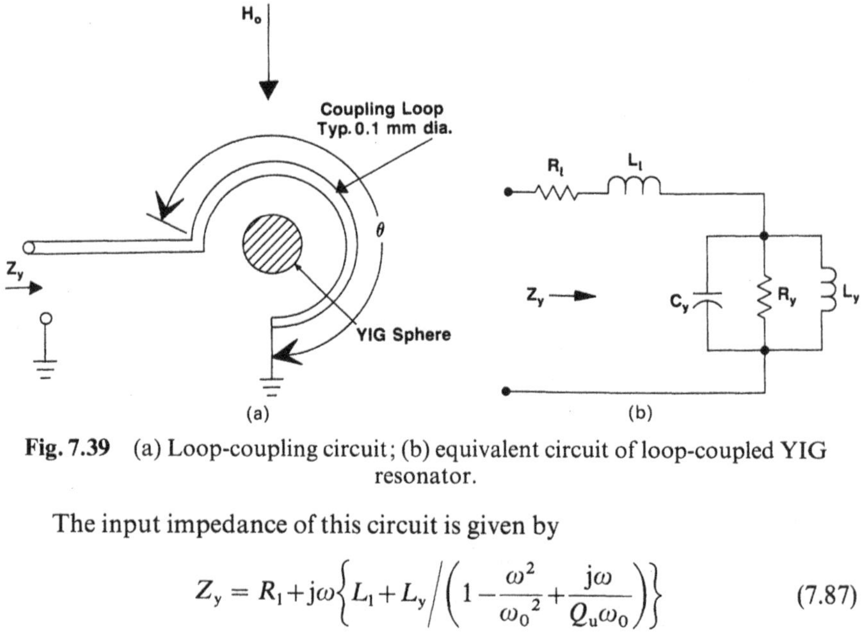

Fig. 7.39 (a) Loop-coupling circuit; (b) equivalent circuit of loop-coupled YIG resonator.

The input impedance of this circuit is given by

$$Z_y = R_l + j\omega\left\{L_l + L_y\Big/\left(1 - \frac{\omega^2}{\omega_0{}^2} + \frac{j\omega}{Q_u\omega_0}\right)\right\} \tag{7.87}$$

where ω_0 is the resonant frequency of the YIG and

$$Q_u = \frac{R_y}{\omega_0 L_y} = \frac{H_0}{\Delta H}$$

The equivalent circuit parameters are related to the basic YIG resonator and loop parameters as follows:

$$L_y = \mu_0 V_y K^2 \omega_m \tag{7.88}$$

$$R_y = \omega_0 L_y Q_u \tag{7.89}$$

$$C_y = \frac{1}{\omega_0{}^2 L_y} \tag{7.90}$$

where μ_0 is the permeability of free space, K is the coupling coefficient between the YIG sphere and the coupling loop and $\omega_m = 4\pi M_s \gamma$. The volume V_y of a YIG sphere of diameter D_y is given by

$$V_y = \frac{4\pi}{3}\left(\frac{D_y}{2}\right)^3 = 0.523\, D_y{}^3$$

The coupling coefficient K is a geometric coupling factor and is related to the loop angle θ and the loop diameter D_l as follows:

$$K = \frac{1}{D_l}\frac{\theta}{360} \tag{7.91}$$

The closeness of the coupling between the loop and the YIG can also be expressed in terms of Q_e, the external Q of the resonator. If an external resistance R_0 is connected to the terminals of the loop, Q_e is defined as

$$Q_e = \frac{R_0}{\omega_0 L_y} \tag{7.92}$$

and

$$R_y = \omega_0 L_y Q_u = \frac{R_0 Q_u}{Q_e} \tag{7.93}$$

Thus the impedance R_y of the parallel resonant circuit at resonance is proportional to two quantities which are easily obtained from a single bandwidth measurement. Similarly, the inductance L_y is related to Q_e as follows:

$$L_y = \frac{R_0}{\omega_0 Q_e} \tag{7.94}$$

YIG equivalent circuit parameters can thus be estimated rapidly by measuring Q_e. Figure 7.40 shows theoretical values of Q_e as a function of the diameter of the YIG sphere with the loop diameter as a parameter for a 360° loop and a pure YIG sphere.

Magnetic tuning circuit The design of the magnetic tuning circuit forms an important aspect of the design of any YIG device because the value of the magnetic field determines the resonance frequency of the YIG resonator. For the frequency to be maintained within 0.1% requires that the magnetic field also be maintained within 0.1%. Likewise, a 0.1% tuning linearity requires that the magnetic field is linearly related to a control input with a maximum deviation of 0.1%.

The magnetic circuit most commonly used is shown in Fig. 7.41. This structure is known as a re-entrant self-shielding magnetic circuit. It is made of soft magnetic material (e.g. Ni–Fe alloy) with a high initial permeability and is fully annealed after machining to ensure minimum hysteresis. The pole tips are

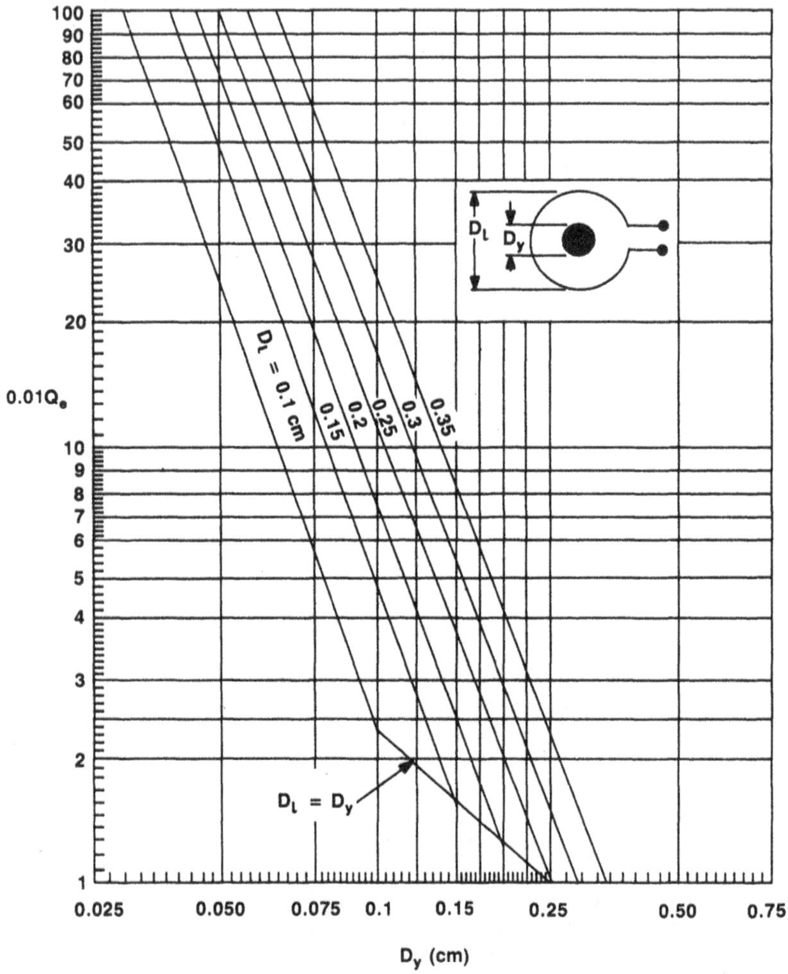

Fig. 7.40 Theoretical quality factor for a YIG resonator (after ref. 7.37).

tapered in order to improve the magnetic design and to allow the location of a larger heat sink close to the active device. Because of the high permeability of the material, the reluctance of the magnetic circuit is primarily in the air gap. Thus temperature-dependent changes in the permeability of the magnetic material have little effect on the magnetic field.

YIG magnetic tuning dissipates large d.c. power. Typically, it is about 5 W at 12 GHz and the tuning sensitivity lies between 15 and 25 MHz mA^{-1}. The latter implies that the current supply must have very low noise and ripple to minimize FM noise. For comparison, if the inherent spectral linewidth is expected to be better than 10 kHz, the magnet power ripple should not exceed 0.5 μA.

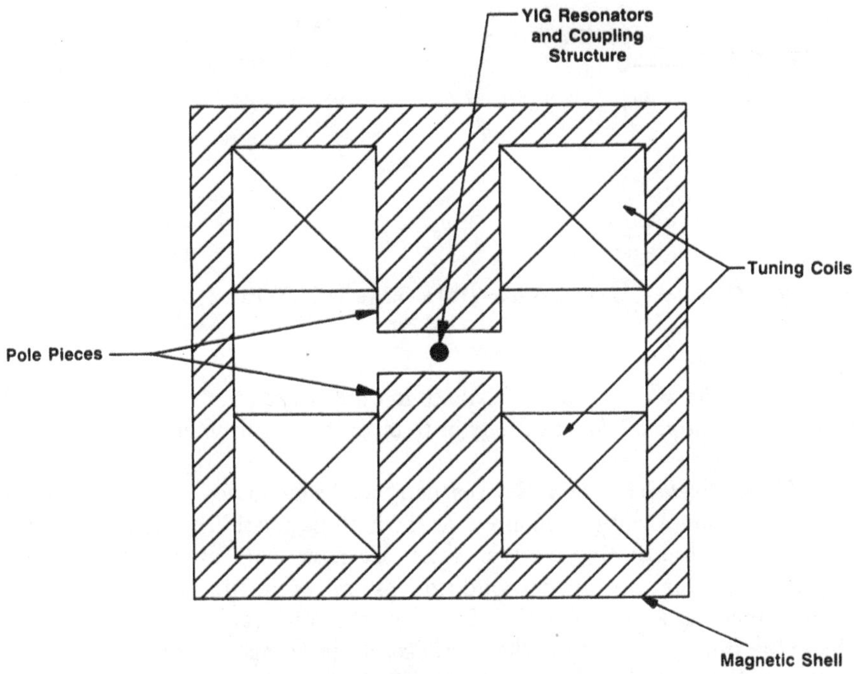

Fig. 7.41 Re-entrant magnetic structure.

In addition to the standard magnetic structure described above, two complementary types of structure are possible: permanent magnet circuits and laminated magnet circuits. The permanent magnet type is attractive because it reduces the power required to tune the resonator to the low end or middle of its tuning range and the electronic sweep adds or subtracts a magnetic field sufficient to tune the entire range of the resonator. A permanent magnet design also provides considerably faster switching speed and owing to its low drive coil dissipation, especially for a coverage of less than one octave, is particularly suitable for temperature stabilization in a thermal enclosure. The laminated magnet circuit has an even faster switching configuration. This design is particularly suitable for digital control where the current is switched from zero for each change in frequency.

7.6.1.2 Configurations of YIG-tuned oscillators

Figure 7.42 shows the most common YIG-tuned oscillator configurations using GaAs MESFETs (Fig. 7.42(a)) and bipolar transistors (Fig. 7.42(b)). Since the YIG sphere is a tunable resonant circuit with a very high Q, the oscillation phase condition $\angle S'_{11} = -\angle \Gamma$ (Fig. 7.42(a)) can easily be satisfied and stable oscillation can be achieved in the negative-resistance frequency band of the device. Under large-signal operation condition (7.34) should also

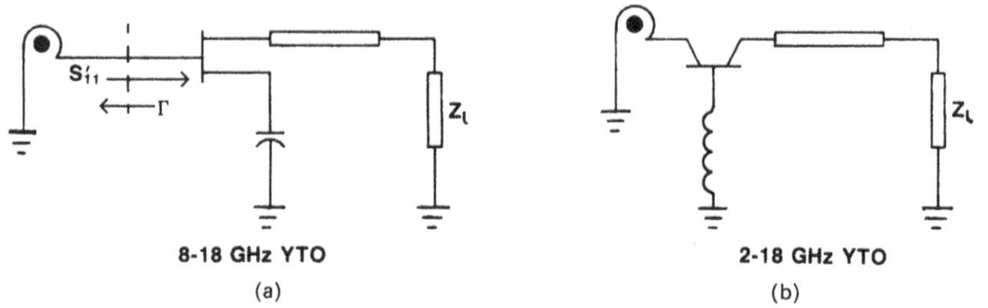

Fig. 7.42 YIG-tuned oscillators using (a) a GaAs MESFET and (b) a silicon bipolar transistor.

be satisfied. The high Q characteristic or in other words the very high value of the reactance slope $dX/d\omega$ again helps to meet this requirement.

Tuning linearity YIG devices can be designed for excellent tuning linearity. It is important to ensure that the high permeability shell and center core pieces of the biasing electromagnet are never saturated in the desired tuning range. This unsaturated condition minimizes the tuning nonlinearity in a YIG-tuned oscillator. Another factor in the YIG-tuned oscillator which affects the tuning linearity is the presence of matching circuits. It should be noted that in the oscillator design approach used here the YIG sphere does not work exactly at its resonant frequency. In fact this component serves as a variable reactance with very high Q which compensates the reactance presented by the transistor and the associated load and feedback circuits. Hence deviations from linearity of the tuning curve depend on the load circuit design. Obviously, the tuning linearity improves with increasing Q.

Parasitic and spurious oscillations YIG-tuned oscillators are susceptible to parasitic oscillations for various reasons. There are two types of parasitic or spurious oscillations: those which have a fixed frequency and do not tune with the resonator, and those which do tune and originate from the YIG sphere itself. The fixed-frequency oscillations are due to the coupling loop and may be present even when the magnetic field is not applied. These oscillations can be minimized by suitable design of the matching circuit and loop so that the fixed-frequency oscillations fall outside the desired frequency band, and parasitic oscillations should not appear at all.

Tunable parasitic oscillations due to the YIG sphere are the result of higher-order modes of the resonator. Thus, for some applied magnetic field values, the oscillator can jump from the principal resonant curve to a parasitic curve and thereby provoke the phenomenon of tuning hysteresis. It is necessary to characterize the YIG sphere loop coupled to a microstrip line before using it in an oscillator circuit. Tuning hysteresis can also result from nonmonotonic behavior of the phase of the active circuit with frequency.

Performance of YIG-tuned oscillators Initial performance results for YIG-tuned transistor oscillators were reported in 1976 [7.42] and 1977 [7.43]. Since then a number of publications dealing with these devices have appeared. An 8–18 GHz MESFET YIG-tuned oscillator in which a capacitive feedback in the source was used for the first time was described in 1976 [7.44]. A common YIG resonator coupled to both the source and the gate ports of a MESFET has been used to obtain a tuning range of 3.5–14 GHz [7.45], and two separate YIG spheres at the gate and source have been used to obtain a 2–20 GHz YIG-tuned MESFET oscillator [7.46]. More recently, Leung *et al.* [7.1] have realized a 2–18 GHz YIG-tuned bipolar transistor oscillator using a single YIG sphere in the emitter of a bipolar transistor as shown in Fig. 7.43. This high frequency performance was made possible by using silicon bipolar transistors (Avantek) with an emitter width of 0.5 μm and an emitter–emitter pitch of 2 μm. The oscillator has superior phase noise as shown in Fig. 7.44.

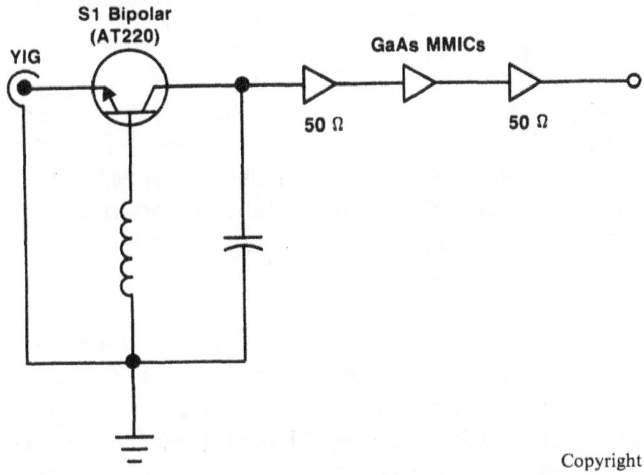

Fig. 7.43 YIG oscillator constructed by buffering the oscillating output of the transistor with a series of GaAs MMIC amplifiers (after ref. 7.1).

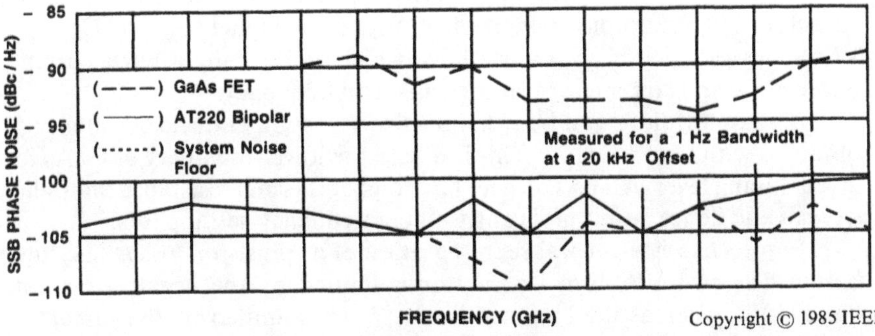

Fig. 7.44 Single-sideband (SSB) phase noise of the AT200 bipolar oscillator (after ref. 7.1).

Much higher frequencies have been reported with GaAs MESFETs. YIG MESFET oscillators with frequency ranges of 26.5–40 GHz and 33–50 GHz have been reported by Zensius *et al.* [7.47]. The magnetic field saturation limitation is likely to limit YIG transistor oscillators to a maximum of 60 GHz.

7.6.2 Varactor-tuned oscillators

A varactor is a diode which, when reverse biased, acts like a voltage-controlled capacitance. Varactors are commonly used to achieve high speed tuning over a moderate frequency range (up to an octave). In this device the applied reverse bias causes the depletion layer to extend into the semi-conductor active layer which acts as a capacitor. In the case of a flat-profile (abrupt) Schottky barrier varactor the capacitance–voltage relationship is given by

$$C_j(V) = \frac{C_j(0)}{(1 - V/V_b)^\gamma} \tag{7.95}$$

where V is the applied voltage, V_b is the built-in potential of the diode, $C_j(0)$ is the capacitance for zero bias voltage and the elastance γ is approximately 0.5. The frequency–voltage relationship for a simple resonant circuit composed entirely of an inductance and the varactor junction capacitance is of the form

$$f(V) = \frac{1}{2\pi(LC_0)^{1/2}}\left(1 - \frac{V}{V_b}\right)^{\gamma/2} \tag{7.96}$$

This simple analysis suggests that there is an inherent nonlinearity between the output frequency and the turning voltage when this type of varactor is used. However, in almost all microwave circuits the varactor is not the only capacitance in the resonator. Parasitic capacitance and purposely designed decoupling capacitors influence the varactor voltage–frequency relationship significantly. The value of γ can also be controlled to achieve better linearity. Hyperabrupt junction varactors ($1 \leqslant \gamma \leqslant 2$) are frequently used to realize VCOs with better linearity than is obtained with abrupt varactors. Hyperabrupt varactors offer lower Q but higher values of C_{max}/C_{min}. VCOs using hyperabrupt diodes have wider tuning ranges but also higher phase noise and larger temperature drifts with frequency.

A varactor diode can be either a silicon or a GaAs device. A higher Q is obtained with GaAs devices. This is due to the lower resistivity of GaAs for a given doping level N. GaAs devices have higher thermal resistance than silicon devices and hence have significantly higher frequency-settling times.

Figure 7.45 is a general representation of a transistor VCO. The tuning bandwidth of a YIG-tuned oscillator is limited by the negative-resistance bandwidth, whereas the bandwidth of a VCO is limited by the susceptance ratio $C_{min}\omega_{max}/C_{max}\omega_{min}$. Let the admittance of the transistor and feedback circuit transferred in the plane of the varactor be $Y(\omega) = G(\omega) + jB(\omega)$ with both $G(\omega)$ and $B(\omega)$ less than zero. From the plots of $-B(\omega)$, $C_{min}\omega$ and $C_{max}\omega$ shown in Fig. 7.46 it can be seen that the electronic bandwidth is limited by the frequencies at which

$$C_{min}\omega_{max} = -B(\omega_{max}) \tag{7.97}$$

$$C_{max}\omega_{min} = -B(\omega_{min}) \tag{7.98}$$

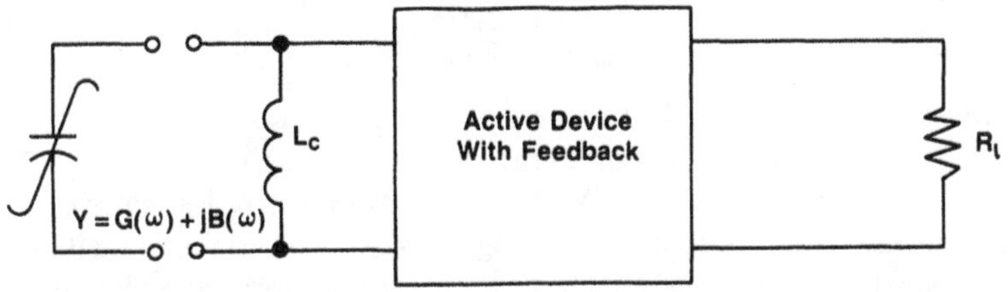

Fig. 7.45 Reactance compensation in a varactor-tuned oscillator.

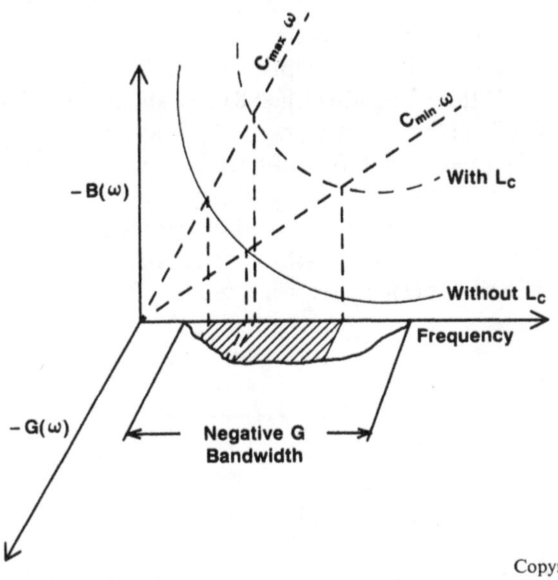

Fig. 7.46 Frequency limits of a varactor-tuned oscillator (after ref. 7.52).

This limitation holds even if the negative-conductance bandwidth is much wider as shown in Fig. 7.46.

To increase the electronic tuning bandwidth for a given C_{max}/C_{min} ratio it is necessary to reduce the ratio $B(\omega_{min})/B(\omega_{max})$ and modify the variation in $B(\omega)$ to that of the susceptance of a lumped inductance. In order to do this an inductance L_c is added in parallel to the transistor terminals (Fig. 7.45). In this case the total susceptance ratio seen by the varactor is given by [7.48]

$$\frac{B_t(\omega_{min})}{B_t(\omega_{max})} = \frac{B(\omega_{min}) + 1/(L_c\omega_{min})}{B(\omega_{max}) + 1/(L_c\omega_{max})} \tag{7.99}$$

If L_c is small, this ratio is large and approaches $\omega_{max}/\omega_{min}$. Equating this ratio to the susceptance ratio presented by the varactor we obtain

$$\frac{\omega_{max}}{\omega_{min}} = \left(\frac{C_{max}}{C_{min}}\right)^{1/2} \tag{7.100}$$

The value of L_c normally has to be optimized to maximize the bandwidth because of the parasitic inductances present in the circuit.

Figure 7.47 shows a typical VCO circuit using a bipolar transistor and a varactor diode. This configuration, using an NEC 567 bipolar transistor and a GaAs hyperabrupt varactor, covers a bandwidth of more than an octave from 3 to 9 GHz [7.49]. Increased bandwidth can be realized by using two varactors, one in the source and the other in the gate, as shown in Fig. 7.48 for a GaAs MESFET VCO. Bandwidths of greater than an octave have been realized using the above configuration. A low noise VCO configuration using a bipolar transistor and a silicon abrupt diode [7.50] is shown in Fig. 7.49. This oscillator has an 8% tuning bandwidth at 8 GHz and has excellent phase noise of −103 dBc at 100 kHz off the carrier.

Wideband frequency operation of a transistor VCO can be extended by using a configuration of the push–push type as shown in Fig. 7.50. This circuit cancels the fundamental frequency and optimizes the second harmonic at the output point. Hyperabrupt diodes with a capacitance variation of 3.2–0.16 pF and a GaAs MESFET device have been used to fabricate a VCO covering 20–30 GHz [7.51].

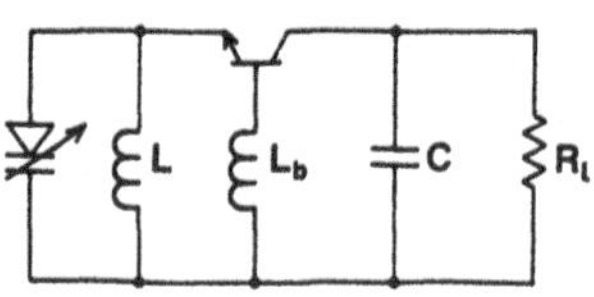

Fig. 7.47 Varactor-tuned bipolar transistor oscillator.

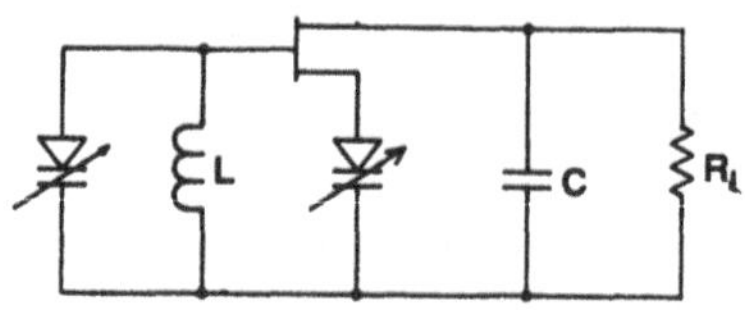

Fig. 7.48 Varactor-tuned GaAs MESFET oscillator.

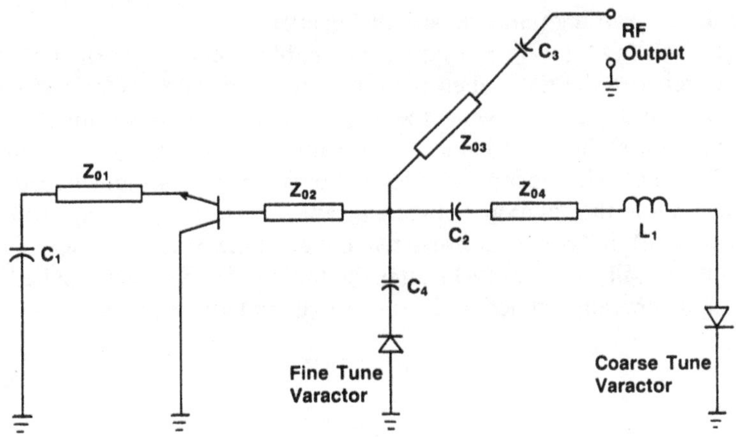

Fig. 7.49 Wideband varactor-tuned bipolar transistor oscillator (after ref. 7.50).

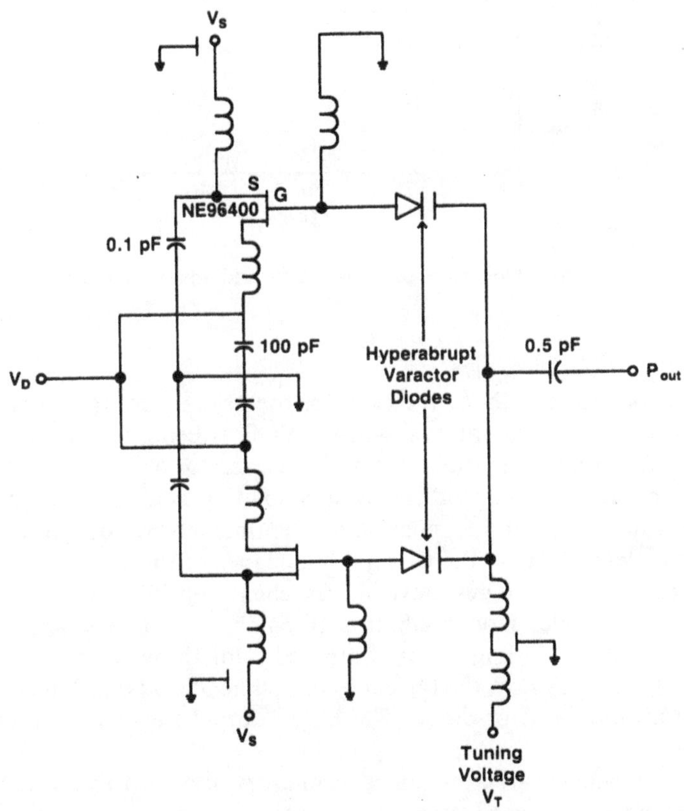

Fig. 7.50 Schematic diagram of a push–push VCO circuit (after ref. 7.51).

7.6.2.1 Settling time and post-tuning drift

The settling time in a frequency-tunable oscillator is defined as the interval between the time when the input tuning drive waveform reaches its final value and the time when the VCO frequency falls within a specified tolerance of a stated final value. The settling time and post-tuning drift are defined in Fig. 7.51. The input drive waveform reaches its final value at t_r. The settling time is the interval between t_{st} and t_r. The post-tuning drift is the frequency drift that occurs between two defined times t_1 and t_2 where t_1 can be specified typically as 10 μs to 1 s after the tuning step has been applied to the VCO; t_2 is generally defined in the range 10 μs to 1 s after t_1.

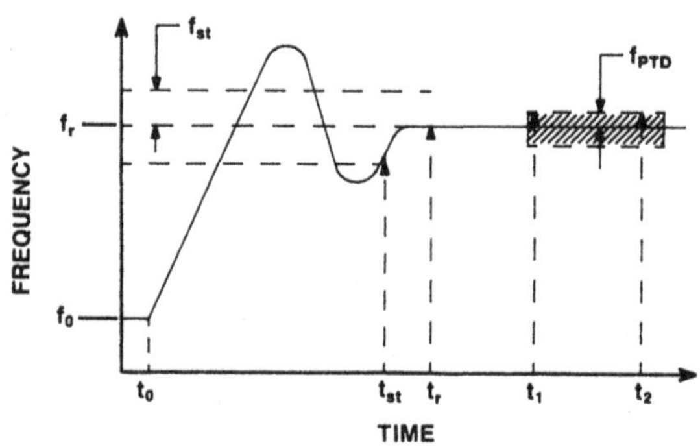

Fig. 7.51 Definition of settling time and post-tuning drift.

Bias voltages and thermal effects are the primary contributors to post-tuning drift. During the interval when a VCO is being tuned, the junction temperatures of both the transistor and the varactor are changing owing to changes in the r.f. circuit efficiency and loading. This causes impedance changes which result in frequency shift. The time interval during which this happens is dependent on the thermal impedance of the device. Abrupt and hyperabrupt silicon diodes have an excellent settling-time performance. Abrupt GaAs diodes have medium settling times and hyperabrupt GaAs diodes have large settling times compared with those of silicon varactor diodes. Figure 7.52 shows the typical settling time of a bipolar Ku band VCO using an abrupt varactor diode [7.52]. The output frequency settles in less than 2 μs.

The YIG- and varactor-tuning techniques discussed above are significantly different from each other. Their major parameters are compared in Table 7.2.

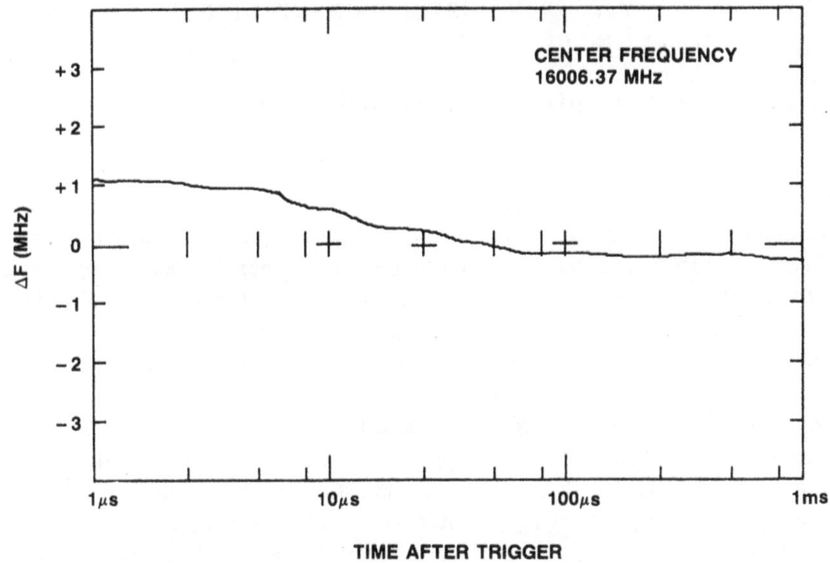

Fig. 7.52 Typical settling time of a bipolar Ku band VCO (after ref. 7.52).

Table 7.2 Varactor-controlled oscillator versus YIG-tuned oscillator

Parameter	*YIG tuning*	*Varactor tuning*
Bandwidth	Wide (octave and multi-octave)	Narrow (octave or less)
Phase noise	Excellent	Good performance in narrow bandwidth
Tuning	Current	Voltage
Linearity	±1%	±10%–15%
Slew rate	<1 MHz ms^{-1}	1–10 GHz μs^{-1}
Step response time	1–3 ms	<0.1 μs
Post-tuning drift constant	Seconds	Microseconds to milliseconds
Temperature stability	50–100 ppm K^{-1}	100–300 ppm K^{-1}
Frequency pushing	5 MHz V^{-1}	70 MHz V^{-1}
Oscillator Q	High (>1000)	Low (<100)
Power consumption	High (10 W maximum)	Low (1 W maximum)
Size	Large (1 in^3 minimum)	Small (typically 0.25 in^3)

7.7 Oscillator measurements

7.7.1 Measurements using a network analyzer

The techniques for measuring oscillator characteristics like power, frequency, harmonics, external Q and FM noise are well known. However, these measurements require such expensive apparatus as power meters, frequency counters, a spectrum analyzer, a signal generator, circulators, variable attenuators, sliding shorts etc. In this section we shall discuss the measurement of major oscillator characteristics using a standard network analyzer and signal generator [7.53, 7.54].

The measurement principle is based on injection locking the TDRO under test with the signal available at the "unknown" port of the network analyzer. An injection-locking polar diagram (ILPD) is thus obtained and displayed on the polar display. The network analyzer is used to measure the magnitude and phase of the injection gain under locked conditions. The resulting ILPD resembles ordinary impedance measurements and can be used to measure the major characteristics of the oscillator. Moreover, the ILPD can be used to optimize key oscillator operating parameters in a manner analogous to using the network analyzer in its conventional role.

The Hewlett-Packard HP8410 network analyzer was used for the system described here. The configuration shown in Fig. 7.53 was used for low power oscillators (up to about +10 dBm). Figure 7.54 shows a higher power configuration that is useful up to an output of about 30 dBm. To increase the dynamic range of the gain measurement and to protect the sensitive harmonic converter from inadvertent overload, attenuators of 30 dB and 3 dB are inserted in the test channel and reference channel respectively. With an r.f. power range of −16 to −43 dBm at the reference channel input of the harmonic frequency converters, the use of a signal generator with a power output of −10 to +17 dBm allows complete coverage of the range of possible power variations encountered. For example, an injection gain range of 13–40 dB can be measured using an oscillator with an output of +10 dBm (Fig. 7.53).

The equipment can be calibrated for either a transmission or a reflection coefficient mode using a through-line and a short circuit. The phase calibration can be accomplished with the circuit injection-locked with very high injection gain (above 30 dB). In this case the phase difference between the locked signal and the locking source at the center frequency can be assumed to be zero at the parallel tuned-circuit oscillator plane. The reference plane extension control is adjusted to bring the ILPD into the desired plane. The free-running oscillator frequency can be measured using a counter to locate the θ_0 point (when $\Delta\omega = 0$) on the injection-locking diagram (Fig. 7.55).

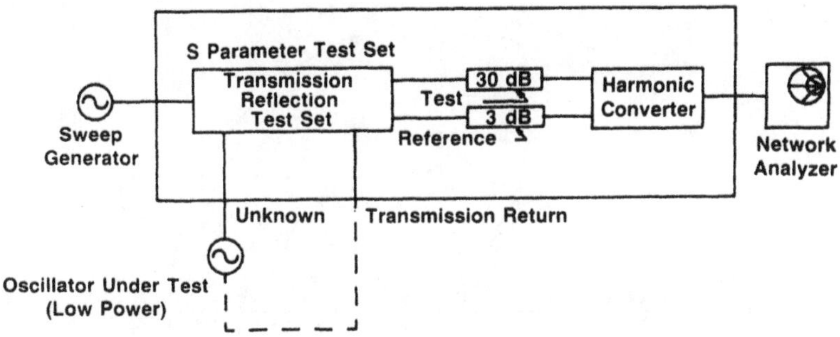

Fig. 7.53 Low power oscillator measurement set-up.

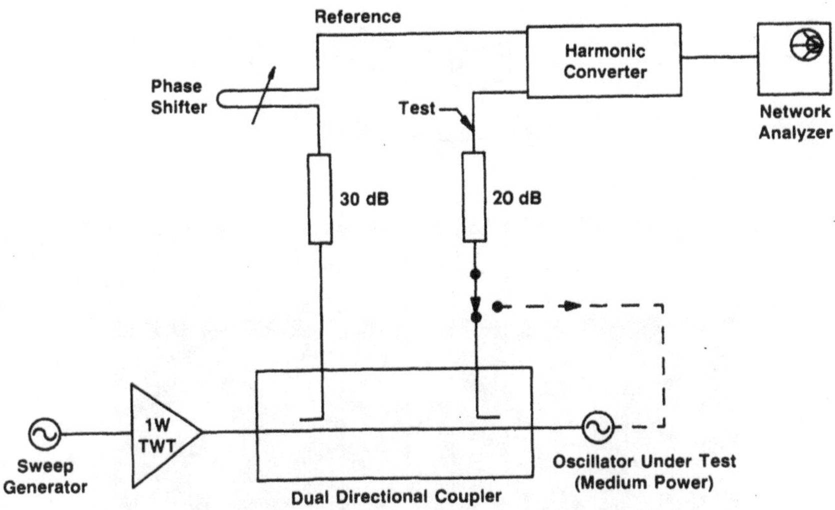

Fig. 7.54 Medium power oscillator measurement set-up.

When the oscillator is connected to the unknown port of the *S* parameter test set and the sweep generator is set to sweep across about 10% of the known oscillator frequency, noise can be observed on both the polar and rectangular display. This noise is in addition to the desired signal peak on the rectangular display and the ILPD on the polar display. Next, the sweep band is gradually reduced and the sweep generator power and the test channel gain are adjusted to obtain the desired ILPD on the polar display for various values of the injection gain. The injection gain is thus directly measurable. Reducing the sweep bandwidth to a usable minimum helps to eliminate the noise and improve the ILPD. Figure 7.55 shows the ILPD for various values of the sweep generator power output and hence the injection gain. Figure 7.56 shows the ILPD of a MESFET DRO at 9.5 GHz.

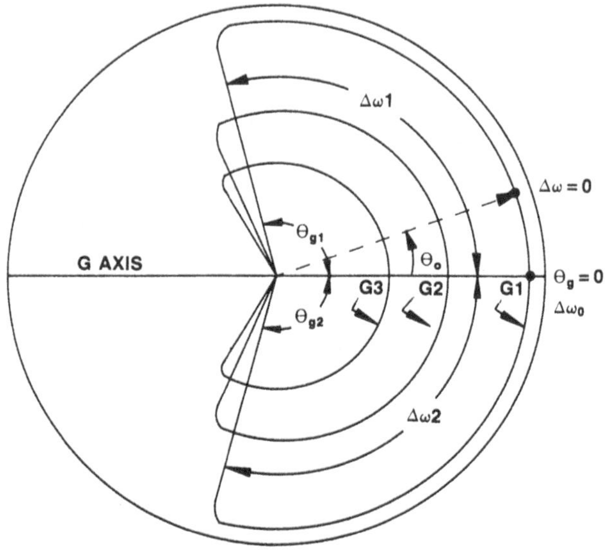

Fig. 7.55 Typical ILPD and definition of various parameters (after ref. 7.53).

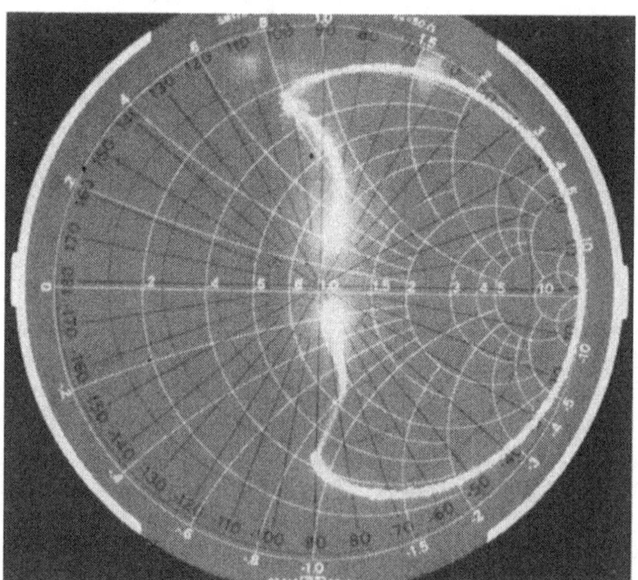

Fig. 7.56 ILPD of a dielectric resonator oscillator at 9.5 GHz (after ref. 7.54).

For r.f. power measurements, the test channel gain is set to correspond to an injection gain of 20 dB. The signal generator output power level control is used to move the $G = 0$ point on the ILPD to the outer edge of the polar display. The oscillator power can then be determined to the accuracy of the signal generator output controls without using a power meter. For the low power test configuration shown in Fig. 7.53 the output power of the oscillator is equal to the output power of the signal generator.

Once the injection gain G has been established on the ILPD, the remaining parameters can easily be measured. The injection bandwidth $\Delta\omega$ is given by $|\Delta\omega_1|+|\Delta\omega_2|$, and the external quality factor Q_e can be determined from

$$Q_e = \frac{2\omega_0}{\Delta\omega}\frac{1}{G^{1/2}} \tag{7.101}$$

Significantly, the measurement is made without using a spectrum analyzer, circulator, directional coupler or power meter, which are usually required. Moreover the display presented by the network analyzer is generally more informative than that obtained from a spectrum analyzer because it also gives injection gain phase information which is not provided by a spectrum analyzer.

The nonlinear constants α and K [7.55] of the oscillator can be determined from the measured values of $\Delta\omega_0$, θ_0 and G using the following relations [7.54]:

$$\alpha = \frac{G \sin \theta_0}{1 - G \cos \theta_0} \tag{7.102}$$

$$K = \frac{\Delta\omega_0(G-1)}{2\alpha} \tag{7.103}$$

This approach provides a rapid means of determining α and K and graphically representing the effects of parameters such as biasing voltage on these nonlinear constants. For low values of injection gain, the elliptical power variation in the injection-locked frequency range as well as in the locking range asymmetry (which is a function of α) can be read directly from the ILPD.

The ILPD also yields valuable insight into the oscillator output matching circuit. Frequency and phase jumps within the locking range are readily displayed and recognized (Fig. 7.57). The network analyzer thus provides a valuable tool for testing and aligning the output load circuit. Since the injection gain represents the oscillator power output (for constant oscillator signal input), the oscillator can easily be tuned for the desired output. The method employed is comparable to using a network analyzer for tuning a filter or amplifier. This results from the comprehensive display that immediately shows the effect of changes to load or circuit impedances.

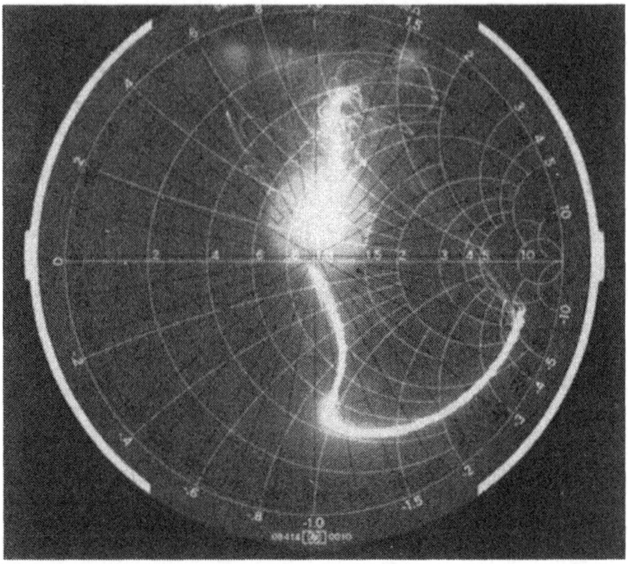

Fig. 7.57 Phase jumping by a TDRO (after ref. 7.54).

7.7.2 Measurement of the pulling figure

Pulling of an oscillator is defined as the total angular frequency deviation $\Delta\omega_t$ due to a known load perturbation. The load is characterized by a reflection coefficient with a constant modulus and a phase varying from zero to 2π rad. The schematic diagram of an oscillator for pulling-figure derivation is shown in Fig. 7.58. Frequency pulling can be shown to be given by [7.55]

$$\Delta\omega_t = \frac{S^2 - 1}{S} K (\alpha^2 + 1)^{1/2} \tag{7.104}$$

where

$$K = \frac{\partial R_t}{\partial I} R_0 \bigg/ \left(\frac{\partial R_t}{\partial I} \frac{\partial X_t}{\partial \omega} - \frac{\partial R_t}{\partial \omega} \frac{\partial X_t}{\partial I} \right) \tag{7.105}$$

$$Z_t = R_t + jX_t \tag{7.106}$$

$$\alpha = \frac{\partial X_t / \partial I}{\partial R_t / \partial I} \tag{7.107}$$

K and α are the nonlinear constants of the oscillator and S is the voltage standing-wave ratio (VSWR) of the load. Neglecting the nonlinear terms $\partial R_t/\partial I$ and $\partial X_t/\partial I$, we obtain the simplified pulling expression

$$\Delta\omega_t = \frac{\omega_0}{2Q_e}\left(S - \frac{1}{S}\right) \tag{7.108}$$

In practice frequency pulling is measured on a spectrum analyzer when the oscillator is connected to a load with a constant reflection coefficient modulus and variable phase. The load phase is varied from zero to 2π rad with the help of a sliding short. Figure 7.59 shows the pulling-figure set-up for a commonly used VSWR of 1.67:1. For example, if a typical 10 GHz DRO has a frequency pull of 4 MHz, Q_e can be calculated from (7.108) to be

$$Q_e = \frac{10000}{2\times 4}\left(1.67 - \frac{1}{1.67}\right) = 1339 \tag{7.109}$$

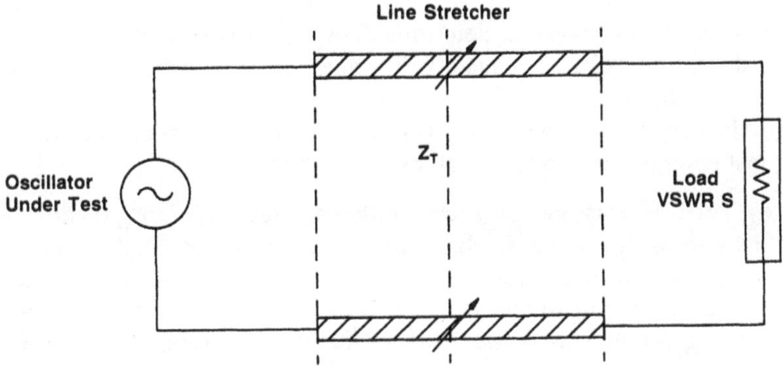

Fig. 7.58 Schematic diagram of an oscillator for pulling-figure derivation.

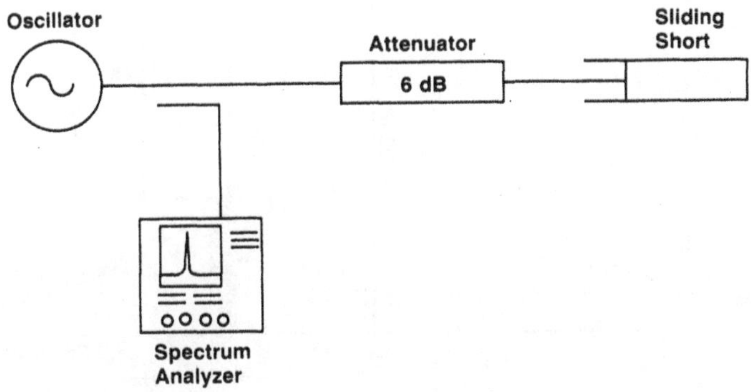

Fig. 7.59 Set-up for measuring oscillator pulling figures.

7.7.3 Measurement of frequency modulation noise

FM noise is an important characteristic of the oscillator. It represents the phase jitter or, in other words, the short-term stability of the device. The oscillator power is distributed in spectral distributions known as noise sidebands on opposite sides of the carrier. Noise can be discussed as a modulation phenomenon, and these sidebands can be separated into AM and FM components. Since the AM components are much smaller than the FM components, we shall mainly consider FM noise in the following discussion. FM noise also represents short-term stability in the time domain. Various terminologies are used to specify the FM noise of an oscillator and this causes significant problems for oscillator engineers. The oscillator noise performance results often reflect the measurement equipment employed, but it is often not possible to compare the results directly without additional information. FM noise can be represented in the time domain or the frequency domain. In the time-domain method the r.m.s. frequency amplitude deviation from the mean square value of the noise is determined. A probability density function is assumed and the time-averaged value gives little information about the spectral distribution.

In the frequency domain the FM noise is represented in various ways. The two most commonly used terminologies are the following.

(a) The FM noise power is represented as the ratio of the power in some specified bandwidth in one sideband to the carrier power. Such power ratios are usually quoted in decibels below carrier amplitude and require specification of the bandwidth and frequency of the carrier at which the measurement is made (Fig. 7.60). This is the most frequently used representation of FM noise.

(b) The FM noise can also be represented as the equivalent r.m.s.

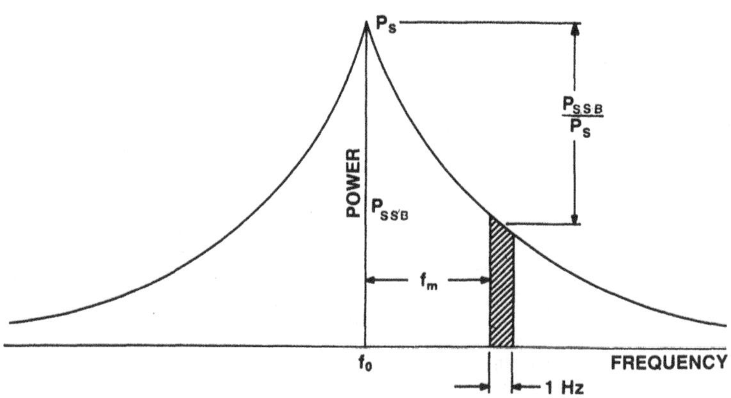

Fig. 7.60 R.f. spectrum of an oscillator.

frequency deviation of a noise-free carrier modulated at a single frequency and producing the same output as the noisy carrier in a narrow band centered at the frequency of interest. This representation is related to the earlier one by the following relation and requires specification of the bandwidth and frequency of the carrier:

$$\text{single sideband (dBc)} = 20\log\left(\frac{\Delta f_{\mathrm{rms}}}{\sqrt{2}f_{\mathrm{rm}}}\right) \tag{7.110}$$

Figure 7.61 shows the conversion nomogram relating these two representations of the FM noise.

In theory an idealized spectrum analyzer with a dynamic range of more than 150 dB and a bandwidth of 1 Hz operating at microwave frequencies would facilitate most noise measurements. In practice such performances are not possible and a variety of techniques have been developed for the measurement of oscillator FM noise. The two most common techniques are discussed below.

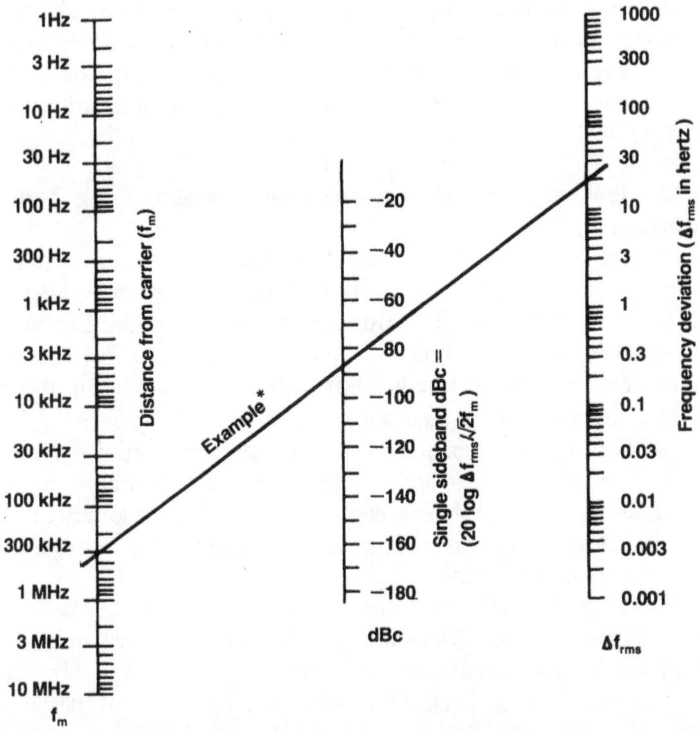

Fig. 7.61 FM noise conversion nomograph. When a straight edge is placed on the frequency deviation (in Hz) (right-hand column) and on the distance from the carrier at which that deviation occurs (left-hand column), the single-sideband noise power ratio (in dBc) is shown in the center column. Example*: 20 Hz deviation in a 1 kHz band at 300 kHz from the carrier ≡ −87 dBc in a 1 kHz bandwidth.

7.7.3.1 Frequency modulation noise measurement using a cavity discriminator [7.10]

A block diagram of the simplified set-up is shown in Fig. 7.62. This diagram represents the frequency discriminator which serves to convert FM to AM using the frequency-response variation of a tuned circuit. The signal from the oscillator to be tested is divided into two paths: the local oscillator (LO) path and the r.f. signal path. An adjustable phase shifter and attenuators are placed in the LO path while a high Q carrier-suppression filter (a tunable cavity and a circulator) is placed in the r.f. path. When the cavity is used in this mode, a reduced r.f. power corresponding only to the FM sidebands of the oscillator signal is reflected towards the balanced mixer. The phase shifter zero is adjusted in order to re-introduce the carrier phase advanced by 90°. FM noise sidebands are thus converted to AM sidebands and are easily detected by mixer diodes, while the AM noise sidebands are converted to FM and are not detected by the mixer. The mixer output is detected by a highly sensitive adjustable-bandwidth selective voltmeter.

The system is calibrated by applying modulation to the oscillator under test. The modulation is set using the microwave spectrum analyzer at a known level (generally at a modulation index of 2.405 corresponding to a carrier null) and the baseband measurement is used to determine the calibration constants of the system.

7.7.3.2 Frequency modulation noise measurement using a delay-line discriminator [7.56]

The block diagram of a delay-line FM discriminator is shown in Fig. 7.63. This set-up has less sensitivity than a cavity discriminator but is wideband and easier to automate. The principle of operation of the delay-line discriminator is similar to that of a cavity discriminator. The length of the delay line is adjusted to place the zero crossing at the oscillator carrier frequency. In practice this means setting up the system to obtain zero d.c. output from the balanced mixer. This minimizes the sensitivity of the system to amplitude variations. The length of the delay line also controls the sensitivity of the system to the FM noise measurement. This can be increased by increasing the length of the delay line at the cost of the bandwidth. The calibration procedure is similar to that used for the cavity discriminator.

Hewlett-Packard have proposed using a test set (HP11729C) with the delay-line–mixer-frequency discriminator technique for phase noise measurement on microwave sources at frequencies from 10 MHz to 18 GHz. The block diagram is shown in Fig. 7.64. This noise test uses an internal low noise microwave reference signal to down-convert the test signal to an intermediate frequency below 1.28 GHz. The intermediate frequency signal is amplified and is then applied to a delay-line frequency discriminator where the phase noise is demodulated and made available for analysis. HP11729C is General Purpose Interface Bus (GPIB) controllable, making automatic phase-noise measure-

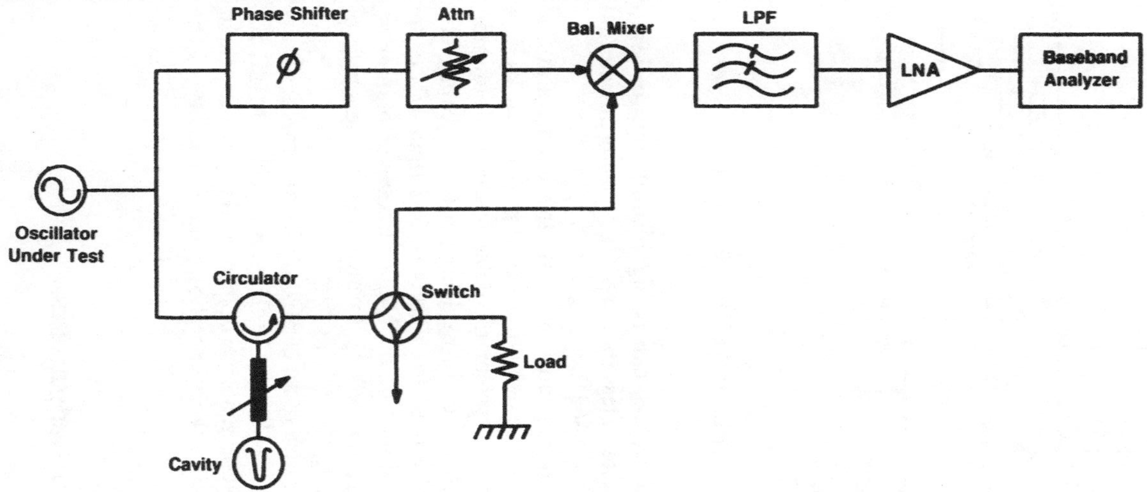

Fig. 7.62 FM noise measurement set-up using the cavity discriminator method.

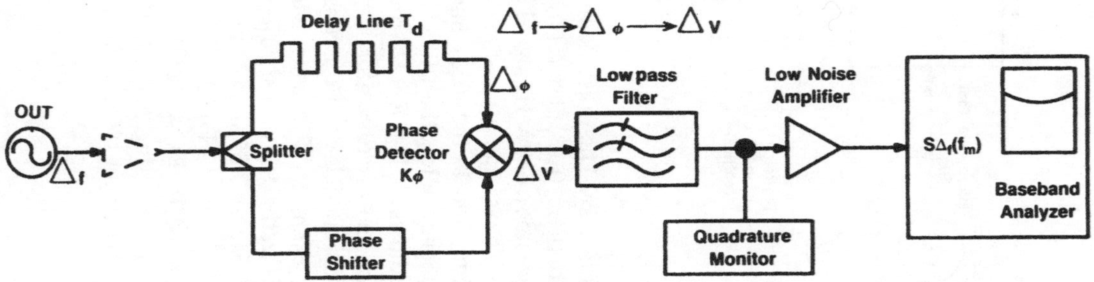

Fig. 7.63 FM noise measurement set-up using the delay-line discriminator method.

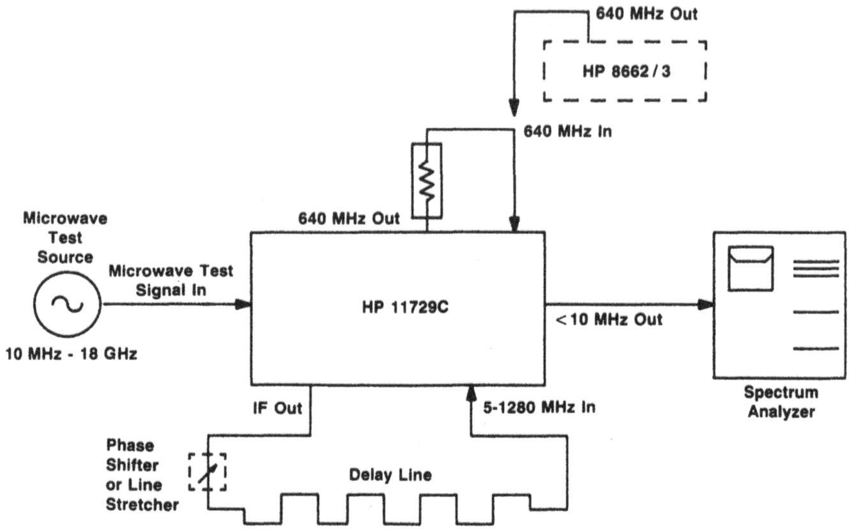

Fig. 7.64 Phase noise set-up using the HP11729C test set.

ments easy. The noise floor of the test set can be significantly improved (to more than 10 dB) at offset frequencies of less than 50 kHz by using an external lower noise 640 MHz signal from HP 8662/3.

The sensitivity of the system depends on both the length and the attenuation of the delay line. Since the delay-line discriminator works at a frequency of less than 1.28 GHz in this system, the delay line can be common coaxial cable like RG223 providing about 1.5 ns delay per foot of cable. The delay-line discriminator is typically used for offsets of less than $1/2\tau_d$; the maximum delay to be used is determined by the highest frequency of interest. For example, if measurements are desired up to an offset frequency f_m of 1 MHz, the delay must be less than $f_m/2 = 500$ ns.

Figure 7.65 shows the phase noise of an electronically and mechanically tuned bipolar TDRO (Avantek DSO 4100) at 10 GHz and a GaAs MESFET low noise TDRO at 18 GHz. Measurements were performed on an automated test set-up using HP11729C. The broken line shows the sensitivity of the system using a 100 ns delay line.

Acknowledgments

Most of the work reported in this chapter was performed at the University of Limoges, France, and Avantek Inc., Santa Clara, CA. The author is grateful for their support.

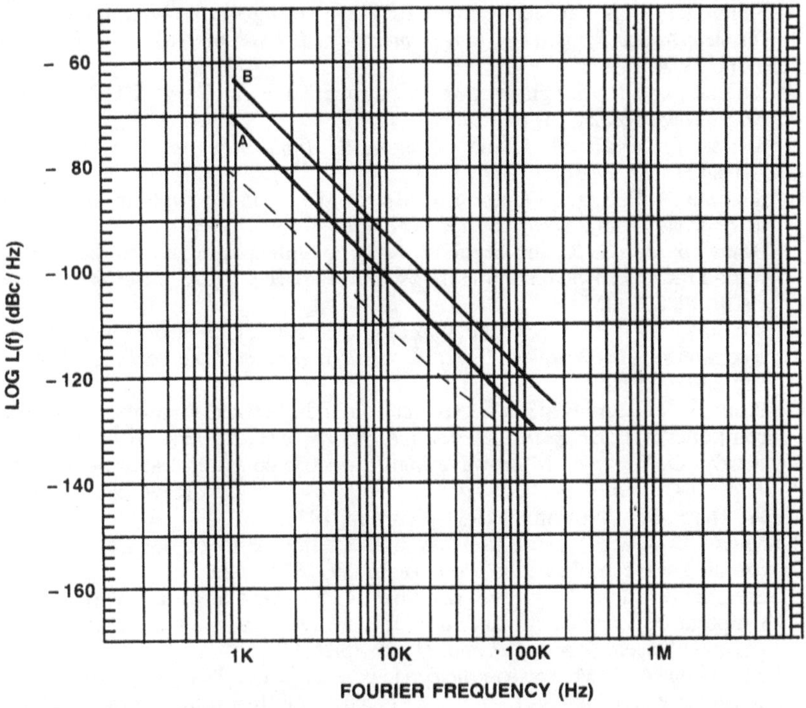

Fig. 7.65 Typical phase noise plot for a bipolar X band dielectric resonator oscillator (line A) and an 18 GHz low noise GaAs MESFET oscillator (line B) (Avantek).

Special thanks are due to Dr Christos Tsironis of SPAR Canada and Bob Oyafuso of Avantek Inc. for fruitful discussions and critical reviews of the chapter.

References

7.1 Leung, C., Snapp, C. and Grande, V. A 0.5 μm silicon bipolar transistor for low phase noise oscillator applications up to 20 GHz, *IEEE Int. Microwave Symp. Dig., June 1985*, pp. 383–385, IEEE, New York, 1985.

7.2 Camiade, M. and Bert, A. Wide tuning bandwidth Ku band varactor FET oscillators, *Proc. 15th European Microwave Conf., Paris, 1985*, p. 413, Microwave Exhibitions and Publishers, Tunbridge Wells, 1985.

7.3 Boyles, J. W. The oscillator as a reflection amplifier: an initiative approach to oscillator design, *Microwave J.*, **29**, 83–98, June 1986.

7.4 Bodway, G. E. Circuit design and characterization of transistor by means of three-port scattering parameters, *Microwave J.*, **11** (5), 55–63, 1968.

7.5 Khanna, A. P. S. Parallel feedback FETDRO using 3-port *S*-parameters, *IEEE Int. Microwave Symp. Dig., San Francisco, CA, 1984*, pp. 181–183, IEEE, New York, 1984.

7.6 Khanna, A. P. S. Three port *S*-parameters ease GaAs FET designing, *Microwaves RF*, **24**, 81–84, 1985.

7.7 Soares, R., Graffeuil, J. and Obregon, J. *Applications of GaAs MESFETs*, Artech House, Dedham, MA, 1983.

7.8 Khanna, A. P. S. and Obregon, J. Microwave oscillator analysis, *IEEE Trans. Microwave Theory Tech.*, **29**, 606–607, 1981.

7.9 Basawapatna, G. R. and Stancliff, R. B. A unified approach to the design of wide-band microwave solid state oscillators, *IEEE Trans. Microwave Theory Tech.*, **27**, 379–385, 1979.

7.10 Khanna, A. P. S. Oscillateurs microondes stables intégrés à transistors et résonateurs diélectriques, *Thèse Docteur-Ingenieur*, University of Limoges, September 1981.

7.11 Plourde, J. K. and Ren, C. L. Application of dielectric resonators in microwave components, *IEEE Trans. Microwave Theory Tech.*, **29**, 754–770, 1981.

7.12 Ready, D. W. *et al.* Microwave high dielectric constant materials, *Quarterly Rep. 1–4, Final Rep., Contract DAAB07–69–C–0455*, Raytheon Company, Waltham, MA, November 1969–February 1971.

7.13 Masse, D. J. *et al.* A new low-loss high-*K* temperature compensated dielectric for microwave applications, *Proc. IEEE*, **59**, 1628–1629, 1971.

7.14 Simonet, W. and Mage, J. C. (Zr, Sn)TiO_4 dielectric materials with adjustable temperature coefficients and low microwave loss, presented at *American Ceramic Society Meet., Chicago, IL, 29 April 1980.*

7.15 Montvala, A. J. Microwave dielectric constant materials, *Quarterly Rep. 1–3, Contract DAAB07–69–C–0402*, ITT Research Institute, Chicago, IL, July 1969–January 1970.

7.16 Kajfez, D. and Guillon, P. *Dielectric Resonators*, Artech House, Dedham, MA, 1986.

7.17 Guillon, P., Mekerta, S. and Garault, Y. Microstrip bandstop filter using a dielectric resonator, *Proc. Inst. Electr. Eng.*, **128** (3), 151–154, 1981.

7.18 Khanna, A. P. S. and Garault, Y. Determination of loaded, unloaded and external quality factors of a dielectric resonator coupled to a microstrip line, *IEEE Trans. Microwave Theory Tech.*, **31**, 261–264, 1983.

7.19 Guillon, P. Personal communication, 1986.

7.20 Gil, F. H. and Martinez, J. P. Analysis of dielectric resonators with tuning screw and supporting structure, *IEEE Trans. Microwave Theory Tech.*, **33**, 1453–1457, 1985.

7.21 Golio, J. M. and Krowne, C. M. New approach for FET oscillator design, *Microwave J.*, **21**, 59–61, October 1978.

7.22 Khanna, A. P. S., Obregon, J. and Garault, Y. Efficient low noise three port X-band FET oscillator using two dielectric resonators, *IEEE Int. Microwave Symp. Dig., Dallas, TX, 1982*, pp. 277–279, IEEE, New York, 1982.

7.23 Wagner, W. Oscillator design by device line measurement, *Microwave J.*, **22**, 43–48, 1979.

7.24 Poulin, D. Load-pull measurements help you meet your match, *Microwaves*, **19**, 61–65, 1980.

7.25 Galani, Z., Bianchini, M. J., Waterman, R. C., Dibiase, R., Laton, R. W. and Cole, J. B. Analysis and design of a single-resonator GaAs FET oscillator with noise degeneration, *IEEE Trans. Microwave Theory Tech.*, **32**, 1556–1565, 1984.

7.26 Ishihara, O., Tetsurou, M., Sawano, H. and Nakatani, M. A highly stabilized GaAs FET oscillator using a dielectric resonator feedback circuit in 9–14 GHz,

IEEE Trans. Microwave Theory Tech., **28**, 817–824, 1980.

7.27 Lan, G., Kalokitis, D. and Mykietan, E. Ultrastable low-noise GaAs FET oscillator with dielectric resonator, *RCA Rev.*, **47**, 472–486, 1986.

7.28 Tsironis, C. and Lesartre, P. Temperature stabilization of GaAs FET oscillators with dielectric resonators, *Proc. 12th European Microwave Conf., Helsinki, 1982*, pp. 181–186, Microwave Exhibitions and Publishers, Tunbridge Wells, 1982.

7.29 Lee, J. *et al.* Digital and analog frequency–temperature compensation of dielectric resonator oscillators, *IEEE Int. Microwave Symp. Dig., 1984*, pp. 277–279, IEEE, New York, 1984.

7.30 Khanna, A. P. S. Review of dielectric resonator oscillator technology, *Proc. Frequency Control Symp., Philadelphia, PA, May 1987*, pp. 478–486, IEEE, New York, 1987.

7.31 Farr, A. N., Blackie, G. N. and Williams, D. Novel technique for electronic tuning of dielectric resonators, *Proc. 13th European Microwave Conf., Nürnberg, 1983*, pp. 791–796, Microwave Exhibitions and Publishers, Tunbridge Wells, 1983.

7.32 Watanabe, S. *et al.* Very high *Q* dielectric resonator voltage controllable oscillator, *Proc. 8th European Microwave Conf., Paris, 1978*, pp. 269–273, Microwave Exhibitions and Publishers, Tunbridge Wells, 1978.

7.33 Christ, S. and Horowitz, P. Miniature X-band transmitter uses nine-volt battery, *Microwave Syst. News*, **12,** 84, December 1982.

7.34 Hercsfeld, P. R., Daryoush, A. S., D'Ascenzo, C., Contarino, V. M. and Rosen, A. Optically tuned and FM modulated X-band dielectric resonator oscillator, *Proc. 14th European Microwave Conf., Liège, 1984*, pp. 268–273, Microwave Exhibitions and Publishers, Tunbridge Wells, 1984.

7.35 Hercsfeld, P. R., Daryoush, A. S., Contarino, V. M., Rosen, A., Turski, Z. and Khanna, A. P. S. Optically controlled microwave devices and circuits, *IEEE Int. Microwave Symp. Dig., St Louis, MO, 1985*, pp. 211–214, IEEE, New York, 1985.

7.36 Pengelly, R. S. *Microwave Field Effect Transistors—Theory, Design and Applications*, Chapter 7, Research Studies Press, Letchworth, 1982.

7.37 Osbrink, N. K. YIG-tuned oscillator fundamentals, *Microwave Syst. News*, **13**, 207–225, March 1983.

7.38 Matthaei, G. L., Young, L. and Jones, E. M. T. *Microwave Filters, Impedance Matching Networks and Coupling Structures*, McGraw-Hill, New York, 1964.

7.39 Lax, B. and Button, K. J. *Microwave Ferrites and Ferromagnetics*, pp. 145–196, McGraw-Hill, New York, 1962.

7.40 Ollivier, P. M. Microwave YIG-tuned transistor oscillator amplifier design application to C-band, *IEEE J. Solid State Circuits*, **7**, 54–60, 1972.

7.41 Trew, R. J. Design theory for broad-band YIG-tuned FET oscillator, *IEEE Trans. Microwave Theory Tech.*, **27**, 8–14, 1979.

7.42 Heyboer, T. and Emery, F. YIG tuned GaAs FET oscillators, *IEEE Int. Microwave Symp. Dig., 1976*, pp. 48–50, IEEE, New York, 1976.

7.43 Ruttan, T. X-band GaAs FET YIG tuned oscillators, *IEEE Int. Microwave Symp. Dig., 1977*, p. 264, IEEE, New York, 1977.

7.44 Oyafuso, R. An 8–18 GHz FET YIG-tuned oscillator, *IEEE Int. Microwave Symp. Dig., 1979*, pp. 183–184, IEEE, New York, 1979.

7.45 Le Tron, Y., Barvet, S. and Obregon, J. Multioctave FET oscillators double tuned by a single YIG, *ISSCC Dig. Tech. Pap.*, 162–163, 1979.

7.46 Obregon, J., Le Tron, Y., Funk, R. and Barvet, S. Decade bandwidth FET functions, *IEEE Int. Microwave Symp. Dig., 1981*, pp. 141–142, IEEE, New York, 1981.

7.47 Zensius, D., Hauptman, J. and Osbrink, N. GaAs FET YIG oscillator tunes from 26 to 40 GHz, *Microwaves RF*, **22,** 129–139, October 1983.
7.48 Marechal, P. and Obregon, J. 1.5 to 4.5 GHz varactor-tuned transistor oscillators, *Proc. 9th European Microwave Conf., Brighton, 1979*, pp. 621–624, Microwave Exhibitions and Publishers, Tunbridge Wells, 1979.
7.49 Obregon, J. *et al.* Ultrabroad band electronically tunable oscillators, *Proc. 11th European Microwave Conf., 1981*, pp. 475–479, Microwave Exhibitions and Publishers, Tunbridge Wells, 1981.
7.50 Nieherke, E. and Hess, R. D. A microstrip low noise X-band voltage controlled oscillator, *IEEE Trans. Microwave Theory Tech.*, **27** (12), 1075–1079, 1979.
7.51 Winch, R. G. K-band FET doubling oscillator, *Electron. Lett.*, **18** (22), 946–947, 1982.
7.52 Khanna, A. P. S. Fast settling low noise Ku band fundamental bipolar VCO, *IEEE Int. Microwave Symp. Dig., 1987*, pp. 579–581, IEEE, New York, 1987.
7.53 Khanna, A. P. S. and Obregon, J. Network analyzer doubles as oscillator diagnostician, *Microwaves RF*, **23**, 106–112, 1984.
7.54 Khanna, A. P. S. and Obregon, J. Direct measurement of the non-linear MIC oscillator characteristics using injection locking polar diagram, *IEEE Int. Microwave Symp. Dig., Boston, MA, 1983*, pp. 501–503, IEEE, New York, 1983.
7.55 Obregon, J. and Khanna, A. P. S. Exact derivation of the non-linear negative resistance oscillator pulling figure, *IEEE Trans. Microwave Theory Tech.*, **30**, 1109–1111, 1982.
7.56 Phase noise characterization of microwave oscillators, *Product Note 11729C*, Hewlett-Packard.

Appendix Conversion of two-port *S* parameters to three-port *S* parameters

Equations (7.4) and (7.5) represent six equations in terms of nine three-port S parameters to be determined. Other equations necessary to determine all nine S parameters of the three-port can be found as follows.

From (7.3) we can write

$$b_1 = S_{11}a_1 + S_{12}a_2 + S_{13}a_3 \tag{A1}$$

$$b_2 = S_{21}a_1 + S_{22}a_2 + S_{23}a_3 \tag{A2}$$

$$b_3 = S_{31}a_1 + S_{32}a_2 + S_{33}a_3 \tag{A3}$$

If port 3 of the transistor shown in Fig. 7.3 is connected to ground, making it a common-source configuration, we have

$$\Gamma_3 = \frac{a_3}{b_3} = -1 \tag{A4}$$

Using (A4) in (A1)–(A3) and eliminating a_3 and b_3, we have

$$\begin{bmatrix} b_1 \\ \\ b_2 \end{bmatrix} = \begin{bmatrix} S_{11} - \dfrac{S_{13}S_{31}}{1+S_{33}} & S_{12} - \dfrac{S_{13}S_{32}}{1+S_{33}} \\ \\ S_{21} - \dfrac{S_{31}S_{23}}{1+S_{33}} & S_{22} - \dfrac{S_{23}S_{32}}{1+S_{33}} \end{bmatrix} \begin{bmatrix} a_1 \\ \\ a_2 \end{bmatrix} \tag{A5}$$

which represents the reduced two-port S matrix of the transistor with terminal pair 3 connected to ground. This relation represents four equations which, combined with the six equations from (7.4) and (7.5), provide the necessary 10 equations to be solved to determine all nine three-port S parameters. The final relations obtained are in simplified form [7.6]

$$S_{33} = \sum_{\substack{i=1,2 \\ j=1,2}} S_{ijT} \Big/ 4 - \sum_{\substack{i=1,2 \\ j=1,2}} S_{ijT} \tag{A6}$$

$$S_{32} = \frac{1+S_{33}}{2}(1 - S_{12T} - S_{22T}) \tag{A7}$$

$$S_{23} = \frac{1+S_{33}}{2}(1 - S_{21T} - S_{22T}) \tag{A8}$$

$$S_{22} = S_{22T} + \frac{S_{23}S_{32}}{1+S_{33}} \tag{A9}$$

$$S_{11} = 1 - S_{21} - S_{31} \tag{A10}$$

$$S_{13} = 1 - S_{23} - S_{33} \tag{A11}$$

$$S_{31} = 1 - S_{33} - S_{32} \tag{A12}$$

$$S_{12} = 1 - S_{22} - S_{32} \tag{A13}$$

$$S_{21} = 1 - S_{22} - S_{23} \tag{A14}$$

Chapter 8

GaAs Digital Integrated Circuits

MAURICE GLOANEC AND ERNESTO H. PEREA

8.1 Introduction

The first GaAs integrated circuit (IC) was reported about 14 years ago and opened exciting perspectives for the realization of very high speed digital circuits [8.1]. A few years later, the potential of this newborn technology for the fabrication of high performance functional MSI circuits operating at gigabit data rates was demonstrated [8.2, 8.3]. The first GaAs IC of LSI complexity with a speed performance about 10 times higher than the best equivalent circuit available in silicon technology at the same power consumption was announced in 1980 [8.4]. This remarkable evolution of high performance GaAs ICs is the result of continuous progress in materials, circuit design and processing technology which has led to an extremely fast growth rate, close to a threefold increase in circuit complexity each year (Fig. 8.1).

The principal features which have allowed the emergence of GaAs digital IC technology and which will dictate its potential field of application are reviewed in this chapter with the aid of various examples. We shall first briefly review the potential advantages of GaAs for the realization of very fast digital ICs and the various GaAs MESFET IC fabrication approaches in use today (other approaches such as IGFET, JFET, E-HMT and HBT are beyond the scope of this chapter). Next, the GaAs Schottky diode FET characteristics which are most relevant to the design of logic ICs will be summarized, and the principal logic families in existence will be presented with a discussion of their main features in terms of speed–power performance, packing density, degree of parameter control required and integration levels attainable. Then, three design approaches—full custom, standard cell and cell array—will be described in order to demonstrate the ability offered by this new technological family. Finally an overview of analog integrated circuits together with some aspects of the testing and packaging of GaAs ICs are given.

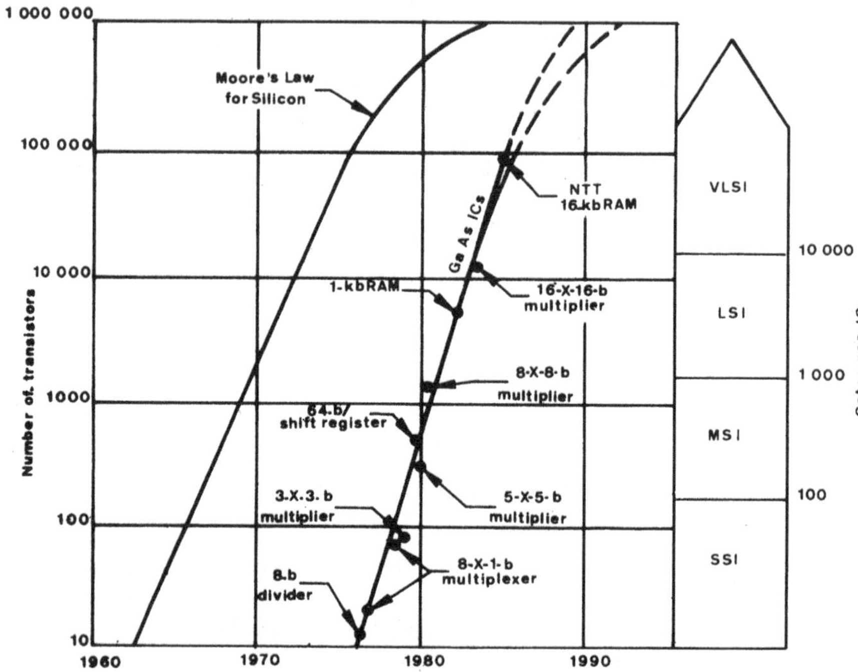

Fig. 8.1 Evolution of GaAs IC complexity.

8.2 Advantages and constraints of GaAs in integrated circuit implementations

8.2.1 Material and electrical properties

Table 8.1 summarizes, for both silicon and GaAs, the main characteristics of relevance for electron devices and IC realization. The emergence of GaAs as a material for high performance ICs results not only from attractive fundamental semiconductor properties but also from a favorable trade-off between a number of features including metallurgical characteristics, possibilities of device implementation, operating temperature range and radiation tolerance.

As far as high speed digital IC implementation is concerned, GaAs offers the following main advantages.

(1) High Schottky barriers can be realized on GaAs with a large variety of metals (e.g. aluminum, platinum, titanium) leading to high quality Schottky junctions with excellent ideality factors ($n < 1.1$) and fairly low reverse currents ($J_s < 1\ \mu\text{A cm}^{-2}$). This feature allows the realization of metal–semiconductor

Table 8.1 Principal properties of GaAs and silicon at 300 K

Properties	*GaAs*	*Silicon*
Electron mobility at $N_d = 10^{17}$ cm^{-3} (cm^2 V^{-1} s^{-1})	5000	800
Maximum electron drift velocity (cm s^{-1})	2×10^7	1×10^7
Critical field for high field transport (V cm^{-1})	3×10^3	1×10^4
Hole mobility at $N_d = 10^{17}$ cm^{-3} (cm^2 V^{-1} s^{-1})	250	350
Energy gap (eV)	1.43	1.12
Type of gap	Direct	Indirect
Density of states in conduction band (cm^{-3})	5×10^{17}	3×10^{19}
Maximum resistivity (Ω cm)	$\approx 10^9$	$\approx 10^5$
Minority carrier lifetime (s)	$\approx 10^{-8}$	$\approx 10^{-3}$
Breakdown field (V cm^{-1})	$\approx 4 \times 10^5$	$\approx 3 \times 10^5$
Schottky barrier height (V)	0.7–0.8	0.4–0.6
MIS interface charge Q_{ss}/q (cm^{-2})	$\approx 10^{12}$–10^{13}	$\approx 10^{10}$

field-effect transistors (MESFETs) with well-controlled characteristics and thus compensates the poor properties of GaAs with regard to minority carriers and the dielectric–semiconductor interface which make the fabrication of bipolar transistors and MOSFETs with stable characteristics very difficult. Nevertheless some encouraging results on GaAs E/D MOSFET ring oscillators have been reported [8.5, 8.6].

(2) n-type GaAs exhibits very favorable transport properties. In lightly doped material the low field electron mobility can reach values as high as 8000–9000 cm^2 V^{-1} s^{-1} at room temperature (50 000–100 000 cm^2 V^{-1}s^{-1} at 77 K). At normal doping levels the electron mobility and maximum drift velocity values in GaAs are respectively about six times higher than in silicon and twice as high as in silicon. However, high field transport, and thus saturation velocity, is achieved at electric fields about four times higher than in silicon. As a consequence, transit times as short as about 5–10 ps, corresponding to current gain–bandwidth products in the range $f_t = 15$–25 GHz, are obtained for GaAs MESFETs with typical gate lengths (1.0–0.5 μm). Such a figure of merit is three to five times better than that of silicon FETs with similar design parameters. It should be pointed out that the high mobilities quoted above have also been observed in modulation-doped GaAs–GaAlAs superlattices in the direction parallel to the multilayer at low temperatures [8.7] and resonant tunnelling is observed in the direction perpendicular to the multilayer; these properties lead to some useful microwave and high speed logic applications [8.8–8.10].

(3) The availability of "semi-insulating" GaAs material (resistivity in the 10^7–10^9 Ω cm range at room temperature) is a very important advantage for both technology simplification and performance optimization. Device isolation can be achieved very simply and with only a negligible influence on speed performance. Also, parasitic capacitances to the ground plane are

reduced to values comparable with those of coplanar capacitances between source–drain contacts or interconnection lines.

Compensation of residual levels in horizontal Bridgmann crystals by selected deep impurities (the most widely used being chromium) has been utilized for a long time as a conventional method of obtaining semi-insulating GaAs substrates. The electrical yields for ion implantation by Bridgmann-grown chromium-doped GaAs crystals tend to be low and unpredictable, mostly because of the phenomenon of thermal conversion during the post-implant anneal [8.11–8.14] which is due to chromium out-diffusion as observed earlier in epitaxial layers grown on chromium-doped GaAs substrates [8.15].

The introduction of refined purification techniques has allowed the fabrication of "intrinsic" GaAs crystals with a low nonintentional background doping level (about $10^{14}\,cm^{-3}$). More recently, the growth of high purity bulk semi-insulating GaAs by the liquid-encapsulated Czochralski (LEC) method has produced thermally stable crystals of large size (diameter, 3 in; weight, up to 5 kg) with a low dislocation density (about $10^4\,cm^{-2}$) and high resistivity which are suitable for use in ion implantation. Large ⟨100⟩-oriented circular wafers compatible with modern semiconductor processing equipment have become available. The lowest residual silicon concentrations (about $10^{14}\,cm^{-3}$) and highest mobilities are observed on ingots pulled from pyrolytic boron nitride (PBN) crucibles [8.16, 8.17].

8.2.2 Approaches to GaAs device structure fabrication

Schematic cross-sections of a MESFET and a Schottky diode for the principal GaAs IC fabrication processes are shown in Fig. 8.2. The main features of these alternative processing technologies are as follows.

(1) "Mesa-isolation" technology, which is widely used in discrete microwave GaAs FETs, was the first approach to the fabrication of digital GaAs ICs. Two options, which differ in that the active layer is either epitaxially grown (vapor phase epitaxy [8.18, 8.19], low pressure organometallic chemical vapor deposition [8.20, 8.21]) or ion implanted [8.22] have been used.

In order to avoid substrate problems (e.g. back-gating effects, leakage currents, light sensitivity) which were rather common in the early days of GaAs technology [8.23, 8.24], a nonintentionally doped epitaxial buffer layer 2.5 μm thick is generally grown prior to deposition of the active layer. MESFETs have also been fabricated with a recessed-gate structure (typically 500–1000 Å deep) in order to improve transistor characteristics by reducing the influence of parasitic resistances and surface effects (Fig. 8.3) and at the same time to allow a fine adjustment of the pinch-off voltage.

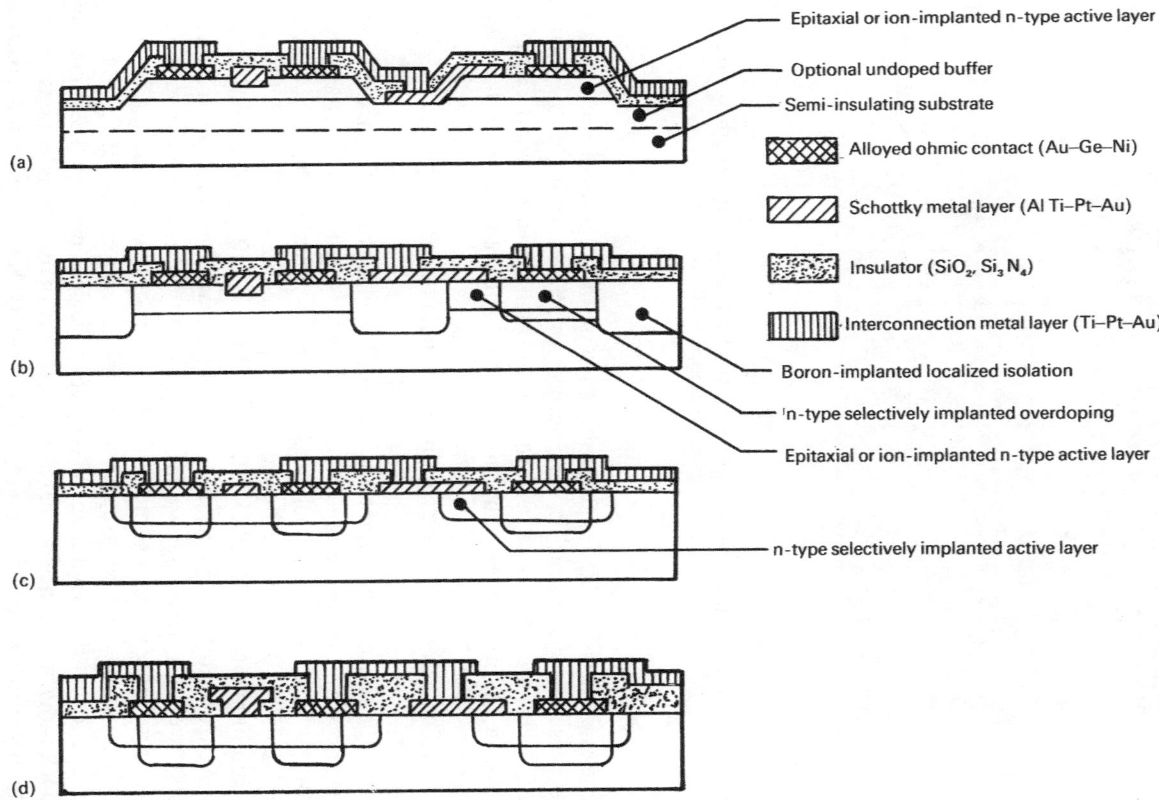

Fig. 8.2 Schematic cross-section of MESFETs and Schottky diodes for three alternative approaches to the fabrication of GaAs ICs: (a) mesa isolation; (b) planar/localized isolation; (c) planar/localized active layer; (d) self-aligned gate.

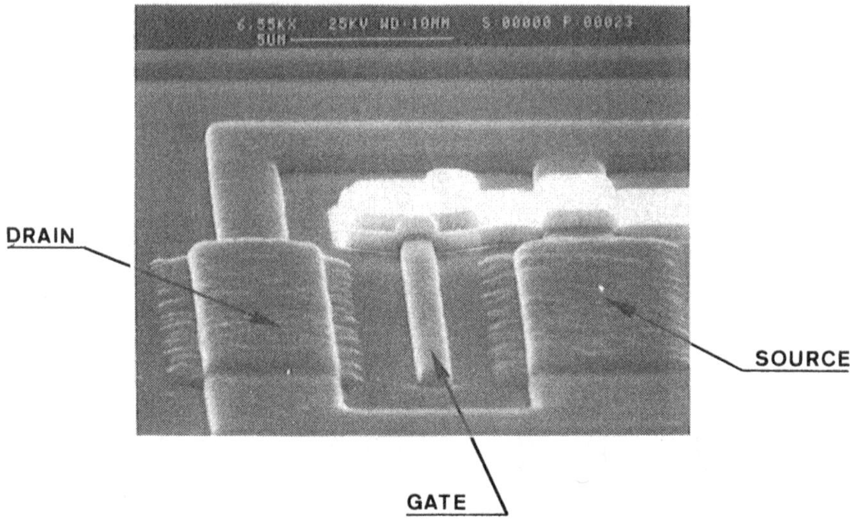

Fig. 8.3 Scanning electron micrograph of a MESFET with a gate length of 1.0 μm and a source–drain spacing of 4.0 μm.

The main disadvantages and problems of mesa-isolation technology are (a) the presence of a parasitic capacitance associated with the mesa-step coverage that limits the speed performance at small transistor widths and (b) that crossings of metal interconnection lines with mesa edges can lead to high resistance or open lines and thus to reduced fabrication yields and reliability problems. These drawbacks are eliminated by planar technologies.

(2) In the processing technology of Fig. 8.2(b) planar isolation is achieved by creating semi-insulating regions in an overall n-type layer (epitaxial or implanted). This is usually achieved by proton bombardment (140 keV, $5 \times 10^{14}\,\text{cm}^{-2}$) [10.25], boron implantation (150 keV, $2 \times 10^{13}\,\text{cm}^{-2}$) [10.26] or oxygen implantation (100 keV, $10^{13}\,\text{cm}^{-2}$) [10.27, 10.28] of silicon-doped layers. Apart from the benefits associated with planar technology, the main advantages of this technique are (a) the possibility of utilizing active layers prepared by a large number of methods (e.g. vapor or liquid phase epitaxy, ion implantation, molecular beam epitaxy) and (b) the easy control of active-layer characteristics at the beginning of the technological process. Its principal drawback is the occurrence of free-carrier recovery in insulating regions under high temperature annealing treatments (900–1100 K). Various ion species (H^+, B^+, O^+,...) have been proposed as being able to create compensated insulating regions in GaAs, but currently boron seems to offer the best trade-off in terms of junction depth–implantation energy and acceptable annealing temperatures [8.29]. A partial view of a planar digital IC fabricated by this

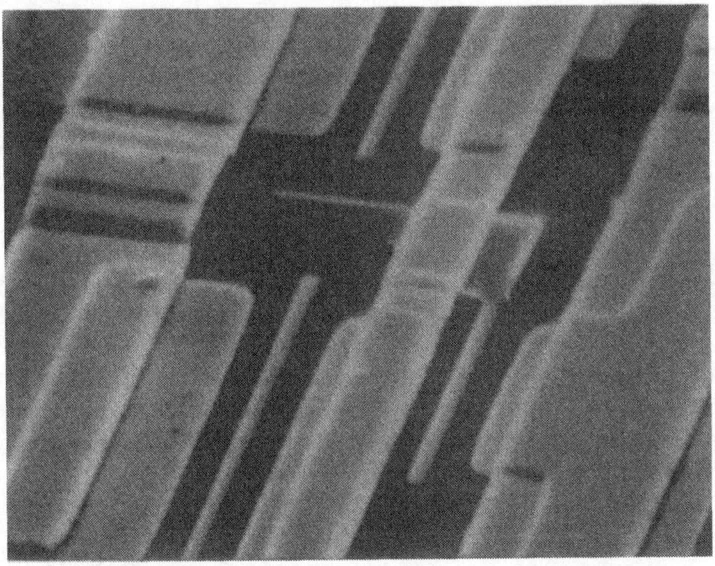

Fig 8.4 Scanning electron micrograph of a planar digital GaAs IC fabricated by localized implantation of boron to form insulating regions. (Courtesy of Thomson-Semiconductors.)

technique is shown in Fig. 8.4. Unlike mesa-isolated circuits, the contours of the devices are no longer apparent.

(3) Planar IC fabrication by multiple selective implantation of active areas and overdoped regions (Fig. 8.2(c)) is another attractive processing approach allowing the development of silicon-like GaAs IC technology. In fact, the variety of options offered by selective implantation allows a dramatic improvement in GaAs IC performance, packing density, integration level and reliability. Reported results on a 1000-gate circuit [8.30] have proved the capability of GaAs technology to reach the LSI level.

Multiple selective implantation offers a number of present or potential advantages including (a) the adjustment of FET channel doping profiles by a suitable choice of ion species and implantation conditions, or by use of a multidose multi-energy implantation scheme, (b) the reduction of parasitic resistances in MESFETs and Schottky diodes by forming localized overdoped regions and (c) the implementation of active loads in enhancement-mode circuits (see Section 8.4) or, more generally, the possibility of IC performance improvement by multiple pinch-off voltage technology [8.31]. Multiple implantation technology also allows the fabrication of GaAs junction FETs, which have been proposed as an alternative to enhancement-mode MESFETs for IC implementation [8.32].

Despite its very attractive features, selective implantation technology requires a close control of GaAs substrate quality in order to obtain reproducible results. In particular, GaAs ingot characteristics, such as dislocation density, residual impurities and electrically active defects, resistivity and thermal stability, have to be checked before wafer processing. High uniformity and good reproducibility of basic FET parameters can be achieved (for instance, a long-range dispersion of threshold voltage lower than 100 mV [8.33]) provided that a good substrate and a good annealing technique are available.

Recent comparisons of the two planar processing technologies have shown that the selective implantation process appears to be less favorable in terms of both electrical isolation between devices and back-gating effects. For instance, an improvement by a factor of 4 in the isolation voltage has been demonstrated for proton isolated circuits in comparison with a selective implant process [8.25]. Simultaneously, the level of the back-gating effect was reduced by at least a factor of 3.

(4) The most recently introduced fabrication process is self-aligned gate MESFET technology (Fig. 8.2(d)). This extremely promising approach has been introduced by Fujitsu in order to overcome the problem of reproducibility in fabricating LSI/VLSI enhancement-mode GaAs MESFET ICs [8.34]. The self-aligned GaAs technology, which is analogous to the polysilicon gate self-alignment technique of silicon LSI technology, has been made possible by the utilization of TiW silicide film to form Schottky contacts on GaAs. This silicide is extremely stable even at temperatures of up to 1100 K and thus allows annealing of self-aligned n^+-overdoped source and drain regions formed by using the TiW gate as an implantation mask.

The self-aligned gate process can be combined with either of the two planar processes previously described. It has been demonstrated to be usable over a wide range of threshold voltages and to improve the pinch-off voltage uniformity of normally-off MESFETs substantially (a typical standard deviation of 40 mV cm^{-2} has been reported [8.34]). In addition, the FET characteristics are greatly improved compared with those of conventional recessed-gate normally-off transistors, since the transconductance appears to be three times greater and the source series resistance five times smaller [8.34]. As a result, the propagation delay of self-aligned MESFET logic circuits is improved by a factor of about 3. For all these reasons, self-aligned MESFET technology should be considered as the key process for fabrication of high performance high reliability GaAs LSI/VLSI circuits.

8.3 GaAs MESFET characteristics

Micron or submicron geometries are needed for integrated devices in order to benefit fully from the potential speed advantages of GaAs. This is why

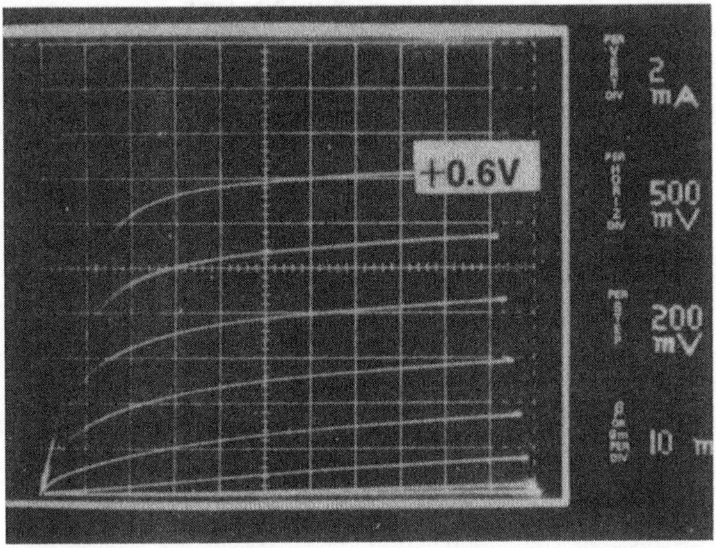

Fig. 8.5 Typical I_{DS}–V_{DS} characteristics of a normally-on GaAs MESFET with gate geometry 0.8 μm × 100 μm.

present-day integrated GaAs MESFETs have gate lengths ranging from 0.5 to 1.0 μm and source–drain intervals of 2.0–4.0 μm (Fig. 8.3). The basic simplicity of MESFET structures facilitates control of such small geometries, but achievement of satisfactory pattern delineation, reproducible linewidths and adequate alignment accuracy requires sophisticated microlithography techniques such as reduction projection photolithography with direct on-wafer stepping, direct-writing E-beam pattern generation or X-ray replication.

Figure 8.5 shows typical drain current I_{DS} versus drain–source voltage V_{DS} characteristics for a GaAs MESFET with a gate geometry of 0.8 μm.

MESFET channels require dopant profiles with a peak concentration of about $(1–3) \times 10^{17}$ cm^{-3} and a depth of about 500–2000 Å. Channel implantation is usually performed directly into the semi-insulating GaAs substrate (or through a dielectric layer (generally Si_3N_4) about 500–1000 Å thick) at doses of $(1–5) \times 10^{12}$ cm^{-2} using silicon (100–200 keV) or selenium (300–400 keV). Deep sulfur implantation (about 1×10^{13} cm^{-2} at 300–400 keV typically) is often preferred for n^+ overdopings. Annealing of implanted layers at temperatures of about 1100 K is needed to achieve high values (70%–100%) of activation efficiency in GaAs. However, free-atmosphere thermal treatments lead to decomposition of GaAs for temperatures exceeding about 873 K, and implantation annealing must be performed with extreme care (under Si_3N_4 or SiO_2 encapsulation or by capless annealing under arsenic pressure) in order to preserve material integrity [8.36].

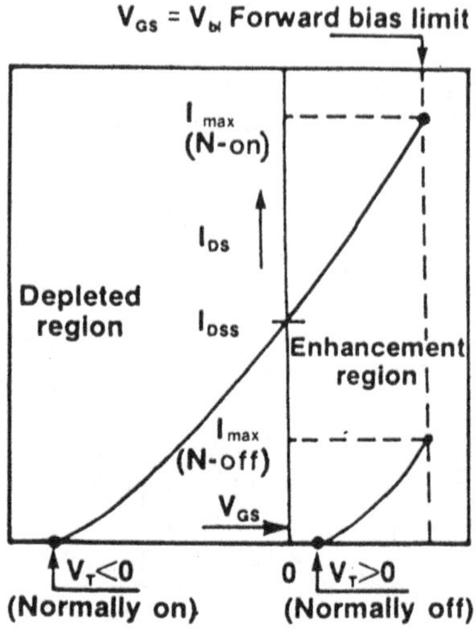

Fig. 8.6 Transfer characteristics in the saturation regime for normally-on and normally-off MESFETs.

The GaAs MESFET characteristics can be approximated by the Shockley relationship [8.35] which is valid for a long uniformly doped channel of carrier concentration N_d, thickness a and mobility μ. The measured saturation current exhibits an almost quadratic dependence on the gate–source voltage V_{GS} (Fig. 8.6),

$$I_{DS} = \frac{Z}{L}\gamma(V_{GS} - V_t)^2$$

which allows two important parameters of the device to be defined: the threshold voltage V_t and the current driving factor γ. Their ideal expressions are

$$V_t = V_{bi} - \frac{qN_d a^2}{2\varepsilon}$$

$$\gamma = \frac{2\varepsilon\mu}{3a}\alpha_c K_s$$

where α_c and K_s are reduction factors ($0 < \alpha_c$, $K_s < 1$) which depend in particular on the actual gate length and the parasitic source resistance respectively.

Depending on whether the device is operating with reverse ($V_p < V_{GS} < 0$) or forward ($0 < V_{GS} < V_{bi}$) gate–source bias, the mode of operation is usually called depletion or enhancement respectively (Fig. 8.6). Although exactly the same physical principles are involved, it is also usual to define two types of

MESFETs, normally-on ($V_t<0$) and normally-off ($V_t>0$), which lead to important differences in circuit implementation and IC performance.

The principal design parameters and characteristics of MESFETs currently utilized for the fabrication of digital GaAs ICs are summarized in Table 8.2.

Table 8.2 Current characteristics of integrated GaAs MESFETs

Channel doping level N_d	$(1–3) \times 10^{17}$ cm^{-3}
Electron mobility μ in channel	4000–5000 cm^2 V^{-1} s^{-1}
Channel thickness a	500–2000 Å
Gate built-in voltage V_{bi}	0.7–0.8 V
Gate length L	0.5–1.0 μm
Gate width Z	5–20 μm
Source–drain interval X_{SD}	2–3 μm
Transistor length X	10–12 μm
Threshold voltage V_t (normally-on)	-2.5 to -0.5 V
Threshold voltage V_t (normally-off)	0 to $+0.2$ V
Current driving factor γ	50–100 μA V^{-2}
Drain current I_{DSS} (at $V_p = -1.5$ V)	100–150 mA mm^{-1}
Transconductance g_m (at $V_p = -1.5$ V)	100–150 mS mm^{-1}
Drain conductance g_d (at $V_p = -1.5$ V)	10–20 mS mm^{-1}
Source–gate capacitance C_{GS}	0.5–1.5 pF mm^{-1}
Gate–drain capacitance C_{GD}	0.1–0.2 pF mm^{-1}
Drain–source capacitance C_{DS}	0.05–0.1 pF mm^{-1}
Current gain–bandwidth product f_t	15–25 GHz

8.4 Main features of the principal GaAs logic families

Since GaAs ICs were first developed, various logic circuit approaches have been proposed exhibiting different trade-offs in speed–power performance, logic versatility, power supply requirements, fabrication constraints and yield, and achievable integration level. Since normally-on FETs require a negative voltage to be switched off, depletion-mode logic circuits, buffered FET logic (BFL) and Schottky diode FET logic (SDFL) require a dual-supply voltage ($+V_{DD}$ and $-V_{SS}$) and a two-stage circuit design. However, enhancement-mode logic circuits and direct-coupled FET logic (DCFL) operate with a single supply ($+V_{DD}$) and can use either linear or saturated resistors or depletion-mode devices as pull-up loads or current sources. More recently, a new approach called source-coupled FET logic (SCFL) has been developed by Japanese companies.

Other features and detailed characteristics of the various logic approaches are discussed below. Transfer characteristics are usually measured on test gates, while power–delay performances are inferred from measurements performed on ring oscillator (RO) circuits.

8.4.1 Buffered FET logic (BFL)

BFL (Fig. 8.7) was the pioneering GaAs logic family [8.1, 8.2, 8.37] and it is widely utilized by many European, American and Japanese companies involved in the manufacture of GaAs ICs.

It is considered to be the most mature approach to GaAs logic but is highly power consuming. In fact, propagation delays t_{pd} as short as 34 ps at $P = 42$ mW have been reported for BFL circuits with 0.5 μm gate geometry [8.38]. In order to obtain these outstanding performances, the pinch-off voltage usually ranges between −1.5 V and −2.5 V, and three level-shifting diodes are used. This approach is called "high speed BFL" (hs-BFL) and is capable of multigigabit operation, but at the expense of a rather high power consumption which limits the achievable circuit complexity to the MSI level. Recent improvements in processing technology (in particular, achievable pinch-off voltage control) now allow the definition of a second BFL subclass offering a very different speed–power trade-off [8.39]. This is the "low power BFL" (lp-BFL) approach, which is obtained by a simultaneous optimization of the threshold voltage and the gate design ($V_T \approx -0.6$ V, two-diode level shifting). In this approach the power consumption can be reduced to about 1–3 mW per gate without significantly sacrificing the speed. This optimization in power consumption has led to 80 ps at 3 mW per gate and 108 ps at 1.5 mW per gate (0.16 pJ per gate) with widths of 20 μm and 10 μm respectively [8.40].

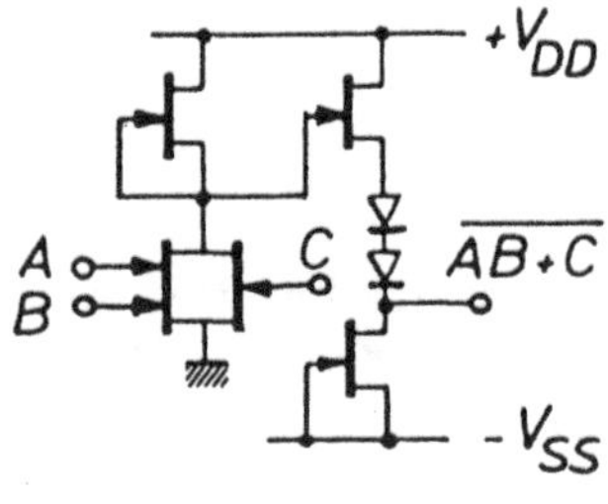

Fig. 8.7 Basic schematics of a BFL logic gate.

Figure 8.8 shows an extremely dense lp-BFL RO circuit including 29 inverters implemented with MESFETs of gate width 5 μm, gate length 0.5 μm and source–drain distance 2.5 μm. The effective inverter area is 500 μm^2 which corresponds to a packing density of 2000 gates per square millimeter [8.40] (the unwired gate density would reach a value of about 3000 gates per square millimeter). This value is similar to packing densities expected from ultra-dense VHSIC silicon technology.

Figure 8.9 shows a detailed scanning electron micrograph of an RO like the one in Fig. 8.8 but with a transistor 10 μm wide. It has been shown [8.41]

Fig. 8.8 Magnified view of an lp-BFL RO test circuit (29 integrated inverters). The MESFETs are 5 μm wide. The effective density is 2000 gates per square millimeter. (Courtesy of Thomson-Semiconductors.)

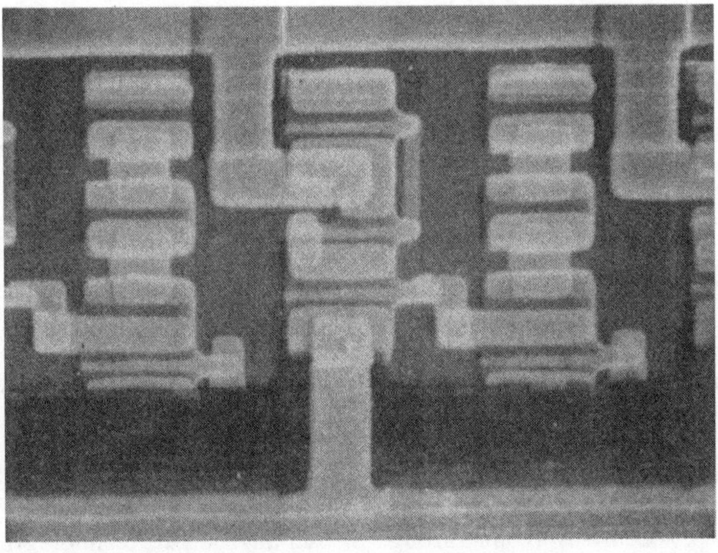

Fig. 8.9 Detail of an RO test circuit designed with a MESFET geometry of 10 μm. (Courtesy of Thomson-Semiconductors.)

that under certain fan-in–fan-out (FI–FO) conditions the elimination of the common-drain load driver results in a power–delay product optimization.

8.4.2 Schottky diode FET logic (SDFL)

SDFL (Fig. 8.10) uses both MESFETs and Schottky diodes to perform logic functions and has been proposed as an alternative approach to BFL in LSI circuit implementation [8.3, 8.4, 8.30, 8.42–8.44]. Very attractive speed–power trade-offs have been obtained with SDFL. Multilevel logic can be achieved using this logic approach [8.44].

The performances reported by Rockwell also include the benefits of a highly sophisticated multiple selective implantation technology [8.45, 8.46]. A more realistic performance comparison between BFL and SDFL is given in ref. 8.47 where it is shown that the speed advantage of SDFL over BFL does not exceed about 20% under the same conditions of power consumption, circuit design, FI–FO and process refinements.

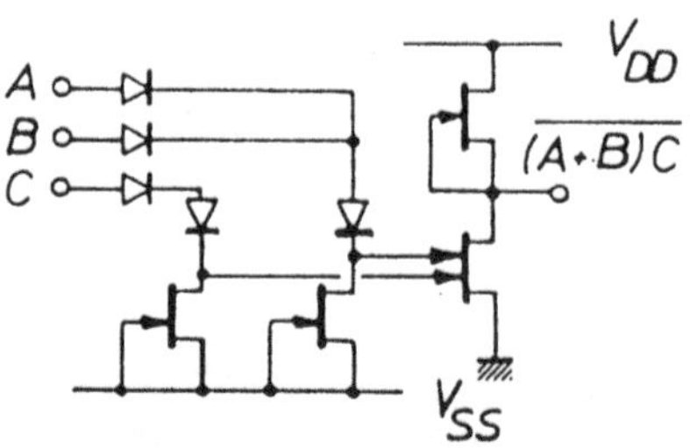

Fig. 8.10 Basic schematics of an SDFL logic gate.

8.4.3 Direct-coupled FET logic (DCFL)

DCFL (Fig. 8.11) offers the advantage of both extremely simple circuit design and very low power consumption for enhancement-mode logic. Thus it has been presented as an unrivalled approach for future LSI/VLSI GaAs ICs. The rather slow emergence of DCFL is mainly a result of the degree of control required for the pinch-off voltage ($\Delta V_t = \pm 100$ mV typically) in order to avoid decaying noise margins [8.34]. Thus DCFL suffers from much more severe constraints in IC processing than hs-BFL and SDFL (or equivalently lp-BFL) which tolerate V_T deviations five and three times wider respectively. This drawback is being overcome through the self-aligned gate technology, but it has delayed the emergence of DCFL circuits compared with normally-on logic approaches.

Apart from its technological difficulties, the enhancement-mode or normally-off FET approach is the key to GaAs LSI circuits. Although not as fast as depletion-mode devices at high FOs, enhancement-mode FETs dissipate much less power, occupy less space and can be operated with a single power supply voltage. Moreover, the enhancement-mode devices plus depletion-mode loads are similar to the relatively simple NMOS logic designs used in silicon. The self-aligned planar process has been shown to yield E–D mode inverters (FI/FO = 1/1) with propagation delays as low as 50 ps with a 287 fJ power–delay product at V_{DD} = 5 V [8.31]. The lowest power–delay product measured on the same devices was 14.5 fJ with a 98 ps delay time at V_{DD} = 0.6 V.

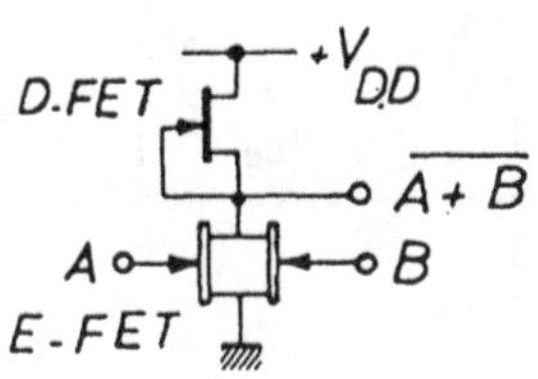

Fig. 8.11 Basic schematics of a DCFL logic gate.

8.4.4 Source-coupled FET logic (SCFL)

The circuit shown in Fig. 8.12 was recently proposed by NTT. It belongs to the current mode logic (CML) family. Several reference voltages are necessary in order to keep the transistors in the saturated regime and thus improve the speed. This configuration is highly tolerant with respect to pinch-off voltage dispersion and inherits all the known CML advantages. Some simulations yield propagation delays of 25 ps [8.48, 8.49].

8.4.5 Other approaches

Other approaches different from those presented in Figs 8.7 and 8.10–8.12 have been proposed. Capacitor-coupled logic (CCL) [8.50] is an example where interstage coupling is obtained by means of a capacitor. Another example is Schottky-barrier-coupled Schottky barrier gate FET logic (SSFL) [8.51] where Schottky diodes are used for the same purpose. Since these gates do not need a level-shifting buffer stage, the power–delay product is reduced. However, the circuits will not function properly at very low frequencies since the gates are coupled through a high-pass filter. This approach does not need very close control of the pinch-off voltage, and the typical measured performance is 59 ps for 3.8 mW per gate on an SSFL RO (FI/FO = 1/1) [8.52].

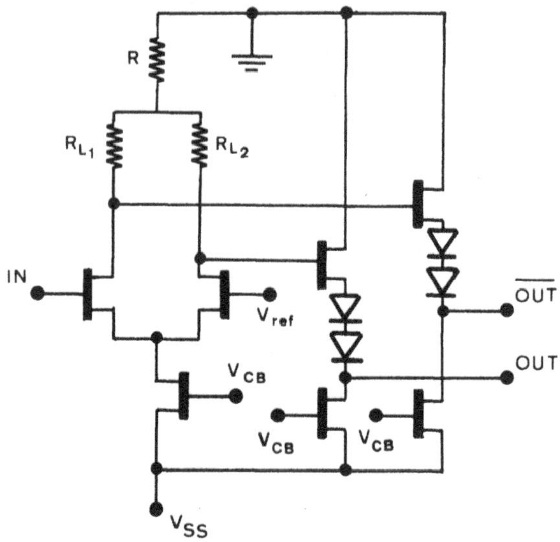

Fig. 8.12 Basic schematics of an SCFL logic gate.

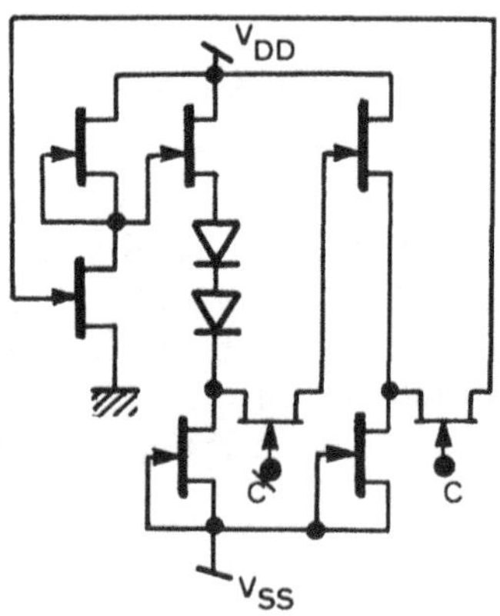

Fig. 8.13 Frequency divider by 2 using dynamic logic.

GaAs dynamic logic, which is well known in MOS technology, was proposed by LEP/Philips [8.53]. It is based on the use of MESFET transmission gates. This approach allows a drastic reduction in the complexity of the circuit, thus improving its performance. As an example, Fig. 8.13 shows the schematics of a dynamic frequency divider by 2 which uses a single BFL gate, two transmission gates and a buffer. With regard to the speeds attainable using this approach, a 10.2 GHz fixed-range frequency divider [8.54] and a power consumption as low as 0.25 mW for a DCFL-gate-based T flip-flop with a maximum clock frequency of 1.9 GHz [8.55] have been reported.

8.4.6 Summary

We can summarize Section 8.4 by presenting in Table 8.3 typical RO performances for each of the logic families described. Gate lengths are in the range 0.5–1 μm.

Table 8.4 shows the typical speed–power products for various logic families with 1 × 20 μm transistors and FI/FO = 2/2. The threshold voltage is the key parameter for circuit performance and optimum power supply voltages. For example, for a threshold voltage variation from −2.5 to +0.1 V, the typical propagation delay increases from 75 to 200 ps. We can then see that,

Table 8.3 Ring oscillator speed–power performances for various GaAs logic families

Logic family	*Company*	*Gate length × width* (μm)	*Stages*	T_{pd} (ps)	*P* (mW)	PT_{pd} (pJ)	*Reference*
hs-BFL	Hughes	0.5 × 50	5	34	41	1.4	8.38
hs-BFL	TH-SC	0.8 × 20	11	46	18	0.83	8.40
hs-BFL	Tektronix	1 × 10	7	62	15	0.93	8.56
lp-BFL	TH-SC	0.6 × 10	5	108	1.5	0.16	8.39
lp-BFL	TH-SC	0.6 × 20	5	80	3.1	0.25	8.39
SDFL	Rockwell	1 × 5	9	156	0.17	0.03	8.43
SDFL	Rockwell	1 × 20	9	75	2.3	0.17	8.42
SDFL	Honeywell	1 × 20	25	150	1.5	0.22	8.57
DCFL	Hughes	0.6 × 11	5	15	5.4	0.08	8.58
DCFL	Hughes	1 × 10	11	45	0.4	0.02	8.59
DCFL	TH-SC	0.8 × 20	15	55	4.3	0.24	8.39
DCFL	TH-SC	0.8 × 20	15	173	0.07	0.01	8.40
DCFL	Fujitsu	1.5 × 30	—	50	5.7	0.29	8.6
DCFL	NTT	0.6 × 20	15	30	1.9	0.06	8.60
DCFL	NTT	0.3 × 10	15	17	7.3	0.12	8.61
DCFL	Bell Labs	0.7 × 10	11	21	3.4	0.07	
DCFL	Oki	0.7 × 10	21	25	2.6	0.07	8.62

Table 8.4 Typical performances of GaAs logic families for FI/FO = 2/2

Logic family	*Threshold voltage*	T_{pd} (ps)	*Power* (mW)	PT_{pd} (pJ)	V_{DD} (V)	V_{SS} (V)
hs-BFL	−2.5	75	20	1.5	+4.0	−4.0
SDFL	−1.0	100	2	0.2	+2.5	−1.5
lp-BFL	−0.7	125	2	0.25	+2.5	−1.5
DCFL	+0.1	75–200	0.1–2.0	0.02–0.15	+1.5	—

if we exclude hs-BFL for which a high price in consumption is paid for its excellent speed performance, all other families present a power consumption compatible with LSI. The results presented in this table are the averages of published data. Typical standard deviations are about 20%–30% for the propagation delays, and as much as 100% for the power consumption.

8.5 GaAs digital integrated circuit design

In this section we summarize the field of computer-aided design, which includes all operations leading to mask definition.

8.5.1 Computer-aided design: work stations

The concept of a work station has evolved considerably in recent years owing to innovations in computer technology. With the advent of VLSI, a work station is generally based on a 32 bit minicomputer which includes at least a 1 Mbit random-access memory (RAM) and a 20 Mbit hard disk. The database is hierarchically organized in such a way that a close degree of coherence between different levels of circuit description is maintained.

Software packages typically include schematics capture, electrical simulation, logic fault simulation, test vector generation, graphic editors, design rule check, electrical rule check and gate array/standard cell placement and routing. Even though these software packages have been developed for the design of silicon ICs, they can easily be adapted for GaAs.

8.5.2 Logic simulation

Logic simulation is carried out at the gate level. It can be applied to either combinatorial or sequential circuits. Its purpose is to verify that the circuit satisfies its Boolean definition. Most logic simulators accept variable signal delays and some even allow the description of rising and falling edges. These delays can be a function of FO. Capacitive loading can sometimes also be included.

8.5.3 Electrical simulation

Electrical simulation is carried out at the component description level. The node equations are solved and the static or dynamic behavior is obtained. These techniques are only useful if a realistic model of the various components

(MESFETs, diodes etc.) is available and is coupled to the use of an efficient node equation algorithm in order to minimize central-processing unit time and memory.

Today, several electrical simulation programs are available. The most popular is Spice [8.63], which was originally developed at Berkeley and has been substantially improved in recent years. Other programs such as CIRCEC, ASTEC etc. are currently used in Europe. These programs only differ in their execution speed, their convergence and the possibility they offer of introducing user-defined models. The last feature is very important for GaAs IC simulation since it is then possible to simulate MESFET and Schottky diode devices correctly.

8.5.4 MESFET modelling

Figure 8.14 shows the equivalent circuit of a GaAs MESFET: R_g, R_d and R_s are the gate, drain and source access resistances; J_{DS} is the drain–source current generator. The value of the gate-controlled current generator is given by

$$J_{DS} = \gamma^{**}\frac{Z}{L_g}V_p^2\,\Omega\left[S\left\{u_d+\left(\frac{1}{\Omega}-1\right)g(u_{DS})\right\}-S\left\{u_s+\left(\frac{1}{\Omega}-1\right)\right\}g(u_{SD})\right]+$$
$$+Z\,g_c\,g(V_{GST}-V_Q)V_{D_1S_1}$$

where

$$\Omega=\frac{\lambda}{1-S(u_{GST})}\left(\frac{1}{2}-\left(1+\frac{2\lambda}{3}\right)^{1/2}\sin\left[\frac{1}{3}A\sin\left\{\frac{\lambda-1+2S(u_{GST})}{(1+2\lambda/3)^{3/2}}\right\}\right]\right)$$

$$\lambda=\frac{3L_gE_c}{V_p}$$

$$V_p=V_{bi}-V_t$$

$$V_Q=V_t-0.15$$

$$g_cZ=\frac{1}{2}\frac{\partial I_d}{\partial V_{DS}}\frac{1}{V_{GS}-V_t-V_Q}$$

$$g(x)=\begin{cases}x \text{ for } x>0\\ 0 \text{ for } x\leqslant 0\end{cases}$$

$$S(x)=\begin{cases}0 \text{ for } x\leqslant 0\\ x(3-2x^{1/2}) \text{ for } 0<x<1\\ 1 \text{ for } x\geqslant 1\end{cases}$$

$$u_d=\frac{V_{bi}-V_{G_1D_2}}{V_p}$$

$$u_s = \frac{V_{bi} - V_{G_1S_2}}{V_p}$$

$$u_{DS} = u_d - u_S = \frac{V_{D_2S_2}}{V_p}$$

$$u_{SD} = -u_{DS}$$

$$u_{GST} = \frac{V_{bi} - V_{GST}}{V_p}$$

$$V_{GST} = \begin{cases} V_{G_1S_2} \text{ if } V_{D_1S_1} > 0 \\ V_{G_1D_2} \text{ if } V_{D_1S_1} < 0 \end{cases}$$

These are nonlinear equations and the solution of a system of such equations is only possible if a very efficient alogrithm is used.

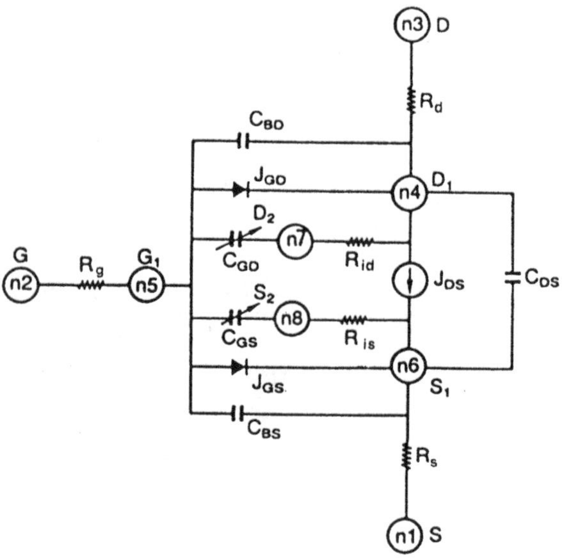

Fig. 8.14 Equivalent circuit of a GaAs MESFET. (Courtesy of Thomson-CSF.)

8.5.5 Design rules for digital integrated circuits

Table 8.5 presents a set of design rules commonly used for the design of GaAs ICs. Design rules are always specific to a given technology since they depend on the details of the fabrication process. Their definition and evolution are a result of constant analysis of the fabrication yields, as measured using specific test structures. The values shown in Table 8.5 are conservative in order

Table 8.5 GaAs design rules

Parameter	*Minimum dimensions* (μm)
Metal interconnections	2
Metal line separation	2
Ohmic contact size	2
Active zone separation	3
Source–drain spacing	2
Transistor width	5
Gate length	0.5

to maximize the fabrication yield. They do not represent absolute minima but are the result of a trade-off between performance, integration level and yield.

Careful layout is necessary to achieve top performance. Interconnections deserve special care, since they will determine the maximum performance of a circuit. They are usually considered from a purely architectural point of view. Most silicon automatic routers simply minimize the interconnection length and complexity in order to reduce capacitive loading and area. In GaAs technology the interconnection problem is more difficult to solve since the rise and fall times of the signals involved are of the same order of magnitude as the gate propagation delays. Furthermore, crosstalk also becomes important. Transmission line effects are not usually included for on-chip interconnections since the total length is still much smaller than a quarter-wavelength which corresponds to the highest signal harmonic.

The orders of magnitude of the various capacitances are shown in Fig. 8.15 for 2 μm lines separated by a distance of 4 μm [8.64]. The gate–source

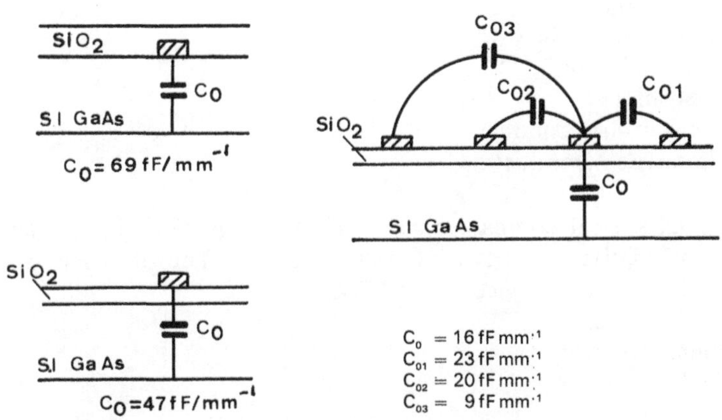

Fig. 8.15 GaAs IC interconnection line capacitances.

capacitance of a 10 μm transistor is close to 12 fF while the crossover capacitance between two 2 μm intercommunication lines is 1.3 fF.

Stray capacitances degrade circuit performance. In the case of the DCFL 16 × 16 multiplier developed by Fujitsu, as much as 55% of the critical path delay is caused by stray interconnection capacitance.

8.6 Manufacture of GaAs digital integrated circuits

Figure 8.16 shows a typical GaAs IC pilot line fabrication process (Thomson-Semiconductors). This process involves 23 operations and seven mask levels. Its main features are the following:

(1) localized active-zone ion implantation through an Si_3N_4 cap layer 500 Å thick;
(2) contact optical lithography ($L_g = 1$ μm);
(3) two levels of metallization;
(4) recessed-gate transistors;
(5) Si_3N_4 passivation 5000 Å thick.

8.7 Examples of GaAs digital integrated circuits

The development of GaAs ICs has been an important event for the catalog product market. In 1986 six companies were selling GaAs digital ICs:

Gigabit Logic
Harris Semiconductors
Honeywell
Matsushita
Thomson-Semiconductors
Triquint Semiconductors

In addition, several companies were offering GaAs foundry services, thus proving the relative maturity of GaAs technologies. The following companies were offering foundry services in the USA:

Adams–Russel
Gigabit Logic
Honeywell
Triquint Semiconductors

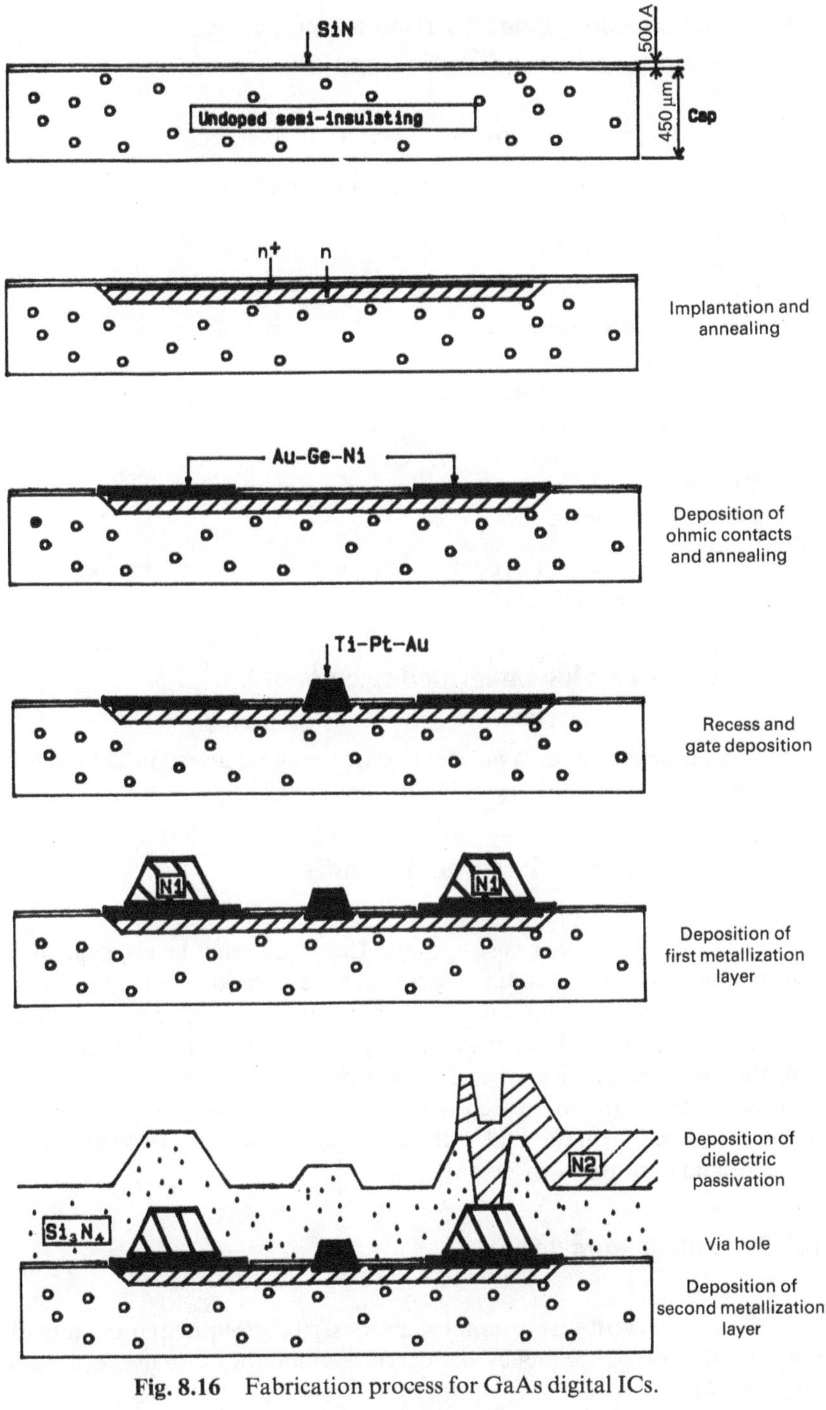

Fig. 8.16 Fabrication process for GaAs digital ICs.

The companies offering foundry services in Europe were

GEC
STC
Thomson-Semiconductors

Furthermore, several other companies announced their start-up in 1986 and 1987:

Anadigics
Celeritek
Ford Microelectronics
Gain Electronics
Microwave Monolithics
Microwave Technology
Plessey III–V Group
Tachonics
Vitesse Electronics

All these announcements confirm the expectations of industry for the emerging GaAs technology.

8.7.1 GaAs catalog integrated circuits

The catalog products available to date are summarized in Table 8.6. Most of these products use BFL gates for maximum yield.

8.7.2 GaAs custom integrated circuits

GaAs foundries offer customer services at different levels, depending on the demand and the foundry. Design can be carried out using any of the following approaches: full custom design; standard cell design; gate array design. These approaches are differentiated by the design turn-around time and the expected performance. This fact is shown in Fig. 8.17 for circuits of MSI complexity. It can be seen that, for a given technology, the best design approach depends on the skill of the designer, the acceptable design time and the expected performance.

8.7.3 Full custom design

Three representative examples are described: sequential circuits (fixed- and variable-range frequency dividers); combinatory circuits (4 bit adder); static RAMs.

Table 8.6 Typical GaAs catalog products

Product	*Gigabit Logic*	*Harris Semiconductors*	*Triquint Semiconductors*	*Honeywell*	*Matsushita*	*Thomson-Semiconductors*
OR–NOR gate	×	×				×
AND–NAND gate		×				×
XOR–XNOR gate		×				×
D-flip-flop	×	×				×
Fixed-range divider	×	×				×
Program divider	×	×	×		×	×
Comparator	×	×				
Counter	×		×			
Multiplexer	×		×			
Demultiplexer	×		×			
Register	×					×
Gate array		×	×	×		×

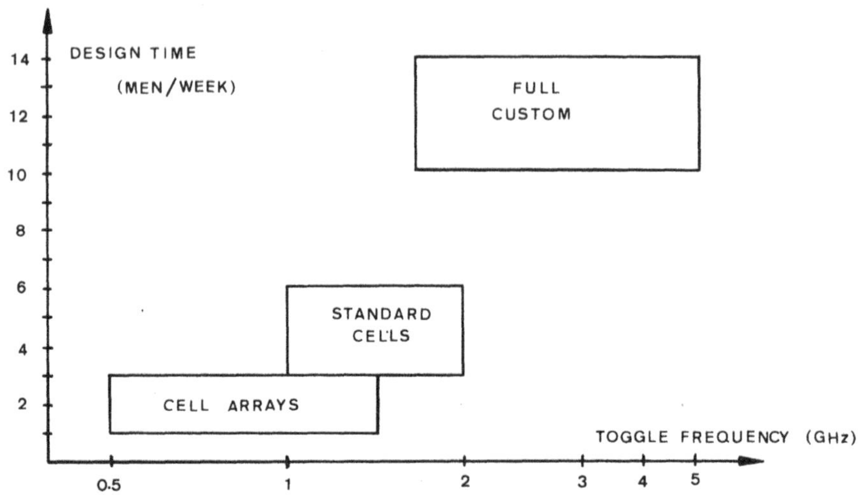

Fig. 8.17 Design time versus performance for various design approaches.

8.7.3.1 Fixed-range frequency dividers

These circuits are used as prescalers at microwave frequencies in order to bring the frequency down to the range where further treatment can be taken over by conventional silicon emitter-coupled logic (ECL). The divider's maximum clocking frequency is the most important parameter since the wideband characteristics of digital frequency dividers largely simplify the necessary circuitry, thus reducing the cost of the system.

The most commonly used divider structures are shown in Fig. 8.18 [8.65]. The maximum clock frequency is given by

$$f_m = \frac{1}{nt_{pd}}$$

where n is a factor depending on the frequency-divider configuration and t_{pd} is the average gate propagation delay. In Figs 8.18(a), 8.18(b) and 8.18(c) n is equal to 4, 2 and 4.85 respectively. The fastest structure is the RST$\bar{T}$ but it needs two complementary clocks. The logic schematics presented in Fig. 8.18 are based on NOR–OR BFL gates. The same functions could have been implemented using AND–NAND gates. Figure 8.19 shows typical transistor-level schematics for these logic functions.

In order to carry out a full custom design, the following items are necessary:

(1) very precise MESFET and Schottky diode models in order to be able to perform realistic simulations;

(2) electrical and geometrical design rules.

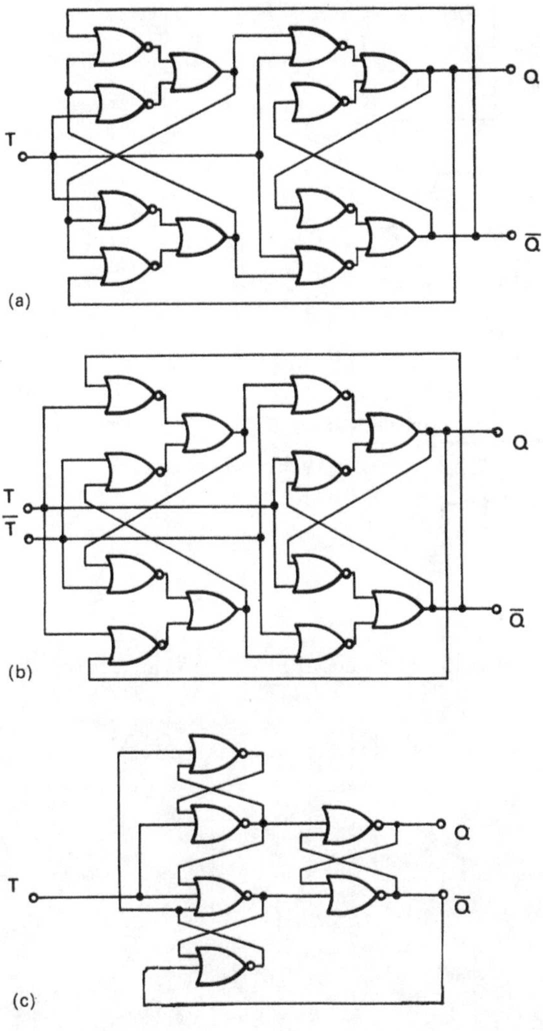

Fig. 8.18 Flip-flop structures for frequency dividers: (a) RST flip-flop used as a divider by 2; (b) RST$\bar{T}$ master–slave flip-flop used as a divider by 2; (c) D-flip-flop used as a divider by 2.

The full custom design of a frequency divider is a simple matter in principle. Figure 8.20 shows, as an example, a frequency divider by 2 and by 4 realized as a catalog product by Thomson-Semiconductors using 1 μm hs-BFL technology. The maximum clock frequency is 4.5 GHz for a total power consumption of 300 mW. Figure 8.21 shows the input and output waveforms for a frequency divider by 2 based on master–slave RST$\bar{T}$ flip-flops at a clock frequency of 4 GHz.

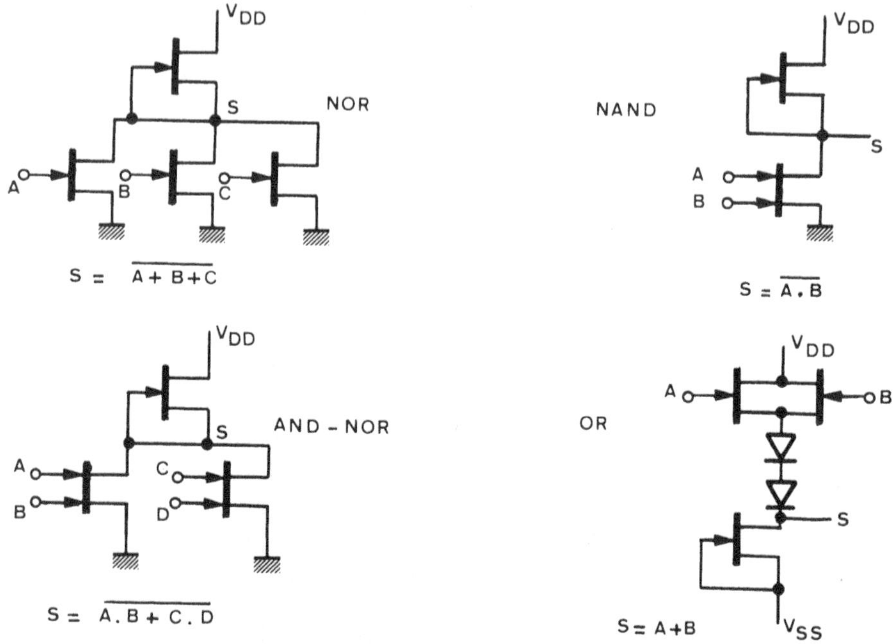

Fig. 8.19 D-mode MESFET GaAs logic functions.

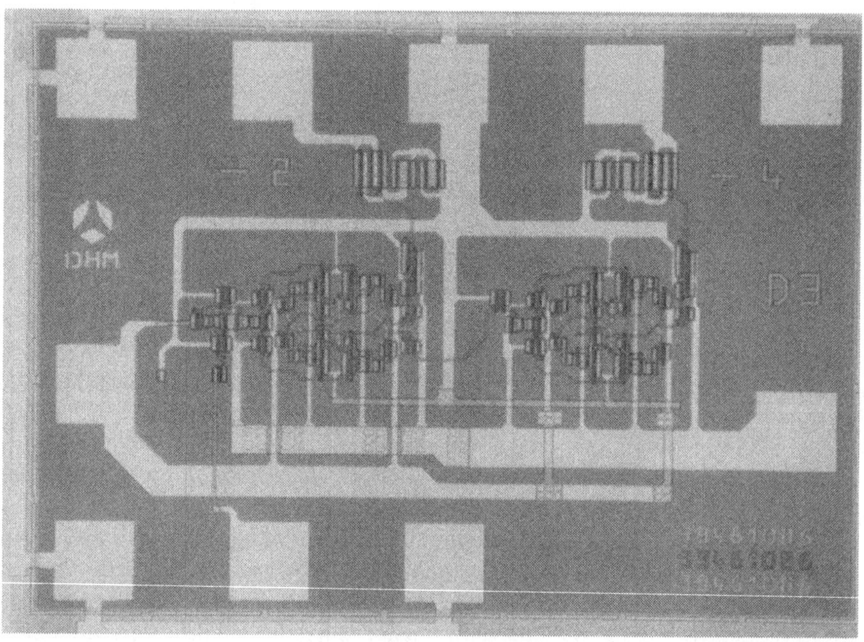

Fig. 8.20 Frequency divider by 2 and by 4. (Courtesy of Thomson-Semiconductors.)

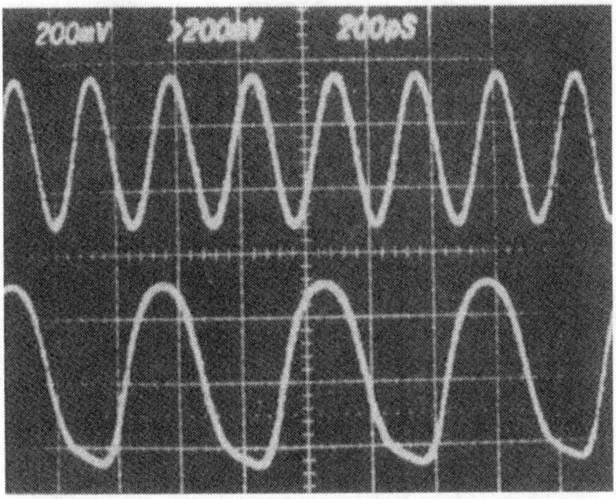

Fig. 8.21 Input and output waveforms for a frequency divider by 2 running at 4 GHz. (Courtesy of Thomson-Semiconductors.)

GaAs frequency dividers have been proposed by several companies. The reported laboratory and commercial product performances [8.66] are summarized in Fig. 8.22.

8.7.3.2 Variable-range frequency dividers

Variable-range frequency division is a very useful function for frequency synthesis applications. In general, a variable-range frequency divider (modulo N, $N+1$) consists of several flip-flops and feedback loops. The output signal period is an exact submultiple of the input signal period, depending on the feedback loop selected. The counter flip-flops can be either synchronous or asynchronous. The maximum clock frequency at which the circuit will still function properly will depend on the propagation delay t_{ff} of the flip-flops:

$$f_{\max} = 1/t_{\mathrm{ff}}$$

for synchronous counters;

$$f_{\max} = 1/nt_{\mathrm{ff}}$$

for asynchronous counters, where n is the number of stages of modulo N counter given by

$$n = \log_2(N+1)$$

It is evident that the synchronous counters are the fastest but at the price of a higher complexity and power consumption.

Figure 8.23 shows the typical architecture of a programmable frequency divider. It consists of a counter, a decoder and an initialization circuit.

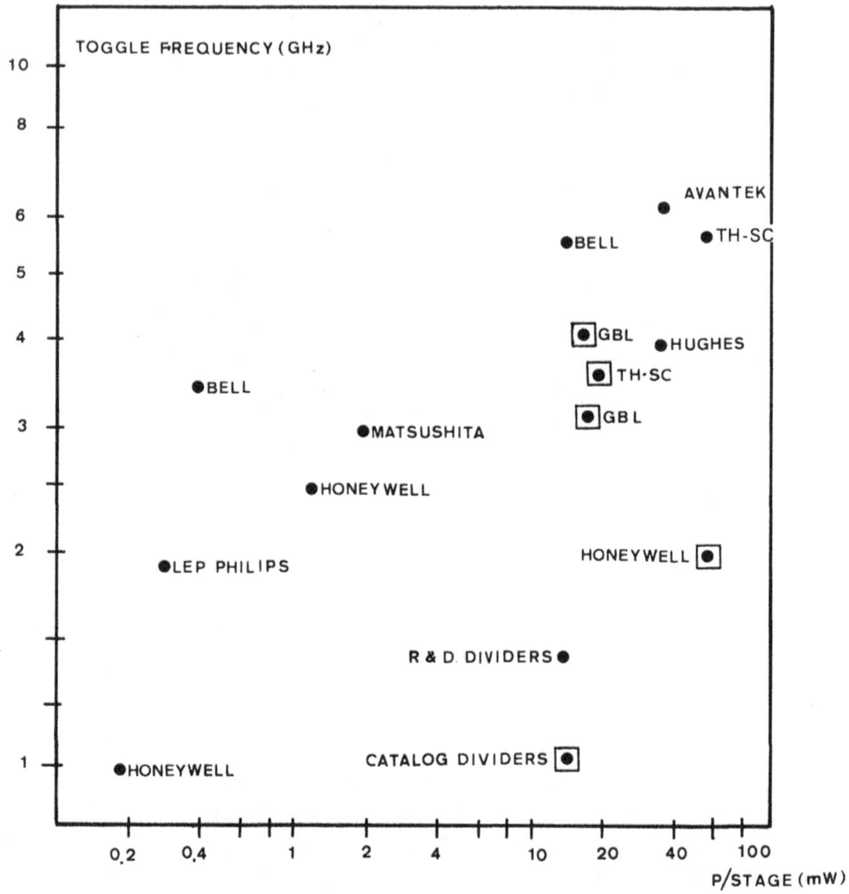

Fig. 8.22 Performances of frequency dividers.

Figure 8.24 shows another possible configuration for a variable-range frequency divider. It consists of a GaAs $N, N+1$ prescaler coupled to two ECL programmable counters. The output frequency is given by

$$F_o = \frac{F_{in}}{AN + B(N+1)}$$

In this case the clock frequency of the ECL counters is a factor of N lower than the input frequency of the GaAs prescaler.

Figure 8.25 shows an example of a commercial variable-range frequency divider by 5,6 fabricated by Thomson-Semiconductors using standard pilot-line 1 μm BFL technology. F_{max} is 4 GHz for a total power consumption of 600 mW.

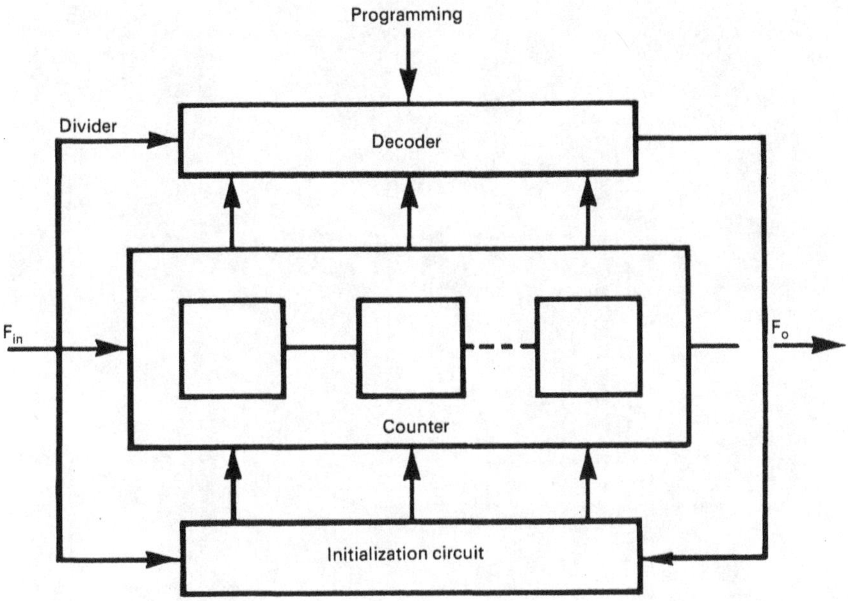

Fig. 8.23 Block diagram of a variable-range frequency divider.

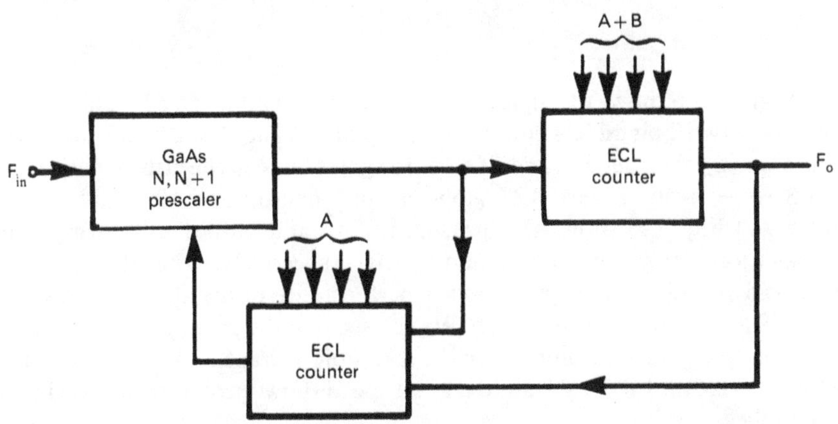

Fig. 8.24 Variable-range frequency divider using a programmable prescaler.

8.7.3.3 Four bit adder

The following gives a good example of the performances that can be attained by an MSI GaAs lp-BFL IC. We recall that the lp-BFL approach allows a drastic reduction of the power consumption per gate, and thus higher complexities are possible [8.40]. Typical pinch-off voltages for lp-BFL are in the range -0.9 to -0.6 V, and so only two level-shifting diodes are necessary.

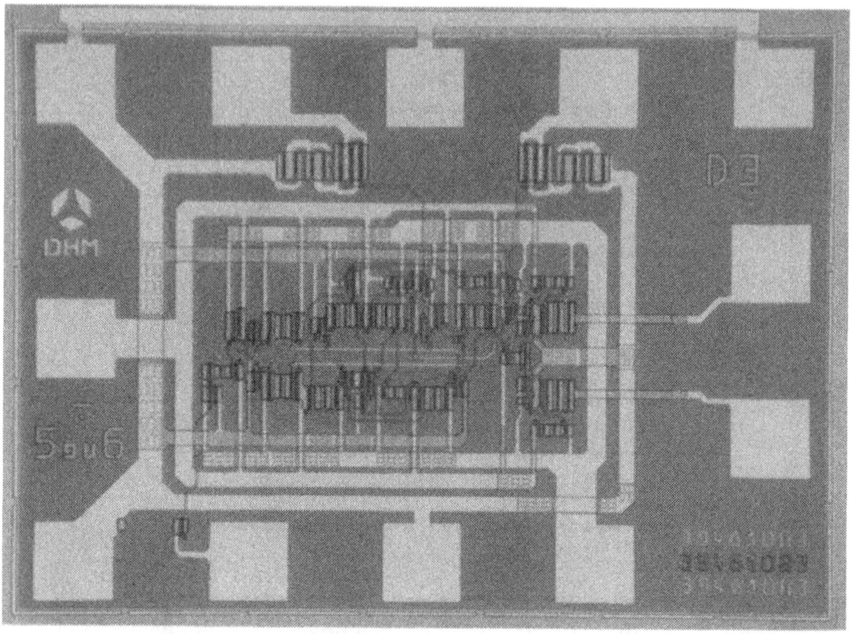

Fig. 8.25 GaAs programmable frequency divider by 5, 6. (Courtesy of Thomson-Semiconductors.)

Figure 8.26 shows the logic schematics of a 4 bit adder [8.67]. This circuit consists of two 2 bit adders with ripple carry in cascade. The critical path goes from C_0 to S_4. Average FIs and FOs for the gates lying in the critical path are 5 and 3 respectively. Figure 8.27 shows a photomicrograph of the fabricated adder and Fig. 8.28 shows the measured input and output waveforms. The propagation delays are 1.5 ns and 2.5 ns respectively. The effect of the threshold voltage on the speed–power performance of the adder is shown in Fig. 8.29 for low power dissipation. We can see that the propagation delay per gate is a rapidly varying function of the threshold voltage and that values of V_t between -0.6 and -0.9 V lead to a good speed–power trade-off for MSI and LSI applications.

This example clearly shows the very promising prospects of GaAs ICs for very high speed arithmetic applications. Efforts are increasing very rapidly in this field and highly encouraging results have already been reported. For instance, an SDFL-implemented 8×8 parallel multiplier including 1008 gates has been reported [8.4]. The chip size is $2.25\,\text{mm} \times 2.7\,\text{mm}$. This NOR gate implementation forms the 16 bit product of two 8 bit input numbers in 5.25 ns with a total power consumption of 1 W.

More recently Fujitsu has developed a 10.5 ns 16×16 multiplier. The circuit includes 3168 DCFL NOR gates (10625 active devices) and the chip

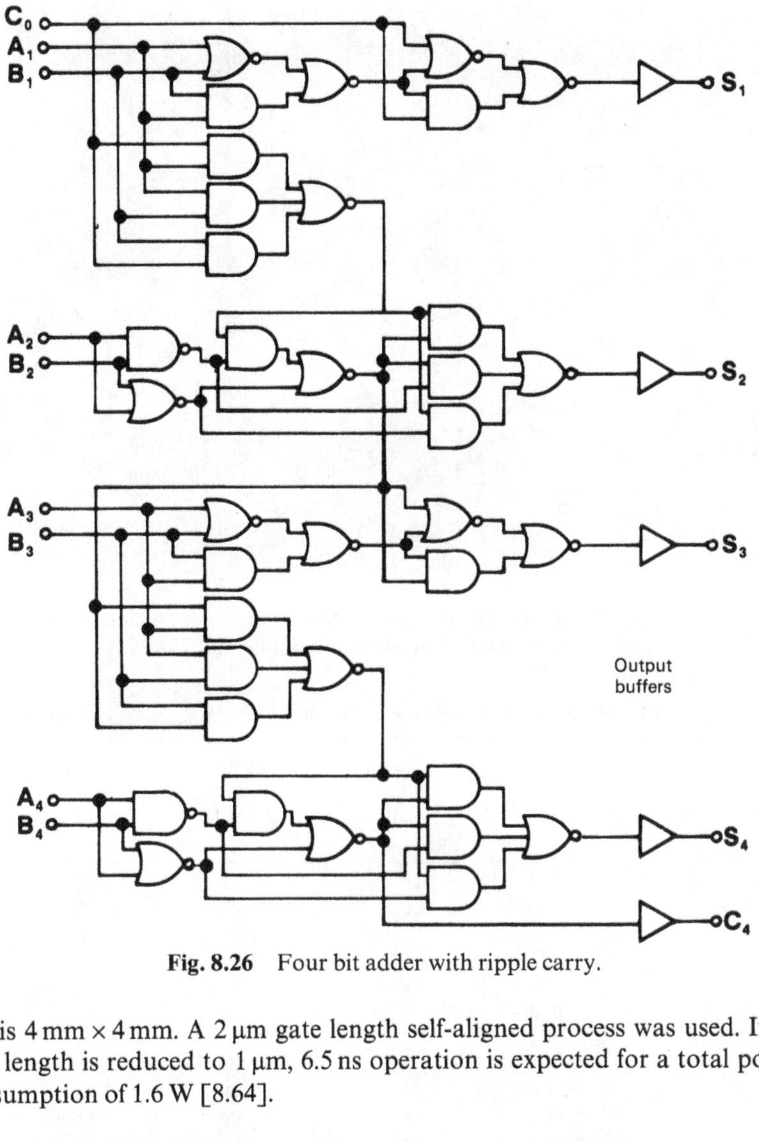

Fig. 8.26 Four bit adder with ripple carry.

size is 4 mm × 4 mm. A 2 μm gate length self-aligned process was used. If the gate length is reduced to 1 μm, 6.5 ns operation is expected for a total power consumption of 1.6 W [8.64].

8.7.3.4 High speed GaAs static random-access memories

Most applications of GaAs ICs in future high performance digital systems, such as computers and detection or communication systems, will require memory circuits with performances compatible with those of GaAs digital ICs. The principal applications of these circuits could be buffer or cache memories, and the complexity of some kilobits permitted by the present status of GaAs digital IC technology appears to be acceptable for many high speed digital-processing requirements. The development of GaAs RAM circuits is at

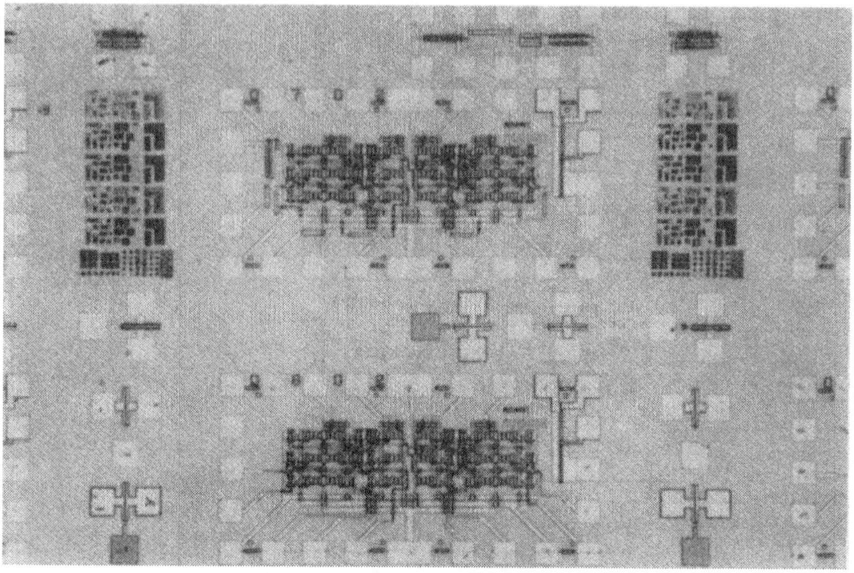

Fig. 8.27 Photomicrograph of a GaAs wafer containing a 4 bit adder circuit and test patterns. (Courtesy of Thomson-Semiconductors.)

Table 8.7 Reported performances of GaAs random-access memories

Company	*Size*	*Access time* (ns)	*Total power* (mW)	*Reference*
Thomson-SC	4 × 2	0.6	80	8.68
Rockwell	256 × 1	1.0	267	8.69
Toshiba	256 × 4	4.0	175	8.70
Hitachi	256 × 4	2.3	550	8.71
Rockwell	1024 × 1	2.8	180	8.72
Mitsubishi	1024 × 1	3.8	38	8.73
LEP/Philips	1024 × 1	3.0	700	8.74
Texas Instruments	1024 × 1	2.5	1200	8.75
NTT	4096 × 1	2.8	1200	8.76
Toshiba	4096 × 1	2.6	1800	8.77
NTT	16384 × 1	4.1	2520	8.78
Fujitsu	4096 × 1	3.0	700	8.79

present being carried out by a number of companies. Several results over a wide range of performances are presented in Table 8.7.

Owing to the complexity of GaAs RAMs a circuit approach using enhancement-mode MESFETs is preferred in order to optimize the power consumption. This is why all the GaAs RAM designs reported to date are

(a)

(b)

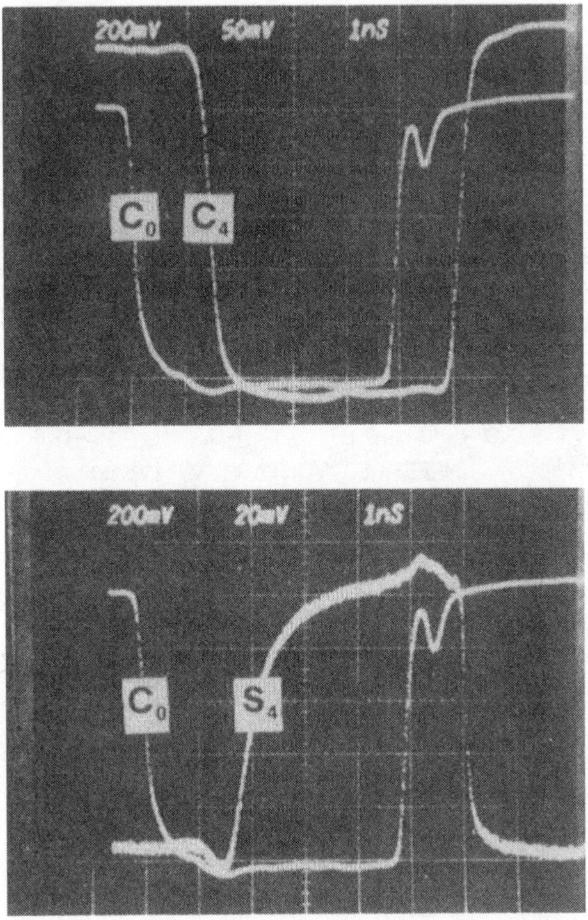

Fig. 8.28 (a) Sampling scope C_0 and C_4 traces for $V_{DD} = +3.5$ V and $V_{SS} = -2.5$ V; (b) sampling scope C_0 and S_4 traces for $V_{DD} = +3.5$ V and $V_{SS} = -2.5$ V.

based on the DCFL circuit approach. In RAMs all memory cells are random addressable for either a read or a write operation. In sequential memory the cells can only be addressed in a predefined order. RAM cells are usually matrix organized. Access to cells is obtained by activation of a "word line" and either read or write on a "bit line". The position of the word and bit lines activated corresponds to the coordinates of the addressed cell, as shown in Fig. 8.30. GaAs RAMs are usually designed using the full custom approach, since each function block has to be carefully optimized for maximum performance and minimum power consumption.

In 1984 NTT reported the realization of a 4 kbit static RAM using self-aligned refractory gate technology. The design is based on the DCFL

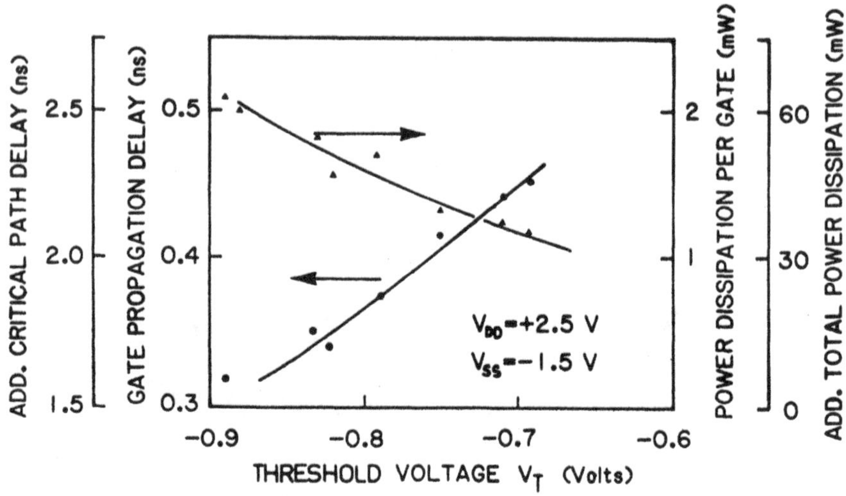

Fig. 8.29 Propagation delays and power dissipation as a function of the pinch-off voltage V_t.

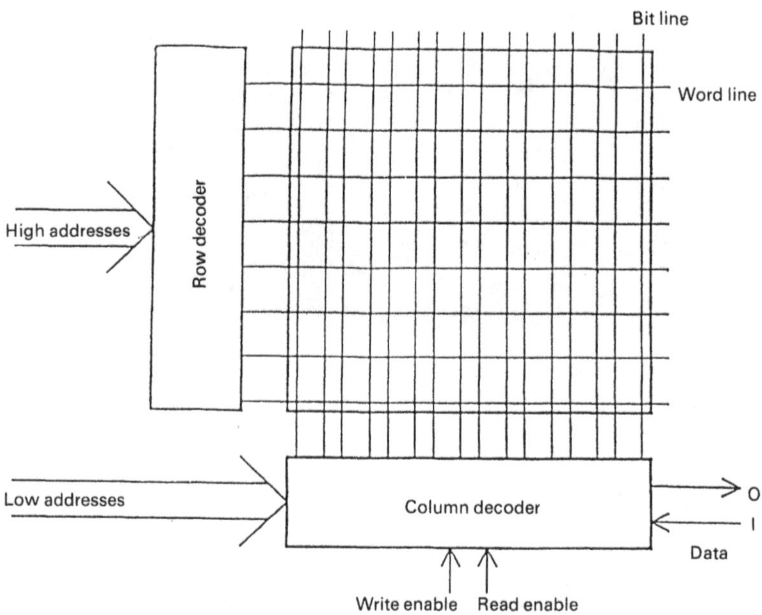

Fig. 8.30 Organization of a RAM.

approach and integrates 26031 MESFETs in a 4.6 mm × 3.4 mm chip. The memory cell size is 47 μm × 32 μm. The measured access time is 3 ns and the total power consumption is 700 mW. Simulation shows that an access time of

3 ns and a power consumption of 900 mW can be attained for 1 μm MESFETs and 1.5 μm design rules [8.80].

8.7.4 Standard cell design

Standard cell IC design reduces the cost of the design phase since use is made of optimized cells which have already been verified. However, performance is somewhat degraded (by 10%–15%) owing to the constraints of the approach.

In general, a standard cell library includes a set of computer-aided design (CAD) tools which allow a symbolic as well as a mask-level representation of the layout. Since the cells have been previously designed and characterized, the layout phase involves placement and routing. Very efficient algorithms have been developed for this operation, mostly for silicon applications. The topology is such that power rails become interconnected by simple placement of the cells in rows. The rows of cells are separated by routing channels whose width is optimized by the placement-and-routing CAD tool.

Let us consider, as an example, the 4 bit universal register fabricated by Thomson-Semiconductors using their standard cell library. The operations performed by this circuit are as follows: 4 bit parallel-in–parallel-out register; shift left; shift right; parallel load. Figures 8.31 and 8.32 show respectively a

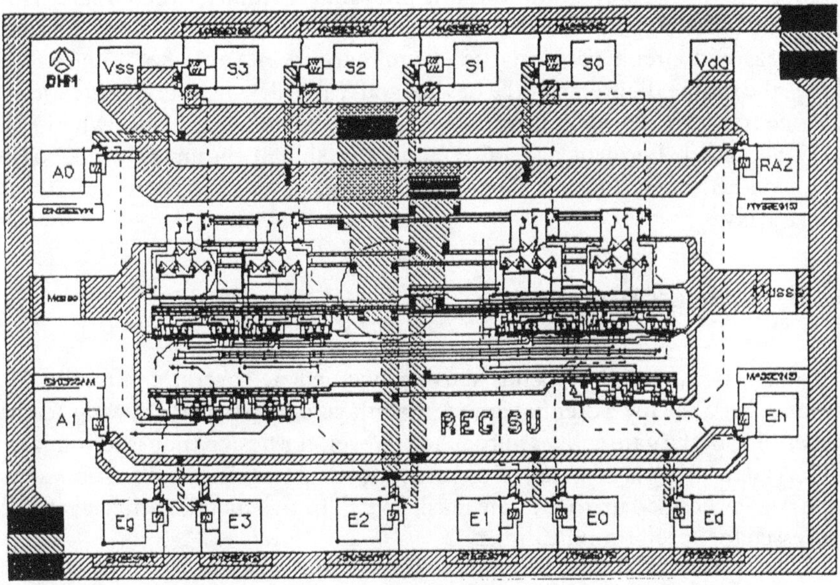

Fig. 8.31 Symbolic layout of a 4 bit universal register. (Courtesy of Thomson-Semiconductors.)

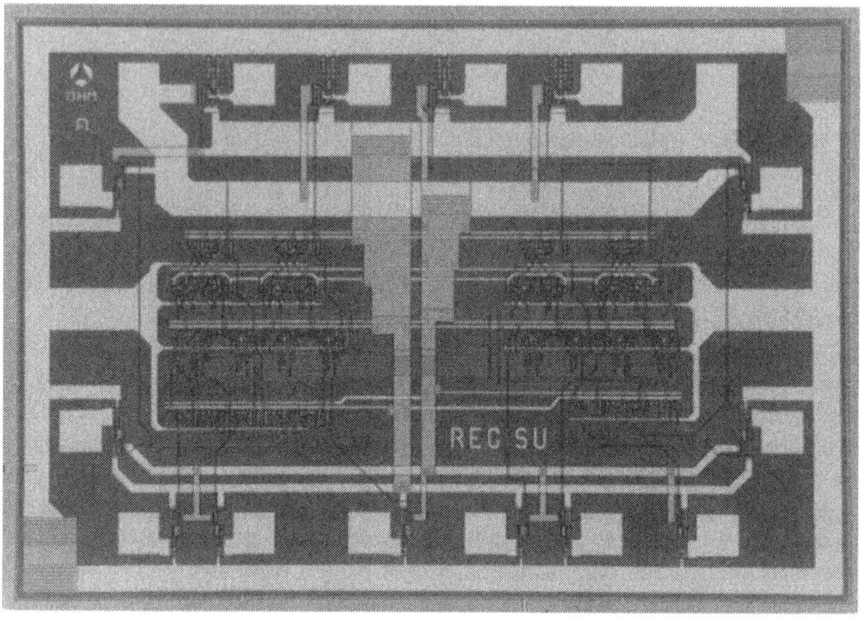

Fig. 8.32 GaAs 4 bit universal register. (Courtesy of Thomson-Semiconductors.)

symbolic layout representation and a photomicrograph of the circuit. This circuit includes 60 gates (450 active devices). The chip size is 2 mm × 1.5 mm. The measured circuit performance is summarized in Table 8.8. Figure 8.33 shows the timing diagrams of the device, and Table 8.9 gives its specifications. A scope trace of a working universal register in the shift mode is shown in Fig. 8.34. The clock frequency is 1 GHz and the shifted datum is a 500 MHz synchronous signal. Output signal rise and fall times are 150 ps and 200 ps respectively.

8.7.5 Gate array design

Gate array design is being actively pursued by several companies. Its critics argue that the potential speed advantage of GaAs is lost owing to the constraints of the gate array approach; however, the design time is minimized. A GaAs gate array is an array of unwired identical transistor cells. The wafers have been processed up to the gate mask level. The two metallization levels are processed for specific applications.

In Europe Thomson-Semiconductors have developed the 1000-equivalent-gate BFL gate array with 32 ECL-compatible input–outputs shown in Fig. 8.35. Typical propagation delays are 130 ps for FI/FO = 2/2

Table 8.8 Measured performance of a GaAs 4 bit universal register

Positive power supply V_{DD}	+3 V
Negative power supply V_{SS}	−2.5 V
Power consumption P_t	900 mW
Maximum clock frequency f_m	1.2 GHz

Table 8.9 Specifications of a GaAs universal register

Parameter	*Chip*	*Packaged device*
Maximum clock frequency f_m (GHz)	0.8–1.2	0.8–1.2
Reset pulse width T_1 (ps)	650	650
RES–S_n delay (CK = 0) T_2 (ps)	1230	1280
RES–S_n delay (CK = 1) T_3 (ps)	880	1010
Mode set-up T_4 (ps)	280	280
Mode hold time T_5 (ps)	−80	−80
Data set-up ($E = 0$) T_6 (ps)	390	390
Data set-up ($E = 1$) T_7 (ps)	230	230
Data hold time ($E = 0$) T_8 (ps)	10	10
Data hold time ($E = 1$) T_9 (ps)	170	170
Data pulse width T_{10} (ps)	400	400
CK–S delay ($E = 0$) T_{11} (ps)	1070	1200
CK–S delay ($E = 1$) T_{12} (ps)	880	1010
Internal cycle ($E = 0$) T_{13} (ps)	1080	1080
Internal cycle ($E = 1$) T_{14} (ps)	740	740
Clock low time T_{15} (ps)	400	400

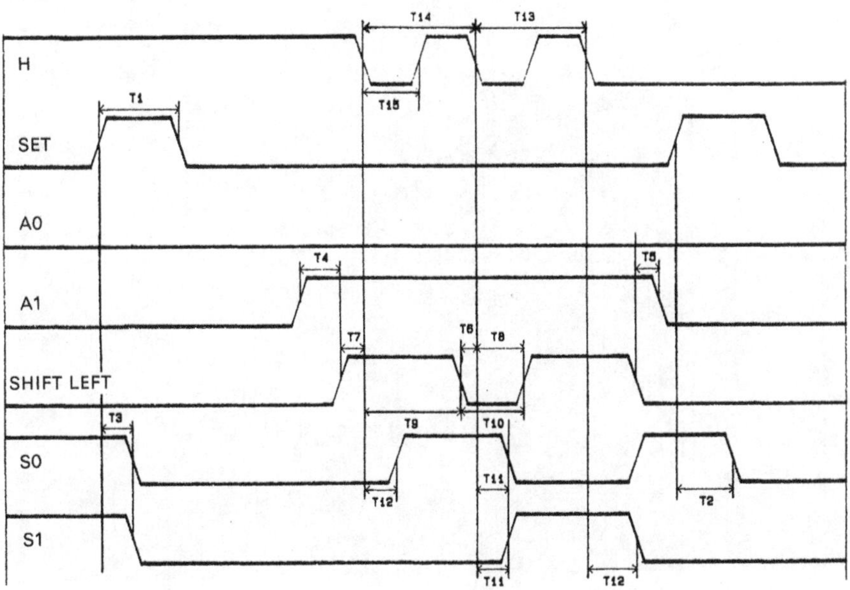

Fig. 8.33 Timing diagrams of the 4 bit universal register.

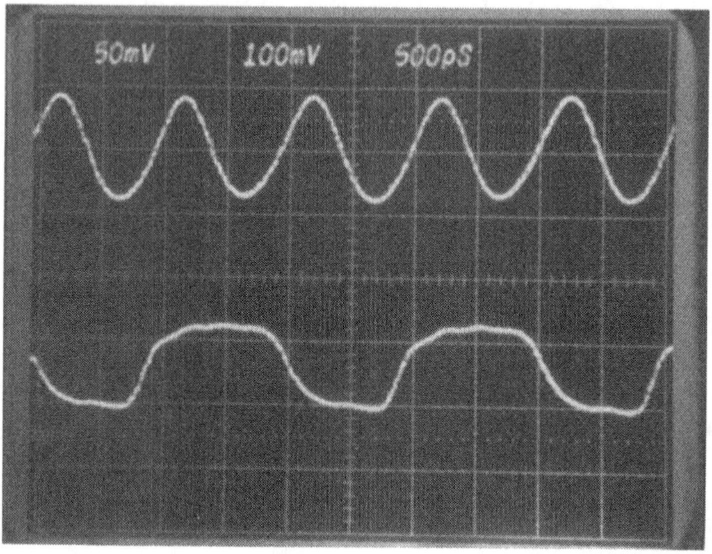

Fig. 8.34 Waveforms of the GaAs universal register.

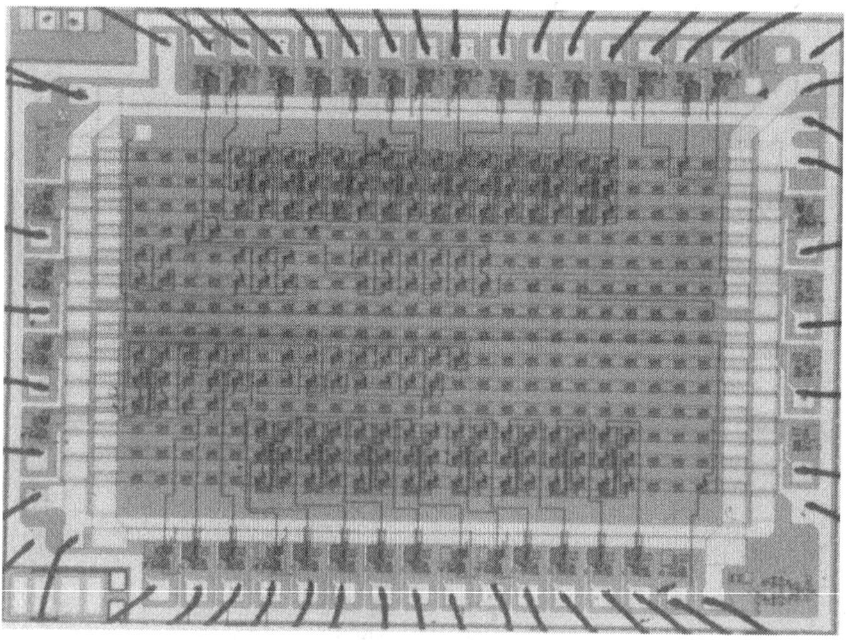

Fig. 8.35 GaAs gate array. (Courtesy of Thomson-Semiconductors.)

and 6 mW per gate. The typical clock frequency is 1.2 GHz. Other companies are currently offering gate arrays (BFL or DCFL) and cell arrays where the cell function has been carefully optimized and the cells have been preplaced. This is the approach adopted by Harris Semiconductors. The main features of available GaAs gate arrays are summarized in Table 8.10.

Table 8.10 Available GaAs gate arrays

Company	*Logic family*	*No. of cells*	*Cell power* (mW)	*Gate delay* (ps)	*Reference*
Harris Semiconductors	BFL	170	35	130	8.81
Thomson-Semiconductors	BFL	320	6.5	140	8.82
Honeywell	SDFL	432	1.5	110	8.83
Triquint Semiconductors	BFL	280	20	190	8.84
Toshiba	DCFL	1050	0.2	350	8.85
Ford Microelectronics	FDCL	500	—	—	8.86
Oki	DCFL	1000	2.25	390	8.87
Rockwell	BFL	306	2	200	8.88

8.8 GaAs analog integrated circuits

The definition "analog integrated circuits" usually includes a large number of families other than those already presented and monolithic microwave ICs. Analog ICs can include small-signal and large-signal circuits and, depending on the application, signal frequencies going from d.c. to a few gigahertz.

GaAs is expected to be the key material for high speed signal-processing devices. This application requires great precision and speed, and these often contradictory requirements are the source of the difficulty in developing analog GaAs IC technologies since a high material quality and good reproducibility are necessary. This implies extremely low parameter dispersions, usually a factor of 6 less than those acceptable for current digital ICs. The following main parameters need to be closely controlled:

(1) the threshold voltage V_t;
(2) the g_m/g_{DS} ratio, indicating the intrinsic MESFET gain;
(3) the observed hysteresis in $I_{DS} = f(V_{GS}, V_{DS})$;
(4) parasitic effects such as side-gating or back-gating;
(5) the thermal noise of the MESFET which is usually high;
(6) temperature drift of the parameters;
(7) the gain–bandwidth product of the MESFET.

The optimization criteria for analog ICs differ substantially from those for digital ICs. Gate lengths for digital applications are as short as possible (less than 1 μm) whereas for analog circuits they range between 1 and 2 μm which increases the drain conductance. Despite the difficulties associated with parameter nonuniformity, interesting results have been reported in the fields of analog ICs for signal processing and analog ICs for digital communications.

8.8.1 GaAs analog circuits for signal processing: analog-to-digital converters

Signal processing has evolved from analog to digital during the last 20 years. High rate analog-to-digital conversion is the main motivation behind the efforts of most companies developing GaAs analog ICs. Figure 8.36 shows the block diagram of a typical digital signal-processing system. The fastest silicon analog-to-digital converters (ADCs) have 6 bit precision at about 100 Msamples s^{-1} (Matsushita, Sony, TRW, Siemens). The main initial aim for GaAs ADCs is to attain 6 bit precision at 1 Gsample s^{-1}.

In order to improve the accuracy and speed, the following problems will have to be solved.

Speed Since gigasample per second rates are necessary, very high performance components will have to be integrated.

Precision The offset of the comparator input must be less than a quarter of the converter's resolution. This means that in order to obtain 20 mV resolution the offset must be less than 5 mV, which is incompatible with present GaAs MESFET technology.

Complexity Speed constraints impose flash converter architectures leading to integration levels in the MSI–LSI range.

The development effort must be concentrated in the following areas.

(1) The material and the processing must be improved in order to reduce parasitic effects such as side-gating, hysteresis and defect density.

(2) New converter architectures must be explored in order to tolerate the higher parameter dispersions current in GaAs. Dynamic comparators are an example.

Table 8.11 presents experimental results obtained in the high speed analog-to-digital domain. Some results for digital-to-analog conversion are also included. Heterojunction bipolar transistor technology holds a great potential for analog ICs, as the first realizations have shown [8.94]. Rockwell

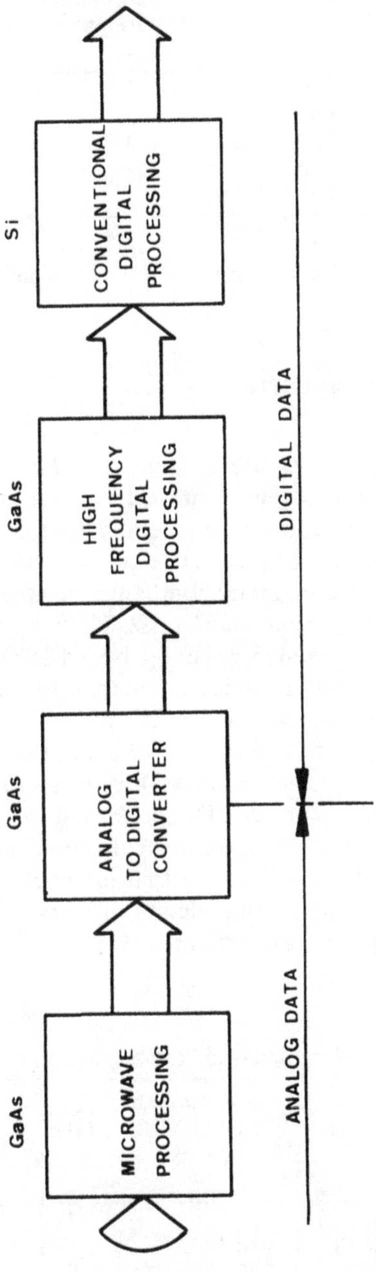

Fig. 8.36 Digital signal-processing system.

Table 8.11 High speed analog-to-digital and digital-to-analog conversion results

Company	*Product*	*Performance* (Gsamples s^{-1})	*Reference*
RCA	3 bit ADC	1–2	8.89
Raytheon	3 bit ADC	1–2	8.90
LEP/Philips	4 bit ADC	1–2	8.91
Tektronix	4 bit ADC	1	8.92
Hewlett-Packard	12 bit DAC	1	8.93

has reported comparator offsets smaller than 4 mV and hysteresis of less than 1 mV.

8.8.2 Digital link regenerator

Signal-restoring circuits are a familiar component in high-data-rate digital links. They reshape signals degraded by the transmission.

Silicon IC data regenerators are generally limited to data rates lower than 600 Mbit s^{-1}. The use of GaAs leads to data rates as high as several gigabits per second. For this reason, many companies engaged in the telecommunications business have developed GaAs ICs. These companies include AT&T [8.95], NTT [8.96], Thomson [8.97] and CNET [8.98].

A regenerator circuit includes a comparator, a flip-flop and a low impedance output interface as shown schematically in Fig. 8.37. The main specifications of a signal regenerator are the input sensibility, the maximum clock frequency and the transition time. Table 8.12 shows the specifications of the various realizations reported. The performance of these devices should increase in the years to come to allow the realization of optical data links of up to 4 Gbit s^{-1}. It should also be noted that many companies are developing high-data-rate multiplexers and demultiplexers. The first commercial multiplexer–demultiplexer products at 1.6 Gbits s^{-1} have been proposed by Gigabit Logic [8.99].

Table 8.12 Reported specifications for signal regenerators

Company	*Logic family*	*Sensibility* (mV)	$f_{CK\,max}$ (GHz)	*Rise time* (ps)	*Reference*
Thomson-SC	hs-BFL	75	3	115	8.97
NTT	SCFL	80	2	150	8.96
Bell	DCFL	70	2.4	150	8.95
Thomson-SC	hs-BFL	80	4.4	100	8.98

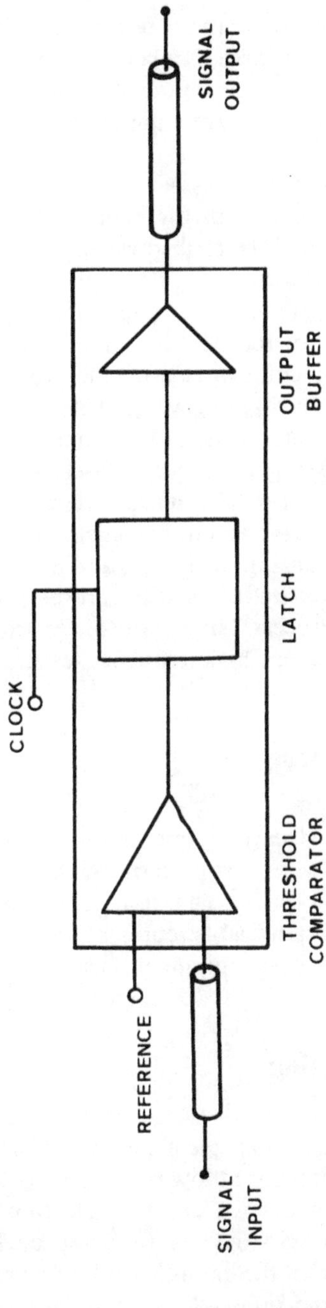

Fig. 8.37 Schematics of a signal regenerator.

8.9 Tests of GaAs digital integrated circuits

GaAs digital ICs are difficult to test because of their ever increasing complexity, which requires a large number of test vectors, and their extremely high speed. Silicon test equipment is not well suited for GaAs ICs because it is almost impossible to apply a large number of gigahertz frequency signals through a probe card.

This technological problem has been solved in some special cases. These solutions involve improvement of the signal interconnection between the measurement equipment and the probed wafer. The electronic test equipment is also improved so that the old manual sampling techniques are replaced by completely automatic systems. The main difficulty stems from the non-existence of adequate commercial test equipment for this purpose. This is why GaAs foundries are often confronted with the problem of developing GaAs ICs and at the same time having to develop their own test equipment.

High frequency on-wafer testing of the functionality of GaAs circuits is extremely difficult to carry out in most cases. A GaAs IC wafer usually includes, in addition to the specific application circuit, various test patterns for the following purposes: measurement of some important parameters, such as the threshold voltage V_t, during the fabrication process and extraction of the complete parameters after the fabrication process in order to evaluate the quality of the wafer and process. Three types of test are carried out: parametric tests; functional tests; dynamic high frequency tests.

8.9.1 Parametric testing

Parametric testing is carried out automatically after the fabrication process using special on-wafer test patterns. A wafer probe is used to contact the various test pattern pads. The test results are stored and treated statistically in order to obtain usable representations of the various parameter distributions. Commercial test equipment is used.

8.9.2 Functional testing

Owing to the limitations of present commercial automatic test equipment functional testing is carried out at low frequencies (less than 50 MHz). These measurements do not give an accurate indication of what the different electrical parameters of the circuit will be at full speed. For this reason the tests are limited to verification of the circuit's truth table and measurement of the power supply currents and input–output interface signal levels. Experience has shown that most circuits which work properly at low frequencies also function correctly at high frequencies.

8.9.3 Dynamic testing

8.9.3.1 On-wafer dynamic testing

The correct transmission of a microwave signal requires special care since transmission has to take place with minimum losses through impedance-matched lines to minimize reflections. The probe-card high frequency IC test is only possible over a limited number of probes. Signals have to reach and leave the probe card through coaxial lines. Probe cards have to be carefully designed using strip line techniques in order to avoid impedance mismatching. It can clearly be understood that, as the number of high frequency probes increases, difficulties also increase owing to higher crosstalk. This technique is thus limited to about 20 high frequency probes.

Because of these difficulties a number of designers are considering autotesting as a solution. This approach, which is well known to silicon IC designers, involves about 20% of the chip area and has the following advantages: it reduces the number of test vectors to be applied and verified, and in some cases a fail-safe operation mode is possible. Figure 8.38 shows a high frequency probe card.

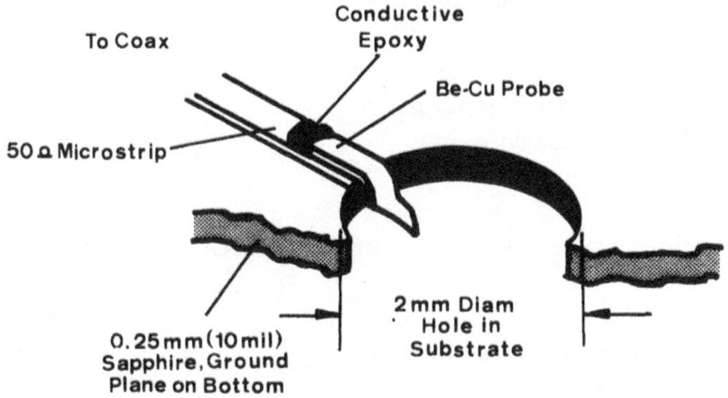

Fig. 8.38 Detail of one probe of a high frequency probe card. (Courtesy of Hewlett-Packard.)

The high frequency probe-card technology allows low complexity circuit testing at frequencies as high as 4–5 GHz [8.100] and MESFET discrete device testing up to 18 GHz [8.101].

8.9.3.2 Packaged device test

High performance GaAs IC packaging requires special consideration.

The intrinsic performances of the circuit must be preserved, which means that the electrical link to the external world must be as perfect as possible.

Microwave techniques are used in the package design to minimize crosstalk and improve impedance matching. In general, a GaAs IC input impedance consists of a 10 fF capacitance in parallel with a 0.5–5 kΩ resistance. Input and output signal levels are compatible with the logic family to be interfaced, in general a few hundreds of millivolts over a 50–75 Ω impedance. When sampling techniques are used, the input impedance is considered to be that of an open circuit (total reflection). Since GaAs ICs are not impedance matched to input–output transmission lines, external matching resistances are added to the package. Special test jigs are used to measure the performance of packaged devices. These jigs allow interfacing between the package and the coaxial lines going to the test equipment.

8.10 Packaging of GaAs digital integrated circuits

The performance of GaAs ICs requires high performance packaging and very low crosstalk levels, which cannot be met with commercial silicon circuit packages [8.102]. GaAs packages must comply with the following requirements.

Hermeticity: protection against the environment
Small size: high printed-circuit-board density and performance
Thermal dissipation: low case temperature
Good signal quality: low parasitics
Crosstalk: low cross-coupling
Cost: as low as possible

The options chosen by the various GaAs IC manufacturers are quite different.

Triquint: a coplanar ceramic including power supply decoupling capacitances [8.103].
Harris: flat-pack packages [8.104].
Gigabit Logic: commercial chip carriers with a silicon substrate where the GaAs IC is mounted and including power supply decoupling capacitances and transmission lines to reduce the length of the bonding wires [8.105].

Thomson-Semiconductors have developed a family of very high performance packages for GaAs ICs. Figure 8.39 shows 8 pin and 16 pin packages belonging to this family. These are ceramic–metal flat packages. The basic

Fig. 8.39 Eight pin and 16 pin GaAs packages. (Courtesy of Thomson-Semiconductors.)

structure is made of high purity (more than 96%) alumina on which the Mo–Mg 50 Ω outputs are silk-screened. The back plane is made of Cu–Ag alloy. The alumina frame is laser machined. The ceramic cap ensures hermeticity. This package family offers an excellent mechanical–electrical–cost trade-off. The 8 pin package bandwidth is 8 GHz and the crosstalk isolation between two lines with a grounded line between is greater than 40 dB up to 6 GHz [8.106].

References

8.1 Van Tuyl, R. and Liechti, C. *IEEE J. Solid State Circuits*, **9**, 269–276, 1974.

8.2 Van Tuyl, R. and Leitchi, C. *IEEE J. Solid State Circuits*, **12**, 485–496, 1977.

8.3 Long, S. I., Lee, F. S., Zucca, R., Welch, B. M. and Eden, R. C. *IEEE Trans. Microwave Theory Tech.*, **28**, 466–472, 1980.

8.4 Lee, F. S., Shen, E., Kaelin, C. R., Welch, B. Y., Eden, R. C. and Long, S. I. *1980 GaAs IC Symp., Las Vegas, NV, November 1980*, Paper 3.

8.5 Mimura, T. and Fukuta, M. *IEEE Trans. Electron Devices*, **27**, 1147–1155, 1980.

8.6 Yokoyama, N., Mimura, T. and Fukuta, M. *IEEE Trans. Electron Devices*, **27**, 1124–1128, 1980.

8.7 Dingle, R., Stormer, H. L., Gossard, A. C. and Wiegmann, W. *Appl. Phys. Lett.*, **33**, 665–667, 1978.
8.8 Change, L. L., Esaki, L. and Tsu, R. *Appl. Phys. Lett.*, **24**, 593–598, 1978.
8.9 Mimura, T., Hiyamizu, S., Fyjii, T. and Nanbu, K. *Jpn. J. Appl. Phys.*, **19**, 225–227, 1980.
8.10 Delagebeaudeuf, D., Delescluse, P., Etienne, P. and Laviron, M. *Electron. Lett.*, **16**, 667–668, 1980.
8.11 Zucca, R. *Inst. Phys. Conf. Ser.*, **33**, 20–22, 1976.
8.12 Asbeck, P., Tandon, J., Babcock, E., Welch, B. and Evans, C. *IEEE Trans. Electron Devices*, **26**, 1853, 1979.
8.13 Huber, A. M., Morillot, G. and Linh, N. T. *Appl. Phys. Lett.*, **34**, 358, 1979.
8.14 Evans, C. S., Deline, V., Sigmon, T. W. and Lidow, A. *Appl. Phys. Lett.*, **35** (3), 291, 1979.
8.15 Tuck, B., Adegoboyega, G. A., Jay, P. R. and Cardwell, M. J. *Inst. Phys. Conf. Ser.*, **45**, 114–124, 1978.
8.16 Fairman, R. D., Chen, R. T., Olivier, J. R. and Ch'en, D. R. *IEEE Trans. Electron Devices*, **28**, 135–140, 1981.
8.17 Hobgood, H. Y., Eldridge, G. W., Barett, D. L. and Thomas, R. N. *IEEE Trans. Electron Devices*, **28**, 140–149, 1981.
8.18 Komeno, J., Nogami, N., Shibatomi, A. and Ohkawa, S. *Conf. on Gallium Arsenide and Related Compounds, Vienna, 1980*. pp. 9–18.
8.19 Mitsutami, T., Kato, N., Ida, M. and Ohmori, M. *IEEE Trans. Microwave Theory Tech.*, **28**, 479–486, 1980.
8.20 Duchemin, J. P., Bonnet, M., Koelsh, F. and Huyghe, D. *J. Electrochem. Soc.*, **126**, 1134–1142, 1979.
8.21 Gloanec, M., Nuzillat, G., Arnodo, C. and Peltier, M. *IEEE Trans. Microwave Theory Tech.*, **28**, 472–478, 1981.
8.22 Dobratz, B. E., Ho, N., Krumm, C. F. and Greiling, P. T. *IEEE Trans. Microwave Theory Tech.*, **28**, 486–490, 1980.
8.23 Rossel, P., Nuzillat, G., Tranduc, H., Bert, G., Graffeuil, I. and Azizi, C. *Rev. Phys. Appl.*, **13**, 1978.
8.24 Houng, Y. M. and Pearson, G. L. *J. Appl. Phys.*, **49** (6), 3348, 1978.
8.25 D'Avanzo, D. *1981 GaAs IC Symp., San Diego, CA, October 27–29, 1981*, pp. 21–23.
8.26 Berth, M., Cathelin, M. and Durand, G. *Proc. 1977 Int. Electron Devices Meet., Washington, DC, December 1977*, pp. 201–204.
8.27 Favennec, P. N. *J. Appl. Phys.*, **47**, 2532–2536, 1976.
8.28 Berth, M., Venger, C. and Martin, G. M. *Electron. Lett.*, **17**, 873–874, 1981.
8.29 Rao, E. V., Duhamel, N., Favennec, P. N. and L'Azidon, H. *J. Appl. Phys.*, **49**, 3898–3905, 1978.
8.30 Eden, R. C., Welch, B. M. and Zucca, R. *IEEE J. Solid State Circuits*, **13**, 419–426, 1978.
8.31 Yokoyama, N., Mimura, T., Fukuta, M. and Ishikawa, M. *IEEE ISSCC Dig. Tech. Pap.*, pp. 218–219, February 1981.
8.32 Lehovec, K. and Zuleeg, R. *IEEE Trans. Electron Devices*, **27**, 1074–1091, 1980.
8.33 Nuzillat, G., Bert, G., Phan Ngu, T. and Gloanec, M. *IEEE Trans. Electron Devices*, **27**, 1102–1109, 1980.
8.34 Yokoyama, N., Ohnishi, T., Odani, K., Onodera, M. and Abe, M. *IEEE Trans. Electron Devices*, **29** (10) , 1541–1547, 1982.
8.35 Shockley, W. *Proc. IRE*, **40** (11), 1365, 1952.
8.36 Croset, M., Ecole, J. and Perrocheau, J. *Rev. Tech. Thomson-CSF*, **12** (4), 827–852, 1980.

8.37 Van Tuyl, R. and Lietchi, C. *1979 IEEE Int. Solid State Circuit Conf.*, pp. 20–21, IEEE, New York, 1979.
8.38 Greiling, P. T., Lundgren, R. E., Krumm, C. F. and Lomr, R. F. *Microwave Syst. News*, 48–60, January 1980.
8.39 Damay-Kavala, F., Gloanec, M., Peltier, M., Nuzillat, G. and Arnodo, C. *Int. Symp. on GaAs and Related Compounds, Oiso, September 20–23, 1981*, Paper 15.
8.40 Pam Ngu, T., Gloanec, M. and Nuzillat, G. *Proc. 7th European Specialist Workshop on Active Microwave Semiconductor Devices, Greece, 1981.*
8.41 Barna, A. and Liechti, C. *IEEE J. Solid State Circuits*, **14** (4), 708–715, 1979.
8.42 Eden, R. C., Welch, B. M. and Zucca, R. *IEEE J. Solid State Circuits*, **14**, 299–317, 1979.
8.43 Long, S. I., Welch, B. M., Eden, R. C., Lee, F. S. and Zucca, R. *Proc. Int. Conf. on Solid State Devices, Tokyo, Japan, August 1979, Jpn J. Appl. Phys.*, **19** (Suppl. 19-1), 319–324, 1980.
8.44 Eden, R. C., Lee, F. S., Long, S. J., Welch, B. M. and Zucca, R. *1980 IEEE Int. Solid State Circuits Conf., San Francisco, CA, February 13–16, 1980*, pp. 122–123, IEEE, New York, 1980.
8.45 Welch, B. M., Shen, Y. D., Zucca, R. and Eden, R. C. *1979 IEDM Tech. Dig.*, pp. 493–496, 1979.
8.46 Welch, B. M., Shen, Y. D., Zucca, R., Eden, R. C. and Long, S. I. *IEEE Trans. Electron Devices*, **27** (6), 1116–1124, 1980.
8.47 Barna, A. *European Solid State Circuits Conf., Southampton, September 18–21, 1979*, pp. 144–146.
8.48 Idda, M. and Takada, T. *IEEE Trans. Microwave Theory Tech.*, **32** (1), 5–10, 1984.
8.49 Katsu, S., Nambu, S., Shimano, A. and Kano, G. *IEEE Trans. Electron Devices*, **32** (6), 1114–1118, 1985.
8.50 Mellor, P. and Livingstone, A. *Electron. Lett.*, **16** (19), 749–750, 1980.
8.51 Hashizume, N., Yamada, H. and Tomisawa, K. *Electron. Lett.*, **17** (1), 51–52, 1981.
8.52 Hashizume, N., Yamada, H., Kosima, T. and Tomizawa, K. *Gallium Arsenide and Related Compounds, 1981.*
8.53 Rocchi, M. *Electron. Lett.*, **17** (4), 168–169, 1981.
8.54 Rocchi, M. and Gabillard, B. *IEEE J. Solid State Circuits*, **18** (3), 1983.
8.55 Frey, P., Gadillard, B. and Rocchi, M. *1983 GaAs IC Symp., Phoenix, AZ, November 1983*, pp. 62–65.
8.56 Rode, A., MacCormack, A., MacCormack, G. and Vetanen, B. *1982 IEDM Tech. Dig.*, pp. 162–165, December 1982.
8.57 Vu, T., Rogers, P. *et al. IEEE J. Solid State Circuits*, **19** (1), 1984.
8.58 Sadler, R. A. and Eastmann, L. F. *IEEE Trans. Electron Devices Lett.*, **4** (7), 215–217, 1983.
8.59 Levy, M. M. and Lee, R. E. *IEEE Trans. Electron Devices Lett.*, **4** (4), 1983.
8.60 Mizutami, T., Kato, N., Osafune, K. and Ohmori, M. *IEEE Trans. Electron Devices*, **29** (2), 199–204, 1982.
8.61 Yamasaki, K., Asai, K. and Kurumada, K. *14th Conf. on Solid State Devices, Tokyo, August 1982*, pp. 201–208.
8.62 Nakamura, M., Sano, Y., Nukada, T., Ishida, T. and Kaminishi, K. *1983 GaAs IC Symp., Phoenix, AZ, November 1983*, pp. 134–137.
8.63 Nagel, L. and Spice, Z. *Memo UCB/ERL M 75/520*, ERL, May 1975.
8.64 Nakayama, Y., Suyana, K., Shimizu, M., Yokoyama, N., Ohnishi, M., Shibatami, A. and Ishikawa, M. *IEEE J. Solid State Circuits*, **18** (5), 1983.
8.65 Gloanec, M., Jarry, J. and Nuzillat, G. *Journées Nationales Microondes*,

Toulouse, June 28–30, 1982.
8.66 Gloanec, M. *1985 European Solid State Devices Conf., Aachen, September 1985*, Panel session.
8.67 Perea, E. H., Kaszinski, A., Nuzillat, G. and Arnodo, C. *GaAs IC Symp., New Orleans, LA, November 1982*, pp. 104–107.
8.68 Bert, G., Morin, J. P., Nuzillat, G. and Arnodo, C. *IEEE Trans. Microwave Theory Tech.*, **30** (7), 1014, 1982.
8.69 Lee, C. P., Hou, D., Lee, S. I., Miller, D. L. and Anderson, R. J. *1983 GaAs IC Symp., Phoenix, AZ, November 1983*, pp. 162–165.
8.70 Toyoda, N., Kanazawa, K., Terada, T., Mochizuki, M., Hirose, M., Ikawa, Y. and Hojo, A. *1983 GaAs IC Symp., Phoenix, AZ, November 1983*, pp. 86–89.
8.71 Hayashi, T., Masaki, A., Tanaka, H., Yamashita, H., Masuda, N. and Doi, T. *1984 GaAs IC Symp., Boston, MA, October 1984*, pp. 111–114.
8.72 Lee, S., Vahremkamp, R., Deming, R. and Chang, F. *1984 GaAs IC Symp., Boston, MA, October 1984*, pp. 115–117.
8.73 Takano, S., Tamino, N., Yoshihara, T., Mitsui, Y. and Nishitani, K. *IEEE Trans. Electron Devices*, **32** (6), 1135–1139, 1985.
8.74 Rocchi, M. *Journées Electriciens, Electroniciens, Automaticiens, Lyon, February 1985.*
8.75 MacLevige, W. V. and Chang, C. T. *1985 GaAs IC Symp., Monterey, CA, November 1985*, pp. 203–206.
8.76 Hirayama, M., Ino, M. and Matsuoka, Y. *1984 IEEE Int. Solid State Circuits Conf., February 1984*, pp. 46–47, IEEE, New York, 1984.
8.77 Mizoguchi, T., Toyoda, N., Kanazawa, K., Ikawa, Y., Terada, T., Mochizuki, M. and Wojo, A. *1984 GaAs IC Symp., Boston, MA, October 1984*, pp. 117–120.
8.78 Ishii, Y., Ino, M., Ida, M., Hirayama, M. and Ohmori, M. *1984 GaAs IC Symp. Boston, MA, October 1984*, pp. 121–124.
8.79 Yokoyama, N., Onodera, H., Shinoki, T., Ohnishi, N. and Nishi, N. *IEEE Trans. Electron Devices*, **32** (3), 1797–1801, 1985.
8.80 Yokoyama, N., Onodera, M., Shinoki, T. and Ohnishi, M. *1984 IEEE Int. Solid State Circuits Conf., February 1984*, pp. 44–45, IEEE, New York, 1984.
8.81 Harris Semiconductors, *Data Sheet*, 1985.
8.82 Gobbi, J. M. *Journées Electriciens, Electroniciens, Automaticiens, Bordeaux, March 1986.*
8.83 Lee, G., Steeves, T., Schroederer, M., Nelson, R., Helix, M., Jamison, S., Hanka, S., Brown, J., Vu, T., Hansai, B. and Roberts, P. *1983 GaAs IC Symp., Phoenix, AZ, November 1983*, pp. 174–177.
8.84 Pengue, L., MacCormack, G., Strid, E., Smith, D. and Bowman, T. *1984 GaAs IC Symp., Boston, MA, November 1984*, pp. 27–30.
8.85 Toyoda, N., Uchitomi, N., Kitoura, Y., Mochizuki, M., Kanasawa, K., Terada, T. and Ikana, Y. *1985 IEEE Int. Solid State Circuits Conf., February 1985*, IEEE, New York, 1985.
8.86 Laude, D. P. *IEEE 1986 Custom ICs Conf., Rochester, NY, May 1986*, p. 805, IEEE, New York, 1986.
8.87 Nakamura, H., Tanaka, K., Tsunotani, M., Kanakami, Y., Akiyama, M. and Kaminishi, K. *1985 IEEE Int. Solid State Circuits Conf., February 1985*, pp. 204–205, IEEE, New York, 1985.
8.88 Deming, R. N., Zucca, R., Vamrenkamp, R., Hou, L. D., Naused, B. A. and Gilbert, B. K. *IEEE J. Solid State Circuits*, **19** (5), 732–738, 1984.
8.89 Upadhayula, L. C., Curtice, W. R. and Smith, R. *IEEE Trans. Electron Devices*, **30** (1), 2–10, 1983.
8.90 Namordi, M. R., O'Connell, M. R. and Newman, P. F. *1985 IEEE Int. Solid*

State Circuits Conf., February 1985, pp. 18–21, IEEE, New York, 1985.
8.91 Ducourant, M., Binet, M., Baelde, J. C., Rocher, C. and Gibereau, J. C. *1986 GaAs IC Symp. Grenelefe, FL, October 28–30, 1986*, pp. 209–212.
8.92 Larue, G. S. *1983 GaAs IC Symp., Phoenix, AZ, November 1983*, pp. 70–73.
8.93 Hsiem, K. C., Knotts, T. and Baldwin, G. *1985 GaAs IC Symp., Monterey, CA, November 1985*, pp. 187–190.
8.94 Wang, K., Asbeck, P., Chang, M., Miller, D. and Eisen, F. *1985 GaAs IC Symp., Monterey, CA, November 1985*, pp. 99–102.
8.95 O'Connor, P., Flahive, P. G., Clemetson, W., Panock, R. L., Wemple, S. H., Shunk, S. C. and Takanashi, D. P. *IEEE Trans. Electron Devices Lett.*, **5** (7), 226–227, 1984.
8.96 Ohta, N. and Takada, T. *Electron. Lett.*, **19** (23), 1983.
8.97 Peltier, M., Nuzillat, G. and Gloanec, M. *1980 IEEE MTT/S Int. Microwave Symp., Washington, DC, May 1980*, pp. 472–473, IEEE, New York, 1980.
8.98 Goloubkoff, M., Henry, M. and Klein, J. P. *L'Echo des Recherches*, No. 116, pp. 59–68, 1984.
8.99 Hickling, R. M., Argyroudis, P. Lai, M., Chow, J., Lee, F. S. and Eden, R. C. *1985 GaAs IC Symp., Monterey, CA, November 1985*, pp. 79–82.
8.100 Van Tuyl, R., Kumar, V., D'Avanzo, D. C., Taylor, T. W., Peterson, V. E., Hornbuckle, D. P., Fisher, R. A. and Estreich, D. B., *IEEE Trans. Microwave Theory Tech.*, **30** (7), 935–942, 1982.
8.101 Gleason, K. R., Reeder, T. M. and Strid, E. W. *Microwave Syst. News*, 55–65, May 1983.
8.102 Wilson, D., Frick, W., Kwiat, J., Lo, S., Churchill, J. and Barrera, J. *1982 GaAs IC Symp., New Orleans, LA, November 1982*, pp. 13–16.
8.103 Smith, D. H., Bowman, T. G., Lind, R. and Riley, T. S. *1985 GaAs IC Symp., Monterey, CA, November 1985*, pp. 151–154.
8.104 Koophan-Larson, D. A., Frisco, J. A. and Heckman, D. E. *1985 GaAs IC Symp., Monterey, CA, November 1985*, pp. 155–158.
8.105 Gheewala, T. R. *1985 GaAs IC Symp., Boston, MA, October 1985*, pp. 143–146.
8.106 Jarry, J., Gloanec, M. and Camps, P. *Journées Nationales Microondes, Lannion, June 1984*, pp. 162–163.

Chapter 9

Microwave Monolithic Integrated Circuits

C. KERMARREC AND C. RUMELHARD

9.1 Introduction

The emergence of GaAs monolithic microwave integrated circuit (MMIC) technology, which is based on the integration on the same substrate (GaAs) of both passive and active circuit elements, is revolutionizing the design and fabrication of microwave components. It is the result of a long-term effort aimed at developing miniaturized microwave circuits with low cost and high performance.

From 1930 to about 1960, microwave circuits were manufactured using waveguide technology. The design was generally experimental and the fabrication was long and costly. At that time the microwave engineer was often referred to as a "plumber". Around 1960, the introduction of cheap low loss dielectric materials (plastic, glass, alumina etc.) and microwave silicon planar devices resulted in the development of microwave integrated circuit (MIC) technology. This technology, which is based on discrete devices mounted on a dielectric substrate that acts as a microwave transmission medium, is also known as *hybrid* microwave integrated circuit technology (HMIC) since "integrated" now usually implies "monolithically integrated". The vast majority of microwave circuitry produced at present still uses HMIC technology.

The concept of MMICs dates back to the mid-1960s. At that time it became obvious that a fully integrated microwave technology process which would allow all the passive and active devices to be formed within an insulating crystalline material would appear. Silicon-based integrated circuits in which silicon is used as the dielectric and the active-layer material (silicon MMICs) or as semiconducting material deposited on a sapphire dielectric substrate (silicon-on-sapphire (SOS) technology) have been investigated. However, the inability of silicon to retain its high resistivity during high temperature processes has restricted the applicability of silicon MMICs, and SOS technology has effectively been abandoned because of critical problems involved with the sapphire–silicon interface.

By the late 1960s GaAs had become the material of choice for high performance microwave devices. Two unique properties of this material that make it particularly suitable for the fabrication of microwave devices are high electron mobility and high peak velocity (see Chapter 1). These translate into high frequency operation. In addition, the high resistivity semi-insulating GaAs substrate reduces crosstalk between devices and permits the fabrication of MMICs which are composed of active and control devices, transmission lines and lumped elements, all integrated together on the same substrate.

In comparison with other technologies, GaAs MMICs offer the following advantages:

(a) size and weight reduction;
(b) cost reduction for medium and high production volumes as a result of the batch-processing capability;
(c) enhanced reproducibility due to uniform processing and integration of all parts of the circuit;
(d) the enhanced reliability inherent in integration;
(e) broadband performance owing to the reduction of parasitic elements and optimization of device size;
(f) circuit design flexibility and a single-chip multifunction capability.

Many possible areas of application (e.g. phased-array radar systems, global positioning systems etc.) will not be technologically or financially feasible unless the promise of cheap high volume GaAs MMICs is realized. However, it should be noted that some particular applications, such as very low noise receivers or high power transmitters, are still best served by HMIC technology since the integration of high electron mobility transistors (HEMTs) or high power transistors (> 10 W) is difficult.

Unlike HMICs [9.1], GaAs MMICs cannot be easily trimmed by adjusting lumped or distributed elements (laser trimming, wire bonding etc.). The design of these components must therefore be based on accurate physical and electrical models (of both passive and active elements) including dispersion effects due to manufacturing process tolerances. Furthermore, powerful interactive software programs for the synthesis, analysis and layout of linear and nonlinear circuits should be available to minimize the time-consuming interactive processes needed to develop integrated circuits. To some extent, these are already available.

The first GaAs MMICs developed [9.2, 9.3] were narrowband amplifiers, but applications now cover almost all the functions of microwave systems. We can even say that GaAs MMICs are changing system concepts by allowing new circuit capabilities such as microwave signal processing.

The capability of GaAs MMICs is discussed in this chapter together with design considerations and circuit applications. A typical technological process and the methodology of characterization are described in Section 9.2. Electrical models of active and passive elements used in GaAs MMICs, including active devices (MESFETs), transmission lines, spiral inductors, capacitors and resistors, are presented in Section 9.3. Various wideband and medium power amplifiers are analyzed in Section 9.4. The performance of switching and control functions by GaAs MMICs with MESFETs as variable resistors is dealt with in Section 9.5. The applications include variable attenuators, switches and phase shifters. Oscillators and frequency-processing functions are addressed in Section 9.6 which covers the application of GaAs MMICs to free-running oscillators, voltage-controlled oscillators, mixers and analog frequency dividers. Multifunction MMICs are discussed in Section 9.7, and a single-chip X-band down-converter is described. Finally, future developments in the applications of GaAs MMICs are discussed in Section 9.8.

9.2 Technology

9.2.1 A typical technological process

The process described in this section should be considered as a typical process that might be used for the fabrication of GaAs MMICs. It does not represent any single current GaAs MMIC foundry process. However, it provides a good example of the chain of operations needed for complete MMIC fabrication including depletion-mode MESFETs, resistors, capacitors, inductors, interconnections and via holes. Other technological options will be outlined during the following description of the process.

The various steps in the fabrication are summarized in Fig. 9.1. It should be noted that the cross-sections illustrating the process are not to scale. The steps are as follows.

Formation of the active layer A uniform n-doped layer is implanted (the doping density is typically 2×10^{17} atoms cm^{-3}) into a thick semi-insulating GaAs substrate (400 μm thick for example) and this is followed by an n^+ implant (the doping density is typically 10^{18} atoms cm^{-3}). The n layer acts as the active layer for the MESFETs or as the bulk resistive layer for the resistors, and the n^+ layer acts as a low resistive region to reduce contact resistance. Epitaxy could be considered for the formation of the active layer, but ion implantation is preferred for the production of GaAs MMICs because of its superior uniformity and reproducibility, its unique capability for enabling multiple selective implants to be performed, its higher throughput and finally its lower cost.

Isolation Isolation is achieved by mesa etching down to the semi-insulating GaAs. It can also be achieved by ion or proton implantation. Basically, implantation produces defect centers which trap electrons and holes, thus reconverting n-GaAs to semi-insulating GaAs.

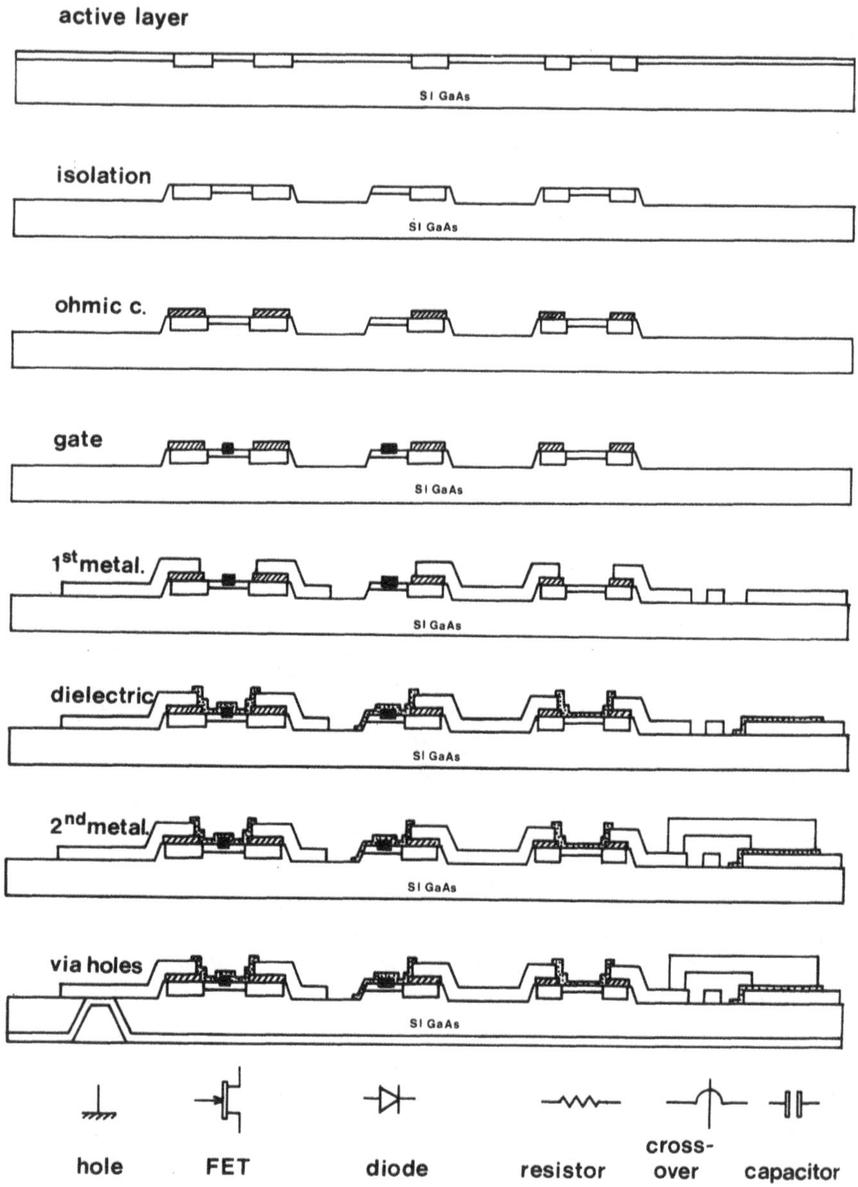

Fig. 9.1 Summary of a typical technological process.

Ohmic contacts Nonrectifying or ohmic contacts consist of Au–Ge, nickel and gold layers deposited on the n^+ contact areas and alloyed at high temperature (typically 400 °C).

Gate metallization Once the gate channel is recessed to provide the required pinch-off voltage, the gate metal is deposited. It consists of a three-layer metallization system—titanium, platinum and gold. Titanium associated with n-GaAs forms the Schottky contact, gold reduces the resistivity of the gate metal and platinum acts as a diffusion barrier for gold.

First metallization level A Ti–Au first metallization level is deposited to form the passive elements and the bottom electrodes of the metal–insulator–metal (MIM) capacitors, and to thicken ohmic contact metallization.

Dielectric layer Plasma-deposited Si_3N_4 is used as a passivation layer for exposed n-doped regions (MESFETs and resistors) and as a dielectric layer for MIM capacitors. Si_3N_4 is selected mainly from reliability considerations [9.4] and for its excellent dielectric properties [9.5]. Typically, the square capacitance is 600 pF mm^{-2} (for a thickness of 1200 Å) and the breakdown voltage is greater than 30 V depending on the capacitor size.

Second metallization level, air bridges This metallization level provides top plate capacitors and interconnections between elements to be achieved by air bridges. It is usually produced by a selective gold-plating process to obtain metallization as thick as 3–5 μm.

Via holes After lapping the back-side of the wafer to the required thickness (typically 100–200 μm) the via holes are chemically etched (wet etching or dry etching) through the GaAs substrate. Then the wafer is back-metallized using gold deposition followed by gold plating. The via holes interconnect MESFET sources or any required top-side element to the ground plane.

In general, each mask level requires a photolithographic process which can be summarized as follows: deposit a photoresist over the entire wafer; expose the photoresist to ultraviolet or deep ultraviolet radiation through a patterned mask (for very short gate definition, direct-write electron beam lithography is used); develop the photoresist; perform metallization or dielectric deposition and achieve wafer patterning using the lift-off technique (the etching technique is used when layer deposition is made before photoresist definition); remove the remaining photoresist.

More details of the technological process can be found in ref. 9.6.

9.2.2 Process-control monitor

The process-control monitor (PCM) is a group of test patterns, included in the general pattern, which is used to characterize each step of the technological process (in-line process monitor), to perform final tests and statistical analysis, and to provide yield information. It usually includes patterns for measurement of metal-layer sheet resistances, contact resistances, crossover isolation, characteristics of the dielectric layer, doping profiles and MESFET parameters. Typical elements such as MESFETs, varactors, resistors, capacitors and inductors can also be included in order to evaluate the validity of the design model using d.c. and microwave measurements.

The wafer is accepted or rejected on the basis of PCM results. All the parameters should be measured within a predefined range of a mean value for a PCM to be classified as good. A wafer is validated when the yield is more than a specified value (e.g. 60%). The yield is defined as the ratio of the good PCMs to the PCMs inside the wafer (typically ten).

PCM results are key parameters for MMIC designers. PCMs provide information on the mean value and standard deviation of parameters for a given technology prior to circuit design. The designer can then perform sensitivity and statistical analyses to ensure that the designed circuits meet the specifications, taking into account the expected deviations in the parameters. After the process, PCM results can be used to analyze any discrepancy between the measured and simulated performances of the circuit and to determine whether it is due to the fabrication process or the design procedure.

Figure 9.2 shows a typical PCM which includes the following patterns to be tested at each relevant step of the fabrication process:

(a) a series of long-gate MESFETs (2, 5, 8, 12, 20 and 45 μm) which allow the disclosure of surface effects, the assessment of variation in the transconduc-

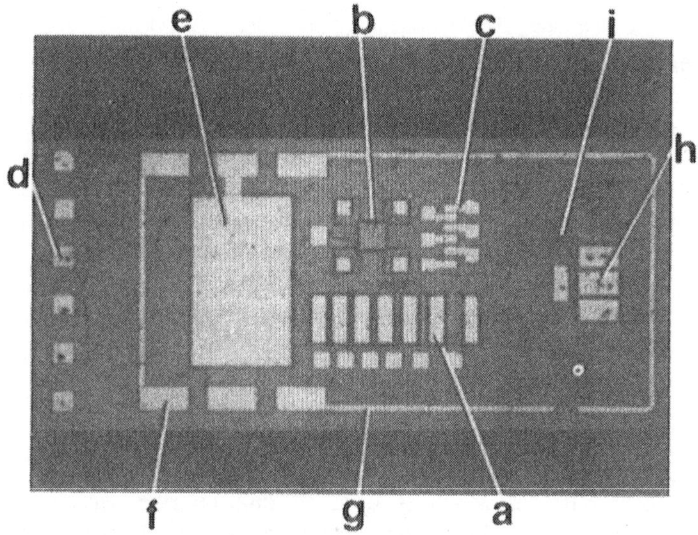

Fig. 9.2 Test pattern used to control circuit fabrication processes.

tance g_m as a function of the gate length, and finally the measurement of the drift mobility and doping profile using the longest gate [9.7];

(b) a Hall clover to measure Hall mobility [9.8];

(c) a test pattern to measure contact resistances and the sheet resistance of the active layer [9.9];

(d) a test pattern to measure the gate resistance;

(e) a MIM capacitor to measure the thickness of the dielectric layer and the dielectric loss;

(f), (g) lines to measure the attenuation and resistivity of the second metallization level;

(h) a MESFET to measure S parameters and the mobility using a magnetotransconductance method [9.10, 9.11];

(i) spot marks to check the alignments of the various mask levels.

These test patterns show an example of what is needed to characterize the material, the technological process and the active element (the MESFET) simultaneously.

9.3 Monolithic circuit components

A key factor in the GaAs MMIC approach is the availability of accurate electrical models to characterize active and passive components. These components include GaAs MESFETs, Schottky diodes, transmission lines, inductors, capacitors and resistors. Surprisingly, rigorous usable mathematical models do not exist for most of the above components owing to the difficulty of taking the fundamental physical mechanism into account in a three-dimensional analysis. This lack of mathematical models combined with the need for fast computer-aided design (CAD) software has encouraged design engineers to develop semi-experimental models based on a particular well-defined reproducible technology.

We do not intend to analyze the existing models in detail here but just to present an overview of the problems involved with component modelling together with references to basic studies.

9.3.1 Active element: the MESFET

A typical recessed-gate MESFET is shown in Fig. 9.3. Modelling of GaAs MESFETs has been discussed in Chapter 2. Despite a better understanding of fundamental physical mechanisms and the use of powerful numerical analysis to compute MESFET parameters as functions of geometrical and technolog-

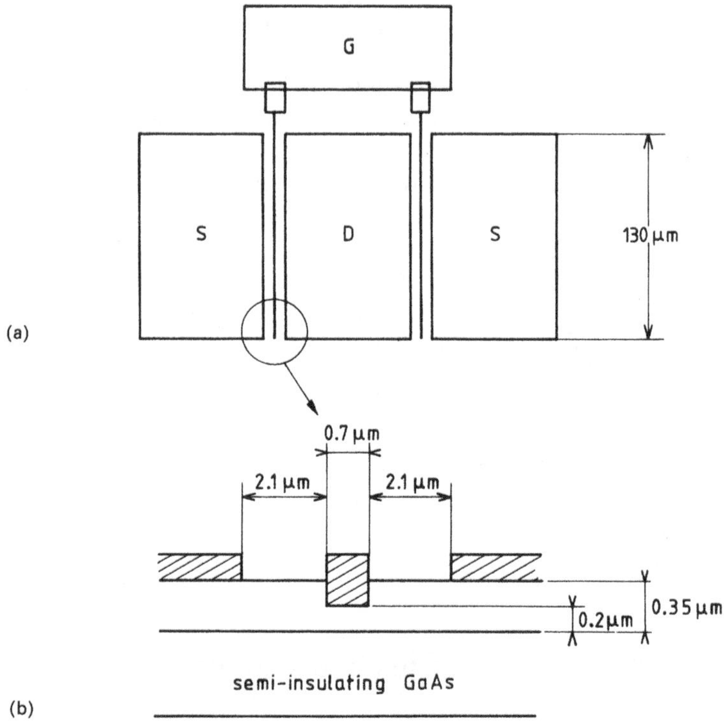

Fig. 9.3 Structure of a GaAs MESFET: (a) top view; (b) cross-section showing the unit cell structure.

ical data, MESFET models are not yet accurate enough to be implemented in general high speed CAD software for GaAs MMIC design.

In general, the following models should be supplied to the MMIC designer: a small-signal model which consists of an equivalent circuit including noise sources or S and noise-figure parameters; a large-signal model consisting of the MESFET characterization as a function of the input and output voltages; a thermal model consisting of the MESFET characterization as a function of temperature.

The semi-empirical approach is preferred for MESFET characterization. Accurate measurement data can be collected to characterize MESFETs for a specific application and implemented in CAD software in tabular or polynomial form for fast computation. Measurements should be performed on discrete devices fabricated using the same process as for their MMIC counterparts. The equivalent circuit (Fig. 2.35) and noise-figure parameters for low noise MESFETs (see Chapters 2 and 3) and the output power and linearity characteristics for power MESFETs (see Chapter 4) are derived from these measurements. Furthermore, the collected data can be scaled up or down according to the MESFET gate width when different geometries are required.

9.3.2 Passive elements

The passive components used in GaAs MMICs can generally be considered as lumped elements because of their small size. Only spiral inductors or particular structures may have lengths approaching a quarter wavelength. However, the prime lumped-element value is modified by parasitic effects due to the proximity of the coplanar ground plane or back-side metallization. Finite metallization thickness and finite material conductivity also introduce parasitic resistances. Finally, inductive or capacitive parasitic coupling influences the behavior of the elements.

It is therefore more convenient to use distributed models to include parasitic effects for even those lines which are short relative to a quarter wavelength. These models are composed of single or coupled microstrip or coplanar lines with or without losses. An overview of line models and of results obtained using numerical analysis for modelling coplanar and microstrip lines is presented next.

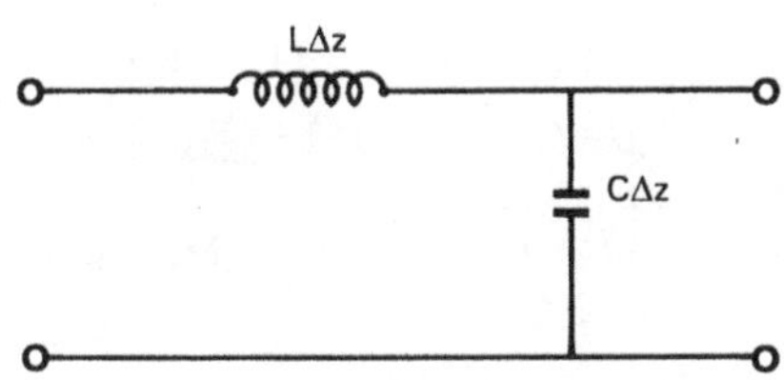

Fig. 9.4 Equivalent circuit of an infinitely small lossless transmission line element: L is the inductance per unit length, C is the capacitance per unit length and Δz is an infinitely small line length.

9.3.2.1 Transmission lines

A lossless TEM mode transmission line can be modelled by an infinite number of series inductors and shunt capacitors defined by their inductance and capacitance per unit length (Fig. 9.4). The characteristic impedance Z_0 of the line and the phase velocity are given by the relations

$$Z_0 = \left(\frac{L}{C}\right)^{1/2} \qquad v_p = \frac{1}{(LC)^{1/2}}$$

so that

$$Z_0 = \frac{1}{v_p C}$$

For an inhomogeneous line $v_p = c/\varepsilon_{eff}^{1/2}$, where c is the speed of light and ε_{eff} is the effective dielectric constant which is related to the linear capacitance by the equation

$$\varepsilon_{\text{eff}} = \frac{C}{C_0} \tag{9.1}$$

C_0 is the linear capacitance of the line when the dielectric has been replaced by air. These relations give

$$Z_0 = \frac{1}{c(CC_0)^{1/2}} \tag{9.2}$$

For a set of coupled lines such as those in Fig. 9.5 the capacitive influence coefficients c_{ij} are defined by the equations

$$\begin{aligned} Q_1 &= c_{11}V_1 + c_{12}V_2 \\ Q_2 &= c_{21}V_1 + c_{22}V_2 \end{aligned} \tag{9.3}$$

where the Q_i are the charges on each conductor and the V_i are their potentials with respect to ground. If we rewrite the charges Q_i as a function of the potentials V_i but using the linear capacitances C_{ij} we obtain

$$\begin{aligned} Q_1 &= C_{11}V_1 + C_{12}(V_1 - V_2) \\ Q_2 &= C_{21}(V_2 - V_1) + C_{22}V_2 \end{aligned} \tag{9.4}$$

This gives the expressions

$$\begin{aligned} c_{11} &= C_{11} + C_{12} \\ c_{12} &= -C_{12} \\ c_{21} &= -C_{21} \\ c_{22} &= C_{22} + C_{21} \end{aligned} \tag{9.5}$$

or vice versa

$$\begin{aligned} C_{11} &= c_{11} + c_{12} \\ C_{12} &= -c_{12} \\ C_{21} &= -c_{21} \\ C_{22} &= c_{22} + c_{21} \end{aligned} \tag{9.6}$$

Calculation of either the influence coefficients or the linear capacitances involves onerous numerical methods: conformal mapping (in cases presenting some symmetry) [9.12], finite-difference methods [9.13] or image methods using Green functions [9.13]. The results of these calculations are usually given in the form of influence coefficients, which are less familiar to the engineer than are linear capacitances. Hence the set of backward and forward relationships given above is very useful.

Microstrip lines in the presence of a coplanar ground plane Numerical

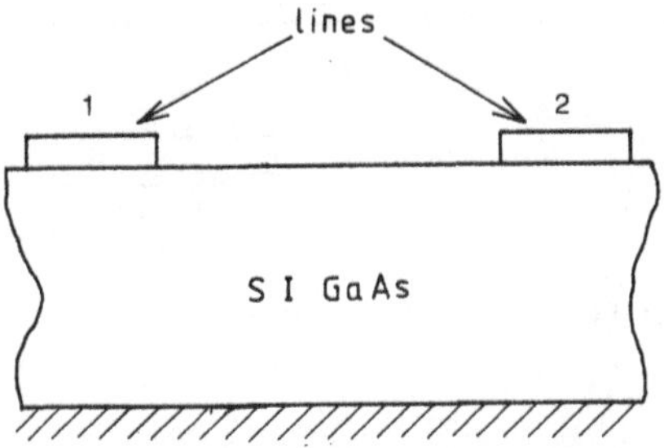

Fig. 9.5 Coupled lines on GaAs.

analysis [9.13] can be used to calculate the influence coefficients of coupled lines. If the potential of conductor 2 in Fig. 9.5 is set to zero (thus simulating the coplanar ground plane) eqns (9.1), (9.2) and (9.6) for a single line can be used to calculate ε_{eff} and Z_0:

$$\varepsilon_{\text{eff}} = \frac{c_{11}}{c_{11}{}^{0}} = \frac{C_{11}+C_{12}}{C_{11}{}^{0}+C_{12}{}^{0}} \tag{9.7}$$

$$Z_0 = \frac{1}{c(c_{11}c_{11}{}^{0})^{1/2}} = \frac{c^{-1}}{\{(C_{11}+C_{12})(C_{11}{}^{0}+C_{12}{}^{0})\}^{1/2}} \tag{9.8}$$

where the superscript 0 indicates a coefficient calculated assuming that the dielectric medium is air. Figures 9.6 and 9.7 show the results of the application of this type of calculation to realistic microstrip–coplanar ground configurations on GaAs substrates [9.14, 9.15].

Ohmic losses start to become significant for narrow lines (10–30 μm) and it is very difficult to estimate them from measurements. Figure 9.8 shows the results of a finite-element loss calculation [9.16] compared with results obtained by Pucel *et al.* [9.17]. In this figure α is the attenuation coefficient per unit length (dB cm^{-1}) and R_s is the surface skin resistivity. It can be shown that, although the approach used in ref. 9.17 is valid for lines wider than 100 μm, it significantly underestimates the losses of narrow lines. For example, ref. 9.17 predicts $\alpha R_s = 21$ dB cm^{-1} compared with the 27 dB cm^{-1} predicted by a finite-element calculation [9.16]. For a gold conductor $R_s = 0.03\ \Omega$, giving $\alpha = 0.63$ dB cm^{-1} in the first case and 0.71 dB cm^{-1} in the second case.

Line characteristics can also be affected by their proximity to the chip edge, but tedious calculations are required to account for this factor accurately. A detailed treatment of the effect can be found in ref. 9.18.

9.3.2.2 Spiral inductors

The modelling of multiturn round or rectangular inductors for GaAs MMICs is not trivial. The prime inductance value of such inductors was predicted a long time ago by Grover [9.19]. However, models of spiral inductors should include capacitive coupling between adjacent turns, parasitic

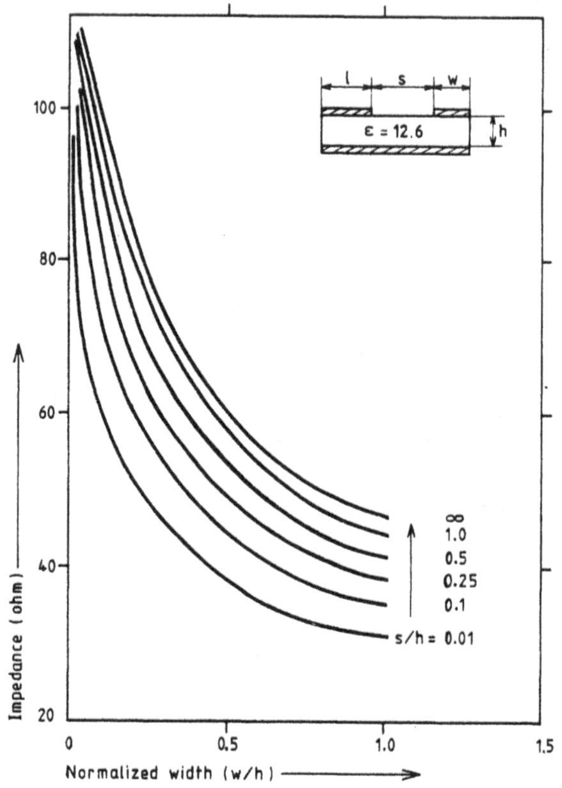

Fig. 9.6 Characteristic impedance of GaAs microstrip lines in the presence of a coplanar metal plane.

capacitance to the ground plane to take into account the distributed nature of the inductor and the mutual inductance of a mirror inductor present as a result of ground-plane effects. In addition, the contribution of any crossover, discontinuous or dielectric layer should be determined if the models are to be accurate.

Figure 9.9 shows the distributed-element model of a square spiral inductor. Each lumped reactive component can be derived using a three-dimensional magnetostatic–electrostatic approach as proposed in ref. 9.20. Jansen *et al.* [9.21] have developed a theoretical hybrid-mode transmission method to derive an equivalent circuit for square spiral inductors. This approach can include the contribution of a multilayer dielectric structure and predict inductor losses.

The modelling of circular spiral inductors has been achieved by dividing the inductor into circular lines connected together [9.22]. A

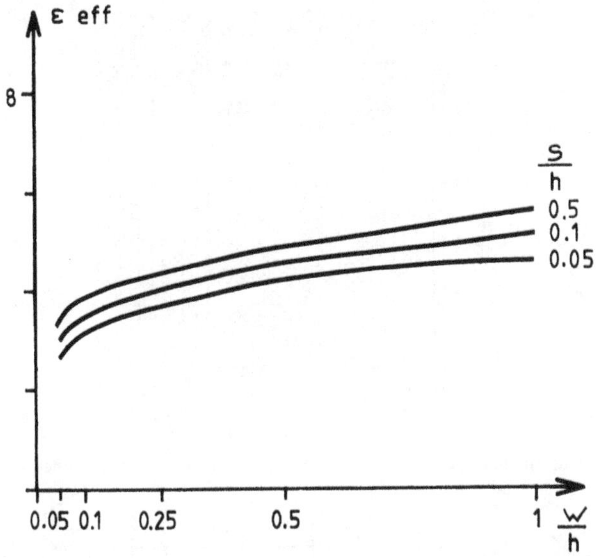

Fig. 9.7 Effective dielectric constant of GaAs microstrip lines in the presence of a coplanar metal plane.

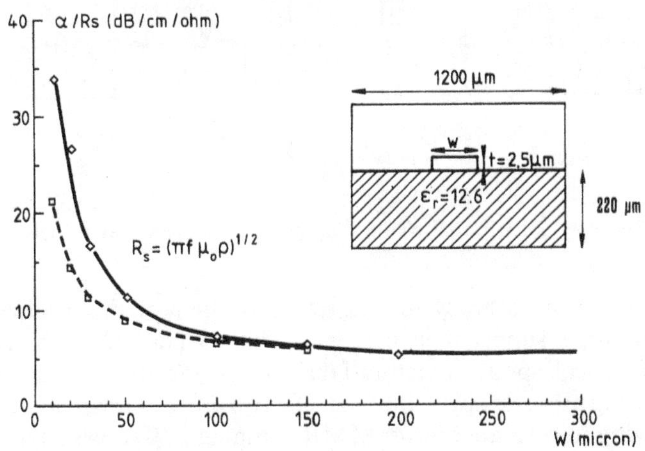

Fig. 9.8 Losses in GaAs microstrip lines: α, line attenuation constant; f, frequency; μ_0, vacuum magnetic susceptibility; ρ, bulk resistivity; ◇, finite elements; □, Pucel *et al.* [9.17].

magnetostatic–electrostatic approach has been used to calculate the inductance and the capacitive effects of a single-turn inductor.

Despite improvements, the above models are only valid up to an inductor impedance of about 100–150 Ω and can predict the first parallel resonance

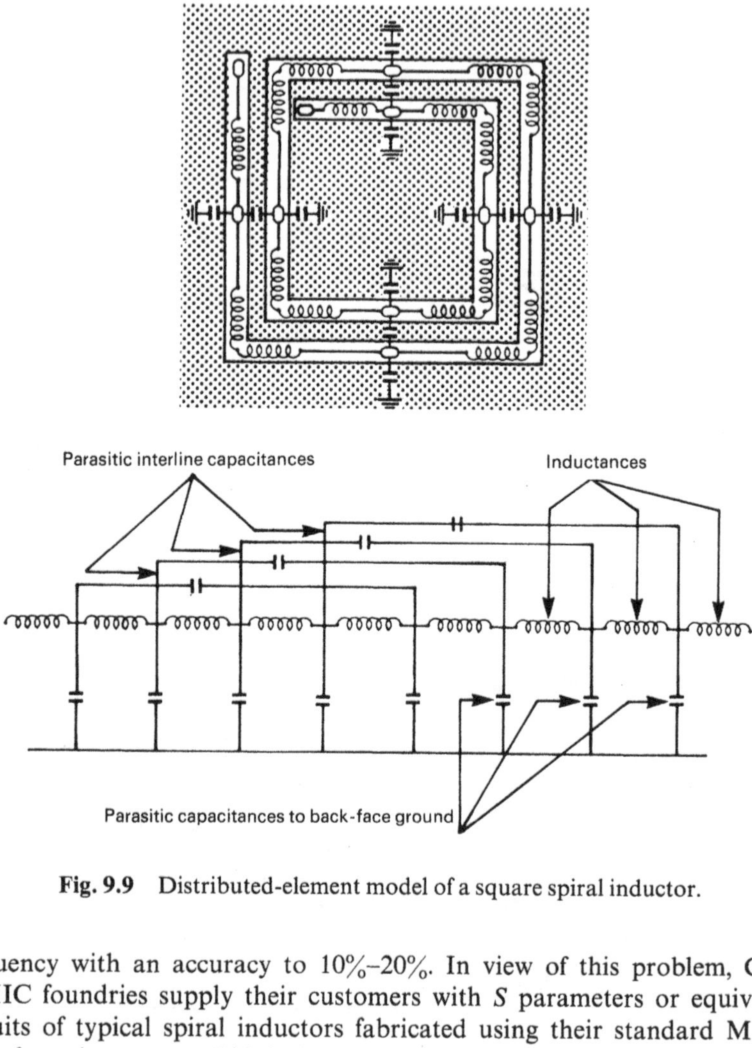

Fig. 9.9 Distributed-element model of a square spiral inductor.

frequency with an accuracy to 10%–20%. In view of this problem, GaAs MMIC foundries supply their customers with *S* parameters or equivalent circuits of typical spiral inductors fabricated using their standard MMIC manufacturing process. This is the most efficient and accurate method so far of characterizing spiral inductors for MMICs. Figure 9.10 shows a typical round spiral inductor including an air bridge for connecting the inner termination.

9.3.2.3 Metal–insulator–metal capacitors

Overlay or MIM capacitors are preferred for GaAs MMICs mainly because of their high capacitance per unit area and low parasitics compared with interdigital capacitors, and their linear behavior and high breakdown voltage compared with Schottky capacitors. The dielectric materials used for MIM capacitors are generally SiO_2 ($\varepsilon_r \approx 5$), Si_3N_4 ($\varepsilon_r \approx 7$) and Ta_2O_5 ($\varepsilon_r \approx 24$) [9.23–9.25]. The dielectric thickness is chosen by a trade-off analysis

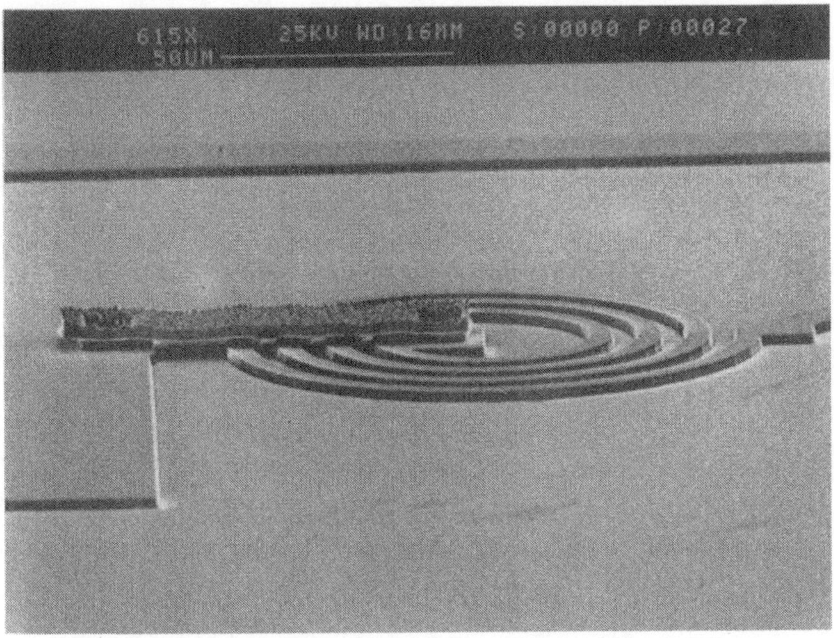

Fig. 9.10 Scanning electron micrograph of a round spiral inductor. (Courtesy of Thomson-Semiconductors.)

Fig. 9.11 Three-layer MIM capacitor. (Courtesy of Thomson-Semiconductors.)

between the need for a high capacitance per unit area (thin layer) to increase the integration density and the requirement for a minimum breakdown voltage (typically 30–40 V). This usually results in a dielectric thickness in the range 1000–4000 Å.

Care should be taken in the design to avoid parasitic breakdown spots in the capacitor structure. Figure 9.11 shows a capacitor based on a simple metal–SiO_2–metal structure. Potential breakdown spots at the periphery of the lower electrode and at the crossover level owing to the thinness of the dielectric layer make this structure unacceptable for high yield circuits.

The use of a second thick dielectric layer (e.g. SiO_2) to cover the edge of the capacitor to provide a bridge over the second metallization level has

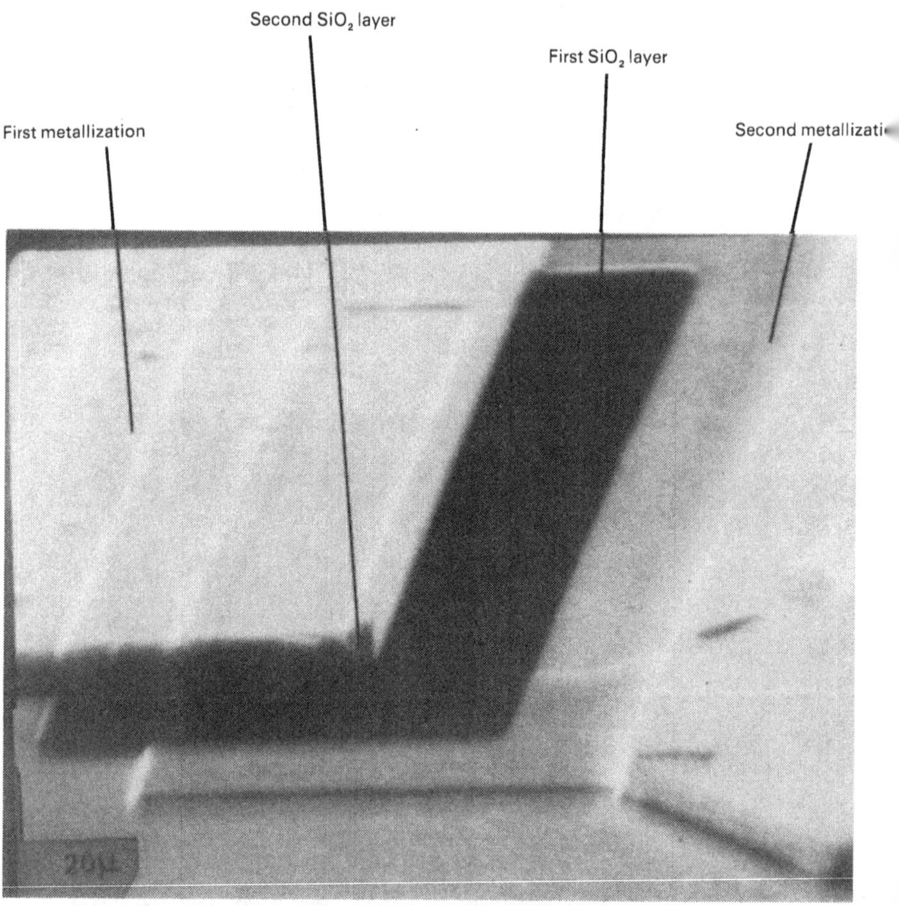

Fig. 9.12 Microphotograph of a MIM capacitor showing a second SiO_2 layer near the edges. (Courtesy of Thomson-Semiconductors.)

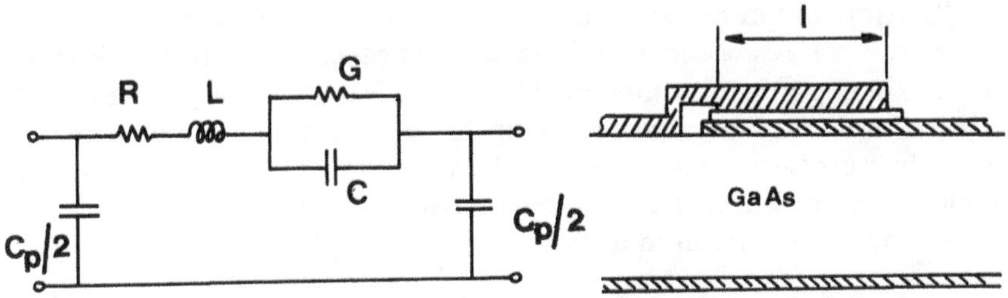

Fig. 9.13 Equivalent circuit of a MIM capacitor: $R = (2/3)rl$, $C = C_{\square}A$, $C_p = C_p l$, $L = l_s l$ and $G = C\omega \tan \delta$, where l is the length of the top electrode, r is the electrode resistance per unit length, l_s is the electrode inductance per unit length, C_p is the electrode capacitance per unit length, $C_{\square}$ is the area capacitance of the dielectric film, A is the area of the top electrode and $\tan \delta$ is the loss tangent of the dielectric film.

dramatically improved the breakdown voltage and the yield of MIM capacitors. This improved structure is shown in Fig. 9.12. Air bridges can also be used to interconnect the top electrode, thus improving the breakdown voltage and reducing the crossover parasitic capacitance.

Models for MIM capacitors have been developed using the lumped [9.26] or distributed [9.27] approach. Figure 9.13 shows a typical lumped-element equivalent circuit together with the element values.

9.3.2.4 Resistors

Resistors can be fabricated using a GaAs doping layer (the MESFET active layer for instance) or by depositing resistive metallic material on the wafer. The latter method involves an additional mask level but increases the design flexibility. For example, metallic resistors are not limited in size by the electron saturation velocity which exists in GaAs resistors and limits the electric field across the resistor to about 500 V cm^{-1} for linear operation [9.28]. Furthermore, as shown in Table 9.1, the temperature coefficient of a metallic resistor is lower and can be adjusted by controlling the technological process.

Table 9.1 Properties of typical resistive layers

Material	*Sheet resistance* ($\Omega/\square$)	*Temperature coefficient* (ppm K^{-1})
Cr	15	+3000
Ti	10–100	+2500
Ta	30–200	−100 to +500
Ni–Cr	60–600	+200
TaN	300	−500 to −1000
Active layer	30–300	+3000

Resistors are modelled as a lossy microstrip transmission line or as a lumped resistor associated with two grounded capacitors at its terminations. As the skin depth in lossy material is much greater than the thickness of the resistive layer (a few micrometers compared with tenths of micrometers in X band), the d.c. resistance is used in models. Care should be taken to include the ohmic contact resistance for GaAs bulk resistors in the resistance value.

Figure 9.14 shows an equivalent circuit for a typical thin film resistor.

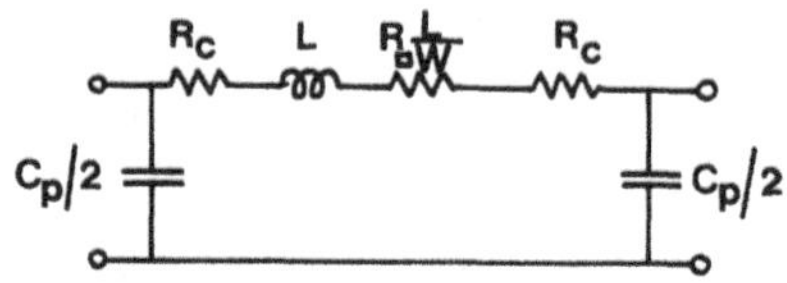

Fig. 9.14 Equivalent circuit of a thin film resistor: $L = l_s l$, $C_p = C_p l$, R_c is the electrode contact resistance, $R_{\square}$ is the sheet resistance of the resistive layer, l and w are the length and width respectively of the resistive layer, and l_s and C_p are respectively the inductance and capacitance per unit length.

9.4 Amplifiers

9.4.1 Amplifier topologies

The design of small-signal and large-signal amplifiers has been intensively discussed in Chapters 3 and 4. The advantages of the GaAs MMIC approach over the HMIC approach have been summarized in Section 3.7, and consequently the basic design of GaAs MMIC amplifiers will not be addressed in this section. Five amplifiers covering medium-band, broadband and medium power applications are presented below. They include the following: a small-signal medium-bandwidth amplifier in X band using reactive lumped elements as matching circuits; two broadband amplifiers (0.5–7 GHz), one using a lossy match approach and the other a distributed structure; a broadband distributed amplifier (2–18 GHz); a medium power C-band amplifier (400 mW) using a tree structure. These amplifiers have been developed at the Laboratoire Central de Recherche (LCR), Thomson CSF [9.29], except for the 2–18 GHz distributed amplifier which was developed at the Laboratoires d'Electronique et de Physique appliquée (LEP).

9.4.2 Small-signal X-band cascadable amplifier

The small-signal X-band cascadable amplifier, which is used as a gain block for phased-array antennas, was developed to meet the following

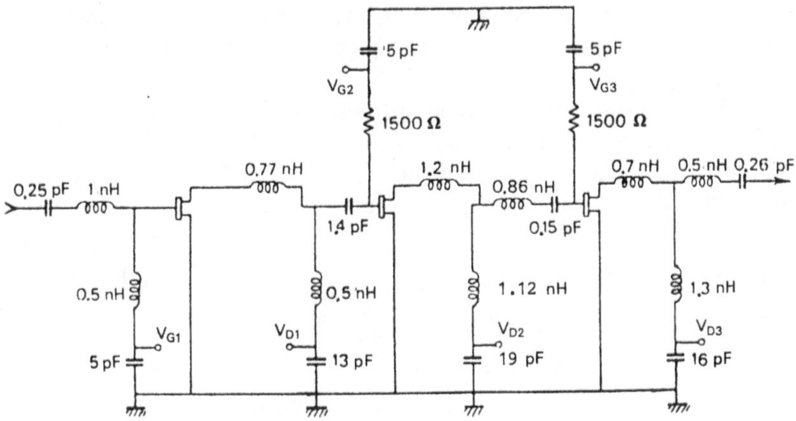

Fig. 9.15 Circuit diagram of an 8–12GHz amplifier.

specifications: frequency band, 8–12 GHz; gain, 14 ± 1 dB; input–output VSWR, <2. Figure 9.15 shows the electrical circuit of this three-stage amplifier. Transistors are built using an interdigitated structure based on two gate fingers of dimensions 150 μm × 1 μm. The matching circuits are composed of inductors and capacitors arranged in a bandpass topology [9.30]. The second and third MESFET gates are biased through a 1500 Ω GaAs bulk resistor (no current flows through the gate). The drains and the first gate bias are bypassed by a high value (low impedance) MIM capacitor. It should be

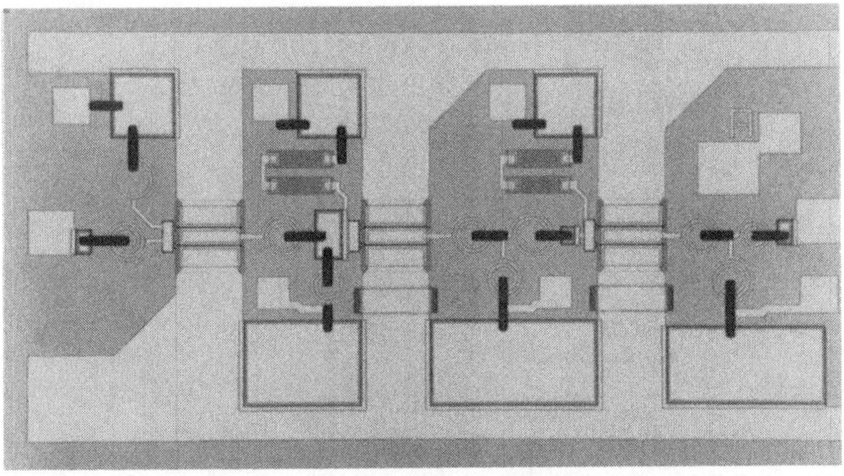

Fig. 9.16 Microphotograph of an 8–12 GHz amplifier. (Courtesy of Thomson-Semiconductors.)

noted that the input–output matching circuit structure includes a series capacitor which allows the amplifier to be cascadable without additional components.

Figure 9.16 shows the amplifier chip which is of dimensions 2 mm × 1 mm, and a wafer of diameter 2 in, for example, contains 1000 amplifiers. Figure 9.17 shows the envelopes of measurements made on 15 amplifiers and the simulated results obtained using MESFET parameters derived from the measurement of

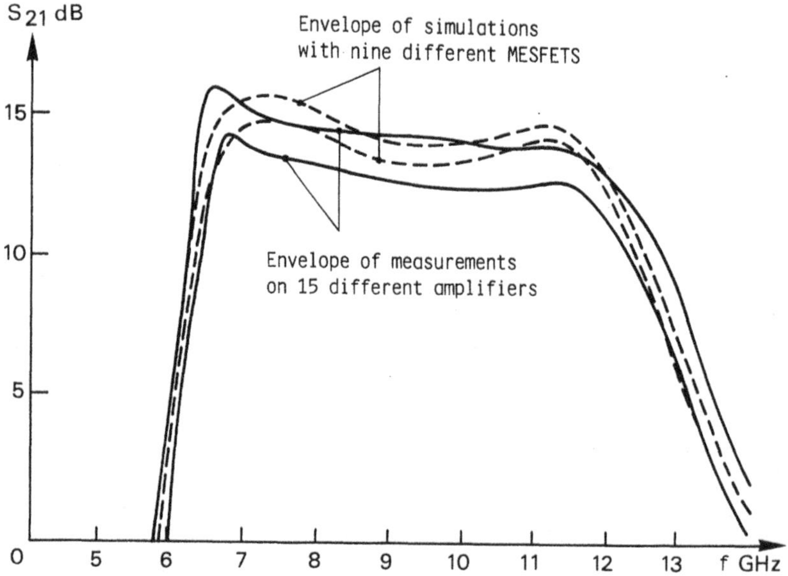

Fig. 9.17 Gain responses of 8–12 GHz amplifiers.

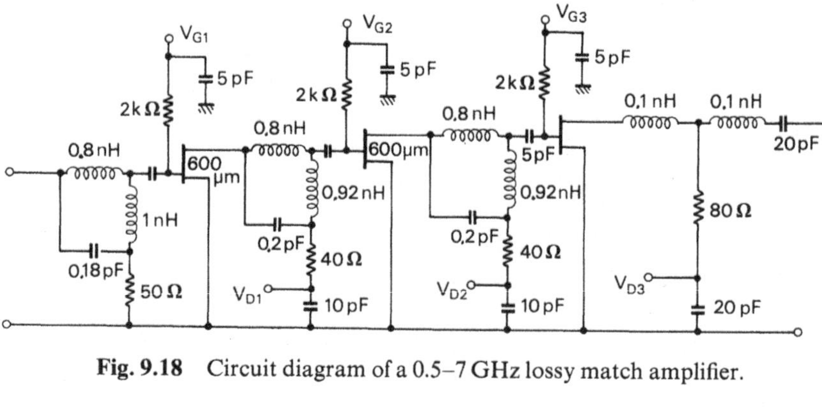

Fig. 9.18 Circuit diagram of a 0.5–7 GHz lossy match amplifier.

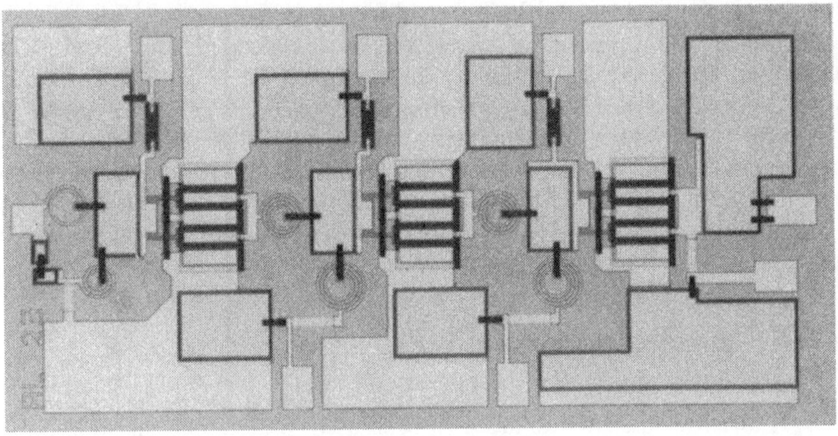

Fig. 9.19 Microphotograph of a 0.5–7 GHz lossy match amplifier. (Courtesy of Thomson-Semiconductors.)

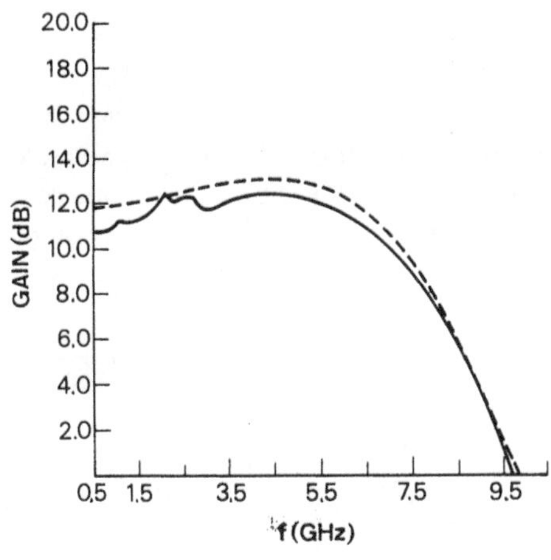

Fig. 9.20 Gain response of a 0.5–7 GHz lossy match amplifier: ——, measured; – – –, simulated.

nine transistors processed on the same wafer. These curves demonstrate the validity of the passive element model [9.20] in this frequency bandwidth.

9.4.3 Broadband lossy match amplifier (0.5–7 GHz)

Figure 9.18 shows the electrical circuit of a lossy match amplifier [9.31, 9.32] designed to meet the following objectives: frequency bandwidth,

0.5–7 GHz; gain, > 10 ± 1 dB. This amplifier is based on three MESFETs with gate widths of 600 μm and lossy matching circuits derived from a bridged-tee structure. The gate bias is supplied through a 2000 Ω GaAs resistor and the drain bias through a 40 or 80 Ω titanium resistor. Figure 9.19 shows the amplifier chip which is of dimensions 2.2 mm × 1.1 mm, and Fig. 9.20 demonstrates the excellent agreement between simulated and measured results. The gain is 11 ± 1 dB and the noise figure varies from 7.5 to 9 dB. This rather high value is due to the contribution of the matching circuit resistors to the overall output noise. The power at the 1 dB compression point is 20 mW.

9.4.4 Distributed amplifier (0.5–7 GHz)

A distributed structure has also been used to design an amplifier in the 0.5–7 GHz frequency bandwidth [9.33–9.36]. The objective of the design was a gain of 10 ± 1 dB. Figure 9.21 shows the electrical circuit and the element values of a distributed amplifier providing a 6 dB gain in the required bandwidth. Two identical amplifiers have been cascaded to meet the specifications.

Figure 9.22 shows a two-stage version of the circuit. Each stage is composed of four MESFETs of dimensions 200 μm × 0.5 μm. The gate and drain biases are supplied through the gate and drain line termination resistance. This approach allows the wideband space-consuming drain bias tee to be avoided. However, the size of the 65 Ω drain line resistor should be large enough for it to dissipate the d.c. power due to drain currents.

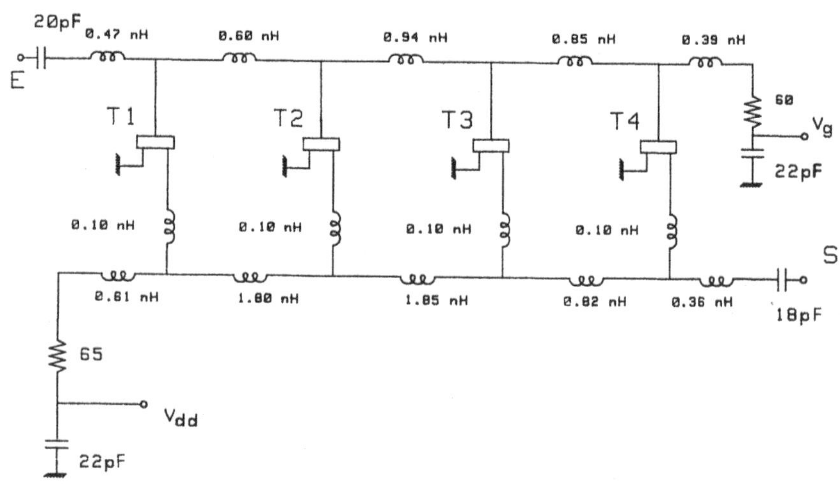

Fig. 9.21 Circuit diagram of a 0.5–7 GHz distributed amplifier.

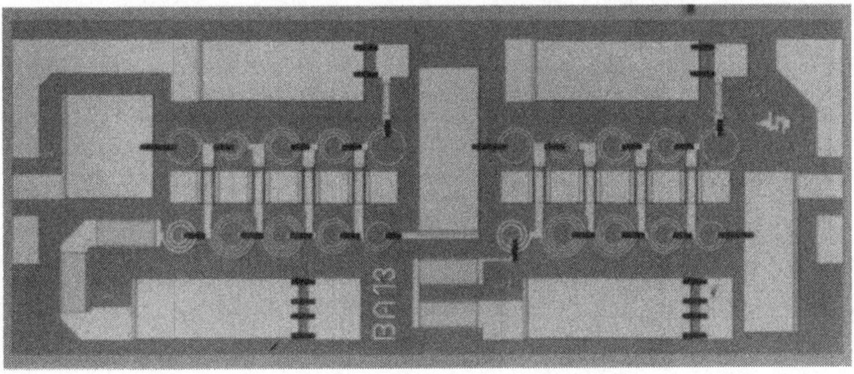

Fig. 9.22 Microphotograph of a 0.5–7 GHz distributed amplifier. (Courtesy of Thomson-Semiconductors.)

The dimensions of the chip shown in Fig. 9.22 are 1 mm × 2.5 mm. MESFET sources are connected to the back-side of the 100 μm thick chip through chemically etched via holes. The gain and the input and output return loss measured for three amplifiers are plotted in Fig. 9.23. The gain is 11.5 ± 1 dB and the return loss is less than −10 dB. The noise figure is lower than 6.5 dB over the entire bandwidth. These curves demonstrate the excellent technological uniformity and design accuracy.

In comparison with the lossy match amplifier the gain is almost the same, the noise figure is lower and the chip is slightly larger. The total gate width is twice that of the lossy match amplifier, resulting in a higher d.c. power consumption.

9.4.5 Broadband distributed amplifier (2–18 GHz)

Another example of a distributed amplifier is shown in Fig. 9.24 [9.37]. The amplifier is based on four MESFETs (four gate fingers each of dimensions 50 μm × 0.6 μm) as part of the gate and drain artificial lines. It should be noted that the drain bias circuit is included on the chip. Source and termination resistor grounding is achieved by plated via holes. The chip dimensions are 2.8 mm × 2.3 mm. The gain is 5.7 ± 0.6 dB in the band 2–18 GHz, the VSWRs are less than 2.2 and the noise figure varies from 6 to 8 dB.

9.4.6 Medium power C band amplifier using the tree structure

A medium power C band amplifier has been designed to provide a 400 mW output power associated with a 14 dB gain between 5.9 and 6.4 GHz.

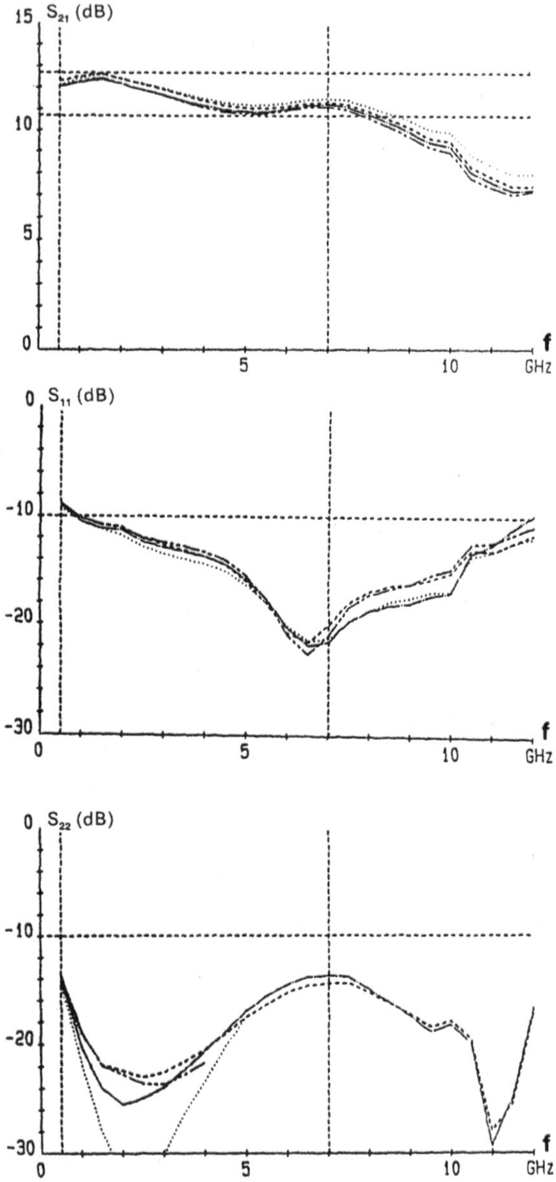

Fig. 9.23 Measured response of three amplifiers of the type shown in Fig. 9.21 showing the effect of spread.

The tree structure chosen for the application is shown in Fig. 9.25. The MESFETs, which were fabricated using ion implantation, have a gate length of 1 μm and a gate width of 600 μm. The pinch-off voltage is −2.5 V.

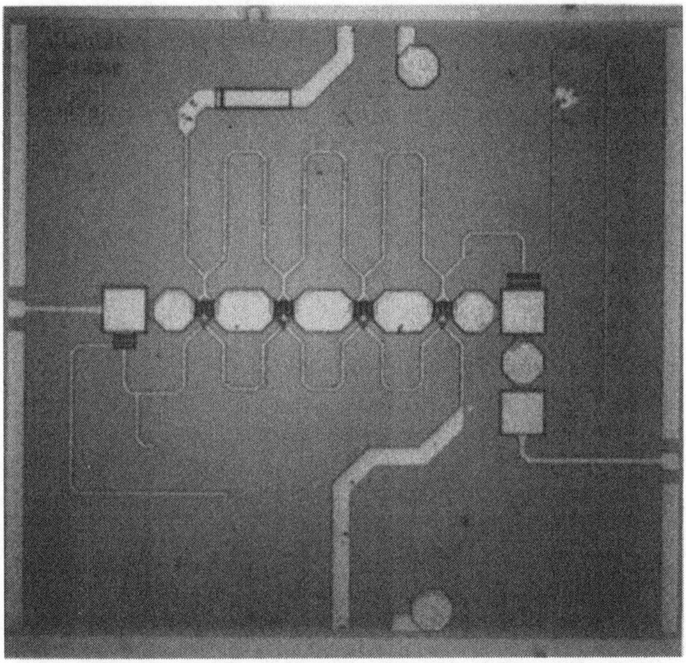

Fig. 9.24 Microphotograph of a 2–18 GHz distributed amplifier. (Courtesy of P. Gamand, LEP.)

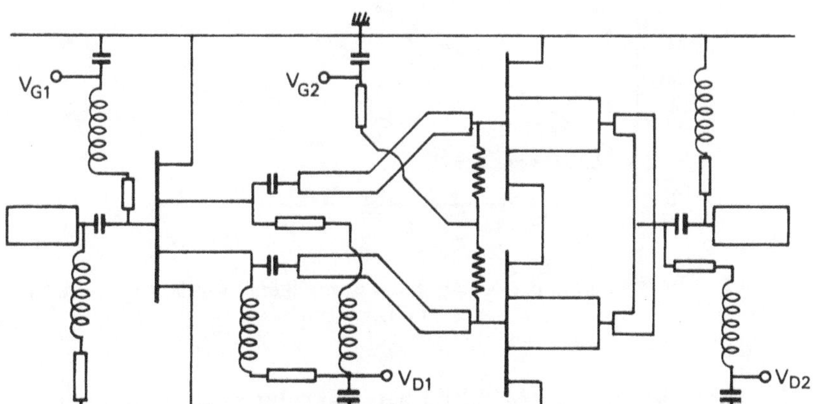

Fig. 9.25 Circuit diagram of a tree-structure amplifier (5.9–6.4 GHz).

Figure 9.26 shows a photograph of the chip. The same transistor is used once in the first stage and twice in the second stage. As a result of this tree structure, the interstage matching circuits are easier to design and introduce

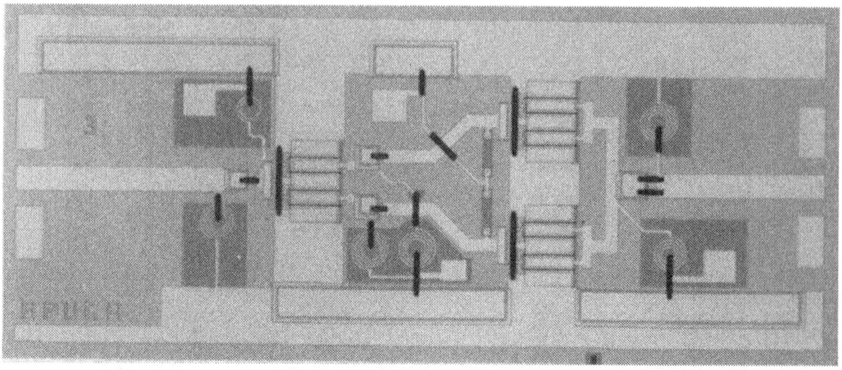

Fig. 9.26 Microphotograph of a tree-structure amplifier (5.9–6.4 GHz). (Courtesy of Thomson-Semiconductors.)

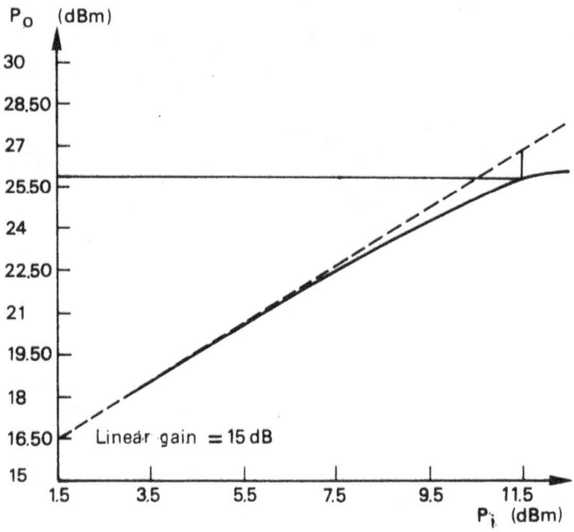

Fig. 9.27 Power output performance of the amplifier of Fig. 9.25 measured at 6 GHz.

low losses. At the input the MESFET acts as a splitter. At the output the two MESFETs are simply connected in parallel. Figure 9.27 shows a typical in-band power response. The amplifier presents a 15 dB linear gain and a 380 mW output power at the 1 dB compression point.

This tree structure could be extended to increase the output power, provided that each output feeds another similar cell with a single input and two outputs, i.e. the power level per cell remains more or less constant but the number of cells increases.

9.5 Switching and control functions

9.5.1 Introduction

Microwave functions are no longer limited to amplification or up- and down-conversion because signal processing at microwave frequencies is becoming increasingly common in commercial and military systems. Therefore microwave control circuits such as variable attenuators, switches and phase shifters are now used extensively. The operating principles of these three components, their suitability for monolithic integration, their applications and some examples of their realization are described next. The general modelling of MESFETs used in GaAs MMICs as switching devices is introduced in Section 9.5.2 which is concerned with variable-attenuator considerations.

Planar pin diodes [9.38], which have long suffered from a high forward series resistance due to surface recombination in the intrinsic zone, are now used as switching devices in GaAs MMIC applications [9.39]. However, resistive MESFETs, which do not need a d.c. current to be driven in the on state and require much simpler technology compared with pin diodes, are preferred. Consequently, only MESFET-based circuits are considered below.

9.5.2 Attenuators

Variable attenuators are widely used in microwave subsystems since they can achieve critical automatic level control functions, automatic gain control functions, temperature gain control functions and, in general, any amplitude control function. GaAs MMICs offer the unique possibility of producing very wideband variable attenuators. One approach, based on GaAs MESFETs used as variable resistors, can cover the frequency range from d.c. to 18 GHz [9.40, 9.41]. The principle and examples of GaAs integrated variable attenuators using MESFETs are described here.

9.5.2.1 Principles

MESFET modelling It is well known that MESFETs exhibit two regions of operation as shown in Fig. 9.28. In region 1 ($V_{DS} < 1$ V) the electrons flowing under the depletion zone have not reached saturation velocity and the MESFET behaves as a variable resistor controlled by the gate voltage. Region 1 is used in attenuators, switches and mixers, and region 2 is suitable for amplifier and oscillator applications. Figures 9.29 and 9.30 show physical and electrical models of a MESFET in ohmic region 1. All parameters are essentially constant with the exception of the channel resistance R and the gate capacitance C.

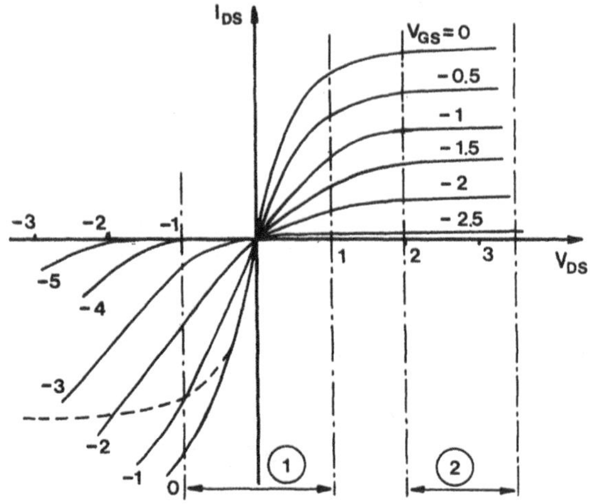

Fig. 9.28 D.c. characteristics of a MESFET.

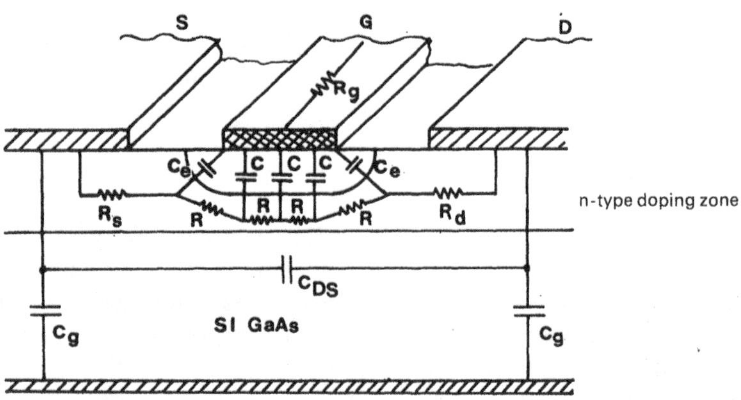

Fig. 9.29 Physical model of a resistive MESFET: R_s, source resistance; R_d, drain resistance; R_g, gate resistance; C_e, edge capacitance; C_g, ground capacitance; C_{DS}, drain-to-source capacitance.

These two components are more adequately described by a distributed *RC* network (the common electrode of the network is the gate metallization). *R* and *C* can be obtained as a function of the gate potential if the depth of the depletion region is known. This, in turn, is obtained by integrating the one-dimensional Poisson equation. The volume density of the charge underneath the gate is given by the doping profile. In the case of an ion-implanted doping profile, satisfactory results are generally obtained by using an equivalent flat profile of constant-volume free-charge density N_d over a thickness A. The values of N_d and A of this equivalent profile are chosen to give the same pinch-

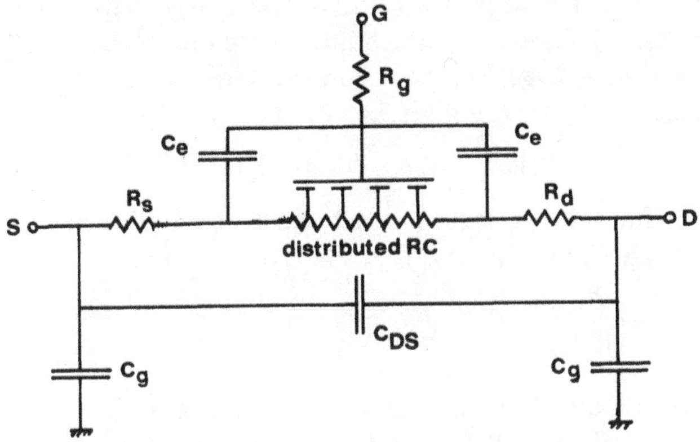

Fig. 9.30 Electrical circuit of a resistive MESFET.

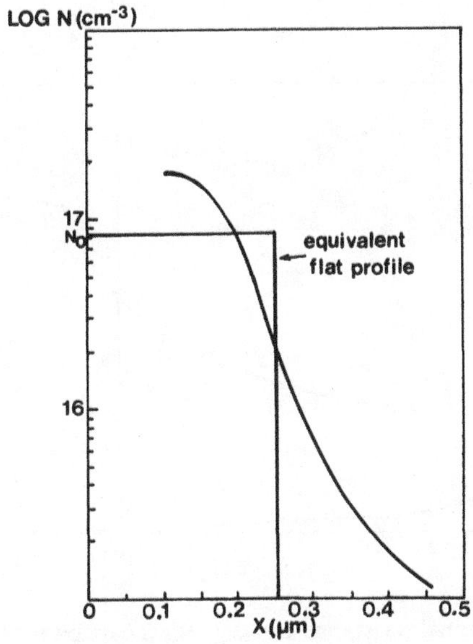

Fig. 9.31 Ion-implanted profile and equivalent flat profile.

off voltage and sheet resistance at $V_{GS} = 0$ as the true profile (Fig. 9.31).

A detailed derivation of $R(V_{GS})$ and $C(V_{GS})$ is given in ref. 9.6, and we shall given only a brief summary here. The depletion depth is obtained from Poisson's equation (see eqn (1.4)):

$$y_d(V_{GS}) = \left\{\frac{2\varepsilon(V_{bi} - V_{GS})}{qN_d}\right\}^{1/2}$$

where q is the electron charge, N_d is the volume charge density, $\varepsilon = \varepsilon_0\varepsilon_r$ is the dielectric constant of GaAs, V_{bi} is the built-in potential of the Schottky barrier and V_{GS} is the gate voltage. Since the current flows through the neutral zone under the depleted region, R is given by

$$R(V_{GS}) = \{zq\mu_0 N_d(A - y_d)\}^{-1} L_g$$

where z is the gate width, μ_0 is the low field electron mobility and L_g is the gate length, and the gate capacitance C is given by

$$C = \frac{\varepsilon_0\varepsilon_r zL_g}{y_d}$$

Typical curves of R and C versus V_{GS} are shown in Fig. 9.32.

More exact values of $R(V_{GS})$ and $C(V_{GS})$ can be obtained by numerical integration of Poisson's equation using the real doping profile. Other parameters appearing in Fig. 9.30 can be calculated as described in refs 9.42 and 9.43 or from d.c. or low frequency measurements [9.44].

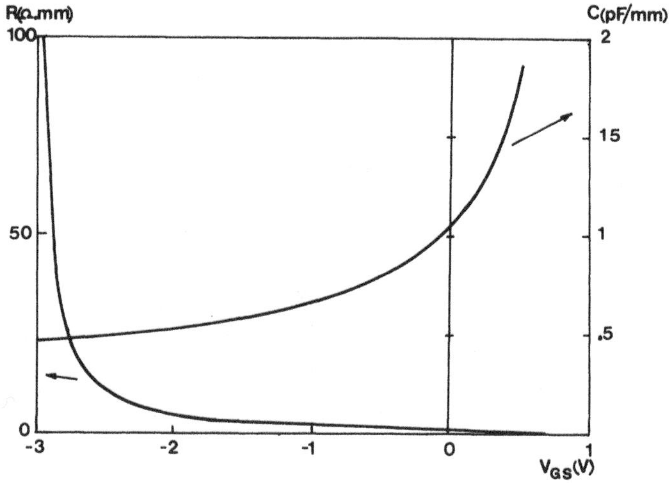

Fig. 9.32 Typical channel resistance and depletion capacitance as functions of V_{GS}.

Attenuator configurations Table 9.2 shows the basic structure of three types of MESFET attenuators (tee, π and bridged-tee) together with the impedance-matching conditions and attenuation as functions of the resistance values R_1 and R_2. If the range and absolute values of the MESFET resistance were the only important parameters, the bridged-tee attenuator would be the best choice. However, series and parallel parasitic capacitances are present in MESFETs (see Fig. 9.30). In both the π and the bridged-tee attenuators the

Table 9.2 Characteristics of tee, π and bridged-tee attenuators

Type	Tee		π		Bridged-tee	
Configuration	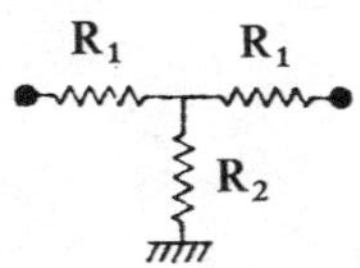		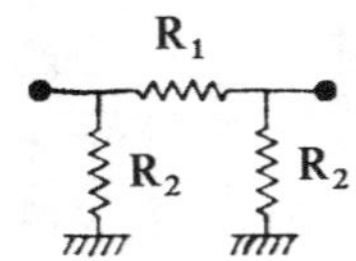		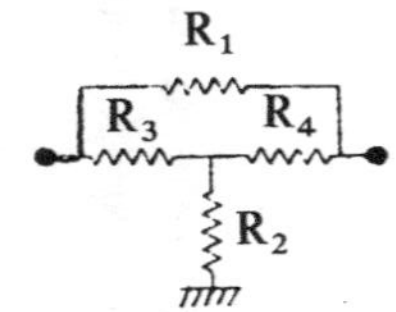	
Matching conditions	$Z_0{}^2 = R_1{}^2 + 2R_1R_2$		$Z_0{}^2 = \dfrac{R_2{}^2}{1+2R_2/R_1}$		$R_3 = R_4 = Z_0$ $Z_0{}^2 = R_1R_2$	
Attenuation at matching	$A = \dfrac{V_0}{V_1} = \dfrac{R_2}{R_1+R_2+Z_0}$		$A = \dfrac{V_0}{V_1} = \dfrac{R_2Z_0}{R_1R_2+Z_0(R_1+R_2)}$		$A = \dfrac{R_2}{Z_0+R_2}$ $A = \dfrac{Z_0}{Z_0+R_1}$	
Attenuation (dB)	R_1 (Ω)	R_2 (Ω)	R_1 (Ω)	R_2 (Ω)	R_1 (Ω)	R_2 (Ω)
0.5	1.4	868	2.9	1737	3	843
1	2.9	433	5.8	869	6	409
3	8.5	142	17.6	292	20	120
5	14	82	31	179	39	64
10	26	35	71	96	108	23
15	35	18.4	136	72	231	10.8

parallel capacitances C_{DS}, C_E and C_g associated with the series resistance R_1 reduce the high frequency performance drastically at high attenuation. This factor is important enough to ensure that the largest frequency band and dynamic range are generally provided by tee attenuators.

These design considerations have been used to develop "untuned" variable attenuators as described in Section 9.5.2.2.

9.5.2.2 Circuit applications

Many laboratories and GaAs foundries have developed wideband (d.c. to 18 GHz) variable attenuators, some of which are commercially available [9.45, 9.46].

For example, Tajima *et al.* [9.40] have developed a tee-type variable attenuator as shown in Fig. 9.33. By using a control circuit, a single control voltage can vary the attenuation electrically from 2 to 12 dB in the frequency range 2–18 GHz. The return loss is better than 10 dB under all conditions. Dual-gate MESFETs, in which the knee voltage determining the ohmic behavior and the breakdown voltage are substantially increased, were used in a high power version (600 mW) of the attenuator.

Figure 9.34 shows two tee-type attenuators developed at Spar Aerospace and CRC (Canada) [9.47] which cover frequency bands of 1–18 GHz and 1–15 GHz and exhibit a dynamic attenuation range of 10 dB. As shown in Fig. 9.35 the insertion losses are 2.5 dB and 1 dB respectively. The return loss is better than 15 dB under all conditions. This example illustrates the trade-off

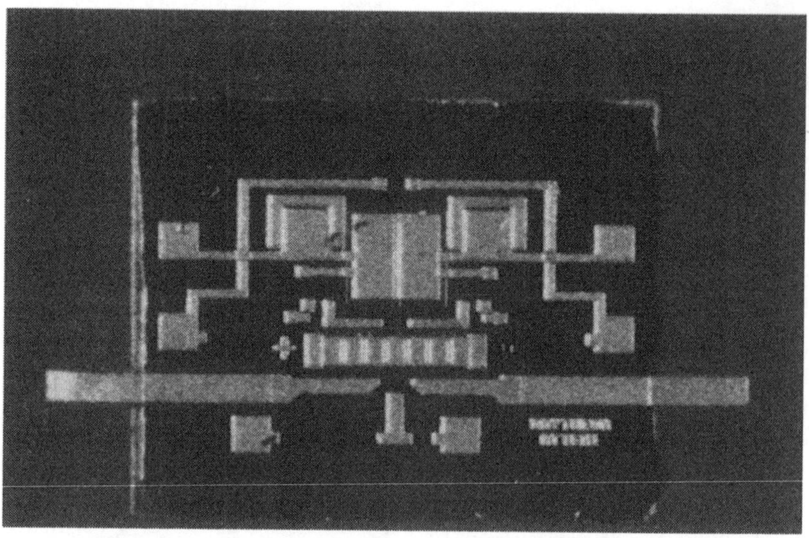

Fig. 9.33 Monolithic variable attenuator. (Courtesy of Raytheon Co.)

analysis applied to the size of series MESFETs. The first attenuator (bottom circuit in Fig. 9.34) achieves a 1 dB insertion loss by using large series MESFETs, while the second attenuator (top circuit in Fig. 9.34), which uses a smaller MESFET, has an insertion loss of about 2.5 dB but a larger bandwidth (up to 18 GHz) owing to the reduction in the ground capacitance C_{GS} (see Fig. 9.29). The overall chip size is 1 mm × 1 mm. Such a small size associated with a very simple technological process leads to integrated attenuators with high yields and low production costs. These low costs and the unique wideband capabilities will probably ensure that GaAs MMIC variable attenuators will dominate the market for wideband low power applications.

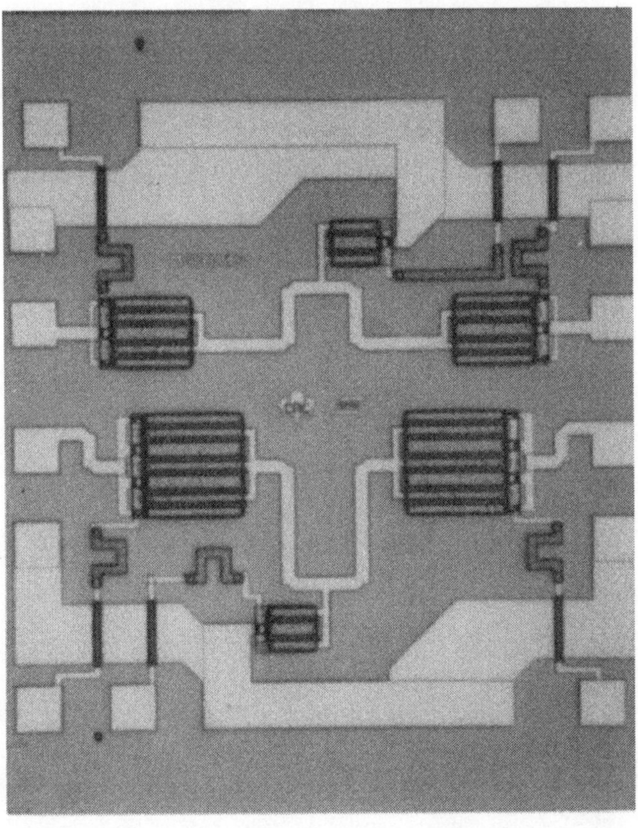

Fig. 9.34 Attenuator chip of dimensions 1 mm × 1 mm. (Courtesy of Spar Aerospace.)

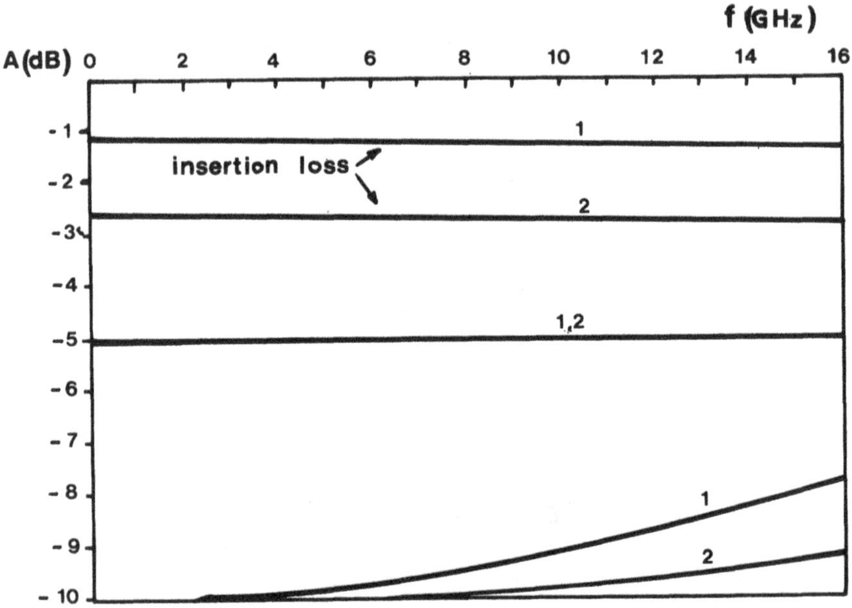

Fig. 9.35 Attenuation versus frequency for the two attenuators shown in Figs 9.33 and 9.34.

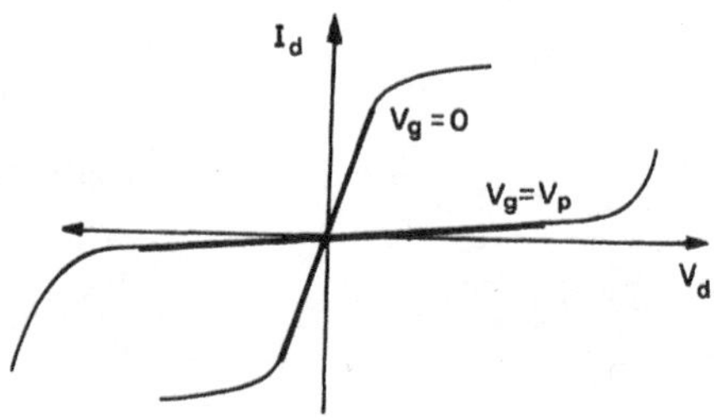

Fig. 9.36 MESFET characteristics in the on and off states.

9.5.3 Switches

One of the most suitable circuits for monolithic integration is the switching circuit, which has many applications and thus requires a large production technology capability. Typical applications include the following:

transmit–receive modules for phased-array antennas
basic switching blocks for phase shifters and phase modulators
switching matrices for TDMA communication systems

switching blocks for highly reliable systems based on the redundancy concept
multiple-switching spot-beam coverage of territory by communication satellites
multiple-switching frequency oscillators for military applications
multiple access for test and measurement equipment

An overview of the principles of GaAs MMIC switching together with an example of an application is presented next.

9.5.3.1 Principles

Switching device The basic device is a resistive MESFET reverse biased at its gate terminal. The MESFET model described in Section 9.5.2 in relation to the design of MESFET attenuators can also be applied to the design of MESFET switches for both the on and off states. As seen in Fig. 9.36 the on state (low resistance) corresponds to a gate voltage close to zero ($V_{GS} \approx 0$), and the off state corresponds to a gate voltage close to the pinch-off voltage ($V_{GS} < V_p$). It should be noted that no d.c. current is required to drive the MESFET switch.

As shown in Fig. 9.37, the power rating of the switch in the off state is limited by the difference between the gate-to-drain diode breakdown voltage V_{GB} and the pinch-off voltage V_p whereas in the on state it is limited by the maximum drain-to-source current at which the MESFET still behaves as a resistor. If the gate terminal is connected to a high impedance source, the instantaneous gate potential V_g is equal to the instantaneous value of the average drain and source potentials owing to the symmetry of the structure.

A trade-off analysis [9.48] addressing maximum power handling,

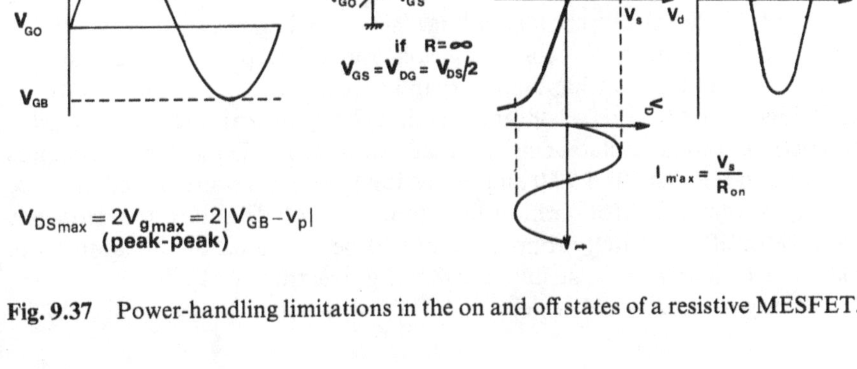

Fig. 9.37 Power-handling limitations in the on and off states of a resistive MESFET.

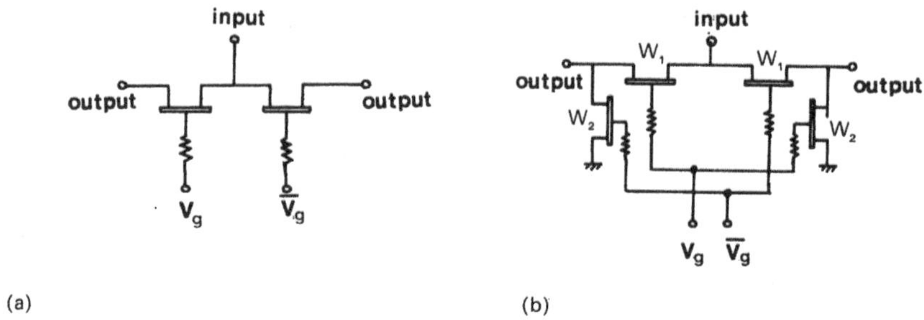

Fig. 9.38 Equivalent circuit of SPDT switches using (a) two MESFETs and (b) four MESFETs.

insertion loss, isolation and frequency bandwidth requirements has to be performed to optimize the switching device for a given application. For instance, the insertion loss can be reduced by increasing both the pinch-off voltage V_p and the electron doping concentration N_d at the expense of the frequency bandwidth and maximum power handling (see Fig. 9.37).

Basic single-pole double-throw configurations Two configurations are used for single-pole double-throw (SPDT) switch applications. One is based on a wideband "untuned" concept and the other on a frequency-dependent transformation concept (quarter-wavelength transformer).

Figure 9.38(a) shows the equivalent circuit of an SPDT switch based on two MESFETs and Fig. 9.38(b) shows an enhanced version using four transistors to improve isolation. The series configuration (Fig. 9.38(a)) provides an inherently low loss broadband response. However, while the off-state resistance is high, isolation is limited at high frequencies by the drain-to-source capacitance which tends to short-circuit the transistor. One method of improving switch isolation, while keeping the broadband properties, is to use a series–shunt configuration as shown in Fig. 9.38(b). Isolation is improved by adding a shunt MESFET which presents a low impedance (on state) when isolation is required. The control inputs V_g and $\overline{V}_g$ of the series–shunt SPDT switch are networked to the gate of each MESFET via resistors and are arranged so that complementary voltages set the switch in one of its two states. Optimization of insertion loss, isolation and other performances such as return loss and power-handling capability can be carried out, following models proposed in ref. 9.48 and Section 9.5.2, by varying the transistor geometries and manufacturing process parameters. Typical performances reported in refs. 9.49 and 9.50 and in Section 9.5.3.2 are summarized in Table 9.3 for convenience. Monolithic integration will probably allow the frequency bandwidth of the switch proposed above to be increased by the addition of matching elements or a distributed switching structure [9.51].

The performance of wideband SPDT switches becomes ineffective for

Table 9.3 Typical performances of series–shunt switches

Frequency range	*Insertion loss* (dB)	*Isolation* (dB)	*Return loss* (dB)	*Switching time* (ns)	*Power* (W)
D.c.–8 GHz	0.5–1.5	50–20	25–15	<1	0.5–1

very high frequency and/or high power applications. The principle of an SPDT switch suitable for multiwatt power handling in X band and above is shown in Fig. 9.39. Two complementary voltages V_g and $\bar{V}_g$ are applied to the MESFET gates such that one MESFET is in the on state and the other is in the off state. As a result of quarter-wave transformation the on-state low resistance appears as a very high impedance at the input terminal, thus enabling the signal to flow into the other arm. The off-state high impedance does not perturb signal propagation. As in the series–shunt switch, frequency limitation of the isolation level by the capacitance of the series MESFET is eliminated because isolation is achieved by the low on-state resistance of a shunt MESFET associated with a quarter-wavelength transformer. Furthermore, the off-state parallel capacitance can be compensated by series inductances to improve the switch insertion loss as shown in Fig. 9.40.

Of course this type of switch is inappropriate for monolithic integration at low frequencies because of the space-consuming quarter-wavelength transformer. However, at X band frequencies and above, MMIC technology allows the MESFET to be integrated into the transmission line where it is most

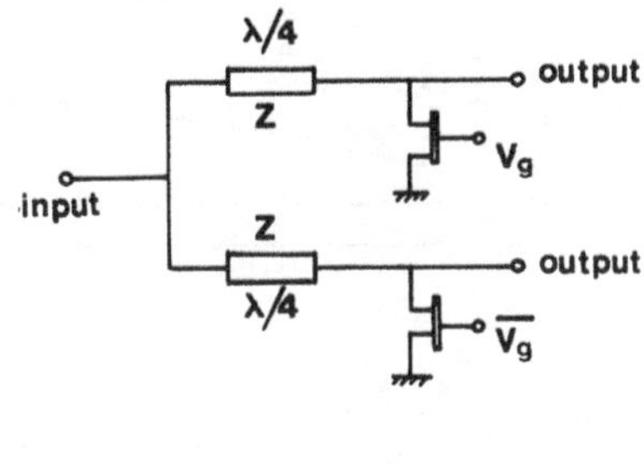

Fig. 9.39 Electrical circuit of a quarter-wavelength switch.

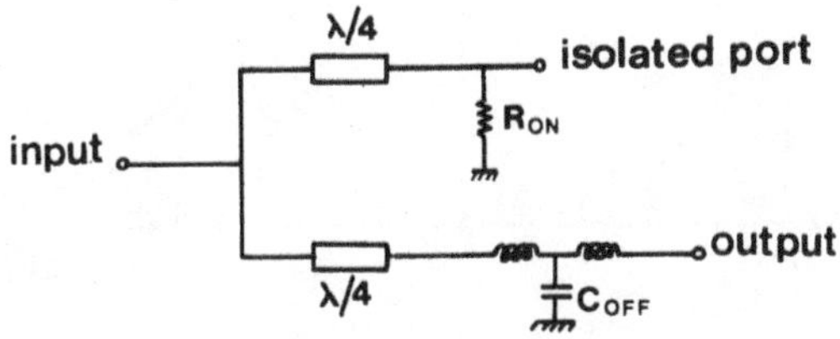

Fig. 9.40 Electrical circuit of a compensated switch in the on state.

effective and also allows additional inductive elements (see Fig. 9.40) to be inserted in the drain-to-source spacing of each FET finger, thus leading to optimum performance. An X band monolithic switch based on the quarter-wavelength transformer principle has been developed for transmit–receive module applications [9.52]. This switch is capable of switching more than 10 W with an insertion loss of about 1 dB, an isolation higher than 25 dB and a return loss better than 10 dB in the 8.5–10.5 GHz frequency range.

9.5.3.2 Circuit applications

SPDT switches are commercially available from most GaAs foundries [9.46, 9.53]. Figure 9.38(b) shows the electrical circuit of a series–shunt SPDT switch developed at Spar Aerospace [9.54]. The design was based on ion-implanted MESFETs with a gate length of 1 μm, a pinch-off voltage of −3 V

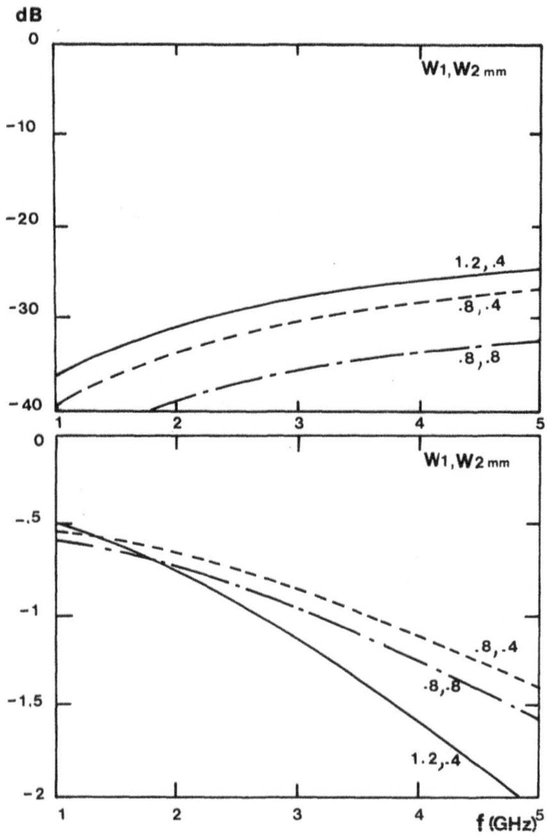

Fig. 9.41 Simulated performances for several combinations of transistor gate widths.

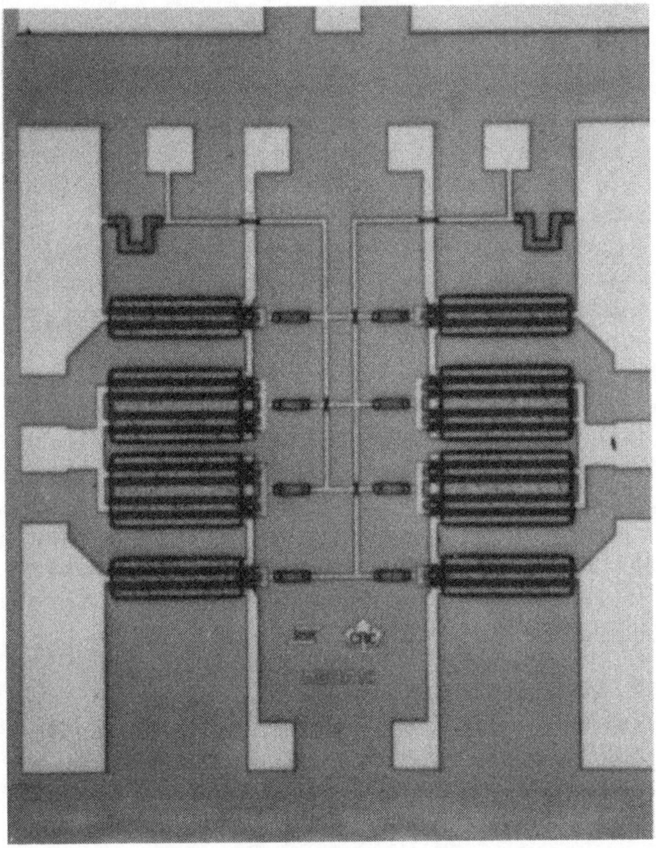

Fig. 9.42 Two SPDT switches of dimensions 1 mm × 1 mm. (Courtesy of Spar Aerospace.)

and a peak doping concentration of 2×10^{17} atoms cm^{-3}. Figure 9.41 shows computed performances for several combinations of transistor gate widths. As expected, good isolation is obtained for devices with large shunt and small series MESFETs, and a low insertion loss is obtained for devices with a shunt and a series size resulting from a trade-off analysis between low and high frequency losses. A 0.8 mm–0.4 mm combination is a reasonable solution for applications up to 4 GHz. Figure 9.42 shows two 0.8 mm–0.4 mm SPDT switches sharing the same control voltages. The insertion loss, isolation and return loss capabilities are better than 1.5 dB, 25 dB and 10 dB respectively, while power handling is better than 0.5 W and the switching time is in the nanosecond range (determined mainly by the external gate circuit). The chip size is 1 mm × 1 mm.

9.5.4 Phase shifter

A key component of transmit–receive modules (phased-array antennas), phase linearizers or phase modulators is a programmable phase shifter. The performance of the systems depends strongly on the characteristics of this component. MMIC technology is suitable for microwave phase shifters to be used in phased-array antenna systems because of its small size, light weight, reliability, reproducibility, low cost and high-volume production capabilities. Some applications such as space-based radar (systems of several low orbit satellites designed to measure the position and speed of flying and nonflying objects) will require up to 100 000 transmit–receive modules [9.55].

As the cost of monolithic circuits is directly related to the number of circuits per wafer, the chip area must be reduced as much as possible. This is a challenge for a microwave designer because phase shifting is almost always achieved by time delay of the signal propagation through a transmission line, which is generally unsuitable for monolithic integration. For example, a 180° phase shift at 10 GHz corresponds to a GaAs microstrip line about 5 mm long. Other methods must be used to reduce the size of the phase shifter. Various phase-shifter configurations, including passive and active phase shifters, and some application circuits are described next.

9.5.4.1 Phase-shifter configurations

Phase shifters can be divided into two main classes: lossy reciprocal passive and unidirectional active. They can be further subdivided into digitally controlled fixed phase angle shifters and continuously variable phase angle shifters.

Passive phase shifters The most common configuration uses the loaded-line approach (Fig. 9.43) to realize circuits with phase shifts of up to 45° because it offers a simple solution with a good performance. The theory of the loaded-line phase shifter is well known [9.56] and is widely applied in MIC as well as MMIC technology [9.57, 9.58]. If B_{off} is the susceptance value loading the line when the MESFETs are set in the off state and B_{on} is the susceptance in the on state, the phase change of the network between the two states, for small perturbations, is

$$\Delta\phi = \tan^{-1}\{(B_{off} - B_{on})Z_0\} \tag{9.9}$$

The line presents an equivalent characteristic impedance for both states:

$$Z_0' = \frac{Z_0}{\{1-(BZ_0)^2 + 2BZ_0 \cot\theta\}^{1/2}} \tag{9.10}$$

where B is either B_{on} or B_{off} and θ is the electrical length defined in Fig. 9.43. The maximum achievable phase shift is about 90°. However, as the magni-

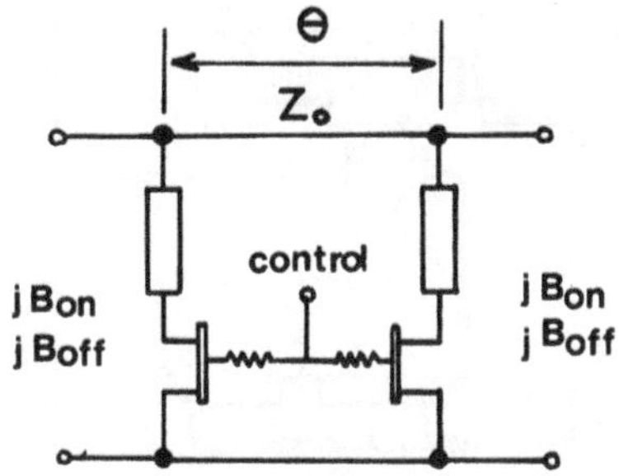

Fig. 9.43 Electrical circuit of a phase shifter based on the loaded-line approach.

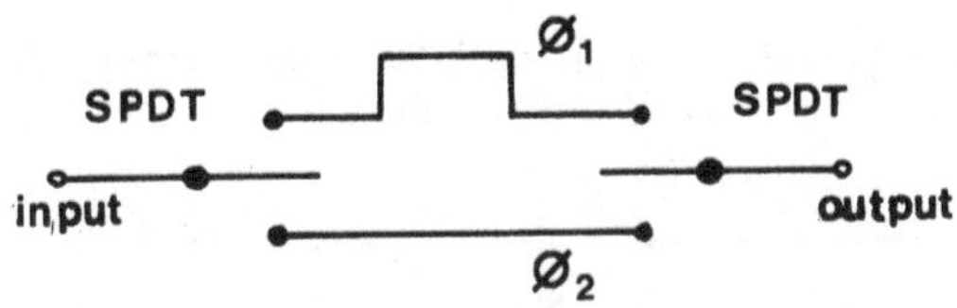

Fig. 9.44 Principle of the switched-line phase shifter.

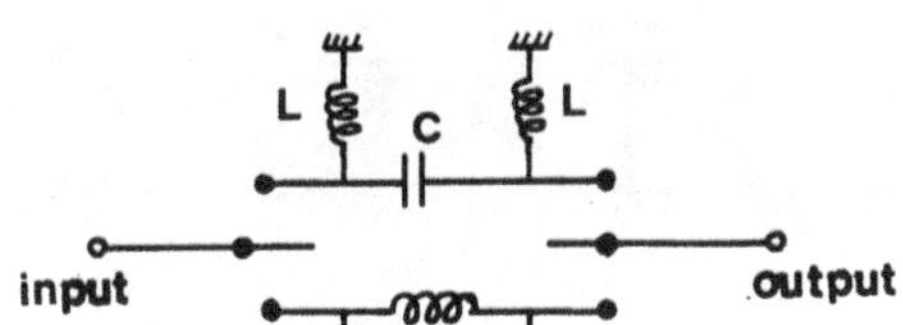

Fig. 9.45 Highpass–lowpass phase-shifter circuit.

tudes of the susceptances B_{on} and B_{off} must be kept small for a good match (see eqn (9.10)), loaded-line phase shifters are generally limited to maximum phase shifts of 45°. The two devices providing the switching susceptances B_{on} and B_{off} are resistive MESFETs as described in Section 9.5.2.

Switched-line phase shifters are composed of two SPDT switches (see Section 9.5.3) which are used to choose between two transmission paths of different phase lengths ϕ_1 and ϕ_2 as shown in Fig. 9.44. Thus the net phase shift is $\phi_1 - \phi_2$. As the value of the phase shift is limited only by the chip size, the switching approach is appropriate for the design of 90° and 180° phase shifters. The two delay lines can be defined as either distributed lines or lumped-element highpass–lowpass filters (Fig. 9.45). This approach allows smaller circuits to be built and larger frequency bandwidths to be achieved because of the opposite phase slopes of the two filter types [9.59]. An example of this

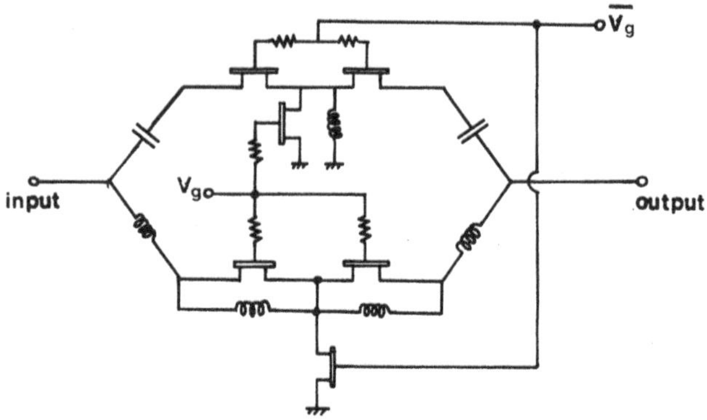

Fig. 9.46 Highpass–lowpass phase-shifter circuit with switching devices incorporated in the filters.

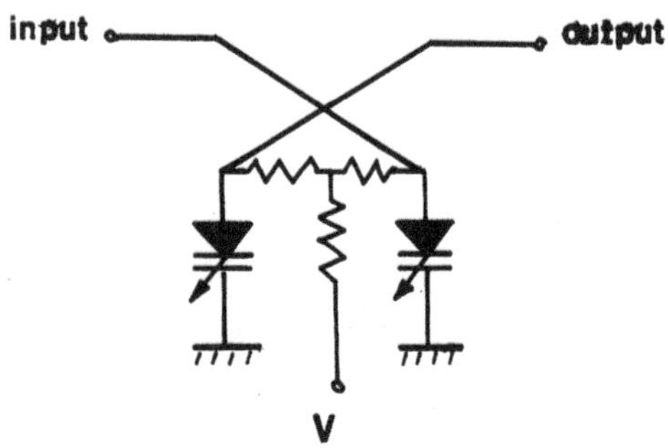

Fig. 9.47 Electrical circuit of an analog phase shifter.

switch configuration is given in Section 9.5.4.2. Another attractive solution is to eliminate the two SPDT switches and to use switching MESFETs as part of the highpass and lowpass filters [9.60] as shown in Fig. 9.46. This leads to an improved frequency bandwidth and a miniaturized phase shifter. The MESFETs behave as a small resistance in the on state and as a capacitance in the off state (see Section 9.5.2).

In contrast with the digital phase shifters discussed above, analog phase shifters can provide variable phase shifts. They have the advantage that fine phase tuning can be used to compensate for manufacturing deviations and thermal or aging effects. In addition, in multibit phase shift systems a single analog phase shifter can replace several digital phase shifters and thus minimize the circuit area required. The electrical circuit of an analog phase shifter [9.61) is shown in Fig. 9.47. A quadrature 3 dB coupler is terminated by two GaAs Schottky diodes. The transfer characteristic of the phase shifter determines the reflection coefficient of the diodes. Ideally a 180° phase shift could be achieved. However, the finite capacitance ratio C_{max}/C_{min} and the

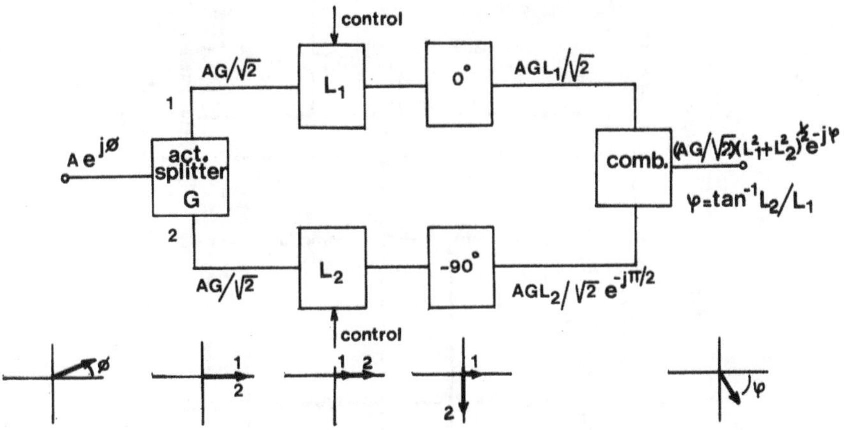

Fig. 9.48 Block diagram and vector transformation of a typical 90° vector generator.

parasitic elements limit the phase shift to typically 120°. The monolithic integration of a high-Q GaAs Schottky diode is discussed in Sections 9.6.2 and 9.6.4. The 3 dB coupler can be designed using a folded Lange coupler or a hybrid coupler based on the lumped-element approach [9.62] to minimize the size.

Active phase shifters These are so called because an active device used in a conventional amplifier configuration performs a basic function of the phase shifter. Analog phase shifters based on a dual-gate MESFET have been demonstrated [9.63, 9.64], but the most promising configuration for monolithic integration is probably the vector generator based on a MESFET active splitter and/or active combiner.

The vector generator is a circuit that allows continuous phase and amplitude control within a constrained complex plane [9.65]. Figure 9.48 shows the block diagram and vector transformation of a typical 90° vector generator based on an active splitter–attenuator, two lines in phase quadrature and a combiner. In order to ensure a constant output voltage magnitude, independent of the phase shift ϕ, the following conditions must be obeyed by attenuators L_1 and L_2:

$$L_1 = \cos\phi \qquad L_2 = \sin\phi \tag{9.11}$$

where ϕ is the required phase shift. The output voltage is given by

$$|V| = \frac{AG}{\sqrt{2}} \tag{9.12}$$

The minimum loss of a passive vector generator used as a continuous 90° phase shifter is 3 dB (obtained by setting $G = 1$ in eqn (9.12)) which is often

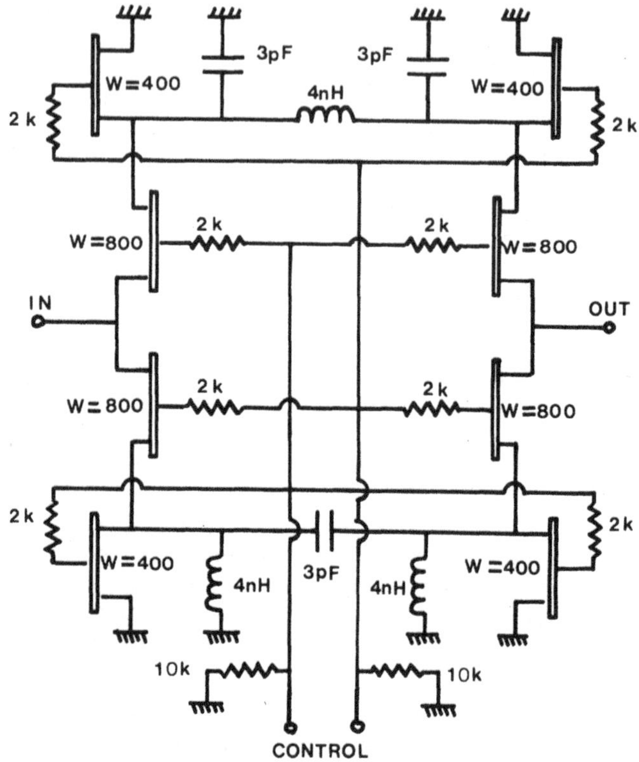

Fig. 9.49 Topology of a phase shifter.

unacceptable for system requirements. An X band GaAs integrated vector generator based on the use of a dual-gate MESFET gain-controlled splitter has been developed [9.66]. The dual-gate MESFET consists of a split structure (gate 2 and drain) sharing the same input terminal (gate 1). Monolithic integration of the dual-gate active splitter produces a device which is almost perfectly symmetrical in terms of technological processes and thus provides excellent amplitude balance between the two states of operation. By inserting transmission lines with a 90° differential phase shift at the operating frequency and combining the two split signals in a passive combiner a 3 dB gain has been achieved over a 10% bandwidth in X band on a chip of dimensions 2.5 mm × 3 mm. The same concept has been used to develop a GaAs integrated vector generator in the v.h.f., L, S and C bands [9.67, 9.68].

9.5.4.2 Circuit applications

The design of a 180° L band phase shifter based on the highpass–lowpass approach [9.51] is described here. As shown in Fig. 9.49, the circuit is constructed using two SPDT switches as described in Section 9.5.3.1 and two

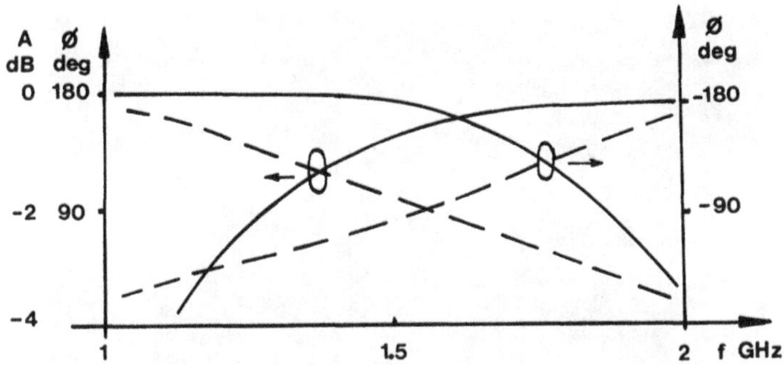

Fig. 9.50 Insertion loss and phase response of lowpass and highpass filters.

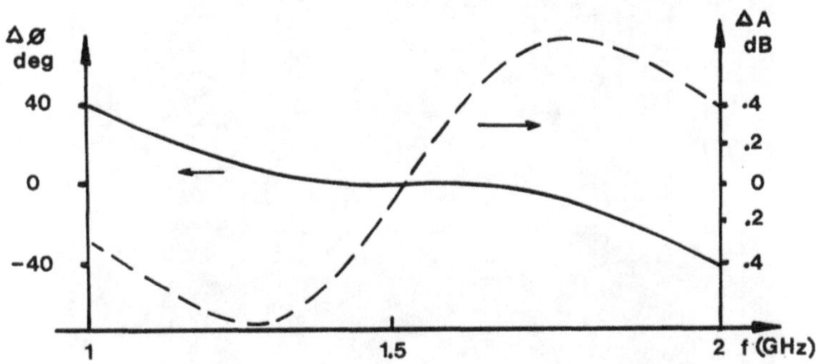

Fig. 9.51 Phase shift and amplitude imbalance of a lowpass–highpass phase shifter.

lumped-element highpass–lowpass filters. The highpass structure, which comprises two shunt inductors and a series capacitor, generates a $+90°$ phase shift and the lowpass structure, which comprises two shunt capacitors and a series inductor, generates a $-90°$ phase shift. Thus the total phase shift difference between the two paths is 180°. Furthermore, owing to the opposite signs of the phase slope of the two circuits, a 20% bandwidth is achievable (assuming a phase shift error of $\pm 5°$) as shown in Fig. 9.50. A spiral inductor and a MIM capacitor are used to obtain the required 4 nH inductance and 3 pF capacitance (see Fig. 9.49).

The phase shift and amplitude imbalance are plotted against the frequency of the complete phase shifter in Fig. 9.51. It exhibits a 300 MHz bandwidth (20%) corresponding to a $\pm 5°$ phase error and an amplitude imbalance of ± 0.7 dB. Within this bandwidth the loss is lower than 1.5 dB and the input and output return losses are better than 10 dB. The circuit chip is illustrated in Fig. 9.52. Despite the low frequency operation, a 180° phase-shifter chip as small as 1 mm × 1 mm has been designed by using lumped-element filters and optimizing the layout.

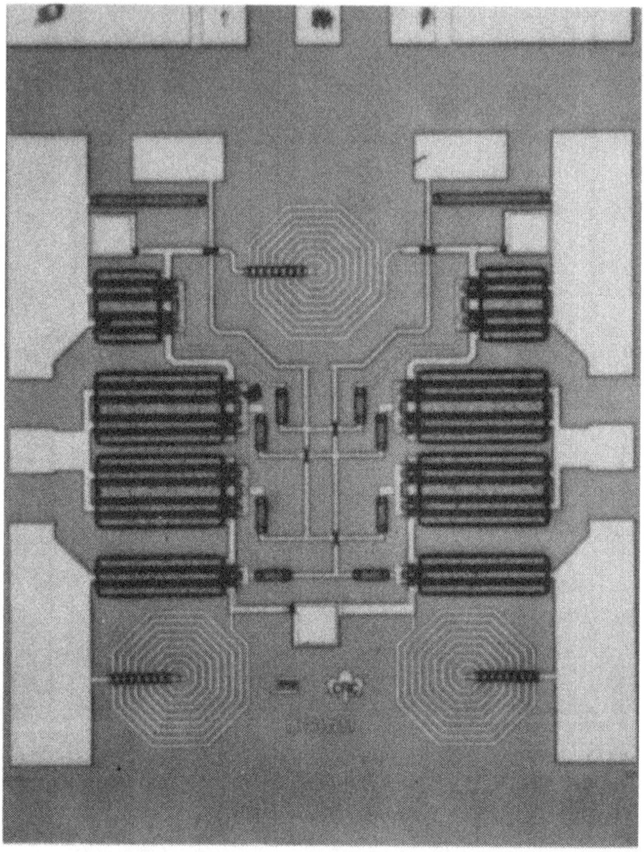

Fig. 9.52 Phase-shifter chip of dimensions 1 mm × 1 mm. (Courtesy of Spar Aerospace.)

9.6 Oscillators and frequency-processing functions

9.6.1 Introduction

Although GaAs MMIC technology has made an impressive contribution to improvements in the performance of amplifiers, switching circuits and control circuits, this is not yet the case for nonlinear analog circuits. This is probably because the demand for increased complexity in mixers and oscillators, although emerging, is still low. The lack of accurate nonlinear models and associated CAD software (which is still in its infancy) limits the progress which could be made by using MMIC technology. The increasing efforts of MMIC designers to model nonlinear devices will bring benefits to the design of MIC components. MICs, in contrast with MMICs, are still often fabricated using the "cut-and-try" experimental method. However, as will be shown in this section, GaAs MMICs have already made significant advances in this field and are expected to play a major role in the future by improving the performances of known structures and introducing new circuit capabilities.

In this section we shall discuss the principles of operation using as examples nonlinear analog circuits based on a GaAs MESFET or a Schottky diode as the nonlinear device (see also Chapter 5). The circuits considered are monolithic fixed and voltage-controlled oscillators, analog frequency dividers and mixers. Although this list is not complete (frequency multipliers, limiters and modulators could also have been included), it provides a representative picture of the field.

9.6.2 Oscillators

The emergence of micron and submicron GaAs MESFET technology combined with monolithic integration of passive components on GaAs substrates has allowed the development of microwave and millimeter-wave fixed-frequency oscillators [9.69–9.71] and voltage-controlled oscillators (VCOs) which are competitive with Gunn and IMPATT diode oscillators [9.72, 9.73]. The increasing demand for e.h.f. oscillators, frequency-agile wideband sources and FM continuous-wave (CW) radar in commercial or military applications has attracted the attention of systems designers to GaAs technology. ICs permit an increase in the oscillation frequency to close to the maximum MESFET oscillation frequency f_{max} [9.74] or allow an increase in the bandwidth of wideband VCOs by reducing parasitic elements [9.75]. The potential reproducibility of MMIC technology from chip to chip and wafer to wafer can benefit VCO requirements in, for example, FM CW radar where reproducible voltage-controlled oscillation from chip to chip allows computerized linearization of VCOs, thus relaxing intrinsic linearity requirements.

Oscillator design considerations and examples of applications are described next.

9.6.2.1 Oscillator design

The design of MESFET oscillators has been extensively discussed in Chapter 7. As the design principles for MIC and MMIC oscillators are essentially the same, no further fundamental considerations are needed here. However, it should be noted that most oscillators, in particular VCOs, are designed using a semi-empirical approach. Despite efforts in VCO modelling [9.76–9.78] output power is still difficult to predict and important parameters such as frequency bandwidth or tuning linearity are predicted using small-signal analysis.

This situation is unacceptable for a MMIC designer because design and process iterations are long and costly. Therefore the implementation of design tools able to predict oscillator parameters such as power, bandwidth, harmonic level, linearity, settling time etc. is vital. The key activities will be the development of user-friendly nonlinear analysis programs, as proposed in Chapter 5, and accurate device models suitable for high operating frequencies and high power dynamic ranges. One of the major problems faced by designers of MMIC VCOs is the definition of varactor characteristics. To illustrate this point, the electrical and equivalent circuits of a typical VCO are shown in Fig. 9.53. A voltage-controlled capacitor with a capacitance C_V and a resistance R_V is connected via an inductor to the transistor gate. The feedback capacitor C provides gate and drain port impedances with a negative real part while the inductor L ensures that the gate circuit resonates at the operating frequency. The oscillation conditions are

$$f = \frac{1}{2\pi\{LC_T C_V/(C_T + C_V)\}^{1/2}} \tag{9.13}$$

$$R_V = R_T \tag{9.14}$$

where R_V is the varactor series resistance which is a function of the varactor biasing voltage, C_V is the varactor capacitance which is voltage controllable from a minimum value $C_{V\,min}$ to a maximum value $C_{V\,max}$, and $-R_T$ and C_T are the equivalent resistance and capacitance presented by the MESFETs, the load impedance and the feedback capacitor C at the gate terminal.

The minimum and maximum oscillation frequencies f_{min} and f_{max} can be calculated by substituting the equivalent capacitance C_T (generally $C_T \ll C_{V\,max}$) and the minimum varactor capacitance $C_{V\,min}$ (generally $C_{V\,min} \ll C_T$) respectively in eqn (9.13):

$$f_{min} \approx \frac{1}{2\pi(LC_T)^{1/2}} \tag{9.15}$$

$$f_{max} \approx \frac{1}{2\pi(LC_{V\,min})^{1/2}} \tag{9.16}$$

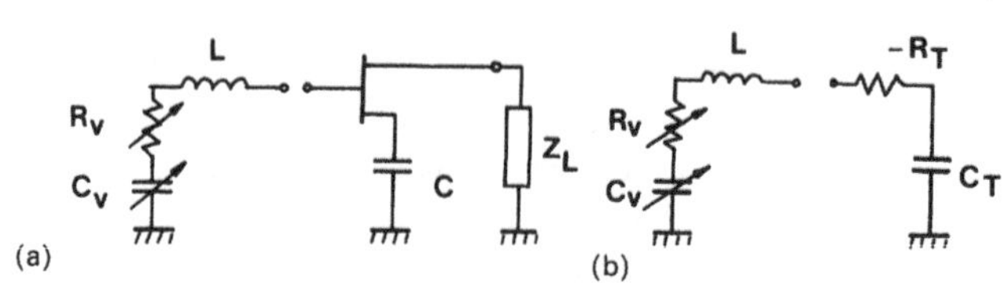

Fig. 9.53 (a) Electrical and (b) equivalent circuits of a gate varactor VCO.

Although the complete model is more complicated, the main conclusions relating to varactor characteristics can be obtained from eqns (9.14) and (9.15): the series resistance R_V and the capacitance $C_{V\,min}$ must be low to ensure wideband frequency operation. The linearity of the VCO and the levels of the harmonic and spurious signals also depend strongly on the variation in the capacitance as a function of the applied voltage.

The design procedure for a varactor for wideband VCO applications

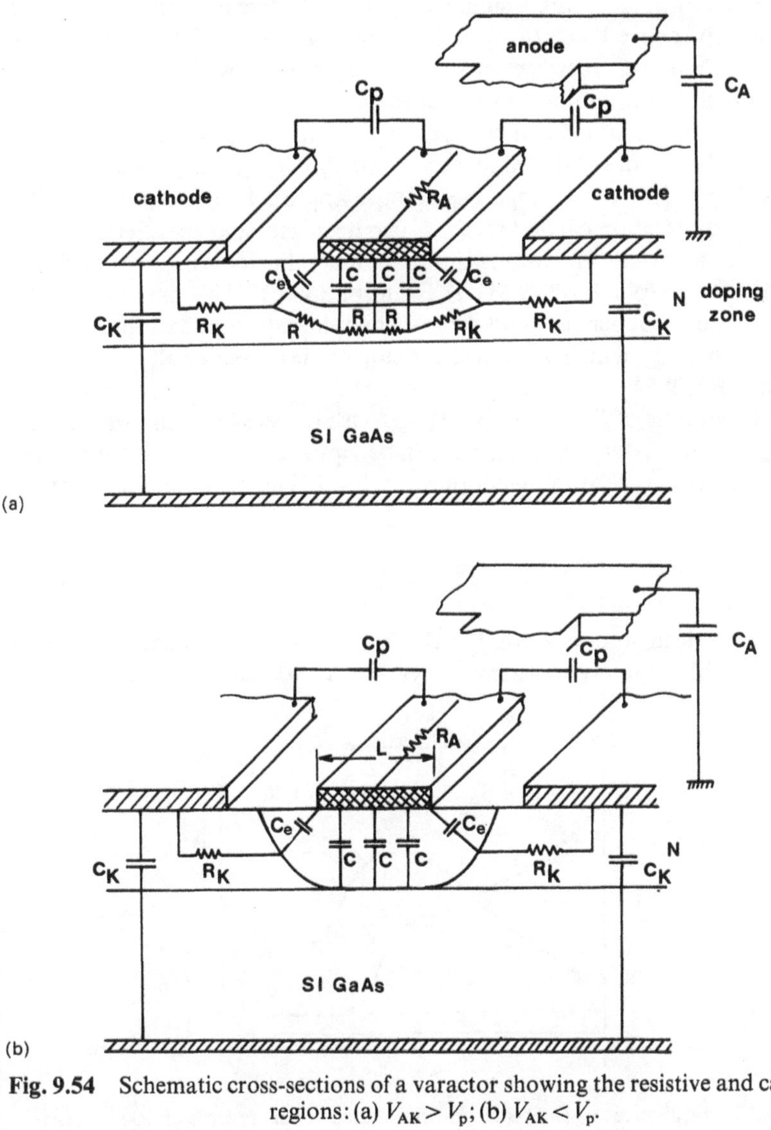

Fig. 9.54 Schematic cross-sections of a varactor showing the resistive and capacitive regions: (a) $V_{AK} > V_p$; (b) $V_{AK} < V_p$.

which makes use of a single implanted active layer is described as an example in the following section.

9.6.2.2 Design of a varactor for wideband applications

As explained in Section 9.6.2.1 wideband VCOs require varactors with a small value of $C_{V\,min}$, a high $C_{V\,max}/C_{V\,min}$ ratio and a low value of the parasitic series resistance. Our discussion is focused on a lateral implanted Schottky varactor suitable for monolithic integration as shown in Fig. 9.54. The design is undertaken by considering a single n layer (a selective n^+ contact layer can be included to reduce the cathode access resistance) [9.79, 9.80]. Figure 9.54 shows two schematic cross-sections of such a varactor with its resistive and capacitive regions for two biasing conditions.

Elements of the model can be calculated using the approach proposed in ref. 9.81. The behaviors of the equivalent resistance R_V and capacitance C_v are shown in Fig. 9.55. As the negative voltage applied to the anode terminal decreases, the depletion capacitance C_V also decreases and the effective series resistance R_V increases. R_V attains its maximum value for a bias voltage close to punch-through (pinch-off in MESFET terminology). This maximum value $R_{V\,max}$ is frequency dependent and occurs when the capacitance approaches its minimum value C_{min} which is equal to the sum of the two sidewall capacitances as shown in Fig. 9.54.

Since the value of $R_{V\,max}$ can be critical with respect to oscillator start-up, the design of the lateral varactor must be optimized taking into account the geometrical and technological parameters. It has been demonstrated in ref. 9.81 that

$$R_{V\,max} \leqslant \frac{A}{WF} \tag{9.17}$$

where W is the total anode width, F is the frequency and A is a constant which depends on the doping profile (and is typically 50 Ω mm GHz). Figure 9.56

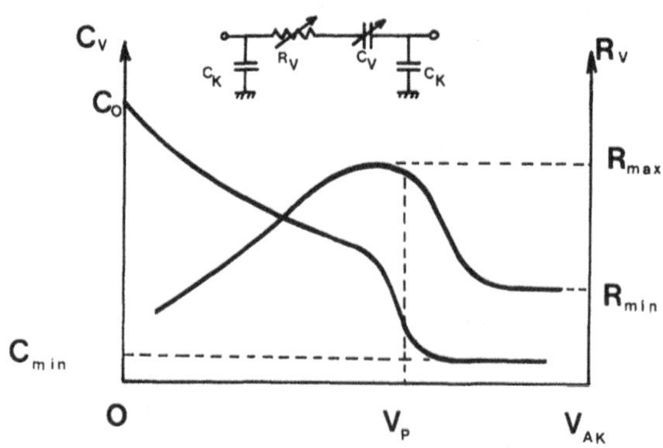

Fig. 9.55 Equivalent capacitance and resistance versus reverse anode voltage.

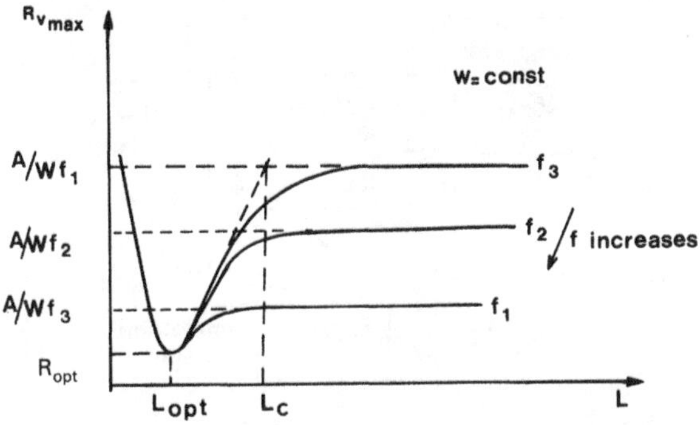

Fig. 9.56 Maximum series resistance versus the anode length with frequency as parameter for a constant anode width.

shows the variation in $R_{V\,max}$ as a function of the anode length L for a constant anode width W. $R_{V\,max}$ reaches a minimum R_{opt} when the loss due to the anode series resistance R_A equals the loss in the n layer (R_K and R in Fig. 9.54). Generally, the corresponding anode length L_{opt} is small (1–4 μm). $R_{V\,max}$ is limited to an upper value defined by eqn (9.17). The above calculation procedure can also be applied to multifinger varactor structures.

Given the maximum acceptable value of C_{min}, W can be calculated (the value of C_{min} is essentially independent of the anode length L) and the acceptable value of $R_{V\,max}$ can be used to specify L between L_{opt} and L_c (see Fig. 9.56). For example, a five-finger varactor using two anodes of dimensions 3 μm × 100 μm, an implanted layer with a sheet resistance of 800 Ω/□ at $V_{AK} = 0$ and a punch-through voltage of −3 V (corresponding to a typical FET n layer) exhibits a 1.2 pF capacitance at 0 V, a maximum resistance $R_{V\,max}$ of 6 Ω near punch-through and a minimum capacitance C_{min} of 0.09 pF.

This approach has been used to design the VCO described in the next section.

9.6.2.3 Circuit applications

Two of the many GaAs MMIC oscillators developed so far are described here: a stable fixed-frequency oscillator in X band and a wideband VCO in Ku band.

Stable oscillator An 11 GHz stable oscillator has been developed at the Laboratoires d'Electronique et de Physique appliquée (LEP) using a capacitive series feedback and an off-chip dielectric resonator [9.82]. Figure 9.57 shows its electrical circuit. The source series feedback capacitance C_F and the drain tuning circuit (L_d, C_d) were determined using an experimental design method [9.83] to maximize the output power.

Figure 9.58 shows the oscillator chip, the dimensions of which are

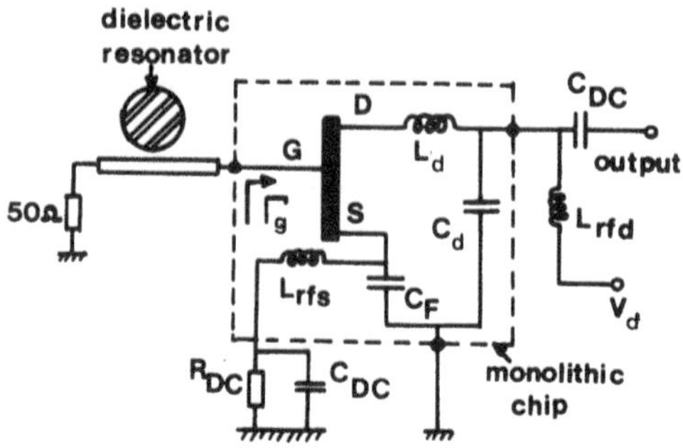

Fig. 9.57 Electrical circuit of an 11 GHz stable oscillator.

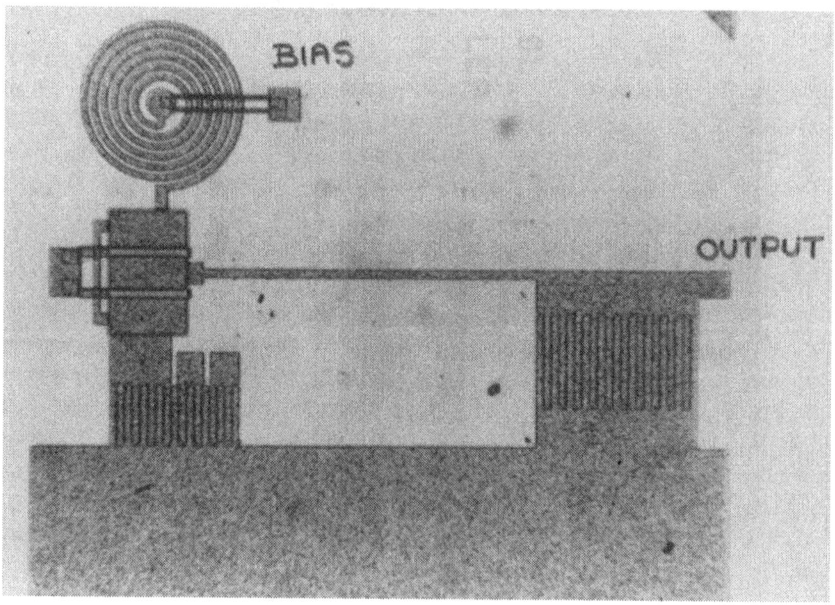

Fig. 9.58 Monolithic oscillator chip (1.2 mm × 1.4 mm).

1.2 mm × 1.4 mm. The MESFET uses two gates of dimensions 0.8 μm × 150 μm. The feedback and output capacitors C_F and C_d are interdigitated capacitors designed to have a minimum finger inductance and therefore a series resonance at a frequency much higher than 11 GHz. The spiral bias inductor is designed to present a maximum impedance (parallel resonance) at 11 GHz. This chip exhibits an impedance with a negative real part in the frequency range 6–14 GHz, thus enabling an oscillator operating between 6 and 14 GHz to be built using the appropriate off-chip dielectric resonator.

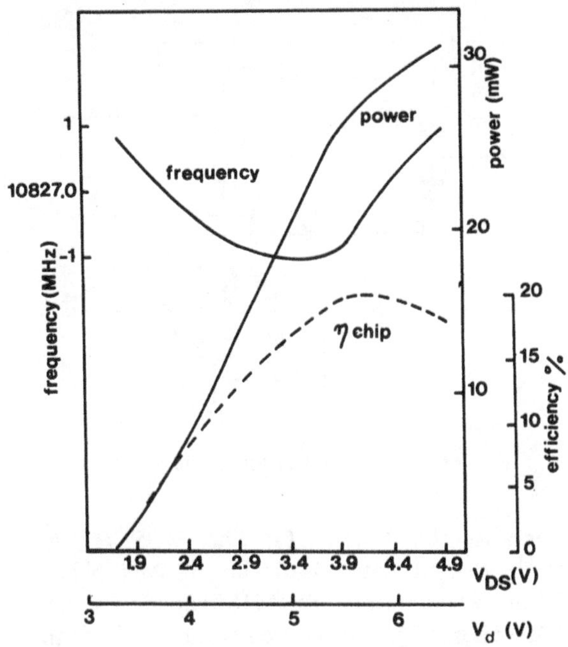

Fig. 9.59 Performance of the monolithic chip.

The performance of the 11 GHz oscillator is illustrated in Fig. 9.59. The output power exceeds 30 mW and the frequency push related to a drain voltage variation from 2 to 5 V is less than 1 MHz. The chip efficiency reaches 20%, but the oscillator efficiency is 14% because of the external bias resistance R_{DC} (Fig. 9.57). The frequency drift is better than 1 ppm K^{-1} and is due to compensation of the device parameter drift and the dielectric resonator characteristics (see Chapter 7).

Voltage-controlled oscillator in the Ku band A wideband VCO operating in the Ku band which uses two varactor diodes and covers the 11–16 GHz frequency bandwidth has been designed at LEP and processed at Radiotechnique Compélec (RTC) using the available GaAs foundry capabilities. Figure 9.60 shows the electrical circuit of the VCO. The variable source series feedback varactor essentially provides the gate terminal impedance with a negative real part in the required bandwidth, and the gate varactor enables the VCO to be tuned over the oscillation frequency bandwidth. The chip includes microwave as well as biasing circuits.

The gate varactors consist of four anode fingers of dimensions 75 μm × 5.1 μm, and the source varactors consist of two anode fingers of dimensions 150 μm × 3.8 μm. They use the same implanted doping layer and

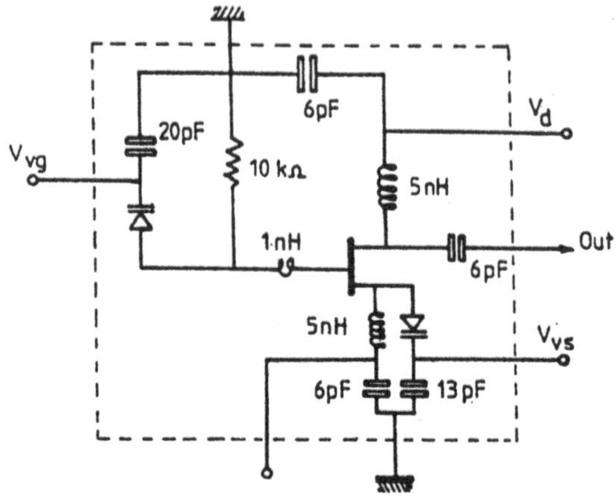

Fig. 9.60 Electrical circuit of the VCO chip.

gate metallization as the MESFETs. The MESFET gate geometry is 0.7 μm × 300 μm divided into four parallel fingers. An SiN layer 1500 Å thick is deposited as the dielectric material for MIM capacitors resulting in a square capacitance of 450 pF mm^{-2}. A negative resistance of less than −15 Ω at the gate port in the frequency range 9–18 GHz has been predicted using the modified small-signal approach presented in ref. 9.81.

Figure 9.61 shows a VCO chip of dimensions 0.7 mm × 0.7 mm. This chip has demonstrated an oscillation capability in the frequency range

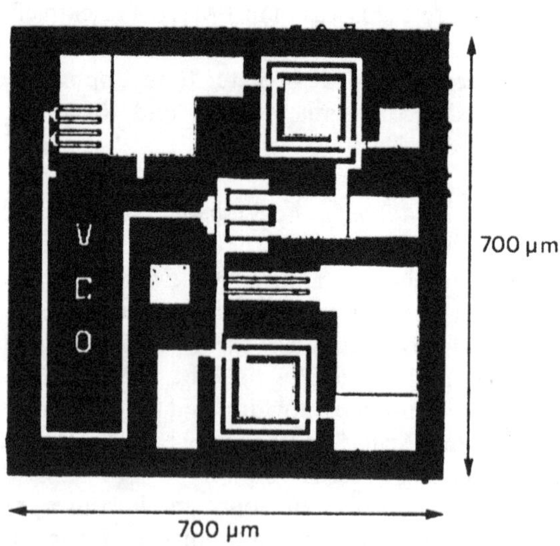

Fig. 9.61 VCO chip of dimensions 0.7 mm × 0.7 mm. (Courtesy of LEP and RTC.)

11–16.5 GHz with an output power of 7–10 dBm and a typical efficiency of 10%. Excellent MMIC technology reproducibility has been confirmed since a typical wafer dispersion for the tuning bandwidth of less than 5% has been obtained.

9.6.3 Dividers

The need for frequency dividers operating at microwave frequencies becomes extremely important in applications like instrumentation, microwave communication systems and electronic warfare (frequency counters, frequency synthesizers and instantaneous frequency measurements).

High speed GaAs logic dynamic dividers have demonstrated a maximum operating frequency of 20 GHz [9.84], and therefore analog frequency dividers are still the only circuits able to operate at extremely high frequencies. This is mainly because logic circuits divide the frequency on a cycle-to-cycle basis limited by the delay time of the gate arrays. In contrast, analog dividers are based on either the regenerative modulation principle, which is limited by the ability of a feedback loop to achieve gain, or the parametric oscillator principle, which is limited by the quality factor of a nonlinear reactive device.

An overview of the principles is given next together with examples of GaAs integrated dividers based on regenerative and parametric division concepts.

9.6.3.1 Regenerative modulation dividers

Principles The fundamental ideas utilized in regenerative modulation were described by Miller [9.85] as early as 1939. This general theory can be summarized as follows (Fig. 9.62). The divider comprises an amplifier and a feedback loop with a mixer and a bandpass filter to select the required output

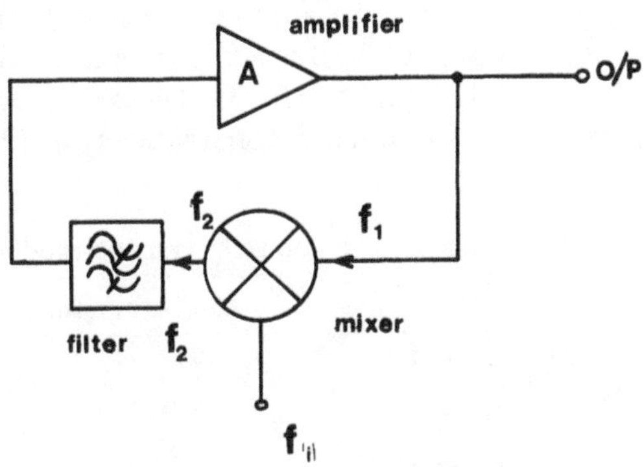

Fig. 9.62 Schematic diagram of regenerative modulation.

frequency. When a signal of frequency f_i is applied to the input of the mixer, the following frequencies are generated:

$$\begin{aligned} f_2 &= nf_i - mf_1 \qquad & nf_i > mf_1 \\ f_2 &= mf_1 - nf_i \qquad & nf_i < mf_1 \end{aligned} \tag{9.18}$$

where m and n are integers. To ensure coherence in the loop f_1 and f_2 must be equal, leading to the relation

$$f_1 = \left(\frac{n}{m+1}\right) f_i \tag{9.19}$$

Equation (9.19) shows that both integer and fractional division can be achieved with regenerative dividers. Table 9.4 gives the frequency division for the first four mixing products.

The feedback loop (Fig. 9.62) can sustain an oscillation at frequency $f_1 = f_i/2$ provided that the transfer function $\mu\exp(-j\phi)$ corresponding to the amplifier and filter cascade satisfies the following conditions [9.86]:

$$\mu = \frac{1}{L_c} \tag{9.20}$$

$$\phi = \frac{\pi}{2}(2k+1) \tag{9.21}$$

where L_c is the conversion loss of the mixer and k is an integer between zero and infinity. These two equations show that the amplifier should compensate the mixer and filter losses and that the electrical length of the loop should be adjusted to satisfy eqn (9.21). When no power is applied to the input of the mixer the feedback loop is stable provided that the combined mixer isolation and filter losses are greater than the amplifier gain.

The maximum operating frequency of regenerative dividers by 2 can be determined from eqn (9.20). If the mixer loss is independent of frequency and is flat at 8 dB, the maximum operating frequency is about 50 GHz if 0.5 μm MESFET technology is used for the amplifier or about 70 GHz if 0.25 μm technology is used since a gain of 8 dB is achievable at 25 GHz and 35 GHz respectively by 0.5 μm and 0.25 μm MESFETs. These frequencies are three or four times higher than the maximum division frequency of digital dividers.

Table 9.4 Frequency division versus mixing products

Mixing products		*Frequency division*
Second order	$m = 1, n = 1$	$f_1 = f_i/2$
Third order	$m = 1, n = 2$	$f_1 = f_i/3$
Fourth order	$m = 1, n = 3$ $m = 2, n = 2$	$f_1 = f_i/4$ $f_1 = 2f_i/3$

The dynamic range of regenerative dividers is limited on the one hand by the power required to drive the mixer into the low loss (or high gain for an active mixer) region (usually 5–10 dBm) and on the other hand by the maximum power permitted to keep the divider free from spurious frequency generation (usually 15–20 dBm). Finally, despite their high frequency capability, regenerative dividers offer a modest instantaneous frequency bandwidth [9.86, 9.87] since operating conditions are frequency dependent as shown in eqn (9.20).

Two examples illustrating the properties of GaAs MMIC regenerative dividers are given below.

Circuit applications Relatively few dividers based on regenerative modulation have been designed using MMIC technology. Stubbs *et al.* [9.88] have designed a frequency halver at 10 GHz based on a GaAs MESFET mixer with gain, thus combining mixer and amplifier functions in one device. The feedback loop is achieved by a selective reactive network as shown in Fig. 9.63. Consequently, the divider works using only one MESFET. This circuit

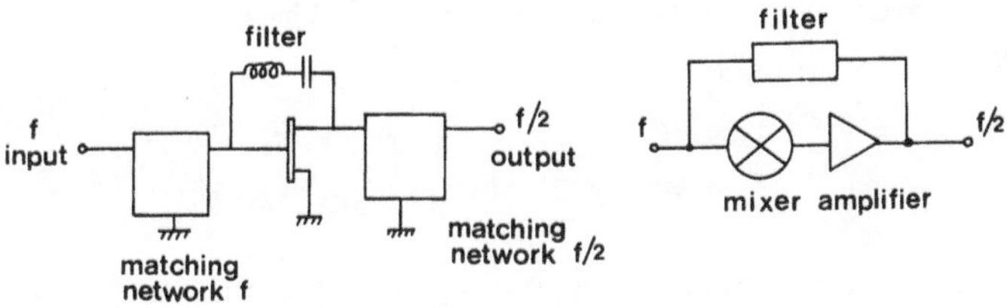

Fig. 9.63 Single MESFET divider and its equivalent block diagram.

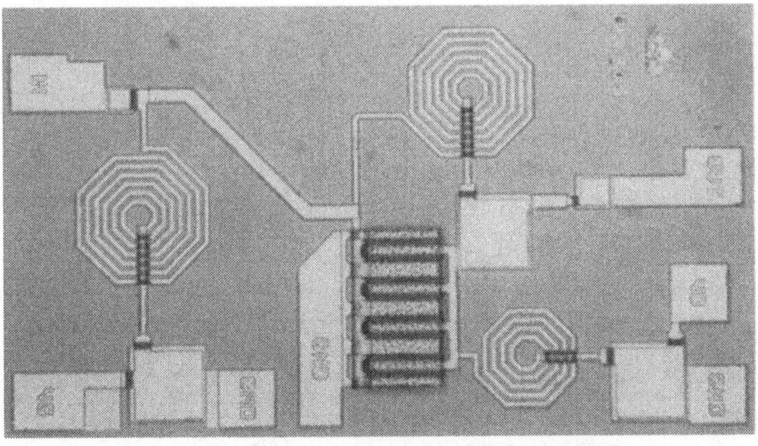

Fig. 9.64 GaAs MESFET regenerative divider chip of dimensions 1 mm × 0.65 mm. (Courtesy of CRC.)

presented a frequency bandwidth of 800 MHz at 10 GHz with an input power of 16 dBm and a conversion loss of 12 dB. Figure 9.64 shows the corresponding chip which has dimensions of 1 mm × 0.65 mm. Coupling this circuit to a VCO will facilitate the construction of phase-locked sources synchronized by low frequency reference sources.

Research workers at NEC have developed a GaAs MMIC divider by 4 at 11 GHz [9.89] which is based on a dividing loop including a MESFET amplifier, a dual-gate MESFET mixer, a MESFET loop amplifier and a delay line to satisfy the phase condition for minimum divider loss. This divider has demonstrated a conversion loss of 6 dB (11–2.75 GHz). The input power is 10 dBm and the frequency bandwidth is 1.5 GHz.

9.6.3.2 Parametric dividers

Principles The theory of parametric amplification and oscillation dates back to the 1930s [9.90]. The general theory can be understood by reference to Fig. 9.65. A power pump of frequency f_p and an incoming signal of frequency f_s are applied to a nonlinear variable reactor (in this case a variable capacitor). Consequently, a third frequency f_i, which is usually called the idler frequency, is generated through the nonlinear capacitor C. These three frequencies are related by the following equations:

$$\begin{aligned} f_i &= nf_p - mf_s \qquad & nf_p > mf_s \\ f_i &= mf_s - nf_p \qquad & nf_p < mf_s \end{aligned} \tag{9.22}$$

where m and n are integers. Depending on the matching and filtering conditions, the basic configuration shown in Fig. 9.65 could be an amplifier, an oscillator, a multiplier, a divider, a down-converter or an up-converter. For $f_i = f_s$ the above equations become

$$f_s = \frac{nf_p}{m+1} \tag{9.23}$$

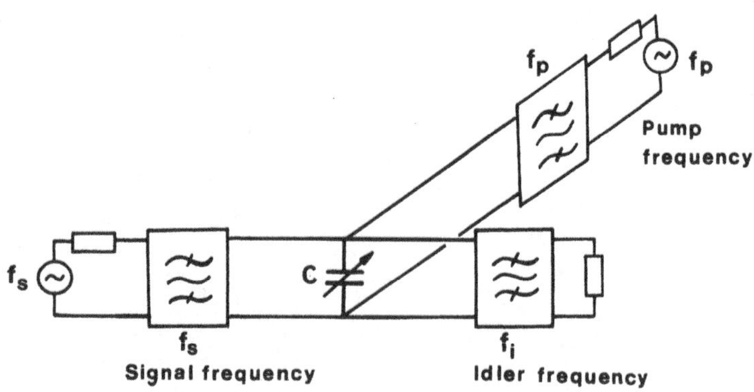

Fig. 9.65 General schematic diagram of parametric circuits.

This means that Table 9.4 is also valid for parametric dividers and, as in the case of regenerative dividers, integer or fractional parametric dividers can be obtained by designing appropriate bandpass filters for different frequencies.

The simplified approach proposed in ref. 9.91 enables the key parameters determining the working conditions of the frequency halver to be identified. It is assumed that a capacitor C_0 vibrating at a pump frequency f_p with a small amplitude $\Delta C = \varepsilon C_0$ is part of a resonant RLC circuit as shown in Fig. 9.66. The condition for the loop to sustain an oscillation at frequency f_s is given by the relation

$$\varepsilon Q = 2 \tag{9.24}$$

where $Q = \frac{1}{2}r\, C_0 f_s$ and $\varepsilon = \Delta C/C_0$. This expression shows that a variable capacitor suitable for frequency dividers is characterized by a high quality factor Q and/or a high C/C_0 ratio. More sophisticated models have been developed [9.92] to derive equations describing the behavior of parametric circuits of the type shown in Fig. 9.65. However, condition (9.24) is still valid.

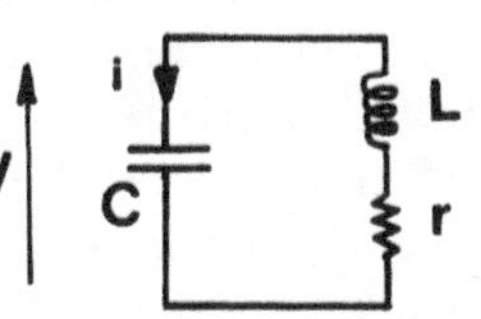

$$C = C_0\{1+\varepsilon \sin(2\pi f_p t)\}$$

$$LC_0(2\pi f_S)^2 - 1 = 0$$

Fig. 9.66 Resonant RLC circuit (frequency f_s) including a capacitor pumped at f_p.

Parametric dividers based on Schottky diodes [9.93], silicon bipolar transistors [9.94] and GaAs MESFETs [9.95] have already been developed using MIC and MMIC technologies. The present state of GaAs MMIC technology enables frequency halvers to be developed at more than 100 GHz using GaAs Schottky diodes with cut-off frequencies f_c of several hundred gigahertz (f_c is defined as the frequency where $Q = 1$). Dividers at up to 70 GHz can be obtained using 0.3 μm GaAs MESFETs exhibiting an input circuit cut-off frequency of about 60 GHz and a transition frequency of about 35 GHz. The dynamic range of parametric dividers is limited by the minimum power required to satisfy the operating conditions defined by eqn (9.24) (a high-Q varactor leads to low input power) and by the maximum power that keeps the divider free from spurious frequency generation.

The following example illustrates the properties of the GaAs MMIC parametric divider.

Circuit applications Stapleton *et al.* [9.96] have used a parametric approach to develop a GaAs monolithic analog frequency halver at 12 GHz. The divider utilizes the nonlinear behavior of the gate-to-source capacitance of two MESFETs of gate length 1 μm arranged in a balanced configuration at the

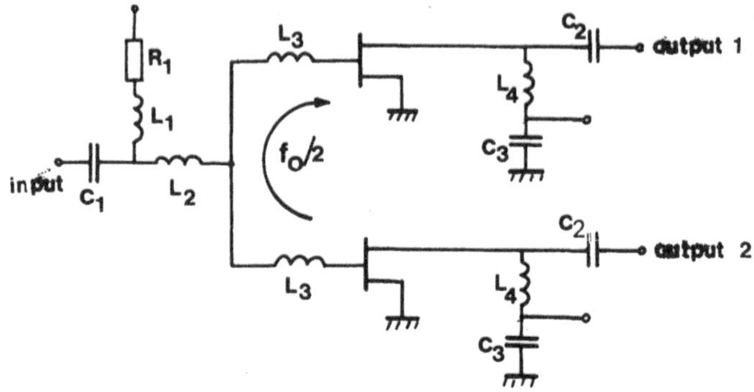

Fig. 9.67 Circuit diagram of a parametric divider by 2.

output to extract the divided signals but cancel the input signal. Figures 9.67 and 9.68 show the circuit diagram of this divider and a photograph of the divider chip (the chip size is 0.5 mm × 1.5 mm). The minimum input power for division to occur is 10 dBm. A bandwidth of 1.2 GHz and a conversion loss of 9 dB have been obtained with an input power of 18 dBm.

MESFETs with a gate length of 1 μm are sufficient for 12 GHz parametric or regenerative analog division, but a digital divider at the same frequency requires devices with gate lengths of 0.5 μm.

9.6.3.3 Future developments

Microwave designers have just started to exploit the ability of GaAs MMICs to produce analog dividers. The bandwidth is still small compared with that obtainable with digital dividers, although octave bandwidths are feasible [9.97]. Furthermore, MMIC technology can produce millimeter-wave dividers, and this frequency range is not accessible to the digital counterpart using current technology. Since the demand for Ku band and millimeter-wave systems is increasing, analog frequency dividers are assured of an interesting future.

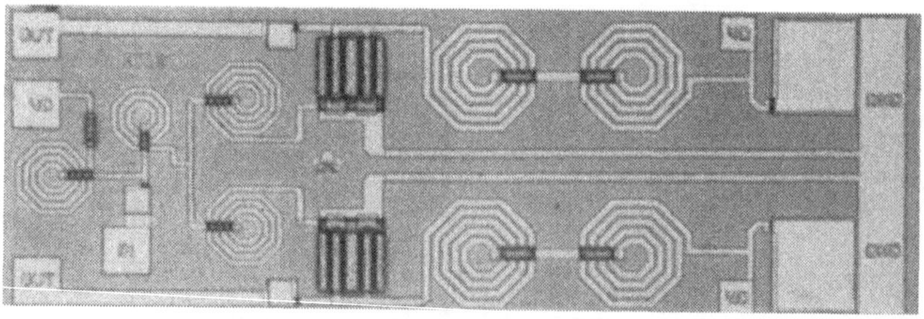

Fig. 9.68 Photograph of a parametric divider chip of dimensions 1.5 mm × 1.5 mm. (Courtesy of CRC.)

9.6.4 Mixers

Mixers generally determine the dynamic range requirements as well as the spurious and harmonic signal levels of microwave receivers. They can also define the spectral purity requirements for local oscillators. Furthermore, when used without front-end amplification, as for example in millimeter-wave applications, the mixer also defines the receiver sensitivity. Consequently, sophisticated mixers are increasingly required. The performance of a frequency conversion unit is dependent on the type of mixer used: single-ended, single- or double-balanced, image rejection, image recovery or sub-harmonically pumped etc. [9.98].

The use of GaAs MMIC technology for mixer fabrication allows the design of mixers with improved performance based on balanced configurations and dramatically reduces the fabrication complexity of millimeter-wave mixers. Mixing is usually achieved using diodes or transistors as nonlinear devices, with the choice depending on operating frequency or the required performance. The two devices can be integrated in GaAs MMICs in the form of GaAs Schottky diodes and GaAs MESFETs.

The principles of GaAs integrated mixers are described together with an example of their application.

9.6.4.1 Operating principles

Device considerations The MESFETs and diodes used to provide mixing in GaAs MMICs require two different technological processes as shown in Figs 9.69(a) and 9.69(b). The MESFET is essentially a planar device since the current flowing parallel to the surface from the drain to the source terminals is modulated by the gate voltage, whereas the diode is a transverse device because the current flows perpendicularly to the surface of the crystal.

The thick buried n^+ layer (2–5 μm) in the Schottky diode provides a low resistance path for the diode current leading to a high cut-off frequency ($f_c = 1/2\pi C_{j0} r_s$ where r_s is the diode series resistance and C_{j0} is the diode capacitance at zero voltage), thus allowing high frequency operation and low conversion loss. A simpler construction can be achieved by eliminating the n^+ layer for compatibility with MESFET technology. However, this results in a reduction in operating frequencies owing to the increased diode series

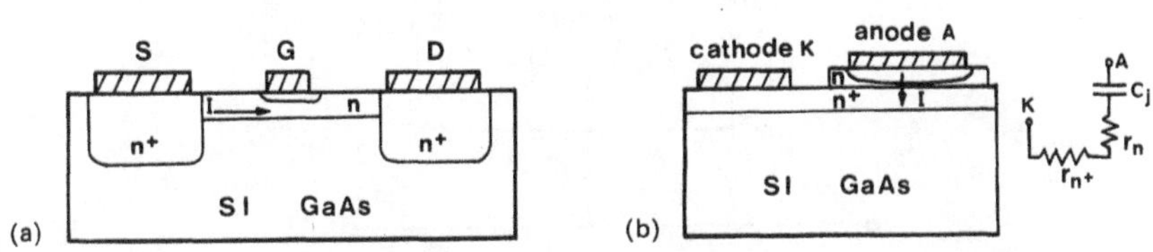

Fig. 9.69 Cross-section of (a) a MESFET and (b) a Schottky diode.

resistance. It may be possible to overcome this problem using dot matrix diode structures [9.99].

Technological processes have been developed which allow Schottky diodes with high cut-off frequencies to be integrated on semi-insulating GaAs substrates. Earlier processes were based either on a selective epitaxial growth of n^+- and n-GaAs layers [9.100] or on a uniform growth followed by ion implantation or mesa isolation [9.101]. More recently, selective ion implantation has been used to form n^+ and n layers [9.102]. This technique enables diodes and MESFETs to be fabricated on the same wafer using the flexible and reproducible ion implantation process which is well suited for high-volume production applications. Cut-off frequencies in excess of 1000 GHz and 500 GHz have been achieved using epitaxial and implanted layers respectively.

GaAs MESFET technology has been described in Chapter 1, and so it will not be discussed further in this section.

Design considerations An important factor for MMIC mixers is the reduction of chip size. This can be achieved by using single- or dual-gate MESFETs because the intrinsic isolation between the r.f., local oscillator (LO) and intermediate-frequency (i.f.) signals is provided inherently by these three- or four-port devices without the necessity for space-consuming filters or couplers. Figure 9.70 shows three different devices which are candidates for monolithic integration. Mixer properties depend strongly on the choice of port

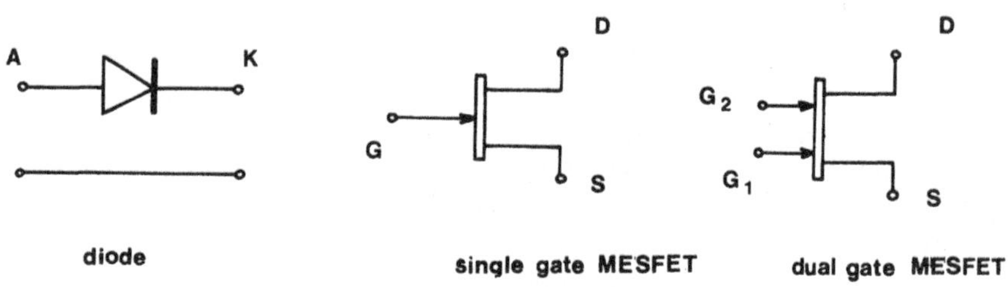

Fig. 9.70 Three candidate devices for mixer circuits.

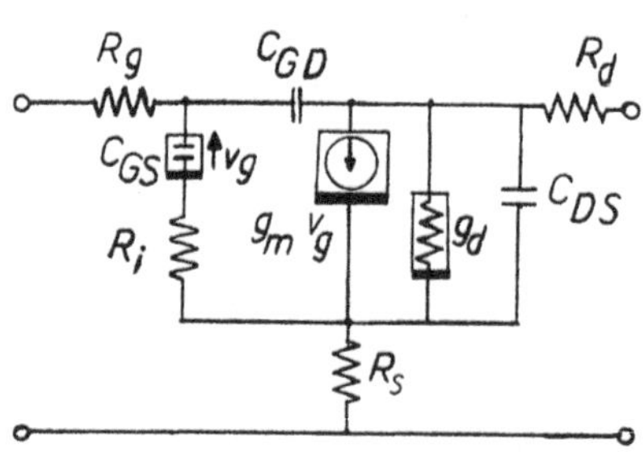

Fig. 9.71 MESFET equivalent circuit and associated nonlinear elements.

to be used for r.f. and LO signal input and for i.f. signal output. Furthermore, MESFET mixers can provide conversion gain while diode mixers exhibit conversion loss. However, in contrast with MESFET mixers, diode mixers do not generally need a biasing current, provide an excellent noise figure (5 dB or less) in the microwave region and are suitable for operation up to the submillimeter waveband.

The mixing process in Schottky diodes is mainly due to the nonlinear behavior of the diode junction resistance [9.103], whereas in GaAs MESFETs the nonlinear elements are the gate capacitance C_{GS}, the drain resistance g_d and the transconductance g_m, all of which depend on bias conditions [9.104]. Figure 9.71 shows the MESFET equivalent circuit with its main nonlinear elements.

Mixer diode design is well documented and MMIC designers can benefit from extensive work on the topic [9.105, 9.106]. The design of GaAs MESFET mixers is more complicated. In particular, prediction of the noise figures of MESFET mixers is still experimental despite some tentative approaches to modelling [9.107]. The fact that a given device produces a higher noise figure as a mixer than as an amplifier is not completely understood.

Mixing mechanisms for various single-gate MESFET mixers are given in Table 9.5. MESFET mixers can be classified as gate, source or drain mixers depending on which port is used for the LO input. The mixing process in a gate mixer is determined by g_m and C_{GS}, in a source mixer by g_m, g_d and C_{GS}, and in

Table 9.5 Various types of MESFET nonlinearities

Type of mixer	*Nonlinear element*	*Behavior*	*Bias point*
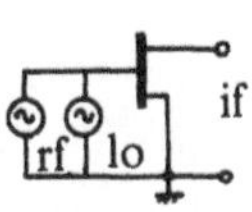 Gate mixer	g_m, c_{GS}	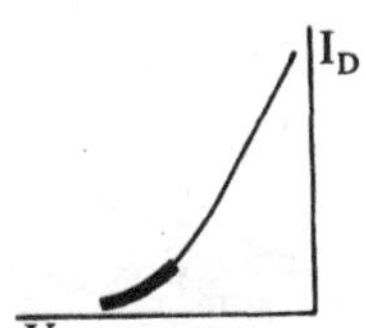	Near pinch-off
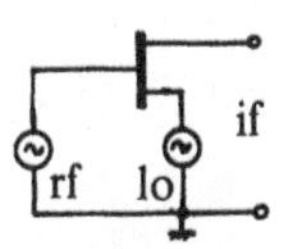 Source mixer	g_m, g_d, c_{GS}	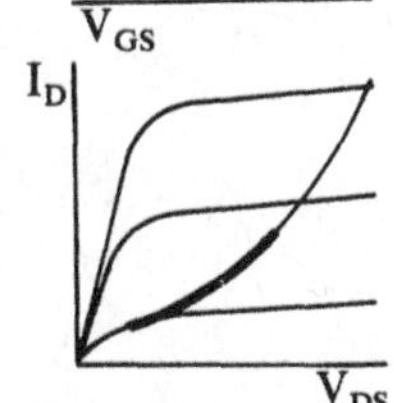	Middle of characteristics
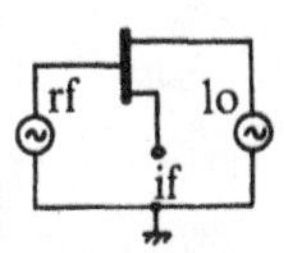 Drain mixer	g_m, g_d	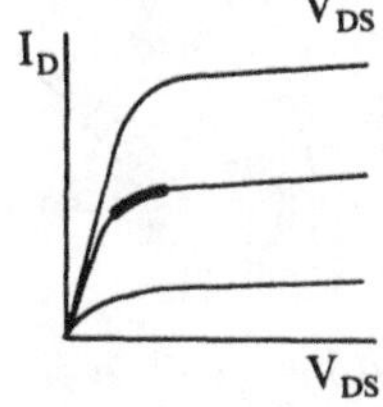	Knee of characteristics

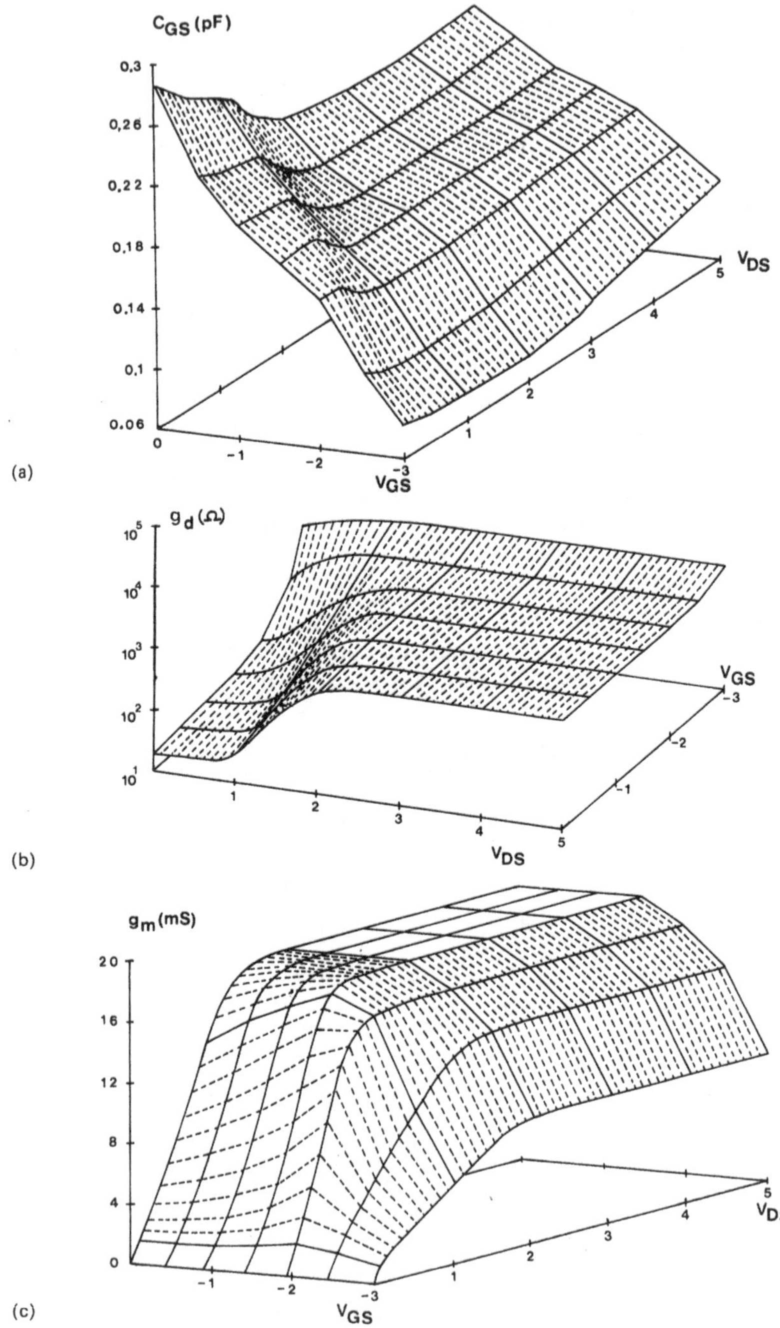

Fig. 9.72 Three-dimensional plots of (a) C_{GS}, (b) g_d and (c) g_m as functions of V_{DS} and V_{GS}.

a drain mixer by g_m and g_d. Consequently, modelling of g_m, g_d and C_{GS} is very important. The behavior of these three nonlinear elements mainly responsible for mixing is shown in Fig. 9.72 which gives a three-dimensional plot of C_{GS}, g_d and g_m as functions of the drain and gate voltages V_{DS} and V_{GS} for a typical gate MESFET of dimensions 0.7 μm × 150 μm.

Harrop [9.108, 9.109] has demonstrated that, with appropriate impedance terminations, gate mixers exhibit the highest conversion gain and require the lowest LO power compared with other structures. A simplified expression has been derived [9.98] to estimate the conversion gain of MESFET gate mixers under appropriate load conditions (matching or short-circuit impedances):

$$G = \frac{g^2_{m\,max}\,Q_i^{\,2}\,(R_s + R_i + R_g)\,\mathrm{Re}\{Z_L(\omega_{if})\}}{16} \tag{9.25}$$

where

$$Q_i = \{\omega_s C_{GS\,0}\,(R_s + R_i + R_g)\}^{-1}$$

and $g_{m\,max}$ is the maximum d.c. transconductance. R_s, R_i and R_g are defined in Fig. 9.71, $C_{GS\,0}$ is the gate capacitance at $V_{GS} = 0\,\mathrm{V}$, $Z_L(\omega_{if})$ is the drain impedance at intermediate frequency and ω_s is the angular frequency of the signal. It can be shown that MESFET mixers require high transconductance and high input quality factor Q_i to achieve a high conversion gain at a given frequency. With the technology available at present for GaAs MMICs on the production line (gate length of 0.5 μm) conversion gain is achievable up to about 25 GHz.

Dual-gate MESFET: an extension of the single-gate MESFET Injection of an LO signal into the source or drain port of a single-gate MESFET is an attractive solution since no coupler is required, as in the gate mixer, to inject signal and LO powers into the gate terminal. However, the drain mixer demands a high LO power and the source mixer requires stabilization circuitry because of the potential instability created by the feedback circuit located in the source path.

A drain mixer can be built using a second MESFET connected to the drain and acting as a device for LO power injection. As shown in Fig. 9.73 the

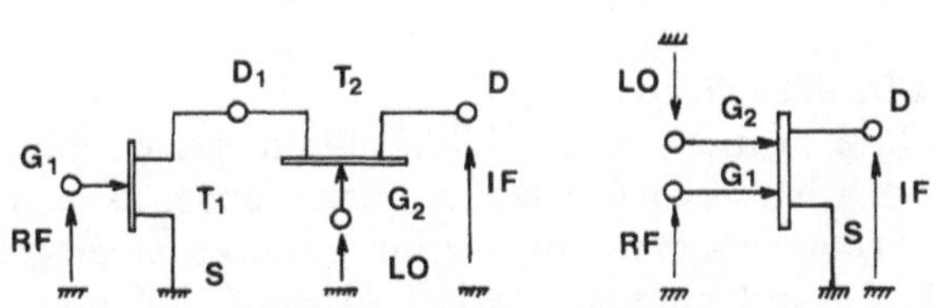

Fig. 9.73 Equivalent structure of a dual-gate MESFET.

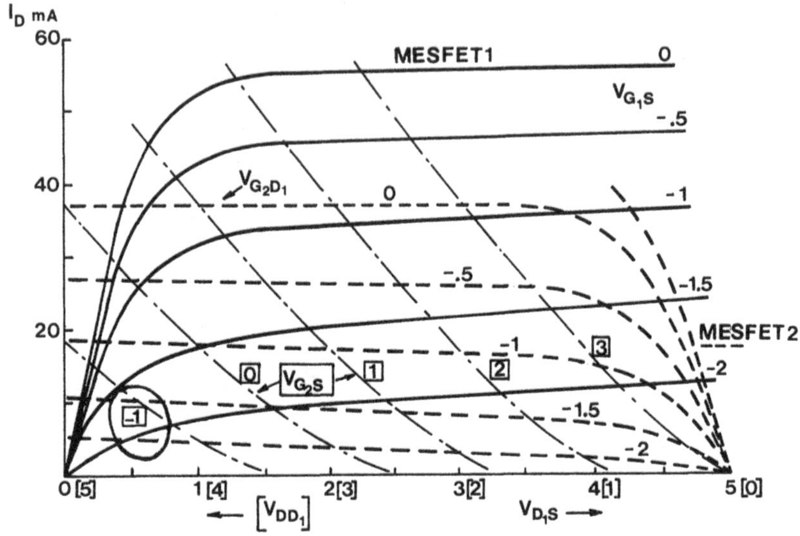

Fig. 9.74 Static characteristics of a dual-gate MESFET. The circled area on the left-hand side corresponds to low noise operation.

dual-gate MESFET structure corresponds to this two-transistor arrangement without allowing independent control of the intermediate voltage V_{D_1S}.

The static characteristics of the dual-gate MESFET in the region of low noise operation are given in Fig. 9.74. This figure shows that under the proposed operation mode MESFET 1 acts as a drain mixer while MESFET 2, working in its saturated region, acts as an amplifier. The dual-gate MESFET mixer is mainly used because of its compactness and its gain capability.

The dual-gate MESFET can be modelled as two series transistors characterized by their linear and nonlinear elements (see Fig. 9.71). The three-dimensional plots shown in Fig. 9.72 are applicable to the dual-gate MESFET once the internal potential V_{D_1} of the device is known. Modelling of the dual-gate MESFET has been achieved using the quasi-static approach [9.110] and dual-gate MESFET mixer performance has been analyzed using the Fourier series approach [9.111, 9.112] or Volterra series analysis [9.113].

9.6.4.2 Circuit applications

Intensive investigations of monolithically integrated mixer circuits for application from v.h.f. to e.h.f. bands are being undertaken. The following examples give an overview of the applications of GaAs integrated mixers.

A MESFET-based integrated mixer for u.h.f. TV tuner applications which exhibits a flat gain of 10 dB between 50 MHz and 1 GHz with a power consumption as low as 50 mW has been developed at LEP [9.114]. Researchers at LEP have also demonstrated that, in contrast with silicon ICs which have so far failed to meet u.h.f. TV tuner requirements, GaAs MMICs offer attractive solutions for complete TV tuner integration including the wideband mixer as well as the VCO. This is mainly because of the superior capabilities of GaAs high gain and low noise-figure devices and the inherent good isolation between the different functions.

At frequencies about two decades higher workers at Hughes Aircraft have demonstrated the superiority of GaAs MMIC technology over MIC technology in the construction of a 94 GHz cross-bar Schottky diode balanced mixer using integrating diodes, matching circuits and cross-bar probes on the same GaAs substrate [9.115]. Consequently parasitic connecting elements have been eliminated and improved reproducible performances have been obtained by simplifying assembly procedure. A conversion loss as low as 7.5 dB has been measured.

Tang *et al.* [9.116] have proposed extending the concept of the distributed amplifier to distributed mixers in order to make a wideband MESFET mixer. This approach provides a method of achieving wideband mixers in monolithic form without using space-consuming baluns or broadband couplers. Titus *et al.* [9.117] have applied this concept to the fabrication of a distributed monolithic image rejection mixer. The circuit, which is based on four dual-gate MESFETs acting as elements of two artificial transmission lines, down-converts a 14–20 GHz frequency band to a 2–8 GHz i.f. band using a fixed 12 GHz LO. A 10–20 dB image rejection, 6 dB conversion loss, 30 dB LO to i.f. isolation and 20 dB LO to r.f. isolation have been obtained over the whole bandwidth.

Figure 9.75 shows an example of a dual-gate MESFET integrated mixer developed at LEP for 12 GHz direct broadcasting satellite applications [9.118]. The signal is converted from the 11.7–12.5 GHz frequency band down to the 0.95–1.75 GHz band using a 10.75 GHz 10 mW oscillator. The mixer comprises a dual-gate MESFET using two 0.8 μm × 150 μm gates and appropriate circuits connected to gate 1, gate 2 and the drain ports for matching conditions at signal, LO and intermediate frequencies respectively. The interdigitated capacitor connected to the drain has been designed to work at its series resonant frequency mode at 11 GHz in order to reflect the LO signal back into the mixer. This mixer achieves a conversion gain of 2 dB and a noise figure of 8 dB. The chip size is 1.2 mm × 0.8 mm.

For the same application Watanabe *et al.* [9.119] have developed a balanced mixer–i.f. amplifier chip based on Schottky diodes. Schottky diode mixing has been chosen to avoid the sensitivity of dual-gate MESFET mixers to bias conditions. The diode has been processed using a single n implantation layer for MESFET technology compatibility resulting in a fairly high series resistance. However, owing to the moderately high operating frequency (12 GHz), the mixer conversion loss remains at an acceptable value of 8 dB.

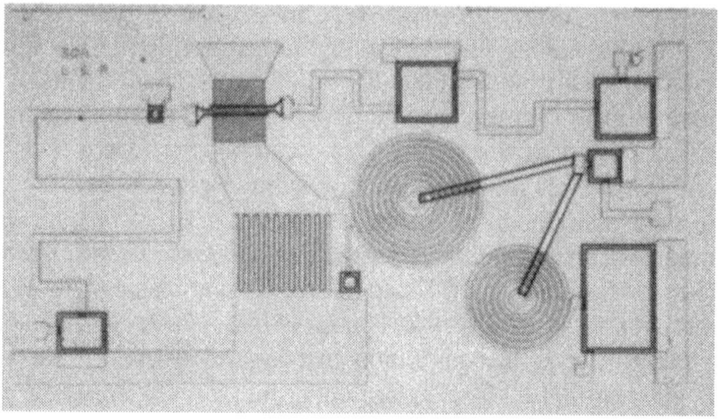

Fig. 9.75 Dual-gate MESFET integrated mixer chip of dimensions 1.2 mm × 0.7 mm. (Courtesy of LEP.)

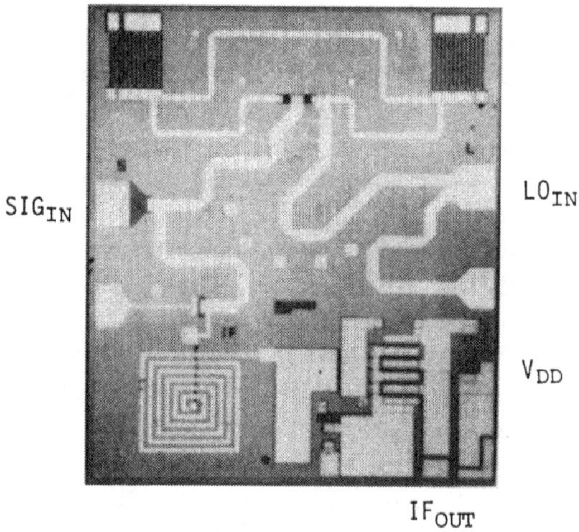

Fig. 9.76 Diode mixer–amplifier chip of dimensions 2.2 × 2.2 mm. (Courtesy of Toshiba.)

Figure 9.76 shows the mixer–amplifier chip. The overall chip, including a balanced mixer and a feedback i.f. amplifier at 1 GHz, exhibits a conversion gain of 2 dB and a noise figure of 10 dB. The chip size is 2 mm × 2.25 mm.

9.7 Multifunction MMICs

9.7.1 General

As described in previous sections, most of the research and development work performed on GaAs MMICs so far has been aimed at individual single-chip circuit cells such as amplifiers, mixers, oscillators, switches, phase shifters etc. This first step in monolithic integration provides cost-effective functional building blocks for use in system integration in the frequency range 1–18 GHz. However, attention should be paid to methods of interconnecting basic MMIC cells without significant performance degradation. Furthermore, basic cells designed for general-purpose applications (covering a wide variety of application areas to increase chip volume production) rarely offer optimum performances.

With this in mind, the integration of multiple functions on a single chip appears desirable. Although the concept of a complete system on a single chip is still at an early stage, MMIC chips integrating multiple functions have been demonstrated [9.120, 9.121]. GaAs MMIC macrocells have been developed on a commercial basis by Pacific Monolithic [9.122]. They offer a cell library covering the 0.5–18 GHz range and containing the most widely used microwave building blocks for signal processing in receivers and transmitters.

One of the first multifunction MMICs was an integrated receiver chip developed at the LEP for 12 GHz direct broadcasting satellite applications [9.120, 9.123]. As the concept of the system is to make direct broadcasting satellite reception available to all users of small dish antennas, a large market volume and low-cost requirements well suited to GaAs MMIC technology are implied. Multifunction MMICs were used to minimize cost further and to improve performance to a certain extent.

Details of the physical and electrical design and processing technology of this multifunction MMIC are presented in the following section together with the overall performance. When first designed in 1984, this circuit represented the state of the art in GaAs MMICs.

9.7.2 Single-chip receiver for 12 GHz direct broadcasting satellite applications

9.7.2.1 Circuit design

Figure 9.77 shows the diagram of the receiver which requires a low noise amplifier covering the r.f. band (11.7–12.5 GHz), a filter rejecting the image frequency band (9.0–9.8 GHz), a stable LO (10.75 GHz), a mixer circuit and an i.f. amplifier (0.95–1.75 GHz). Monolithic integration of these different func-

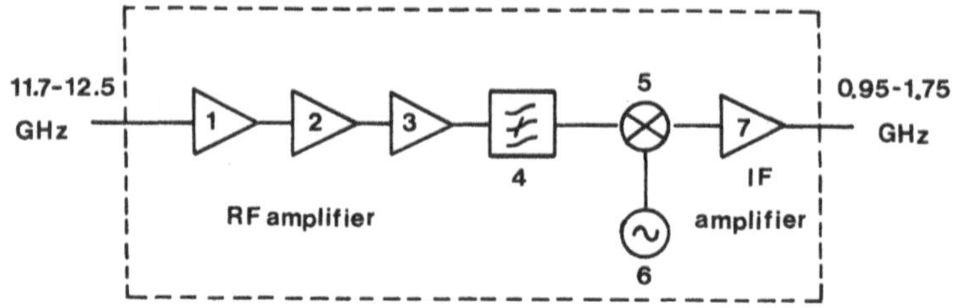

Fig. 9.77 Direct broadcasting satellite receiver: 4, image frequency rejection filter; 5, mixer; 6, local oscillator.

tions allows optimum matching regardless of the 50 Ω reference impedance, leading to simpler circuits, lower losses and minimized gain ripple.

In particular, the three-stage low noise amplifier has been designed using bandpass and highpass matching circuits in the different interstage matching networks in order to fulfill both matching conditions in the useful bandwidth (11.7–12.5 GHz) and enable distributed image frequency filtering (9.0–9.8 GHz). In this way the requirement for the space-consuming image rejection filter in front of the mixer to reject noise contained in the image band is avoided.

The mixer has been designed using a dual-gate MESFET as described in Section 9.5.4 for gain and space-saving considerations. The i.f. amplifier was coupled directly to the mixer output since considerable gain can be achieved by using unmatched MESFETs occupying the same area as the usual i.f. matching circuit (see Fig. 9.75). I.f. output matching has been achieved by using a MESFET in the common-drain configuration (see Section 9.6.4).

The LO–mixer coupling circuit has been designed to satisfy appropriate oscillator operating conditions. A second version using a buffer stage between the oscillator and the mixer to keep the oscillator insensitive to load changes (pulling) has been designed without significantly increasing the active chip space. The LO is described in detail in Section 9.6.2.

The equivalent circuit of the complete receiver is shown in Fig. 9.78.

9.7.2.2 Circuit topology

The topology of multifunction MMICs is very important with respect to minimizing the intermodule coupling caused by their physical proximity. Therefore the receiver is organized around a gold-plated area with three via holes etched through the substrate. This area acts as a ground plane for the circuits and as an electrical wall shielding the low noise amplifier from the oscillator located on the opposite side. Therefore any parasitic crosstalk which could saturate the amplifier is avoided.

A detailed description of the technological process used to realize the

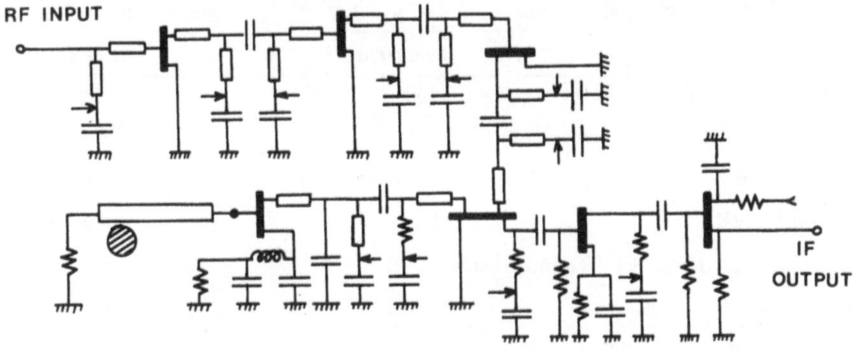

Fig. 9.78 Equivalent circuit of the complete receiver.

receiver chip is given in ref. 9.20. This process, or a very similar approach, is used by most GaAs MMIC foundries. However, the two-step procedure which has been developed to etch holes of diameter 100 μm through a GaAs substrate 300 μm thick is still unique. This particular technique consists in first opening a tub 150 μm deep and then producing holes of diameter 100 μm to meet the top ground plane.

The receiver chip is shown in Fig. 9.79. It has dimensions of 2.5 mm × 2.5 mm.

9.7.2.3 Circuit performance

A summary of the circuit performance together with the predicted values

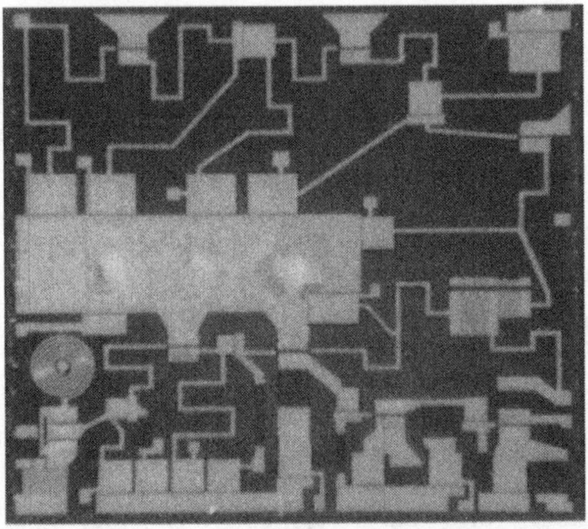

Fig. 9.79 Fully integrated receiver chip of dimensions 2.5 mm × 2.5 mm. (Courtesy of LEP.)

Table 9.6 Predicted and measured performances of the single-chip receiver

Parameter	*Predicted*	*Measured*
Conversion gain (dB)	34	35[a]
Noise figure (dB)	<4	<4.5
Input VSWR	<2	<1.8
Output VSWR	<1.5	<1.4
Output power (dBm) at 1 dB compression	0	0
D.c. power (W)	0.45	0.45

[a] Slightly frequency shifted (about 1 GHz).

is given in Table 9.6. Problems associated with the slight frequency shift can probably be solved in the production stage by better control of the fabrication process. Noise figures as low as 2.5 dB are achievable using technological processes currently available on production lines. This type of integrated multifunction chip is promising for cost-effective complex GaAs MMICs.

9.8 Future developments

A major problem which must be solved in the near future is the development of low-cost GaAs MMIC production facilities. Many factors may contribute to the production of "affordable" GaAs MMICs, and some are listed below:

(a) the development of a stable mature process by increasing automation, standardizing the process and controlling the quality of the GaAs;
(b) reduction of the development cost by improving the design accuracy (improved models) and producing interactive CAD software (design, layout, design-rule checking);
(c) improvement of integration density for a more efficient use of the GaAs area;
(d) development of a fast and accurate testing procedure;
(e) development of low-cost high performance packages.

Trimming procedures or "prediffused" GaAs MMIC concepts similar to those used in the silicon industry may emerge in the near future as a cost-reducing approach.

The availability of a mature low-cost technology should hasten the development of complex multifunction MMICs like transmit–receive modules, complete front-ends or switching matrices. Functions combining

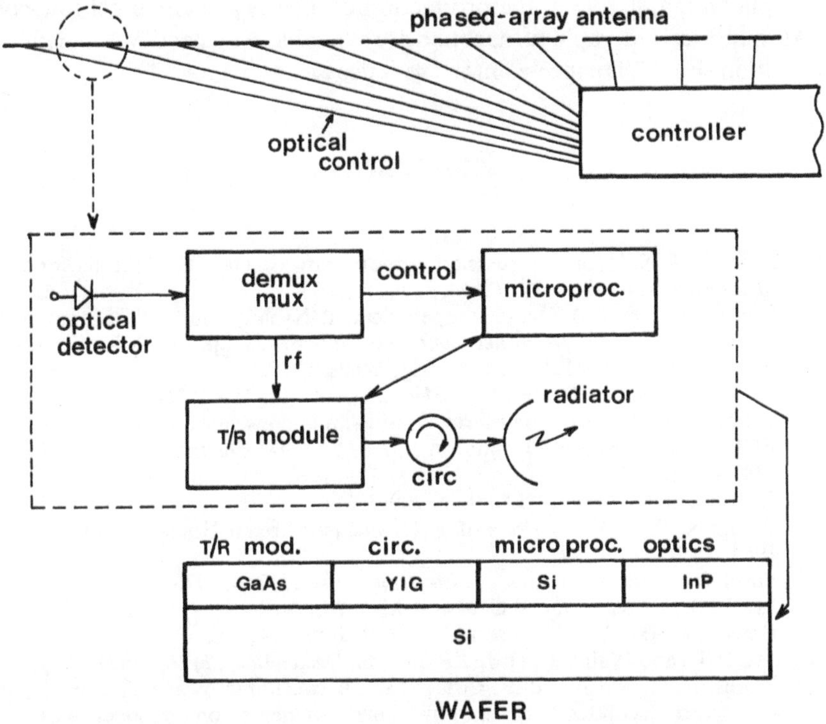

Fig. 9.80 Block diagram of an optically controlled phased-antenna system showing the interest for integration of several crystalline materials on the same substrate (e.g. silicon).

digital, microwave and optical technologies, such as optical receivers, may also be produced. The need for high performance circuits will probably lead to the integration of heterostructure devices such as HEMTs on a production basis.

In the millimeter-wave region, the antenna and the microwave transmit–receive functions of emitters and/or transmitters at 94 GHz and above could be integrated on the same substrate. The demonstration of excellent MESFETs fabricated from a GaAs epitaxial layer grown on a silicon substrate [9.124] could eventually lead to the production of integrated circuits combining silicon-based digital circuits and GaAs-based microwave circuits. Perhaps they will be called silicon MMICs. This idea could be extended to the integration of a number of crystalline materials on the same wafer. The dream of system design engineers would become a reality as illustrated in Fig. 9.80 which represents the block diagram of an optically controlled phased-array antenna together with an integrated module including various crystalline materials deposited on a silicon substrate.

In conclusion, although the market is slow to emerge, GaAs MMICs could play a major role in the production of microwave circuits if low-cost GaAs foundries with rapid turnaround are available. Further developments in integration should be expected in the next decade.

References

9.1 Pengelly, R. S. Hybrid versus monolithic microwave circuits – a matter for cost, *Microwave Syst. News*, **13**, 77–14, January 1983.

9.2 Pengelly, R. S. Broadband lumped-element X-band GaAs FET amplifiers, *Proc. 5th European Microwave Conf., September 1975*, pp. 301–305, Microwave Exhibitions and Publishers, Tunbridge Wells, 1975.

9.3 Pengelly, R. S. and Turner, J. A. *Electron. Lett.*, **12**, 251, 1976.

9.4 Wemple, S. H. *et al.* Long term and instantaneous burnout in GaAs power FETs: mechanisms and solutions, *IEEE Trans. Electron Devices*, **28** (7), 834–840, 1981.

9.5 Jackson, G. N. *Thin Solid Films*, **5**, 209, 1970.

9.6 Williams, R. E. *GaAs Processing Techniques*, Artech House, Dedham, MA, 1984.

9.7 Pucel, R. A. and Krumm, C. F. *Electron. Lett.*, **12**, 240, 1976.

9.8 Jay, P. R., Crossley, I. and Cardwell, M. J. *Electron. Lett.*, **14**, 190, 1978.

9.9 Dewulf, P. *Rep. 151*, Thomson-CSF LCR, June 1981.

9.10 Jay, P. R. and Wallis, R. H. *IEEE Electron Device Lett.*, **2**, 265, 1981.

9.11 Wallis, R. H. and Jay, P. R. Comparison between microwave parameters and magnetotransconductance mobility measurements on submicron GaAs MESFETs, *Proc. 9th Int. Symp. on Gallium Arsenide and Related Compounds, Oiso, September 1981*, Institute of Physics, London, 1981.

9.12 Akhtarzad, S., Rowbotham, T. R. and Johns, P. B. *IEEE Trans. Microwave Theory Tech.*, **23**, 486, 1975.

9.13 Daumas, R., Pompei, D., Rivier, E. and Ros, A. *Ann. Telecommun.*, **7**, 325, 1973.

9.14 Zamanski, P. *Rev. Tech. Thomson-CSF*, **5**, 505, 1973.

9.15 Le Brun, M., Jay, P. R. and Rumelhard, C. Coupling and impedance between line and ground electrodes on GaAs: implications for MMIC design, *Proc. 11th European Microwave Conf., Amsterdam, September 1981*, Microwave Exhibitions and Publishers, Tunbridge Wells, 1981.

9.16 Aubourg, M., Villotte, J. P. and Godon, F. *Rapport d'Etude*, LEM-Université de Limoges, January 1982.

9.17 Pucel, R. A., Masse, D. J. and Hartwig, C. P. *IEEE Trans. Microwave Theory Tech.*, **16**, 342, 1968.

9.18 Smith, C. E. and Chang, R. S. *IEEE Trans. Microwave Theory Tech.*, **28**, 90, 1980.

9.19 Grover, F. W. *Inductance Calculations*, Van Nostrand, Princeton, NJ, 1946.

9.20 Parisot, M., Archambault, Y., Pavlidis, D. and Magarshack, J. Highly accurate design of spiral inductors for MMICs with small size and high cut-off frequency characteristics, *1984 IEEE MTT Symp. Dig.*, pp. 106–110, IEEE, New York, 1984.

9.21 Jansen, R. H., Wiemer, L., Finlay, H. J., Suffolk, J. R., Roberts, B. D. and Pengelly, R. S. Theoretical and experimental broadband characterization of multiturn square spiral inductors in sandwich type GaAs MMICs, *Proc. 15th*

European Microwave Conf., Paris, 1985, pp. 946–948, Microwave Exhibitions and Publishers, Tunbridge Wells, 1985.

9.22 Wolff, I. and Kapusta, H. Modeling of circular spiral inductors for MMICs, *1987 IEEE MTT Symp. Dig.*,pp. 123–126, IEEE, New York, 1987.

9.23 Danto, Y., Barriere, A. S., Girma, T., Pichon, J., Jay, P. R. and Rumelhard, C. Physical characterisation of different dielectrics for thin film capacitors in GaAs MMICs, *Proc. 1st Int. Conf. on Conduction and Breakdown in Solid Dielectrics, Toulouse, July 1983.*

9.24 Elta, M. E., Chu, A., Mahoney, L. J., Cerretani, R. T. and Courtney, W. E. Tantalum oxide capacitors for GaAs monolithic integrated circuits, *IEEE Electron Device Lett.*, **3** (5), 127–129, 1982.

9.25 Girma, T. Conception, réalisation de condensateurs en couches minces pour circuits intégrés hyperfréquences sur arséniure de gallium, *Thesis*, University of Bordeaux I, October 1985.

9.26 Pucel, R. A. Design considerations for monolithic microwave circuits, *IEEE Trans. Microwave Theory Tech.*, **29**, 513–534, 1981.

9.27 Mondal, J. P. An experimental verification of a simple distributed model of MIM capacitors for MMIC applications, *IEEE Trans. Microwave Theory Tech.*, **35**, 403–408, 1987.

9.28 Lee, C. P., Welch, B. M. and Zucca, R. Saturated resistor load for GaAs integrated circuits, *IEEE Trans. Microwave Theory Tech.*, **30** (7), 1007–1013, 1982.

9.29 Rumelhard, C., Camiade, M., Chaplart, J., Le Brun, M., Pavlidis, D. and Parisot, M. Quelques amplificateurs monolithiques microondes sur GaAs, *Rev. Tech. Thomson-CSF*, 1986.

9.30 Dao, T. H. and Ha, T. T. Explicit formulas for GaAs FET amplifier interstage matching networks, *Proc. Inst. Electr. Eng., Part G*, **128** (1), 1981.

9.31 Niclas, K. B. On design and performance of lossy match GaAs MESFET amplifiers, *IEEE Trans. Microwave Theory Tech.*, **30** (11), 1900–1907, 1982.

9.32 Niclas, K. B. Reflective match, lossy match, feedback and distributed amplifiers: a comparison of multioctave performance characteristics, *1984 IEEE MTT Symp. Dig.*, pp. 215–217, IEEE, New York, 1984.

9.33 Niclas, K. B., Wilser, W. T., Kritzer, T. R. and Pereira, R. R. On theory and performance of solid state microwave distributed amplifiers, *IEEE Trans. Microwave Theory Tech.*, **31** (6), 447–456, 1983.

9.34 Niclas, K. B. and Tucker, B. A. On noise in distributed amplifiers at microwave frequencies, *IEEE Trans. Microwave Theory Tech.*, **31** (8), 661–668, 1983.

9.35 Strid, E. W. and Gleason, K. R. A DC–12 GHz monolithic GaAs FET distributed amplifier, *IEEE Trans. Electron Devices*, **29** (7), 1065–1071, 1982.

9.36 Kenna, W. T., Andrade, T. and Hwang, C. A miniature 2–18 GHz monolithic GaAs distributed amplifier, *1984 IEEE Microwave and Millimeter-wave Monolithic Circuits Symp.*, pp. 41–44, IEEE, New York, 1984.

9.37 Gamand, P., Rabinzohn, P., Varin, C. and Baudet, P. Amplificateur distribué monolithique 2–18 GHz sur arséniure de gallium, *LEP Revue Annuelle*, pp. 24–26, Laboratoires d'Electronique et de Physique appliquée, Paris, 1985.

9.38 Summers, J. G. and Ng Thow Hing, L. A. An overview of GaAs MMIC processes, *GEC J. Res.*, **4** (2), 104–113, 1986.

9.39 Nesbit, G. H., Wong, D. W., Li, D. and Chen, J. C. A W-band monolithic GaAs pin diode switch, *1986 IEEE Microwave and Millimeter-wave Monolithic Circuits Symp.*, pp. 51–55, IEEE, New York, 1986.

9.40 Tajima, Y., Tsukii, T., Mozzi, R., Tong, E., Hares, L. and Wrona, B. GaAs monolithic wideband (2–18 GHz) variable attenuators, *1982 IEEE MTT Symp. Dig.*, pp. 479–481, IEEE, New York, 1982.

9.41 Barta, G. S., Jones, K. E., Herrick, G. C. and Strid, E. W. A 2 to 8 GHz leveling loop using a GaAs active splitter and attenuator, *1986 IEEE Microwave and Millimeter-wave Monolithic Circuits Symp.*, pp. 75–79, IEEE, New York, 1986.

9.42 Pucel, R. Signal and noise properties of GaAs microwave FETs, *Adv. Electron. Electron Phys.*, **38**, 193–265, 1975.

9.43 Rocchi, M. Outil CAO pour circuits intégrés numériques sur GaAs, *Acta Electron.*, **23**, 3, 1980.

9.44 Rocchi, M. Post-caractérisation des transistors à effet de champ et CI GaAs sur motif de test, *Acta Electron.*, **23** (1), 63–80, 1980.

9.45 *MMIC Catalog*, MA/COM, 100 Chelmsford Street, Lowell, MA 01851.

9.46 *Cell Library Catalog*, Pacific Monolithics, 245 Santa Ana Court, Sunnyvale, CA.

9.47 Landry, P. P. *Internal Report*, Spar Aerospace Ltd, February 1987.

9.48 Ayasli, Y. Microwave switching with GaAs FETs, *Microwave J.*, **25**, 61–74, November 1982.

9.49 Bedard, B. A high performance monolithic GaAs SPDT switch, *Technical Note*, Adams Russell, 1985.

9.50 Ayasli, Y. *et al.* A monolithic X-Band SPDT bidirectional GaAs FET switch, *GaAs Integrated Circuits Symp., October 1980.*

9.51 Schindler, M. and Harris, A. M. D.c.–40 GHz and 20–40 GHz MMIC SPDT switches, *1987 IEEE MTT Symp. Dig.*, pp. 85–88, IEEE, New York, 1987.

9.52 Ayasli, Y., Mozzi, R., Hanes, L. and Reynolds, L. D. An X-band 10 W monolithic transmit–receive GaAs FET switch, *1982 IEEE Microwave and Millimeter-wave Monolithic Circuits Symp. Dig.*, pp. 42–45, IEEE, New York, 1982.

9.53 *MMIC Catalog*, Tachonics Corporation, 107 Morgan Lane, Plainsboro, NJ.

9.54 Meierer, R. *Internal Rep.*, Spar Aerospace, February 1987.

9.55 Brookner, E. *et al.* Derivation of a satellite radar architecture for air surveillance, *Microwave J.*, **29**, 173–191, February 1986.

9.56 White, J. Phase shifters for array antennas, *IEEE Trans. Microwave Theory Tech.*, **22**, 658–674, 1974.

9.57 Yahara, T. A note on designing digital diode loaded line phase shifters, *IEEE Trans. Microwave Theory Tech.*, **20**, 703–704, 1972.

9.58 Ayasli, Y., Platzker, A., Vorhaus, J. and Reynolds, L. D. A monolithic single chip X-band four bit phase shifter, *IEEE Trans. Microwave Theory Tech.*, **30**, 2201–2205, 1982.

9.59 Garver, R. Broadband diode phase shifters, *IEEE Trans. Microwave Theory Tech.*, **20**, 314–323, 1972.

9.60 Ayasli, Y., Miller, S., Mozzi, R. and Hanes, L. Wideband S–C band monolithic phase shifters, *1984 IEEE Microwave and Millimeter-wave Monolithic Circuits Symp. Dig.*, pp. 11–13, IEEE, New York, 1984.

9.61 Dawson, D., Conti, A., Lee, S., Shade, G. and Dickens, L. An analog X-band phase shifter, *1984 IEEE Microwave and Millimeter-wave Monolithic Circuits Symp. Dig.*, pp. 6–10, IEEE, New York, 1984.

9.62 Gupta, R. K. and Getsinger, W. J. Quasi-lumped-element 3 and 4 port networks for MIC and MMIC applications, *1984 IEEE MTT Symp. Dig.*, pp. 409–411, IEEE, New York, 1984.

9.63 Tsironis, C. Dual gate GaAs MESFET phase shifter with gain at 12 GHz, *Electron. Lett.*, **16** (14), 553–554, 1980.

9.64 Pengelly, R. Broad and narrow band frequency discriminators using dual gate FETs, *Proc. 9th European Microwave Conf., 1979*, pp. 326–330, Microwave Exhibitions and Publishers, Tunbridge Wells, 1979.

9.65 Meierer, R. Optimum design of amplitude modulators, *Proc. 14th European*

Microwave Conf., 1984, pp. 443–448, Microwave Exhibitions and Publishers, Tunbridge Wells, 1984.

9.66 Vorhaus, J. L., Pucel, R. A. and Tajima, Y. Monolithic dual gate GaAs FET digital phase shifter, *IEEE Trans. Microwave Theory Tech.*, **30**, 982–992, 1982.

9.67 Chen, Y. A GaAs multi-band digitally controlled 0–360° phase shifter, *1985 IEEE GaAs Integrated Circuits Symp.*, pp. 125–128, IEEE, New York, 1985.

9.68 Suckling, S. S-band phase shifter using monolithic GaAs circuits, *IEEE Solid-State Circuits Conf., 1982*, pp. 134–135, IEEE, New York, 1982.

9.69 Joshi, J. S., Cockrill, J. R. and Turner, J. A. Monolithic microwave GaAs FET oscillators, *IEEE Trans. Electron Devices*, **28**, 158–162, 1981.

9.70 Maki, D. W., Schellenberg, J. M., Yamasaki, H. and Liu, L. C. T. A 69 GHz monolithic FET oscillator, *1984 IEEE Microwave and Millimeter-wave Monolithic Circuits Symp. Dig.*, pp. 62–66, IEEE, New York, 1984.

9.71 Tserng, H. Q. and Macksey, H. M. Performance of monolithic GaAs FET oscillators at J band, *IEEE Trans. Electron Devices*, **28**, 163–165, 1981.

9.72 Brehm, G. E., Scott, B. N. and Doerbeck, F. H. Fabrication technique for X band monolithic VCOs, *1982 IEEE Microwave and Millimeter-wave Monolithic Circuits Symp. Dig.*, pp. 57–60, IEEE, New York, 1982.

9.73 Patout, G. VCO en bande X intégrés sur GaAs, *Journ. Nationales Microondes Dig.*, Lannion, 1984.

9.74 Tserng, H. Q. A 115 GHz monolithic GaAs FET oscillator, *1985 IEEE GaAs Integrated Circuits Symp. Dig.*, pp. 11–13, IEEE, New York, 1985.

9.75 Scott, B. N., Wurtele, M. and Cregger, B. B. A family of four monolithic VCO MICs covering 2–18 GHz, *1984 IEEE Microwave and Millimeter-wave Monolithic Circuits Symp. Dig.*, pp. 58–61, IEEE, New York, 1984.

9.76 Rauscher, C. Generalized technique for designing broadband varactor-tuned negative resistance oscillators, *Circuit Theory Appl.*, **7**, 313–320, 1979.

9.77 Camiade, M. Wide tuning bandwidth Ku band varactor FET oscillators, *Proc. 15th European Microwave Conf., Paris, 1985*, Microwave Exhibitions and Publishers, Tunbridge Wells, 1985.

9.78 El Kamali, W., Grimm, J. P., Meierer, R. and Tsironis, C. New design approach for wide-band FET VCOs, *IEEE Trans. Microwave Theory Tech.*, **34**, 1059–1063, 1986.

9.79 Brehm, G. E., Scott, B. N., Seymour, D. J., Duncan, W. N. and Doerbeck, F. H. High capacitance ratio monolithic varactor diode, *1981 Cornell Microwave Conf. Dig.*, pp. 53–63.

9.80 El Kamali, W. *et al.* Physical model for planar Schottky varactor diode, *IEEE Trans. Electron Devices*, in the press.

9.81 El Kamali, W. Limitations de la bande de fréquence des oscillateurs monolithiques microondes accordables par capacitance variable, *Ph.D. Thesis*, ENST 87E002, Paris, February 1987.

9.82 Tsironis, C., Kermarrec, C., Faguet, J. and Harrop, P. Stable monolithic GaAs FET oscillator, *Electron. Lett.*, **18** (8), 345–347, 1982.

9.83 Tsironis, C. and Lesartre, P. X band and Ku band dual gate MESFET oscillators stabilized using dielectric resonators, *Proc. 11th European Microwave Conf., Amsterdam, 1981*, pp. 469–474, Microwave Exhibitions and Publishers, Tunbridge Wells, 1981.

9.84 Osafune, K. and Ohwada, K. 20 GHz dynamic frequency divider with GaAs advanced saint and air-bridge technology, *Electron. Lett.*, **23** (6), 300–302, 1987.

9.85 Miller, R. Fractional frequency generator utilizing regenerative modulation, *Proc. IRE*, 446–457, 1939.

9.86 Rauscher, C. Regenerative frequency division with GaAs MESFET, *IEEE Trans. Microwave Theory Tech.*, **32**, 1461–1468, 1984.

9.87 Kamiski, D., Goussu, P., Funck, R. and Bert, A. G. A dual gate GaAs FET analog frequency divider, *1983 IEEE MTT Symp. Dig.*, pp. 352–354, IEEE, New York, 1983.

9.88 Stubbs, M. and Stapleton, S. A single stage monolithic regenerative 1/2 analog frequency divider, *1986 IEEE GaAs Integrated Circuits Symp.*, pp. 199–201, IEEE, New York, 1986.

9.89 Madihian, M. and Honjo, K. 11 GHz band GaAs monolithic VCO with 1/4 analog divider, *1985 IEEE GaAs Integrated Circuits Symp.*, pp. 133–136, IEEE, New York, 1985.

9.90 Mumford, W. Some notes on the history of parametric transducers, *Proc. IRE*, **48**, 848, 1960.

9.91 Bernard, M. Y. *Circuits*, Vol. 2, Ribert, Paris, 1973.

9.92 Burq, P. The degenerate and quasi-degenerate mode of parametric amplification, *IRE Trans. Circuit Theory*, 200–210, 1960.

9.93 Harrison, R. A broadband frequency divider using microwave varactors, *IEEE Trans. Microwave Theory Tech.*, **25**, 1055–1059, 1977.

9.94 Wroblewski, R. Personal communication, 1986.

9.95 Stapleton, S. A microwave frequency halver with conversion gain, *Proc. 16th European Microwave Conf., 1986*, Microwave Exhibitions and Publishers, Tunbridge Wells, 1986.

9.96 Stapleton, S. and Stubbs, M. GaAs monolithic frequency halver, *Electron. Lett.*, **22** (15), 173–174, 1986.

9.97 Harrison, R. Varactor frequency halvers with enhanced bandwidth and frequency range, *1986 IEEE MTT Symp. Dig.*, pp. 305–308, IEEE, New York, 1986.

9.98 Maas, S. *Microwave Mixers*, Artech House, Dedham, MA, 1986.

9.99 Bauhahn, P. *et al.* 30 GHz monolithic balanced mixers using ion implanted FET-compatible 3-inch GaAs wafer process technology, *1986 IEEE Microwave and Millimeter-wave Monolithic Circuits Symp. Dig.*, pp. 45–50, IEEE, New York, 1986.

9.100 Chu, A., Courtney, W. E. and Sudbury, R. W. A 31 GHz monolithic GaAs mixer/preamplifier circuit for receiver applications, *IEEE Trans. Electron Devices*, **28**, 149–154, 1981.

9.101 Chao, C., Contolatis, A., Jamison, S. and Butter, C. 94 GHz monolithic GaAs balanced mixers, *1983 IEEE Microwave and Millimeter-wave Monolithic Circuits Symp. Dig.*, pp. 50–51, IEEE, New York, 1983.

9.102 Barker, G. 60 GHz monolithic GaAs front-end circuit for receiver applications, *Electron. Lett.*, **20** (8), 334–335, 1984.

9.103 Egami, S. Nonlinear, linear analysis and computer-aided design of resistive mixers, *IEEE Trans. Microwave Theory Tech.*, **22**, 270–275, 1974.

9.104 Meierer, R. Modelling of single and dual gate MESFET mixers, *Electron. Lett.*, **20** (2), 97–98, 1984.

9.105 Cox, R. Diode calculations for mixers, *Solid-State Electron.*, **10**, 1213, 1967.

9.106 Held, D. N. and Kerr, A. R. Conversion loss and noise of microwave and millimeter-wave mixers: Part 1. Theory, *IEEE Microwave Theory Tech.*, **26**, 49–61, 1978.

9.107 Tie, G. K. and Aitchison, C. S. Noise figure and associated conversion gain of a microwave MESFET gate mixer, *Proc. 13th European Microwave Conf., 1983*, pp. 579–584, Microwave Exhibitions and Publishers, Tunbridge Wells, 1983.

9.108 Harrop, P. Modelling of an FET mixer, *Electron. Lett.*, **14** (12), 369–370, 1978.

9.109 Harrop, P. GaAs FET mixers: theory and applications, *Acta Electron.*, **23** (4), 291–297, 1980.

9.110 Tsironis, C. and Meierer, R. Microwave wideband model of a GaAs dual gate

MESFET, *IEEE Trans. Microwave Theory Tech.*, **30**, 243–251, 1982.
9.111 Tsironis, C., Meierer, R. and Stahlmann, R. Dual gate FET mixers, *IEEE Microwave Theory Tech.*, **32**, 248–255, 1984.
9.112 Stahlmann, R. Dual gate MESFET self oscillating X band mixers, *Electron. Lett.*, **3** (17), 524–526, 1979.
9.113 Taleb, L. Etude de trois types de mélangeurs utilisant des TEC GaAs, *Ph.D. Thesis*, Lille, 1984.
9.114 Dautriche, P., Pauker, V. and Collet, A. GaAs monolithic circuits for TV tuners, *1985 IEEE GaAs Integrated Circuits Symp.*, pp. 165–168, IEEE, New York, 1985.
9.115 Yvan, L. A W-band monolithic GaAs crossbar mixer, *1986 IEEE Microwave and Millimeter-wave Monolithic Circuits Symp. Dig.*, pp. 67–69, IEEE, New York, 1986.
9.116 Tang, O. S. A. and Aitchison, C. S. A practical microwave travelling wave MESFET gate mixer, *1985 IEEE MTT Symp. Dig.*, pp. 605–608, IEEE, New York, 1985.
9.117 Titus, W., Tajima, Y., Pucel, R. and Morris, A. Distributed monolithic image rejection mixer, *1986 IEEE GaAs Integrated Circuits Symp.*, pp. 191–194, IEEE, New York, 1986.
9.118 Kermarrec, C., Faguet, J., Collet, A. and Mayousse, C. Récepteur d'émissions de télévision diffusées par satellite à 12 GHz sur une seule puce de GaAs, *Acta Electron.*, **27** (1–2), 65–76, 1985.
9.119 Watanabe, S., Ozaki, J. Miyauchi, M., Tatematsu, M., Hori, S. and Kamei, K. GaAs monolithic MIC mixer–IF amplifiers for DBS receivers, *1984 IEEE Microwave and Millimeter-wave Monolithic Circuits Symp. Dig.*, pp. 19–23, IEEE, New York, 1984.
9.120 Kermarrec, C., Faguet, J., Vançon, B., Mayousse, C., Collet, A., Kaikati, P. and Beaufort, D. The first fully integrated microwave receiver for DBS application at 12 GHz, *Proc. 14th European Microwave Conf., 1984*, pp. 749–754, Microwave Exhibitions and Publishers, Tunbridge Wells, 1984.
9.121 Bastida, E. A monolithic 800 MHz bandwidth DBS front-end for mass production, *Proc. 14th European Microwave Conf., 1984*, pp. 755–760, Microwave Exhibitions and Publishers, Tunbridge Wells, 1984.
9.122 Podell, A. GaAs MMICs designed for EW applications, *Microwave Syst. News Commun. Technol.*, **17** (4), 8–19, April 1987.
9.123 Kermarrec, C., Faguet, J., Harrop, P. and Tsironis, C. Monolithic circuits for DBS reception, *1982 IEEE Microwave and Millimeter-wave Monolithic Circuits Symp. Dig.*, pp. 5–10, IEEE, New York, 1982.
9.124 Fischer, R. J., Chand, N., Kopp, W. F., Peng, C. K., Morkoc, H., Gleason, K. R. and Scheitlin, D. A d.c. and microwave comparison of GaAs MESFETs on GaAs and Si substrates, *IEEE Trans. Electron Devices*, **33** (2), 206–213, 1986.

Chapter 10

An Aspect of GaAs MESFET Reliability: Surface-induced Degradation Mechanisms

J. M. DUMAS, J. PAUGAM AND C. VUCHENER

10.1 Introduction

The electrical properties of GaAs, in particular (a) the significantly increased low field electron mobility in n-type layers, (b) the semi-insulating character of GaAs which allows electrical isolation and (c) the existence of high Schottky barriers with metals commonly used in microelectronics such as aluminum, titanium, platinum and chromium, associated with advances in photolithography [10.1] and the empirical optimization of ohmic contacts [10.2], led Mead to propose the GaAs MESFET as a novel microwave device [10.3]. At the beginning of the 1970s, on the basis of its electrical performance, this device was described as a promising "laboratory curiosity" [10.4]. However, its extreme brittleness caused problems for the first users. The emergence in 1976 of commercially available FETs [10.5] opened the way to the "MESFET revolution" [10.6]. Consequently, the GaAs MESFET found widespread application throughout the microwave spectrum.

One area that has been of concern is reliability. This is because of (i) the introduction of this new GaAs technology itself and (ii) the stringent reliability requirements applied to communication equipment. The reliability of GaAs MESFETs has been reviewed comprehensively elsewhere [10.7, 10.8]. Therefore we shall only discuss some recently investigated degradation mechanisms in this chapter. Throughout the chapter we shall use the terms defined below.

Instantaneous degradation (time and temperature independent): this appears in the first few minutes after connecting the d.c. power supply and it affects both the static and microwave parameters dramatically.

Catastrophic degradation (time and temperature dependent): this occurs suddenly and affects both the static and microwave parameters.

Gradual degradation (time and temperature dependent): the microwave performance drifts slowly, but this r.f. degradation is not necessarily accompanied by significant changes in the static parameters usually measured during a reliability investigation.

Figure 10.1 illustrates these types of degradation.

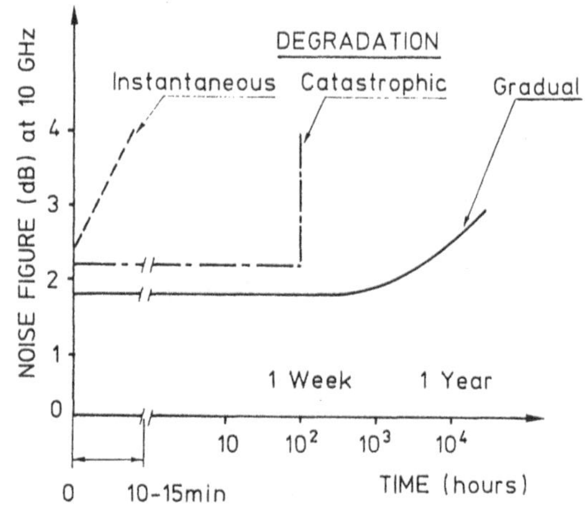

Fig. 10.1 The degradation modes of a GaAs MESFET shown in terms of changes in the microwave parameter with time.

Instantaneous degradation and catastrophic degradation have been reported, discussed and summarized elsewhere [10.7, 10.8]. Handling, lithographic defects and dirt are mainly responsible for instantaneous degradation, whereas catastrophic degradation depends on the metallization systems. The mechanism of gradual degradation involves the GaAs itself and particularly its surface. The role played by the surface and protection layers in the degradation of large-signal devices, i.e. power MESFETs and MESFETs developed for digital integrated circuits, is discussed in this chapter.

10.2 Power MESFETs

10.2.1 Static parameters for power capability [10.9]

Figure 10.2 is a schematic diagram of the I–V characteristics. It can be seen that the maximum output power P_0 is obtained for the device parameters I_{sat}, BV_{DGO}, V_p and V_{sat}. However, the pinch-off voltage V_p and the saturation voltage V_{sat} are lower than the gate-to-drain breakdown voltage BV_{DGO}. In other words, P_0 depends most strongly on I_{sat} and BV_{DGO}.

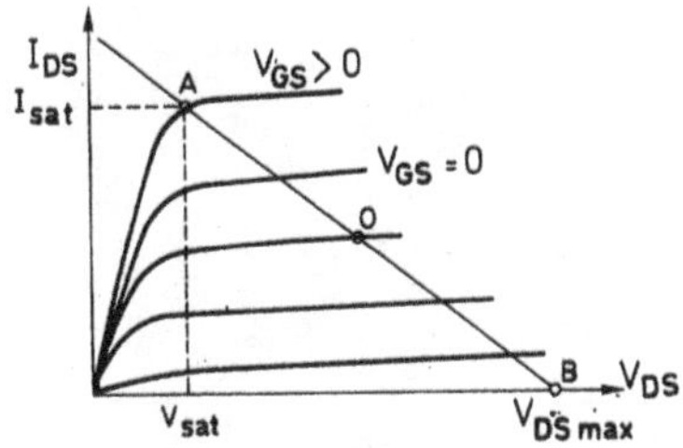

$$V_{DS\,max} = BV_{DGO} - |V_p|$$

$$P_O = \frac{1}{8} I_{sat} (BV_{DGO} - |V_p| - V_{sat})$$

Fig. 10.2 Schematic representation of the *I–V* characteristics of a power MESFET showing the parameters for power capability.

10.2.2 Gradual degradation of output power

10.2.2.1 Experimental procedure

Commercially available power MESFETs (0.5–1.5 W, X band) were purchased from six manufacturers. The devices tested are listed in Table 10.1. Types A, B and C_1 are unprotected, type C_2 is passivated with a plasma-enhanced chemical-vapor-deposited (PECVD) Si_3N_4 layer, and types D, E and F are protected with a low temperature (473–573 K) chemical-vapor-deposited (CVD) SiO_2 layer. In type E the SiO_2 layer is grown before gate processing (the windows are then selectively open), whereas in types D and F the SiO_2 layer is grown after gate processing. Sets of devices have been life tested under the following conditions: (a) d.c. operation; (b) gate reverse bias; (c) r.f. drive at 8 GHz at channel temperatures in the range 393–413 K (for four types of devices); (d) high temperature storage. A set of d.c. and microwave parameters was measured for the whole aging experiment.

Whatever the life-test conditions, a gradual degradation is observed during aging (Fig. 10.3). We report only the results of the gate reverse bias test, which are the most representative. Two types of degradation occur: (a) the first affects the unprotected devices (A, B and C_1); (b) the other occurs on SiO_2-protected MESFETs (D, E and F).

10.2.2.2 Unprotected devices

A gradual degradation of these components is observed during aging as illustrated in Fig. 10.4 for the output power at 8 GHz. (For clarity devices from only one manufacturer are represented here, but similar behavior is found for devices from the other two.) Two types of degradation occur.

The first, which occurs during storage at 523 K, is due to the failure of ohmic contacts. This is revealed by scanning electron microscopy (SEM)

Table 10.1 Structures of seven types of device purchased from six manufacturers (each type from the same wafer)

Manufacturer	A	B	C_1[a]	C_2[a]	D	E	F
Metal gate	Ti–W–Au	Al	Al	Al	Al	Al	Ti–Al
Recess gate	Yes	Yes	Yes	Yes	Yes	Yes	Yes
Active layer	VPE on buffer layer	VPE on buffer layer	VPE on buffer layer	VPE on buffer layer	VPE on buffer layer	VPE on buffer layer	Ion implantation in a Cr-doped substrate
Surface protection	Unprotected	Unprotected	Unprotected	PECVD Si_3N_4 layer	CVD SiO_2 layer	CVD SiO_2 layer	CVD SiO_2 layer

VPE, vapor phase epitaxy.
[a] C_1 and C_2 were both obtained from manufacturer C.

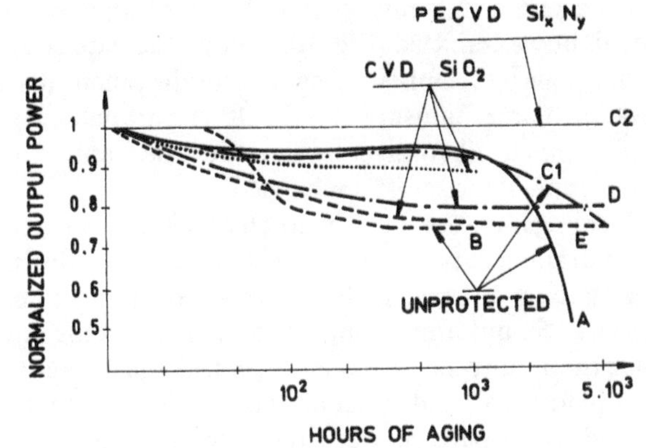

Fig. 10.3 Normalized output power at 1 dB gain compression measured at 8 GHz for the gate reverse bias test and averaged for each type of device.

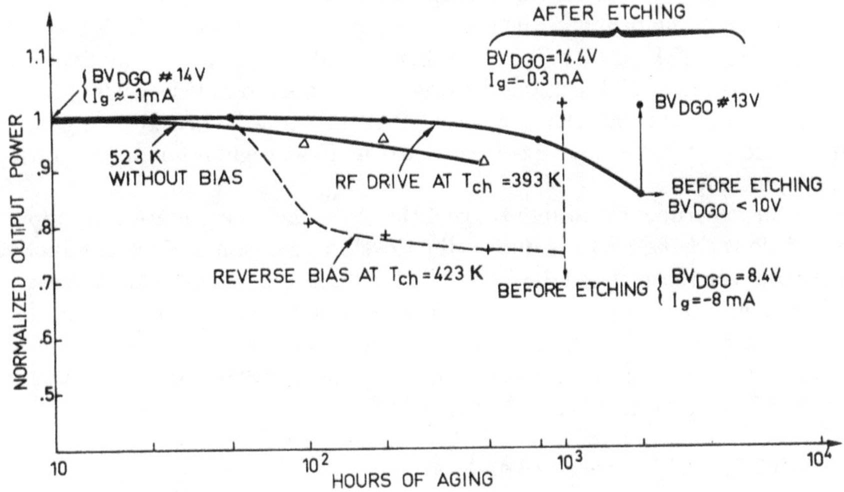

Fig. 10.4 Normalized output power at 8 GHz (aging characteristics averaged on 15 devices from one manufacturer): Δ, storage at 523 K; ●, r.f. operation at 393 K; +, reversed biased at 423 K near the pinch-off voltage.

examination and by an increase in the contact resistance and a decrease in the drain saturation current. This degradation was expected from previous results and is negligible at temperatures below 448 K [10.10].

The other type of degradation occurs at lower temperatures (a channel temperature of 423 K) and is bias dependent. A 25% decrease in the output power was observed after 200 h when the device was biased at the pinch-off voltage or after 2000 h of r.f. operation. At the same time BV_{DGO}, which was measured at 100 μA per millimeter of gate width, decreased from 14 to 8.5 V for the reverse biasing test and from 14 to 10 V for r.f. operation. During r.f. operation white light was emitted (Fig. 10.5) along the gate edge between the gate and the drain. This light emission occurs near the pinch-off condition and is due to an avalanche mechanism [10.11]. Other parameters, such as drain saturation current, pinch-off voltage, transconductance and ideality factor of the Schottky gate, remained almost unchanged.

A marked change in the *I–V* characteristics of the Schottky diode between gate and drain occurs at low currents and voltages during aging (Fig. 10.6). Before aging the *I–V* curve is that of an almost perfect Schottky diode ($\Phi_B = 0.65$ V, $n = 1.15$), but after aging, if the ideality factor and the barrier height do not exhibit any changes, a large leakage current appears for $V_g < 0.2$ V in both forward and reverse bias. Such a current involves the creation of a conduction path between the gate and the drain as found by Ozeki *et al.* [10.12] for passivated devices. This path is responsible for the decrease in the breakdown voltage and then, since the drain voltage swing is reduced, for the decrease in the output power.

To check the origin of the conductive path, we etched some MESFETs with FNH_4:HF:H_2O (6:1:7), rinsed them in distilled water and finally baked them for 10 min at 393 K under flowing nitrogen. As found elsewhere [10.12] the gate excess current decreases (Fig. 10.6) and both the breakdown voltage and microwave performance are restored (Fig. 10.4). Light is no longer emitted at the gate edge.

Our experimental results suggest that long-term degradation in unprotected power MESFETs is due to the development of a surface conductive layer between the gate and the drain which reduces the breakdown voltage and subsequently the output power. A possible failure mechanism is the gradual release of elemental metallic arsenic as the result of a reaction, probably enhanced by the electric field, between GaAs and native oxides formed during MESFET processing or aging [10.13, 10.14].

10.2.2.3 SiO_2-protected devices

The following parameters were measured: the zero-gate-bias current I_{DSS}, the gate-to-source voltage V_{GS} at $V_{DS} = 5$ V and $I_{DS} = 24$ mA (i.e. 10 mA per millimeter of gate width), the near-zero-gate-bias transconductance g_m, the gate-to-drain breakdown voltage BV_{DGO} at a current of 250 μA (i.e. about 100 μA per millimeter of gate width) and the output power $P_s(-1)$ at 1 dB gain compression at 8 GHz. The drift of each of these parameters is shown in Fig. 10.7. For clarity, only the first type of component is represented. A slight increase in the total drain-to-source resistance $R_{DS\,on}$ from 2.1 to 2.3 Ω (an increment of 0.2 Ω) was observed. However, the ideality factor, the barrier height and the gate-to-source and gate-to-drain leakage currents were unchanged. This indicated that there was no deterioration in the gate properties, and this was confirmed by SEM examination. In addition, there was no failure of the ohmic contacts. However, the ohmic contacts did degrade

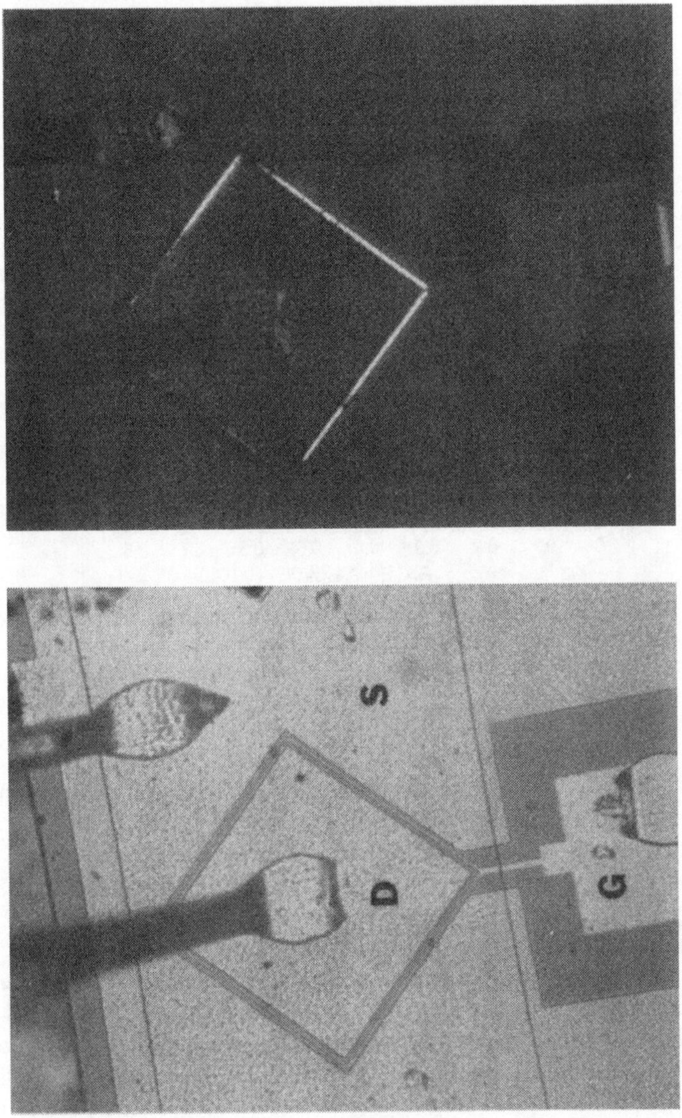

(a) (b)

Fig. 10.5 Light emission: (a) photograph of one cell (D, drain pad; S, source pad; G, gate pad); (b) gate light emission from the cell.

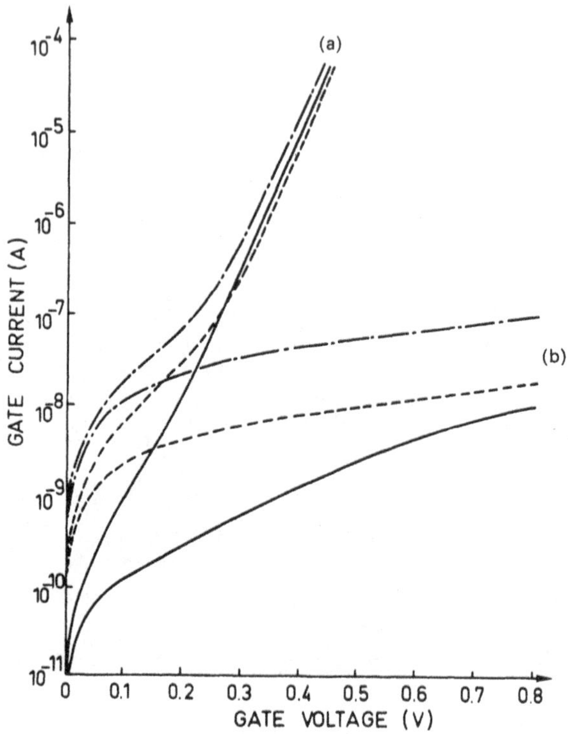

Fig. 10.6 Typical *I–V* characteristics between gate and drain for devices before and after aging and after etching: (a) forward bias; (b) reverse bias. Measurements were performed at 300 K: —, before aging, –·–, after aging; – – –, after etching.

after 650 h at 573 K without bias; $R_{DS\,on}$ increased by 0.5–0.6 Ω and I_{DSS} decreased. In this case SEM investigation showed structural changes in the ohmic contacts and overlayers.

Some of the samples subjected to life testing were analyzed using SEM. The first observation showed "particles" in the channel area inside the SiO_2 dielectric layer (Fig. 10.8) for samples prepared using both the chemical vapor deposition and the evaporation deposition techniques. These "particles" did not exist on the set of devices which had not been aged and which were used as control samples during the experiment. The dielectric layers of the failed and control devices were then sequentially etched. The "particles" appeared to be debris which had originated on the GaAs surface and grown through the SiO_2 layer. Figure 10.9 shows scanning electron micrographs of the devices after etching. Figure 10.9(a) shows a typical channel area of a control device: the recess region appears clean. Figure 10.9(b) shows a typical channel area of a degraded device. It can be seen that (i) the recessed channel is severely blocked by debris and (ii) debris is also visible in the nonrecessed channel regions.

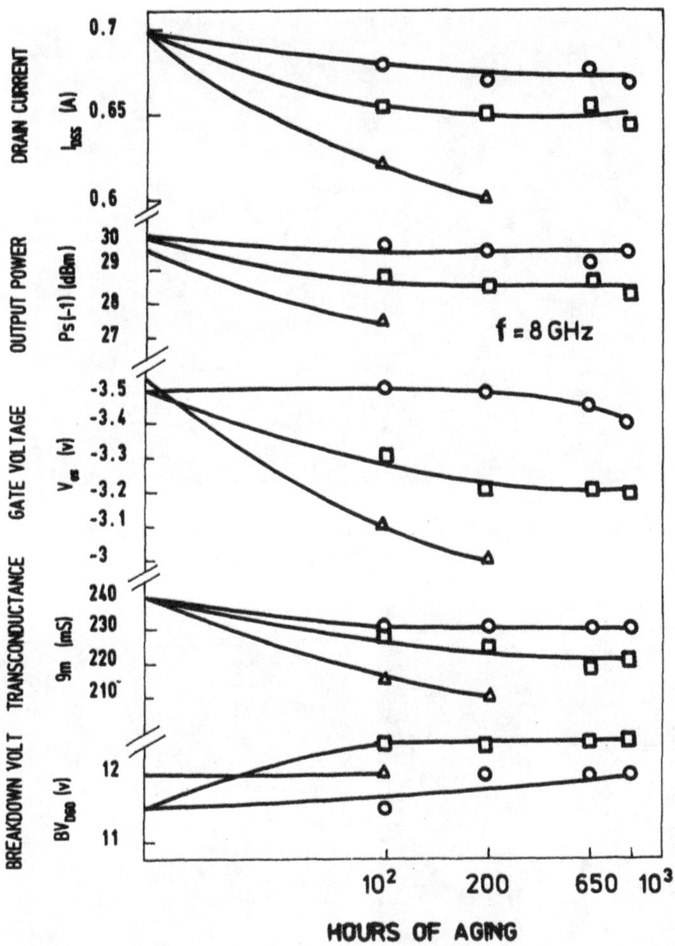

Fig. 10.7 Aging characteristics averaged on 30 devices from one manufacturer: ○, gate reverse bias at 423 K; □, d.c. operation at 478 K; △, d.c. operation at 538 K.

These observations suggest that the gradual degradation of the devices during biased aging tests could be due to detrimental GaAs surface effects.

Another group of degraded and control samples were then analyzed using a JEOL JAMP 10 Auger microprobe. This apparatus is equipped with beam brightness modulation which improves the signal-to-noise ratio at low currents [10.15]. This permits a reduction in the primary electron beam intensity with a corresponding increase in the spatial resolution. Beam currents in the range 5–10 nA with a corresponding beam size of about 0.1–0.2 μm were used for the experiments. The beam energy was set at 5 keV. These conditions result in the $E\,N(E)$ mode of operation instead of the usual $E\,dN(E)/dE$ mode. Debris in the channel is located using the SEM facilities of

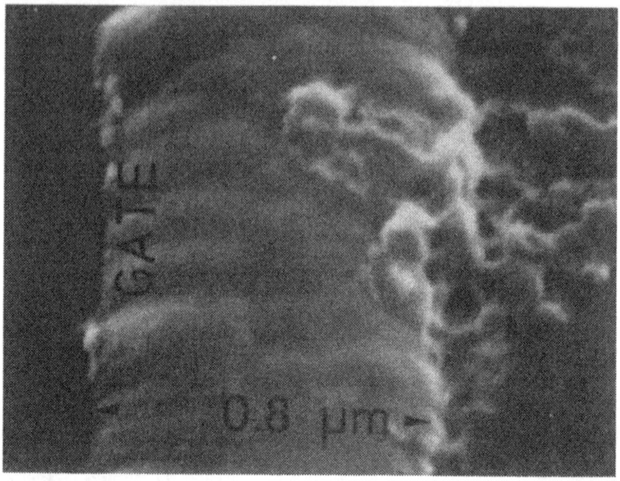

Fig. 10.8 Scanning electron micrograph of a "particle" inside the SiO_2 layer and along the drain edge of the gate.

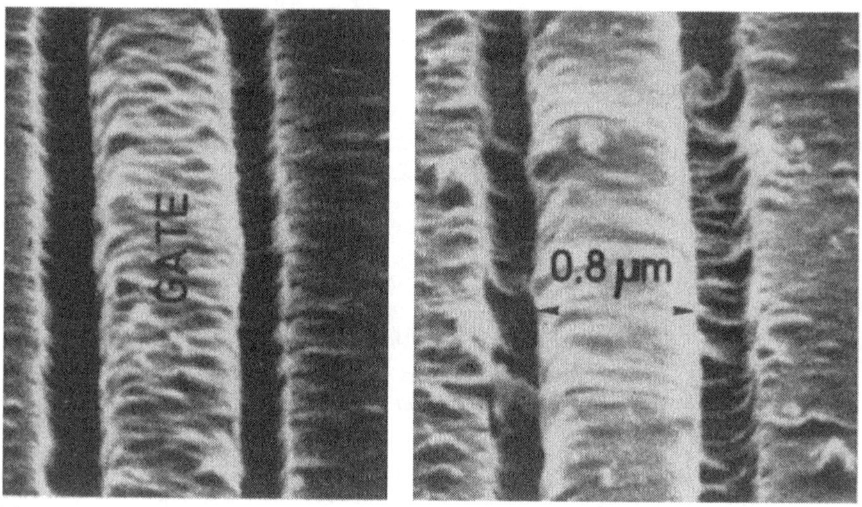

Fig. 10.9 Scanning electron micrographs of channel areas after removal of the surface coating: (a) control device, (b) degraded device.

the apparatus. An *in situ* ion-sputtering gun allows the SiO_2 layers to be etched over a few hundred ångströms for MESFET analysis. This permits a better access to debris. Figure 10.10 shows spectra taken in the low energy range (0–100 eV) under the same experimental conditions.

Spectrum (a) was obtained for a GaAs sample after *in situ* cleaning.

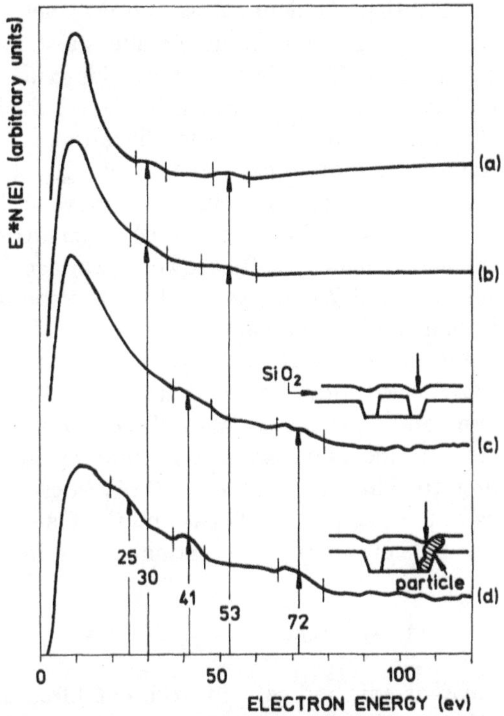

Fig. 10.10 Auger spectra from (a) a clean GaAs sample, (b) an oxidized GaAs sample, (c) an untested MESFET and (d) a failed MESFET.

Arsenic and gallium peaks were found around 30 eV and 53 eV respectively, as expected [10.16]. Spectrum (b) was obtained for a wafer which had been oxidized in boiling water for a few minutes to produce a thin oxide layer a few nanometers thick [10.17]. As previously reported [10.18, 10.19] peak broadening due to oxidized species (46 and 51 eV for gallium) is observed, mainly for gallium. Spectrum (c) was obtained from the SiO_2 layer located between the gate and drain of a control MESFET. The silicon peak at 72 eV associated with the presence of SiO_2 appears with another peak centered around 41 eV. Spectrum (d) was determined for debris (dimensions about 0.5 μm) from a failed MESFET. In addition to the two previous peaks (with an increase in the 41 eV peak), another broad peak centered around 25 eV is present. The charging effect of an insulating film under an electron beam produces an energy shift [10.20] towards lower values in the case of the SiO_2 CVD layer [10.21]; because of this we can attribute the 25 eV peak to arsenic. Comparison with the 30 eV peak in spectrum (b) of Fig. 10.10 reveals broadening, suggesting that the area analyzed also contains oxidized arsenic. Application of a similar energy shift to the 41 eV peak leads to a peak at 46 eV. This energy value is attributed to strongly oxidized gallium [10.18, 10.19].

Our results suggest that the debris observed on failed devices contains excess arsenic and oxidized gallium. Gallium oxide is also found in the SiO_2 coatings on the control MESFETs. The presence of this oxide can be explained by the occurrence of an out-diffusion mechanism during the SiO_2 deposition [10.22]. Thus the technique induces surface modification.

The gradual degradation observed in the biased aging tests is not due to the metallurgy (Schottky gate, ohmic contacts, overlayers). Parameter drifts similar to those shown in Fig. 10.7 have previously been attributed to in-bulk deterioration of the active layer [10.10]. However, surface effects have been proposed by Irvin [10.8] and Wemple *et al.* [10.23]. It should be noted that similar behavior is found for both implanted layers and active layers deposited by vapor phase epitaxy, and this leads us to eliminate in-bulk defects. However, we have observed surface changes leading to the accumulation of debris in the recessed and nonrecessed areas of the channel. This debris was identified as excess arsenic and oxidized gallium by Auger microprobe analysis. According to Thurmond *et al.* [10.14] such products can be explained in terms of the oxidation of GaAs and the Ga–As–O equilibrium phase diagram. This indicates that As_2O_3 should react with GaAs according to

$$As_2O_3 + 2GaAs \rightarrow Ga_2O_3 + 4As$$

This reaction involves arsenic and gallium oxides and free elemental arsenic and behaves like GaAs surface erosion, as can clearly be seen in Fig. 10.9(b). Moreover, it can modify the electrical properties of the surface, leading to an increase in the surface space charge layers in the gate-to-drain and gate-to-source access regions. The narrowing of the active layer in the access regions explains the electrical parameter drifts (access resistances).

Suitable surface preparation and passivation are required to overcome gallium out-diffusion during SiO_2 deposition and the oxidation mechanism during operation.

10.2.2.4 A possible remedy

With this problem in mind, some transistor manufacturers (C_1, E, F) have deliberately underrated the microwave performances of their devices. Subsequently manufacturer C developed a PECVD Si_3N_4 layer on the same transistor (C_2 type). This method, which excludes or minimizes oxidation, has been established as a suitable passivation process for III–V devices [10.13] and increases the median lifetime of the MESFETs [10.23]. Thus the drifts measured on C_2 devices after 5000 h are within our margin of error (Fig. 10.3). A comparative study of the effects of various surface protection layers on the stability of low noise MESFET amplifiers from turn-on through 168 h of operation confirmed that PECVD Si_3N_4 eliminates the time-dependent drift of the r.f. gain [10.24].

10.2.2.5 The gate lag effect

We now report the results of life tests showing a decrease in P_0 which is not accompanied by drifts in I_{DSS}, BV_{DGO} or any other basic static parameter. We describe the electrical behavior of the degradation and then propose a new method for characterization.

Life tests were conducted on commercially available power MESFETs (1 W, 12 GHz). Fifty devices fabricated from the same wafer were supplied by one manufacturer. The active layer was obtained by silicon implantation on an undoped substrate. They were coated with a PECVD Si_3N_4 layer 3000 Å thick deposited at low temperature (523 K) after fabrication of the gate and ohmic contacts. Sets of 10 devices were submitted to the following life tests: gate reverse bias at 423 K; d.c. operation at $V_{DS} = 8$ V and $I_{DS} = I_{DSS}/2$ at channel temperatures of 443 K (DC 443) and 493 K (DC 493); storage at 523 K without bias.

A set of 12 basic static parameters was measured during the entire aging experiment. We also measured a "maximum" linear output power P_{01} at 11 GHz on a fixed tuned amplifier. The sample studied was homogeneous; P_{01} was about 30 dBm and was obtained with a static drain current of 350 mA ($\approx I_{DSS}/2$) for all the devices.

A gradual degradation of P_{01} during aging was observed for about two-thirds of the devices, independently of the life-test conditions (Fig. 10.11). However, the static parameters remained almost unchanged. (A slight decrease in I_{DSS} associated with an increase in the drain-to-drain resistance $R_{DS\,on}$ was observed on some devices at 523 K without bias. We attributed these drifts to a stabilization of the ohmic contacts.) The decrease in P_{01} was not due to changes in the fixed tuned amplifier: the output power of a set of devices which had not been aged and which were used as control samples did not vary during the experiment.

However, we noted that during the microwave characterization the degradation in the output power was accompanied by a decrease in the drain current I_{DS} (Fig. 10.12) and a marked increase in the positive gate current I_{GS} (5–10 mA) under large-signal conditions. These values of I_{GS} clearly demonstrate that the gate was forward biased. This behavior can be explained as follows. As the quiescent level was maintained at the same drain current, the dynamic drain current decreased owing to a waveform-clipping effect (when the input power was abruptly switched off I_{DS} returned to its static value of 350 mA). Such behavior was not observed before aging: I_{DS} remained stable or increased slightly (see Fig. 10.12 for $P_0 > 26$ dBm), indicating that avalanche regions were reached, and I_{GS} stayed close to zero with a negative value.

The output power loss appeared to be due to a decrease in I_{DSS} which occurred only when the device operated under large r.f. signal conditions. In other words, the current I_{DSS} available for the microwave operation is lower than the static current I_{DSS}. This phenomenon can be explained by the development of a gate lag effect during aging. The gate lag effect [10.25]

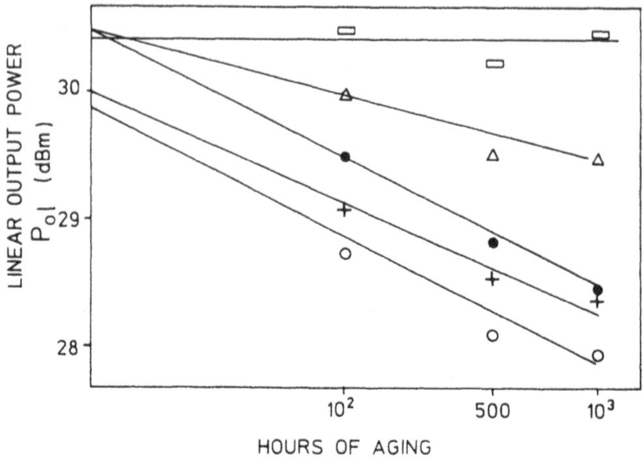

Fig. 10.11 Typical aging characteristics for each life test: △, gate reverse bias at 423 K; +, d.c. operation at 443 K; ●, d.c. operation at 493 K; ○, high temperature storage at 523 K; □, control sample.

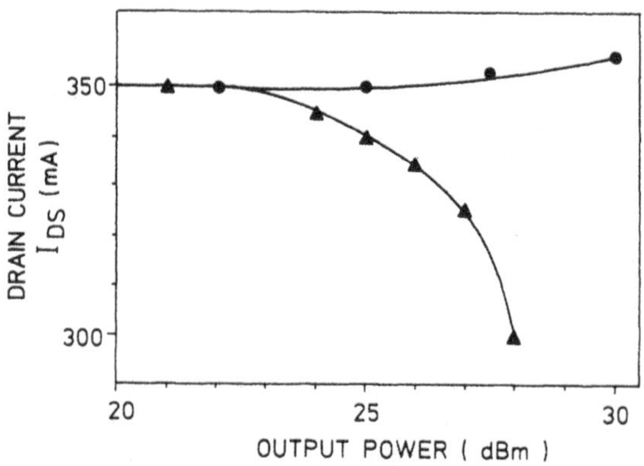

Fig. 10.12 Microwave characterization: drain current versus output power before (●) and after (▲) aging.

corresponds to a drain current transient responding to a voltage pulse on the gate with a small variation in the drain-to-source voltage V_{DS}. (A large variation in V_{DS} can trigger a kinetic effect attributed to the formation of a negative trapped space charge at the channel–substrate interface region [10.26].) Although several research groups are studying this parasitic effect, which limits GaAs device performance, few results have been published so far [10.25].

Figure 10.13(a) shows the electrical scheme for the measurement we performed on the MESFETs. The gate is switched from the pinch-off voltage V_p to $V_{GS} = 0$ in the saturation region of the I–V characteristics ($V_{DD} = 7$ V).

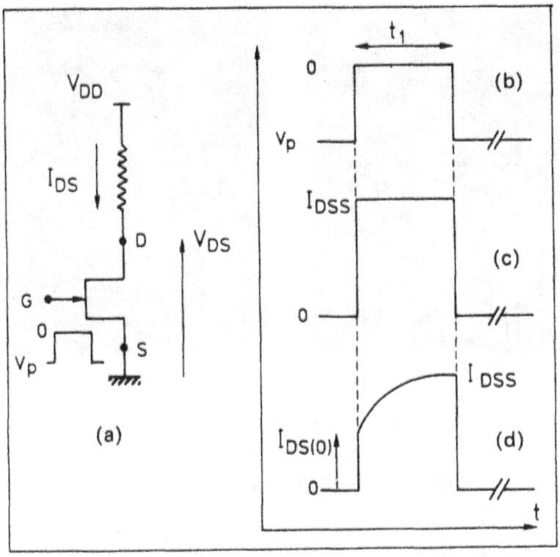

Fig. 10.13 Gate lag measurements: (a) electrical scheme; (b) gate pulse; (c) ideal drain current response; (d) gate lag effect.

The drain current is measured through a 1 Ω load or using a Hall current probe. (The variation in V_{DS} is 0.7 V as I_{DSS} is typically 0.7 A.) The gate pulse width t_1 can vary from 1 μs to several seconds with a 10% repartition rate (Fig. 10.13(b)). Figure 10.13(c) shows the response of the ideal drain current to the pulse: the static value of I_{DSS} is reached instantaneously. If the gate lag effect is present, as shown in Fig. 10.13(d), the drain current switches from $I_{DS} = 0$ to $I_{DS}(0)$ and increases to the static value of I_{DSS} with a time constant ranging from 1 μs to several seconds. It is obvious that $I_{DS}(0)$ is the maximum drain current available for microwave operation.

This effect has been attributed to the time-dependent (or frequency-dependent) modulation of the access resistances R_s and R_d. When it is observed immediately after processing, it has been shown to be linked to slow surface states. These states could be induced by free ions in the dielectric coating or deep traps in the disordered region close to the GaAs surface in the access regions [10.25].

We have observed the same effect on aged devices that initially exhibited no lagging. Figure 10.14(a) shows the typical drain current response of a MESFET before aging ($V_p = -3$ V at $I_{DS} = 3$ mA, i.e. 1 mA per millimeter of gate width). There is no gate lag effect. Figure 10.14(b) shows a MESFET after 1000 h in the DC 493 life test. P_{01} decreased by 1.5 dBm and the difference between I_{DSS} and I_{DS} was about 100 mA, with I_{DSS} remaining equal to its static value of 700 mA. In other words the maximum high frequency available drain current is reduced to 600 mA during aging. As we were limited by the rise time (200 ns) of our current probe, $I_{DS}(0)$ is certainly lower than this value.

In order to investigate this degradation more extensively we made the

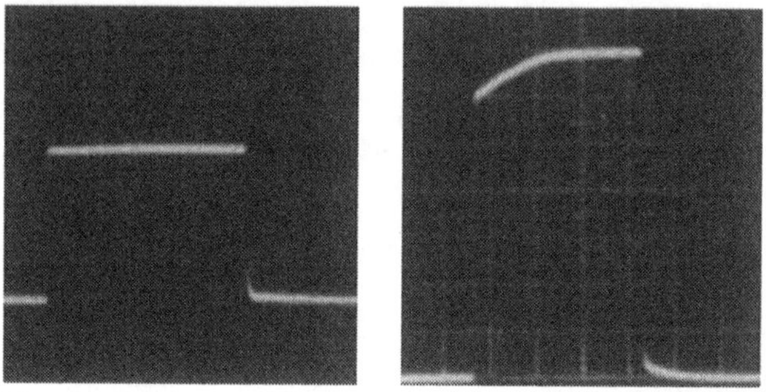

(a) (b)

Fig. 10.14 Typical drain current responses: (a) before aging (vertical scale, 200 mA per division; horizontal scale, 2 μs per division); (b) after aging (vertical scale, 100 mA per division; horizontal scale, 5 μs per division).

following measurements: the I–V characteristics were vertically scanned from V_p to -2 V, V_p to -1 V and V_p to 0 V, and the difference δI_{DS} between the static and transient values of I_{DS} was measured each time. The transient I–V characteristics can then be deduced. Figure 10.15 is a schematic representation of the results obtained for the MESFET shown in Fig. 10.14(b). These results clearly show that almost all the lag effect (70 mA) occurs between $V_{GS} = -1$ V and $V_{GS} = 0$. This confirms the reduction in the drain current in the upper part of the I–V characteristics. There is no gate lag effect before aging and the high frequency excursion therefore follows the static device characteristics shown by the solid curves in Fig. 10.15. Thus AB represents the dynamic load line. The static characteristics remain unchanged after aging, but a gate lag effect is observed which reduces the dynamic excursion available to the r.f. signal (broken curves in Fig. 10.15). As the static bias point zero is maintained and the measurements are made in the same fixed tuned amplifier, the gate will be forward biased and a waveform-clipping effect will appear when the device operates beyond A′. This is the first observation of the gate lag phenomenon in power MESFETs during aging.

When the gate lag effect is observed immediately after processing, it has been shown to be linked to slow surface states. These states could be induced by free ions in the dielectric coating or deep traps in the disordered region close to the GaAs surface in the access regions [10.25]. We have verified that the gate lag effect disappears after etching SiO_2 or r.f. sputtered Si_3N_4 layers on various devices supplied by different manufacturers. The same result is obtained after 100–200 h at 423 K without bias with all transistor connections short-circuited [10.27]. This last experiment, which demonstrates that the effect can be reversible, strongly suggests that mobile ions are involved.

However, the effect is irreversible with PECVD Si_3N_4 and after aging, although the measured time constants (10–50 μs) agree with those observed after processing [10.25, 10.27]. Owing to the complexity of the device pattern

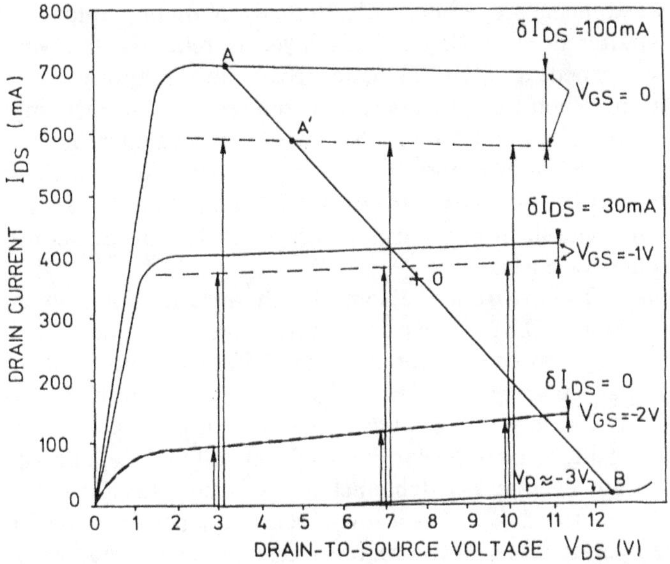

Fig. 10.15 *I–V* characteristics: —, static device characteristics before and after aging; – – –, deduced from gate lag measurements. The gate lag effect appeared during aging.

(multigate geometry, a second dielectric level and crossover metallization), we were unable to etch the Si_3N_4 layer and take the failure analysis further; however, the degradation mechanism is certainly correlated with the surface treatment and Si_3N_4 deposition conditions.

We can thus conclude that (i) the absence of a gate lag effect after processing does not guarantee reliability for the MESFET user and (ii) gate lag measurement is a recommended method for characterizing MESFETs during reliability investigations.

10.3 MESFETs developed for digital integrated circuits

After more than 10 years of research and development in many laboratories, GaAs logic integrated circuits (ICs) are becoming available for high speed logic application. In order to determine the degradation mechanisms and establish the reliability of such devices, we have carried out life tests on drop-in chips.

Drop-in chips are processed on the same wafers as high frequency dividers. The basic logic gate configuration is buffered FET logic. A Czochralski undoped semi-insulating substrate is used for the fabrication, and the active channel is prepared by selective silicon ion implantation through a thin r.f. sputtered Si_3N_4 layer. This layer is retained and windows are selectively opened to etch the gate recess and evaporate a Ti–Pt–Au metallization. When this process is used the recess flanks are unpassivated. The ohmic contacts are of Au–Ge–Ni. Two Ti–Pt–Au interconnection levels, which are isolated by a second PECVD Si_3N_4 layer 5000 Å thick, are used. A drop-in contains various test patterns (MESFETs, diodes, side-gate pattern, isolation and metallization patterns etc.) developed for process control and for studying parasitic effects.

About 100 chips from the same batch were mounted in dual-in-line ceramic packages (DIL 24); the pattern bond pads were connected to the pins. Such packaging allowed us to conduct biased life tests appropriate to each test pattern. In particular, sets of MESFETs with gate lengths of 1 and 5 μm and gate widths of 20 μm were aged under the following conditions: gate reverse bias at 423 K; d.c. operation near the pinch-off voltage $V_{GS\,off}$ and at the zero-gate-bias drain current I_{DSS} (channel temperature, 423 K); gate switching between $V_{GS} = 0$ and $V_{GS\,off}$ at a low clock frequency (F = 100 kHz) and a channel temperature of 423 K. A set of 35 d.c. parameters was measured on the drop-in chips over the whole aging experiment. Automatic test equipment was used.

No major degradation of the metallization patterns, ohmic contacts or Schottky gate was observed. This means that the improvements made to the metallurgy of GaAs discrete devices have been successfully transferred to GaAs IC technology. The isolation provided by the PECVD Si_3N_4 layer 5000 Å thick between the two Ti–Pt–Au metallization levels did not change during aging. A slight increase in the side-gating threshold voltage was measured.

However, the following failure was identified and investigated. High gate-to-drain and/or gate-to-source leakage currents occurred in the MESFETs under biased life tests and induced short-circuits. Some MESFETs exhibited such leakages before aging. The leakages were due to the development of conductive paths (about 0.5 μm wide) between the gate and the ohmic contacts at the interface between the r.f. sputtered Si_3N_4 layer and the GaAs in the access regions. These paths were localized using a laser beam microscope (λ = 442 nm) operating (i) in the reflected mode and (ii) in the optical beam induced current (OBIC) mode (Fig. 10.16). They cannot be seen using classical SEM examination and the associated electron beam induced current (EBIC) mode is sensitive to charging problems.

High resolution Auger analysis was conducted as described in Section 10.2.2.2. After *in situ* etching of the r.f. sputtered Si_3N_4 layers the paths were analyzed (Fig. 10.17) using a beam spot size of about 0.1–0.2 μm. Spectrum 1 was obtained at the Si_3N_4–GaAs interface of a control device outside the access region. Excess oxygen (1.5 at.%) is clearly seen together with a significant amount of arsenic (peak at 38 eV). This means that oxygen is

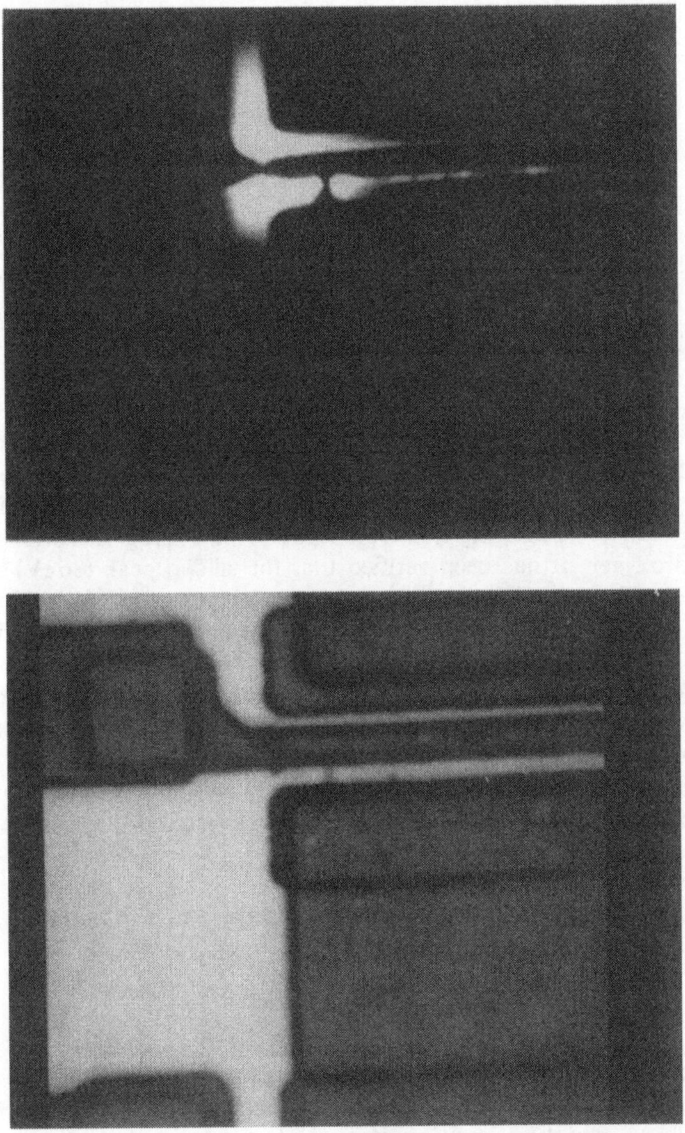

(a) (b)

Fig. 10.16 Laser scan microscopy images ($G = 3500$): (a) reflected mode; (b) OBIC mode.

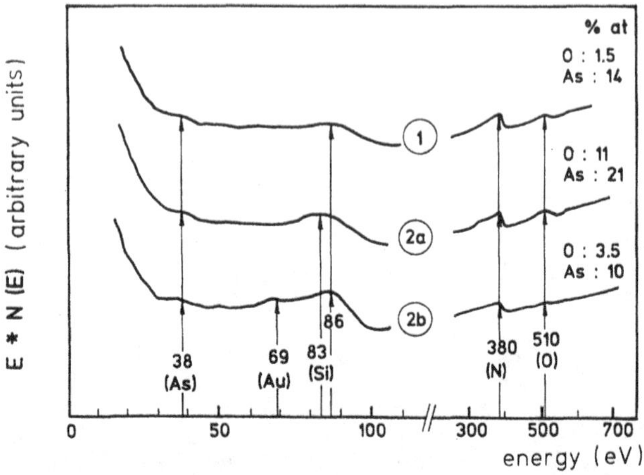

Fig. 10.17 Auger spectra: 1, spectrum taken outside the access regions in a control device; 2a, spectrum observed on a conductive path in a control device; 2b, spectrum observed on a conductive path after aging.

introduced during the Si_3N_4 deposition process and may induce oxidation of GaAs [10.14]. Spectrum 2a was obtained from the path of a control device: 21 at.% As and 11 at.% O are found. Spectrum 2b was measured on the path of a failed device in a biased aging test. Gold is found (69 eV) together with some arsenic and oxygen. It has been verified that the silicon peak (86 eV) shifts toward lower values (83 eV) when the oxygen content increases [10.28].

These results lead us to postulate the following degradation mechanisms.

(1) Before aging a path is created by the gradual release of arsenic through a GaAs oxidation mechanism [10.14] and conduction takes place through this elemental arsenic [10.29].

(2) After aging the path is created by electric-field-induced migration of electrode material (gold from the ohmic contacts or gate) which depends strongly on the amount of As_2O_3 present [10.30].

Again, suitable GaAs surface passivation is required.

10.4 Conclusion

We have investigated the gradual degradation mechanisms of large-signal GaAs MESFETs. It has been found that chemical changes in the GaAs surface, such as oxidation, may be responsible for these degradations on uncoated devices and on devices coated with SiO_2 or r.f. sputtered Si_3N_4. A PECVD Si_3N_4 layer can solve this problem if the surface treatment and deposition conditions are well controlled.

References

10.1 Hewitt, B. S., Cox, H. M., Dilorenzo, J. Y., Schlosser, W. O. and Iglesias, D. E. Low-noise GaAs MESFETs: fabrication and performance, *Inst. Phys. Conf. Ser.*, **33a**, 246–254, 1977.

10.2 Rideout, V. L. A review of the theory and technology for ohmic contacts to group III–V compound semiconductors, *Solid-State Electron.*, **18**, 541–550, 1975.

10.3 Mead, C. A. Schottky barrier gate field effect transistor, *Proc. IEEE*, **54**, 307–308, 1966.

10.4 Loriou, B., Bellec, M. and Le Rouzic, M. Performances à basse température d'un transistor MESFET, *Electron. Lett.*, **6**, 122–123, 1970.

10.5 Sekido, K. and Arden, J. A. Recent advances in FET device performance and reliability, *Microwave Syst. News*, **6** (2), 71–81, 1976.

10.6 Special Issue on Microwave Field Effect Transistors, *IEEE Trans. Microwave Theory Tech.*, **24** (6), 1976.

10.7 Davey, J. E. and Christou, A. Reliability and degradation of active III–V semiconductor devices. In Howes, M. J. and Morgan, D. V. (eds), *Reliability and Degradation*, Chapter 5, pp. 237–300, Wiley, Chichester, 1981.

10.8 Irvin, J. C. The reliability of GaAs FETs. In Dilorenzo, J. V. and Khandelwal, D. D. (eds), *GaAs FET Principles and Technology*, Chapter 6, pp. 349–400, Artech House, Dedham, MA, 1982.

10.9 Dilorenzo, J. V. and Wisseman, W. R. GaAs power MESFETs: design, fabrication and performance, *IEEE Trans. Microwave Theory Tech.*, **27** (5), 367–378, 1979.

10.10 Fukui, H., Wemple, S. H., Irvin, J. C., Niehaus, W. C., Hwang, J. C. M., Cox, H. M., Schlosser, W. O. and Dilorenzo, J. V. Reliability of power GaAs field-effect transistors, *IEEE Trans. Electron Devices*, **29**, 395–401, 1982.

10.11 Yamamoto, R., Higashisaka, A. and Hasegawa, F. Light emission and burnout characteristics of GaAs power MESFETs, *IEEE Trans. Electron Devices*, **25**, 567–573, 1978.

10.12 Ozeki, M., Kodama, K., Takikawa, M. and Shibatomi, A. Analysis of electrical and optical properties of insulating film GaAs interfaces using MESFET-type structures, *J. Vac. Sci. Technol.*, **21**(2), 438–441, 1982.

10.13 Capasso, F. and Williams, G. F. A proposed hydrogenation–nitridization passivation mechanism for GaAs and other III–V semiconductor devices, including InGaAs long wavelength photodetector, *J. Electrochem. Soc.*, **129**, 821–824, 1982.

10.14 Thurmond, C. D., Schwartz, J. P., Kammlot, G. W. and Schwartz, B. GaAs oxidation and Ga–O equilibrium phase diagram, *J. Electrochem. Soc.*, **127**, 1366–1371, 1980.

10.15 Mogami, A. Analytical techniques with the Auger scanning electron microscope, *Thin Solid Films*, **57**, 127–139, 1979.

10.16 *Handbook for Auger Electron Spectroscopy*, JEOL, Tokyo, 1982.

10.17 Moritomo, H. and Takusagawa, M. Accelerated facet degradation of InGaAs/InP double heterostructure lasers in water, *J. Appl. Phys.*, **53**(6), 4028–4037, 1982.

10.18 Ranke, W. and Jocabi, K. Electron stimulated oxidation of GaAs studied by quantitative Auger electron spectroscopy, *Surf. Sci.*, **47** (2–3), 525–542, 1975.

10.19 Bartels, F. and Mönch, W. Chemisorption of oxygen at cleaved GaAs(110) surfaces: photon stimulation and chemisorption states, *Surf. Sci.*, **143** (2–3), 315–341,1984.

10.20 Le Gressus, C. Personal communication, 1984.

10.21 Arnaud d'Avitaya, F. and Gourrier, S. Personal communication, 1984.

10.22 Gyulai, J., Mayer, J. W., Mitchell, I. V. and Rodriguez, V. Outdiffusion through silicon oxide and silicon nitride layers on gallium arsenide, *Appl. Phys. Lett.*, **17**(8), 332–334, 1970.

10.23 Wemple, S. H., Niehaus, W. C., Fukui, H., Irvin, J. C., Cox, H. M., Hwang, J. C. M., Dilorenzo, J. V. and Schlosser, W. O. Long term and instantaneous burnout in GaAs power FETs: mechanisms and solutions, *IEEE Trans. Electron Devices*, **28**, 834–840, 1981.

10.24 Tenedorio, J. G. and Terzian, P. A. Effects of Si_3N_4, SiO, and polyimide surface passivations on GaAs MESFET amplifier RF stability, *IEEE Electron Device Lett.*, **5**(6), 199–202, 1984.

10.25 Rocchi, M. Status of the surface and bulk parasitic effects limiting the performances of GaAs ICs. In Noblanc, J. P. and Zimmermann, J. (eds), *Proc. European Solid State Devices Research Conf.*, pp. 119–138, North-Holland, Amsterdam, 1985.

10.26 Makram-Ebeid, S. and Minondo, P. The role of the surface and bulk of the semi-insulating substrate in low frequency anomalies of GaAs integrated circuits, *IEEE Trans. Electron Devices*, **32**(3), 632–642, 1985.

10.27 Dumas, J. M. Unpublished results, 1986.

10.28 Von Oostrom, A., Augustus, L., Habraken, F. H. P. M. and Kuiper, A. E. T. Quantitative Auger microanalysis of the silicon–oxygen–nitrogen system, *J. Vac. Sci. Technol.*, **20**(4), 953–956, 1982.

10.29 Folkes, P. A., Chang, C. C. and Lane, E. Process-induced leakage current in a low noise GaAs MESFET, *J. Electrochem. Soc.*, **132**(6), 1417–1423, 1985.

10.30 Kretschmer, K. H. and Hartnagel, H. L. XPS analysis of GaAs-surface quality affecting interelectrode material migration, *23rd Ann. Proc. IEEE Int. Reliability Physics Symp.*, pp. 45–49, IEEE, New York, 1985.

Index

www.ingramcontent.com/pod-product-compliance
Lightning Source LLC
LaVergne TN
LVHW101451200926
848483LV00001B/107
* 9 7 8 0 8 9 0 0 6 2 6 7 8 *